World Geomorphological Landscapes

Series Editor

Piotr Migoń, Institute of Geography and Regional Development, University of Wrocław, Wrocław, Poland

Geomorphology – 'the Science of Scenery' – is a part of Earth Sciences that focuses on the scientific study of landforms, their assemblages, and surface and subsurface processes that moulded them in the past and that change them today. Shapes of landforms and regularities of their spatial distribution, their origin, evolution, and ages are the subject of geomorphology. Geomorphology is also a science of considerable practical importance since many geomorphic processes occur so suddenly and unexpectedly, and with such a force, that they pose significant hazards to human populations. Landforms and landscapes vary enormously across the Earth, from high mountains to endless plains. At a smaller scale, Nature often surprises us creating shapes which look improbable. Many geomorphological landscapes are so immensely beautiful that they received the highest possible recognition – they hold the status of World Heritage properties. Apart from often being immensely scenic, landscapes tell stories which not uncommonly can be traced back in time for millions of years and include unique events. This international book series will be a scientific library of monographs that present and explain physical landscapes across the globe, focusing on both representative and uniquely spectacular examples. Each book contains details on geomorphology of a particular country (i.e. The Geomorphological Landscapes of France, The Geomorphological Landscapes of Italy, The Geomorphological Landscapes of India) or a geographically coherent region. The content is divided into two parts. Part one contains the necessary background about geology and tectonic framework, past and present climate, geographical regions, and long-term geomorphological history. The core of each book is however succinct presentation of key geomorphological localities (landscapes) and it is envisaged that the number of such studies will generally vary from 20 to 30. There is additional scope for discussing issues of geomorphological heritage and suggesting itineraries to visit the most important sites. The series provides a unique reference source not only for geomorphologists, but all Earth scientists, geographers, and conservationists. It complements the existing reference books in geomorphology which focus on specific themes rather than regions or localities and fills a growing gap between poorly accessible regional studies, often in national languages, and papers in international journals which put major emphasis on understanding processes rather than particular landscapes. The World Geomorphological Landscapes series is a peer-reviewed series which contains single and multi-authored books as well as edited volumes.

World Geomorphological Landscapes – now indexed in Scopus® !

More information about this series at https://link.springer.com/bookseries/10852

Milan Lehotský • Martin Boltižiar
Editors

Landscapes and Landforms of Slovakia

Springer

Editors
Milan Lehotský
Department of Physical Geography
Geomorphology and Natural Hazards
Institute of Geography
Slovak Academy of Sciences
Bratislava, Slovakia

Martin Boltižiar
Department of Geography
Geoinformatics and Regional Development
Faculty of Natural Sciences
Constantine The Philosopher University in Nitra
Nitra, Slovakia

Institute of Landscape Ecology
Slovak Academy of Sciences
Nitra, Slovakia

ISSN 2213-2090 ISSN 2213-2104 (electronic)
World Geomorphological Landscapes
ISBN 978-3-030-89292-0 ISBN 978-3-030-89293-7 (eBook)
https://doi.org/10.1007/978-3-030-89293-7

This Springer imprint is published by the registered company Springer Nature Switzerland AG
The registered company address is: Gewerbestrasse 11, 6330 Cham, Switzerland

Series Editor Preface

Landforms and landscapes vary enormously across the Earth, from high mountains to endless plains. At a smaller scale, nature often surprises us creating shapes which look improbable. Many physical landscapes are so immensely beautiful that they received the highest possible recognition—they hold the status of World Heritage properties. Apart from often being immensely scenic, landscapes tell stories which not uncommonly can be traced back in time for tens of million years and include unique events. In addition, many landscapes owe their appearance and harmony not solely to the natural forces. For centuries, and even millennia, they have been shaped by humans who have modified hillslopes, river courses and coastlines, and erected structures which often blend with the natural landforms to form inseparable entities.

These landscapes are studied by geomorphology—'the science of scenery'—a part of Earth Sciences that focuses on landforms, their assemblages, surface and subsurface processes that moulded them in the past and that change them today. To show the importance of geomorphology in understanding the landscape, and to present the beauty and diversity of the geomorphological sceneries across the world, we have launched a book series *World Geomorphological Landscapes*. It aims to be a scientific library of monographs that present and explain physical landscapes, focusing on both representative and uniquely spectacular examples. Each book will contain details on geomorphology of a particular country or a geographically coherent region. This volume presents the geomorphology of Slovakia—a Central European country, which contains an enormous diversity of landforms within its rather small territory. The spectrum includes the high-mountain scenery of the Tatras—the highest massif within the entire Carpathians Mountains system, lower ranges on crystalline, volcanic and sedimentary bedrock, spectacular karst, intriguing intramontane basins and lowlands finely shaped by fluvial and aeolian processes. This is also an area of protracted history of human–landscape interactions, which is reflected in many chapters within the volume.

The World Geomorphological Landscapes series is produced under the scientific patronage of the International Association of Geomorphologists (IAG)—a society that brings together geomorphologists from all around the world. The IAG was established in 1989 and is an independent scientific association affiliated with the International Geographical Union (IGU) and the International Union of Geological Sciences (IUGS). Among its main aims are to promote geomorphology and to foster dissemination of geomorphological knowledge. I believe that this lavishly illustrated series, which sticks to the scientific rigour, is the most appropriate means to fulfil these aims and to serve the geoscientific community. To this end, my great thanks go to the editors of the volume, Milan Lehotský and Martin Boltižiar, who agreed to add this publication project to their busy agendas and did their best to show the fascinating geomorphology of Slovakia to the world. I am also grateful to all individual contributors who accepted the invitation to contribute to the book and shared their expertise and experience. Being long fascinated by landforms of Slovakia myself, and still recalling my illuminating first visit to the 'textbook karst' of the Slovak Karst back in 1990, I am particularly pleased to introduce this excellent reference source about geomorphology of Slovakia to the global readership.

Wrocław, Poland

Piotr Migoń
Series Editor

Introduction

The forms of the land can be well seen only through the eye of understanding, which shows them to be the product of agencies that have operated for long periods of past time.

W. M. Davis (1900)

The book *Landscapes and Landforms of Slovakia* is not intended to act as textbook describing all geomorphological regions. Rather, considering geomorphological particularities and diversity of the country the book aims at providing an overview of the unique landscapes of Slovakia from both scientific and scenic viewpoint. The volume is divided into three parts. In the first part, it provides background of geomorphological diversity of Slovakia, short history of landforms research, maps of geomorphological division and landscapes presented in this book (Chap. 1). Then, it deals with geological and tectonic development (Chap. 2), the past and recent climatic aspects (Chap. 3) and geomorphological history of the Slovak landscape (Chap. 4). Part II includes 16 chapters, among which 14 illustrate different landscapes ranging from the glacially shaped high mountains, through the symmetrical mountain ridge, relicts of volcanic terrain to aeolian and inland delta lowlands, the unique braided-meandering river patterns, fluvial imprints on the Flysch valley bottoms, passing through hilly lands of intra-montane basins, highlands of limestone Klippen belt and karst plateaux. Two last chapters (Chaps. 19 and 20) of part II deal with gullies and landslides, which are relatively small landforms, but very widespread across the country. In this way, they complete the images of regions susceptible to the occurrence of such natural hazards. Part III is concerned with and provides to the readers the main aspects of geoheritage, historical cultural landscapes, geo-protection and geotourism in Slovakia. They show the history of landscape protection in the country, main protected areas and policy to promote geodiversity and geoheritage to people and accelerate tourism focused on valuable landforms.

This book is the result of collaboration of geomorphologists, geologists and geographers, and also included the participation of experts from climatology. Forty-one authors from 13 research centres, including the Slovak Academy of Sciences, universities, governmental institutes, and emeritus scientists, have contributed to the book.

I am very grateful to Prof. Piotr Migoń, Series Editor, for offering us to participate in the editorial work of this book and for his continuous support. My special thanks go to the individual authors for the enthusiasm with which they responded to my invitation, and for the outstanding efforts made for the success of this important editorial initiative.

I also acknowledge the technical support of Dr. Miloš Rusnák and Dr. Anna Kidová. I would like to warmly thank all contributors. It was a real pleasure to cooperate on such an exceptional book describing geomorphological and landscape peculiarities and beauties of Slovakia.

Milan Lehotský
geogleho@savba.sk

Reference

Davis VM (1900) Physical geography in the high school. The School Review 8(7): 388–404

Contents

Editors and Contributors

About the Editors

Milan Lehotský is a physical geographer and fluvial geomorphologist at the Institute of Geography of the Slovak Academy of Sciences. He was many years head of the Department of Physical Geography, Geomorphology and Natural Hazards. His research topics are responses of fluvial systems to environmental changes, sedimentological connectivity, evolution trajectories, hydromorphology and GIS and remote sensing applications in rivers and landforms research. He is also working as an external lecturer at the Department of Physical Geography and Geoecology, the Faculty of Natural Sciences of the Comenius University in Bratislava.

Martin Boltižiar is a Professor of Geography and head of Department of Geography, Geoinformatics and Regional Development of the Constantine the Philosopher University in Nitra. He is working also as a senior scientist at the Institute of Landscape Ecology of the Slovak Academy of Sciences. Vice-president of the Slovak Geographical Society and scientific secretary of the Slovak Ecological Society. Member of the Slovak National Geographical Committee. He is specializing in physical and regional geography, landscape ecology, high-mountain geomorphology, environmentalistics and geoinformatics, researching land-use/land cover changes based on remote sensing data and historical maps using the GIS technology.

Contributors

Dušan Barabas Institute of Geography, Faculty of Natural Sciences, Pavol Jozef Šafárik University in Košice, Košice, Slovak Republic

Pavel Bella Slovak Cave Administration, Liptovský Mikuláš, Slovakia;
Department of Geography, Faculty of Education, Catholic University in Ružomberok, Ružomberok, Slovakia

Martin Boltižiar Department of Geography, Geoinformatics and Regional Development, Faculty of Natural Sciences, Constantine the Philosopher University in Nitra, Nitra, Slovakia; Institute of Landscape Ecology, Slovak Academy of Sciences, Bratislava, branch Nitra, Nitra, Slovakia

Ján Bóna Institute of Geography, Faculty of Natural Sciences, Pavol Jozef Šafárik University in Košice, Košice, Slovak Republic

Libor Burian Department of Physical Geography and Geoinformatics, Faculty of Natural Sciences, Comenius University in Bratislava, Bratislava, Slovakia

Ľudovít Gaál Slovak Cave Administration, Liptovský Mikuláš, Slovakia

Bohuslava Gregorová Department of Geography and Geology, Matej Bel University, Banská Bystrica, Slovakia

Zdenko Hochmuth Institute of Geography, Faculty of Natural Sciences, Pavol Jozef Šafárik University, Košice, Slovakia

Juraj Holec Slovak Hydrometeorological Institute, Bratislava, Slovakia

Mikuláš Huba Department of Human and Regional Geography, Institute of Geography, Slovak Academy of Sciences, Bratislava, Slovakia;
Institute of Management, Slovak University of Technology, Bratislava, Slovakia

Jozef Hók Department of Geology and Palaeontology, Faculty of Natural Sciences, Comenius University, Bratislava, Bratislava, Slovakia

Jozef Jakál Department of Physical Geography, Geomorphology and Natural Hazards, Institute of Geography, Slovak Academy of Sciences, Bratislava, Slovakia

Kacper Jancewicz Institute of Geography and Regional Development, University of Wrocław, Wrocław, Poland

Marián Jenčo Department of Physical Geography and Geoinformatics, Faculty of Natural Sciences, Comenius University in Bratislava, Bratislava, Slovakia

Anna Kidová Department of Physical Geography, Geomorphology and Natural Hazards, Institute of Geography, Slovak Academy of Sciences, Bratislava, Slovakia

Štefan Koco Department of Geography and Applied Geoinformatics, Faculty of Humanities and Natural Sciences, University of Presov, Presov, Slovakia

Michal Kováč Department of Geology and Palaeontology, Comenius University, Bratislava, Slovakia

Peter Labaš Department of Physical Geography, Geomorphology and Natural Hazards, Institute of Geography of the Slovak Academy of Sciences, Bratislava, Slovakia

Ján Lacika Department of Geography, Geoinformatics and Regional Development, Faculty of Natural Sciences, Constantine the Philosopher University in Nitra, Nitra, Slovakia

Milan Lapin Comenius University, Bratislava, Slovakia

Milan Lehotský Department of Physical Geography, Geomorphology and Natural Hazards, Institute of Geography, Slovak Academy of Sciences, Bratislava, Slovakia

Pavel Liščák State Geological Institute of Dionýz Štúr, Bratislava, Slovakia

Juraj Littva Slovak Cave Administration, Liptovský Mikuláš, Slovakia

Juraj Maglay State Geological Institute of Dionýz Štúr, Bratislava, Slovakia

Marián Melo Faculty of Mathematics, Physics and Informatics, Department of Astronomy, Physics of the Earth, and Meteorology, Comenius University, Bratislava, Slovakia

Piotr Migoń Institute of Geography and Regional Development, University of Wrocław, Wrocław, Poland

Jozef Minár Department of Physical Geography and Geoinformatics, Faculty of Natural Sciences, Comenius University in Bratislava, Bratislava, Slovakia

Mário Molokáč Faculty of Mining, Ecology, Process Control and Geotechnologies, Institute of Earth Resources, Technical University of Kosice, Košice, Slovakia

Ján Novotný Department of Physical Geography, Geomorphology and Natural Hazards, Institute of Geography, Slovak Academy of Sciences, Bratislava, Slovakia

Pavol Papčo Department of Geography, Faculty of Education, Catholic University in Ružomberok, Ružomberok, Slovakia

Peter Pauditš State Geological Institute of Dionýz Štúr, Bratislava, Slovakia

Jozef Pecho Slovak Hydrometeorological Institute, Bratislava, Slovakia

Dušan Plašienka Department of Geology and Palaeontology, Faculty of Natural Sciences, Comenius University, Bratislava, Slovakia

Wioleta Porębna Institute of Geography and Regional Development, University of Wrocław, Wrocław, Poland

Juraj Procházka Department of Physical Geography and Geoinformatics, Faculty of Natural Sciences, Comenius University, Bratislava, Slovakia

Miloš Rusnák Department of Physical Geography, Geomorphology and Natural Hazards, Institute of Geography, Slovak Academy of Sciences, Bratislava, Slovakia

Ján Sládek Department of Physical Geography, Geomorphology and Natural Hazards, Institute of Geography, Slovak, Academy of Sciences, Bratislava, Slovakia

Miloš Stankoviansky Department of Physical Geography and Geoinformatics, Faculty of Natural Sciences, Comenius University in Bratislava, Bratislava, Slovakia

Ján Urbánek Department of Physical Geography, Geomorphology and Natural Hazards, Institute of Geography, Slovak Academy of Sciences, Bratislava, Slovakia

Ladislav Vitovič Department of Physical Geography and Geoinformatics, Faculty of Natural Sciences, Comenius University in Bratislava, Bratislava, Slovakia;
State Geological Institute of Dionýz Štúr, Bratislava, Slovakia

Rastislav Vojtko Department of Geology and Palaeontology, Comenius University, Bratislava, Slovakia

Ladislav Šimon State Geological Institute of Dionýz Štúr, Bratislava, Slovakia

Ľubomír Štrba Faculty of Mining, Ecology, Process Control and Geotechnologies, Institute of Earth Resources, Technical University of Kosice, Košice, Slovakia

Short History of Geomorphological Research and Geomorphological Division of Slovakia

1

Milan Lehotský, Miloš Rusnák, and Ján Novotný

Abstract

Slovakia is a small landlocked country in Central Europe. Its position is characterized by the intersection of major Eurasian tectonic and climatic domains, which predetermines general features of the Slovak landscape and the occurrence of specific landforms. The first work on the landscape, history and people of Slovakia, "*Notitia Hungariae Novae Historico Geographica*" (1735–1742) and its abridged version "*Compendium Hungariae geographicum*" (1753) was written by Matej Bel. The pioneer study dealing with the geomorphological development of Slovakia is the book by Jan Hromádka, "*Všeobecný zemepis Slovenska*" (Hromádka in Všeobecný zemepis Slovenska. Slovenská akadémia vied a umení, Bratislava, 256 pp, 2013). The strong input to the research of the physical landscape of Slovakia was made in 1950–1960s. A history of the geomorphological division of Slovakia is shortly described and the map of geomorphological division of Slovakia and 14 landscapes portrayed in individual chapters of the book are presented.

Keywords

Slovakia • Geomorphology • Research • Division

1.1 Geomorphological Research

Although Slovakia is a small landlocked country in Central Europe, it is a territory of outstanding variety of landforms, where, apart from scenic coastlines and smoking volcanos, rugged alpine mountains, narrow gorges, floodplains of large rivers, karst areas can be seen during one-day journey. The evolution of the Slovak landscape cannot be understood without taking into account differences between the large morphostructural regions, directly controlled by the geology and tectonics. The Central European position of Slovakia is characterized by the intersection of major Eurasian tectonic and climatic domains, which predetermines general features of the Slovak landscape and the occurrence of specific landforms. The Cenozoic fold-and-thrust belt of the Western Carpathians with intermontane basins meets here the Pannonian basin. The Tatra Mts. with glacial landforms are in contrast to the periglacial aeolian region in Western Slovakia and the Danube plain, crossed by the second largest European river. Landscape diversity in many regions of the country is tightly connected with human presence since ancient times and the variety of customs and traditions. Original landforms of some regions are parts of landscape heritage and are protected as geosites and geomorphosites.

The altitude of Slovakia ranges from 94.3 m a.s.l. (the Bodrog River floodplain near to the Slovak-Hungarian boundary) to 2655 m a.s.l. (Gerlachovský štít/Gerlach peak in the High Tatra Mts.). Hilly lands in intra-mountainous basins and margins of the plains dominate in Slovakia in terms of area (39%). Uplands and highlands occupy 26% and 22% of the Slovak area, respectively. Plains belonging to the Pannonian basin occupy 11% and the glacially shaped high mountains occur over 2% of the total area of Slovakia.

The first work on the landscape, history and people of Slovakia, "*Notitia Hungariae Novae Historico Geographica*" (1735–1742) and its abridged version "*Compendium Hungariae geographicum*" (1753), was written by Matej Bel, with the contribution of a number of important personalities of the Kingdom of Hungary, which Slovakia belong to at that time. The pioneer study dealing with the geomorphological development of Slovakia is the book by Jan Hromádka, "*Všeobecný zemepis Slovenska*" (1943). Hromádka acknowledged the theory of geographical cycle by W.M. Davis, and studied relief in association with other geographic factors, including the role of humans. During the study stay in Paris (1931–1932, with A. Demangeon and E.

M. Lehotský (✉) · M. Rusnák · J. Novotný
Department of Physical Geography, Geomorphology and Natural Hazards, Institute of Geography Slovak, Academy of Sciences, Bratislava, Slovakia
e-mail: geogleho@savba.sk

M. Lehotský and M. Boltižiar (eds.), *Landscapes and Landforms of Slovakia*, World Geomorphological Landscapes,
https://doi.org/10.1007/978-3-030-89293-7_1

de Martonne) he attempted regionalization of the territory, which resulted in the first geomorphological division of the Slovakia. The strong input to the research of the physical landscape of Slovakia was made in 1950–1960s by Slovak geomorphologists, mainly by Michal Lukniš, Emil Mazúr, Jozef Kvitkovič, Jozef Jakál and Ján Urbánek, and resulted in the series of geomorphological maps published in the "Atlas of Slovak Socialist Republic" (1980). Their research activity was followed by a younger generation of Slovak and Czech geomorphologists and stimulated the launch of "*Geomorphologia Slovaca et Bohemica*" journal in 2001.

1.2 Geomorphological Division of Slovakia

The geomorphological delimitation of the territory of Slovakia has a relatively long history and results from the general human need to name the observed objects, connect them with similar ones and distinguish them from different objects. From the point of view of geomorphology as a science, it is then both the subject of the research itself, as well as one of its basic tools, which allows the localization and characterization of research sites.

According to Mazúr and Lukniš (1978), we can distinguish three basic periods in the development of the geomorphological delimitation of Slovakia. The first was the laic distinction and naming of the most distinctive shapes of the relief, resulting from the need of the orientation in the living space of human society. The origin of traditional names such as the Carpathians, Pieniny, Tatra, Matra, Fatra, Beskyd, etc. dates back to this period. In the second phase, which took place in the second half of the nineteenth and in the beginning of the twentieth century, scholars working in the field of geography and geology tried to organize the shapes of the earth's surface into chains, but without their precise spatial definition. Among several, the work of the geologist Štúr (1862) is especially worth mentioning. The very delimitation of orographic units, the drawing of their boundaries into maps and their systematic classification is the content of the third phase. Its foundations in Slovakia were laid mainly in the works of Hromádka (1943, 1956), in which several principles of division were presented, valid in a modified form to this day. It was based on older knowledge, but mainly on knowledge of the field and analyses of the topographic maps. Among the most significant innovations brought by J. Hromádka are the introduction of a multi-level hierarchical classification (at the level of today's provinces, sub-provinces and regions), perception of mountain ranges and inter-mountain basins as equal units, drawing boundaries between units in places of the greatest change of gradient (i.e. foothill lines), as well as introduction of the typology of relief, both in the definition of units and in their nomenclature. In his regional works, Lukniš (1972) and Lukniš and Plesník (1961) followed the progressive conception of Hromádka's division of the relief of Slovakia.

On the basis of works of Hromádka, the most comprehensive *Regional Geomorphological Delimitation of Slovakia* (Mazúr and Lukniš 1980; Mazúr 1980; Mazúr and Lukniš 1986) was subsequently created, which is still valid today and is one of the basic and widely used works of Slovak geography. Its primary goal was to define surface units in Slovakia as unique individuals, so it was not a typological or genetic division, but individual regionalization (Mazúr and Lukniš 1978). Therefore, in addition to the typology of relief, the authors also relied on knowledge about the geological structures and morphostructures of the area, as well as their own knowledge based on detail field mapping. The manuscript maps corresponded to a scale of 1:200 000 and the work was published in various versions at a scale of 1:500 000 (Mazúr and Lukniš 1980; Mazúr 1980; Mazúr and Lukniš 1986).

The criteria for location of the geomorphological boundaries had been contacting lines or boundaries between the different relief types such as foothill lines, terrain edges, saddles, morphostructural and morphosculptural lines, etc. The resulting delimitation has eight hierarchical levels. At the highest level, the entire territory of Slovakia belongs to the Alpine-Himalayan system, within which two sub-systems are distinguished—the Carpathians and the Pannonian Basin. Within the Carpathians, the territory of Slovakia is covered by two provinces—the Western Carpathians and the Eastern Carpathians. The Western Carpathians have the largest extent and are further divided into the Outer Western Carpathians and the Inner Western Carpathians at the level of sub-provinces. Within the territory of Slovakia, each of these sub-provinces is then divided into five regions (Fig. 1.1). The Eastern Carpathians are represented in Slovakia by two sub-provinces and three regions. The Pannonian Basin is divided into two provinces—the West Pannonian Basin and the East Pannonian Basin, and a total of three sub-provinces and four regions cover the territory of Slovakia (Fig. 1.1).

The fundamental part of the geomorphological delimitation is the sixth hierarchical level—geomorphological units, which represent individual mountain ranges, highlands, uplands, basins, etc. These are the most traditionally perceived elements of the relief, the identification and spatial definition of which are usually the least problematic and largely follow the older systems of classification of geomorphological units in Slovakia. A total of 84 units were allocated, which are divided into subunits (263) and parts (237) at the lowest levels.

The geomorphological delimitation of Mazúr and Lukniš corresponded to the state of knowledge and the requirements of the time in which it originated. As stated by Urbánek et al. (2009a), it is an issue that needs to be continuously updated

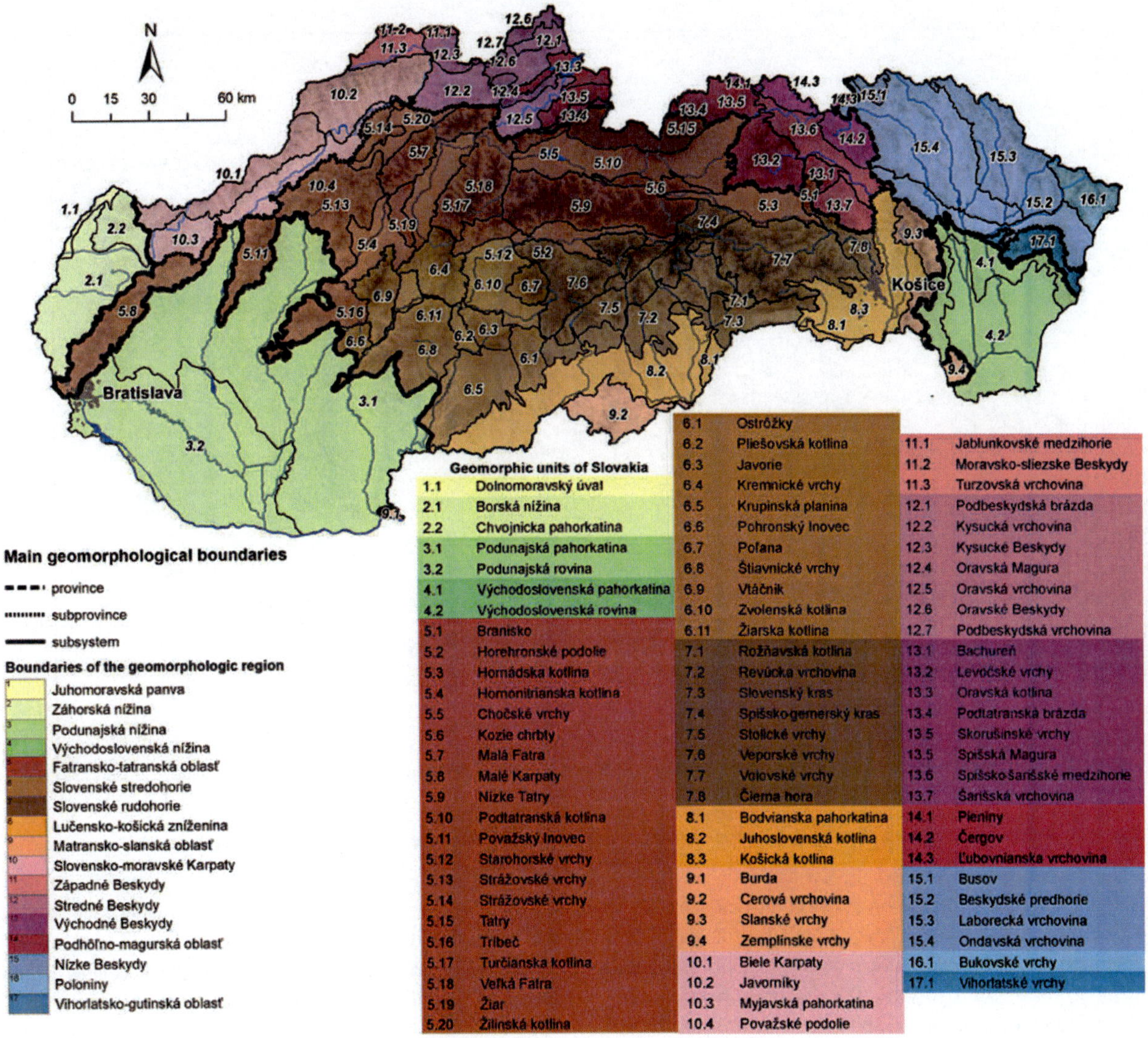

Fig. 1.1 Geomorphological division of Slovakia, based on Mazúr and Lukniš (1978), with 17 geomorphic region ("*geomorfologická oblasť*" in Slovak) and 84 geomorphological unit ("*geomorfologická jednotka*" in Slovak) represented by individual mountain ranges, highlands, uplands, basins, etc

in the context of expanding knowledge in the field of geology and geomorphology, as well as in relation to technological progress and the resulting possibilities of mapping and cartographic expression. The development of geographic information systems has resulted in the need to adapt the geomorphological delimitation for the needs of its display and analyses in the digital environment. Urbánek et al. (2009a) characterize several areas that represent challenges for such updates. As they state, the method used for the production of the Regional geomorphological delimitation map remained uncertain and the criteria for location of the boundaries in a particular area are not clearly defined. In several cases, the identification and interpretation of boundaries of different hierarchical levels, especially at the fifth level of the regions, is also problematic. The authors also point out the uneven and inhomogeneous division of units at the lowest hierarchical levels. In addition, many geomorphological units lie on the borders with neighboring states and the harmonization of individual national delimitation systems is complicated (e.g., Bandura et al. 2019).

These facts motivated several authors to try to update the map of Mazúr and Lukniš and to create a digital database of geomorphological delimitation of Slovakia. The most important work can be considered the digital database *Regional Geomorphological Division of Slovakia, 1:50 000* (Urbánek et al. 2009b), which was created at the Institute of

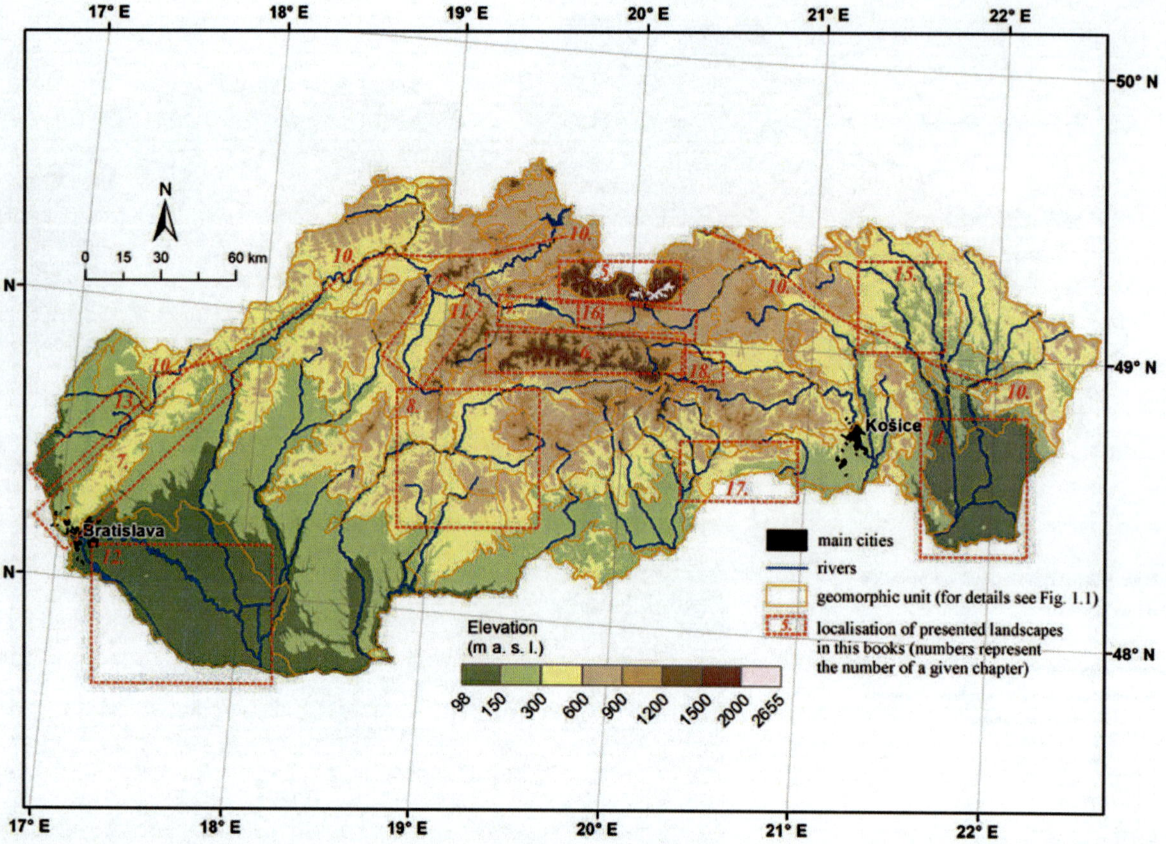

Fig. 1.2 Location of landscapes presented in this book, with numbers of respective chapters

Geography of the Slovak Academy of Sciences. It was elaborated by interpreting the boundaries of the original delimitation so that their course corresponds to topographic maps at a scale of 1:50 000. The database therefore corresponds in content to the delimitation of Mazúr and Lukniš, but the course of the boundaries is more accurate in detail. The resulting course of the border lines is thus still subject to expert evaluation of all relevant knowledge. In addition to this database, digital maps of geomorphological division were created, applying different methodological procedures and semi-automatized approach, based on a DMR (e.g. Kočický and Ivanič 2011; Bandura et al. 2019).

The simplified and generally accepted scheme of geomorphological division according to Mazúr and Lukniš (1980) is presented in Fig. 1.1, where geomorphic regions and geomorphological units are shown.

The book, apart from chapters dealing with general Slovak landscape descriptions and descriptions of kind of dispersed landforms, presents 14 individual landscapes which according geomorphological division represent either geomorphologic regions or geomorphologic units and their localities are depicted in Fig. 1.2.

References

Bandura P, Minár J, Drăguţ L (2019) Morphometrical-morphostructural subdivision of the Western Carpathians by Object-based image analysis. Geomorphologia Slovaca Et Bohemica 19(1):1–101

Hromádka J (1943) Všeobecný zemepis Slovenska. Slovenská akadémia vied a umení, Bratislava, 256 pp

Hromádka J (1956) Orografické třídení Československé republiky. Sborník Československé Společnosti Zeměpisné 11(3):161–299

Kočický D, Ivanič B (2011) Geomorfologické členenie Slovenska. https://apl.geology.sk/temapy/. Accessed 27 May 2021
Lukniš M (1972) Reliéf. In: Lukniš, M (ed). Slovensko 2 – Príroda, Obzor, Bratislava, pp 124–202
Lukniš M, Plesník P (1961) Nížiny, kotliny a pohoria Slovenska. Osveta, Martin, 134 pp
Mazúr E (1976) Morphostructural features of the West Carpathians. Geografický Časopis 28(2):101–111
Mazúr E (ed) (1980) Atlas slovenskej socialistickej republiky. Slovenská Akadémia Vied, Slovenský úrad geodézie a kartografie, 296 pp
Mazúr E, Lukniš M (1978) Regionálne Geomorfologické Členenie SSR. Geografický Časopis 30(2):101–125
Mazúr E, Lukniš M (1980) Regionálne geomorfologické členenie (1:500 000). Geografický ústav SAV, Bratislava
Mazúr E, Lukniš M, Balatka B, Loučková J, Sládek J (1986) Geomorfologické členenie SSR a ČSSR. Mapa 1:500 000. Slovenská kartografia, Bratislava
Štúr D (1862). Geologicko-geografická osnova polohopisu Slovenska. Sokol (reedícia 1960 vydavateľstvo SAV, Bratislava)
Urbánek J, Beták J, Jakál J, Lacika J, Novotný J (2009a) Regional geo morphological division of Slovakia: old problem in new perspectives. Geographia Slovaca 26:237–259
Urbánek J, Jakál J, Lacika J, Beták J, Novotný J (2009b) Digitálna databáza Regionálne geomorfologické členenie Slovenska, 1 : 50 000. Geografický ústav SAV

Milan Lehotský is a physical geographer and fluvial geomorphologist at the Institute of Geography of the Slovak Academy of Sciences. He was many years head of the Department of Physical Geography, Geomorphology and Natural Hazards. His research topics are responses of fluvial systems to environmental changes, sedimentological connectivity, evolution trajectories, hydromorphology and GIS and remote sensing applications in rivers and landforms research. He is also working as an external lecturer at the Department of Physical Geography and Geoecology, the Faculty of Natural Sciences of the Comenius University in Bratislava.

Miloš Rusnák is a fluvial geomorphologist at the Institute of Geography of the Slovak Academy of Sciences (Department of Physical Geography, Geomorphology and Natural Hazards). His research topics are fluvial geomorphology, spatial data processing in GIS, UAV data acquisition and processing, fluvial processes and sediment connections in gravel-bed rivers and remote sensing applications in rivers and landforms research. He is the author and co-author of several papers dealing with fluvial system evolution in the Outer Western Carpathians.

Ján Novotný is a geomorphologist at Institute of Geography, Slovak Academy of Sciences,Bratislava. His research interests focus especially on morphostructures, fluvial geomorphology, geomorphic division of the Western Carpathians and geodiversity.

Part I
Physical Environment

2 Outline of Geology and Cenozoic Evolution of Slovakia

Rastislav Vojtko, Dušan Plašienka, and Michal Kováč

Abstract

The territory of Slovakia is located in Central Europe and is built up by the Western Carpathian mountain range to the north and the overstepping Cenozoic Basin system in the central and southern parts. The Western Carpathians are an integral part of the northern branch of the Carpathian arc and the whole European Alpine orogenic system, which was formed by the closure of the Tethys and Alpine Atlantic oceans during the Mesozoic and Cenozoic times. The Western Carpathians can be longitudinally subdivided into three major tectonic zones, namely, the External, Central, and Internal Western Carpathians. The External Western Carpathians (the Carpathian Flysch Belt) cover the northern territory of Slovakia and represent the Paleogene to Neogene accretionary complex. The Central Western Carpathians, covering the majority of the Slovak territory, include numerous Late Jurassic to Cretaceous nappe units, which are divided from the Carpathians Flysch Belt by the narrow Pieniny Klippen Belt. The Central Western Carpathians include, from the bottom to the top, the Tatric, Veporic, and Gemeric units composed of predominantly Variscan basement and its Upper Paleozoic to Mesozoic sedimentary cover. The thick-skinned thrust sheets are covered by detached thin-skinned nappes of the Fatric, Hronic, and Silicic thrust systems. In the south, the Internal Western Carpathians are represented by the Jurassic accretionary wedge composed of the Meliata, Turňa, and Silica nappes. The Alpine nappe system of the Western Carpathians is covered by sediments deposited in various Upper Cretaceous, Paleogene, and Neogene Basins and volcanic formations. The present landscapes and landforms of Slovakia are a result of many forces, like the active tectonics, selective erosion and denudation of the elevated parts of the Earth surface, but also the transport and deposition of sediments into depressions. This process can be traced back during the Quaternary time (up to 2.6 Ma), and perhaps very weakly during the Neogene period (up to 23 Ma). For the older Cenozoic and Mesozoic periods, it is not possible to document landscape, but we prepared palaeogeographical models that lack data and a strong generalisation increase with the shift to the past. Such palaeogeographical maps or 2.5D models can than inform us about the occurrence of mountain ranges, lowlands, but also rivers, lakes and seas or oceans in these ancient periods from a geological record.

Keywords

Western Carpathians • Pannonian Basin • Geological structure • Mountain building processes • Palaeogeography

2.1 Introduction

The territory of Slovakia is entirely contained within the Western Carpathians, which are the northernmost segment of the European Alpine orogenic system (Fig. 2.1). They form a northward-convex arc, which is approximately 500 km long in the W–E direction and 300 km wide across. Taking into consideration the pre-Cenozoic units, the Western Carpathian area includes not only the mountainous regions mostly in its northern part, but also the subsurface of wide flat lowlands of the Pannonian Basin system to the south. The western limit of the mountains next to the Eastern Alps is conventionally placed in the basement of the Vienna

R. Vojtko (✉) · D. Plašienka · M. Kováč
Department of Geology and Palaeontology, Comenius University, Ilkovičova 6, 84215 Bratislava, Slovakia
e-mail: rastislav.vojtko@uniba.sk

D. Plašienka
e-mail: dusan.plasienka@uniba.sk

M. Kováč
e-mail: kovacm@uniba.sk

M. Lehotský and M. Boltižiar (eds.), *Landscapes and Landforms of Slovakia*, World Geomorphological Landscapes,
https://doi.org/10.1007/978-3-030-89293-7_2

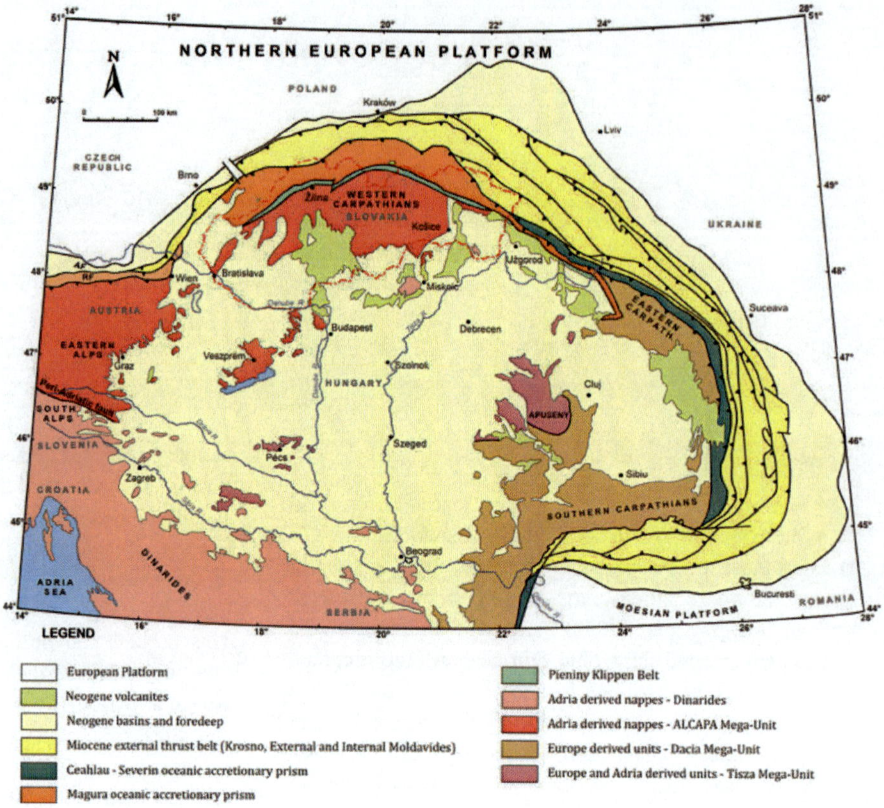

Fig. 2.1 Tectonic scheme with major tectonic units of the Alps, Carpathians, and Dinarides (according to Kováč 2000 and Schmid et al. 2008; modified). *Note* Red dotted line represents the state boundary of Slovakia

and Danube Basins; the geographical boundary is placed in the so-called Carnuntum Gate between the Leitha and Hundsheim Mountains. North of Vienna this boundary roughly corresponds to the border of the Rhenodanubian Flysch and the Waschberg Zone. The arcuate northern limit of the Western Carpathians follows the boundary between the most external Western Carpathian units (including the foredeep) and the foreland of Northern European Platform composed of various pre-Alpine units and their Meso-Cenozoic sedimentary cover (Figs. 2.2 and 2.3).

The eastern boundary of the Western Carpathians with respect to the Eastern Carpathians is not so clear, because there is a distinct division from the geographical and geological points of view. Conventionally, the boundary approximately follows the north–south trending Polish and Slovak border with the Ukraine. However, the official geographic boundary of the Western and Eastern Carpathians is located further to the west. The extensive Rhenodanubian-Magura Unit of the External Western Carpathians and the Central Western Carpathian system wedge out in this area. In contrast, the outermost, typically Eastern Carpathian Stebnik, Sambor-Rozniatov, and Borislav-Pokuty units first appear and these widen to the southeast. However, some other large Western Carpathian units (Skole-Skiba-Tarcău, Silesian-Chornohora, and Dukla units, Pieniny Klippen Belt) continue southeastwards without any considerable changes. The conventional southern boundary of the pre-Cenozoic Western Carpathian units follows the SW–NE trending, broad Mid-Hungarian Fault Zone, which is linked with the Peri-Adriatic Fault in the Eastern and Southern Alps. This complex Cenozoic fault zone juxtaposes the southernmost Western Carpathian elements against the largely subsurface Tisza (Tisia) lithospheric fragment in the southern part of the Pannonian Basin. However, all along this fault the contact is buried beneath a thick Neogene sedimentary cover (Fig. 2.1).

Geomorphologically, the Western Carpathians are divided into two principal parts—the mountainous Western Carpathians proper and the lowlands of the northern portion of the Pannonian Basin; i.e. the Východoslovenská a Podunajská nížina lowlands and the Great Hungarian Plain. The mountainous Western Carpathians are subdivided into the Outer Western Carpathians (mainly the Cenozoic Flysch Belt) and the Inner Western Carpathians (mainly formed by pre-Cenozoic complexes and Neogene volcanic structures), but their geographic boundary only partly corresponds to the major tectonic divide in the Western Carpathians, the Pieniny Klippen Belt. The Inner Western Carpathians are split into numerous horsts (so-called core mountains) separated by small intramontane basins of the Neogene age, and

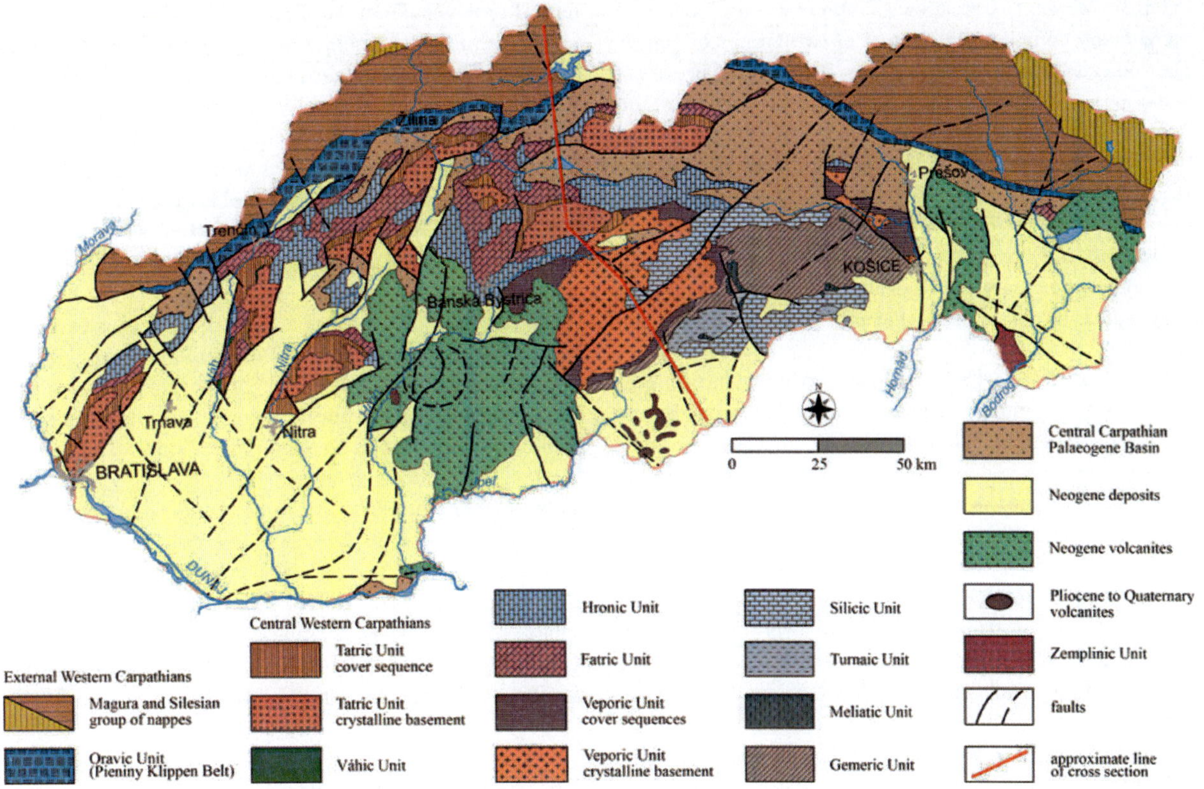

Fig. 2.2 Tectonic division of Slovakia (base map according to Biely et al. 1996)

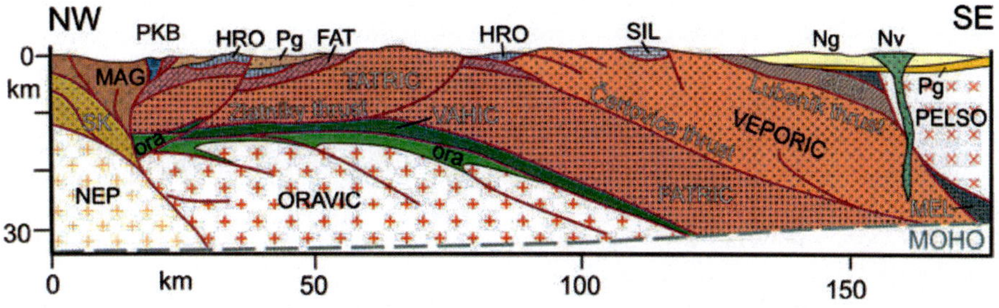

Fig. 2.3 Schematic cross section showing mutual relationships of the principal Western Carpathian tectonic units (according to Plašienka 2019, modified). Explanations: NEP—North European Platform; SK—Silesian–Krosno units of the External Western Carpathians; MAG—Magura Unit; ora—Oravic cover rocks; FAT—Fatric nappe system; HRO—Hronic nappe system; MEL—Meliatic accretionary wedge; SIL—Silicic nappes; GEM—Gemeric Unit; PZ—Internal Western Carpathian Paleozoic complexes; Pg—Paleogene overstepping sediments; Ng—Neogene deposits; Nv—Neogene volcanics

embayments of the Pannonian Basin System along the southern margin of the orogen (Fig. 2.2).

2.2 Tectonic Subdivision

Apart from the overstepping Cenozoic formations, the Western Carpathians can be longitudinally subdivided into three major tectonic zones, namely, the External, Central, and Internal Western Carpathians (Froitzheim et al. 2008; Plašienka 2018).

The External Western Carpathians cover the territories of northeasternmost Austria, southeastern Moravia (Czech Republic), northwestern and northeastern Slovakia, and southeastern Poland. They include the Carpathian foredeep, which is the eastern prolongation of the Alpine Molasse Basin, and the Carpathian Flysch Belt. Both continue along the entire Carpathian arc. The Carpathian Flysch Belt of the

External Western Carpathians represents the Cenozoic accretionary complex and consists of two groups of nappes. The outer, Moldavide tectonic system (Silesian-Krosno units) has no connections westwards into the Alps, but include a substantial part of the Eastern Carpathian Flysch Belt with the exception of the Outer Dacides (e.g. Tomek 1993; Picha et al. 2006). The inner, Magura Unit is considered a direct prolongation of the Rhenodanubian Flysch of the Eastern Alps. It wedges out at the Western/Eastern Carpathian boundary. Palaeogeographically, the Magura Unit represents the northern Penninic realm (Figs. 2.2 and 2.3).

The Central Western Carpathians, covering the majority of the Slovak territory, include various pre-Cenozoic units and the overstepping Upper Mesozoic to Cenozoic sedimentary deposits and volcanic complexes (Fig. 2.1). The former is distributed in several longitudinal morphotectonic belts. The narrow Pieniny Klippen Belt represents the boundary between the External and Central Western Carpathians. It includes units of the (Middle) Penninic provenance (Oravic Unit) and units probably derived from the Central Western Carpathian tectonic system (e.g. Andrusov 1974; Birkenmajer 1986; Plašienka 2018). Areas to the south of the Pieniny Klippen Belt are composed of numerous Upper Jurassic to Cretaceous nappe units, which represent the principal structural elements of the Central Western Carpathians. The Tatra-Fatra Belt of core mountains includes the lowermost Tatric Unit composed of predominantly Variscan basement and its Upper Paleozoic to Mesozoic sedimentary cover and detached thin-skinned nappes of the Fatric and Hronic thrust systems. The Vepor–Gemer Belt to the south is composed of a stack of basement-dominated north-verging thick-skinned Veporic and Gemeric basement/cover thrust sheets and the Jurassic accretionary wedge composed of the Bôrka, Meliata, Turňa, and Silica nappes. The Veporic and Gemeric units gradually merge towards the south under the Cenozoic sediments and Miocene volcanic structures, making it difficult to follow the boundary between the Central and Internal Western Carpathians (Figs. 2.1 and 2.2).

The so-called Pelso Mega-Unit (e.g. Kovács 1992; Haas et al. 1995) exposed mostly in isolated mountains (inselbergs) in N Hungary (Transdanubian Range, the Bükk and surrounding mountains and the Aggtelek-Rudabánya Mountains) represents the Internal Western Carpathians (Fig. 2.2). The oceanic complexes of the Meliata Unit represent the supposed Central/Internal Western Carpathian boundary, but superimposed nappe units and unconformable cover rocks mostly obliterate this suture. The Internal Western Carpathians mainly comprise unmetamorphosed or very low grade metamorphosed Paleozoic and Mesozoic complexes that form a south-directed fold-and-thrust belt. In terms of palaeogeography, the Pelso Mega-Unit is closely related to the southern Austroalpine or the Southern Alpine facies realm (Transdanubian Range), or even to that of the Dinarides (Bükk Mountains; e.g. Schmid et al. 2008).

In the plate tectonic terms, the External Western Carpathians correspond to a Cenozoic accretionary complex related to the southward subduction of the Northern Penninic (Magura), and possibly also the Moldavide, oceanic realms. The Pieniny Klippen Belt forms a narrow steep transpressional zone between the accretionary wedge and the Central Western Carpathians representing the wedge backstop (Plašienka et al. 2020). It is also the boundary between the Penninic and Austroalpine-related units in the Western Carpathians (Figs. 2.1 and 2.2). Accordingly, it is a fossil plate boundary (i.e. suture) although surface evidence for ophiolite complexes is lacking. The Central and Internal Western Carpathians represent the stacks of crustal-scale units within the late Mesozoic collisional pro- and retro-wedge, respectively, shaped by the latest Jurassic to mid-Cretaceous subduction-collision processes related to the closure of the Triassic–Jurassic Meliata Ocean. The ophiolite-bearing Meliata suture is located approximately in a central position between the pro- and the retro-wedge of the Western Carpathian orogenic system.

2.3 Palaeogeographic Subdivision

From the point of view of Mesozoic–Cenozoic palaeogeography, two principal evolutionary periods have to be distinguished in the Western Carpathians. During the first one (Triassic-Jurassic), units of the External and Central Western Carpathians were located on the northern, rifted, passive European margin of the Meliata Ocean (Schmid et al. 2008). In contrast, the Internal Western Carpathian elements formed the southern (in present coordinates) passive (Triassic) and later active (Jurassic) margin of the ocean, corresponding to the northeastern part of Apulia-Adria (Plašienka 2018). The Central and Internal Western Carpathians were welded together by Jurassic/Cretaceous boundary time as a result of closure of the Meliata Ocean. From this time onward, the Western Carpathian Cretaceous palaeogeography generally corresponds to that of the Alps, including the following zones from north to south:

- European continental margin, largely overridden by the Western Carpathian units during the Cenozoic;
- Moldavide Basins (partly oceanic?) in the north to northeast only, and widening eastwards;
- Elongated continental fragment (the Silesian Ridge);
- The Northern Penninic-Magura Oceanic Basin;
- Rifted continental fragment in a Middle Penninic position (the Oravic or Czorsztyn Ridge);

- The Southern Penninic-Vahic Oceanic Zone;
- Broad and dissected Adria-related continental margin including Central Western Carpathian (Austroalpine) and Internal Western Carpathian (Upper Austroalpine-Southern Alpine-Dinaridic) units.

2.4 Outline of the Palaeogeographic and Palaeotectonic Evolution

2.4.1 Palaeoalpine Tectonic Evolution

The Alpine edifice of the Western Carpathians was formed on the remnants of the Variscan orogenic belt of Central Europe (e.g. Bielik et al. 2004). The Variscan orogeny culminated in the late Early to early Late Carboniferous collisional processes. Post-Variscan unconformable successions include marine Pennsylvanian and continental Permian clastic sedimentary successions. During the Permian, several narrow rift basins were formed, which were filled with continental red beds. Some of these were accompanied by alkaline to calc-alkaline, acid to intermediate (locally also mafic) volcanism (Vozárová and Vozár 1988). Permian to Lower Triassic granitoid intrusions include late orogenic S-type granites in the Gemeric Unit and several small, post-orogenic A-type intrusions in the southern zones of the Western Carpathians (e.g. Poller et al. 2002). In the Early Triassic, the region had become a peneplain and most of the Western Carpathian area was covered by mature continental siliciclastics subsequently overlain by lagoonal or sabkha deposits. However, the site of the later Meliata rift was already marked by the presence of much thicker Lower Triassic deposits.

Gradual subsidence of the European shelf during the Anisian is reflected by the deposition of widespread carbonate ramp and platform deposits, with intervening narrow intra-shelf basins. Early Mesozoic Neo-Tethyan rifting culminated in the opening of the Meliata Ocean in the early Middle Triassic (as indicated by the Pelsonian break-up unconformity in adjacent areas; Kozur 1991). The Meliata Ocean has been interpreted as a back-arc basin related to the northward subduction of Palaeo-Tethys (Stampfli and Kozur 2006). This ocean subsequently separated the stable European shelf in the north from the mobile Adria-related continental fragments in the southwest. The broad northern shelf exhibited only restricted subsidence during the Triassic. Epicontinental successions similar to those of the German Basin were deposited (e.g. the Upper Triassic Carpathian Keuper Formation). Along the edge of the northern shelf; however, extensive, reef-cored Middle-Upper Triassic carbonate platforms developed (Kovács et al. 2011). These reflect the thermal subsidence of the Meliata Ocean flanks. In contrast, the southern Meliata margin records Tethyan type, mostly deep-water pelagic sedimentation, since deposition did not keep pace with subsidence. In the axial zones of the Meliata Ocean, deep-marine sedimentation below the calcite compensation depth predominated. Ladinian and Upper Triassic radiolarites today occur as blocks within the Jurassic olistostromes and mélanges, together with dismembered ophiolite fragments (Mock et al. 1998; Plašienka et al. 2019).

From the Triassic/Jurassic boundary onwards, important palaeogeographic changes occurred. The northern shelf experienced several strong rifting phases that led to the disintegration of the Upper Triassic carbonate platform and ultimately resulted in the opening of new, Penninic-related oceanic basins. The Central Carpathian domain was separated from the Northern European Platform by the Middle Jurassic break-up of the Southern Penninic-Vahic Ocean, and was dissected into several longitudinal, wide, subsiding basins, floored by extended lithosphere and filled with deep-water, predominantly calcareous pelagic sediments. These Jurassic-Lower Cretaceous basins were separated by narrow, occasionally emergent ridges with shallower and/or condensed sedimentation (Plašienka 2018).

The southward subduction of the Meliata Ocean probably commenced during the late Early Jurassic and its closure occurred in the latest Jurassic. Subsequent shortening migrated from the collision zone to both sides. During the Early Cretaceous, the Internal Western Carpathian retro-wedge (Pelso Mega-Unit) was formed by generally southward thrusting and obduction of a small Upper Jurassic back-arc oceanic basin, as indicated by the presence of the Szarvaskő ophiolites of the Bükk Mountains and by back-thrusting and formation of a retro-arc, ophiolite detritus-bearing deep-marine clastic basin in the Transdanubian Range. Subsequently, the units of the Pelso Mega-Unit were unconformably overlain by the Senonian (Gosau Group) and by Eocene to Lower Miocene epicontinental deposits of the Northern Hungarian (Buda) Basin (Kováč et al. 2016).

2.4.2 Eoalpine and Neoalpine Tectonic Evolution

The Central Western Carpathian pro-wedge grew significantly during the late Early and early Late Cretaceous. Considerable shortening of the Central Western Carpathian crust, attenuated by previous Jurassic rifting, took place and the principal thick- and thin-skinned nappe sheets were thrust northward (e.g. Putiš et al. 2009). Mid-Cretaceous deep-marine clastic basins developed in front of the advancing basement thrust wedges. The Cretaceous nappe stack in the Central Western Carpathians was completed by

Fig. 2.4 A winter view of Mt. Kráľova hoľa from Mt. Fabova hoľa. The foreground mountains of the Veporic Unit are represented by a gently modelled landscape

the Turonian and is unconformable overlain by the Senonian deposits, although these are only locally preserved. The thickened crust collapsed in the southern zones of the Central Western Carpathians in the Late Cretaceous, when the Veporic metamorphic core complex was exhumed by an orogen-parallel extensional unroofing (Janák et al. 2001; Jeřábek et al. 2007; Vojtko et al. 2016; Fig. 2.4).

North of the Central Carpathian domain, Penninic rifting occurred in three main phases (Plašienka 2012, 2018). The first Early Jurassic phase affected broad areas, but did not yet lead to the continental break-up. The Middle Jurassic localised asymmetric rifting led to the opening of the Southern Penninic-Vahic Ocean Basin, while the onset of oceanic crust production in the Northern Penninic-Magura Ocean probably began as late as in the Early Cretaceous. There is, however, no direct evidence for the oceanic nature of these basins, since no ophiolites are found. Southward-directed subduction of the Vahic oceanic crust commenced in Senonian times. The Oravic continental fragment began to collide with the Central Carpathian orogenic wedge at the Cretaceous-Paleogene boundary. Only detached Vahic, Oravic (Pieniny Klippen Belt), and Magura sediments were frontally accreted to the orogenic wedge, while the entire Penninic lithosphere, either oceanic or continental, was subducted (Plašienka and Soták 2015). Subduction of the Magura Ocean probably began during the Eocene and lasted until the earliest Miocene. The eastward-migrating subduction of the basement of the Moldavian units, followed by underthrusting of the European margin below the External Western Carpathian accretionary wedge, commenced in the Late Oligocene and ended by the Early Miocene in the west and by the Middle Miocene farther to the east (e.g. Sperner et al. 2002; Kováč et al. 2017a).

The Middle to Late Miocene retreat (rollback) of the subduction zone below the External Western Carpathians had crucial consequences for the evolution of the Central and Internal Western Carpathians in the upper plate position. From the Middle Miocene onwards, the Pannonian Basin System formed in a back-arc position (Kováč et al. 2017a). Initially, small pull-apart basins opened, followed by widespread rifting and thermal subsidence that unified numerous small depocentres into the extensive Lake Pannon (Magyar et al. 1999). Marine conditions in the Central Paratethys were replaced by brackish and then by freshwater conditions during the Late Miocene to Pliocene, as the connections with the Mediterranean and Black Seas were closed. Lithospheric stretching and asthenospheric upwelling were associated with voluminous calc-alkaline andesitic volcanism, which culminated during the Middle and Late Miocene and was terminated by minor basanitic extrusions during the Pliocene and Quaternary. In the Western Carpathians, this back-arc rifting was also associated with the uplift of small horsts, forming mountains from the Late Miocene up until the present (Králiková et al. 2016). The peaks of the High Tatra Mountains are the highest in the entire Carpathians, though the highest point (Gerlachovský štít) is only 2665 m above the sea level.

2.5 Paleogene Evolution

2.5.1 Paleocene

Geodynamic processes influencing the evolution of Western Carpathian orogenic system, often in a close relationship with the development of Alpine and Dinaric systems controlled the changes of Cenozoic landscape. After the Alpine crustal thickening and burial due to basement nappe stacking, a tectonic collapse represented by an orogen-parallel extensional unroofing of the Gemeric, Veporic, and partly Tatric units occurred during the Late Cretaceous to early Eocene. In the Veporic and Gemeric units (Fig. 2.2), the Late Cretaceous to early Eocene tectonic collapse and exhumation of the units from the depth of ≈10 up to 2.5 km is related to eastward unroofing of the underlying Veporic Unit (Fig. 2.4). The unroofing finished at approximately 80 Ma and was followed by an en-block exhumation of the western portion of Veporic Unit (western part of the Slovenské Rudohorie Mts., northeastern part of the Tribeč Mts. -Rázdiel Mts., and some small occurrences from beneath the Neogene volcanites). In the eastern Veporic Unit (Branisko and Čierna Hora Mts.), the gradual rise, tectonic and erosional denudation exhumed the basement to depth of ≈6 km during the latest Late Cretaceous–Early Eocene. In general, the exhumation of the Tatric Unit took place during the mid-Cretaceous to early Eocene as was documented by ZFT and ZHe data from the crystalline basements of the Veľká Fatra, Žiar, Tribeč, Malé Karpaty, Považský Inovec, eastern part of Lúčanská Fatra and Tatry Mts (Fig. 2.5; Králiková et al. 2016). However, the exhumation phase was replaced by the middle Eocene to early Miocene burial beneath sedimentary sequences of the Central Carpathian Paleogene Forearc Basin.

With respect to the topographic and structural axis of an evolving orogen we distinguish, besides these uplifted/exhumed parts of the Central Western Carpathians Palaeoalpine wedge, also basins in two broad-scale tectonic settings: (i) the fore-axis and (ii) back-axis basin system (*sensu* Kováč et al. 2016).

Paleogene palaeogeography of the Western Carpathians (≈65–35 Ma) is characterised by (i) uplifted, mostly axial part of the Palaeoalpine consolidated accretion wedge of the Central Western Carpathians; (ii) basins in front of this structural axis of the forming orogen (fore-axis basin system) to which are ranked residual oceanic basins on the subducted/lower plate and basins located above the edge of the thrust plate, represented by wedge-top and fore-arc basin; (iii) retro-arc type basins in the hinterland of the orogen, i.e. basins behind the structural axis of the orogen (back-axis basin system).

In the Paleocene (≈66–55 Ma), an extensive domain of residual oceanic basins of the External Western Carpathian domain appeared between the southern edge of the Northern European Platform and the forming Central Western Carpathian orogenic wedge (Fig. 2.6). The basement of these basins gradually subducted below the rising orogen, with the orogenic front oriented nearly the N–S direction and with active deep-sea trench located in the same direction. This was in contrast to the Alps, where the orientation of the trench was in the W–E direction (Csontos and Vörös 2004). The Szolnok-Kričevo-Iňačovce domain, located within the

Fig. 2.5 A view to the main ridge of the Malá Fatra Mts. looking east. Peaks are composed of the Fatric nappe (predominantly Jurassic and Cretaceous marlstones; smooth landscape) and the Hronic nappe (Triassic shallow-water carbonates; rocky landforms)

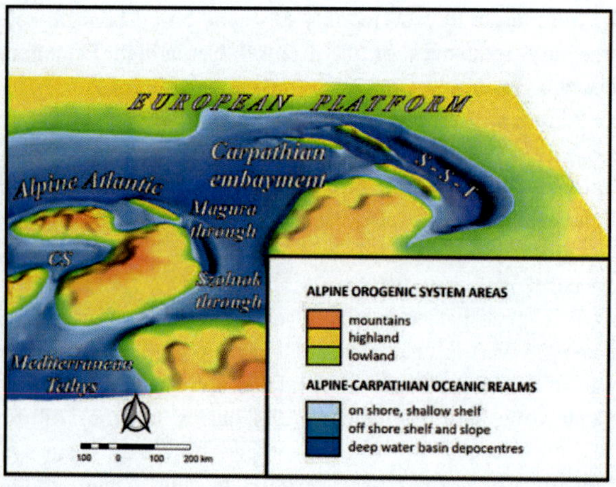

Fig. 2.6 Paleocene palaeogeography of oceanic Alpine and Mediterranean domains (≈65–50 Ma). In the front of growing (Cretaceous) Eastern Alpine–Western Carpathian orogenic wedge subduction of the Pennine oceanic crust (Alpine Atlantic) started and led to evolution of the Cenozoic fold-and-thrust belt of the External Alpine and Western Carpathian tectonic zones. In the west, the Alpine Atlantic in front of the Alps continued eastward into the Magura flysch Basin connected towards the south-east with the Szolnok–Kričevo troughs. The Szolnok flysch belt is presently located on the western slopes of the Tisza–Dacia microplate (see Fig. 2.1). In the north to northeast, the Carpathian embayment with Penninic type basins was located, where the internal Silesian and external Skole–Skyba–Tarcau flysch troughs (S-S-T) were divided by continental ribbons—cordilleras, which were emerged above the sea level at that time. The hinterland of the Eastern Alpine–Western Carpathian orogenic wedge was flooded by the Carinthian Sea (CS; according to Kováč et al. 2016, modified)

Magura oceanic realm in the northern and southeastern part of the front of Central Western Carpathians, belonged to the group of back-axis basin system (sensu Kováč et al. 2016). The eastern boundary of this originally oceanic domain was formed by the continental edge of the Tisza–Dacia Mega-Unit (Csontos et al. 1992).

2.5.2 Eocene

In the Eocene (≈56–34 Ma), subduction of the Magura domain led to the separation of the Magura-related Biele Karpaty Unit belonging to the External Western Carpathians from its substratum. Its sedimentary complex was subsequently shortened and incorporated to the front of the accretionary wedge of the Central Western Carpathians in the west. Compressional regime in the frontal part of the orogen migrating to the east and later to the northeast led to further out-of-sequence thrusting, backthrusting, and thickening of the accretionary prism. During this time, the accretionary wedge was formed by the Oravic units (Pieniny Klippen Belt s.s.). In the west, the accretion wedge was flooded by the Paleocene (Danian–Ypresian) sea, after which the wedge-top Paleogene basins occurring along the Pieniny Klippen Belt subsided to bathyal depths (Plašienka and Soták 2015). The uplifted edge of the Central Western Carpathians provided huge masses of carbonate slope breccias from the Triassic complexes predominantly of the Hronic Unit (Marschalko and Samuel 1993), which, together with fine-grained sediments, were in a part deposited below the calcite compensation depth. Already in the middle Eocene (late Lutetian–Bartonian), in the area of the Central Western Carpathians the compression regime prevailed again, causing uplift and consequent disintegration and erosion of the wedge-top Paleocene-Middle Eocene basins (Fig. 2.7).

In the eastern part of the Central Western Carpathian front, there was a gradual closing of the Iňačovce-Kričevo oceanic domain from west to east during the Paleocene and Eocene. Sedimentary complexes of the Iňačovce-Kričevo oceanic domain, which was related to the Pennine realm to the west and the Szolnok-Sava oceanic zones to the south-southeast, were partially underthrust beneath the frontal part of the Central Western Carpathians and underwent low temperature and medium pressure metamorphosis (Soták et al. 2000). Moreover, the NW–SE direction of the Czorsztyn continental margin formed by the Oravic Unit (Pieniny Klippen Belt) influenced formation of the frontal part of the accretionary wedge during the dextral transpressional tectonic regime. Although transpressive tectonic regime prevailed in the prograding Alpine-Western Carpathian orogenic system, the different movements of its individual segments also generated extension.

Thus, between the Eastern Alps moving northwards and the Central Western Carpathians moving east- to north-eastwards, a subsiding area flooded by the sea likely formed above the weakened crust. This sea linked the Magura oceanic domain with the rest of the Penninic Ocean in the west and south during the Paleocene and nearly all of the Eocene (Kováč et al. 2016). The area of Carinthia Basin in the hinterland of the Alpine orogenic system (back-axis basin system) was flooded from the west only in the Late Paleocene (Kázmér et al. 2003). The depocentres of the Eocene sea were formed in the area of Paleogene Northern Hungarian Basin in a retro-arc position (Kováč et al. 2016).

The Lutetian carbonate platform was deposited above the pre-Cenozoic sequences, often with bodies of bauxites and coal complexes beneath the Northern Pannonian domain in the hinterland of the Central Western Carpathians (Kázmér et al. 2003; Tari et al. 1993). The similar palaeo-bio-provincial character of fossil assemblages in sediments suggests that this basin was also connected with the Dinaride foredeep (Kecskeméti and Vörös 1975; Drobne et al. 1977; Less et al. 2000).

In the Late Eocene (≈41–34 Ma), the trench, i.e. subduction zone, was moved to the edge of the Magura domain

Fig. 2.7 The Suľovské vrchy Mts. The Paleogene carbonate conglomerates forming castellated landforms are among the most valuable parts of the mountains

also in the eastern portion of the Central Western Carpathians (Fig. 2.8). The subduction movement allowed the ocean domain to be shortened in front of the whole orogen, with the trend of process rejuvenation eastwards. At the same time, an accretionary prism was formed between the eastern frontal edge of the Central Western Carpathians and the edge of the Tisza–Dacia Mega-Unit, consisting of future units of the Iňačovce–Kričevo zone (partly also Szolnok zone), whose origin according to Kováč et al. (2016) can be linked with the Sava zone in the Dinarides.

The pull of the subducting Magura plate caused extensive collapse of the margin of thickened Central Western Carpathian crust, on which the basin began to form. Stretching caused by subduction under the edge of the Central Western Carpathians resulted in the beginning of deposition in the Central Carpathian Paleogene Basin in fore-arc position (Figs. 2.2 and 2.8). Continental siliciclastic rocks and shallow marine bioclastic carbonates were rapidly replaced by deep-marine sedimentation of several thousand metres thick, deep-water flysch deposits (Soták et al. 2001). These indicate a rapid collapse of the northern part of the Central Western Carpathians following the eastward-migrating transgression in time span between the Eocene and the end of Oligocene. The Paleogene deposits on the Central and Internal Western Carpathians are generally not deformed and this orogenic wedge functioned as a relatively rigid buttress for the developing accretionary wedge of the External Western Carpathians during the Oligocene and Neogene.

2.5.3 Oligocene

Oligocene palaeogeography (≈34–23 Ma) was efficiently influenced by the early stage of collision of the Alpine-Western Carpathian orogenic system with the southern margin of the European Platform. The fore-axis basin system (cf. Kováč et al. 2016) of the Western Carpathians was represented by the diminishing space of the

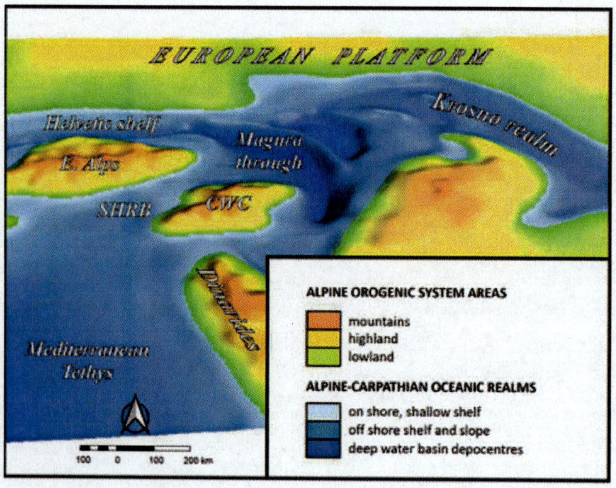

Fig. 2.8 Late Eocene palaeogeography of the Alpine–Western Carpathian chains (≈40–35 Ma). Partial uplift of central zones of the orogenic wedges in a form of an archipelago (Eastern Alps, Central Western Carpathians—CWC) between the Alpine Atlantic and Mediterranean Tethys is documented. In the north, along the slopes of the European Platform, the Helvetic shelf continued towards the southwest in the residual basins of the Alpine Atlantic. In the east, the internal Magura flysch troughs and the external basin system of the Krosno realm subsided. The hinterland of the Eastern Alpine–Western Carpathian orogenic wedge was flooded by the sea (Slovenian–Hungarian retro-arc basin—SHRB) from the Mediterranean domain (according to Kováč et al. 2016, modified)

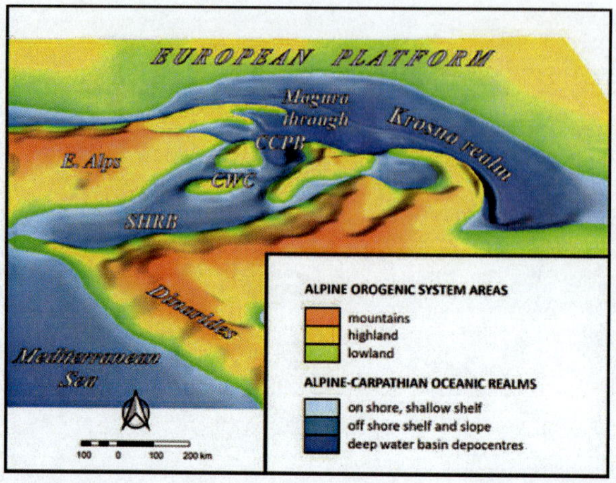

Fig. 2.9 Chattian palaeogeography of the Alpine–Western Carpathian orogenic wedge and adjacent basin systems (≈30–25 Ma). In the north, along the slopes of the European Platform, the Alpine foredeep basin continued towards the east, into the flysch troughs of the Krosno realm. The incipient Magura fold-and-thrust belt in the west, and the Szolnok–Kričevo fold-and-thrust belt in the east started to be uplifted and eroded. Transport of coarse clastic material into the Central Carpathian Paleogene Basin (CCPB) in a fore-arc position, which subsided along the northern Central Western Carpathians margin, is documented from both directions. In the hinterland of the Eastern Alpine–Western Carpathian orogenic system, a wide Slovenian-Hungarian retro-arc basin (SHRB) developed. Besides the front of the Alps, communication of the Central Paratethys Sea with the Mediterranean is supposed also along the northern tip of the Dinarides (according to Kováč et al. 2016, modified)

Magura and Silesian-Krosno domains. As a result of increasing compression, the Krosno Basin was isolated from the "world ocean" (van Couvering et al. 1981; Oszczypko-Clowes 2001), resulting in a decrease of the water mass circulation and anoxic conditions at the basin bottom and deposition of dark deep-water claystones. To the contrary, the syntectonic turbiditic sandstones were deposited in the Magura Basin at the same time (Picha and Straník 1999; Oszczypko-Clowes and Żydek 2012). During the Rupelian, but also the Chattian, the accretionary prism of the External Western Carpathians grew gradually, while the major part of it belonged already to the folded sediments of the future Magura Group of nappes (Oszczypko and Oszczypko-Clowes 2006, 2009).

The Central Carpathian Paleogene Basin was filled with clastic material derived not only from the uplifted portions of the Central Western Carpathian crystalline complexes and nappes (Sliva 2005), but also from the uplifted areas of the accretionary prism (Fig. 2.9). From the northwest, sand material was transported mainly during the Rupelian (Orava megabeds; Starek et al. 2013). The source of siliciclastic deposits was apparently an uplifted part of the accretionary prism of the Flysch Belt located between the European Platform and the axial part of the orogenic system (Rhenodanubian and Magura units). In the east, i.e. on the opposite side of the Central Western Carpathian Basin, the direction of the turbidite current systems, as well as the composition of the Chattian siliciclastics, point to another source, located in the east or southeast. In addition to the increasing content of silici- and calciclastics, the presence of ophiolite-derived material is typical (Soták et al. 1996). It follows that the source area had to contain obducted ophiolite melange-like complexes.

At the end of the Oligocene, the fore-arc basin was characterised by shallowing of the sedimentary environment on the eastern margin of the orogenic wedge (Kováč et al. 2018b). The progressive rock and surface uplift of the individual parts of the consolidated Central Western Carpathians was accompanied by the activity of the NE–SW sinistral strike-slip faults (Fig. 2.10), which at the same time controlled the continental and shallow marine sedimentation of intramontane depressions (Marko and Vojtko 2006; Pešková et al. 2009; Vojtko et al. 2010; Sůkalová et al. 2012).

The Oligocene evolution of the topographic and structural axis of the evolving orogen led to exhumation and surface uplift of the Eastern Alpine and Central Western Carpathian junction. The Penninic complexes of the lowermost orogen pattern appeared near the surface (Kováč et al. 1994; Dunkl and Demény 1997; Fügenschuh et al. 1997; Hejl 1997; Dunkl et al. 2005; Danišík et al. 2004, 2008; Kuhlemann

Fig. 2.10 A southwestward view of the Muráň fault zone from the Predná hora saddle. There is a fault contact of the Triassic shallow-water carbonates of the Silica Unit (right side) with Lower Paleozoic orthogneiss complex of the Veporic Unit (left side)

2007; Králiková et al. 2014a, b, 2016; Tari et al. 2021). This uplift was also associated with uplift of the Transdanubian Unit in the orogenic wedge, leading to widespread erosion of the Upper Eocene back-arc basin in the Rupelian (≈34–28 Ma), as well as a shift of the depocentres in the hinterland of orogen to the east (Kázmér et al. 2003, Kováč et al. 2017b).

The depocentre of the residual retro-arc basin subsided along the dextral strike-slip fault system (Klučiar et al. 2016). This basin, located far to the east than the original sedimentation area of the North Hungarian Paleogene Basin, was spread not only above the Northern Pannonian domain (Transdanubicum), but also above the Central Western Carpathian units. The basin is filled by firstly anoxic, later by calcareous marls, and on the margins by siliciclastic formations, which were derived from the eroded Eastern Alps and Western Carpathians (Tari et al. 1993). The basin reached the maximum extent in the Late Oligocene (28–23 Ma). During the Late Oligocene to Early Miocene, the evolution of the residual retro-arc basin was influenced by the increasing role of faults (e.g. Mid-Hungarian Fault).

2.6 Neogene Evolution

2.6.1 Early Miocene

The early Miocene (≈23–16 Ma) collision of the Western Carpathian orogen with the European Platform led to a further rising on the accretion wedge of the Flysch Belt. In front of the raised Magura nappe pile, the residual oceanic Krosno Basin occurred. The Krosno Basin was also the deepest part of the sea, on the northern shore, where on the slopes of the platform were formed depocentres of the internal zone of foredeep basin (Kováč et al. 2017a, b).

Compressional tectonic regime along the eastern edge of the orogenic wedge led to disintegration of the remnants of the Central Carpathian Paleogene Basin (fore-arc basin). To the contrary, during this period, new piggy back or wedge-top basins were formed in the west (Vienna Basin, Považie), partly located on the flysch nappes of the accretion prism, as well as on the edge of the Eoalpine consolidated portion of the Central Western Carpathian margin (Kováč et al. 2004).

On the triple junction of the Eastern Alps, Central Western Carpathians, and Transdanubicum, the orogenic collision led to the uplift and subsequent erosion of rock complexes that provided a rich source of clastic materials to the wedge-top basins. Oligocene to early Miocene exhumation of the western part of the Central Western Carpathians continued to the middle Miocene, resulting in denudation of the future Danube Basin basement built by the crystalline rocks, since the Mesozoic and Paleogene sedimentary complexes are absent there (Biela 1978; Fusán et al. 1987).

The rollback effect of the subduction of the submerged orogenic ocean-bottom domain together with push of the Apulian plate has influenced the individualization of the ALCAPA Mega-Unit (Fig. 2.1). Later, the process allowed the tectonic escape of this lithospheric fragment northeastwards (e.g. Ratschbacher et al. 1991a, b; Csontos et al. 1992; Kováč et al. 1994; Fodor et al. 1998, Konečný et al. 2002).

Movement of the ALCAPA Mega-Unit led to a collision with the edge of the Tisza-Dacia Mega-Unit, on which the Szolnok flysch formations were gradually thrust along the Mid-Hungarian fault system (Csontos and Nagymarosy 1998). Tectonic escape of lithospheric fragment of the ALCAPA Mega-Unit northeastwards was associated with a counter-clockwise rotation at ≈50° (Márton et al. 2013).

In the early Miocene, basins depocentres located to the north of the Mid-Hungarian Zone (in present) began to open along the southern edge of the ALCAPA Mega-Unit in transtensional tectonic regime. Formation of these depocentres is related to the initial rifting of the hinterland basin system (Kováč et al. 2018a). In the eastern part of the overriding microplate, a deeply submerged part of the accretion wedge of the Iňačovce-Kričevo zone was exhumed from under the frontal part of the Central Western Carpathians (Iňačovce Unit; Soták and Bebej 1996). Finally, in order to compensate for the rotational movement, at the end of the early Miocene (≈17 Ma), depocentres of the Eastern Slovak Basin (Východoslovenská nížina lowland) were opened by the pull-apart mechanism in the dextral transtensional tectonic regime.

A significant tectonic event associated with the oblique collision of the orogen with the edge of the European Platform is recorded around the boundary between the early and middle Miocene (≈17–15 Ma; cf. Kováč et al. 2018b). In the west, compression resulted in the disintegration of wedge-top basins and their inversion and surface uplift. Moreover, back thrust faults were also documented, with the movement vergence into the hinterland of the orogen (e.g. Marko et al. 2005; Pešková et al. 2009; Vojtko et al. 2010; Hók et al. 2016). The inversion of the basins took place between the rising and uplifted accretion wedge and the exhumed area of the axial portion of the mountain range (Kováč et al. 2016). Denudation of the exhumed parts provided a large amount of material transported into the remaining and newly forming back-arc basins. The magnitude of erosion is well documented by the angular discordance of the sedimentary formations, for example, in the northern part of the Vienna Basin (e.g. Kováč et al. 2004). The movement of the ALCAPA Mega-Unit to the northeast in this period is also associated with additional ≈ 30° counter-clockwise rotation of the peripheral Ždánice Unit of the flysch accretion wedge of the External Western Carpathians (Márton et al. 2009).

2.6.2 Middle Miocene

In the middle Miocene (≈16–11.6 Ma), the External Western Carpathian nappe system was folded, thrust, and emplaced in the final position. The foredeep external basin reached a surface area considerably larger than in the early Miocene, and the large areas of the European Platform were flooded by the sea. Maximum subsidence, as well as uplift of accretionary wedge of the orogen migrated over time eastwards (Meulenkamp et al. 1996). Based on the relatively short duration of the island-arc type volcanic activity in the Eastern Slovak Basin, extending from the Late Badenian to Early Sarmatian, the length of the submerged Krosno oceanic plate was estimated to 120–150 km (Konečný et al. 2002). The deposition of evaporites, particularly in the northern and eastern parts of the foredeep, shows basin isolation from the surrounding marine areas in the Badenian period (Fig. 2.11). Sedimentation in the Western Carpathian foredeep terminated before the late Miocene.

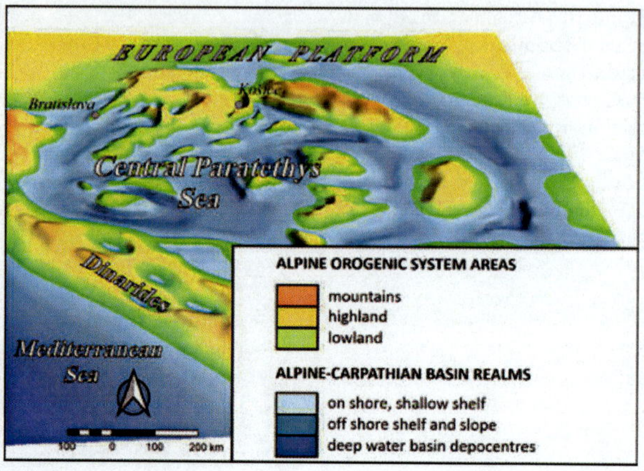

Fig. 2.11 Serravallian palaeogeography of the Central Paratethys Sea and adjacent areas (≈13–12 Ma). In the north and northeast, the Carpathian foredeep covers the European Platform margin situated in front of the External Western Carpathian accretionary wedge, which was rising above the sea level. The uplifted Central Western Carpathian units represent the northern border of an extensive back-arc basin system. From left to right the Vienna, Danube, Novohrad, and Eastern Slovakia basins are figured (according to Kováč et al. 2017a, modified)

Stretching of ALCAPA Mega-Unit resulted in the synrift stage of basins evolution (Šujan et al. 2021). The formation of the back-arc basin system in the rear of the orogeny was accompanied by huge calc-alkaline and acid volcanism (e.g. Konečný et al. 2002). The maximum of the middle Miocene subsidence was gradually shifted eastwards in time (Kováč 2000). The oldest (Early Badenian) depocentre was located in the western portion of the Danube Basin, followed by the younging towards the eastern depressions (Sarmatian). In the Eastern Slovak Basin, the subsidence has reached its peak in the Sarmatian (Kováč et al. 2017a). The Middle Miocene Basins were partly opened under the conditions of the simple shear tectonic regime (Tari et al. 1993; Lankreijer et al. 1995), but later, approximately from the Sarmatian/Pannonian boundary pure shears were often activated (Šujan et al. 2016; Sztanó et al. 2016).

According to fission track ages, the middle Miocene collision of the Central Western Carpathian orogenic wedge with the European continental margin led to final exhumation of the most external horsts formed by the Tatric Unit (Považský Inovec, Strážovské vrchy, Malá Fatra, Tatry Mts.).

2.6.3 Late Miocene to Pliocene

The period of the late Miocene (≈11.6–5.6 Ma) partly represents the late synrift, but predominantly the postrift stage of the back-arc basin system evolution. First, there is a rifting of the Lake Pannon sedimentation areas in the hinterland of the orogen (Magyar et al. 2013). The evolution of the upper Miocene depocentres of the Danube Basin was most probably controlled by pure shear tectonics (Majcin et al. 2015), as opposed to the middle Miocene depocentres in the Pannonian Basin System (Fig. 2.12). During the late Miocene, the intramontane basins in the Central Western Carpathians were formed (e.g. the Turiec and Žiar basins; Fig. 2.13).

The origin of the Pannonian Basin System was related to the stretching of both ALCAPA and Tisza-Dacia Mega-Units, probably due to the subduction pull at the front of the Eastern Carpathians. After filling of deep depocenters (hinterland basins) of the northern part of the Lake Pannon around 9 Ma, a thermal subsidence began (Magyar et al. 1999; Šujan et al. 2016). From the Pliocene, the basin system has been in the tectonic inversion stage and is accompanied by erosion of the basin margins. The final destruction of the volcanic structure, exhumation, and denudation of the mountains began in the late Miocene. However, the relatively fast exhumation most probably occurred during the Pliocene and Quaternary period, as inferred from preserved mid-mountain planation surfaces approximately 1000 m asl. (Králiková et al. 2016; Vojtko et al. 2016, 2017). At the same time, the Central Western Carpathian area is characterised by the onset or acceleration of subsidence of an intramontane depression controlled by extension in the NW–SE direction. The last phase of subsidence was accompanied by a clockwise rotation of the Tisza-Dacia Mega-Unit at about 30° (Dupont-Nivet et al. 2005). Since the Pliocene, the contemporary shape of the Western Carpathian area has started to be formed as it is known in present (e.g. Vojtko et al. 2008, 2011a, b). The highest mountainous part of Slovakia was modelled by Quaternary glaciation (Tatry and Nízke Tatry Mts.). On the contrary, sedimentation continues in the axial parts of the Vienna, Danube, and Eastern Slovak Basins.

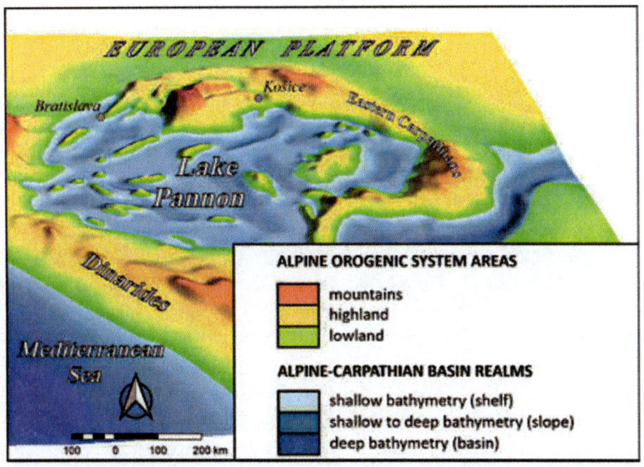

Fig. 2.12 Tortonian palaeogeography of the Lake Pannon and adjacent areas (≈11–9 Ma). In the north and northeast, the Carpathian foredeep deposits covering the European Platform rimmed the uplifting External Carpathian arc. The uplifted Central Western Carpathian mountain ranges represent the northern border of the wide Pannonian basin system. From left to right the Vienna, Danube and Eastern Slovak basins are shown (according to Kováč et al. 2017a, modified)

2.7 Summary and Conclusions

The Western Carpathians, as a part of the European Alpine orogenic system, form an arc located in its northeastern portion. The mountain range is approximately 500 km long in the W–E direction and 300 km wide across. The Western Carpathian area includes not only the mountainous regions, but also the fundament of wide flat lowlands of the Pannonian Basin System (e.g. Danube Lowlands, Great Hungarian Plain, Eastern Slovak Basin). The External Western Carpathians, situated in the north, include the sediments of the Carpathian foredeep and the Carpathian Flysch Belt, representing the Cenozoic accretionary complex consisting of the outer Silesian-Krosno and the inner Magura units. The Pieniny Klippen Belt consisting of predominantly Mesozoic to Paleogene complexes represents a boundary between the External and Central Western Carpathians. The Central Western Carpathians include various pre-Cenozoic units and the unconformable Upper Cretaceous, Paleogene, Neogene, and Quaternary sedimentary deposits. They are divided into morphotectonic belts, from the north to the south (i) the Tatra-Fatra Belt that includes the lowermost Tatric Unit composed of predominantly Variscan basement and its Upper Paleozoic to Mesozoic sedimentary cover and detached nappe pile of the Fatric and Hronic thrusts; (ii) the Vepor–Gemer Belt, which is composed of a stack of basement-dominated nappes. It is formed by the north-verging thick-skinned Veporic and Gemeric basement/cover thrust sheets and the Jurassic Meliata-Turňa-Silica nappe system. The Internal Western Carpathians are exposed mostly in isolated mountains outside of Slovakia (Transdanubian Range, Bükk and

Fig. 2.13 Intramontane basins are typical geomorphological features of the Western Carpathians. The photo shows the Turiec Basin located between the Fatra mountain ranges, with the snowy Tatra Mts. on the horizon

surrounding mountains, and the Aggtelek-Rudabánya Mountains). The tectonic units of the Central and Internal Western Carpathian complexes are often buried by the Paleogene forearc and retro-arc basins deposits, as well as by the Neogene wedge top, intramontane and back-arc basins fill, often associated with the Neogene volcanic edifices. In today's geomorphological picture is clearly visible for example, the Central Carpathian Paleogene Basin, or the dominant Neogene basins like the Vienna, Danube, Novohrad, or Eastern Slovak basins.

The lithologically very variable composition of the Western Carpathian units and their complex tectonic evolution resulted in a range of different morphological landforms and often-picturesque landscapes. Comparatively soft flysch rocks of the External Carpathians conditioned hilly relief with smooth ridges and narrow incised valleys, enriched by scattered peculiar rocky cliffs in the Pieniny Klippen Belt. Owing to the Neogene extensional tectonic movements, the Cretaceous nappe units of the northern Central Carpathian zones were dissected by normal faults into elevated horst structures separated by grabens filled with soft sediments. The southern Central Western Carpathians (Vepor–Gemer area) were less affected by the extension and partly preserve the pre- Neogene morphology of a peneplain character. In contrast, extensive lowlands of the Pannonian Basin system to the south exhibit a monotonous flat landscape, but eroded remnants of Neogene stratovolcanoes in central and eastern Slovakia provide another distinct type of landscape occurring in Slovakia. The landforms were strongly affected also by the Quaternary climate and tectonics, which controlled the deposition of slope and alluvial sediments, river terraces, aeolian sand dunes, loess, continental and mountain glacier deposits, or travertine mounds.

Acknowledgements This work was supported by the Slovak Research and Development Agency under the contract No. APVV-17-0170, APVV-16-0121, APVV-20-0120 and the Science Grant Agency—project VEGA 1/0346/20.

References

Andrusov D (1974) The Pieniny Klippen Belt. In: Maheľ M (ed) Tectonics of the Carpathian-Balkan regions. Dionýz Štúr Geological Institute, Bratislava, pp 145–158

Biela A (1978) Hlboké vrty v zakrytých oblastiach vnútorných Západných Karpát. Regionálna geológia Západných Karpát 11, Geological Institute of Dionýz Štúr, Bratislava, Slovakia, 224 pp

Bielik M, Šefara J, Kováč M, Bezák V, Plašienka D (2004) The Western Carpathians—interaction of Hercynian and alpine processes. Tectonophysics 393:63–86. https://doi.org/10.1016/j.tecto.2004.07.044

Biely A, Bezák V, Elečko M, Kaličiak M, Konečný V, Lexa J, Mello J, Nemčok J, Potfaj M, Rakús M, Vass D, Vozár J, Vozárová A (1996) Geological map of Slovakia. Ministerstvo životného prostredia SR, Geologická služba SR, Bratislava

Birkenmajer K (1986) Stages of structural evolution of the Pieniny Klippen Belt, Carpathians. Studia Geologica Polonica 88:7–32

van Couvering JA, Aubry MP, Berggren WA, Bujak JP, Naeser CW, Wieser T, (1981) The terminal Eocene event and the Polish connection. In: Paleogene paleogeography and the geological events at the Eocene/Oligocene boundary. Palaeogeogr

Palaeoclimat Palaeoecol 36:321–362. https://doi.org/10.1016/0031-0182(81)90111-5
Csontos L, Nagymarosy A (1998) The Mid-Hungarian line: a zone of repeated tectonic inversions. Tectonophysics 297:51–71. https://doi.org/10.1016/S0040-1951(98)00163-2
Csontos L, Vörös A (2004) Mesozoic plate tectonic reconstruction of the Carpathian region. Palaeogeogr Palaeoclimat Palaeoecol 210:1–56. https://doi.org/10.1016/j.palaeo.2004.02.033
Csontos L, Nagymarosy A, Horváth F, Kováč M (1992) Tertiary evolution of the Intra-Carpathian area: a model. Tectonophysics 208:221–241. https://doi.org/10.1016/0040-1951(92)90346-8
Danišík M, Dunkl I, Putiš M, Frisch W, Kráľ J (2004) Tertiary burial and exhumation history of basement highs along the NW margin of the Pannonian Basin—an apatite fission track study. Austrian J Earth Sci 95(96):60–70
Danišík M, Kohút M, Dunkl I, Frisch W (2008) Thermal evolution of the Žiar Mountains basement (Inner Western Carpathians, Slovakia) constrained by fission track data. Geol Carpath 59:19–30
Drobne K, Pavlovec R, Drobne F (1977) Paleogene larger foraminifera from the area between Mežica and Slovenj Gradec (NW Yugoslavia). Razprave IV Razred SAZU 20:1–88
Dunkl I, Demény A (1997) Exhumation of the Rechnitz Window at the border of the Eastern Alps and Pannonian Basin during Neogene extension. Tectonophysics 272:197–211. https://doi.org/10.1016/S0040-1951(96)00258-2
Dunkl I, Kuhlemann J, Reinecker J, Frisch W (2005) Cenozoic relief evolution of the Eastern Alps—constraints from apatite fission track age-provenance of Neogene intramontane sediments. Austrian J Earth Sci 98:92–105
Dupont-Nivet G, Vasiliev I, Langereis Cor G, Krijgsman W, Panaiotu C (2005) Neogene tectonic evolution of the southern and eastern Carpathians constrained by paleomagnetism. Earth Planet Sci Lett 236:374–387. https://doi.org/10.1016/j.epsl.2005.04.030
Fodor L, Jelen B, Márton E, Skaberne D, Čar J, Vrabec M (1998) Miocene-Pliocene tectonic evolution of the Slovenian Periadriatic fault: implications for Alpine-Carpathian extrusion models. Tectonics 17:690–709. https://doi.org/10.1029/98TC01605
Froitzheim N, Plašienka D, Schuster R (2008) Alpine tectonics of the Alps and Western Carpathians. In: McCann T (ed) The geology of Central Europe, vol 2. Mesozoic and Cenozoic. Geological Society Publishing House, London, pp 1141–1232
Fügenschuh B, Seward D, Mancktelow N (1997) Exhumation in a convergent orogen: the western Tauern window. Terra Nova 9:213–217. https://doi.org/10.1111/j.1365-3121.1997.tb00015.x
Fusán O, Biely A, Ibrmajer J, Plančár J, Rozložník L (1987) Podložie terciéru Vnútorných Západných Karpát. GÚDŠ, Bratislava, 100 pp
Less Gy, Kecskeméti T, Ozsvárt P, Kázmér M, Báldi-Beke M, Kollányi K, Fodor L, Kertész B, Varga I (2000) Middle–Upper Eocene shallow water benthos in Hungary. In: Bassi D (ed) Field-trip guide-book. Shallow water benthic communities at the Middle–Upper Eocene boundary. Southern and North-Eastern Italy, Slovenia, Croatia, Hungary. 5th Meeting of the IUGS-UNESCO IGCP 393—July 18th–31st, 2000, Annali Universitá di Ferrara, Scienze della Terra, Ferrara, 8, pp 151–181
Haas J, Kovács S, Krystyn L, Lein R (1995) Significance of Late Permian-Triassic facies zones in terrane reconstruction in the Alpine-Nord Pannonian domain. Tectonophysics 242:19–40. https://doi.org/10.1016/0040-1951(94)00157-5
Hejl E (1997) "Cold spots" during the Cenozoic evolution of the Eastern Alps: thermochronological interpretation of apatite fission-track data. Tectonophysics 272:159–173. https://doi.org/10.1016/S0040-1951(96)00256-9
Hók J, Kováč M, Pelech O, Pešková I, Vojtko R, Králiková S (2016) The Alpine tectonic evolution of the Danube Basin and its northern periphery (southwestern Slovakia). Geol Carpath 67:495–505. https://doi.org/10.1515/geoca-2016-0031
Janák M, Plašienka D, Frey M, Cosca M, STh, Schmidt, Lupták B, Méres Š (2001) Cretaceous evolution of a metamorphic core complex, the Veporic unit, Western Carpathians (Slovakia): P-T conditions and in situ $^{40}Ar/^{39}Ar$ UV laser probe dating of metapelites. J Metamorp Geol 19:197–216. https://doi.org/10.1046/j.0263-4929.2000.00304.x
Jeřábek P, Stünitz H, Heilbronner R, Lexa O, Schulmann K (2007) Microstructural-deformation record of an orogen-parallel extension in the Vepor Unit, West Carpathians. J Struct Geol 29:1722–1743. https://doi.org/10.1016/j.jsg.2007.09.002
Kázmér M, Dunkl I, Frisch W, Kuhlemann J, Ozsvárt P (2003) The Palaeogene forearc basin of the Eastern Alps and Western Carpathians: subduction erosion and basin evolution. J Geol Soc 160:413–428. https://doi.org/10.1144/0016-764902-041
Kecskeméti T, Vörös A (1975) Biostratigraphische und paläoökologische Untersuchungen einer transgressiven eozänen Schichtserie (Darvastó, Bakony-Gebirge). Fragm Min Et Paleont 6:63–93
Klučiar T, Kováč M, Vojtko R, Rybár S, Šujan M, Králiková S (2016) The Hurbanovo-Diósjenő fault: a crustal-scale weakness shear zone at the boundary of the Central Western Carpathians and Northern Pannonian Domain. Acta Geol Slovaca 8:59–70
Konečný V, Kováč M, Lexa J, Šefara J (2002) Neogene evolution of the Carpatho-Pannonian region: an interplay of subduction and back-arc diapiric uprise in the mantle. EGS Stephan Mueller Spec Publ Ser 1:105–123
Kováč M, Kráľ J, Márton E, Plašienka D, Uher P (1994) Alpine uplift history of the Central Western Carpathians: geochronological, paleomagnetic, sedimentary and structural data. Geol Carpath 45:83–96
Kováč M, Baráth I, Harzhauser M, Hlavatý I, Hudáčková N (2004) Miocene depositional systems and sequence stratigraphy of the Vienna Basin. Cour. Forsch. – Inst. Senkenberg, Frankfurt Am Main 246:187–212
Kováč M, Plašienka D, Soták J, Vojtko R, Oszczypko N, Gy L, Ćosović V, Fügenschuh B, Králiková S (2016) Paleogene palaeogeography and basin evolution of the Western Carpathians, Northern Pannonian domain and adjoining areas. Glob planet Change 140:9–27. https://doi.org/10.1016/j.gloplacha.2016.03.007
Kováč M, Hudáčková N, Halásová E, Kováčová M, Holcová K, Oszczypko-Clowes M, Báldi K, Less G, Nagymarosy A, Ruman A, Klučiar T, Jamrich M (2017) The Central Paratethys palaeoceanography: a water circulation model based on microfossil proxies, climate, and changes of depositional environment. Acta Geol Slovaca 9:75–114
Kováč M, Márton E, Oszczypko N, Vojtko R, Hók J, Králiková S, Plašienka D, Klučiar T, Hudáčková N, Oszczypko-Clowes M (2017) Neogene palaeogeography and basin evolution of the Western Carpathians, Northern Pannonian domain and adjoining areas. Glob Planet Change 155:133–154. https://doi.org/10.1016/j.gloplacha.2017.07.004
Kováč M, Márton E, Klučiar T, Vojtko R (2018) Miocene basin opening in relation to the north-eastward tectonic extrusion of the ALCAPA Mega-Unit. Geol Carpath 69:254–263. https://doi.org/10.1515/geoca-2018-0015
Kováč M, Rybár S, Halásová E, Hudáčková N, Šarinová K, Šujan M, Baranyi V, Kováčová M, Ruman A, Klučiar T, Zlinská A (2018) Changes in Cenozoic depositional environment and sediment provenance in the Danube Basin. Basin Res 30:97–131. https://doi.org/10.1111/bre.12244

Kováč M (2000) Geodynamický, paleogeografický a štruktúrny vývoj Karpatsko-panónskeho regiónu v miocéne Nový pohľad na neogénne panvy Slovenska. VEDA, Bratislava, 202 pp

Kovács S (1992) Tethys "western ends" during the Late Paleozoic and Triassic and their possible genetic relationships. Acta Geol Hung 35:329–369

Kovács S, Sudar M, Grădinaru E, Gawlick H-J, Karamata S, Haas J, Cs P, Gaetani M, Mello J, Polák M, Aljinović D, Ogorelec B, Kolar-Jurkovšek T, Jurkovšek B, Buser S (2011) Triassic evolution of the tectonostratigraphic units of the Circum-Pannonian region. Jahrb Geol Bundesanst 151:199–280

Kozur H (1991) The evolution of the Meliata-Hallstatt ocean and its significance for the early evolution of the Eastern Alps and Western Carpathians. Paleogeogr Palaeoclimatol Palaeocol 87:109–135. https://doi.org/10.1016/0031-0182(91)90132-B

Králiková S, Vojtko R, Sliva Ľ, Minár J, Fügenschuh B, Kováč M, Hók J (2014) Cretaceous—quaternary tectonic evolution of the Tatra Mts. (Western Carpathians): constraints from structural, sedimentary, geomorphic and fission track data. Geol Carpath 65:307–326. https://doi.org/10.2478/geoca-2014-0021

Králiková S, Vojtko R, Andriessen P, Kováč M, Fügenschuh B, Hók J, Minár J (2014) Late Cretaceous-Cenozoic thermal evolution of the northern part of the Central Western Carpathians (Slovakia): revealed by zircon and apatite fission track thermochronology. Tectonophysics 615–616:142–153. https://doi.org/10.1016/j.tecto.2014.01.002

Králiková S, Vojtko R, Hók J, Fügenschuh B, Kováč M (2016) Low-temperature constraints on the Alpine thermal evolution of the Western Carpathian basement rock complexes. J Struct Geol 91:144–160. https://doi.org/10.1016/j.jsg.2016.09.006

Kuhlemann J (2007) Paleogeographic and paleotopographic evolution of the Swiss and Eastern Alps since the Oligocene. Glob Planet Change 58:224–236. https://doi.org/10.1016/j.gloplacha.2007.03.007

Lankreijer A, Kováč M, Cloetingh S, Pitoňák P, Hlôška M, Biermann C (1995) Quantitative subsidence analysis and forward modelling of the Vienna and Danube basins: thin-skinned versus thick-skinned extension. Tectonophysics 252:433–451. https://doi.org/10.1016/0040-1951(95)00099-2

Magyar I, Geary DH, Müller P (1999) Paleogeograhic evolution of the Late Miocene Lake Pannon in Central Europe. Palaeogeogr Palaeoclimat Palaeoecol 147:151–167. https://doi.org/10.1016/S0031-0182(98)00155-2

Magyar I, Radivojević D, Sztanó O, Synak R, Ujszászi K, Pócsik M (2013) Progradation of the paleo-Danube shelf margin across the Pannonian Basin during the Late Miocene and Early Pliocene. Glob Planet Change 103:168–173. https://doi.org/10.1016/j.gloplacha.2012.06.007

Majcin D, Bilčík D, Klučiar T (2015) Thermal state of the lithosphere in the Danube Basin and its relation to tectonics. Contrib Geophys Geodesy 45:193–218. https://doi.org/10.1515/congeo-2015-0020

Marko F, Vojtko R (2006) Structural record and tectonic history of the Mýto-Tisovec fault (Central Western Carpathians). Geol Carpath 57:211–221

Marko F, Plašienka D, Fodor L (1995) Meso-Cenozoic tectonic stress fields within the Alpine-Carpathian transition zone: a review. Geol Carpath 46:19–27

Marschalko R, Samuel M (1993) Sedimentológia východnej vetvy súľovských zlepencov. Západné Karpaty. Sér. Geol. 17:7–38

Márton E, Rauch-Włodarska M, Krejčí O, Tokarski AK, Bubík M (2009) An integrated palaeomagnetic and AMS study of the Tertiary flysch from the Outer Western Carpathians. Geophys J Int 177:925–940. https://doi.org/10.1111/j.1365-246X.2009.04104.x

Márton E, Grabowski J, Plašienka D, Túnyi I, Krobicki M, Haas J, Pethe M (2013) New paleomagnetic results from the Upper Cretaceous red marls of the Pieniny Klippen Belt, Western Carpathians: evidence for general CCW rotation and implications for the origin of the structural arc formation. Tectonophysics 592:1–13. https://doi.org/10.1016/j.tecto.2013.01.027

Meulenkamp JE, Kováč M, Cicha I (1996) On late Oligocene to Pliocene depocenter migrations and the evolution of the Carpathian-Pannonian system. Tectonophysics 266:301–317. https://doi.org/10.1016/S0040-1951(96)00195-3

Mock R, Sýkora M, Aubrecht R, Ožvoldová L, Kronome B, Reichwalder P, Jablonský J (1998) Petrology and stratigraphy of the Meliaticum near the Meliata and Jaklovce Villages, Slovakia. Slovak Geol Mag 4:223–260

Oszczypko N, Oszczypko-Clowes M (2006) Evolution of the Magura Basin. In: Oszczypko N, Uchman A, Malata E (eds) Paleotectonic evolution of the Outer Carpathian and Pieniny Klippen Belt basins. Geological Institute of Jagellonian University, Krakow, pp 133–164

Oszczypko N, Oszczypko-Clowes M (2009) Stages in the Magura Basin: a case study of the Polish sector (Western Carpathians). Geodin Acta 22:83–100. https://doi.org/10.3166/ga.22.83-100

Oszczypko-Clowes M (2001) The nannofossil biostratigraphy of the youngest deposits of the Magura Nappe (East of the Skawa river, Polish Flysch Carpathians). Ann Soc Geol Pol 71:139–188

Oszczypko-Clowes M, Żydek B (2012) Paleoecology of the Upper Eocene-Lower Oligocene Malcov Basin based on the calcareous nannofossils: a case study of the Leluchów section (Krynica Zone, Magura Nappe, Polish Outer Carpathians). Geol Carpath 63:149–164. https://doi.org/10.2478/v10096-012-0012-8

Pešková I, Vojtko R, Starek D, Sliva Ľ (2009) Late Eocene to quaternary deformation and stress field evolution of the Orava region (Western Carpathians). Acta Geol Polon 59:73–91

Picha FJ, Stráník Z, Krejčí O (2006) Geology and hydrocarbon resources of the Outer Western Carpathians and their foreland, Czech Republic. In: Golonka, Picha FJ (eds) The Carpathians and their foreland: Geology and hydrocarbon resources. AAPG Memoir 84:49–175

Picha FJ, Straník Z (1999) Late Cretaceous to early Miocene deposits of the Carpathian foreland basin in southern Moravia. Int J Earth Sci 88:475–495. https://doi.org/10.1007/s005310050280

Plašienka D (2012) Jurassic syn-rift and Cretaceous syn-orogenic, coarse-grained deposits related to opening and closure of the Vahic (South Penninic) Ocean in the Western Carpathians—an overview. Geol Q 56:601–628. https://doi.org/10.7306/gq.1044

Plašienka D (2018) Continuity and episodicity in the early Alpine tectonic evolution of the Western Carpathians: How large-scale processes are expressed by the orogenic architecture and rock record data. Tectonics 37:2029–2079. https://doi.org/10.1029/2017TC004779

Plašienka D, Soták J (2015) Evolution of Late Cretaceous-Palaeogene synorogenic basins in the Pieniny Klippen Belt and adjacent zones (Western Carpathians, Slovakia): tectonic controls over a growing orogenic wedge. Ann Soc Geol Pol 85:43–76. https://doi.org/10.14241/asgp.2015.005

Plašienka D, Méres Š, Ivan P, Sýkora M, Soták J, Lačný A, Aubrecht R, Bellová S, Potočný T (2019) Meliatic blueschists and their detritus in Cretaceous sediments: new data constraining tectonic evolution of the West Carpathians. Swiss J Geosci 112:55–81. https://doi.org/10.1007/s00015-018-0330-7

Plašienka D, Bučová J, Šimonová V (2020) Variable structural styles and tectonic evolution of an ancient backstop boundary—the Pieniny Klippen Belt of the Western Carpathians. Int J Earth Sci 109:1355–1376. https://doi.org/10.1007/s00531-019-01789-5

Poller U, Uher P, Broska I, Plašienka D, Janák M (2002) First permian —early Triassic zircon ages for tin-bearing granites from the Gemeric unit (Western Carpathians, Slovakia): connection to the post-collisional extension of the Variscan orogen and S-type granite magmatism. Terra Nova 14:41–48. https://doi.org/10.1046/j.1365-3121.2002.00385.x

Putiš M, Frank W, Plašienka D, Siman P, Sulák M, Biroň A (2009) Progradation of the Alpidic Central Western Carpathians orogenic wedge related to two subductions: constrained by $^{40}Ar/^{39}Ar$ ages of white micas. Geodin Acta 22:31–56. https://doi.org/10.3166/ga.22.31-56

Ratschbacher L, Frisch W, Linzer HG, Merle O (1991) Lateral extrusion in the Eastern Alps; part 2: structural analysis. Tectonics 10:257–271. https://doi.org/10.1029/90TC02623

Ratschbacher L, Merle O, Davy P, Cobbold P (1991) Lateral extrusion in the Eastern Alps; part 1: boundary conditions and experiments scaled for gravity. Tectonics 10:245–256. https://doi.org/10.1029/90TC02622

Schmid SM, Bernoulli D, Fügenschuh B, Matenco L, Schefer S, Schuster R, Tischler M, Ustaszewski K (2008) The Alpine-Carpathian-Dinaridic orogenic system: correlation and evolution of tectonic units. Swiss J Geosci 101:139–183. https://doi.org/10.1007/s00015-008-1247-3

Sliva Ľ (2005) Sedimentary facies of the Central Carpathians Palaeogene Basin in the area of Spišská Magura Mts. Dissertation, Comenius University, Bratislava, 137 pp

Soták J, Bebej J (1996) Serpentinitic sandstones from the Tertiary collisional zone in the Eastern Slovakia (Šambron-Kamenica Zone). Geol Carpath 47:227–238

Soták J, Bebej J, Biroň A (1996) Detrital analysis of the Paleogene flysch deposits of the Levoča Mts.: evidence for sources and paleogeography. Slovak Geol Mag 3–4:345–349

Soták J, Biroň A, Prokešová R, Spišiak J (2000) Detachment control of core complex exhumation and back-arc extension in the East Slovakian Basin. Slovak Geol Mag 2:130–132

Soták J, Pereszlényi M, Marschalko R, Milička J, Starek D (2001) Sedimentology and hydrocarbon habitat of the submarine-fan deposits of the central Carpathian Paleogene Basin (NE Slovakia). Mar Petrol Geol 18:87–114. https://doi.org/10.1016/S0264-8172(00)00047-7

Sperner B, Ratschbacher L, Nemčok M (2002) Interplay between subduction retreat and lateral extrusion: Tectonics of the Western Carpathians. Tectonics 21:1051, 24 pp. https://doi.org/10.1029/2001TC901028

Stampfli G, Kozur H (2006) Europe from the Variscan to the Alpine cycles. In Gee DG, Stephenson RA (eds) European lithosphere dynamics. Geological Society, London. Memoirs 32:57–82

Starek D, Soták J, Jablonský J, Marschalko R (2013) Large-volume gravity flow deposits in the Central Carpathian Paleogene Basin (Orava region, Slovakia): evidence for hyperpycnal river discharge in deep-sea fans. Geol Carpath 64:305–326. https://doi.org/10.2478/geoca-2013-0022

Šujan M, Braucher R, Kováč M, Bourlès D, Rybár S, Guillou V, Hudáčková N (2016) Application of the authigenic $^{10}Be/^{9}Be$ dating method to Late Miocene-Pliocene sequences in the northern Danube Basin (Pannonian Basin System): confirmation of heterochronous evolution of sedimentary environments. Glob Planet Chan 137:35–53. https://doi.org/10.1016/j.gloplacha.2015.12.013

Šujan M, Rybár S, Kováč M, Bielik M, Majcin D, Minár J, Plašienka D, Nováková P, Kotulová J (2021) The polyphase rifting and inversion of the Danube Basin revised. Glob Planet Change 196. https://doi.org/10.1016/j.gloplacha.2020.103375

Sůkalová Ľ, Vojtko R, Pešková I (2012) Cenozoic deformation and stress field evolution of the Kozie chrbty Mountains and the western part of Hornád Depression (Central Western Carpathians). Acta Geol Slov 4:53–64

Sztanó O, Kováč M, Magyar I, Šujan M, Fodor L, Uhrin A, Rybár S, Csillag G, Tőkés L (2016) Late Miocene sedimentary record of the Danube/Kisalföld Basin: interregional correlation of depositional systems, stratigraphy and structural evolution. Geol Carpath 67:525–542. https://doi.org/10.1515/geoca-2016-0033

Tari G, Báldi T, Báldi-Béke M (1993) Paleogene retroarc flexural basin beneath the Neogene Pannonian Basin: a geodynamic model. Tectonophysics 226:433–456. https://doi.org/10.1016/0040-1951(93)90131-3

Tari G, Bada G, Beidinger A, Csizmeg J, Danišík M, Gjerazi I, Grasemann B, Kováč M, Plašienka D, Šujan M, Szafián P (2021) The connection between the Alps and the Carpathians beneath the Pannonian Basin: selective reactivation of Alpine nappe contacts during Miocene extension. Glob Planet Change 103401. https://doi.org/10.1016/j.gloplacha.2020.103401

Tomek Č (1993) Deep crustal structure beneath the central and inner West Carpathians. Tectonophysics 226:417–431. https://doi.org/10.1016/0040-1951(93)90130-C

Vojtko R, Hók J, Kováč M, Sliva Ľ, Joniak P, Šujan M (2008) Pliocene to quaternary stress field change in the Western Carpathians (Slovakia). Geol Q 52:19–30

Vojtko R, Tokárová E, Sliva Ľ, Pešková I (2010) Reconstruction of Cenozoic paleostress fields and revised tectonic history in the northern part of the Central Western Carpathians (the Spišská Magura and Východné Tatry Mountains). Geol Carpath 61:211–225. https://doi.org/10.2478/v10096-010-0012-5

Vojtko R, Betak J, Hók J, Marko F, Gajdoš V, Rozimant K, Mojzeš A (2011) Pliocene to quaternary tectonics in the Horná Nitra Depression (Western Carpathians). Geol Carpath 62:381–393. https://doi.org/10.2478/v10096-011-0028-5

Vojtko R, Marko F, Preusser F, Madarás J, Kováčová M (2011) Evidence for late quaternary uplift along the Vikartovce Fault (Western Carpathians, Slovakia). Geol Carpath 62:563–574. https://doi.org/10.2478/v10096-011-0040-9

Vojtko R, Králiková S, Jeřábek P, Schuster R, Danišík M, Fügenschuh B, Minár J, Madarás J (2016) Geochronological evidence for the Alpine tectono-thermal evolution of the Veporic Unit (Western Carpathians, Slovakia). Tectonophysics 666:48–65. https://doi.org/10.1016/j.tecto.2015.10.014

Vojtko R, Králiková S, Andriessen P, Prokešová R, Minár J, Jeřábek P (2017) Geological evolution of the western part of the Veporic Unit (Western Carpathians): based on fission track and morphotectonic data. Geol Carpath 68:285–302. https://doi.org/10.1515/geoca-2017-0020

Vozárová A, Vozár J (1988) Late Paleozoic in West Carpathians. Geol Inst D Štúr, Bratislava, 314 pp

Rastislav Vojtko is an Associate Professor at the Department of Geology and Palaeontology, Comenius University in Bratislava. He is engaged in research and teaching of the regional geology and early Alpine tectonic evolution of the Western Carpathians, particularly in the Vepor-Gemer region, Malé Karpaty and Považský Inovec Mts, and in the Pieniny Klippen Belt.

Dušan Plašienka is a Professor at the Institute of Geography, Faculty of Science, Pavol Jozef Šafárik University in Košice. He focuses on the landscape water balance, climate change, fluvial geomorphology and landscape ecology. In the past, he focused on the evaluation of erosion and the effectiveness of hydromelioration in the East Slovakian Lowland.

Michal Kováč is an expert in sedimentology, especially of siliciclastic formations. Solved Western Carpathians palaeogeography and geodynamics during the Neogene and Quaternary within the Central European context. He was many years head of research team within many Slovak and international projects interested in research of Neogene basins (sedimentology, biostratigraphy, sequence stratigraphy, tectonics, interpretation of geophysical data, geomorphology, as well as palaeoenvironmental and palaeoclimatic changes).

3 Climate in the Past and Present in the Slovak Landscapes—The Central European Context

Marián Melo, Milan Lapin, and Jozef Pecho

Abstract

The aim of the chapter is to provide an overview of climatic conditions in the territory that is now Slovakia and the wider Central European region from the Mesozoic Era to the present. The climate for the earliest period can only be reconstructed in the broadest terms. The level of detail possible increases as we approach to the period of the instrumental observations. The regular systematic meteorological measurements began in Slovakia in the middle of the nineteenth century. The mean air temperature in Slovakia has increased by about 2 °C since 1881. The results show that Slovakia's climate, like everywhere on Earth, is constantly changing and evolving.

Keywords

Warm period • Cold period • Glaciers • Air temperature • Precipitation

3.1 Introduction

Through the historical epochs, changes in the Earth's global climatic system (mainly characterised by slow, progressive changes in the global climate) were mainly the result of natural factors such as long-term solar constant fluctuations, changes in the Earth's orbital parameters, tectonic movements of the Earth's core, changes in deep-sea and surface currents in the oceans, volcanic activity and long-term changes in atmospheric chemistry. From time to time there were sudden, catastrophic events such as asteroid impacts, causing drastic and abrupt climate changes. In the most recent period, humans have begun to affect the climate through deforestation, burning of fossil fuels, environmental and air pollution and the like. Scientific research has built up evidence that the growing concentrations of greenhouse gases in the atmosphere caused by human activity are having a significant impact on the Earth's climate system (IPCC 2013). Higher concentrations of greenhouse gases contribute to global warming that produces rapid changes throughout the system.

The territory of Slovakia is situated in a temperate climate zone with a regular seasonal cycle. Continentality effect grows from west to east. Slovakia has a very diverse climate relative to its area thanks to the variations in altitude, considerable terrain segmentation, the variety of natural conditions in its territory and varied conditions of air circulation in the individual regions.

The climate was characterised by very different temperatures and precipitation levels in the past compared to the present. The present paper aims to provide an overview of climatic conditions in the territory that is now Slovakia and the wider Central European region from the Mesozoic Era to the present. While only the broadest outlines can be given for the earliest period, there is more temporal and spatial data available for later periods. The level of detail possible increases as we approach the period of instrumental observations. Much of the description of paleoclimate and historical climate is based on citations from published sources. The description of climatic conditions from the period of instrumental observations is based on the analysis of data provided by Slovak Hydrometeorological Institute.

The Mesozoic Era is the interval of geological time from about 252–65.5 Ma (Ma = million years ago), following the Palaeozoic, and is usually divided into three periods: Triassic, Jurassic and Cretaceous. The following era is called

M. Melo (✉) · M. Lapin
Comenius University, Mlynská dolina F1, 842 48 Bratislava, Slovakia
e-mail: Marian.Melo@fmph.uniba.sk

M. Lapin
e-mail: Milan.Lapin@fmph.uniba.sk

J. Pecho
Slovak Hydrometeorological Institute, Jeséniova 17, 833 15 Bratislava, Slovakia
e-mail: Jozef.Pecho@shmu.sk

M. Lehotský and M. Boltižiar (eds.), *Landscapes and Landforms of Slovakia*, World Geomorphological Landscapes,
https://doi.org/10.1007/978-3-030-89293-7_3

the Cenozoic, which is divided into the following periods: Palaeogene (from 65.5–23 Ma), Neogene (23–2.58 Ma) and the Quaternary (the last 2.58 million years). The Palaeogene itself includes three epochs: the Palaeocene (65.5–55.8 Ma), Eocene (55.8–33.9 Ma) and Oligocene (33.9–23 Ma). The Neogene is divided into two epochs: the Miocene (23–5.33 Ma) and Pliocene (5.33–2.58 Ma). The Quaternary is also divided into two epochs: the Pleistocene (2.58 Ma–11.7 ka) (ka = thousand years ago) and the Holocene (the last 11,700 years BP) (BP = before the present (AD 1950)).

3.2 Climate in the Mesozoic

The geological conditions on Earth in the past were significantly different from what we experience at present. The territory of Slovakia has experienced several periods of marine sedimentation as well as periods of terrestrial development and extensive mountain formation. Towards the end of the Mesozoic, in the Cretaceous period (145–65.5 Ma), the Central European region was strongly affected by the Alpine orogeny, which has continued throughout the Cenozoic until the present, with its peak during the Oligocene. At the end of the Mesozoic, most of Slovakia's territory was dry land.

The climate of the Mesozoic can only be reconstructed in the broadest terms. The Mesozoic climate on Earth was predominantly warm. Between the middle Triassic and the middle Cretaceous, climates were characterised by mean annual temperatures higher than those of the present day, possibly as much as 10 °C warmer on a global basis at some times (Frakes 1979). The middle and higher latitudes in particular were warmer than they are at present (the polar regions had no ice-caps) and therefore climatic zones were not as sharply differentiated as they are today, especially at the beginning of the Mesozoic. The temperature gradient between the polar and tropical regions gradually became steeper towards the end of the Mesozoic. During the Mesozoic era temperatures ranged from 10–20 °C at the poles to 25–30 °C at the equator (Oliver 2008). Mesozoic climates also featured arid conditions which were more widespread than at present, and which locally and temporally extended 10° of latitude farther poleward (Frakes 1979).

3.3 Climate in the Palaeogene and Neogene

An important source of information on the pre-Quaternary Cenozoic climate is the development of plant and animal life, which includes forms that we know today. The finer details of Earth's climate since 65 Ma can be deciphered from investigations of deep-sea sediment cores (Zachos et al. 2001). The cooling trend that began at the end of the Cretaceous continued throughout the Palaeogene and the Neogene, although there were warmer oscillations within these periods. During the most prominent and best-studied warm period of the Cenozoic, the Palaeocene–Eocene Thermal Maximum (about 55 Ma), the global temperature increased by more than 5 °C in less than 10,000 years (Zachos et al. 2008). The middle Miocene Climatic Optimum between 17 and 14.7 Ma is a generally warm time interval punctuating the long-term Cenozoic cooling trend (Zachos et al. 2008; Greenop et al. 2014). The global cooling process was most noticeable in the mid-latitudes and especially in the higher latitudes. Subtropical flora and fauna ended in the polar regions, which experienced a sharp decrease in precipitation. At high latitudes, a new climatic zone appeared during the Cenozoic. Its meteorological regime resembled one that is now found in the mid-latitudes. The winter temperature dropped below freezing and snow cover was formed. There was greater thermal zonation, with greater contrasts in temperature during the year and greater differentiation of climate zones all over the Earth.

The Cenozoic experienced various marked short-term cooling episodes: at the mid-Palaeocene (ca. 60 Ma), in the Middle Eocene (ca. 45 Ma); at the Eocene–Oligocene boundary (ca. 38 Ma); and in the Middle Oligocene (ca. 28 Ma) (Oliver 2008). The crucial event for cooling of Antarctica was the opening of Drake Passage between Antarctica and South America about 35 Ma (Wilson et al. 2000). Around 5 Ma, the already substantial ice sheets in Antarctica underwent rapid growth and attained their present volume (Robinson and Henderson-Sellers 1999). The closure of the gap between North and South America between 4.6 and 2.5 Ma may have initiated Northern Hemisphere glaciation (Wilson et al. 2000). The climatic deterioration of the later Neogene eventually led to the glacial events of the Pleistocene (Oliver 2008).

Continental climate change in Central Europe during the last 45 million years is primarily characterised by the decrease of cold month temperature means, nearly 20 °C between the Eocene and Pliocene. In contrast, mean annual temperature and warm month means did not undergo as significant a change from the Eocene to the Pliocene. Only near the end of the Pliocene did these variables show a notable decrease. Change in precipitation was also small, with over 1,000 mm during most of this interval. Cold month means and warm month means trends clearly diverged during the Cenozoic, implying an increase of seasonality in Central Europe during this time (Mosbrugger et al. 2005).

Terrestrial development on the territory of present-day Slovakia continued in the early part of the Cenozoic. At that time climate was tropical, with one dry season and one wet season. Until the mid-Eocene, Central Europe received heavy rainfall in an annual summer monsoon that came from the sea to the mainland. At that time, what is now the Mediterranean Sea was the western end of the Tethys Ocean, so warm currents could flow in from where the Indian Ocean is now. Eastern Europe, especially the territory of present-day Ukraine, was rich in rainfall. Evergreen tropical and subtropical rain and marsh forests grew here. The period of terrestrial development in the territory of present-day Slovakia was interrupted by the progressive transgression of the sea, which flooded this territory from both the north and the south. Progress was particularly intensive in the Eocene. It is very likely that even when sea levels were at their highest in the Oligocene, the whole territory of Slovakia was not under water. There is abundant fossil evidence of this (mainly Foraminifera and Bivalvia, to a less extent Gastropods) (Paturi 1995). During the Oligocene, drier climate was again succeeded by damper climate in Central Europe, with the air temperature slightly decreasing. The tropical swamp forests that had been typical of Central Europe in the previous Eocene were gradually replaced by subtropical mixed forests. Increasing precipitation meant that these also took on a character of swamp forests or alder forests, especially in large valleys. At the same time, the forests became thinner and acquired a savannah character. Palm trees still grew in the forests, though later they disappeared from the deciduous forests. After uplift of the Carpathian block, the sea again withdrew from north and south Slovakia. The formerly contiguous water bodies ceased to exist. They became bays and lagoons reaching deep into the mainland and towards the end of the Oligocene they began to dry up (Paturi 1995).

The climate continued to cool unevenly in the Miocene. The temperature drop was more noticeable at higher latitudes. One effect of falling temperature was the creation of the ice sheet that has covered most of Antarctica since 20 million years ago. Continental glaciers had not yet developed in the northern hemisphere at this time. There were only alpine glaciers in Iceland. The large quantity of water bound up as ice in Antarctica and the orogenic processes that produced the Alps, Carpathians and Himalayas led to a worldwide decline in sea levels. By the period between 15 and 13 Ma, the only flooded areas in the territory of present-day Slovakia were the Vienna Basin and East Slovak Basin, though there were also large bodies of water in the Danube Basin and the intra-Carpathian basins. The sea left thousands of metres of sedimentary deposits containing many fossilized animals (the sands of Sandberg Hill near Bratislava have yielded evidence of more than 300 species of animals, including fossilised gastropods, bivalves, oysters, sea urchins, sponges and foraminifera, while for larger species there are sharks' teeth, the vertebrae of a whale and seal bones). After the sea retreated from the lowlands and the Carpathian basins between 13 and 11.5 Ma, the territory of Slovakia remained dry. Rainfall amounts in Europe were very variable during the Miocene. The dry beginning of the Miocene was followed by a wet period, then another period of aridity and at the end of the Miocene the climate became once again wet. In the drier periods, some of the forest areas of Europe became steppes, including the Vienna Basin and other previously humid parts of Eastern Europe such as Ukraine. During the Miocene, massive seams of brown coal were formed from peat bogs in Central Europe. The main locations in the territory of present-day Slovakia were the South Slovak Basin (Pôtor, Modrý Kameň) in the early Miocene and the Upper Nitra Basin (Handlová and Nováky basins) in the late Miocene.

Climate cooling continued in the Pliocene. Around 5 Ma, the first large ice sheets appeared on land in the Arctic, in Alaska, Greenland and Iceland. The whole Arctic became ice-covered around 3 Ma. The European steppes moved south (Paturi 1995). At the Kostná valley site near Hajnáčka (southern part of central Slovakia), rare fauna fossils have been found, including tapirs, monkeys and pandas alongside the flora that grew in this area around 3 Ma. These deposits were buried by volcanic ash from an eruption of one of the active volcanoes in the region, ensuring their perfect preservation. Many of the tree species that occur in Central Europe today began to populate the region's forests. Subtropical flora continued to grow alongside. The warm climate of this period is also evidenced by old relict or fossil soils, such as terra rossa. This soil developed in limestone areas in warmer climates at the end of the Neogene. In Slovakia fragments of it have been preserved, mainly as soil sediment in various karst areas, often as cavity fillings (Mičian 1972). Several species of flora and fauna have survived from this period in Slovakia. These pre-Quaternary relics are most richly represented in the Carpathian Mountains, such as *Campodea augens*—an insect species living in mountain soil and Red-eyed Apollo (*Parnassius apollo*)—an alpine butterfly. Plant species include *Daphne arbuscula*, which grows as a paleoendemic species only on the Muránska Planina Plateau, and Smoke tree (*Cotinus coggygria*), which grows mainly in southern Europe and has its northern limit in Slovakia (Korbel 1972; Futák 1972).

3.4 Climate of the Pleistocene

3.4.1 Period until the Last Glacial Maximum

The Earth's global cooling reached its peak in the Pleistocene. The basic outline of the Pleistocene was an alternating sequence of cold events (glacials) and warm intervals (interglacials). There were also warmer and cooler episodes within the glacials and interglacials, the former known as interstadials and the latter considered as stadials. In the areas not covered by ice, dry periods alternated with periods of intense precipitation (pluvial periods). During glacials, ice sheets (continental glaciers) and sea ice expanded to lower latitudes and during interglacials they retreated towards the poles. Similarly, mountain glaciers extended to lower altitudes during glacials and retreated into headwater valleys in interglacials. Terrestrial glaciers were most extensive in humid climates.

Glacials have different designations and different timing based on the origin of the main ice centres (Böse et al. 2012; Engel et al. 2015; Hughes et al. 2013; Lindner et al., 2003; Lundqvist 1986). In the Alpine region, which can be taken to include the Western Carpathians in Slovakia, six glacials are distinguished in the Quaternary: Biber (2.4–2.1 Ma), Donau (1.7–1.38 Ma), Günz (1.20–0.82 Ma), Mindel (0.44–0.32 Ma), Riss (0.18–0.12 Ma), Würm (0.07–0.01 Ma), together with their corresponding interglacials: Biber-Donau (2.1–1.7 Ma), Donau-Günz (1.38–1.20 Ma), Günz-Mindel (0.82–0.44 Ma), Mindel-Riss (0.32–0.18 Ma), Riss-Würm (0.12–0.07 Ma) and Holocene (0.01 Ma to the present).

During the last glacial period (Würm in the Alpine region, Weichselian (Vistulian) in northern parts of Europe), the maximum area of the Northern Hemisphere ice sheets was equal to approximately 90% of the maximum achieved during the last million years of the Pleistocene (Robinson and Henderson-Sellers 1999). The onset of the last glacial was about 70 ka. The cold period following the initial fall in temperature lasted for only a few thousand years and was followed by a relatively clement period (an interstadial) that lasted until about 30 ka (Oliver 2008). A further cold period, the Last Glacial Maximum (LGM), followed. This lasted about 10,000 years and the lowest temperature and the greatest extent of ice sheets of this entire glacial episode were attained at about 20–18 ka (Oliver 2008). According to other various sources, the last Pleistocene ice age reached the LGM in Europe between 37 and 16 ka. The maximum ice sheet limit during the Late Weichselian was time-transgressive in Poland and occurred at 24–19 ka, becoming younger to the east (Marks 2012). During the last glacial cycle (Würmian Stage), the Alpine ice sheet was most extensive between 30 and 18 ka (Hughes et al. 2013). The timing of the local LGM glacier advance in Central Europe was as follows (in brackets: equilibrium line altitude of local glaciers (m a.s.l.)): in Eastern Alps 26.5–21 ka (900–1100), Bohemian Forest 27–22 ka (1000–1160), Krkonoše Mts. 24–21 ka (990–1170), Western Carpathians 26–20 ka (1460–1700), Easthern Carpathians 37–26 ka (1631–1765), Southern Carpathians 26–19 ka (1700–1800), Dinaric Mountains 23–16 ka (1900–2300), Stara Planina and Rhodopes 25–23 ka (2150–2290) (Engel et al. 2015). Two LGM oscillations in the High Tatra Mountains were dated at 26–21 ka and at around 18 ka (Makos et al., 2014). The maximum advance of glaciers on the southern flank of the High Tatra Mts. was around 22.5 ± 2.9 ka (Engel et al. 2015). At the Würm glacial maximum global sea level was lower by at least 120 m (Frakes et al. 1992), though according to Robinson and Henderson-Sellers (1999) the sea level dropped by approximately 85 m. Climate reconstructions from pollen records indicate winter cooling of 5–15 °C across Europe during the LGM, with the greatest cooling in western Europe, and precipitation decreases of as much as 300–400 mm per year in the west (Clark et al. 2012). Sea-surface temperature fell by as much as 10 °C in mid-latitudes of the North Atlantic and 3 °C in the Caribbean in this period (Robinson and Henderson-Sellers 1999). During the LGM, some 20 ka, the temperature of Greenland was lower by up to 20 °C, but the climate was probably only a few degrees colder than normal in the tropics (Rapp 2019). Climate modelling results show that the LGM glaciers (maximum advance) could have advanced in the High Tatras when the mean annual temperature was lower than today by 11–12 °C and precipitation was reduced by 40–60% (Makos et al. 2018).

3.4.2 Ice Sheets

During the last glacial period several large land-based ice sheets as much as 3 km thick existed in the northern hemisphere (Frakes et al. 1992). During the LGM the Fennoscandian Ice Sheet coalesced with the Svalbard-Barents-Kara and the British-Irish Ice Sheets (= the Eurasian ice sheet) (Šibrava 1986; Siegert et al. 2001; Svendsen et al. 2004; Hughes et al. 2013, 2015). The Eurasian ice sheet complex as a whole attained its maximum extent (5.5 Mkm^2) and volume (circa 9.7 Mkm^3, circa 24 m Sea Level Equivalent) between 21 and 20 ka (Hughes et al. 2015). The Fennoscandian ice sheet as part of the Eurasian ice sheet complex reached deep into mainland Central Europe. Based on the evidence of moraines and stray boulders left over from washed-away moraines containing red Scandinavian granite, the ice sheet is known to have reached the foothills of the Carpathians around the Moravian Gate (Ostrava) in the Czech Republic and Krakow in southern Poland. Its closest approach to the Slovak border was 30 km in the Mindel (Elsterian) glacial. It also came almost as close in the Riss (Saalian) glacial. In the last ice age (Würm) (Weichselian) the ice sheet terminus was 270 km from Slovakia (Lukniš 1972).

During the glacial periods the territory of Slovakia was on the margins of the Fennoscandian Ice Sheet and was covered with cold, non-forested steppes. Once ice sheets formed, they created an area of high air pressure around which cold, falling winds gathered dust and fine sand into loess and sand dunes. The cold steppes in Central Europe, whose soil, which was covered only in light vegetation, was more susceptible to wind erosion. In this way, aeolian sediments were formed in several places in Slovakia, such as wind-blown sand (Záhorská Lowland, the Danubian Lowland, the Eastern Slovak Lowland) and loess deposits (in hill lands of the Danubian Lowland, the Eastern Slovak Lowland, South Slovak Basin and Košice Basin). The flora of Central Europe in the Ice Age was typical of subarctic or arctic steppes (tundra vegetation). The land further away from the ice sheets became a subarctic loess steppe, covered by dwarf shrubs and occasional trees. When the ice retreated during the warm intervals, the subarctic steppe was replaced by birch or pine forests, sometimes by mixed deciduous forests. Terra fusca as a relict soil from interglacial periods has been preserved to this day on plateaus such as the Slovak Karst, Muránska Planina and Slovak Paradise (Mičian 1972). There are several locations in Slovakia that have yielded remains of animals that lived in the last ice age and are now extinct such as the woolly mammoth (*Mammuthus primigenius*) and the cave bear (*Ursus spelaeus*). The woolly mammoth lived on open grassland in cold steppes. Most mammoths died out at the end of the last ice age. In Central Europe this extinction happened more than 12 ka, possibly because the grasslands (steppe tundra) significantly decreased and were replaced by forests. The most significant Slovak woolly mammoth finds include those at Banka (Piešťany), Beša, Čaňa, Moravany, Šal'a and Veľký Meder. Another important site is Malé Leváre, where remains were found when dredging gravel sediments from the River Morava. Fragments of mammoths were also found in sediments in the Danube at Nové Košariská. The most complete and best-preserved mammoth skeleton in Slovakia was discovered at Senec in 1961. It includes nearly the whole skeleton of the forelimbs and an almost completely preserved tusk. The find of a mammoth skull (so far the only one in Slovakia) was reported in the vicinity of Borovce. A mammoth tooth (molar) was discovered in the bed of the River Kysuca near the village of Krásno nad Kysucou. At the end of 2003, the tusks and part of the limb of a mammoth were discovered during the construction of a motorway tunnel near the zoo in Mlynská dolina in Bratislava. Remains of an older species, the steppe mammoth (*Mammuthus trogontherii*), have also been found in Slovakia. They lived during the mid-Pleistocene, approximately 700–370 ka. The cave bear is a type of bear that lived in Slovakia during the late Pleistocene, from 270 to 10 ka. They disappeared from most of Eurasia 20–30 ka and completely died out about 10 ka. Cave bear remains have been discovered in several caves in Slovakia, such as Važecká Cave, Bear Cave in the Jánska Valley in the Low Tatras and Jánošík Cave in the Strážovské Vrchy Mountains. The richest finds of cave bear bones come from two caves in the Slovak Paradise—Bear Cave near Letanovce and Stratenská Cave near Stratená.

3.4.3 Mountain Glaciers

During glacial events there were also mountain glaciers in Central Europe, mainly in the Alps but to a lesser extent also in the Carpathians. On the northern side of the Alps, the glaciers were very long and extended far into the Alpine foothills. Compared to the longest Alpine glacier of the present day, the Aletsch Glacier, the Inn Glacier (340 km) was thirteen times its length and the Rhine Glacier (200 km) nearly eight times its length (Paturi 1995). Evidence of glaciers can be found both in the Slovakian and Polish parts of the mountains as glaciofluvial deposits located at different altitudes, and in some cases also as terminal and lateral moraines. There are no moraines for the three oldest glaciations, Biber, Donau and Günz (Lindner et al. 2003). A greater extent of the Tatra Mts. glaciers occurred during the Riss I (Odranian) Glaciation, while they were less extensive during the Riss II (Wartanian) Glaciation. During the Würm (Vistulian) Glaciation the glaciers were surprisingly large. This might have resulted from many factors, including changes in atmospheric circulation responsible for the distribution of precipitation (Lindner et al. 2003).

During glacials, there were mountain glaciers in the Tatras (in the Western Tatras in the Roháče range, the Liptov Tatras, the Red Mountains and the Liptov Hills; in the Eastern Tatras in the High Tatras and less so in the Belianske Tatras) and the Low Tatras, with traces of glacial activity also in the Oravské Beskydy Mountains and the Malá Fatra Mountains. The longest glaciers were in the Tatras (specifically the High Tatras). During the last ice age (Würm) the glacier in the Bielovodská Valley was 13.0 km long and 280 m thick, the glacier in Kôprová Valley was 12.5 km long and 250 m thick, and the glacier in Mengusovská Valley was 10.7 km long and 200 m thick (Lukniš 1972). Another source on the Tatra glaciers during the Würm describes the largest glacier in the Bielovodská Valley as having a total area of 43.6 km^2, a length of 13.4 km, a maximum width of 2.2 km and a maximum thickness of 400 m (Zasadni et al. 2015). There were a total of 55 glaciers in the Tatras in the LGM with a total surface area of 280 km^2 and a volume of 24.6 km^3 (Zasadni et al. 2015). On the southern slopes of the High Tatras, glacial cirques are located at altitudes between 1,600 and 2,200 m a.s.l. The terminal (end) moraines in the High Tatras now form a depositional package around 80 m thick. There are also places in the Tatras with mountain lakes of glacial origin (tarns). In several places in the Tatras, moraine sediments have been found from older glaciers that had a greater extent than those of the last ice age (Würm) and thus testify to a more extensive glaciation in the Tatras in the more distant past. The glaciers in the western parts of the Tatras were short while those in the High Tatras spilled out from the mountains like piedmont glaciers (Zasadni et al. 2015). According to Lukniš (1972), the moraines from the penultimate glaciation (Riss) were strongly affected by water erosion and weathered, whereas the moraines from the last glaciation have not suffered so much from weathering. Based on the remnants of the end moraines and stray boulders, it can be said that the Riss glaciers were about 1 km longer in the main valleys of the Tatras and about 50 m thicker than the Würm glaciers. The even older moraines from the Mindel glacial are severely weathered and were all removed except for a small residue (Lukniš 1972). The Low Tatras had several glaciers in the valleys on the north side of the range, especially in the Ďumbier group. They left behind 31 cirques below the main ridge and on the upper slopes of the northern spurs. The longest glacier was 5 km long and 100 m thick (in the Križianka and Palúčanka valleys). There are only a few, small glacial lakes. In the Malá Fatra Mountains, glaciers created small cirques on the north-facing slopes below the summits of Chleb and Veľký Rozsutec (Lukniš 1972).

Several plant and animal species have survived in the territory of Slovakia from the last ice age to the present as glacial relicts. These are species that were widespread in the ice ages but have largely disappeared along with the retreat of the glaciers and have been preserved only in limited habitats. In the new warmer climate, mountain environments provided a suitable habitat for their survival in Central Europe. Other species gradually withdrew to remote areas and are now found only in northern Europe. In Slovakia, the largest number of glacial relict plant and animal species are in the Tatras, including *Gentiana frigida*, *Ranunculus glacialis* and *Dianthus glacialis*. The fairy shrimp *Branchinecta paludosa* is found in just one habitat in Slovakia—Furkotské pleso in the High Tatras. The copepod *Mixodiaptomus laciniatus* is also found in the Tatra mountain lakes. Glacial relicts found in other regions of Slovakia include the minute snail *Pupilla alpicola*, which lives in swampy habitats in the Liptov Basin, a species of chrysanthemum (*Dendranthema zawadskii*) that grows in the Pieniny area and *Ligularia sibirica*, a plant that grows by the River Hnilec in the Slovak Paradise. Insects have the most glacial relict species, such as the springtail *Tetracanthella arctica*, which lives in the Tatras (Korbel 1972; Futák 1972).

3.4.4 Period after the Last Glacial Maximum

The termination of the LGM was marked by a slight increase in air temperature. The climate of Central Europe was strongly affected by the continental and mountain glaciers even as they retreated (Engel et al. 2015; Stroeven et al. 2016). The Baltic Sea, part of the mainland and the highest mountains were still under continuous ice cover. Most of the climate variability from 19 to 11 ka is associated with the overall deglacial trend with relatively little variance at millennial time scales (Shakun and Carlson 2010). The most prominent deglacial events of the Northern Hemisphere are Heinrich Event (17.5–16 ka) and the associated Oldest Dryas cold period (18–14.7 ka), the Bølling-Allerød warm and moist interstadial period (14.7–12.9 ka) and the Younger Dryas cold event (12.9–11.7 ka) (Shakun and Carlson 2010). In some regions, a cold period known as the Older Dryas can be detected in the middle of the Bølling-Allerød interstadial. In these regions the period is divided into the Bølling oscillation, which peaked around 14.5 ka, and the Allerød oscillation, which peaked closer to 13 ka. Somewhere, the Older Dryas is negligible in the evidence.

In case of the southern flank of the High Tatra Mts. (Veľká and Malá Studená dolina valleys) the period of LGM was around 22.5 ± 2.9 ka (Engel et al. 2015). The first post-LGM glacier re-advance occurred no later than 20.5 ± 1.7 ka when the glacier terminated at the mouth of the Veľká Studená dolina valley trough. The subsequent period of an overall glacier recession was interrupted by a re-advance in the central section of the troughs no later than 15.5 ± 1.1 ka (Engel et al. 2015). During the Younger Dryas cold period, glaciers were restricted to the upper part of the investigated valleys. In the Malá Studená dolina valley, glacier termini descended slightly out of the cirque-in-cirque forms and retreated back over the cirque steps around 11.2–10.7 ka. The retreat of glaciers after 11 ka in the range is in accordance with the onset of the post-Younger Dryas glacier recession in central European mountains (Engel et al. 2015). Glaciers finally retreated from the Tatra Mts. by the end of the pre-optimal part of the Holocene (Lindner et al. 2003).

From the Oldest Dryas period (18.0–14.7 ka) onward, the reconstruction of the climate in the Central Europe region can be assisted by analysing pollen from peat bogs and the analysis of charcoal and plant seeds in archaeological finds (Krippel 1986). The aim of such analyses is to capture the state of vegetation in the broad vicinity of the find sites during the time period concerned. During the Oldest Dryas, much of the land in the present-day Slovakia remained frozen all year round. Average annual air temperatures in Central Europe were 10 °C lower than in the present and there was probably less precipitation (Krippel 1986). Wind conditions in the Oldest Dryas were probably also different than in the present. Especially in the spring and autumn months, Slovakia was exposed to strong westerly and north-westerly winds that picked up sand from dry riverbanks or brought loess sediment to this territory from the foot of the continental glacier. Evidence of aeolian activity is provided by wind-borne sand on the bed of the Cerová-Lieskové peat bog in western Slovakia (Krippel 1986). At this time altitudinal differentiation in vegetation did not exist and the whole territory belonged to one vegetation level. At lower altitudes the territory of Slovakia was covered by cold tundra steppe with thin taiga-type forests of pine and birch in sheltered, relatively warm places. In the Oldest Dryas, there were probably glaciers and permanent snow cover on peaks above 1,600–1,800 m a.s.l. (Krippel 1986).

The subsequent Bølling-Allerød interstadial (14.7–12.9 ka) was warmer and somewhat more humid. Especially the summers were relatively warm. At the onset of the Bølling, temperatures increased rapidly by 3–5 °C across western Europe (Clark et al. 2012). In Slovakia there continued to be frozen wasteland at altitudes above 1,000–1,100 m a.s.l., and even lower than that on the north-facing slopes. Glaciers remained only in smaller areas high in the mountains, above 2,000–2,200 m a.s.l. In this period, the first signs of differences in vegetation coverage based on altitude appeared. At lower altitudes, there were pine and birch forests with corresponding alluvial forests and forest steppe vegetation on rocks, sand and loess. Tundra lichens growing at the vegetation limit made up the higher level (Krippel 1986).

This warming was interrupted by a short cold regression (The Younger Dryas, 12.9–11.7 ka). The Scandinavian Ice Sheet readvanced, depositing an extensive moraine system across southern Norway, Sweden, and Finland. Mountain glaciers also readvanced in the Swiss Alps (Carlson 2013). The Younger Dryas Stadial was associated with a sudden temperature drop of 7 °C at the beginning of the period and a similar increase at the end (Boulton et al. 2004). Cooling at the start of the Younger Dryas ranged between 5 and 10 °C across Western Europe (Clark et al. 2012). Steppe tundra covered most of central and northern Europe again (Stuart 2005). In the territory of Slovakia, there was ice cover on the highest mountains above 1,800–1,900 m a.s.l. but it was not so extensive as in the Oldest Dryas. The signs of stratification of vegetation by altitude that began in the earlier Alleröd period disappeared again and there was practically just one vegetation level covering the whole territory. There was a frozen wasteland without taller trees or plants on all uplands higher than 800–900 m a.s.l., and above 600–700 m a.s.l. on the north-facing slopes. Compared to the previous period, the climate had a slightly drier character (Krippel 1986).

The average annual air temperature on the lowlands and lower lying hills of Slovakia in the Oldest Dryas was around −1 °C to −3 °C. In the Bølling-Allerød interstadial it increased to +2 to +3 °C and in the Younger Dryas stadial, an average annual air temperature fell to a value of around 0 °C (Krippel 1986). For comparison, the present-day average annual air temperature in the Danube, Eastern Slovak and Záhorie Lowlands is around +8.9 °C to +10.0 °C (in the period 1951–1980) and around +10.0 °C to +11.1 °C (in the latest 30-year period 1990–2019). According to Makos et al. (2018), during the Late Glacial stages the temperature decrease in the High Tatras changed from 10 °C during the Oldest Dryas to 6 °C during the Younger Dryas and precipitation lowering decreased from −50% to −30% or even −10%, respectively, comparing to modern conditions.

The Younger Dryas is the last event in the last glacial, followed by a renewed sharp warming trend in Holocene. This stadial ended abruptly in Europe at approximately 11.7 ka, with a rapid temperature increase of approximately 4 °C, but varied latitudinally, with a greater increase in the north (Clark et al. 2012).

3.5 The Climate in the Holocene

The Holocene covers the most recent 11,700 years (Walker et al. 2009) (remark: last 11 700 calendar years before AD 2000, or 11,650 cal. years before present (BP = AD 1950)). Within this period of warm interglacial there have been several significant climate variations. Change was especially rapid during the first few millennnia (Oliver 2008). During this period, ice began to retreat in the continental and mountain glaciers and also on the sea in the Arctic Ocean, leading to a gradual rise in sea level (to within a few metres). During the second half of the Holocene, human impact has become an increasingly important agency in the modification of natural environments (Oliver 2008).

According to the Blytt-Sernander classification based on peat stratigraphy, the Holocene can be divided into the following stages: the Preboreal (11,700–9,000 years BP), Boreal (9,000–7,500 years BP), Atlantic (7,500–5,000 years BP), Subboreal (5,000–2,500 years BP) and Subatlantic (2,500 years BP to the present). These phases may vary slightly in timing depending on the different dating methods and on the locality of research (e.g. Schrøder et al. 2004). Some authors prefer other classification, which includes Postglacial, the Holocene Climatic Optimum and Neoglacial.

Information on the climate in a territory before the start of systematic meteorological measurements can be drawn from various natural or anthropogenic sources. Reconstructions for the Holocene are based on data from climate proxy records such as ice cores, tree rings, pollen grains, marine and lake sediments, corals, speleothems. Valuable information is provided by finds of plant residues in lake and peat bog sediments, which occurred in larger quantities in formerly glaciated areas (Krippel 1986; Bodnariuc et al. 2002; Kalis et al. 2003; Felis and Pätzold 2004; Fischer 2004; KIHZ-Consortium 2004; Helle and Schleser 2004; Andrič et al. 2020). In addition to methods based on the study of preserved paleoclimatic records, the most recent part of the Holocene (approximately the last millennium) is also documented in historical records. The information that historical documents provide on the climate mainly concerns extreme weather phenomena such as intensive precipitation or their long-term deficits, storms, strong winds, heavy snowfall, deep snow cover, severe frosts, high air temperatures, extreme hydrometeorological phenomena such as floods and droughts (Brázdil et al. 2005, 2012; Brázdil and Kundzewicz 2006; Matejovič 2011; Stankoviansky and Pišút 2011; Pekárová et al. 2014; Melo et al. 2014, 2016, 2019; Kiss 2019). Although instrumental meteorological observations in Central Europe started in December 1654 in Warsaw, Poland (Przybylak 2010), the regular systematic measurements (continued until now) cover only about the last 250 years at a Central European scale and only about the last 170 years in Slovakia.

3.5.1 Period to the Beginning of Meteorological Measurements

The Preboreal period is characterised by an already irreversible warming of the cold climate. Precipitation amounts increased as the climate warmed. At the beginning of the period temperatures were on average about 5 °C colder than at present (1951–1980) (average annual air temperature in the lowlands of Slovakia reached around +4 to +5 °C). The abundance of birch pollen grains in the pollen spectrum of peatlands in Slovakia shows that summers in particular were still very cold at the start of the period. This was the result of rapid melting of the continental glaciers and the melting of mountain glaciers in Slovakia, which consumed a great deal of heat (Krippel 1986). Altitudinal differentiation of plant species began again in the Preboreal period. At the end of the Preboreal period, at least three quarters of the territory of present-day Slovakia was covered by forests. An average annual lowland temperature in Slovakia rose to +6 to +7 °C and the tree line was probably at 900–1,000 m a.s.l. The parts of the High Tatras and Low Tatras higher than 2,000 m a.s.l. were probably covered by ice or snow all year round (Krippel 1986).

In the Boreal period, the climate was considerably warmer, but with an abrupt event at ca. 8,200 BP (the so-called "8,200 Cold Period"). This sharp cooling was probably caused by an excessive supply of cold glacial meltwater from the Laurentide ice sheet in the Hudson Bay area and the following changes in the mechanism of ocean currents in North Atlantic (similar event probably also caused the Younger Dryas cold period). This was the most prominent Holocene climatic event in Greenland ice-core proxies, with approximately half the amplitude of the Younger Dryas, and occurred approximately 8.4–8.0 ka. Widespread proxy records from the tropics to the north polar regions show a short-lived cool, dry or windy event of similar age (Alley et al. 1997). An average annual air temperature in the lowland part of Slovakia reached +11 °C to +12 °C in the middle of the Boreal period, which is higher than present-day values (Krippel 1986). Favourable climate conditions enabled for intensive infiltration of thermophilic plant species into the territory of Slovakia from the south and allowed the tree line to climb higher. The period had significantly less precipitation than the present day. The summers were very warm and dry, while winters were quite harsh, with little precipitation. Evidence comes from a smaller number of pollen grains of hydrophytes and hygrophytes in individual peat bogs in Slovakia (Krippel 1986). It is also evidenced by the rapid spread of xerothermic elements in flora and fauna. Steppes and forest steppes were widespread in the dry regions of Europe. The warm and dry climate in the territory of Slovakia also facilitated spreading of shrub communities with hazel, which occurred more frequently at medium-high altitudes. They were gradually pushed out of lowland areas, especially in western Slovakia, by deciduous forests (mixed oaks) in the latter half of the period. There was further development of altitudinal differentiation of vegetation that began in the preceding Preboreal period and by the end of the Boreal at least three significant vegetation levels can be distinguished. Above the tree line (around 1,700 m a.s.l.), there were cold-loving alpine meadow communities with a similar species composition to the present (Krippel 1986).

The transition to the following period, the Atlantic (7,500–5,000 years BP; or 5,550–3,050 BC (BC = years before Christ)), was characterised by strong humidification and continued, albeit slower, warming of the climate. The warming trend culminated in this period. The Atlantic period is commonly referred to as the Climatic Optimum of the post ice age period. During the Climatic Optimum of the mid-Holocene temperatures may have been 1–3 °C higher than now (Oliver 2008) (Note: "now" or "today" is the period of the twentieth century (1901–2000)). Summer temperatures in both Antarctica and Europe were 2–3 °C higher than they are today (Robinson and Henderson-Sellers 1999). Summer precipitation conditions relative to the present were wetter in Central Europe. Rainfall was more plentiful also in the now-arid areas of North Africa and the Middle East (Saltzman 2002). The temperatures were the highest of the Holocene and the humidity of the climate was very favourable for all living organisms. In the first half of the period (the older Atlantic) there are reasons to believe that the lowlands of Slovakia received up to 70% more precipitation than at present. Average annual air temperatures in the lowlands of Slovakia may have reached 13 °C–14 °C (Krippel 1986). The summers were warm and humid, and the winters were mild, with abundant precipitation. In the middle of the period, the climate had a shorter dry episode. In the latter half of the Atlantic period, temperatures fell by 1–2 °C and the quantity of precipitation also decreased. Towards the end of the Atlantic, average annual air temperatures in the lowland regions of Slovakia reached values of around 11–13 °C (Krippel 1986). Short dry and wet episodes alternated, but the trend was towards an overall decrease of precipitation. Several peat bogs in Slovakia have retained pollen that can be analysed to characterise this period (Krippel 1986). The relatively warm and humid climate was very favourable for the rapid spreading of forests. Tree coverage reached its maximum extent in Slovakia at this time. Oaks reached all levels of the mountains, even leaving their pollen grains in peat bogs in the Tatras. The tree line was around 300–400 m higher than at present, at an altitude of 1,900–2,000 m a.s.l. At an altitude of 1,900–2,200 m a.s.l., there was a belt of scrub that was wider than in the previous period. Above 2,200 m a.s.l., there were stands of grass and herb alpine meadows (Krippel 1986). In this period humans made their first contiguous agricultural settlements in Central Europe. There is known evidence of ploughing and systematic cattle grazing at the end of the period. The oldest pollen grains of cereals in Slovakia were found in the peat bogs of Kameničná and Blahová Village in the Danube Lowland and date from this period (Krippel 1986). Around 5,500–5,000 years BP occurred the Piora Cold Period. The event is named after the Val Piora Valley in Switzerland, where it was first detected by using pollen analysis. This abrupt cold and wet period at the end of the Atlantic coincides with glacier advance and tree line decline in the Alps. Palaeoenvironmental and archaeological data from Arbon Bleiche, Lake Constance (Switzerland), give evidence of a rapid rise in lake-level dated by tree-ring and radiocarbon to 5,320 years BP. This west-central European climate change may have favoured the quick burial and the preservation of the Alpine Iceman Ötzi found in September 1991 in the Tyrolean Alps (Magny and Haas 2004). The abrupt cooling towards the end of the Atlantic period is also evident in the pollen diagram from the Kameničná peat bog in the Danube Lowland (Slovakia), with a clear increase in the pollen grains of pine (*Pinus*) on one side and a decrease in the pollen grains of representatives of mixed oaks (*Quercetum mixtum*) (Krippel 1986).

The Climatic Optimum was followed by a period of falling temperatures in the Subboreal period (5,000–2,500 years BP; or 3,050–550 BC). The climate was still relatively warm (average annual air temperatures in the lowlands of Slovakia reached 11–12 °C), but there was considerably less precipitation than in the previous period (Krippel 1986). The summers were very warm and the winters were colder, with lower precipitation amounts. The climate was similar in character to the Boreal period. The tree line was now at an altitude of about 1,800 m a.s.l., and at altitudes above 1,900–2,000 m a.s.l. there were alpine meadow communities (Krippel 1986). In this period the global sea level possibly reached a peak of about 3 m above the present levels at approximately 4,000 years BP, illustrating the lag between the glacial melt and the sea level rise (Robinson and Henderson-Sellers 1999). Palaeohydrological and palaeoecological investigations from five locations in northern Scotland reveal a distinct and large-scale shift to wetter climatic conditions which may reflect a major transition in climate from the mid- to late Holocene. Radiocarbon age estimates place this transition between about 3,900 and 3,500 years BP, although it appears to have occurred abruptly, possibly over a decadal to century time scale. This event appears to be synchronous with changes inferred for other regions, suggesting that it reflects a continental-scale, or possibly even global-scale, change in climate (Anderson et al. 1998). With reference to human cultures, the Subboreal period covers the Bronze Age and the early Iron Age. The climate in the period around 3,250 years BP (1,300 BC), which occurred during the Bronze Age, was relatively warm again (the Minoan Warm Period). At the end of the Subboreal and the beginning of the Subatlantic in the Iron Age (900–300 BC), the climate was generally humid, and the cold phase culminated at the beginning of the Subatlantic (probably around 500–300 BC). There is widespread evidence of regrowth of bogs throughout Europe (Robinson and Henderson-Sellers 1999).

The Subatlantic period began (550 BC) with a cooler climate combined with increasing precipitation. Especially the summers were cooler. Temperature minimum was reached at about 500 BC. It was a time of considerable advance of mountain glaciers (Saltzman 2002). Mean annual temperatures in Europe were 1–2 °C lower than today, caused by cool summers with precipitation and well-known glacier advances (Negendank 2004). Whereas in the previous periods (especially the Atlantic), the humidity conditions in lowland and mountain areas were relatively balanced, in the Subatlantic significant differences developed in climate of these regions, partly as a result of the deforestation of the lowlands while the mountains remained wooded (Krippel 1986). The climate of the Subatlantic period in Central Europe was generally characterised by a decrease in average annual air temperatures by 1–2 °C as it gradually came into its present-day character. The limit for the highest ranging trees, which are conifers in Slovakia, has been at 1,600–1,700 m a.s.l. since around the middle of this period. Above the tree line there is scrub and at altitudes above 1,800–1,900 m a.s.l. a belt of alpine meadows with grass and herbs developed (Krippel 1986).

In the time of the Roman Empire, there was a relatively warm climate and the period is referred to as the Optimum of the Roman time or the Roman Warm Period. This period started quite suddenly around 250 BC and ended around 350–400 AD (AD = after Christ). It was a period warmer than the Medieval Optimum and several Alpine passes were ice-free even in winter. High precipitation is documented also in North Africa (Negendank et al. 2004). Past summer temperatures in northern Scandinavia during Roman times and during the Medieval Warm Period were higher than the twentieth century warmth and the first century AD was the warmest 100-year period over the past two millenia (+0.6 °C on average relative to the 1951–1980 mean) of the Common Era (Esper et al. 2012).

In the early medieval period, there climate conditions deteriorated (Pessimum of the Migration time or Migration Cold Period). It was a cool and wet period, with glacier advances in various areas (Negendank et al. 2004). The drought in Eurasia had two maxima in this period, at about 300 AD, which probably triggered the migrations towards both China and the Roman Empire, and around 800 AD. Especially the winters in Europe were cooler during this period.

The warmer period that followed, lasting from AD 1150 to 1300 in most of Europe is known as the Medieval Warm Period (MWP) or Medieval Optimum (Brázdil and Dobrovolný 2010). Different parts of Europe experienced the start and end of this period at different times. The timing primarily depends on the availability and the time resolution of proxy data, and of the regional setting (Brázdil et al. 2005). In Western Europe and the North Atlantic, the Medieval Warm Period ran from AD 950 to 1200 (Brázdil et al. 2005; Brázdil and Dobrovolný 2010). Ogilvie et al. (2000), after combining documentary evidence with proxy data from natural archives such as ice cores and marine sediments, spoke about favourable conditions in the North Atlantic region during AD 800–1100. The Medieval Optimum was characterised by a warm, relatively dry, and storm-free North Atlantic Ocean (Saltzman 2002). Favourable climatic conditions enabled the Vikings to undertake long voyages that reached as far as North America. Temperatures in this region were higher during AD 800–1100 than in the following centuries, but not more than 1–2 °C (Ogilvie et al. 2000). While according to Negendank et al. (2004) this medieval warm period faced mean annual temperatures of 1–1.5 °C higher than today, allowing vine growth in NW Europe, Brázdil et al. (2005) noted that at the hemispheric scale, temperatures were not higher than during the second part of the twentieth century. Based on documentary data over the period AD 750–1300 for a region of western-central Europe severe winters were somewhat less frequent and less extreme during the MWP, AD 900–1300, than in the ninth century and from 1300 to 1900 (Pfister et al. 1998). At the beginning of the Medieval Optimum it was drier, becoming wetter later on (Negendank et al. 2004).

A period of gradual cooling followed. This cooling took Europe into the Little Ice Age (LIA) in 1550–1850, with the most pronounced phase in the period 1550–1700 (Brázdil et al. 2005). Some authors avoided defining universally applicable dates for the onset and end of the LIA due to great regional variations (Brázdil et al. 2005). Based on documentary evidence, Pfister et al. (1996, 1998) concluded from analyses of winter temperature proxies that the LIA started in Central Europe shortly after AD 1300. A run of cold winters from 1303 to 1328 was followed by a run of 'average' winters up to 1354. Then winter temperatures were extremely variable up to 1375. For the rest of the century they fluctuated somewhat below the average of the twentieth century (Pfister et al. 1996). The international climate history literature differentiates three big cooling episodes during the LIA in Europe: the middle of the fourteenth century, the middle of the seventeenth century and the middle of the nineteenth century (Rácz 2016). LIA is characterised by a mean annual temperature of 1 °C less than today, with severe cold winters and extreme variations (Negendank et al. 2004). The Alpine glaciers were significantly larger in extent in this period. Fluctuations of the longest Alpine glacier, the Great Aletsch Glacier, during the last millennium included a smaller advance at about AD 1120 and three extreme maxima around AD 1350 (this maximum is only radiocarbon dated with a possible dating error in the order of tenths years), 1650 and 1860 (Brázdil et al. 2005). The Gorner glacier (Canton of Valais, Swiss Alps) reached a maximum around 1385 (dendrochronologically dated and confirmed by documentary evidence) (Brázdil et al. 2005). The LIA was the most recent period during which glaciers maintained an expanded position in most parts of the globe (Brázdil et al. 2005). The Late Maunder Minimum (time interval 1675–1710) refers to the coldest phase of the LIA with marked climatic variability (especially in extreme values of temperature and precipitation) over wide parts of Europe. It coincided with a reduced solar and an enhanced volcanic activity, as well as a low number of sunspots. Most of the authors dealing with Late Maunder Minimum report cold winters and mostly cool summers in that period, e.g. the winter of 1685 was the most severe in Europe with an anomaly from −4.5 to −5 °C (KIHZ-Consortium 2004). A cool part of Late Maunder Minimum (1689–1699) is in Tatra Mountains not so extreme as a strong cooling during the next Dalton Minimum (1790–1820) (Niedźwiedź 2010). Volcanic eruptions that occurred during this period also had a significant impact on climate. The Lakagígar eruption in Iceland in 1783 was followed by the unusually cold winter of 1783/1784 in Central Europe. This winter season was characterised by a long cold period with low temperatures and a large amount of snow, strong windstorms and by series of floods across this part of Europe (Melo et al. 2019). During this period Johann Ignaz von Felbiger conducted early instrumental meteorological observations in Bratislava. Based on these observations the exceptionally cold character of the winter 1783/84 (especially in January) in Bratislava can be inferred (Melo et al. 2016). Climatic effects and impacts of the 1815 eruption of Mount Tambora were evaluated for the Czech lands by Brázdil et al. (2016). Extremely cold summer for 1816 (the "Year Without a Summer") in Europe was caused by this eruption. Furthermore, it coincided with a secular phase of low solar activity called the Dalton Minimum (Brönnimann 2015). During the characteristic periods of the Little Ice Age the climate became cool and rainy in the Carpathian Basin. The shifting of the precipitation balance is indicated by that from the fourteenth century onward the water level of Lake Balaton in Hungary was constantly rising, and the settlements on the southern shoreline were continually retreating, whereas several of them even disappeared. The cool and pluvial climate regime supposedly culminated in the seventeenth century, when Lake Balaton's level was metres higher, and there were even three islands at the time (Tihany, Szigliget and Fonyód) (Rácz 2016).

From the second half of the nineteenth century, the northern hemisphere entered a period of warming with a small peak in the 1940s, moderate cooling in the 1970s and accelerated warming after 1980. This continuing warm period (Modern Warm Period) is influenced by increasing atmospheric concentration of greenhouse gases produced by human activity.

The sources described above have been used to reconstruct the approximate course of average annual air temperatures at Hurbanovo from the period of LGM around 22.5 ka until the present (Fig. 3.1), though it should be considered just a theoretical possibility. The temperature course will naturally have to be adjusted and refined in future as new information becomes available based on further research. The present-day (0 years BP = AD 1950 = the twentieth century) average annual air temperature at Hurbanovo is +9.9 °C (in the period 1901–2000).

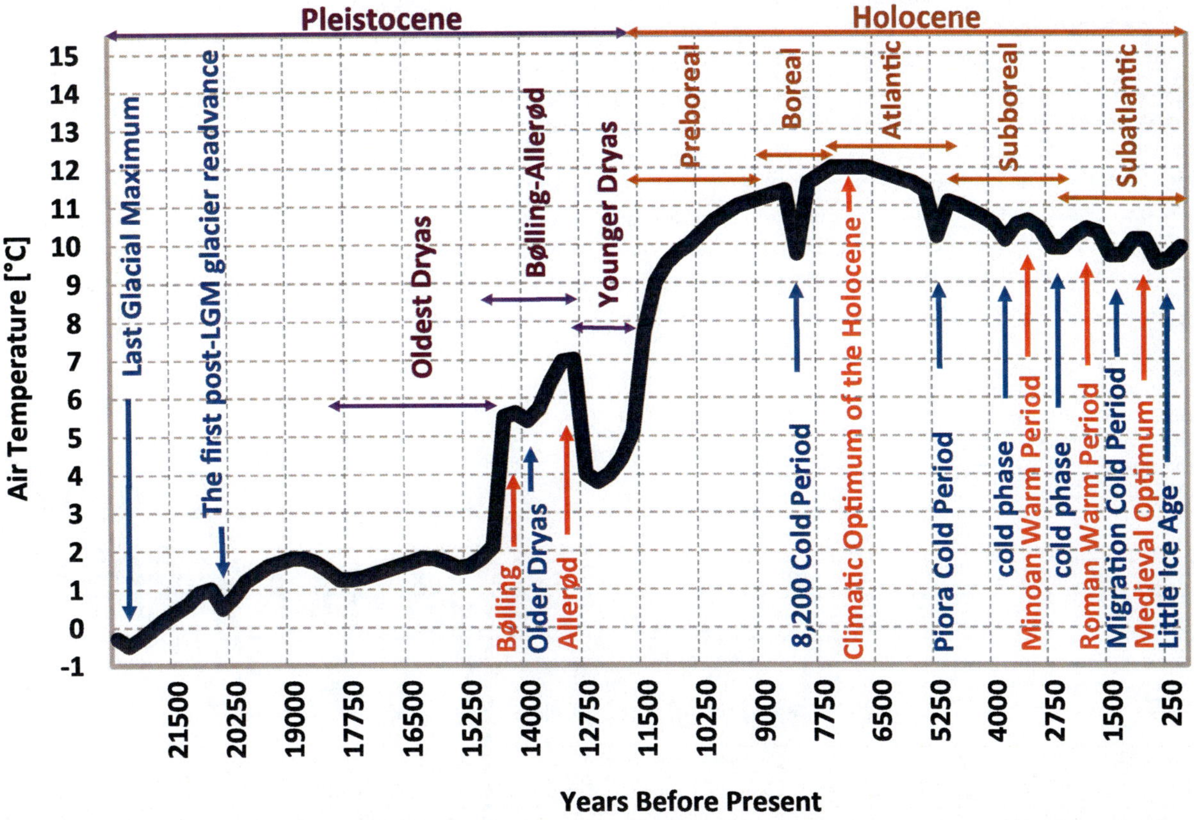

Fig. 3.1 The average annual air temperature at Hurbanovo over the past 22,500 years—only the assumed course based on analysis of the information sources referred to in the text

3.5.2 Period from the Start of Meteorological Measurements to the Present

Meteorological observations cover only a tiny fragment of the Earth's climate history. The basic meteorological instruments like the thermometer and barometer were invented in the first half of the seventeenth century. Systematic meteorological observations began a little later. In Central Europe the first observations were made in Warsaw, Poland, in either December 1654 or at the beginning of 1655 (Przybylak 2010).

The regular systematic meteorological measurements began in Slovakia in the middle of the nineteenth century. Meteorological observations were recorded at some places in Slovakia in earlier times, though only for short periods and after some time they died out completely (Brázdil et al. 2008; Melo et al. 2016). One of the most important meteorological stations in Slovakia is Hurbanovo (115 m a.s.l., SW Slovakia). Its location is typical for the Danube Lowland and its temperature series is currently one of the best in Slovakia. In the period of meteorological measurements from 1881 to 2019 the average annual air temperature at Hurbanovo Observatory increased from 9.3 °C in the oldest 30-year period 1881–1910 to 10.0 °C in the period 1951–1980 and to 11.1 °C in the latest 30-year period 1990–2019 (Fig. 3.2). The graph clearly shows the current warming phase, which has been developing in the world since the second half of the nineteenth century. According to report of IPCC (2007), in the hundred years from 1906 to 2005 the average global annual air temperature had increased by 0.74 °C ± 0.18 °C based on a linear trend, in case of Hurbanovo it was by 1.2 °C in the same century period based on linear trend and even by 2.2 °C in the period of 1881–2019 based on linear trend. Most of this warming in Hurbanovo has occurred in the last 40 years.

Hurbanovo is one of the warmest places in Slovakia according to the average annual air temperature. On the contrary, the coldest place according to the average annual air temperature is the observatory on Mt. Lomnický štít (2,635 m a.s.l., N Slovakia) with a value of −3.9 °C in the period 1951–1980 and −2.9° C in the latest 30-year period 1990–2019 (Fig. 3.3). The average January air temperatures in the case of Hurbanovo increased from −1.5 °C in the period 1951–1980 to 0.0 °C in the period 1990–2019, and in the case of Mt. Lomnický štít there was an increase from −11.3 °C (1951–1980) to −10.1 °C (1990–2019). An even greater growth in the same period was achieved in respect to the average July air temperature, namely, in Hurbanovo it rose from 20.1 °C (1951–1980) to 21.9 °C (1990–2019) and at Mt. Lomnický štít from 3.6 °C (1951–1980) to 5.2 °C (1990–2019).

As already mentioned before, meteorological measurements began in Slovakia in the middle of the nineteenth-century: about 10 meteorological stations performed complete measurements by the year 1881 (including air temperature and air humidity) and about 100 precipitation stations measured daily totals. The number of meteorological and precipitation stations increased continuously up to the year 1950 and then stabilised at about 100 complete meteorological stations and 700 precipitation gauges (including snow cover measurements). When analysing all the data measured, we noted that more than 30 meteorological stations have complete monthly and daily data of all the important elements for the period 1951–2019. About 550 of

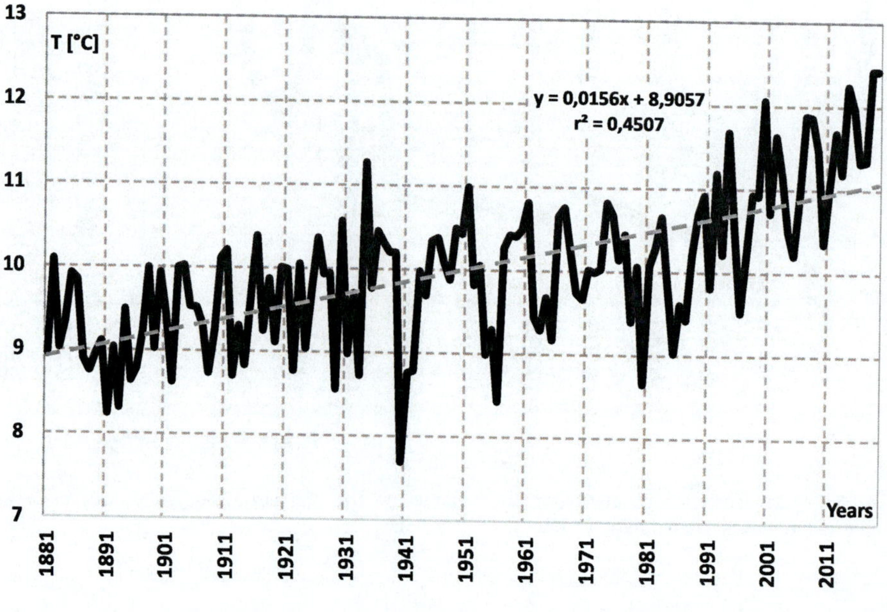

Fig. 3.2 The average annual air temperature (*T*) and the linear trend at the Hurbanovo Observatory during the period of meteorological measurements from 1881 to 2019

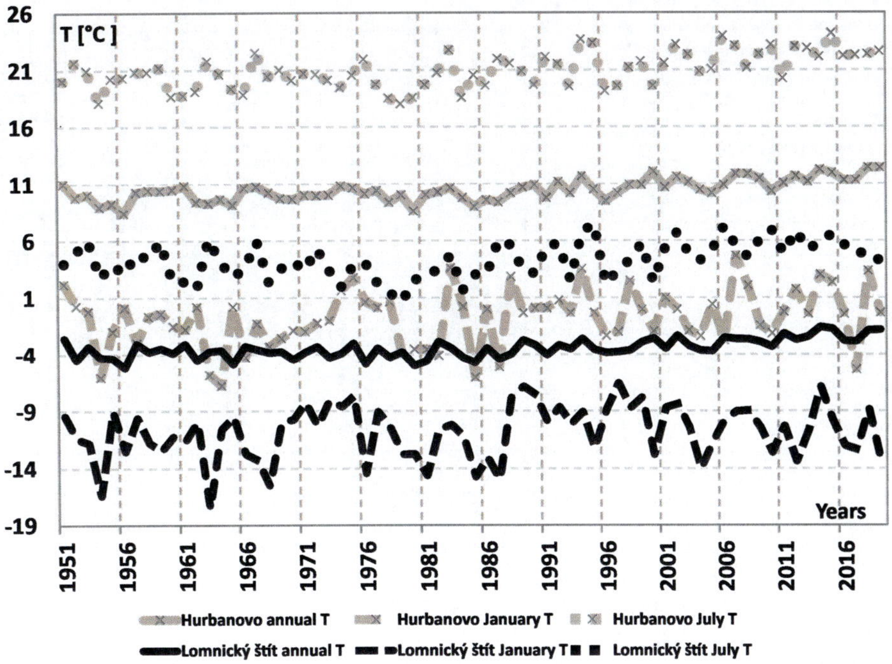

Fig. 3.3 Average annual, January, and July air temperatures (*T*) at the Hurbanovo and Mt. Lomnický štít observatories in the period 1951–2019

the precipitation stations from 1951 to 2019 can be considered as complete. The monthly precipitation data from 203 stations from 1901 to 2019 are complete. Using the monthly data from 203 precipitation stations, the monthly areal precipitation totals in Slovakia were calculated using the Double Weighted Averages method (Šamaj and Valovič 1982). Climatic analyses are periodically published in NCCC (1995, 1997,2001, 2006, 2009, 2014, 2018) and other sources.

In Slovakia, the air temperature has been measured by a classic mercury thermometer in a meteorological shelter 2 m above the ground at 7, 14 and 21 h MLT (mean local time) by the same method since 1851. To calculate the areal deviations of the monthly and seasonal means from the long-term averages of 1881–2019 three stations were selected. They are Hurbanovo, the Košice airport (230 m a.s.l., SE Slovakia) and Liptovský Hrádok (640 m a.s.l., N Slovakia). These or close next stations have been measuring the air temperature since 1881. Comparisons with the mean deviations calculated from more stations from 1981 to 2010 showed only insignificant changes from those calculated by the three stations mentioned. These data are published in the NCCC (1995, 1997, 2001, 2006, 2009, 2014, 2018). The graph (Fig. 3.4) shows the deviations of the mean temperatures and trends in Slovakia for a cold half-year (CHY, October–March) and a warm half-year (WHY, April–September). It is clear that the mean temperature in Slovakia has increased by about 2 °C since 1881 as a linear trend, and nearly the same increase has also occurred since 1981 both in the CHY and WHY. A higher increase in the mean temperature occurred in the months from January to August (Fig. 3.5 for Hurbanovo only).

Daily precipitation totals are measured in Slovakia with the Metra 886 national gauge (1 m above ground and with a 500 cm^2 orifice) by the same method since 1921 (some different methods were used before 1921). Annual and seasonal totals have not exhibited any significant trends since 1881 (Fig. 3.6). In the CHY a decreasing trend was found in southern Slovakia and an increasing trend in northern Slovakia (Fig. 3.7). A greater variability in the annual and seasonal totals and an increasing share of the convective precipitation has been registered since 1995, more information can be found in Lapin et al. (2009) and NCCC (1995,1997, 2001, 2006, 2009, 2014, 2018).

The current temperature and precipitation conditions in Slovakia in the period 1961–2020 are shown in Figs. 3.8 and 3.9. According to the average annual air temperature, the warmest area in Slovakia is the southern part of the Danubian Lowland (at some meteorological stations exceeding 10 °C) and the coldest areas are the highest locations in the Tatras and the Low Tatras (Fig. 3.8). The lowest values of the mean annual precipitation totals are achieved mainly in the Danubian Lowland and the most precipitation amount is recorded in the Tatras (Fig. 3.9).

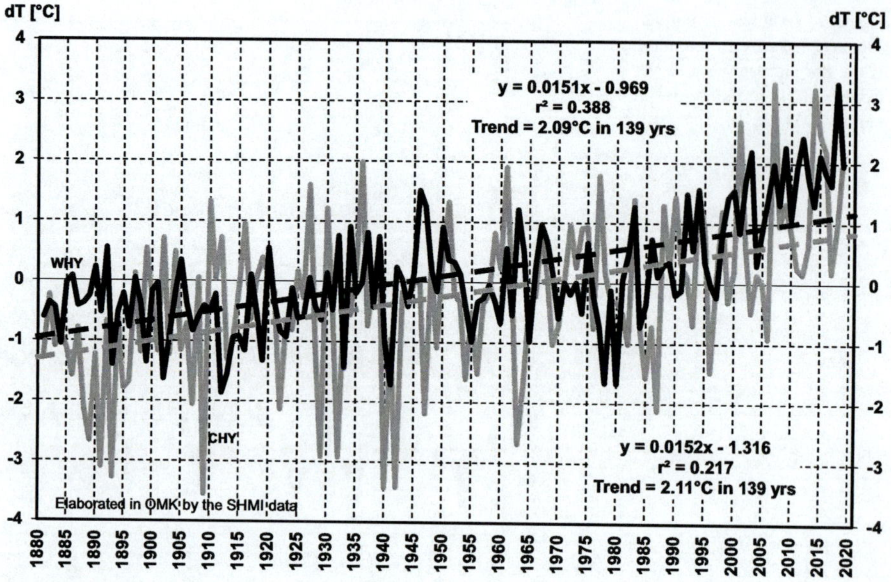

Fig. 3.4 Deviations of mean temperatures (*dT*) and trends in Slovakia for a cold half-year (CHY, October-March) and a warm half-year (WHY, April–September) compared to 1961–1990 in 1881–2019

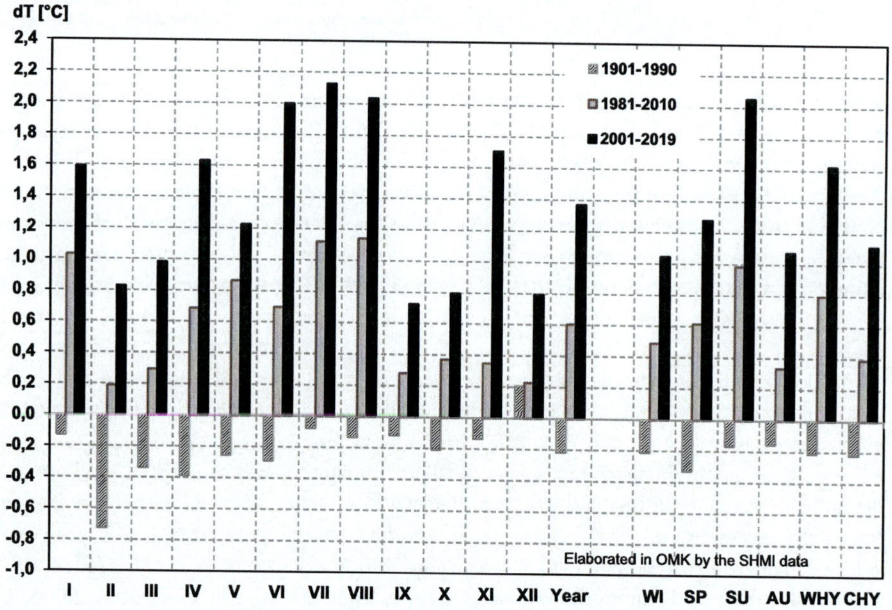

Fig. 3.5 Deviation of mean monthly and seasonal temperatures (*dT*) at Hurbanovo from the 1961–1990 averages in 1901–2019

Based on Konček's climate classification scheme, there are currently 17 climatic areas in Slovakia. There was no very dry region here until 1960. It appeared for the first time at Hurbanovo in the period 1951–1980 and has since spread to the north and east of Slovakia. Konček (1980) classified Slovakia's climate regions based on data for 1901–1950, while Lapin et al. (2002) used data for the period 1961–1990. Since both sources use the same method, they can be used to compare the borders of the climatic areas in Slovakia in these different time periods. Looked at side by side, they show that the most significant climate changes in Slovakia occurred in the Danube Lowland, the Eastern Slovak Lowland and the southern part of central Slovakia, where there was an increase in aridity and thus a change of climate classification. Thus, for example, the Eastern Slovak Lowland and the southern part of central Slovakia were classified as a warm, slightly dry subregion with a cool winter in the period 1901–1950 whereas it became a warm and dry

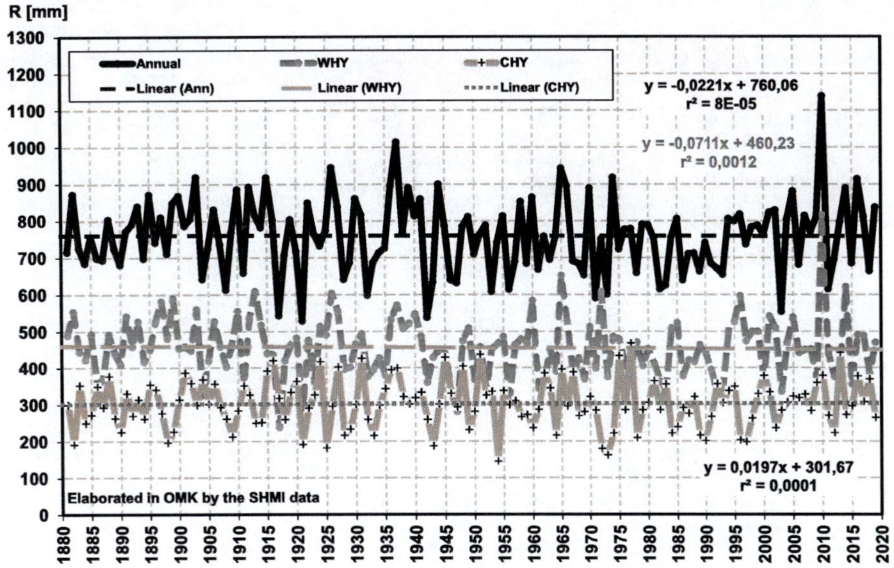

Fig. 3.6 Annual and seasonal (CHY (October–March) and WHY (April–September)) precipitation totals (*R*) in Slovakia in 1881/82–2018/19 (based on 203 stations)

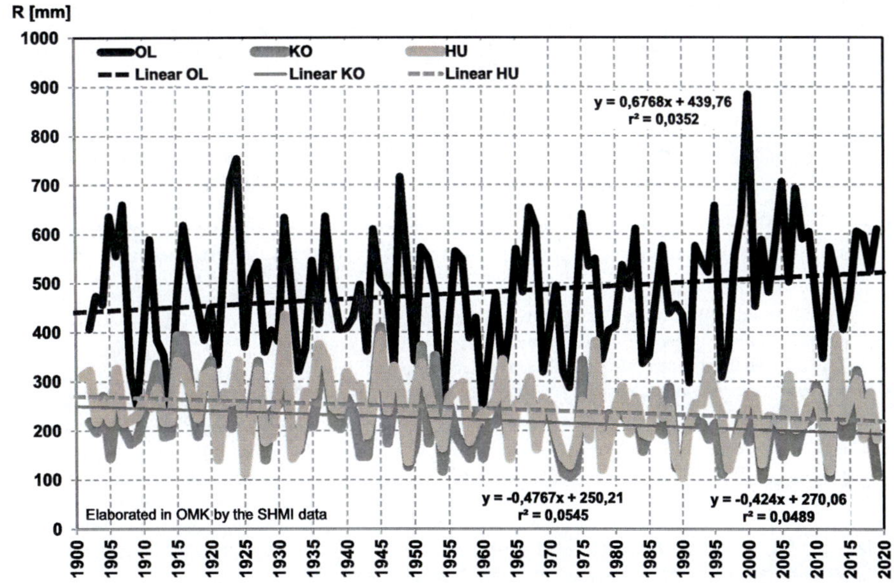

Fig. 3.7 Cold half-year (October–March) precipitation totals (*R*) for 3 stations in Slovakia in 1901/02–2018/19 (Hurbanovo, 115 m a.s.l., SW Slovakia, Košice, 230 m a.s.l., SE Slovakia, Oravská Lesná, 780 m a.s.l., NW Slovakia)

subregion and a cool winter in the more recent period 1961–1990. In the southern part of the Danube Lowland, the climate changed from a warm dry subregion with a mild winter to a warm and very dry subregion with a mild winter. Other changes and shifts in climate have been recorded in other parts of Slovakia (Konček 1980; Lapin et al. 2002). Using the Konček's climate classification, Melo et al. (2013) pointed out changes that took place in the region of the Tatra Mts. and the Poprad Basin between the periods 1951–1980 and 1980–2009. The boundary between the moderately warm and the cold climatic regions shifted about 150 m higher. The moderately cool subregion shifted about 100 m higher and the cool mountainous subregion about 155 m higher between the considered periods (Melo et al. 2013). The scenario for climate change is that the aridity of the Slovak climate will continue to increase (especially in the southern part of the territory) and the climatic areas will move northwards and to higher altitudes.

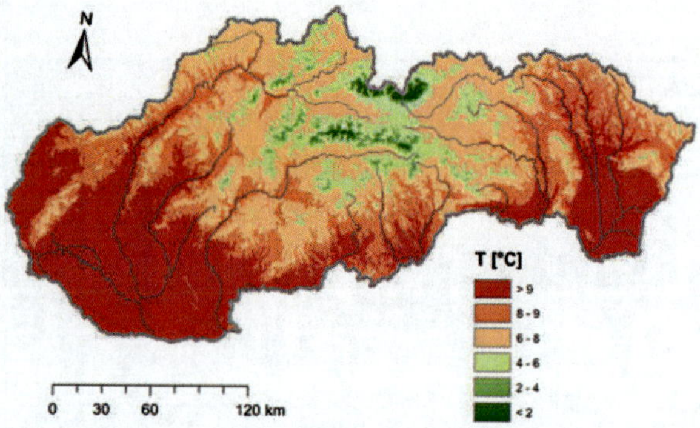

Fig. 3.8 Mean annual air temperature (T) in Slovakia in 1961–2020

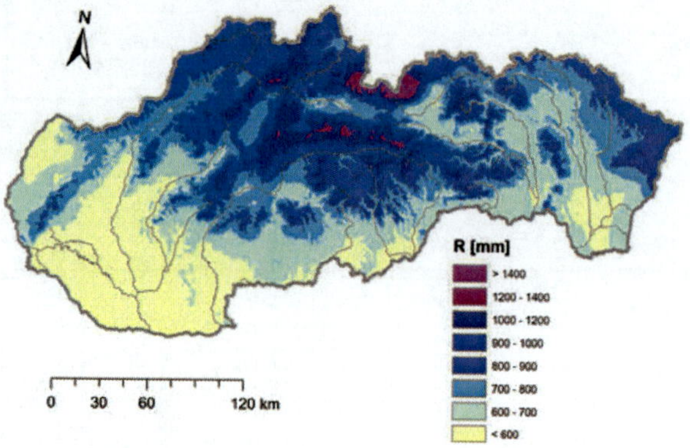

Fig. 3.9 Mean annual precipitation totals (R) in Slovakia in 1961–2020

Acknowledgements This work was supported by the Grant Agency of the Slovak Republic under the Projects VEGA No. 1/0940/17 and No. 1/0781/17. In this chapter, we used measured data from the Slovak Hydrometeorological Institute.

References

Alley RB, Mayewski PA, Sowers T et al (1997) Holocene climatic instability: a prominent, widespread event 8200 yr ago. Geology 25:483–486. https://doi.org/10.1130/0091-7613(1997)025%3c0483:HCIAPW%3e2.3.CO;2

Anderson DE, Binney HA, Smith MA (1998) Evidence for abrupt climatic change in northern Scotland between 3900 and 3500 calendar years BP. The Holocene 8:97–103. https://doi.org/10.1191/095968398669507604

Andrič M, Sabatier P, Rapuc W et al (2020) 6600 years of human and climate impacts on lake-catchment and vegetation in the Julian Alps (Lake Bohinj, Slovenia). Q Sci Rev 227: article 106043. https://doi.org/10.1016/j.quascirev.2019.106043

Bodnariuc A, Bouchette A, Dedoubat JJ et al (2002) Holocene vegetational history of the Apuseni mountains, central Romania. Q Sci Rev 21:1465–1488. https://doi.org/10.1016/S0277-3791(01)00117-2

Boulton GS, Dongelmans P, Punkari M et al (2004) Evidence of European ice sheet fluctuation during the last glacial cycle. In: Ehlers J, Gibbard PL (eds) Quaternary glaciations extent and chronology, part I: Europe. Developments in quaternary sciences, vol 2, pp 441–460. Elsevier

Böse M, Lüthgens C, Lee JR et al (2012) Quaternary glaciations of northern Europe. Q Sci Rev 44:1–25. https://doi.org/10.1016/j.quascirev.2012.04.017

Brázdil R, Dobrovolný P (2010) Historical climate in Central Europe during the last 500 years. In: Przybylak R, Majorowicz J, Brázdil R et al (eds) The polish climate in the European context: an historical overview, pp 41–70. Springer Science + Business Media BV

Brázdil R, Kundzewicz ZW (2006) Historical hydrology—editorial. Hydrolog Sci J 51(5):733–738. https://doi.org/10.1623/hysj.51.5.733

Brázdil R, Pfister C, Wanner H et al (2005) Historical climatology in Europe—the state of the art. Clim Change 70:363–430. https://doi.org/10.1007/s10584-005-5924-1

Brázd, ppil R, Kiss A, Luterbacher J et al (2008) Weather patterns in eastern Slovakia 1717–1730, based on records from the Breslau meteorological network. Int J Climatol 28:1639–1651. https://doi.org/10.1002/joc.1667

Brázdil R, Kundzewicz ZW, Benito G et al (2012) Historical floods in Europe in the past millenium. In: Kundzewicz ZW (ed) Changes in flood risk in Europe. IAHS Special Publication, vol 10, pp 121–166. IAHS Press, Wallingford

Brázdil R, Řezníčková L, Valášek H et al (2016) Climatic effects and impacts of the 1815 eruption of Mount Tambora in the Czech Lands. Clim past 12:1361–1374. https://doi.org/10.5194/cp-12-1361-2016

Brönnimann S (2015) Climatic Changes Since 1700. Advances in global change research, vol 55, p 360. Springer, Switzerland

Carlson AE (2013) The younger dryas climate event. In: Elias SA (ed) The encyclopedia of quaternary science, vol 3. Elsevier, Amsterdam, pp 126–134

Clark PU, Shakun JD, Baker PA et al (2012) Global climate evolution during the last deglaciation. Proc Natl Acad Sci 109(19):E1134–E1142

Engel Z, Mentlík P, Braucher R et al (2015) Geomorphological evidence and 10Be exposure ages for the last glacial maximum and deglaciation of the Velká and Malá Studená dolina valleys in the High Tatra mountains, central Europe. Q Sci Rev 124:106–123. https://doi.org/10.1016/j.quascirev.2015.07.015

Esper J, Büntgen U, Timonen M et al (2012) Variability and extremes of northern Scandinavian summer temperatures over tha past two millenia. Glob Planet Change 88–89:1–9. https://doi.org/j.gloplacha.2012.01.006

Felis T, Pätzold J (2004) Corals as climate archive. In: Fischer H, Kumke T, Lohmann G et al (eds) The climate in historical times. Springer, Towards a synthesis of holocene proxy data and climate models, pp 91–108

Fischer H (2004) The environmental and climate record in polar ice cores. In: Fischer H, Kumke T, Lohmann G et al (eds) The climate in historical times. Springer, Towards a synthesis of holocene proxy data and climate models, pp 149–170

Frakes LA (1979) Climates throughout geologic time. Elsevier Scientific Publishing Company, Amsterdam, Oxford, New York, p 310

Frakes LA, Francis JE, Syktus JI (1992) Climate modes of the Phanerozoic, p 274. Cambridge University Press

Futák J (1972) Vývoj rastlinstva. In: Lukniš M (ed) Slovensko 2, pp 412–421. Príroda, Obzor, Bratislava

Greenop R, Foster GL, Wilson PA et al (2014) Middle Miocene climate instability associated with high-amplitude CO_2 variability. Paleoceanogr Paleoclimatol 29(9):845–853. https://doi.org/10.1002/2014PA002653

Helle G, Schleser GH (2004) Interpreting climate proxies from tree-rings. In: Fischer H, Kumke T, Lohmann G et al (eds) The climate in historical times. Springer, Towards a synthesis of holocene proxy data and climate models, pp 129–148

Hughes ALC, Gyllencreutz R, Lohne ØS et al (2015) The last Eurasian ice sheets—a chronological database and time-slice reconstruction, DATED-1. Boreas 45:1–45. https://doi.org/10.1111/bor.12142

Hughes PD, Gibbard PL, Ehlers J (2013) Timing of glaciation during the last glacial cycle: evaluating the meaning and significance of the 'Last Glacial Maximum' (LGM). Earth Sci Rev 125:171–198. https://doi.org/10.1016/j.earscirev.2013.07.003

IPCC (2007) Climate change 2007: the physical science basis. Contribution of working group I to the fourth assessment report of the IPCC, p 996. Cambridge University Press, Cambridge, New York

IPCC (2013) Climate change 2013: the physical science basis. Contribution of working group I to the fifth assessment report of the intergovernmental panel on climate change, p 1535. Cambridge University Press, Cambridge, New York

Kalis AJ, Merkt J, Wunderlich J (2003) Environmental changes during the Holocene climatic optimum in central Europe—human impact and natural causes. Q Sci Rev 22:33–79. https://doi.org/10.1016/S0277-3791(02)00181-6

KIHZ-Consortium (2004) Evidence for the climate during the late maunder minimum from proxy data and model simulations available within KIHZ. In: Fischer H, Kumke T, Lohmann G et al (eds) The climate in historical times. Springer, Towards a synthesis of holocene proxy data and climate models, pp 397–414

Kiss A (2019) Floods and long-term water-level changes in medieval hungary, p 895. Springer

Konček M (1980) Climatic regions. In: Atlas Slovenskej socialistickej republiky. SAV, Bratislava, Map number 42

Korbel L (1972) Relikty treťohornej a glaciálnej fauny. In: Lukniš M (ed) Slovensko 2. Príroda, Obzor, Bratislava, pp 649–651

Krippel E (1986) Postglaciálny vývoj vegetácie Slovenska. Veda, Bratislava, p 309

Lapin M, Faško P, Melo M et al (2002) Climatic regions. In: Landscape atlas of the Slovak Republic. Ministry of Environment of the Slovak Republic, Bratislava, Slovak Environmental Agency, Banská Bystrica. Map number 27

Lapin M, Gera M, Hrvoľ J et al (2009) Possible impacts of climate change on hydrological cycle in Slovakia and results of observations in 1951–2007. Biologia 64(3):454–459

Lindner L, Dzierżek J, Marciniak B et al (2003) Outline of quaternary glaciations in the Tatra Mountains: their development, age and limits. Geol Q 47(3):269–280

Lukniš M (1972) Reliéf. In: Lukniš M (ed) Slovensko 2 - Príroda, Obzor, Bratislava, pp 124–202

Lundqvist J (1986) Stratigraphy of the central area of the Scandinavian glaciation. Q Sci Rev 5:251–268. https://doi.org/10.1016/0277-3791(86)90191-5

Magny M, Haas JN (2004) A major widespread climatic change around 5300 cal. yr BP at the time of the Alpine Iceman. J Q Sci 19(5):423–430. https://doi.org/10.1002/jqs.850

Makos M, Dzierżek J, Nitychoruk J et al (2014) Timing of glacier advances and climate in the High Tatra mountains (Western Carpathians) during the last glacial maximum. Q Res 82:1–13. https://doi.org/10.1016/j.yqres.2014.04.001

Makos M, Rinterknecht V, Braucher R et al (2018) Last glacial maximum and Lateglacial in the polish high Tatra mountains—revised deglaciation chronology based on the ^{10}Be exposure age dating. Q Sci Rev 187:130–156. https://doi.org/10.1016/j.quascirev.2018.03.006

Marks L (2012) Timing of the Late Vistulian (Weichselian) glacial phases in Poland. Q Sci Rev 44:81–88. https://doi.org/10.1016/j.quascirev.2010.08.008

Matejovič P (2011) Zima A.D. 1500–2010. História a podoby zím v Európe a na Slovensku. VEDA, Bratislava, p 282

Melo M, Lapin M, Kapolková H et al (2013) Climate trends in the Slovak part of the Carpathians. In: Kozak J, Ostapowicz K, Bytnerowicz A et al (eds) The Carpathians: integrating nature and society towards sustainability. Springer-Verlag, Berlin, Heidelberg, pp 131–150

Melo M, Pekárová P, Miklánek P et al (2014) Use of historical sources in a study of the 1895 floods on the Danube River and its tributaries. Geographica Pannonica 18(4):108–116. https://doi.org/10.5937/GeoPan1404108M

Melo M, Pišút P, Matečný I et al (2016) Johann Ignaz von Felbiger and his meteorological observations in Bratislava in the period 1783–85. Meteorol Z 25(1):97–115. https://doi.org/10.1127/metz/2015/0720

Melo M, Pišút P, Melová K et al (2019) Dôsledky extrémnych prejavov počasia v zime 1783/84 na mestá strednej Európy po erupcii vulkánu Lakagígar. In: Bada M, Duchoňová D (ed) Pohromy, katastrofy a nešťastia v dejinách našich miest. Igor Iliť – RádioPrint, Historický ústav SAV, Bratislava, pp 137–182

Mičian Ľ (1972) Pôdy. In: Lukniš M (ed) Slovensko 2. Príroda, Obzor, Bratislava, pp 361–402

Mosbrugger V, Utescher T, Dilcher DL (2005) Cenozoic continental climatic evolution of Central Europe. PNAS 102(42):14964–14969. https://doi.org/10.1073/pnas.0505267102

NCCC (National Communications on Climate change) (1995, 1997, 2001, 2006, 2009, 2014, 2018) Slovak Ministry of the Environment and Slovak Hydrometeorological Institute, Bratislava, vol 1–7. http://unfccc.int/national_reports/annex_i_natcom/submitted_natcom/items/7742.php. Accessed 10 Dec 2019

Negendank JFW (2004) the holocene: considerations with regard to its climate and climate archives. In: Fischer H, Kumke T, Lohmann G et al (eds) The climate in historical times. Springer, Towards a synthesis of holocene proxy data and climate models, pp 1–12

Niedźwiedź T (2010) Summer Temperatures in the Tatra Mountains During the Maunder Minimum (1645–1715). In: Przybylak R, Majorowicz J, Brázdil R et al (eds) The polish climate in the European context: an historical overview, pp 397–406. Springer Science + Business Media BV

Ogilvie AEJ, Barlow LK, Jennings AE 2000 North Atlantic climate c. AD 1000: millennial reflections on the Viking discoveries of Iceland, Greenland and North America. Weather 55:34–45. https://doi.org/10.1002/j.1477-8696.2000.tb04028.x

Oliver JE (ed) (2008) Encyclopedia of world climatology. Springer, Dordrecht, p 854

Paturi FR (1995) Kronika Zeme. Fortuna Print, Bratislava, p 576

Pekárová P, Miklánek P, Melo M et al (2014) Flood marks along the Danube River between Passau and Bratislava. Veda and the Slovak committee for hydrology—national committee for the international hydrological programme of UNESCO, Bratislava, p 104

Pfister C, Schwarz-Zanetti G, Wegmann M (1996) Winter severity in Europe: the fourteenth century. Clim Change 34:91–108. https://doi.org/10.1007/BF00139255

Pfister C, Luterbacher J, Schwarz-Zanetti G et al (1998) Winter air temperature variations in western Europe during the Early and High Middle Ages (AD 750–1300). Holocene 8:535–552. https://doi.org/10.1191/095968398675289943

Przybylak R (2010) Instrumental Observations. In: Przybylak R, Majorowicz J, Brázdil R et al (eds) The polish climate in the European context: an historical overview, pp 129–166. Springer Science + Business Media BV

Rácz L (2016) When did the little ice age end and the recent global warming start in Hungary? Late reflections about a scientific faith debate Historyka. Studia Metodologiczne T 46:197–208

Rapp D (2019) Ice ages and interglacials. Measurements, interpretation, and models, p 346. Springer Nature Switzerland AG

Robinson PJ, Henderson-Sellers A (1999) Contemporary climatology, p 317. Pearson Education Limited

Saltzman B (2002) Dynamical paleoclimatology. Academic Press, Generalized theory of global climate change, p 354

Šamaj F, Valovič Š (1982) Areal precipitation totals in Slovakia in 1881–1980. Meteorologické Zprávy 35:108–112

Schrøder N, Pedersen LH, Bitsch RJ (2004) 10,000 years of climate change and human impact on the environment in the area surrounding Lejre. J Transdiscipl Environ Stud 3(1):1–27

Shakun JD, Carlson AE (2010) A global perspective on last glacial maximum to holocene climate change. Q Sci Rev 29(15–16):1801–1816. https://doi.org/10.1016/j.quascirev.2010.03.016

Šibrava V (1986) Scandinavian glaciations in the Bohemian Massif and Carpathian Foredeep and their relationship to the extraglacial areas. Q Sci Rev 5:373–379. https://doi.org/10.1016/0277-3791(86)90199-X

Siegert MJ, Dowdeswell JA, Hald M et al (2001) Modelling the Eurasian Ice Sheet through a full (Weichselian) glacial cycle. Glob Planet Change 31:367–385. https://doi.org/10.1016/S0921-8181(01)00130-8

Stankoviansky M, Pišút P (2011) Geomorphic response to the little ice age in Slovakia. Geographia Polonica 84(Special Issue Part I):127–146. https://doi.org/10.7163/GPol.2011.S1.9

Stroeven AP, Hättestrand C, Kleman J et al (2016) Deglaciation of Fennoscandia. Q Sci Rev 147:91–121. https://doi.org/10.1016/j.quascirev.2015.09.016

Stuart AJ (2005) The extinction of woolly mammoth (Mammuthus primigenius) and straight-tusked elephant (Palaeoloxodon antiquus) in Europe. Q Int 126–128:171–177. https://doi.org/10.1016/j.quaint.2004.04.021

Svendsen JI, Alexanderson H, Astakhov VI et al (2004) Late Quaternary ice sheet history of northern Eurasia. Q Sci Rev 23:1229–1271. https://doi.org/10.1016/j.quascirev.2003.12.008

Walker M, Johnsen S, Rasmussen SO et al (2009) Formal definition and dating of the GSSP (Global Stratotype Section and Point) for the base of the Holocene using the Greenland NGRIP ice core, and selected auxiliary records. J Q Sci 24:3–17. https://doi.org/10.1002/jqs.1227

Wilson RCL, Drury SA, Chapman JL (2000) The great ice age. Routledge, The Open University, Oxon, New York, Climate change and life, p 267

Zachos J, Pagani M, Sloan L et al (2001) Trends, rhythms, and aberrations in global climate 65 Ma to present. Science 292 (5517):686–693. https://doi.org/10.1126/science.1059412

Zachos JC, Dickens GR, Zeebe RE (2008) An early Cenozoic perspective on greenhouse warming and carbon-cycle dynamics. Nature 451:279–283. https://doi.org/10.1038/nature06588

Zasadni J, Klapyta P, Swiader A (2015) Glaciers of the last glaciation maximum and deposits from the older glaciations. 1:100 000. In: Dabrowska K, Guzik M (eds) Atlas of the Tatra Mountains—abiotic nature. Tatrzanski Park Narodowy, Zakopane, p 42 + maps

Marián Melo is a climatologist at the Department of Astronomy, Physics of the Earth and Meteorology (Faculty of Mathematics, Physics and Informatics) of Comenius University in Bratislava. His research is focused on historical climatology, climate change and variability, climate change scenarios, hydrometeorological extremes and regional climatology.

Milan Lapin is a senior scientist and Professor at the Department of Astronomy, Physics of the Earth and Meteorology, Faculty of Mathematics, Physics and Informatics, Comenius University, Bratislava. He teaches and works in the fields of Physics of the Earth's Climate, Regional Climatology, Applied Climatology, Climate Changes and Variability and he was supervisor of several PhD theses in the Meteorology and Climatology domains.

Jozef Pecho is a climatologist at the Slovak Hydrometeorological Institute in Bratislava. His research is focused on the geomorphometry and theoretical geomorphology, tectonic geomorphology of the Western Carpathians and digital geomorphological mapping.

Geomorphological History of Slovak Landscape

4

Milan Lehotský and Miloš Rusnák

Abstract

The territory Slovakia of reveals exceptionally high diversity of landforms. Geomorphological history of the Western Carpathians can be traced back to the Early/Middle Miocene (10 and 20 Ma) and of the Panonian Basin to the Late Miocene, respectively. The development the Western Carpathians went on in phases, through interaction of tectonic movements and subaerial factors. Tectonic movements manifested themselves in the shape of a general uplift of the Western Carpathians and through differentiated movements of individual blocks within the dome-like structure, which are reflected in the basic internal division of the West Carpathians into two contrasting groups of macroforms: individual mountain ranges and intermountain basins. Terminal Miocene uplift of the Carpathians isolated the Pannonian region from the rest of Paratethys, which allowed for the emergence of the Pannonian Lake. Its extinction at the end of the Pliocene gave the birth to the Slovak lowlands. During the Pleistocene, spatially variable subaerial processes took place epicyclically under the strong influence of tectonic movements and lithologic-structural properties. The variability of processes in space and time led, on the one hand, to re-modelling of macroforms into new patterns of landforms (U-shaped valleys, ridges, moraines, terraces, plateaus, alluvial fans, etc.), and on the other one, to infilling and flattening of subsiding areas to form basins occupied by plains and floodplains. Human impact during the Holocene and Little Ice Age periods is mentioned in brief, too.

M. Lehotský (✉) · M. Rusnák
Department of Physical Geography, Geomorphology and Natural Hazards, Institute of Geography, Slovak Academy of Sciences, Štefánikova 49, 814 73 Bratislava, Slovakia
e-mail: geogleho@savba.sk

M. Rusnák
e-mail: geogmilo@savba.sk

Keywords

Slovakia • Landscape • Geomorphological • History • Drainage basins • Glaciation • River terraces • Periglacial phenomena • Human impact

4.1 Introduction—Turning Points of Geomorphological Evolution

The territory of Slovakia (49,035 km^2) is part of the Alpine–Himalayan mountain system. Its northern and central parts belong to the subsystem of the Carpathians (more than 74% of the area of Slovakia) and its south-western and south-eastern parts to the subsystem of the Pannonian Basin. The predominant section of the Slovak Carpathians (88%) forms the province of the Western Carpathians, the rest is in the Eastern Carpathians. The Western Carpathians are also higher and culminate in the Gerlach in the Tatra Mts. (2,655 m a.s.l.). The highest mountain of the Eastern Carpathians, built mostly of Paleogene flysch and Neogene volcanic complexes, is Kremenec (1,221 m a.s.l.) in the Bukovské vrchy Mountains.

The landforms of Slovakia are a product of complex tectonic evolution and climate changes during the Cenozoic, lithological diversity and also human impact in the Holocene. So, in the geomorphological history of Slovakia, five chronological turning points can be recognized: (1) Miocene–Quaternary uplift and creation of the recent Western Carpathian domal mega-morphostructure, the Pannonian basin and intermountain drainage basins Uplift started in the Early/Middle Miocene (the Styrian neotectonic phase) between 10 and 20 Ma (Kráľ 1977), continued by the Vallachian phase in the Late Pliocene (4–6 Ma) and the Pasadene phase in the Middle Pleistocene; (2) volcanic activity in the Eocene–Pleistocene time span, generating numerous neovolcanic terrains (Fig. 4.1); (3) Pleistocene glaciation/deglaciation and periglacial weathering creating

M. Lehotský and M. Boltižiar (eds.), *Landscapes and Landforms of Slovakia*, World Geomorphological Landscapes,
https://doi.org/10.1007/978-3-030-89293-7_4

Fig. 4.1 The summit of Sninský kameň (1006 m a.s.l.) in the volcanic Vihorlat Mts., which dominate above the eastern Slovakia. The lake below is dammed by landslide (*photo* J. Lacika)

glacial features typical for high mountains as well as systems of terraces/fans and alluvial plains; (4) Little Ice Age (LIA) climatic changes followed by changes in runoff, gravitational and fluvial processes and medieval and early modern age deforestation, miming and settlement development; and (5) recent human intervention in the operation of geomorphic processes. Active and passive morphostructures, which originated during these stages, were sculptured under diverse climatic conditions, involving predominantly humid tropical and subtropical climates in tertiary, cold periglacial and temperate conditions during the Pleistocene as well as temperate climate in the Holocene (Fig. 4.2). The territory of Slovakia also shows a long (>8 ka) history of human activity, which has significantly modified its landforms and landscapes mainly during Anthropocene. Since some chapters of this book are dealing with the Pleistocene glaciation, mainly in the High Tatra Mts. (Chap. 5), volcanic landforms (Chap. 8) and gullies (Chap. 19) in details, in this chapter, we summarize general features of the geomorphological history of the Slovak Republic, i.e. the pre-Pleistocene birth of the Western Carpathians and their drainage basins, briefly Pleistocene development, geomorphic response to Little Ice Age and recent human impact on landscape.

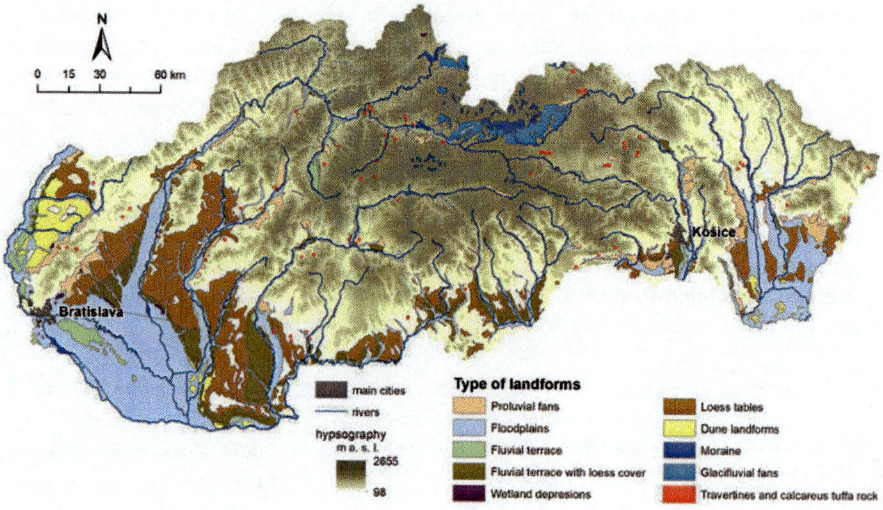

Fig. 4.2 Spatial distribution of depositional landforms of Quaternary age in Slovakia (modified from Maglay and Pristaš 2014)

4.2 Pre-Pleistocene Development

4.2.1 Western Carpathians Dome and The Emergence of the Pannonian Basin

The Western Carpathians represent a morphostructural megaform: a relatively flat elliptical dome, comprising multiple morphostructures of lower hierarchical levels. Mazúr (1965) attributed its rise to repeated vertical movements conditioned probably by sub-crustal magma movement or lateral pressure, in combination with an isostatic response. Along its entire periphery, this elevated dome is bordered by depressions of the Pannonian basin systems and the Vienna basin, the Carpathian Foredeep and the Transcarpathian depression. The surface of the Western Carpathian megaform is irregular, but a mosaic of discrete mountains (mainly horst and dome structures, (Fig. 4.3) and basins (mainly graben and flexure-bounded troughs, Fig. 4.4) creates distinctive patterns. The compactness of the megaform decreases from the NE to SW and, in the same direction, the area occupied by depressions also increases. The SW margin is open to the Pannonian Basin where the extent of the elliptical morphostructure is only indicated by spurs of narrow, low mountain ranges and slightly elevated hilly lands in the Pannonian Basin. The average altitude of the megaform ranges from 300 to 1500 m, with the highest ranges located near the NE focus of the ellipse. In contrast, the altitudinal minimum lies near the SW focus on the gradational boundary with the Pannonian Basin. However, various geological and geomorphological markers indicate the crucial role of Late Miocene, Pliocene and even Quaternary tectonic events conditioned the morphostructural pattern (Minár et al. 2011).

Fission track and radiometric data confirm the very young (post-Middle Miocene) denudational history of many individual mountain ranges (e.g. Kováč et al. 1994; Struzik et al. 2002; Bíl et al. 2004; Danišík et al. 2004, 2008).

The explanation for the dome-like character of the Western Carpathians supra-region with its very young features was mentioned only briefly in older literature. However, the reasons for this phenomenon were either not dealt with or the tectonic explanation was limited by the level of knowledge at that time (c.f. Mazúr 1965; Klimaszewski 1981).

The neotectonic rise of the Western Carpathians dome clearly and continuously deforms the initial planation surface of the "Mid-mountain level" (Lukniš 1962, Mazúr 1963, Fig. 4.5) and uplift should, therefore, be younger. In contrast, exhumation ages derived from fission-track data are very variable within some regions (e.g. Kováč et al. 1994; Kováč 2000; Baumgart-Kotarba and Kráľ 2002; Struzik et al. 2002; Danišík et al. 2004, 2008). This indicates that the detected exhumation history is older than the development of the dome (Minár et al. 2011).

The youngest fission-track data from about 10 Ma determine the maximum age of the "Mid-mountain level" that itself requires a few million years to form. Generally, fine-grained Late Pannonian and Pontian correlative sediments in the Pannonian Basin and intermountain basins of the Western Carpathians are also indicative of the formation of the "Mid-mountain level" planation surface. The general character of sedimentation demonstrably changed in the Pliocene, with coarse sediments replacing fine sediments.

Fig. 4.3 The Western Carpathians "dome" culminates in the High Tatra Mts at the highest mount (Mt. Gerlach - 2655 m a.s.l.) of the Carpathians (*photo* M. Lehotský)

Fig. 4.4 The Liptov basin—one of intra-Carpathians basins. The Chočské vrchy Mts. in the background and the Western Tatra Mts. are on the right (*photo* M. Lehotský)

Fig. 4.5 Mid-mountain level" planation surface—the Sihla plateau in the Veporské vrchy Mts. (*photo* J. Lacika)

Consequently, the dome probably first rose sometime during the last 4–6 million years. However, while the altitude of the "Mid-mountain level" corresponds with the mean altitudes of the Western Carpathians morphostructural regions, the younger "River Level" planation surface (Upper Pliocene–Early Quaternary pediment after Mazúr (1963)) and Quaternary river terraces differ far less between individual regions. This indicates that the main stage of dome formation occurred in the Pliocene and that both the "River level" and the river terraces were formed within the existing dome (Minár et al. 2011).

The projection of the older structural boundaries into new morphostructural regions and the increased abundance of young morpholineament systems (N–S and W–E directions) could be an indication of the gradual spreading of the Western Carpathians into the periphery—surrounding

Fig. 4.6 Elevated structure of the Outer Western Carpathians in north Slovakia—the Mt. Babia hora towers above the Oravské Beskydy Mts. belonging to the Carpathian Flysch Belt (*photo* J. Lacika)

lowlands during the Late Miocene–Quaternary uplift. The abundance of young morpholineaments found frequently in the youngest Neogene sediments in the south, increasing river downcutting since the Middle Pleistocene (based on the height of the river terraces) and possibly also young elevated (Fig. 4.6) and subsided structures of the Outer Western Carpathians (c.f. Zuchiewicz 1998), could indicate that the most recent and more active stage of the morphotectonic development of the Western Carpathians started in the Middle Pleistocene (Minár et al. 2011).

4.2.2 Drainage Basin Development

The main impetus of the transformation of the landscape pattern in the territory of the Slovak Carpathians is morphotectonics. Replacement of older valley and ridge systems by the new ones took place under the direction of morphostructural influences of different hierarchy (Fig. 4.7). Formation of the West Carpathian dome and the tectonic deformation of fault character, both of local and regional importance, played their role as well. Asymmetrical position

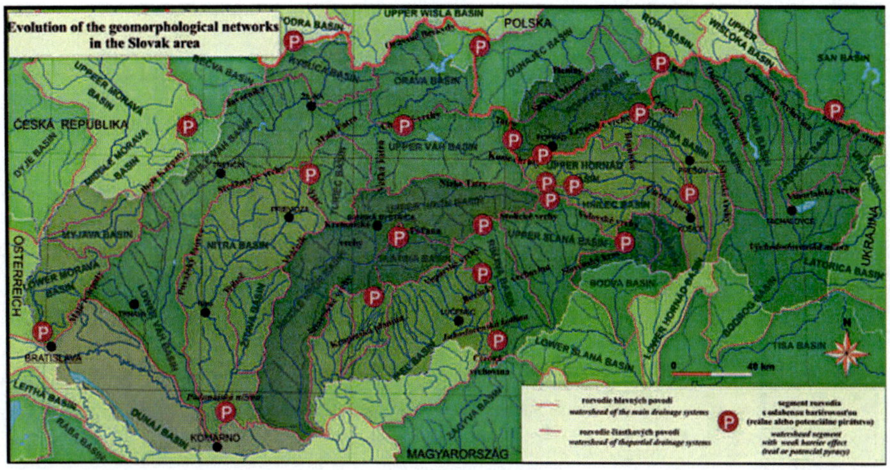

Fig. 4.7 Main drainage basins of Slovakia and their barrier effects showing sites of real or potential piracy (P) (Lacika 2004)

of the top of the dome differentiates morphostructural dynamics of the territory. The basins on the larger western side of the dome and in the centre (the upper Váh, Hron, Poprad and Ipeľ Rivers) develop mostly regressively. Progressively developing partial basins, for instance, in the area of central Považie or in the Hornád basins are affected by the local fault tectonics or are within the reach of large subsidence centres of the Podunajská and Východoslovenská nížina lowlands. The development of basins of the southern and south-eastern wings of the dome of the Western Carpathians is distinctly progressive. They are parts of morphostructurally very dynamic environments between the dome centre and subsiding lowlands in the Hungary.

In the chapter, we show the evolution of geomorphologic networks of two largest basins in the Slovak Carpathians—the Váh and Hron Rivers basins (Lacika 2004).

4.2.2.1 Váh River Basin

The river basin of the Váh River is the largest in Slovakia. It integrates the valleys within an area of 15,755 square kilometres in the territory of the Western Carpathians and the Western Pannonian basin (Podunajská nížina lowland). The Váh traces the longest side of the West Carpathian dome and leaves behind its centre after the longest route. In the Neogene, the Váh mouth followed the regressing lake in the Western Pannonian basin. It responded to uplift of the transversal horsts (the Veľká Fatra and Malá Fatra Mts.) by generation of antecedent gorge-like valleys. The general direction of the Váh valley changes beyond the town of Žilina. Deviation toward the south-west direction can be interpreted as a certain response to the distinct sinking of the Podunajská nížina lowland. Moreover, its course in this reach is linked to the erosion–denudation furrow along the klippen belt, which relatively sank in the Neogene. The Váh connected into the sea in the Late Miocene, which transgressed from the Viennese basin as far as the present territory of Považská Bystrica. In the Pliocene, the Váh prolonged as far as the Ilavská and Trečín basins, where it flowed through freshwater lakes. The lower reach of the Váh flows through the Podunajská nížina lowland. It is its youngest reach as it followed the diminishing Pontian Lake. The aggrading river prevents joining of the lateral streams and the basins taper into the narrow belt of alluvial plain. Aggradation is caused by the young (Quaternary) sinking of the lowland bay between the towns of Nové Mesto nad Váhom and Sereď. It is near Sereď, where the Váh changes its course again, in this case in the south-eastern direction and maintains it as far as its mouth into Dunaj River near Komárno.

4.2.2.2 Hron River Basin

The Hron River can be referred to as the "small Váh". The ground plan of its river basin is similar: nib-like texture of tributaries and deviation of the lowermost part of the basin in the southerly direction. From its spring as far as the town of Banská Bystrica, the Hron follows, as Lukniš (in Lukniš 1972) asserted, the post-Palaeogene megasynclinal riverbed between the Low Tatra Mts. and the Slovenské Rudohorie Mts. The whole upper reach of the Hron is morphologically delimited by the very distinct watershed formed by massive mountain ranges. The comparatively narrow valley widens only in the tectonic Breznianska kotlina basin, with rectangular valley texture. The middle reach of the Hron developed under the strong influence of intensive volcanic activity in the area of the Slovenské stredohorie Mts. In the Badenian the Hron probably entered into the bay of the Neogene Sea, which later retreated to the south, while its course was influenced by volcano-tectonic events. It is difficult to reconstruct the form of the Hron valley in the area of erupting stratovolcanoes and large volcano-tectonic depressions. The dynamic and complicated development of the middle Hron River took place until the Pliocene when three contemporary intra-mountain basins of the Slovenské stredohorie individualized. The Zvolenská and Žiarska kotlina basins were filled by the Pliocene lakes, which were then probably connected by the Hron. Its existence is testified to by the occurrence of the Hron gravel formation (Halouzka 1998), which skirts the present valley of the Hron from the lower Horehronie up to the Žiarska kotlina basin. The present valley of the Hron between Žarnovica and Kozárovce follows the distinct fault system, which further in north separates the mountain range of the Vtáčnik Mts. from the Žiarska kotlina basin. Before the Hron mouths into the Podunajská nížina lowland, it passes through the Slovenská brána gorge and overcomes the south-eastern protuberance of the Štiavnické vrchy Mts. (Kozmálovské vŕšky hills). In the Pliocene, it debouched into the lake and deposited its delta there. The occurrence of the Lower Pleistocene fluvial sediments on the ridges of the northern part of the Hronská pahorkatina hilly-land testifies to the fact that after regression of the Pliocene lake the Hron probably flowed in the south-westerly direction, to the area of what is today the valley of the Žitava River. It created its typical bent to south during the Early Pleistocene. Since then the lower reach of Hron heads to the existing valley between the Hronská and Ipeľská pahorkatina hill lands. The asymmetry of its terrace staircase points to the gradual migration of the riverbed to the east, probably as the result of tectonic tilting of this area (Lacika 2004).

4.2.3 Antecedent and Epigenetic Valleys

Large rivers of the Western Carpathians are older than the Miocene morphostructures and hence, during the Miocene–Pleistocene phase of uplift the rivers incised their channels

Fig. 4.8 The Váh River incised meander (the Domašin meander) in the Strečno gorge as the example of an antecedent valley, passing through the Fatra Mts range (*photo* J. Lacika)

into the rising mountains and so created antecedent valleys. The most famous are the antecedent reaches of the Váh River valley (Kraľovany, Strečno (Fig. 4.8), Nosice gorges); the Orava River valley (the Orava gorge upstream from the confluence with the Váh River); the Hornád River gorge (upstream the city of Košice). The Hron River also connects his wider valley segments by an antecedent valley reach and the Kvačianska and Prosiecka valleys in the Choč Mts. are also of antecedent origin. In addition to the antecedent valleys, epigenetic or epigenetic-antecedent valley reaches have been developed. The Hornád River at the Slovenský raj Mts. (Fig. 4.9), the Dunajec River in the Pieniny Mts., the Nitra River at the edge of the Tribeč Mts. close to the town of Nitra and Brezovský potok near Brezová pod Bradlom through the marginal part of the Little Carpathians Mts.

Fig. 4.9 The Hornád River epigenetic reach cuts a part of the Slovenský raj Mts. (*photo* J. Lacika)

4.2.4 Karst Landscape

Slovakia has karst landscape which covers more than 2700 km^2 (5.5% of the whole area) and the country is well known for an abundance of caves (Fig. 4.10). The first references about them go back to the thirteenth century. Cave maps have been produced since the eighteenth century. The evolution of a platform-like karst (plateau karst) dates back to the Mesozoic (the Cretaceous), whereas the mountain karst is younger (Quaternary up to the Recent). The total number of caves currently known is more than 6000 (Bella et al. 2007). Most of them are small, and only seven caves are longer than 10 km and two of them are longer than 30 km. The cave system of Demänová is the longest (35.358 km). Regarding the cave depth, there are 11 caves deeper than 200 m, while the deepest one reaches 495 m (Hipman`s caves). Sixteen caves are accessible to the public and these are administered by the Slovak Caves Administration (SSJ) seated in the Liptovský Mikuláš town and governed by the State Nature Protection Office (ŠOP). Four caves are under private ownership or owned by a museum. Hochmuth (2008) worked out the most recent regionalization of the karst areas in Slovakia and a list of caves and cave holes was published by Bella et al. (2007), which uses the cadastral area as the main locator of the karst phenomena.

4.3 Pleistocene/Early Holocene Imprints on Landscape

4.3.1 Glaciers, Glacio-Fluvial Fans and Tarns

During the Pleistocene, the territory of Slovakia was located between two large ice masses, namely an ice sheet that stretched across the northern European lowlands from Scandinavia to the northern foothills of the Carpathians and an ice cap that covered the Alps and reached the Danube River. In this period, the snow line on the southern slopes of the Slovak mountains was about 1,400–1,600 m a.s.l. The traces of the last three glacials (Mindel, Riss and Würm according to Alpine chronology) are evident (Fig. 4.11). Zasadni and Kłapyta (2014) inferred that the coalescence of the Tichý and Kôprový pre-Last Glacial Maximum glaciers in the Tatra Mts. created probably the largest Pleistocene glacier system (ca. 50 km^2) in Slovakia. The largest moraines have been preserved at the southern foot of the Tatras, in front of the valley outlets and are up to 100 m high. The exposure ages (Engel et al. 2015; Makos et al. 2014) for the terminal moraine below the Veľká studená and Velická valleys confirm that the Last Glacial Maximum (LGM) occurred no later than 21.5 ka. Braided rivers flowing out of the glaciated Tatra, associated with glacial meltwater, have deposited several generations of glacio-fluvial fans in the

Fig. 4.10 The Demänovka underground stream as the main factor influencing the development of the cave system in the Demänovská dolina valley. In the picture, the Demänovka brook in the Demänovska cave of Liberty (photo M. Lehotský)

Fig. 4.11 The High Tatra Mts landscape with landforms of glacial origin (*photo* M. Lehotský)

Fig. 4.12 Tarns in glacial cirques are typical phenomena in the glacially shaped High Tatra Mts. (*photo* M. Lehotský)

forefield. In total, the altitude of the Tatra was hypothesized to be reduced by glacial activity by about 300 m (Lukniš 1972). The remnants of glacial landforms are also found in the Low Tatra, where 11 glaciers and 31 glacial cirques were expected to occur. Traces of glacial relief can also be found in Mala Fatra, and smaller glacial cirques have also formed on the northern slope of the Babia Hora and Pilsko Mts., on the Slovak–Polish border. In addition to picturesque glacial relief, the Tatra and the Low Tatra Mts. are rich in moraine-dammed lakes and tarns (Fig. 4.12).

4.3.2 River Terraces

By alternating aggradation, lateral migration of channels and deepening of the valleys terrace staircases were created along Slovak rivers in the Quaternary. Generally, the Slovak larger rivers created 4–7 terraces. Their heights above the present floodplain level vary depending on the local tectonic movements. The highest remnants of terraces are located at heights of 100–130 m above the contemporary water level. At the entrance of the Danube River into Slovakia in the Devín Gate seven terraces occur (Mahr and Šajgalík 1979). Three terrace systems of the Hron, Žitava and Danube Rivers were identified in the eastern part of the Danube Basin, different in geometry, layout of the terraces and petrography of the sediment. The Hron terrace system consists of six wide north–south-oriented levels retreating to the east, while terraces of the Žitava are only erosive remnants of four levels retreating to the north-west. Two levels of the Danube terrace system are oriented in west–east direction. Accumulation of the terraces started at the Pliocene–Pleistocene boundary, as indicated by fossil mammals found in the highest terrace levels (Šujan and Rybár 2014). The six levels of Pleistocene terraces of the Hron River can be found in its upper reaches (Škvarček 1973) and four ones in the Žiarska basin (Holec et al. 2015). The Váh River left 13 terraces in the Liptov basin (Vitovič and Minár 2018), 8 in the incised meander of Domašín (Ondrášik and Gajdoš 2011) and 7 in the Žilina basin (Mazúr 1963). Lukniš (1972) and Mazúrova (1978) documented 5 terraces of the Ipeľ River in the Lučenecká Basin and 6 ones of the Slaná River in the Rimavská Basin. The same number is registered along the Dunajec River (Lukniš 1972) and the Topľa River (Harčár 1995).

4.3.3 Sackungen and Pseudokarst Caves

As in many paraglacial environments, also in the Slovak mountains formed by crystalline rocks (the Tatra and the Low Tatra Mts., Malá and Veľká Fatra Mts.) the sackungen landforms (Fig 4.13) can be found (Nemčok 1972; Mahr and Baliak 1973; Mahr 1977; Ondrášik 2002), although they origin is not directly connected with climatic changes. There are two examples provided below. Dating of these typical linear landform assemblages involving primarily expressive uphill-facing scarps and double-crested ridges in the Tatra Mts. revealed that the sackungen occurred between ~7.5 and 4.2 ka BP, representing a 4 ka time lag after the disappearance of glaciers (Pánek et al. 2015). Radiocarbon ages of the sackungen occuring in the Low Tatra Mts. indicate a displacement event of four studied trenches in the late Holocene (shortly after 1,410–1,860 cal yrs. BP) and the longest well-dated record in a single trench contained four inferred displacement events in the past 6 ka, yielding a long-term average recurrence of ca. 1.5 ka (McCalpin et al. 2019). Pseudokarst caves in flysch sandstones and neovolcanic rocks and their morphometric characteristics are briefly mentioned by Gaál (2003) and Hochmuth (2008). Hochmuth

Fig. 4.13 Occurrence of sackungen along the rounded ridge of the Western Tatra Mts. The dissected relief of the High Tatra Mts. in the background (*photo* J. Lacika)

pointed out that even though pseudokarst does not create any "pseudokarst areas" there are still large regions of increased concentration of these landforms. Based on the basic structural differences of the rocks, as well as patterns of spatial occurrence, he recognized five regions with pseudokarst phenomena in Slovakia.

4.3.4 Sand Dunes, Loess Tables and Other Periglacial Landforms

Another phenomena linked with the Pleistocene climate in Slovakia are aeolian landforms and loess plateaus. The largest continuous area with different types of sand dunes, up to 20 m high, extends in the Záhorská lowland. In addition to it, longitudinal sand dunes are located in the south-eastern part of the Danube lowland. They are several kilometres long, following NW–SE direction of the prevailing winds during sand deposition. Barchan type of sand dunes occur in the southern part of the East Slovak lowland. Their height is usually 5–10 m. Generally, sands are non-calcareous and are dated to MIS2 (Fordinál et al. 2013).

Loess in Slovakia covers an area of ca. 7 000 km^2 by a 15–20 m thick layer (Košťálik, 1997). Originally, loess has been horizontally deposited in the form of sheets in MIS2 and MIS3 (Ďurža and Dlapa 2009; Hošek et al. 2017), levelling the undulated relief beneath. However, such flat relief was dissected into separate plateaus due to tectonics and fluvial erosion during the Holocene (Fig. 4.14). So, the loess plateaus in the Podunajská and Východoslovenská nížina lowlands represent dominant landforms in lowland hilly lands.

The response of the Slovak landscape to the alternation of freezing and thawing periods, typical for periglacial climate, was the development of block-fields, frost-riven cliffs, periglacial nivation cirques/hollows and patterned ground. These landforms frequently occur at higher elevations as in the Tatra and Low Tatra Mts., the Babia hora and Pilsko Mts., and in the Malá Fatra Mts. Besides, these landforms can be locally found also in the mountains of lower altitude.

4.4 Holocene History

The first significant human interventions in the Carpathian landscape can be dated to what is referred to as the Great Colonization period in the thirteenth and fourteenth centuries (Stankoviansky and Barka 2007). Towns were founded and central Slovakia even became the land of mining and metallurgy. Another important intervention in the Carpathian landscape was the Wallachian shepherd colonization, which reached the territory of the today's Slovakia in the fifteenth century and peaked in the sixteenth and seventeenth centuries. The following shepherd and so-called "kopanitse" colonization took place in the seventeenth and the first half of the nineteenth centuries. Thus, between the thirteenth and nineteenth centuries, during the above-described colonization waves, gradual deforestation was followed by

Fig. 4.14 The Hron River undercuts the Hronská loess tableland at Bíňa village. In the 12 m high bluff, several layers of different Würm ages are visible (*photo* M. Lehotský)

exploitation of the acquired plots, bringing about an important increase of open areas, affected by accelerated geomorphic processes at a scale incomparable with the entire preceding the Holocene period. In the subalpine belt, the removal of dwarf pine cover, lowering of the present timberline by 280 m and locally even as much as 350–400 m, and overgrazing has led to the higher frequency of snow avalanches, shallow landslides and the occurrence of cryonival processes since the Wallachian colonization.

Not only colonization and land use changes but also climatic changes influenced the operation of geomorphic processes and landform development. The Little Ice Age (LIA) occurred between in 1550 and 1850 and was characterized by an increased frequency of extreme meteorological–hydrological events (Stankoviansky and Pišút 2011; Pišút 2002).

The year 1662 was a memorable one in the series of extreme flood events. During these floods, the village Chmelnica was destroyed by the Poprad River and the channel was shifted by 800 m (Horváthová 2003; Pekárová et al. 2011). Heavy damage was inflicted to the towns of Kežmarok, Levoča and several villages in Spiš country (Réthly 1962). A major flood of the Danube at the turn of November 1787 was possibly the second largest one of the last millennium, with a character of 200–500 yr flood and estimated peak discharge of at least 11,800 m^3. s^{-1} in Bratislava (Pišút 2011). Downstream of Bratislava, channel adjustments eventually resulted in a dramatic planform change, from actively meandering to braid in the 1780s, when a series of large meanders were cut-off (Pišút 1995). Instead, numerous non-vegetated bars appeared as a sign of channel over-enrichment with coarse material, that caused bilateral channel widening and related damage to dikes (Földes 1896; Pišút 2006). Major floods of this period were also responsible for massive gravel accumulations dated to the 1790s–1820s (Pišút and Timár 2007). In the case of the Váh River, the most conspicuous geomorphic fluvial effects were undoubtedly linked to the most catastrophic flood on record in Slovakia, with a character of 500–1000 yr flood on August 23–26, 1813. This flood claimed at least 243 victims, heavily damaged or even obliterated more than 50 villages, pulled down bridges, destroyed and/or damaged important public buildings and initiated many landslides. New bars and alluvial islands appeared in the Váh River channel, whereas extensive areas of the floodplain (fields, meadows, orchards) were covered by gravel and sand. Within this period, conspicuous changes of the Váh channel occurred along its middle and lower reach (Arcanum 2006a, b, 2007). For example, at Trenčín the channel seems to have transformed from an actively meandering one in the 1770s to a wandering channel pattern of the early nineteenth century. Along the lower stretch at Leopoldov, the river changed from an actively meandering, high-sinuosity river in 1775 into a low-sinuosity river, oversaturated with bedload due to the March 1830 flood (Arcanum 2006b). A major flood in the Hron River occurred also on May 8, 1853 (Munkáči and Rigo 1998). Not much lesser in scope than the flood of 1813 was a Váh River flood in August 1854 (Bitara 1998), which destroyed almost all bridges across this river in the Liptov county. It also hit the basins of the Poprad, Hnilec and Torysa rivers (Horváthová 2003). In addition to floods, Kotarba (2004) considered the debris flow event of 1813 among the most intensive ones in the Zelené pleso Valley in the High Tatra Mts. Within the period of the LIA large debris flows occurred not only in the Tatra Mts. Perhaps the most known has been the event in the Malá Fatra Mts in 1848, which destroyed the village of Štefanová under Veľký Rozsutec Mt (1,610 m a.s.l.).

After the World War Two, the most unfortunate terrain adjustment during vast collectivization of agriculture was the levelling of former cultivation terraces. The increased intensity of runoff processes after collectivization is also confirmed by the vertical accretion of colluvium, as well as aggradation in the valley bottoms, reaching approximately 1 m (Stankoviansky 2003). Soil erosion on large agricultural blocks, intensified due to extreme hydro-meteorological events, is frequently accompanied by muddy floods and flash floods. Perhaps the most terrifying event so far was that of May 1, 1996, in the village of Ivanka pri Nitre, where 175 houses were flooded by mud including four that crushed down seriously threatening human lives (Stankoviansky 2002). In 1998, a flash flood on the Malá Svinka Brook killed 44 people in the village of Jarovnice (the Flysch Carpathians) due to bank erosion and houses destruction. Another problem linked with soil erosion in agricultural land generating large amount of suspended load is intensive reservoir silting, mainly in the middle and upper dammed reaches of the Váh River. Considerable amounts of hillslope material are transported by uprooting of trees mostly during wind calamities (in 1941, 1947, 1948, 1949, 1964 and 1976, 2004, Fig. 4.15). According Hreško et al. (2005), there has also been a rise in the frequency of gravitational processes and snow avalanche intensity since the winter of 2000, particularly in the forest belt (Fig. 4.16). At present, the significance of anti-erosion function of woodlands is often weakened by large-scale clear-cuts, inappropriate technology of log skidding, construction of unmetalled roads, ski tracks, ski lifts, etc. Concerning recent fluvial processes, erosion prevails over accumulation as the forested slopes do not release sufficient material so as to reduce the erosive activity of streams. Geomorphologically efficient fluvial processes take place only under high water levels, caused either by heavy rainfalls or snowmelt, with the resulting continuous channel incision and narrowing (c.f. Kidová et al. 2016; Rusnák and Lehotský 2014).

Significant anthropogenic landforms are heaps of tailings. As a result of coal mining, they arose in the vicinity of towns

Fig. 4.15 The consequence of wind disaster in the High Tatra Mts. in 2014 was 2.7 million m^3 of destroyed trees. Since that time several not so serious events occurred (*photo* M. Lehotský)

Fig. 4.16 In June 2014, the Vrátna valley and Terchová village in the north part of the Malá Fatra Mts. were affected by heavy rain, which generated landslides in the subalpine zone. These in turn triggered several debris flows. The valley-station of the cableway at the head of the valley was destroyed and covered by about 3 m thick layer of debris. The village situated at the end of the valley suffered from a large flood (*photo* M. Lehotský)

of Veľký Krtíš, Handlová and Novák. Another occurrence of heaps is related to the extraction of asbestos (Dobšiná) and magnesite (Jelšava and Lubeník). Many heaps have witnessed ore mining in the past (environs of Banská Štiavnica, Špania dolina, Rožňava, Nižná Slaná, Smolník, Švábovce). The largest and newest heaps were created next to large metallurgical factories, especially near towns Žiar nad Hronom and Sereď. Many areas were changed by large quarries for limestone and other solid rocks. Limestone quarries are located mainly in the core mountains and karst areas

(Devínska Kobyle, Rohožník, Trstín, Nové Mesto nad Váhom, Horné Srnie, Lietavská Lúčka, Včeláre, Plešivc). Many large gravel extraction pits, flooded by groundwater, arose at localities rich in gravels and sands and are used for recreational purposes (Zlaté piesky, Rusovce, Štrkovec, Senecké jazerá, Zelená voda near Nové Mesto nad Váhom).

4.5 Conclusions

Geomorphological history of two major geotectonic domains of Slovakia, the Western Carpathians and the Pannonian basin can be traced back to the Miocene. Their landscape features are mainly the result of several phases of tectonic movements and subaerial factors. The Western Carpathians and the Slovak lowlands as parts of Pannonian Basin were significantly reshaped by processes conditioned by a highly fluctuating climate pattern in the Quaternary. Mountain glaciation affected most of Slovak mountains and periglacial processes in the extraglacial zone became significantly imprinted in the topography of Slovakia. Human impact on landscape can be found mainly in old and contemporary mining areas as heaps of tailings, along water courses (damming and training), along motorway constructions and slopes suitable for skiing (ski pists), as well as in arable land (large plots affected by soil erosion) and in suburbanized areas due to house and road construction (trenches, surface levelling, etc.). The geodiversity of Slovakia demonstrates that glacial landforms, karst phenomena, remnants of volcanic landforms as well as rocky relief of the Klippen Belt, can serve as an interesting point accessible to tourists as well as for further scientific research. As far as geomorphological research of the Slovak landscape is concerned, it seems that it is time to apply modern research supported by LiDAR and geochronological data more widely. Such research strategy can bring new insights into several geomorphic problems, including the assessment of age and origin of glaciofluvial fans, valley bottoms development, recent and contemporary tectonic movements, river behaviour and planation surfaces.

Acknowledgements he research was supported by Science Grant Agency (VEGA) of the Ministry of Education of the Slovak Republic and the Slovak Academy of Sciences; 02/0086/21. The authors wish to thank Dr. Ján Lacika for providing photos.

References

Arcanum (2006a) A Magyar Országos Levéltár Térképtára [Map Archives of the Hungarian National Archives]. I. Kamarai Térképek [Maps of Hungarian Royal Chamber, 1747–1882], DVD–ROM, Arcanum Adatbázis Kft. and National Archives of Hungary, Budapest

Arcanum (2006b) A Magyar Országos Levéltár Térképtára [Map Archives of the Hungarian National Archives]. II. Helytartótanácsi Térképek [Maps of the Hungarian Locumtenential Council, 1735–1875], Magyar Országos Levéltár and Arcanum Adatbázis Kft., Budapest

Arcanum (2007) Kéziratos Térképek az Országos Széchényi Könyvtár Térképtárában [Manuscript maps from the map repository of the National Széchényi Library], DVD–ROM, Arcanum Adatbázis Kft. and Országos Széchényi Könyvtár, Budapest

Baumgart-Kotarba M, Kráľ J (2002) Young tectonic uplift of the Tatra Mts. (fission track data and geomorphological arguments). In: Geologica Carpathica 53: Proceedings of the XVII Congress of the Carpathian-Balkan Geological Association Bratislava, September 1th-4th 2002. Eletronic version.

Bella P, Hlaváčová I, Holúbek P (2007) Zoznam jaskýň Slovenskej republiky. Liptovský Mikuláš : Slovenské múzeum ochrany prírody a jaskyniarstva. p 363

Bíl M, Krejčí O, Franců J, Hrouda F, Přichystal A (2004) Estimation of the missing eroded sediments in the Bílé Karpaty unit (Outer West Carpathians). Studia Geomorphologica Carpatho-Balcanica 38:59–66

Bitara E (1998) História povodní v povodí Váhu. In: Bacík M, Podkonický L, Szolgay J (eds) Povodne a protipovodňová ochrana, Dom techniky ZSVTS, Banská Bystrica, pp 16–20

Danišík M, Dunkl I, Putiš M, Frisch W, Kráľ J (2004) Tertiary burial and exhumation history of basement highs along the NW margin of the Pannonian basin—apatite fission track study. Austrian J Earth Sci 95(96):60–70

Danišík M, Pánek T, Matýsek D, Dunkl I, Frisch W (2008) Apatite fission track and (U–Th)/He dating of teschenite intrusions gives time constraints on accretionary processes and development of planation surfaces in the Outer Western Carpathians. Z Geomorphol 52(3):273–289. https://doi.org/10.1127/0372-8854/2008/0052-0273

Ďurža O, Dlapa P (2009) Magnetic susceptibility record of loess/paleosol sequence: Case study from south-west Slovakia. Contribut Geophys Geodesy 39(1):83–94

Engel Z, Mentlík P, Braucher R, Minár J, Léanni L, Team A (2015) Geomorphological evidence and ^{10}Be exposure ages for the Last Glacial Maximum and deglaciation of the Velká and Malá Studená dolina valleys in the High Tatra Mountains, central Europe. Quaternary Sci Rev 12:106–123. https://doi.org/10.1016/j.quascirev.2015.07.015

Fordinál K, Maglay J, Nagy A, Elečko M, Vlačiky M, Moravcová M, Zlinská A, Baráth I, Boorová D, Žecová K, Šimon L (2013) Nové poznatky o stratigrafii a litologickom zložení neogénnych a kvartérnych sedimentov regionu Záhorská nížina. Geologické Práce, Správy 121:47–87

Földes G (1896) Felso—Csallóköz árvédekezésének története. Bratislava

Gaál Ľ (2003) Genetické typy rozsadlinových jaskýň na Slovensku. Slovenský Kras 41:29–45

Halouzka R (1998) Pliocén. In: Konečný V (ed) Vysvetlivky ku geologickej mape Štiavnických vrchov a pohronského Inovca (štiavnický stratovulkán) 1:50 000, II. diel. Geologická služba SR, Bratislava, pp 292–294

Harčár J (1995) Reliéf Nízkych Beskýd. Časť A. Povodie Tople, Časť B. Povodie Ondavy. Geographia Slovaca 8:96

Hochmuth Z (2008) Krasové územia a jaskyne Slovenska. Geographia Cassoviensis 2:209

Holec J, Medveďová A, Vitovič L, Prokešová R (2015) Neotektonický vývoj Žiarskej kotliny indikovaný geomorfologickou analýzou v prostredí GIS. Geografický Časopis 67(2):181–195

Hošek J, Lisá L, Hambach U, Petr L, Vejrostová L, Bajer A, Grygar TM, Moska P, Gottvald Z, Horsák M (2017) Middle

Pleniglacial pedogenesis on the northwestern edge of the Carpathian basin: A multidisciplinary investigation of the Bíňa pedo-sedimentary section, SW Slovakia. Palaeogeogr Palaeoclimatol Palaeoecol 487(1):321–339. https://doi.org/10.1016/j.palaeo.2017.09.017

Horváthová B (2003) Povodeň to nie je len veľká voda. Bratislava, Veda.

Hreško J, Boltižiar M, Bugár G (2005) The present-day development of landforms and landcover in alpine environment — Tatra Mts (Slovakia). Studia Geomorphologica Carpatho-Balcanica 39:23–38

Kartniš J, Kvitkovič J (1970) Prehľad geomorfologických pomerov východného Slovenska. SPN, Bratislava. p 220

Kidová A, Lehotský M, Rusnák M (2016) Geomorphic diversity in the braided-wandering Belá River, Slovak Carpathians, as a response to flood variability and environmental changes. Geomorphology 272:137–149. https://doi.org/10.1016/j.geomorph.2016.01.002

Klimaszewski M (1981) Geomorfologia. PanstwoweWydawnictwo Naukowe, Warszawa

Kotarba A (2004) Zdarzenia geomorfologiczne w Tatrach Wysokich podczas małej epoki lodowej. Prace Geograficzne IG i PZ PAN 197:9–55

Košťálik J (1997) Eolic sediments in Western Carpathians, their characterization and chronostratigraphy. Acta Montanistica Slovaca 2. Special Issue 1:39–41

Kováč M (2000) Geodynamic, paleogeographic and structural development of the Carpathian–Pannonian region in Miocene: new view on Neogene basins of Slovakia. Veda, Bratislava

Kováč M, Kráľ J, Márton E, Plašienka D, Uher P (1994) Alpine uplift history of the Central Western Carpathians: geochronological, paleomagnetic, sedimentary and structural data. Geol Carpath 45 (2):83–96

Kráľ J (1977) Fission track ages of apatites from some granitoids rocks in West Carpathians. Geol Carpath 28:269–276

Lacika J (2004) Neotectonic evolution of the West Carpathian drainage basins in Slovakia. Studia Geomorphologica Carpatho-Balcanica 37:23–47

Lukniš M (1972) Reliéf. In: Lukniš M (ed) Slovensko – Príroda. Obzor, Bratislava. pp 124–202

Maglay J, Halouzka R, Baňacký V, Pristaš J, Janočko J (1999) Neotectonic Map of Slovakia. Scale 1:500000. Ministry of Environment of Slovak Republic; Geological Survey of Slovak Republic, Bratislava

Maglay J, Pristaš J (2014) Quaternary deposits. Geological Survey of Slovak Republic, Bratislava

Mahr T (1977) Deep-reaching gravitational deformation of high mountain slopes. Bull Int Assoc Eng Geol 19:121–127

Mahr T, Baliak F (1973) Regional investigation of slope deformations in the high mountain areas of the West Carpathians. In: Proc. 10th Congress of the Carpathian-Balkan Geolog. Assoc Sect 5:169–178

Mahr T, Šajgalík J (1979) Vývoj, zloženie a vlastnosti štvrtohorných sedimentov západnej časti Bratislavy. Geologické Práce, Správy 73:161–172

Makos M, Dzierżek J, Nitychoruk J, Zreda M (2014) Timing of glacier advances and climate in the High Tatra Mountains (Western Carpathians) during the Last Glacial Maximum. Quat Res 82(1):1–13. https://doi.org/10.1016/j.yqres.2014.04.001

Mazúr E (1963) Žilinská kotlina Basin and adjacent mountains. Vydavateľstvo SAV, Bratislava

Mazúr E (1965) Major features of the West Carpathians in Slovakia as a result of young tectonic movements. In: Mazúr E, Stehlík O (eds) Geomorphological Problems of Carpathians. Vydavateľstvo SAV, Bratislava, pp 9–54

Mazúrová V (1978) Terraces of the Czechoslovak Carpathian rivers and their relation to Danube terraces. Geografický Časopis 30 (4):281–301

McCalpin JP, Liščák P, Jelínek P, Zorba MO, Santacana N (2019) Postglacial deformation history of sackungen on the southern slope of Mount Chabenec, Nízke Tatry Mts. Slovakia. Mineralia Slovaca 52:1–30

Minár J, Bielik M, Kováč M, Plašienka D, Barka I, Stankoviansky M, Zeyen H (2011) New morphostructural subdivision of the Western Carpathians: An approach integrating geodynamics into targeted morphometric analysis. Tectonophysics 502:158–174. https://doi.org/10.1016/j.tecto.2010.04.003

Munkáči J, Rigo F (1998) História povodní a protipovodnovej ochrany v územnej pôsobnosti OZ Povodie Hrona. In Bacík M, Podkonický L, Szolgay J (eds) Povodne a protipovodnová ochrana, Dom techniky ZSVTS, Banská Bystrica, pp 21–28

Nemčok A (1972) Gravitational slope deformation in high mountains. In: Proc. 24th Int. Geol. Congress 13:132–141

Ondrášik R (2002) Landslides in the West Carpathians. In: Rybář J, Stemberk J, Wagner P (eds) Landslides. Routledge, London, pp 75–96

Ondrášik R, Gajdoš V (2011) Riečne terasy Váhu v meandri Domašín v Strečnianskej úžine. Geografický Časopis 63(3):275–285

Réthly A (1962) Idojárási események és elemi csapások Magyarországon 1700-ig. Akadémiai Kiadó, Budapest

Pánek T, Mentlík P, Ditchburn B, Zondervan A, Norton N, Hradecký J (2015) Are sackungen diagnostic features of (de)glaciated mountains? Geomorphology 248:396–410. https://doi.org/10.1016/j.geomorph.2015.07.022

Pekárová P, Svoboda A, Novák V, Miklánek P (2011) Historická hydrológia a integrovaný manažment povodí a krajiny. Vodohospodársky Spravodajca 1–2:4–7

Pišút P, Timár G (2007) História územia ostrova Kopáč. In: Majzlan O (ed) Príroda ostrova Kopáč, Fytoterapia OZ, Bratislava, pp 7–30

Pišút P (1995) Meandrovanie Dunaja pri Bodíkoch pred zmenou charakteru riečiska v 18. storočí. Geografický časopis 47(4):285–298.

Pišút P (2002) Channel evolution of the pre-channelized Danube river in Bratislava, Slovakia (1712–1886). Earth Surf Proc Land 27:369–390. https://doi.org/10.1002/esp.333

Pišút P (2006) Changes in the Danube riverbed from Bratislava to Komárno in the period prior to its regulation for medium water (1886–896). In: Mucha I, Lisický MJ (eds) Slovak-Hungarian environmental monitoring on the Danube, Podzemná voda, Bratislava, 186–190

Pišút P (2011) Dunajská povodeň na Sviatok všetkých svätých roku 1787 a Bratislava. Geografický Časopis 63(1):87–109

Rusnák M, Lehotský M (2014) Time-focused investigation of river channel morphological changes due to extreme floods. Z Geomorphol 58(2):251–266. https://doi.org/10.1127/0372-8854/2013/0124

Stankoviansky M (2002) Bahenné povodne – hrozba úvalín a suchých dolín. Geomorphologia Slovaca 2(2):5–15

Stankoviansky M (2003) Historical and present slope evolution in hilly farmland (on the example of the Myjava Hill Land, Slovakia). Supplementi Di Geografia Fisica a Dinamica Quaternaria, Supplemento 6:91–97

Struzik AA, Zattin M, Anczkiewicz R (2002) Apatite fission track analyses from the Polish Western Carpathians. GeoLines 14:87–89

Stankoviansky M, Pišút P (2011) Geomorphic response to the little ice age in Slovakia. Geographia Polonica 84. Special Issue Part 1:127–146

Stankoviansky M, Barka I (2007) Geomorphic response to environmental changes in the Slovak Carpathians. Studia Geomorphologica Carpatho-Balcanica 41:5–28

Škvarček A (1973) Náčrt kvartérneho vývoja horského úseku doliny Hrona. Geografický Časopis 25:136–145

Šujan M, Rybár S (2014) Vývoj pleistocénnych riečnych terás vo východnej časti Dunajskej panvy. Acta Geol Slovaca 6(2):107–122

Vitovič L, Minár J (2018) Morphotectonic analysis for improvement of neotectonic subdivision of the Liptovská kotlina basin (Western Carpathians). Geografický Časopis 70(3):197–216. https://doi.org/10.31577/geogrcas.2018.70.3.11

Zasadni J, Kłapyta P (2014) The tatra mountains during the Last glacial maximum. J Maps 10(3):440–456. https://doi.org/10.1080/17445647.2014.885854

Zuchiewicz W (1998) Structural geomorphological studies in the Polish Carpathians. Studia Geomorphologica Carpatho-Balcanica 32:31–45

Milan Lehotský is a physical geographer and fluvial geomorphologist at the Institute of Geography of the Slovak Academy of Sciences. He was many years head of the Department of Physical Geography, Geomorphology and Natural Hazards. His research topics are responses of fluvial systems to environmental changes, sedimentological connectivity, evolution trajectories, hydromorphology and GIS and remote sensing applications in rivers and landforms research. He is also working as an external lecturer at the Department of Physical Geography and Geoecology, the Faculty of Natural Sciences of the Comenius University in Bratislava.

Miloš Rusnák is a fluvial geomorphologist at the Institute of Geography of the Slovak Academy of Sciences (Department of Physical Geography, Geomorphology and Natural Hazards). His research topics are fluvial geomorphology, spatial data processing in GIS, UAV data acquisition and processing, fluvial processes and sediment connections in gravel-bed rivers and remote sensing applications in rivers and landforms research. He is the author and co-author of several papers dealing with fluvial system evolution in the Outer Western Carpathians.

Part II

Landscapes and Landforms

5 Unique Glacial Landscape on the Roof of the Carpathians—Tatras Mts.

Martin Boltižiar

Abstract

The Tatra Mountains are the highest mountain range within the Carpathians. They are a mountain range of a corrugated nature, originating from the Alpine orogeny, and therefore characterized by a relatively young-looking lay of the land, quite similar to the landscape of the Alps, although significantly smaller. The Tatra Mts., with their 25 peaks exceeding 2500 m a.s.l., together with the Southern Carpathians, represent the only form of alpine landscape in the entire 1,200 kms length of arc of the Carpathians. Their alpine relief is a final of all glaciations. Before the oldest glaciation the Tatra Mts. had prevailingly smooth relief forms. Transformation of the preglacial erosional–denudational valley relief led to the formation of typical U-shaped valleys and cirques in the upper parts. A considerable part of the Tatra Mts. is found within a periglacial morphoclimatic vertical grade. Both active and relict periglacial landforms occur in the Tatra Mts. Some types of limestone karstify in the Tatra Mts. These mountains are a popular as a summer and winter sports area, with separate resorts such as Štrbské Pleso, Starý Smokovec and Tatranská Lomnica. The Tatra Mts. are protected by law by the establishment of the Tatra National Park (1949) in Slovakia which is jointly entered in UNESCO's World Network of Biosphere Reserves (1993).

Keywords

Tatra Mountains • Geomorphology • Glacial forms • Periglacial forms • Debris flows • Tourism

M. Boltižiar (✉)
Department of Geography, Geoinformatics and Regional Development, Faculty of Natural Sciences, Constantine the Philosopher University in Nitra, Nitra, Slovakia
e-mail: martin.boltiziar@savba.sk; mboltiziar@ukf.sk

Institute of Landscape Ecology, Slovak Academy of Sciences, Bratislava, branch Nitra, Nitra, Slovakia

5.1 Localization of the Tatra Mountains

The Tatra Mountains are the highest mountain range within the Carpathians. They are a mountain range of a corrugated nature, originating from the Alpine orogeny, and therefore characterized by a relatively young-looking lay of the land, quite similar to the landscape of the Alps, although significantly smaller. The Tatra Mts., with their 25 peaks exceeding 2500 m a.s.l., together with the Southern Carpathians, represent the only form of alpine landscape in the entire 1,200 km long arc of the Carpathians.

The Tatra Mountains occupy an area of 790 km^2, of which about 610 km^2 (77.7%) lie within Slovakia and about 175 km^2 (22.3%) within Poland (Fig. 5.1). The highest peak—Gerlachovský peak (2655 m a.s.l.) is located to the north of the town of Poprad, entirely in Slovakia.

The course of the main ridge of the Tatra Mts. is not entirely consistent with the main European watershed. The principal ridge also runs to the north of the highest peaks. Although the Tatra Mts. surpass 2500 m a.s.l., only 6.2% of the massif is above 2000 m a.s.l. and its considerable part (57%) below 1500 m a.s.l. (Balon et al. 2015). As in almost any other alpine mountain range, the topography of the Tatra Mts. is dominated by its principal ridge. Measured from the eastern foothills of the Kobylí vrch (1109 m a.s.l.) to the south-western foot of Ostrý vrch (1128 m a.s.l.), in a straight line, it is 57 km, whereas the actual length of the main ridge is 80 km. The range is only 20 km wide.

The main ridge of the Tatras runs from the village of Huty at the western end to the village of Ždiar at the eastern end (Lukniš 1973). It consists of the internal mountain chains of Eastern Tatra Mts. (Východné Tatry), which in turn consist of the Belianske Tatra Mts. (Belianske Tatry) and the High Tatra Mts. (Vysoké Tatry; Fig. 5.2), and of the Western Tatra Mts. (Západné Tatry).

The Tatra Mts. are protected by law by the establishment of the Tatra National Park (1949) in Slovakia, which is also a member of UNESCO's World Network of Biosphere

M. Lehotský and M. Boltižiar (eds.), *Landscapes and Landforms of Slovakia*, World Geomorphological Landscapes,
https://doi.org/10.1007/978-3-030-89293-7_5

Fig. 5.1 Localization of the Tatra Mountains (Orthophoto: ©Terrametrics, 2021)

Fig. 5.2 Panoramic view of the High Tatra Mountains from the Low Tatra Mountains (Kráľová hoľa peak—1946 m a.s.l.) (*photo* M. Boltižiar)

Reserves under its Man and the Biosphere Programme (1993).

Scientific research of the Tatra's landforms and processes has a long tradition. Two centuries ago Stanisław Staszic described it in a treatise "On the origin in the Carpathians". Much geomorphological research has been carried out in the Tatra Mts. and their foreland in the last 70 years. Geomorphological characteristics of the mountain range provided in this chapter reflect the work of many Slovak, Polish and Czech authors (M. Lukniš, J. Hreško, R. Midriak, M. Stankoviansky, M. Boltižiar, M. Žiak, G. Bugár, F. Kohút, P. Mentlík, P. Mida, M. Křížek, T. Pánek, M. Lehotský, J. Minár, M. Klimaszewski, A. Kotarba, Z. Rączkowska, J. Zasadni, P. Kłapyta, S. Kędzia, K. Krzemień, etc.). Most of this text was created on the basis of previously published material resultings from research of Lukniš (1973a).

Very good collaboration between Slovak and Polish researches allowed us today to present a set of maps covering the entire mountain massif, including both general and detailed maps of relief as well as of various landforms and processes: glacial, periglacial, fluvial, erosional–denudational (debris flows, avalanche tracks), caves and karst phenomena (Rączkowski et al. 2015). Particulary interesting is the map of the extent of glaciers during the Last Glacial Maximum and of the distribution of older glacial deposits (Zasadni and Kłapyta 2014; Zasadni et al. 2015).

According to Lukniš (1973b), the Tatra Mts. rise as a high mountain range among the surrounding basins underlain by the Paleogene flysch of claystone facies, which was originally laid in a position over the crystalline complex of the Tatra's core. Today they occur next to each other along the southern foot and the contact between them is tectonic, along a major W–E trending fault. Analogically, next to each other are mostly limestone–dolomite sediments of the Tatra's Mesozoic Group and the claystone flysch on the northern side of the mountain range, but the contact between them is flexural. Along the mentioned dislocation lines the Tatra Mts. were upthrown high above their surroundings in the late Cenozoic. The sharp delimitation in respect to the surrounding depressions is not only of tectonic nature. It also reflects structural control on the rates of erosion and denudation, since along the dislocations mentioned above the very resistant basement and sedimentary rock of the upthrown Tatra Mts. are juxtaposed with the little-resistant claystone flysch of the Liptov—Poprad Basin in the south and of the Subtatran Furrow in the north (Lukniš 1973a). The flysch was moulded into gently rolling topography, with gravel sheets of glacifluvial origin superimposed or inset into the bedrock-cut relief.

The Tatra Mts. have an asymmetric geological structure. According to Nemčok et al. (1994), they are divided into the Crystalline Tatras and the Limestone Tatras (Figs. 5.3, 5.4 and 5.5). The southern slope and the main ridge are built of crystalline rocks, mainly of granodiorite. The northern interfluves consist mostly of carbonate sediments, folded to recumbent and overthrust folds plunging to the north. The boundary between the two parts runs along longitudinal (subsequent, W–E trending) valleys and saddles within the

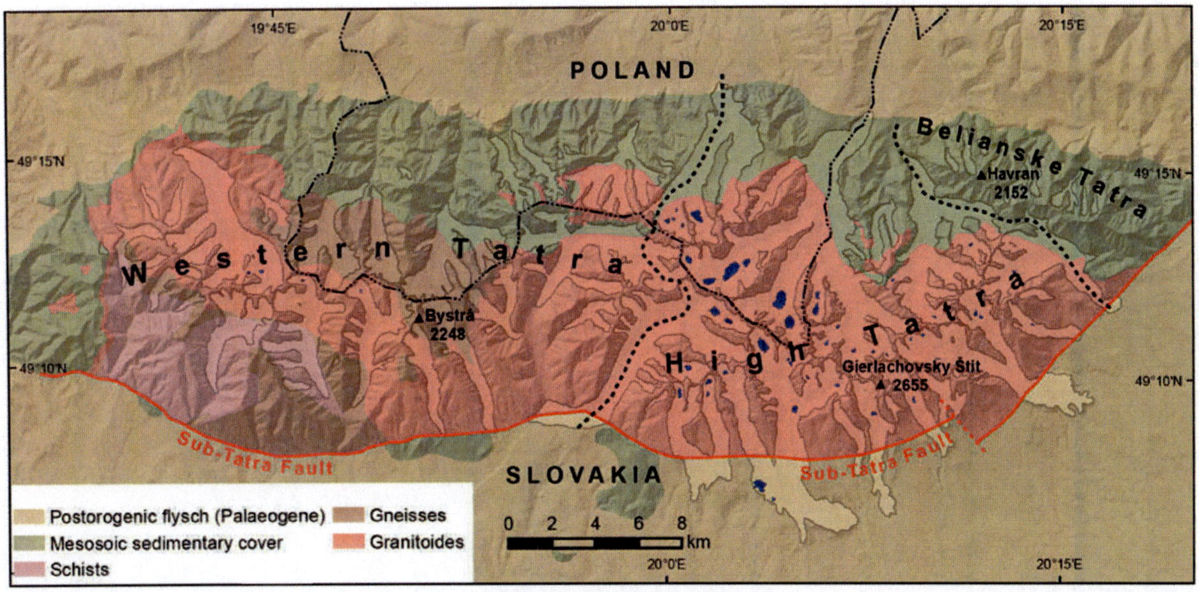

Fig. 5.3 Schematic geological map of the Tatra Mts. (after Zasadni and Kłapyta 2014)

Fig. 5.4 Typical granodiorite rock relief of the High Tatra Mts. Panoramic view from the second highest peak of the Tatra Mts.—Lomnický štít peak (2633 m a.s.l.) (*photo* M. Boltižiar)

northern slopes. Only in the transveral tectonic depression between the arch of the Eastern Tatra Mts. and the arch of the Western Tatra Mts the Limestone Tatra Mts. reach the main ridge (Lukniš 1973b). The Crystalline Tatra Mts. are as a whole formed by more resistant rocks, therefore they are higher. They include the main ridge and the southern interfluves, within which are located the highest mountains such as the Gerlachovský štít peak (2655 m), Lomnický štít peak (2633 m), Bystrá peak (2250 m) and others. A detailed morphological subdivision is influenced by the systems of joints and especially by mylonite zones, with which major saddles and valleys are bound (Figs. 5.4 and 5.6). On crystalline schists, which are abundant in the Western Tatra Mts., the ridges and valley slopes are smoother. On granodiorites, where the slopes were undercut by glaciers so that they have inclination more than 38° (Lukniš 1973a).

The Limestone Tatra Mts. consist of limestones, dolomites, marls, marly slates and exceptionally also quartz sandstones. The relief is characterized by rock defiles and a whole series of sharp-crested ridges, hogbacks and narrow segments of transversal valleys on limestone, dolomite and quartz sandstone (Fig. 5.7). On marls and marly slates the erosive-denudative furrows, saddles and widened segments of valleys have been formed (Lukniš 1973b). Some types of limestone present in the Tatra Mts. are particularly prone to karstification. These are especially the strips of limestone and dolomitic limestone of the Middle Triassic of all tectonic units, the Dogger–Malm–Urgonian limestone of the

Fig. 5.5 Crystalline High Tatra Mts. formed by more resistant granodiorite rocks (in the front) and Limestone Belianske Tatra Mts. with smooth relief (in the back) (*photo* M. Boltižiar)

High-Tatra unit and the strips of the Mural' limestone in a partial nappe of Mt. Havran. Of a large number of the caves known the Belianska jaskyňa cave only has rich sinter decoration. It is 1752 m long. It is easily accessible and is turistically exploited (Lukniš 1973a).

The Tatra Mts. have been divided by a tectonic disturbance into the Western Tatra Mts. (2250 m) and the Eastern Tatra Mts. (2655 m). The Eastern Tatra Mts. were upthrown along that line by about 300 m higher. The difference in height between both parts of the Tatra Mts brought about a more extensive glaciation in the higher Eastern Tatra Mts. Consequently, glacial remodelling was more intense, which is why bare bedrock escarpments prevail over uniformly graded slopes, which dominate in a major part of the Western Tatras (Lukniš 1973b).

From the village of Tatranské Matliare to the village of Podbanské below the southern foot of the Tatra Mts., a more or less expressive step is found, which consists of agglomerated moraines and in some places even flysch and blocks of limestone and dolomite. This is the "piedmont" or foreland of the Tatras, important due to a strip of Tatra recreation-curative settlements linked by the Cesta Slobody way and an electric railway.

The Eastern Tatras, despite hosting the highest peaks, have their mean relative height lower than the Western Tatras, which are lower in terms of absolute altitude. This is due to more expressive glaciation in the Eastern Tatras and more efficient erosion. Here, the main ridge and secondary interfluves were considerably narrowed and steepened by glaciers. Within the sharpened crests, in the consequence of

Fig. 5.6 Typical deglaciated valley with tarns (Žabie plesá) within the cirque floor. Mengusovská dolina valley in the central part of the High Tatra Mts. (*photo* M. Boltižiar)

weathering, glacial undercutting and gravity-driven processes voluminous rock masses lost support and were displaced into the valleys, where they form large extensive boulder fields and taluses. In this way, the crests were lowered more rapidly than in the Western Tatras (Lukniš 1973a).

A curiosity of the Tatras is that the highest summits are located on its southern second-order interfluves rather than on the main ridge. This phenomenon is connected with the fact that the Tatras were being uplifted asymmetrically, especially since the Middle Pliocene, more in the south than in the north, in the form of a tilted block bounded on the south by the Subtatran fault. Some rough estimations have been made concerning the amount of uplift since the emergence of the Tatras in the latest Paleogene. The figures suggested are about 3100 m, of which about 300–400 m goes to the Pleistocene and the remaining 2700 m to the Neogene (Lukniš 1973b; Klimaszewski 1985).

5.2 Glacial Landforms

Before the oldest glaciation, the Tatra Mts. likely had mainly smooth relief forms (Klimaszewski 1960, 1962, 1965; Lukniš 1973a; Vaškovský 1977; Kotarba et al. 1987). In the contemporary morphology, one can recognize remnants of two landform systems, strikingly disharmonious, which may be inherited, preglacial features. The older system, attributed to the Neogene, includes smooth relief that occurs in a reduced form on some summits in the Belianske Tatra Mts. (Fig. 5.7), in the Červene vrchy Mts., and in other places. The younger system developed before the onset of the oldest glaciation in the form of piedmont plateaus and wide-open valley bottoms. Its remnants may be seen mainly along the northern foot of mountain range, but also in the form of rock terraces that even reach the valleys (Mlynická dolina valley, Javorová dolina valley, Ždiarska dolina valley, etc.). In the

Fig. 5.7 Smooth limestone slopes of the Belianske Tatra Mts. View from the Hlúpy peak (2060 m a.s.l.) towards the eastern crest (*photo* M. Boltižiar)

Western Tatra Mts., much of this relief has survived until now, but in the Eastern Tatra Mts., due to more intensive glaciation similar smooth relief features are of very limited extent. Relics of smooth slopes can be observed in terminal places of the southern interfiuves, where they form triangular facets (Fig. 5.8).

High-mountain, alpine relief as is observed today is an outcome of all successive glaciations. Transformation of the preglacial erosional–denudational valley relief led to the formation of typical U-shaped valleys (glacial throughs) and cirques in the upper, non-rejuvenated parts of the valleys (Figs. 5.4 and 5.6). Fluvio-denudational relief occurs in the part of the Tatra Mts. that have not been glaciated (Rączkowska et al. 2015; Klimaszewski 1962, 1988; Kotarba et al. 1987). In the Tatra Mts., traces of three glaciations can be recognized, documented by moraines and intermediary glacifluvial cones. The moraine of the oldest glaciation is found in front of the mouth of the Studené doliny valleys in the area of Pekná vyhliadka (Bellevue) in Horny Smokovec. It is strongly weathered and its surface has been smoothed.

The moraines of the penultimate glaciation are composed of slightly weathered material and are more frequent in the foreland of the Tatras, less so inside the mountain range due to subsequent removal by erosion and denudation. In some places, they have kept a strongly flattened accumulation form as a consequence of protracted denudation, in other places only erratic blocks remain. The glaciers of the penultimate glaciation were by 0.3–2.7 km longer and on an average by 50 m thicker than those of the latest glaciation (Lukniš 1973b).

The latest glaciation has left behind many fresh erosive forms and many fresh moraines (Figs. 5.9, 5.10 and 5.11). Using ramparts of lateral and terminal moraines as the morphological evidence, the extent of the latest glaciation (Würm) as well as transformations of relief that occurred in this interval can be reconstructed (Partsch 1923; Lukniš

Fig. 5.8 Periglacial debris cover/blockfields of crystalline bedrock with dwarf pine stands on the SE triangular facet slope of the Huncovský štít peak (eastern part of the High Tatra Mts.) (*photo* M. Boltižiar)

1959, 1964, 1968; Vaškovský 1977; Halouzka 1992; Baumgart-Kotarba and Kotarba 1997, 2001; Lindner et al. 2003; Kłapyta 2013; Mida and Křížek 2013; Engel et al. 2015). In the period of the greatest extent of the latest glaciation, the glaciers occupied an area of about 150 km^2 in the High Tatra Mts. There were 8470 ha under ice on the northern slope. The southern side was more glaciated, with the catchment areas of snow accumulation in the headwater parts of the valleys being more voluminous and lying, on average, by 200 m higher compared to the cirques of the northern slope (Lukniš 1973a).

Recently, Zasadni and Kłapyta (2014) and Zasadni et al. (2015) have re-evaluated the geomorphological evidence for all glacial systems in the Tatra Mts. identified in the literature, based on new field mapping and remote sensing data analysis (Figs. 5.12, 5.13, 5.14 and 5.15). Utilizing geographic information systems (GIS) and 10 m resolution digital elevation model (DEM) as a base topography, they reconstructed, for the first time, the detailed extent and surface geometry of all Last Glacial Maximum glaciers based on the distribution of glacial erosional and depositional landforms. During this period, 55 glacier systems with a total area ca. 280 km^2 and ice volume of 24.3 km^3 existed in the Tatra Mts. The largest, 13.4 km-long and up to 2.2 km-wide Bielovodský glacier (Fig. 5.16) was nearly 400 m thick and covered 43.6 km^2. (Zasadni et al. 2015).

Fig. 5.9 Lateral moraine ridge (Würm age) in the Skalnatá dolina valley (High Tatra Mts.) is permanently modified by aeolian processes and talus creep. (*photo* M. Boltižiar)

Due to nivation and erosion of the glaciers, the valley heads were gradually re-shaped into wide-floored cirques (Křížek and Mida 2013). They occur above 1500–1650 m above sea level, and many are composite cirques, with individual hollows separated successively by narrow arêtes only. Rock basins with lakes even more than 80 m deep were scoured by glaciers to the floors of stair-like cirque platforms (Šašak et al. 2019). A plenty of roche moutonnées/glaciated knobs rise from the cirque floors (Fig. 5.17). The cirques turn to troughs over rock steps 100–400 m high (Lukniš 1973b). In forked valleys such as the Bielovodská dolina, Javorová dolina, Studená dolina, Mengusovská dolina (Fig. 5.6), Kôprová dolina, Roháčska dolina and others, smaller tributary troughs join the larger troughs over confluent rock steps, on which waterfalls occur. The troughs on the northern slope of the Tatra Mts. are deeper. The cause is not only the uplift of the mountain range in the form of a tilted block, but rock control also plays a part. The northern slope of the mountain range consists of less resistant Mesozoic rocks, which may have facilitated enhanced glacial erosion (Lukniš 1973b).

Voluminous moraines are characteristic for the Tatra Mts., which is causally related to considerable bedrock fracturing due to brittle tectonics. Those on the southern slope of the Tatras are particularly large because here the glaciers descended from the huge cirques to an open slope of the foreland. Different directions of glacier advances resulted in the origin of complex moraine ramparts up to 100 m high,

Fig. 5.10 Glacial accumulation area—late Würm/Holocene moraine plain in the middle of Malá Studená dolina valley (High Tatra Mts.) (*photo* M. Boltižiar)

composed mainly of moraines of the four Würm stadial oscillation, designated by the letters A, B, C and D. They occur in front of major valleys such as the Studená dolina (Lukniš 1955), Velická dolina, Batizovská dolina, Mengusovská dolina, Mlynická dolina, Forkotská dolina, Važecká (Zaťko 1961) and Kôprová dolina. After the blocks of dead ice had melted the depressions were filled with water, so that kettle lakes came into being, and Holocene peat. The glaciers of the late glacial oscillation E remained inside the valleys. They left behind a series of recessional moraines. The last glaciers disappeared in the early Holocene. There occur still four permanent snow patches/fields in the Slovak part of the entire Tatra Mts. (Lukniš 1973a).

The newest published geomorphological map at scale 1:100,000 (Rączkowska et al. 2015; Rączkowski et al. 2015) was compiled from earlier printed maps (Lukniš 1968; Klimaszewski 1985), aerial photographs, field observations, geophysical measurements, digital elevation models and authors' own knowledge (Fig. 5.18).

5.3 Periglacial Landforms

A considerable part of the Tatra Mts., namely from a height of 1600 m above sea level upwards, is located within a periglacial morphoclimatic belt. It is typified by intensive

Fig. 5.11 Moraine of "Senná kopa" in the Slavkovská dolina valley (southern part of the High Tatra Mts.) built by the glacier of the late glacial (Würm) (*photo* M. Boltižiar)

frost weathering and formation of talus below rock walls (Figs. 5.19 and 5.20). In the Slovak part of the High Tatra Mts., the volume of talus is estimated at 350 bil. m^3. They have been deposited after the recession of glaciers, i.e. during last 6–10 thousand years (Pánek et al. 2015, 2016). During this time interval, the ridges have been lowered by about 5 m (Lukniš 1973b; Kotarba 1989; Gądek et al. 2016).

The Little Ice Age (LIA)—a special period during the postglacial evolution—constitutes the coldest period of the last millennia in Europe. A wide range of natural and historical records shows evidence of colder climate conditions between the fourteenth and nineteenth centuries, associated with a higher frequency of extreme hydroclimatic events. During these centuries, temperature and precipitation showed different spatio-temporal patterns across Europe. The LIA in the Tatra Mts. was characterized by both long and short rainy periods (mostly long cold rainy summers), alternated with warmer periods that may have been very dry. Precise identification of the onset and ending of the LIA in the Tatras is not possible. Depending on the criteria adopted, the dates vary slightly. During the LIA, there were no fully developed glaciers in the Tatras, and only glacierettes were present. New rock glaciers did not form while the existing ones did not show any activity. The LIA, in addition to the increased intensity of morphogenic processes (Kotarba 1992, 1996, 2005; Kędzia and Kotarba 2018; Kotarba 2004a; Zasadni and Kłapyta 2009).

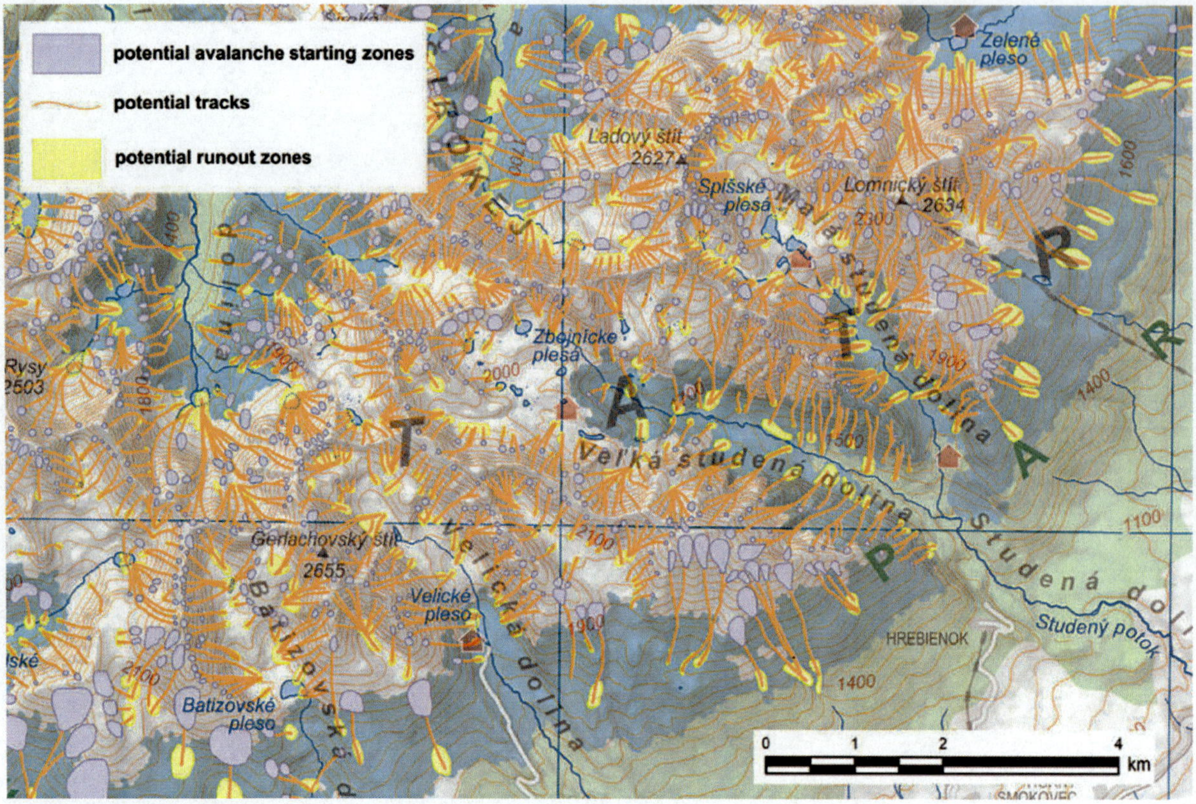

Fig. 5.12 An extract from a map of snow avalanche paths in the Tatra Mts. (original scale 1:50 000) published in Atlas of the Tatra Mts. (Žiak and Čermák 2015)

Both active and relict periglacial landforms occur in the Tatra Mts. (Kaszowski et al. 1988). Relict forms comprise rock glaciers (Figs. 5.21 and 5.22), (Nemčok 1982; Nemčok and Mahr 1974; Kotarba 1986, 1988, 1991, 1992; Kędzia et al. 2004; Kłapyta 2011; Kędzia 2014; Uxa and Mida 2017), nival ramparts (Fig. 5.23) (Kotarba 2007), debris covers (Figs. 5.8 and 5.24), as well as large sorted circles and polygons (Figs. 5.25 and 5.26). Most of these landforms developed in the Late Pleistocene (Kłapyta 2009; 2013). Other landforms—small sorted polygons (Fig. 5.27), periglacial vegetation–soil features (Fig. 5.28) (Plesník 1971), i.e. bald soils (Figs. 5.29 and 5.30) and garland soils (Figs. 5.31 and 5.32), patterned ground (Fig. 5.33), stone-festoons, solifluction sheets (Fig. 5.34) and thufurs are active at present, mostly due to seasonal and diurnal oscillations of ground temperature around 0° C (Midriak 1972, 1996; Baranowski et al. 2004; Kędzia 2004; Rączkowska 2005, 2007a; Gądek and Grabiec 2008; Gądek and Kędzia 2008; Gądek et al. 2009; Orvošová et al. 2013).

5.4 Recent Geomorphic Processes

The present-day geomorphic activity corresponds to the altitudinal climatic and vegetation belts, with the major importance of rapid mass movements like rockfalls, rockslides, debris flows, etc. Permafrost occurrence is possible in the alpine summit zone. Snow avalanches are a special phenomenon (Žiak and Dlugosz 2015; Rączkowska et al. 2016a; Boltižiar et al. 2016; Kędzia 2004).

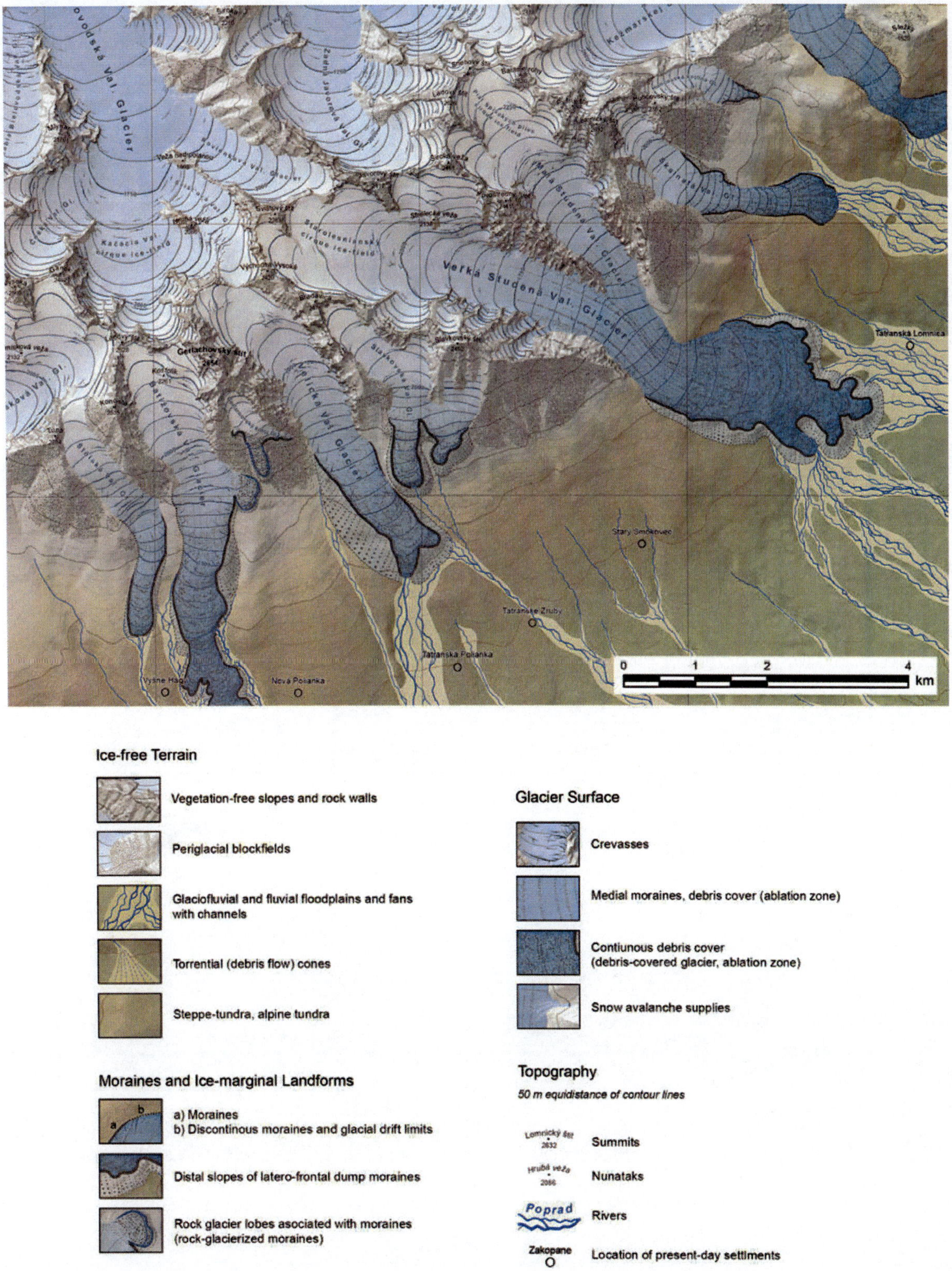

Fig. 5.13 A part of the map of glaciers during the Last Glacial Maximum in the Tatra Mts. (original scale 1:100 000) published in Atlas of the Tatra Mts. (Zasadni et al. 2015) and in Journal of Maps (Zasadni and Kłapyta, 2014)

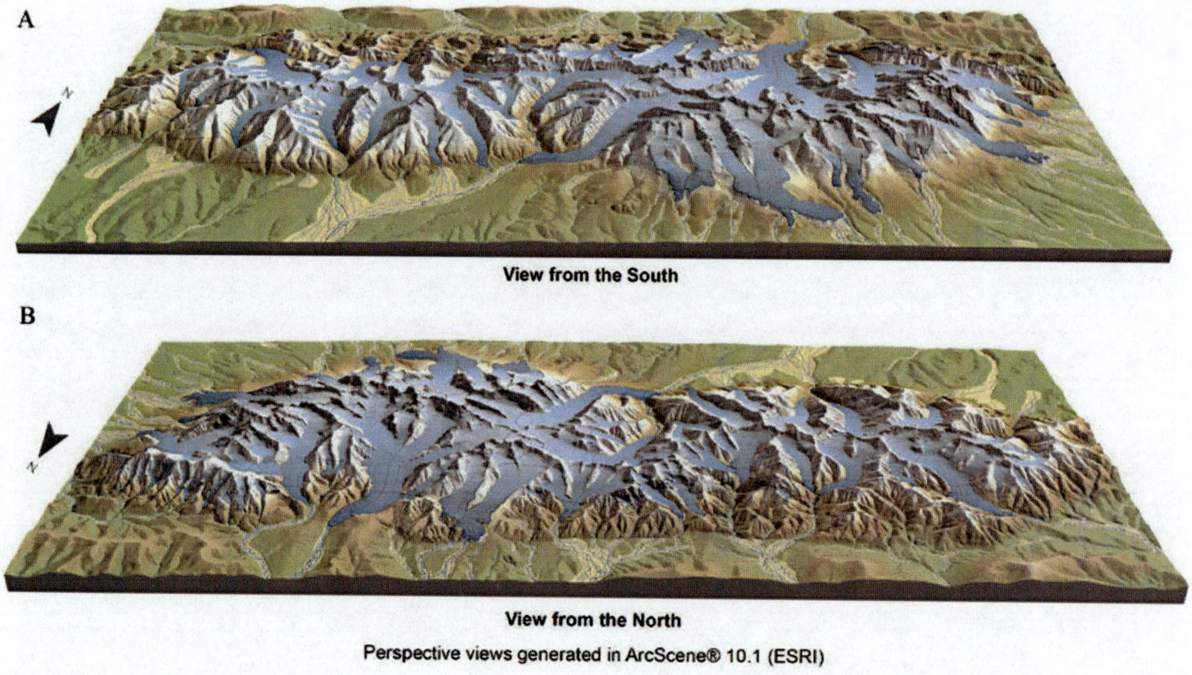

Fig. 5.14 View from the S and N to the Tatra Mts. during the last glacial maximum (after Zasadni et al 2015)

Fig. 5.15 View from the S to the central part of the Tatra Mts. during the last glacial maximum. (after Zasadni et al. 2015; Zasadni and Świąder 2015)

Fig. 5.16 Photorealistic 3D visualization of the Bielovodská dolina valley (High Tatra Mts.), with the terminus of a glacier during the Last Glacial Maximum. Images were generated in Terragen™ 3 software (Planetside software). (after Zasadni and Świąder 2015)

Fig. 5.17 Glaciated knobs (roche moutonnées) on the cirque floor of Malá Studená dolina valley (High Tatra Mts.) (*photo* M. Boltižiar)

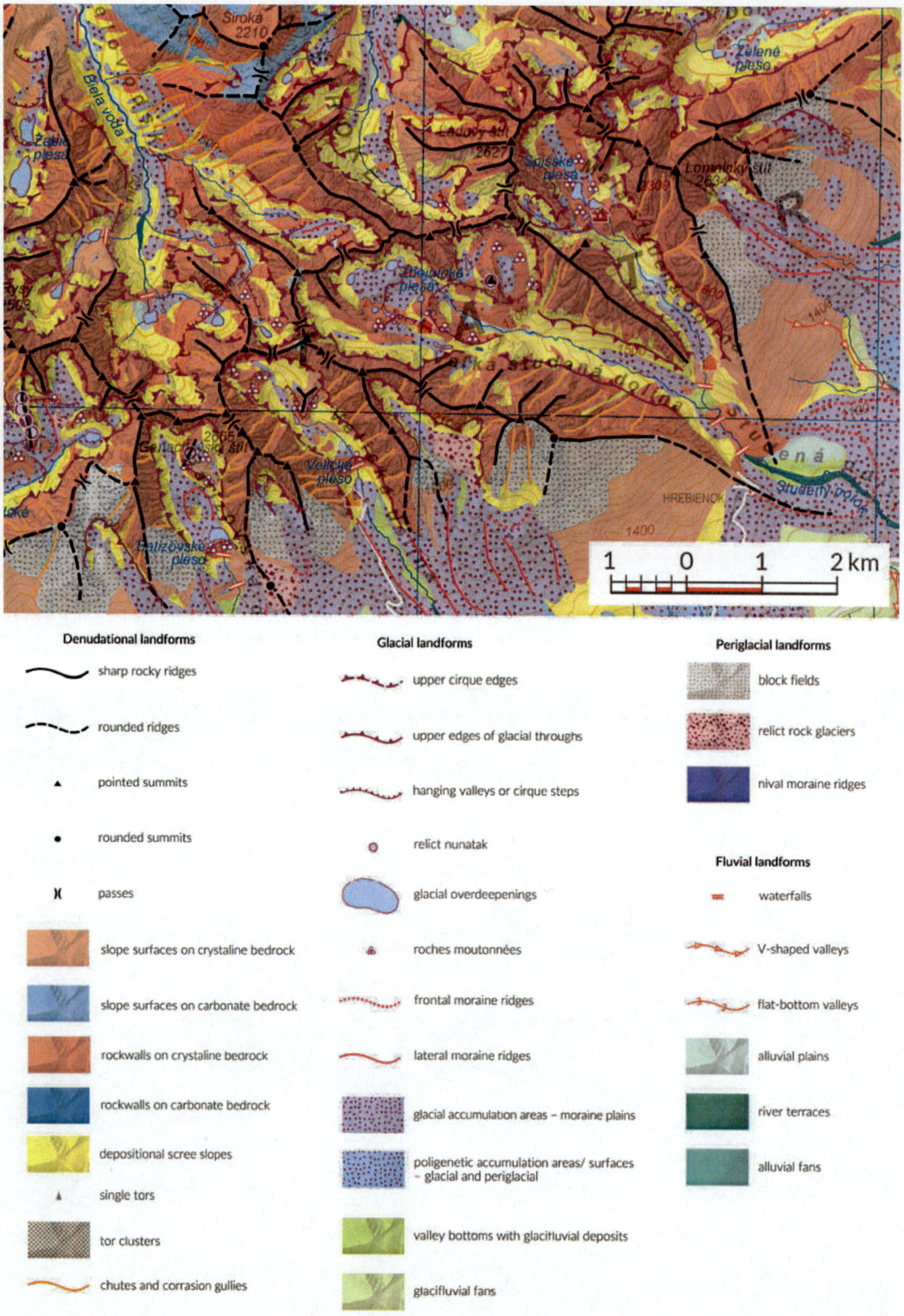

Fig. 5.18 A part of geomorphologic map of the Tatra Mts. (original scale 1:100 000) published in Atlas of the Tatra Mts. (after Rączkowska et al. 2015)

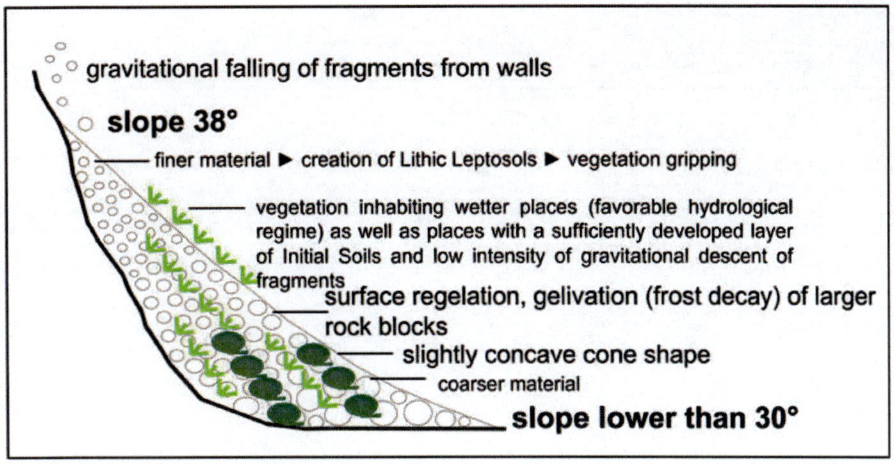

Fig. 5.19 Scheme of the talus cone (Boltižiar 2009)

Fig. 5.20 One of the biggest talus cones (ca 400 m high) in the Tatra Mts. below the Prostredný hrot peak. Malá Studená dolina valley in the High Tatra Mts. (*photo* M. Boltižiar)

Fig. 5.21 Typical lobes of a relict rock glacier covered by dwarf pine stands in the Slavkovská dolina valley (southern part of the High Tatra Mts.) (*photo* M. Boltižiar)

Debris flows are major geomorphic phenomena shaping the present-day slopes of the Tatra Mts. and constitute an important component of their contemporary denudational system. Debris flows can transport a considerable amount of saturated debris-rich regolith, accumulated below crags, in furrows and gullies, often without any noticeable motion for several years or decades (Figs. 5.35, 5.36 and 5.37). The causes of debris flows can vary, but a frequently postulated triggering mechanism is the massive supply of water and the resulting increase in pore-water pressure. The most intensive transport activity is observed each year after the spring melt, as well as at various times during heavy rainstorms with intensities in the range of 35–40 mm per hour and 1 mm. min^{-1} (Kotarba 2002; Kapusta et al. 2010, 2011).

Displacement of a soil profile including weathering mantle and possibly periglacial debris creates debris slides. Unlike the debris flows, these gravitational forms accumulate coarser material in the bottom parts. The generation of debris slides is associated with smooth hillsides with slope angles over 30°, once a waterlogged debris layer becomes separated from the subsoil and begins to move down. This process nevertheless resembles debris flows, being also conditioned by water supply and gravitation. Debris slides mainly affect shallow layers of fine debris (Hreško et al. 2003, 2008).

Geomorphic significance of debris flows is particularly visible in the High Tatra Mts., where they are largest in size and transfer considerable volumes of debris material, from a

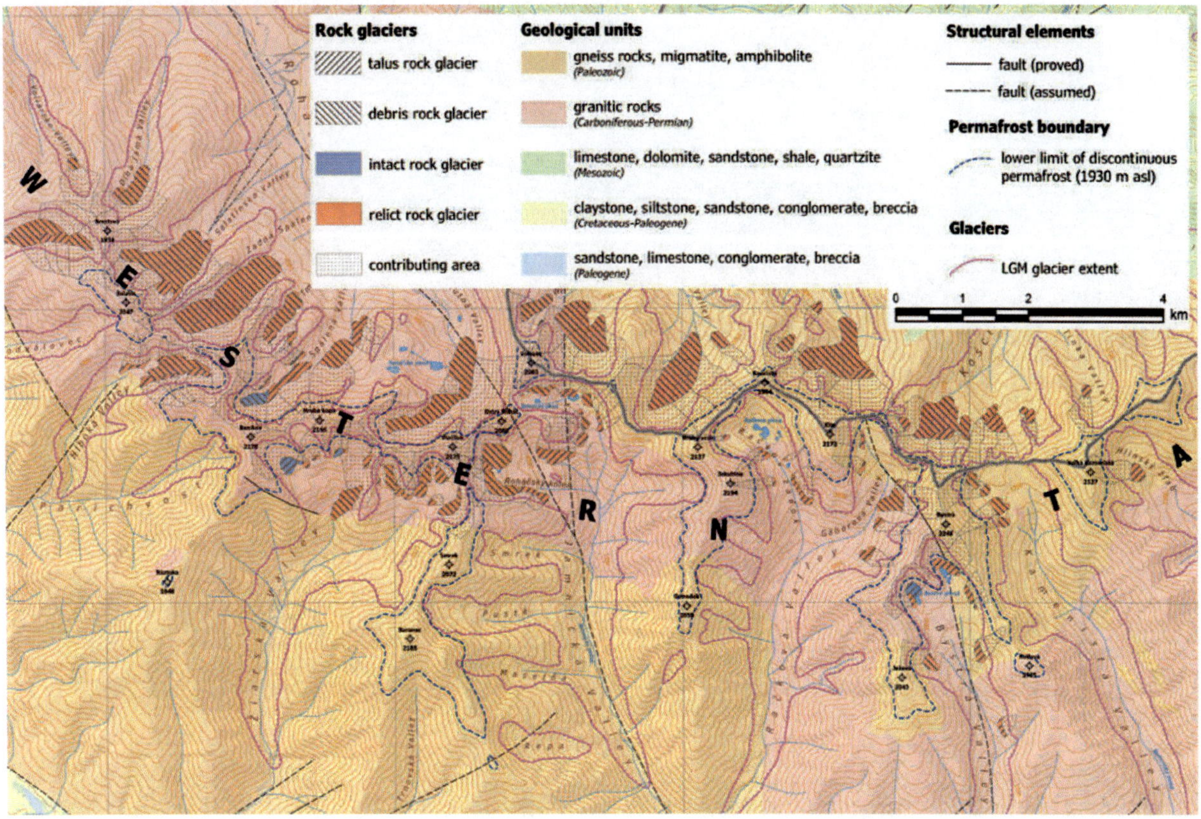

Fig. 5.22 A part of map of rock glaciers in the Western Tatra Mts. (original scale 1:40,000) published in Journal of Maps (after Uxa and Mida 2017)

few to over 25,000 m^3 (Midriak 1983, 1985; Kaszowski et al. 1988; Kotarba 1994; Kotarba et al. 2013).

Wind activity represents a significant phenomenon in an alpine environment. The effect of wind erosion is expressed mainly in deflation that causes winnowing and transport of fine debris particles on (or above) the surface, with larger residual fragments remaining as patches of coarser material. The intensity of the process can increase when ice crystals drift in the wind. Indirect aeolian influence is connected with snow transfer from windward to leeward sides of mountain crests. The resultant snowdrifts may be mobilized as avalanches or create room for nivation processes associated with long-term snow patches. The erosive effect of snow accumulations (nivation hollows) on the surface is particularly significant in areas with discontinuous vegetation cover, though it differs in relation to slope type, in that there is lithologically controlled fragment size composition of vegetation cover. The role of melt water is thus different in various lithological types of hollows (Rączkowska 1995, 2007b; Rączkowska et al. 2015). Nivation hollows and intra-moraine hollows are distinctive for their particular aquatic regime in consequence of the long-term influence of thawed water from snow patches (Kędzia 1993).

Solifluction, broadly defined as slow mass wasting resulting from freeze–thaw action in fine-textured soils, involves several components: needle ice creep (Fig. 5.34) and diurnal frost creep originating from diurnal freeze–thaw action; annual frost creep, gelifluction and plug-like flow originating from annual freeze–thaw action; and retrograde movement caused by soil cohesion. Lukniš (1973a) described solifluction lobes (Fig. 5.38) as elongated steps on slopes inclined by more than 22°, on which turf soil is pushed up into arched ramparts at the front. The lower limit of solifluction activity is to be found at altitudes of about 1700–1800 m.

Fig. 5.23 Relict talus rock glacier in the Kobylia dolina valley (Western Tatra Mts.) (*photo* M. Boltižiar)

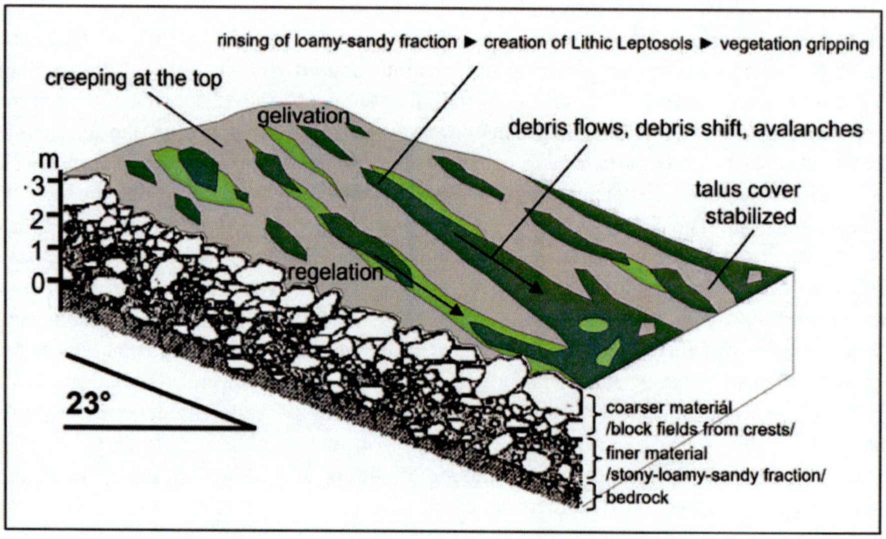

Fig. 5.24 Scheme of the periglacial debris cover/blockfields (according to Lukniš (1973), modified by Boltižiar (2009))

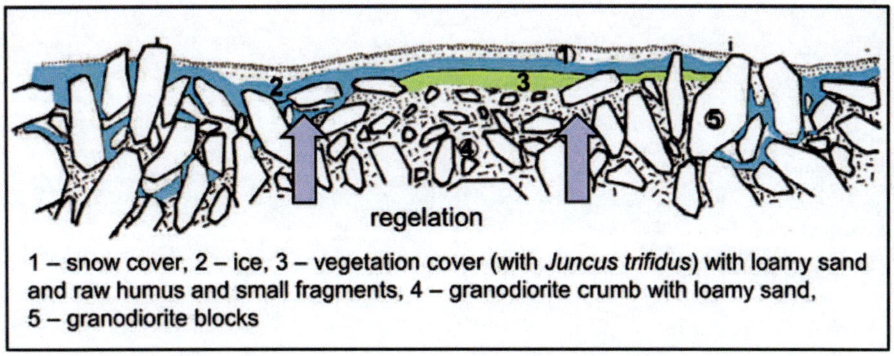

Fig. 5.25 Scheme of sorted polygon (according to Lukniš (1973), modified by Boltižiar (2009))

Fig. 5.26 Relict form of a large sorted polygon on the Lúčne sedlo saddle (central part of the High Tatra Mts.) (*photo* M. Boltižiar)

Fig. 5.27 Active sorted micropolygon on the dry ground of Hincove oko tarn (central part of the High Tatra Mts.) (*photo* M. Boltižiar)

Ploughing boulders represent a relatively common phenomenon in the periglacial environment. They occur in areas of active solifluction, on frost-susceptible soils with low plastic and liquid limits allowing for frost heaving and creeping of blocks. During movement these rotate to adopt an alignment of least resistance. The co-occurrence of ploughing boulders and stone lobes has been demonstrated (Lukniš 1973b; Midriak 1982, 1983; Hreško 1994; Hreško et al. 2005; Boltižiar 2009).

The morphodynamic phenomena of the high-mountain landscape are characterized by high dynamics and frequency of geomorphic processes, extreme disturbance effects on the landscape (Kotarba 2005, 2013; Boltižiar 2007, 2009, 2010; Hreško and Boltižiar 2001; Kędzia 2010; Midriak 1982; Stankoviansky and Midriak 1998; Stankoviansky et al. 2012; Rączkowska 2002, 2004), which are remarkably reflected also by changes of the vegetation patterns.

In terms of snow avalanche activity, the Tatra Mts. represent a dynamic high-mountain terrain with 3000 snow avalanche paths in their Slovak part. The occurrence of both gentle slopes and alpine steep rockwalls with narrow chutes required the use of two mapping methods. Each avalanche path was divided into three sections: starting zone, track and runout zone. The resulting maps (Fig. 5.12) show potential places were avalanches may occur. The total area of the starting zones amounts to 5280 ha, with the largest single starting zone covering 20 ha. Slope angle in the starting zone is the most important morphometric control on the formation

Fig. 5.28 Aeolian patch with sorted patterned soils and vegetation cover formed by wind and regelation processes (Vyšné Kopské sedlo saddle under the Hlúpy peak (2060 m a.s.l., middle part of the Belianske Tatra Mts.) (*photo* M. Boltižiar)

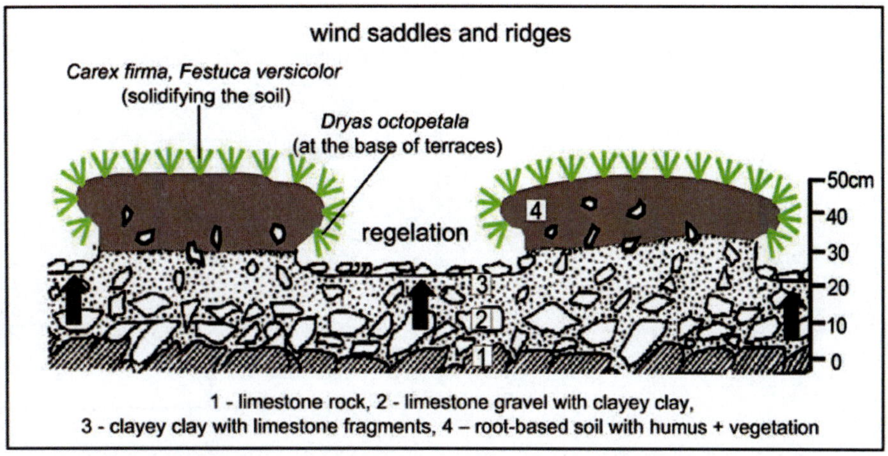

Fig. 5.29 Scheme of "bald" soils (according Lukniš (1973), modified by Boltižiar (2009))

Fig. 5.30 Periglacial vegetation soil forms—"bald" soils, formed by wind and regelation processes. Eastern part of the Belianske Tatra Mts. (*photo* M. Boltižiar)

Fig. 5.31 Scheme of garland soils (according to Lukniš (1973), modified by Boltižiar (2009))

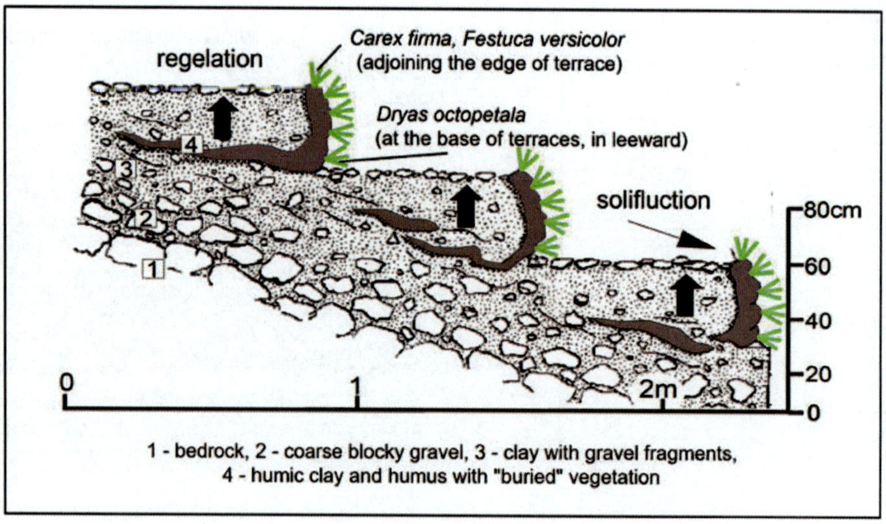

Fig. 5.32 Typical lobes of garland soils on the eastern slope of Hlúpy vrch peak in the Belianske Tatra Mts. (*photo* M. Boltižiar)

of snow avalanches. The average slope gradient in the Tatra Mts. is 42°. The steepest slopes occur in the High Tatra Mts., where fast, new-snow avalanches are commonly generated in the chutes. More gentle relief of the Western Tatra Mts. and the Belianske Tatra Mts. (Fig. 5.39) normally favours formation of large old-snow avalanches. Apart from morphometric features, meteorological conditions have a significant impact on avalanche formation. The total length of avalanche paths is 1860 km, with lengths of individual paths from 500 to 3140 m. The average size of the runout zone is 1 ha, with the largest one, located in the Žiarska valley, covering 26 ha. The combined area of all runout zones reaches 1890 ha. The largest snow avalanche event in the recent history of the Tatra Mts. was recorded on 25 March, 2009, in the Žiarska valley, where several avalanches were released at 11:00 following heavy snowfall and strong wind from the north and north-east that occurred over a short time and at temperatures unusually low for early spring (Žiak and Dlugosz 2015).

5.5 Tourism and Its Geomorphological Impact

According to Rączkowska (2019), human impact in the Tatras shows a variety of types, effects and spatial and temporal organization. Regarding chronology, there was a historical, but already non-existent, human impact associated

Fig. 5.33 Active periglacial patterned ground near Veľké Hincovo pleso tarn (High Tatra Mts.) (*photo* M. Boltižiar)

with mining, metallurgy, grazing and deforestation, whereas the contemporary impact is restricted to tourism and forest management (Rączkowska 2019).

The overall nature of the Tatra Mts., together with their easy accessibility, makes them a favorite destination for tourists. Therefore, the mountains are popular as a summer (mountainering, climbing, hiking, cycling, paragliding, etc.) and winter (skiing, skitouring) sports area, with resorts such as Poprad and the The Town of High Tatras (Mesto Vysoké Tatry), the latter created in 1999 to include former separate resorts of Štrbské Pleso, Starý Smokovec and Tatranská Lomnica, since the second half of the nineteenth century tourism has developed intensively (Rączkowska 2019).

Currently, tourist traffic is not evenly distributed in the area. According to simultaneous monitoring of visitors in 2020 in the Slovak part of the Tatra Mts., almost 30,000 tourists on average entered daily the valleys. On the Slovak side of the mountains there are four places, where the intensity of tourist traffic exceeds 2000 persons/day: the Predné Solisko Mt. in the Mlynická dolina valley (cable car and chair lifts), the Popradské pleso tarn region in the Mengusovská dolina valley (Figs. 5.40 and 5.41), the Hrebienok area (cog railway and chair lifts) and the Mt. Lomnický štit with the Skalnaté pleso tarn (chair lift and cable car; Fig. 5.42). Among the places mentioned above, cable lifts (gondola cars) allow the tourists to access the higher

Fig. 5.34 One of the present-day periglacial processes is regelation. Needle crystals of ice—"pipkrake" formed by groundwater (above right). (*photo* M. Boltižiar)

parts of the mountains while the remaining parts of the alpine zone are characterized by a notably less intensive tourist traffic. Tourist traffic concentrates in summer, from June to September, with the maximum in August (Švajda and Šturcel 2005). More than 70% of tourists entering the valleys reach only mountain huts located usually in the bottom of valley heads (Fidelus et al. 2017). The current hiking routes are largely old pastoral trails and roads (Mirek 1996; Rączkowska 2019).

According to Rączkowska (2019), tourism takes different forms in the Tatras. The main type is hiking along hundreds of kilometres of waymarked paths, leading from the bottoms of the valleys, where they usually have the form of unpaved roads, up to the ridges. Tourist erosion is generally related to hiking and has the most destructive impact on vegetation and soil covers as well as influences the intensity of morphogenetic processes. The main erosional effects of hiking are widening of hiking paths, lowering of their surface and

Fig. 5.35 Debris flow fans with fresh debris flow deposition on their surfaces. (Velická dolina valley, High Tatra Mts.) (*photo* M. Boltižiar)

development of erosional micro-forms within them, as well as deterioration of compact vegetation cover and introduction of synanthropic plant species (Boltižiar 2001; Jodłowski 2011; Rączkowska 2019). The modification of relief within hiking paths results from the activity of natural morphogenetic processes intensified or induced by the touristic use of the area. The intensity of slope degradation is much higher than on slopes not affected by tourism. Midriak (1996) found that soil erosion on slopes protected with vegetation is 0.01–0.03 mm $year^{-1}$, while on the bare surface of hiking paths it increases to 1.7–3.6 mm $year^{-1}$ in the areas with calcareous substrate. The intensity of modification of slope paths differs depending on lithology, slope inclination and orientation, soil structure, vegetation cover, seasons and the number of tourists (Fidelus-Orzechowska et al. 2017). Widening of hiking paths is spatially differentiated.

All tourist trails in the Tatras experience the strongest physical changes during snowmelt in spring, as well as in summer during heavy rainfall events. A distinct asymmetry in slope transformation within the hiking paths was found in the Western Tatra, in the area above the timberline. Northern slopes are transformed to a larger degree (Fidelus 2016; Fidelus-Orzechowska et al. 2017; Rączkowska 2019).

Fig. 5.36 Fresh debris flow track under the Veľké Solisko peak (Mlynická dolina valley in the High Tatra Mts.) (*photo* M. Boltižiar)

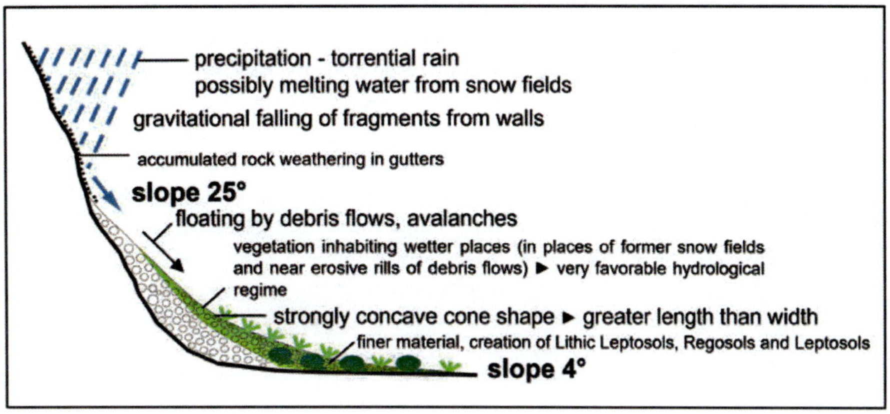

Fig. 5.37 Scheme of debris flow fan (Boltižiar 2009)

Fig. 5.38 Tongue-shaped lobe covered by vegetation and formed by solifluction processes. Northern slope of the Hladký peak (2065 m a.s.l.) (*photo* M. Boltižiar)

Since the middle of the twentieth century, a specific variety of tourism is sporting penetration of karst caves (including the showcave of Belianska jaskyňa). Another type of tourist activity is climbing, with drytooling and bouldering being currenty the most popular (Jodłowski 2015). Also cycling is allowed on roads in the bottoms of a few main valleys. Skiing dates back to the period of the end of nineteenth century. Since that time, downhill skiing became popular in the entire mountain range. In the 1930s, infrastructure for skiers started to be constructed, among them cable cars on the main ridge to reach the Lomnický štit peak (2633 m). Nowadays, skiing concentrates in a few places near chair lifts and cable cars. A rapid increase in popularity of ski-touring has been noted in recent years (Rączkowska 2019).

Landforms and landscape are among the most affected elements of the natural environment. The former due to remnants of historical human activities and effects of present-day erosion by tourism, and the latter because of changes in the altitude and structure of the upper limit of forest and dwarf pine communities. In addition, large forest areas were destroyed by wind or bark beetle (Rączkowska 2019).

Fig. 5.39 Snow avalanches on the smooth slope of Havran peak (2151 m a.s.l.) in the Belianske Tatry Mts. during spring (*photo* M. Boltižiar)

Fig. 5.40 Popular tourist destination—Popradské pleso tarn (*photo* K. Boltižiar)

Fig. 5.41 Tourists in Mengusovská dolina valley in the High Tatra Mts. (*photo* K. Boltižiar)

Fig. 5.42 One of the most popular tourist destination—Skalnaté pleso (tarn) at the altitude of 1751 m. A 4-person cable cars operate from the lower station in Tatranská Lomnica and from there a modern 15-person cable car continues to Lomnický štít peak (2633 m a.s.l.). In front of the image is the cable car station that operates the steepest ski slopes in Slovakia—Lomnické sedlo saddle during winter season. In the middle a new cable car under construction (*photo* M. Boltižiar)

Acknowledgements This work was supported by the Slovak Research and Development Agency under the Contract no. APVV-18-0185.

References

Balon J, Jodłowski M, Krąż P (2015) Położenie Tatr / Lokalizácia Tatier / Localization of the Tatra Mountains. In: Dąbrowska K, Guzik M (eds) Atlas Tatr / Atlas Tatier / Atlas of the Tatra Mountains. Neživá Príroda / Przyroda nieożywiona / Abiotic Nature. Tatrzański Park Narodowy

Baranowski J, Kędzia S, Rączkowska Z (2004) Soil freezing and its relation to slow soil movements on Alpine slopes (of the Tatra Mountains, Poland). Analele Univ. Vest Timis. Geografie 14:169–179

Boltižiar M (2001) Evaluation of vulnerability of high-mountain landscape on example Velická valley in the High Tatras Mts. Ekológia 20(4):101–109

Boltižiar M (2007) Štruktúra vysokohorskej krajiny Tatier. ÚKE SAV, FPV UKF Nitra 248 pp

Boltižiar M (2009) Vplyv georeliéfu a morfodynamických procesov na priestorovú štruktúru vysokohorskej krajiny Tatier. ÚKE SAV, UKF Nitra, p 158

Boltižiar M (2010) Morphogenetic classification of the spatial patterns in the high-mountain landscape structure (on example Tatra Mts.) Ekológia 29(4):373–379

Boltižiar M, Biskupič M, Barka I (2016) Spatial avalanche modelling by application of GIS on the selected slopes of Western Tatra Mts. and Belianske Tatra Mts. (Slovakia) Geographia Polonica 89(1):79–90

Baumgart-Kotarba M, Kotarba A (1997) Würm glaciation in the Biała Woda Valley, High Tatra Mountains. Studia Geomorphologica Carpatho-Balcanica 31:57–81

Baumgart-Kotarba M, Kotarba A (2001) Deglaciation in the Sucha Woda and Pańszczyca Valleys in the Polish High Tatra. Studia Geomorphologica Carpatho Balcanica 35:7–38

Engel Z, Mentlík P, Braucher R, Minár J, Léanni L, Aster team, (2015) Geomorphological evidence and 10Be exposure ages for the Last Glacial Maximum and deglaciation of the Veľká and Malá Studená dolina valleys in the High Tatra Mountains, central Europe. Quat Sci Rev 124:106–123. https://doi.org/10.1016/j.quascirev.2015.07.015

Fidelus J (2016) Slope transformations within tourist footpaths in the northern and southern parts of the Western Tatra Mountains (Poland, Slovakia). Z Geomorphol 60(3):139–162. https://doi.org/10.1127/zfg_suppl/2016/00265

Fidelus-Orzechowska J, Gorczyca E, Krzemień K (2017) Geomorfologiczne skutki gospodarki turystycznej w Tatrach, Tatrzański Park Narodowy, Instytut Geografii i Gospodarki Przestrzennej UJ, Kraków, p 239

Gądek B, Grabiec M (2008) Glacial ice and permafrost distribution in the Medena Kotlina (Slovak Tatras): Mapped with application of GPR and GST measurements. Studia Geomorphologica Carpatho-Balcanica 42:5–22

Gądek B, Grabiec M, Kędzia S, Rączkowska Z (2016) Reflection of climate changes in the structure and morphodynamics of talus slopes (the Tatra Mountains, Poland). Geomorphology 263:39–49. https://doi.org/10.1016/j.geomorph.2016.03.024

Gądek B, Kędzia S (2008) Winter ground surface temperature regimes in the zone of sporadic discontinuous permafrost, Tatra Mountains (Poland and Slovakia). Permafrost Periglac Process 19(3):315–321. https://doi.org/10.1002/ppp.623

Gądek B, Rączkowska Z, Żogała B (2009) Debris slope morphodynamics as a permafrost indicator in the zone of sporadic permafrost, High Tatras. Slovakia. Z Geomorphol 53(2):79–100. https://doi.org/10.1127/0372-8854/2009/0053S3-0079

Halouzka R (1992) Survey of Pleistocene mountain glaciations in the Tatra Mts. (with glacial stratigraphy of the Tatra Mts. In European correlation) and their parallelization with Continental glaciations in the Northmoravian Silesian region. Scripta Fac Sci Univ Masaryka, Brno, Geologia 22:101–107

Hreško J (1994) The morphodynamic aspect of high mountain ecosystem research (Western Tatras—Jalovec valley). Ekológia 13 (3):309–322

Hreško J, Boltižiar M (2001) The influence of the morphodynamic processes to landscape structure in the high mountains (Tatra Mts.). Ekológia 20(3):141–148

Hreško J, Boltižiar M, Bugár G (2003) Spatial structures of geomorphic processes in high-mountain landscape of the Belianske Tatry Mts. Ekológia 22(3):341–348

Hreško J, Boltižiar M, Bugár G (2005) The present-day development of landforms and landcover in alpine environment – Tatra Mts. (Slovakia). Studia Geomorphologica Carpatho-Balkanica 39:23–38

Hreško J, Bugár G, Boltižiar M, Kohút F (2008) Dynamics of recent geomorphic processes in alpine zone of the Tatra Mts. Geogr Pol 81 (1):53–65

Jodłowski M (2015) Nowa koncepcja zarządzania ruchem wspinaczkowym w Tatrzańskim Parku Narodowym jako forma ograniczenia wpływu taternictwa na środowisko. In: Chrobak A, Zwijacz-Kozica T (eds) Nauka Tatrom. Vol 3. Człowiek i środowisko. Wydawnictwo Tatrzańskiego Parku Narodowego, Zakopane, pp 63–72

Kapusta J, Stankoviansky M, Boltižiar M (2010) Changes in Activity and Geomorphic Effectiveness of Debris Flows in the High Tatra Mts Within the Last Six Decades (on the Example of the Velická Dolina and Dolina Zeleného Plesa Valleys). Studia Geomorphologica Carpatho-Balkanica 44:5–35

Kapusta J, Boltižiar M, Stankoviansky M (2011) Changes of geomorphic effectiveness of debris flow in the High Tatra Mts. within the last six decades (on the example of the Kačacia dolina valley, Velická dolina valley and the Dolina Zeleného plesa valley). In: Carpatho-Balcan-Dinaric Conference on Gepmorphology. Hradecký, J, Šilhán, K (Eds.)., Czech Republic, Ostravice, pp 34–35

Kaszowski L, Krzemień K, Libelt P (1988) Postglacjalne modelowanie cyrków lodowcowych w Tatrach Zachodnich. Zeszyty Naukowe Uniwersytetu Jagiellońskiego, Prace Geograficzne 71:121–142

Kędzia S (1993) The climatic conditions of snow patches in the Tatra Mountains as an example glacieret in the Kocioł Mięguszowiecki. Uniwersytet Jagielloński, Zakład Klimatologii Instytutu Geografii, Kraków, p 69

Kędzia S (2004) Klimatyczne i topograficzne uwarunkowania występowania wieloletniej zmarzliny w Tatrach Wysokich (na przykładzie Koziej Dolinki), Instytut Geografii i Przestrzennego Zagospodarowania Polskiej Akademii Nauk, Warszawa, p 124

Kędzia S (2010) The age of debris surfaces on the Żółta Turnia Peak (the Polish Tatra Mts.). Geomorphologia Slovaca et Bohemica 10 (2):29–38

Kędzia S (2014) Are there any active rock glaciers in the Tatra Mountains? Studia Geomorphologica Carpatho-Balcanica 48:5–16

Kędzia S, Kotarba A, Mościcki, J (2004) Lodowiec gruzowy nad Wielkim Hińczowym Stawem w Tatrach Słowackich—wyniki wstępnych badań termicznych. In: A Styczyńska, A A Marsz (Eds.), XXX Międzynarodowe Sympozjum Polarne. Akademia Morska, Polish Polar Studies, Gdynia, pp 167–177

Kędzia S, Kotarba A (2018) The Little Ice Age. The Tatra Mountains. Cuadernos de Investigación Geográfica/Geograph Res Lett 44 (1):47–67. https://doi.org/10.18172/cig.3350

Klimaszewski M (1960) On the influence of pre-glacial relief on the extension and development of glaciation and deglaciation of mountainous regions. Przeglad Geograficzny 32:41–49

Klimaszewski M (1962) Zarys rozwoju rzeźby Tatr Polskich. In: W Szafer (ed) Tatrzański Park Narodowy: Polska Akademia Nauk, Kraków, pp 105–125

Klimaszewski M (1965) On the effect of the preglacial relief on the course and the magnitude of glacial erosion in the Tatra Mountains. Geogr Pol 2:11–21

Klimaszewski M (1985) Mapa Geomorfologiczna Tatry Wysokie, Tatry Zachodnie, 1:30 000. In: Atlas Tatrzańskiego Parku Narodowego, Zakopane—Kraków

Klimaszewski M (1988) Rzeźba Tatr Polskich, PWN Warszawa, p 709

Klimaszewski M (1948) Polskie Karpaty Zachodnie w okresie dyluwialnym. Prace Wrocławskiego Towarzystwa Naukowego, B,7, Wrocław, p 236

Kłapyta P (2009) Glacial and periglacial relief on the southern slopes of the Western Tatra Mts. (Slovakia) – the results of the first detailed geomorphological mapping of the Žiarska, Jamnicka. Račkova and Bystra Valleys. Landform Analysis 10:50–57

Kłapyta P (2011) Relative Surface Dating of Rock Glacier Systems in the Žiarska Valley, the Western Tatra Mountains. Slovakia. Studia Geomorphologica Carpatho-Balcanica 45:89–106

Kłapyta P (2013) Application of Schmidt hammer relative age dating to Late Pleistocene moraines and rock glaciers in the Western Tatra Mountains, Slovakia. CATENA 111:104–121. https://doi.org/10.1016/j.catena.2013.07.004

Kotarba A (1986) Lodowce Gruzowe w Tatrach. Wszechświat 87 (5):97–99

Kotarba A (1988) Fossil rock glaciers in the Polish Tatra Mountains: origin and age. In: M Pecsi, L Starkel (Eds.), Paleography of Carpathian Regions. Geographical Research Institute, Hungarian Academy of Science, Budapeszt, pp 161–169

Kotarba A (1989) On the age of debris slopes in the Tatra Mountains. Studia Geomorphologica. Carpatho-Balcanica 23:139–152

Kotarba A (1991–1992) Reliktowe lodowce gruzowe jako element deglacjacji Tatr Wysokich. Studia Geomorphologica Carpatho-Balcanica 25–26:133–150

Kotarba A (1992) Natural environment and landform dynamics of the Tatra Mountains. Mt Res Dev 12:105–129

Kotarba A (1996) Wspólczesne procesy rzeźbotwórcze. In: Z Mirek et al. (red.) Przyroda Tatrzańskiego Parku Narodowego, Kraków--Zakopane, pp 125–137

Kotarba A (2002) Współczesne przemiany przyrody nieożywionej w Tatrzańskim Parku Narodowym, (red.) W Borowiec, A Kotarba, A Kownacki, Z Krzan, Z Mirek, Przemiany środowiska przyrodniczego Tatr, Wydaw. Instytutu Botaniki PAN, Kraków-Zakopane, pp 13–19

Kotarba A (2004) Rola małej epoki lodowej w przekształcaniu śrdoviska Tatr. IgiPZ PAN, Warszawa, p 116

Kotarba A (2004) Zdarzenia geomorfologiczne w Tatrach Wysokich podczas małej epoki lodowej. Prace Geograficzne IG i PZ PAN 197:9–55

Kotarba A (2005) Geomorphic processes and vegetation pattern changes case study in the Zelene Pleso valley, High Tatra, Slovakia, Studia Geomorph. Carpatho-Balcanica 39:39–48

Kotarba A (2007) Lodowce gruzowe i wały niwalne – efekt późnoglacjalnej ewolucji rzeźby Tatr [Rock glaciers and protalus ramparts—An effect of the Lateglacial evolution of the Tatra Mountains]. Przegląd Geograficzny 79:199–213

Kotarba A (2013) Rzeźba systemu dolinnego Suchej Wody i Pańszczycy [Relief of valley systems of the Suchej Wody and Pańszczyca Valleys]. In: Rączkowska Z, Kotarba A (eds) Dolina Suchej Wody w Tatrach. Środowisko i jego współczesne przemiany, Polish Academy of Sciences, Warszawa, pp 15–33

Kotarba A, Kaszowski L, Krzemień K (1987) High-mountain denudational system of the Polish Tatra Mountains, Geogr. Stud. IGiPZ PAN. Spec Issue 3:1–106

Kotarba A, Rączkowska Z, Dlugosz M, Boltižiar M (2013) Recent debris flow in the Tatra Mountains. In: Loczy D (ed) Geomorphological Impact of Extreme Weather (Case Studies from Central and Eastern Europe). Springer, Berlin, pp 221–236

Křížek M, Mida P (2013) The influence of aspect and altitude on the size, shape and spatial distribution of glacial cirques in the High Tatras (Slovakia, Poland). Geomorphology 198:57–68. https://doi.org/10.1016/j.geomorph.2013.05.012

Lindner L, Dzierzek J, Marciniak B, Nitychoruk J (2003) Outline of Quaternary glaciations in the Tatra Mts.: their development, age and limits. Geol Q 47(4):269–280

Lukniš M (1955) Geomorfologia a kvartér Studenovodskej Doliny w Tatrách. Geograficky Časopis 7:45–67

Lukniš M (1959) Relief a roztriedenie kvartérnych útvarov vo Vysokých Tatrách i na ich predpoli. Geologický Sbornik 10 (1):233–258

Lukniš M (1964) The course of the last glaciation of the Western Carpathians in the relation to the Alps, to the glaciation of northern Europe, and to the division of the central European Würm into periods. Geograficky Časopis 16(2):127–142

Lukniš M (1968) Geomorphological map of the Vysoké Tatry Mts. and their foreland. 1:50 000, GUDŠ, Bratislava

Lukniš M (1973) Reliéf Vysokých Tatier a ich predpolia, SAV Bratislava, p 375

Lukniš M (1973) Reliéf Tatranského národného parku. In: Zborník prác o Tatranskom národnom parku č. 15. Osveta, Martin, pp 89–143

Mida P, Křížek M (2013) New insights into the stratigraphy of moraines in the Studené doliny valleys, the High Tatras. Slovakia. Geomorphologia Slovaca Et Bohemica 13(1):54–55

Midriak R (1972) Deštrukcia pôdy vo vysokohorskej oblasti Belanských Tatier. Lesnícke štúdie 11–12, Príroda, Bratislava, p 207

Midriak R (1982) Súčasné reliéfotvorné procesy a kategorizácia deštruovaných plôch nad hornou hranicou lesa. Lesnícky Časopis 28(4):245–264

Midriak, R (1983) Morfogenéza povrchu vysokých pohorí. Veda, Bratislava, p 516

Midriak R (1985) Debris flows and their occurrence in the Czechoslovak high-mountain West Carpathians. In: Proceedings international symposium on erosion, debris flow and disaster prevention. Japan, Tsukuba, pp 175–180

Midriak R (1996) Present-day processess and micro-landforms evaluation: case of Kopske Sedlo, the Tatra Mountains, Slovakia. Studia Geomorphologica Carpatho-Balcanica 30:39–50

Mirek, Z (1996) Antropogeniczne zagrożenia i przekształcenia środowiska przyrodniczego. In: Mirek Z (ed), Przyroda Tatrzańskiego Parku Narodowego. Tatrzański Park Narodowy, pp 595–617, Kraków-Zakopane

Nemčok A, Mahr T (1974) Kamenné Ladovce v Tatrách. Geografický Časopis 26(4):359–374

Nemčok A (1982) Zosuvy v Slovenských Karpatoch, SAV, Bratislava, p 318

Nemčok J, Bezák V, Biely A, Gorek A, Gross P, Halouzka R, Janák M, Kahan S, Kotański Z, Lefeld J, Mello J, Reichwalder P, Rączkowski W, Roniewicz P, Ryka W, Wieczorek J, Zelman J (1994) Geologicka Mapa Tatier, 1:50 000, GUDŠ, Bratislava

Niedźwiedź T (2003) Extreme precipitation events on the northern side of the Tatra Mountains. Geogr Pol 76(2):15–23

Orvošová M, Deininger M, Milovský R (2013) Permafrost occurrence during the Last Permafrost.Maximum in the Western Carpathian Mountains of Slovakia as inferred from cryogenic cave carbonate. Boreas 43(3):750–758. https://doi.org/10.1111/bor.12042

Partsch J (1923) Die Hohe Tatra zur Eiszeit. F. Hirt & Sohn, Leipzig

Pánek T, Mentlík P, Ditchburn B, Zondervan A, Norton K, Hradecký J (2015) Are sackungen diagnostic features of (de)glaciated mountains? Geomorphology 248:396–410. https://doi.org/10.1016/j.geomorph.2015.07.022

Pánek T, Engel Z, Mentlík P, Braucher R, Břežný M, Škarpich V, Zondervan A (2016) Cosmogenic age constraints on post-LGM catastrophic rock slope failures in the Tatra Mountains (Western Carpathians). CATENA 138:52–67. https://doi.org/10.1016/j.catena.2015.11.005

Plesník P (1971) Horná hranica lesa vo Vysokých a Belanských Tatrach. SAV, Bratislava, p 238

Rączkowska Z (2007) Współczesna rzeźba peryglacjalna wysokich gór Europy, Prace Geograficzne IG i PZ PAN, 212, Wrocław, Warszawa, p 252

Rączkowski W, Boltižiar M, Rączkowska Z (2015) Nowe spojrzenie na mapę geomorfologiczną Tatr. In: Red. Anna Chrobak, Adam Kotarba, Nauka Tatrom. Tom I. Nauki o Ziemi. Materiały V Konferencji Przyroda Tatrzańskiego Parku Narodowego a Człowiek, Zakopane, 24–26 września 2015 roku, Wydawnictwa Tatrzańskiego Parku Narodowego, pp 143–148, Zakopane

Rączkowska Z (1995) Nivation in the High Tatras. Poland. Geogr Ann Ser A 77(4):251–258. https://doi.org/10.1080/04353676.1995.11880445

Rączkowska Z (2002) Morfodynamika stoków dojrzałych w Tatrach na przykładzie Kotła Gąsienicowego i Goryczkowego Świńskiego, (Red.) W Borowiec, A Kotarba, A Kownacki, Z Krzan, Z. Mirek, Przemiany środowiska przyrodniczego Tatr, Wydaw. Instytutu Botaniki PAN, Kraków-Zakopane, pp 49–53

Rączkowska Z (2004) Considerations on periglacial landforms and slope morphodynamic in periglacial zone of Tatra Mountains. Analele Univ Vest Timis Geografie 14:35–50

Rączkowska Z (2005) Morfodynamiczne cechy obszarów z wieloletnią zmarzliną w Tatrach. (Red.) A Kotarba, K Krzemień, J Święchowicz, Współczesna ewolucja rzeźby Polski. VII Zjazd Geomorfologów Polskich, Kraków, 19–22 września 2005, IGiGP UJ, Kraków, pp 385–388

Rączkowska Z (2007) Współczesna rzeźba peryglacjalna wysokich gór Europy. In: Present-day periglacial relief in high mountains of Europe. Polish Academy of Sciences, Warszawa p 212

Rączkowska Z (2019) Human impact in the tatra mountains. Cuadernos De Investigacion Geografica 45(1):219–244. https://doi.org/10.18172/cig.3661

Rączkowska Z, Raczkowski W, Boltižiar M (2015) Plate V.1. Rzeźba/Reliéf/Relief. Map 1. Relief 1:100 000. In: Dąbrowska K, Guzik M (eds) Atlas Tatr/Atlas Tatier/Atlas of the Tatra Mountains. Neživá Príroda/Przyroda nieożywiona / Abiotic Nature. Tatrzański Park Narodowy

Rączkowska Z, Długosz M, Rojan E (2016) Geomorphological conditions of snow avalanches in the Tatra Mountains. Z Geomorphol 60(4):285–297. https://doi.org/10.1127/zfg/2016/0289

Rączkowska Z, Rojan E, Długosz M (2016) The morphodynamics of slopes within the snow avalanche starting zones in the Tatras. Geogr Pol 89(1):17–29

Stankoviansky M, Midriak R (1998) The recent and present-day geomorphic processes in the Slovak Carpathians. Studia Geomorphologica Carpatho-Balkanica 29:69–87

Stankoviansky M, Barka I, Bella P, Boltižiar M, Grešková A, Hók J, Ištok P, Milan Lehotský, Michalková M, Minár J, Ondrášik M, Ondrášik M, Pecho J, Pišút P, Trizna M, Urbánek J (2012) Recent landform evolution in Slovakia. In: Loczy D, Kotarba A, Stankoviansky M (eds) Recent Landform Evolution. Springer, Berlin, pp 141–176

Šašak J, Gallay M, Kaňuk J, Hofierka J, Minár J (2019) Combined use of terrestrial laser scanning and uav photogrammetry in mapping alpine terrain. Remote Sens 11(2154):1–25. https://doi.org/10.3390/rs11182154

Švajda J, Šturcel M (2005) Turystyka w wysokogórskim środowisku Tatr Słowackich. In: Ładygin Z, Chovancová B (eds) Monitoring ruchu turystycznego w Tatrach. Wydawnictwo Tatrzańskiego Parku Narodowego, Zakopane, pp 37–41

Uxa T, Mida P (2017) Rock glaciers in the Western and High Tatra Mountains. Western Carpathians. J Maps 13(2):844–857. https://doi.org/10.1080/17445647.2017.1378136

Vaškovský I (1977) Kvartér Slovenska. Quaternary of Slovakia. GÚDŠ, Bratislava, p 247

Zasadni J, Kłapyta P (2009) An attempt to assess the modern and Little Ice Age climatic snowline altitude in the Tatra Mountains. Landform Anal 10:124–133

Zasadni J, Kłapyta P (2014) The tatra mountains during the last glacial maximum. J Maps 10(3):440–456. https://doi.org/10.1080/17445647.2014.885854

Zasadni P, Kłapyta J, Świąder A (2015) Tatrzańskie zlodovacenia/Tatranské zaľadnenia/Glaciations of the Tatra Mountains. In: Dąbrowska K, Guzik M (eds) Atlas Tatr/Atlas Tatier/Atlas of the Tatra Mountains. Neživá Príroda/Przyroda nieożywiona/Abiotic Nature. Tatrzański Park Narodowy

Zaťko M (1961) Príspevok ku geomorfológii Furkotskej, Suchej a Važeckej doliny v západnej časti Vysokých Tatier. Geografický Časopis 13:271–295

Žiak M, Dlugosz M (2015) Potencjalne obszary lawinowe/Potenciálne lavínové dráhy/Potential avalanches. In: Dąbrowska K, Guzik M (eds), Atlas Tatr/Atlas Tatier/Atlas of the Tatra Mountains. Neživá Príroda/Przyroda nieożywiona/Abiotic Nature. Tatrzański Park Narodowy

Martin Boltižiar is a Professor of Geography and head of Department of Geography, Geoinformatics and Regional Development of the Constantine the Philosopher University in Nitra. He is working also as a senior scientist at the Institute of Landscape Ecology of the Slovak Academy of Sciences. Vice-president of the Slovak Geographical Society and scientific secretary of the Slovak Ecological Society. Member of the Slovak National Geographical Committee. He is specializing in physical and regional geography, landscape ecology, high-mountain geomorphology, environmentalistics and geoinformatics, researching land-use/land cover changes based on remote sensing data and historical maps using the GIS technology.

6 The Longest and the Most Symmetrical Mountain Ridge of Slovakia—Low Tatra Mts.

Milan Lehotský, Bohuslava Gregorová, and Zdenko Hochmuth

Abstract

The Low Tatra Mts. is the most extensive high mountain range of the Western Carpathians. It is approximately 80 km long, 15–30 km wide, covers approximately 1,240 km^2 and culminates in Mt. Ďumbier (2,043 m a.s.l.). As a geomorphological unit of the Inner Western Carpathians, the Low Tatra belong to the Fatra–Tatra region. In terms of the geological structure, it belongs to two different units, the structure of which also conditioned their geomorphological division into two subunits. The western, Ďumbier part, belongs to the core mountains zone—the Tatricum, the eastern, Kráľovohoľská part, belongs to the Veporicum. The Low Tatra Mts. were uplifted into the form of a megaanticline during the post-Paleogene folding in the Miocene. During the Pleistocene, only the uppermost part of the central ridge has been locally shaped by glacial processes. Periglacial climate and processes conditioned, on the one hand, smooth relief on the most parts of the central ridge as well as upper parts of spurs. On the other hand, valleys downcutting, slope mass movement, debris flows and avalanches shaped the spur—ridge system and thus the Low Tatra Mts. roughly got the recent shape. In the north slopes of the Low Tatra Mts., the allogenic karst has been developed in the Fatricum and Hronicum nappes. ^{14}C dating indicate the scarps were formed by episodic displacements over the past 9.6 ka on the south ridge of Mt. Chabenec. The most important part of the Low Tatra Mts. in terms of the avalanche occurrence is the Ďumbier part. The mountain has been legally designated as a national park in 1978, as the largest large-scale protected area in Slovakia with 72,842 ha. Due to the attractiveness of their landscape and suitable climatic conditions, they are intensively used by tourists throughout the year.

Keywords

Low Tatra Mts. • Development • Karst • Glacier landforms • Sackungen • Avalanches shaping

6.1 Introduction

The Low Tatra Mts. is the most extensive high mountain range of the Western Carpathians and the second highest mountain range in Slovakia (Fig. 6.1). It is located in the central part of Slovakia, surrounded by inter-mountain basins from the north and south, and has a west–east elongated shape. It is approximately 80 km long, 15–30 km wide, covers approximately 1,240 km^2 and culminates in Mt. Ďumbier (2,043 m a.s.l.). The Low Tatra spread mainly at altitudes above 550–600 m a.s.l., with more than half of the area lying above 950 m a.s.l. The peaks are located in the subalpine zone and many of them have been glacially shaped during the Pleistocene (Fig. 6.2). The central ridge, apart from rounded relief, displays a variety of tors, thufur hummocks, sackungen, boulder fields and solifluction areas (Fig. 6.3).

In addition to the glacial relief, the Low Tatra abound in gorges and karst phenomena associated with the occurrence of Mesozoic limestones and dolomites. The central ridge forms the water divide between the main Slovak rivers—Váh, Hron, Hornád and Hnilec, and the main European water divide between the Black and Baltic Sea. The borders

M. Lehotský (✉)
Department of Physical Geography, Geomorphology and Natural Hazards, Institute of Geography, Slovak Academy of Sciences, Štefánikova 49, 814 73 Bratislava, Slovakia
e-mail: geogleho@savba.sk

B. Gregorová
Department of Geography and Geology, Matej Bel University, Banská Bystrica, Slovakia
e-mail: bohuslava.gregorova@umb.sk

Z. Hochmuth
Institute of Geography, Faculty of Natural Sciences, Pavol Jozef Šafárik University, Košice, Slovakia
e-mail: hochmuth@upjs.sk

M. Lehotský and M. Boltižiar (eds.), *Landscapes and Landforms of Slovakia*, World Geomorphological Landscapes,
https://doi.org/10.1007/978-3-030-89293-7_6

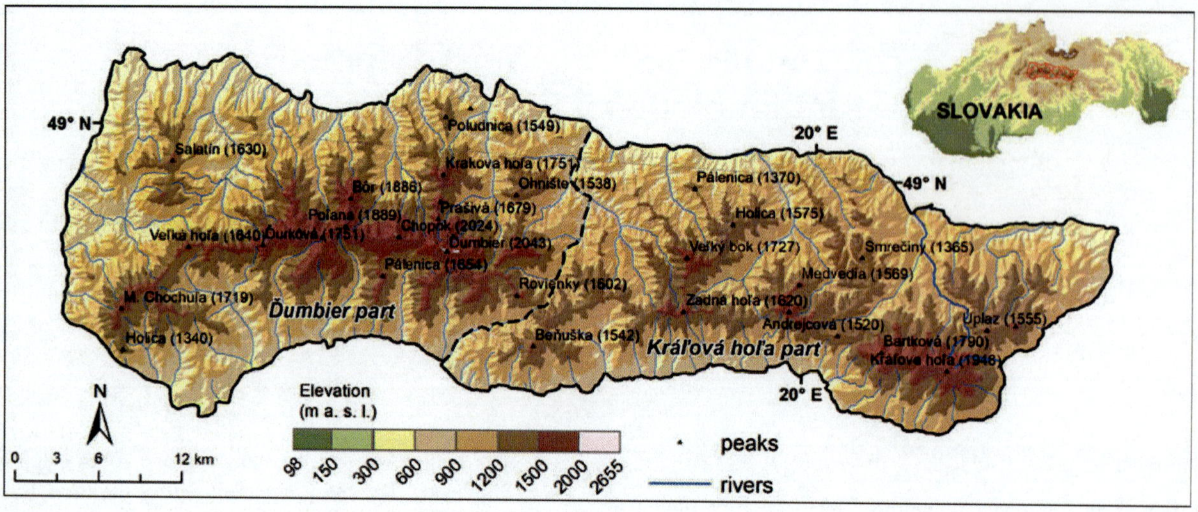

Fig. 6.1 The localization, hypsometry and two parts of the Low Tatra Mts

Fig. 6.2 The central part of the Low Tatra Mts. Rounded main ridge undercut by glacial cirques on the north-facing slopes (*photo* P. Bella)

Fig. 6.3 Tors at the Kráľová hoľa Mt. (1946 m a.s.l.) in the east of the Low Tatra Mts. (*photo* J. Madarás)

of Slovak historical regions have long led through the Low Tatra ridge, separating Zvolenská and Gemerská counties in the south, and Liptovská and partly also Spišská counties in the north. Almost the entire mountain range is protected as a national park due to its exceptional natural richness and values. The Low Tatra Mts. natural resources have been used for mining, charcoal-burning, metallurgy, shepherding, mountain farming and forestry for a long time. These activities have historically shaped the landscape and gave it its distinctive, unique character. The rich relief diversity, heterogeneity and uniqueness of terrestrial and underground phenomena, on the one hand, justify the protection of the Low Tatra nature in the form of a national park, on the other hand, the pressure on nature is continuously increasing by the expansion of new recreational and tourist infrastructure.

6.2 Geology and Development

As a geomorphological unit of the Inner Western Carpathians, the Low Tatra Mts. belong to the Fatra–Tatra region (Mazúr and Lukniš 1978). In terms of geological structure, it belongs to two different units, the structure of which also conditioned their geomorphological division into two subunits. The western, Ďumbier part, belongs to the core mountains zone—the Tatricum, the eastern, Kráľovohoľská part, belongs to the Veporicum (Fig. 6.4).

The division line between these two geotectonic units follows the course of a distinct thrust zone (Čertovica line) composed of mylonitic and phyllonitic rocks (Biely and Fusán 1969), along which the Veporicum was pushed onto the zone of the core mountains during the Cretaceous mountain-forming movements. In the Ďumbier part, the crystalline core consists of the most resistant Palaeozoic igneous and metamorphic rocks (Ďumbier and Prašivá granite types), migmatites, gneisses, phyllites and amphibolites. These types of rocks build mainly the ridge and the southern part of the mountain range. The covering series are formed mainly by Mesozoic sedimentary rocks, such as limestones and dolomites. Their distribution is asymmetric. Mesozoic sediments are less abundant in the southern part of the mountain range; they are prevailingly located in the northern part. The Fatricum builds the western and north-western parts of the Low Tatra Mts. Softer marlstone rocks, which are less resistant to erosion, can be found within the succession, forming a lower and smoothly shaped relief. The more resistant dark Guttenstein limestones and dolomites of Triassic age extend in the area from the Revúcka Valley through Mt. Salatín to the area of the Demänovská Valley. The area from the Iľanovská Valley through Mt. Poludnica and Mt. Ohnište to the Jánská Valley consists of limestones and dolomites of the Hronicum, creating a strongly rugged relief. The most extensive cave systems and the deepest abysses in Slovakia have developed

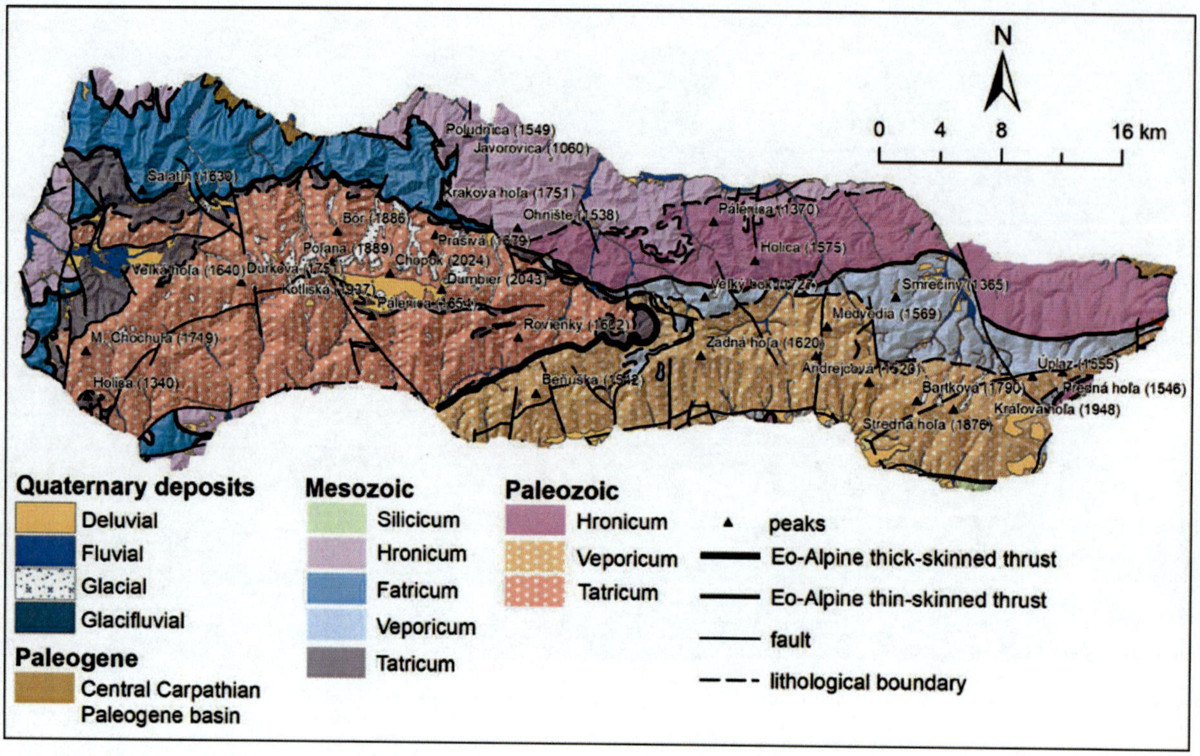

Fig. 6.4 Tectonic division of the Low Tatra Mts. (base map according to Biely et al. 1996)

here thanks to intensive karstification, especially in the area of Demänovská and Jánská valleys. Such rocks are found only in the form of smaller islands on the southwest side of the Low Tatra (between Moštenice and Mýto pod Ďumbierom villages, or in the vicinity of M. R. Štefánik chalet).

In the Kráľovohoľská part of the Low Tatra Mts., which belongs to the Veporicum, the main ridge and the southern slopes are predominantly built of metamorphic rocks (phyllites, gneisses and paragneisses). The northern side of the mountain range has a very complicated structure. It is possible to distinguish a separate cover unit, the so-called Veľký Bok series in the crinkled layers of sedimentary rocks, sloping to the north (slates, quartzites, limestones and dolomites, marls, marly limestones). The Fatricum lies upon the eastern part of the envelope series. Its lowest unit is the melaphyre series. The middle and upper Triassic of the Fatricum is represented by the so-called Čiernovážska series, with a predominance of dolomites. In the south, it is located in the belt between the villages of Boca and Malužiná to Hranovnica. The Bielovážská series of Fatricum contains cherts, sandstones and marls, in addition to limestone, and is located further north from the Malužiná—Svarín line. Limestone–dolomitic siliceous formations cover the eastern edge of the Low Tatra Mts. (Biely et. al. 1997a, 1997b).

The Low Tatra Mts. were uplifted into the form of a mega-anticline in the Miocene. At that time, the mountains were lower and less fragmented. The current relief of the fault-block mountain is the result of positive tectonic movements in the Pliocene and the Pleistocene, which stimulated intense deep erosion. Erosion–denudation processes exposed the crystalline core and cut it into the spur-and-valley system, with local relief around 800 m. So, the Low Tatra Mts. have developed into a form characterized by a central, considerably elongated ridge and a system of spurs alternating with deep valleys at both sides. However, the north side of the mountains, where the Fatricum and Hronicum rock complexes (Mezozoic limestons and dolomites) occur, karst landforms have developed. During the Pleistocene, only the uppermost part of the central ridge was locally shaped by glacial processes. Periglacial climate and processes conditioned, on the one hand, smooth relief on most parts of the central ridge as well as in the upper parts of spurs. On the other hand, valley downcutting, slope mass movement, debris flows and avalanches shaped the spur-ridge system and thus the Low Tatra Mts. roughly got the recent shape.

6.3 Karst Landforms

In the northern slopes of the Low Tatra Mts., the allogenic karst has developed in the Fatricum and Hronicum nappes. Its fundamental feature is that streams, which start in the crystalline core, incise the limestone–dolomite formations of the nappes and form cave systems. The systems comprise old, fossil stream passages but also current underground channels. The evolution of the underground karst landscape was favoured for the chemical properties of waters (unsaturated water, i.e. 'hungry' water) and also transport of solid material (granite pebbles, quartz sand). A high mountain karst evolved in the highest parts of the Mesozoic rock series, where autochthonous rivers generate deep and long, high mountain caves, which belong to the deepest in Slovakia. According to Hochmuth (2008) and Droppa (1973), the karst terrains of the Low Tatra Mts. are classified into twelve spatial units (Figs. 6.5, 6.6).

The maximal areal extent of the karst area typifies the central-north part of the Low Tatra Mts. and is represented by the Demänovská and Jánska valleys karst, of which the Demänovská Valley karst includes around 215 caves of total length more than 55 km. The Demänovská Valley cave system (the system of interconnected caves) presents the longest 37,4 km cave system in Slovakia and the second one in the Carpathians (Bella et al. 2014). The valley/gorge is going through the area of northern Mesozoic cover series and nappes. These geological units are composed of limestones and dolomites prone to karstification, especially the Guttenstein limestone and, in the upper and lower part, the Ramsau dolomites. Oval, river-modelled passages of both larger and smaller dimensions, dome spaces reshaped by collapses and frost weathering dominate in the caves (Droppa 1957, 1972a). The formation of ten individual cave levels as horizontal passages relates to the position of outflow of underground stream to the surface in the past, as well as to the system of river terraces in the adjacent Liptovká basin. Based on Uranium series, paleomagnetic and radiometric dating of the flowstones samples Bella et al. (2011) estimated their age from 0.12 Ma to 2.88 Ma.

The particular position among karst units of the Low Tatra Mts. is occupied by the karst of Mt. Kraková hoľa and Mt. Poludnica and karst of Mt. Ohnište (Fig. 6.7). The Mt. Krakova hoľa and Mt. Poludnica karst unit is stretched on the spur ridge where only headwater drainage system has been developed, whereas perennial streams are active only in underground. The top of Mt. Krakova hoľa (1753 m) is built by dolomites, shaped into monocline structure with front facing by steep cliffs to the Demänovská valley. The ridge leading from the top of Mt. Poludnica to the south is only ten to several tens of meters narrow and has 20–30 m high rock walls on both sides. In this karst unit the deepest cave in Slovakia (495 m), the so-called "The system of Hipman's caves" is located (Fig. 6.8), which originated in 2003 by speleological connection of the previously explored caves Starý hrad and Večná robota. The total length of the cave is currently 7553 m. Droppa (1972b) and Hochmuth (1997) noticed that some caves in the Krakova hoľa unit at locations 600–700 m above the contemporary valley floors have a river character because of the presence of gravels from the packaging series of the Low Tatra Mts. Droppa (1972b), and later Bella (2002) and Orvoš (2005), assumed their Pliocene age (around 5 Ma).

The karst of the Ohnište massif represents a karst plateau towering between the Bocianska and Jánská valleys. It culminates in Mt. Ohnište (Príslop) at 1538 m. The Ohniše plateau represents a planation surface on the Triassic

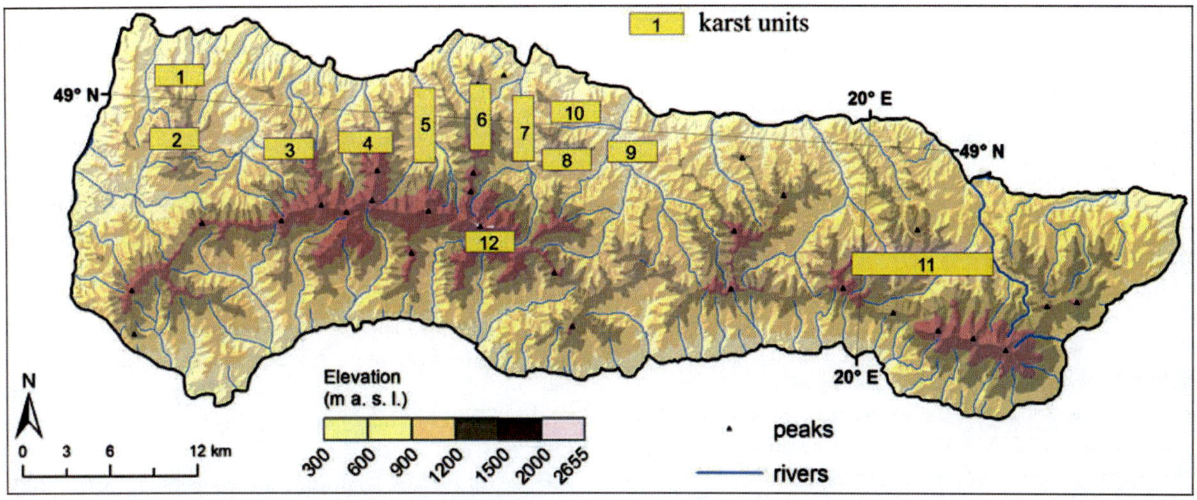

Fig. 6.5 Localization of karst units classified according to Hochmuth (2008). Map legend: 1. karst of Borovinského and Kohúta; 2. Ludrovský karst; 3. Ľupčiansky karst; 4. karst of Mošnica and Kamenica valleys; 5. Demänovská Valley karst; 6. karst of Krakova hoľa and Poludnica; 7. Jánska valley karst; 8. karst of Ohnište; 9. karst of Malužiná; 10. karst of Porúbka; 11. karst of north slopes of Kráľová hoľa; 12. Ďumbier high mountain karst.

Fig. 6.6 The Salatín Mt. (1630 m a.s.l.) as an example of conical karst rocky landform dominates above NW part of the Low Tatra Mts. (*photo* J. Lacika)

limestones and dolomites. The area (about 2 km^2) of the plateau is characterized by low relative relief (up to 50 m), several vertical abysses of the "Aven" type and scarce occurrence of lapies and sinkholes. The relatively low occurrence of karst surface landforms is likely conditioned by anthropogenic interventions (Vajs 1998). Orvošová et al. (2006) proved the existence of allochtonous material in sinkholes located at the head of a "paleo-valley", which suggests that the plateau was significantly larger, reaching as far as the contemporary main ridge of the mountains from where the material was transported by the paleo-river. The touristic attraction here is the 15-m-high rock window of weathering origin, also visible from several places in the Jánska valley (Fig. 6.9).

Another interesting karst unit—the Ďumbier karst—is the carbonate strip 4,5 km long and 100 m wide on average (its max width is 350 m) in the central part of the Low Tatra Mts. at altitudes 1,225–1,740 m a.s.l. According to the stripe karst typology of Lauritzen (2001), the Ďumbier karst presents a type of subvertical unconfined stripe karst attacked by allogenic waters. The very extensive, large multi-level cave system is known in this contact stripe karst. The origin of the Dead Bats Cave (length 18,285 m, denivelation 300 m) is the results of very advanced underground karstification in the subvertical carbonate stripe, enclosed by non-carbonate rocks (karst aquifer was determined and controlled by longitudinal lithological boundaries, parallel structural-tectonic discontinuities and downcutting of valleys). The cave consists of several dominant horizontal epiphreatic passages, phreatic loop segments, epiphreatic good water segments and narrower steep vadose passages, also shafts. The original allogenic stripe structure of the Ďumbier karst was morphologically intensively remodelled and transformed in several places in the consequence of valley downcutting on the southern and northern sides of the Low Tatra Mts. In spite of its recent allogenic–autogenic or autogenic position, more elevated segments of this stripe karst are characterized by remarkable phreatic and epiphreatic features of allogenic paleo-hydrographical structure (Bella 2004).

As far as surface landforms are concerned, rocky walls, cave entrances, karren, dolines, ponors and resurgences occur not only in the Demänovská valley, but they occur in greater or lesser extent in all other karst units developed in

Fig. 6.7 Karst areas of Krakova hoľa, Poludnica and Ohnište situated in the north forks of the central part of the Low Tatra Mts. In the background the Liptov basin and the range of the Tatra Mts. (*photo* J. Lacika)

valleys, such as the Borovinského and Kohúta karst, Ludrovský karst, Mošnica, Kamenica and Jánska valley karst, and the karst of northern slopes of Mt. Kráľová hoľa.

6.4 Glacial Landforms

In the Pleistocene, the whole territory of the Low Tatra Mts. was situated in the periglacial zone. The most intensive glacial shaping of the Low Tatra Mts. occurred in their central part, between Mt. Chabenec and Mt. Ďumbier, while less modelled landscape by glaciers is situated in the east part of Kráľova hoľa (Fig. 6.10). The least elevated cirque floor of the Low Tatra Mts. at an altitude of 1480 m a.s.l. is found on the northern slope of the Mt. Veľký bok (Škvarček 1986, 1989). In the central part of the mountains (Fig. 6.11) the depth of glacial cirques varies from 250 to 350 m (Louček et al. 1960; Škvarček 1980, 1983).

The most abrupt cirque walls are inclined at angle 30–40°. The longitudinal valley profiles often brakes by the rocky steps (30–35°) down the cirque and passing gradually over individual moraines into the outwash fan. The cirque floors are covered by talus cones. Moraines are composed only of large boulder and are devoid of any finer, weathered material deposited by the glaciers. Typical is their semi-circular shape displaying a steeper slope on the outside. The last recessional moraines are to be found sitting on abrupt rock thresholds at the outlets of the cirques. Below these steps there occur moraines dating from the first or second stages of retreat. The oldest retreat stages have left behind much less perfect forms because the parts of the valleys under the cirque steps were as rule affected by strong stream erosion, which washed out some of the moraines and re-deposited their material after a short distance in the form of alluvial fans. The oldest moraines contain well-polished, waterborne boulders and a considerable amount of sandy soil.

Fig. 6.8 Medusa " aven in the Hipman's cave system (*photo* P. Hipman)

Pleistocene glaciation of the Kráľová hoľa part of the Low Tatra Mts. is evidenced by glacier cirques in valley heads and by frontal moraines which occur at different altitudes (Škvarček 1986, 1989). Evidence of Würm glaciation can be observed here only in some valleys dissecting the northern hillsides of the mountains (Fig. 6.12)—in the valleys of Veľký Brunov, Holičná and that of Ždiarsky potok brook which has its heads between Mt. Orlová (1839 m a.s.l.) and Mt. Bartková (1790 m a.s.l.).

Transversal profiles of cirques are slightly asymmetric. Left hillsides are steeper, facing ESE and SE. Asymmetry can be also observed in not perfectly developed parts of

Fig. 6.9 Rock window of weathering origin and the main touristic attraction in the Ohnište karst area. (*photo* E. Hipman)

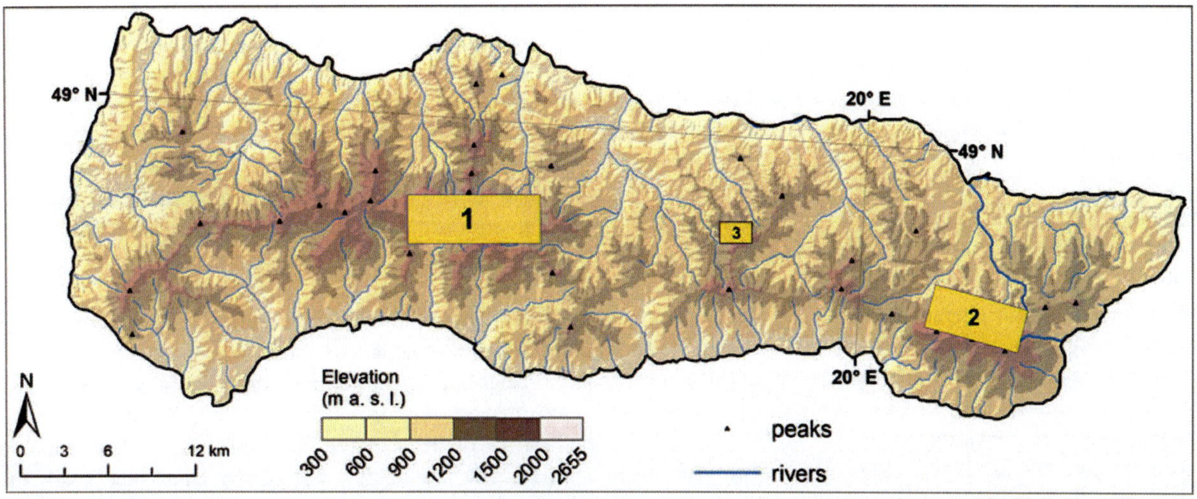

Fig. 6.10 Glacially shaped areas. Map legend: 1. central area between Mt. Chabenec and Mt. Ďumbier; 2. eastern area around Mt. Kráľová hoľa; 3. Mt. Veľký bok

Fig. 6.11 Geomorphological map depicting glacial landforms in the central part of the Low Tatra Mts. (Louček 1960). Rounded central ridge undercut by glacial cirques (left, *photo* P. Bella); northern slopes with glacial cirques of the Dereše Mt. and ski pists in Jasná—the largest ski-resort in Slovakia (middle, *photo* J. Lacika); view of the contemporary relief of a Würm-age glacial cirque in Dereše Mt. (right, *photo* P. Bella). Map legend: 1. recessional moraines; 2. end moraines; 3. outwash moraine material;4. fluvioglacial sediments; 5. talus cones; 6. debris; 7. cirque headwalls; 8. rocky steep slopes; 9. steep slopes; 10. boulder fields; 11. joint-sink-furrows; 12. joint cuesta; 13. thufur hummocks;14. rounded ridge

glacier troughs. This phenomenon is caused by climatic factors. The cirques pass continuously into short and slightly developed through valley parts.

The boundary between these two forms is fuzzy and is indicated only by valley narrowing. The absence of rocky steps under glacier cirques is probably associated with the relatively short duration of glaciation. The longest glacier was in the valley of Velký Brunoy, having the length 2,9 km (Škvarček 1989). The length of glaciers in other valleys did not reach half of that in the valley of Veľký Brunov (Fig. 6.13).

Frontal moraines from the Last Glacial Maximum (LGM) occur in all valleys. The lowest ones are at the altitude 1250 m a.s.l., in the valley of Brunov. The tops of two moraine lines at altitudes 1210 m a.s.l. and 1250 m a.s.l. indicate minuscule stadial oscillations. If we compare the altitudes of cirques and frontal moraines of LGM, we can conclude that the less elevated are cirque floors, the higher are the respective frontal moraines.

After moraines of maximum glaciation were accumulated, the glacier retreated into valley heads and became extinct in the Holičná valley. Frontal moraine of W3 in the valley of Veľký Brunov is at an altitude 1450–1480 m a.s.l. Frontal moraine of the same age in the valley to the north-west of Mt. Orlová is in the bottom of cirque, at an altitude of 1600–1630 m a.s.l. In the Late glacial moraines evolved in cirques (Holičná valley). Two low terraces lying down in the valleys correspond to frontal moraines from the Late glacial. Older terraces pass into glaciofluvial cones being linked with frontal moraines of the LGM. Fragments of a younger terrace from the Late glacial are embedded into an older terrace and also occur over frontal moraines of maximum extent of glaciation. Accumulations of low terraces enlarge at the mouths into main valleys of the Čierny Váh and Žiarsky potok brooks. The older ones, from LGM, end at different heights over levels of water courses, which is associated with headward erosion of pushing back the lower parts of their accumulations. Most frequently, they terminate at the height of 10–15 m above the valley floors. Accumulations of younger cones from the late glacial pass into the bottom fill of the Čierny Váh brook. The valleys on the northern hillsides of this part of mountains, the heads of which end in a close neighbourhood of the snow limit of the Late glacial, are as a rule reworked by nivation processes.

Fig. 6.12 Geomorphological map depicting glacial landforms in the eastern part of the Low Tatra Mts. (Škvarček 1989). Oblique view (GoogleEarth) on the area depicted on the map gives the image about contemporary features of the relief. Map legend:1. glacial cirques; 2. nivation hollow; 3. rock glacier niches; 4. frontal moraines (W2, W3); 5. remnants of ground moraines (W2); 6. ground moraines (W3); 7. moraine translocated by rock glaciers (W2, W3); 8. front of tongue-shaped rock glaciers (W2); 9. frontal part of rock glaciers (W3); 10. tongue-shaped rock glaciers (W2); 11. tongue-shaped rock glaciers (W3); 12. sinkholes; 13. alluvial fans and terraces (W2, W3); 14. alluvial fans and terraces (Riss and older)

Fig. 6.13 View down the glacially shaped Veľký Brunov valley. (*photo* J. Madarás)

Nivation hollows are shallow and have flat bottoms, which become wedge-shaped furrows in the direction of slope. In the valleys, the heads of which are reworked by nivation, accumulations of rock glaciers from the last glacial occurred. Their frontal parts are very steep (20°) and they are linked by alluvial cones. Tongue-like rock glacier accumulations penetrate differently into the valleys. Most frequently, we find only one or rarely two generations of rock glacier accumulations in the valleys, which correspond temporally to frontal moraines of the late Würm, whereas older generations correspond to the moraines from LGM and the younger ones to the moraines from Late glacial.

Apart from two glacially shaped areas the lowest situated glacial cirque in the Low Tatra Mts. from the Late glacial is developed at an altitude of 1480 m a.s.l. in the northern slope of Veľký bok Mt. (Škvarček 1980; Fig. 6.14).

The glaciation of the Low Tatra Mountains took place in two separate phases (Vitásek 1926; Louček et al. 1960; Škvarček 1980, 1983, 1986, 1989). In the first phase, embryonal cirques had been formed and long glaciers reached down far beyond the snow line and deposited a whole series of lateral and terminal moraines. Deposits laid down in these retreat stages date most probably from the early Würm. One presumes that had the Low Tatra Mountains been glaciated also in the Riss stage at all, they must have been together with the other mountain ranges through an effective process of transformation. The fact that the summit area displays rounded surfaces, with thick layers of soil and only slightly indented cirques, shows that climatic conditions were unfavourable and the process of glaciation did not last long (Louček et al. 1960).

In the second phase, the cirques themselves were glaciated. Small retreating glaciers, "névé" fields and snow fields left behind deposits, only here and there transporting them for very short distances. In this stage, terminal and lateral moraines without any stratification were deposited in the cirques. So, the glaciation of the cirques dates from W3 and might have exceeded even into the postglacial period.

After the retreat of the glaciers, the relief was re-shaped again fluvially. Only in places sheltered from stream erosion traces of glaciation are found. In cirques, the effects of glacial activity disappeared under talus cones, which partly covered even the highest situated end and recessional moraines. Consequently, the present Low Tatra relief is the result of polygenetic modelling, with glacial phenomena retreating into the background.

6.5 Sackungen

Characteristic features of the central ridge of the Low Tatra Mts. and to a certain degree also of spurs are numerous, often extensive slope deformations. They are situated mostly in the western part (the Ďumbier part), less often in the eastern one (the Kráľová hoľa part) of the ridge and occur in metamorphic, granitic and Mesozoic sedimentary rocks (Fig. 6.15).

According to Nemčok (1982), slope deformations show different character in each main lithological group. Slope deformations in metamorphic rocks consist primarily of gravitational folds or flexures, cut by shear planes. They can be compared with extensive downslope bending of beds near the slope surface. They affect, for example, slopes of Mt. Struhár, Mt. Skalka (close to the Mt. Kotliská), Mt. Veľká Vápenica, Mt. Medvedia, Mt. Žiarska hoľa, Mt. Žiar and both slopes of the Suchá dolina valley (Stankoviansky 1984).

Fig. 6.14 The lowest situated glacial cirque (circle) in the Low Tatra Mts in the northern slope of Veľký bok Mt. (1727 m a.s.l.). (*photo* M. Slavický)

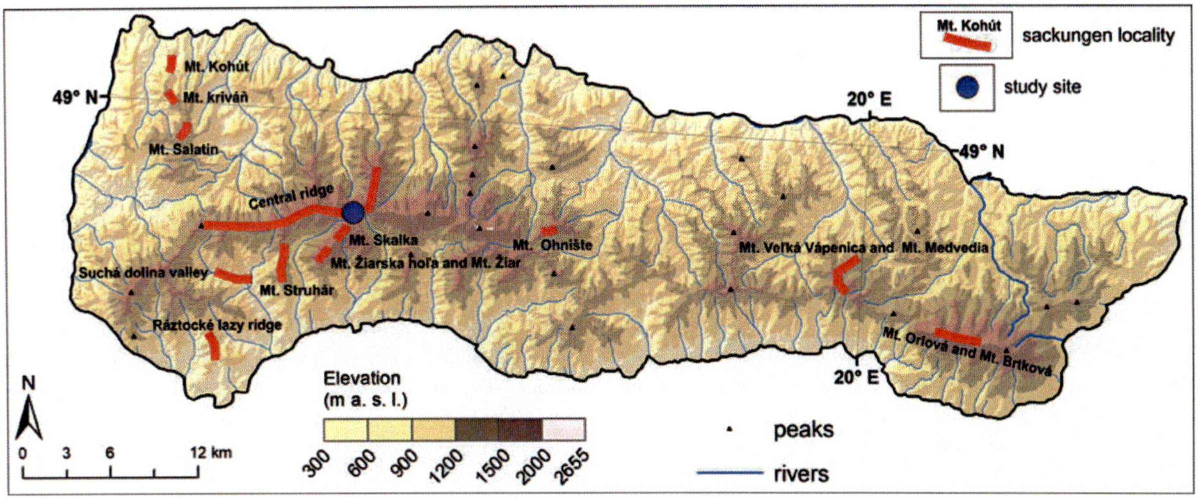

Fig. 6.15 Spatial distribution of sackungen and the study site (according to Stankoviansky 1984; Škvarček 1986, 1989)

Fig. 6.16 Sackungen-deformed ridge at Orlová Mt. (1840 m a.s.l.) (*photo* J. Madarás)

Slope deformations in granitic rocks have the nature of a fan-like disintegration of mountain ridges. They usually loosen along a set of shear planes dipping into the slope. The rock massif is dissected by discontinuity planes into blocks moving downslope and rotating either backward or forward in relation to the slope angle. The general movement of loosened masses is into the valley. The most extensive deformation of this type occurs along the central ridge (the massif of the Mt. Chabenec, Mt. Veľká hoľa, Mt. Latiborská hoľa, Mt. Panská hoľa, Mt. Kotliská, Mt. Poľana, Mt. Ďurková and Mt. Bôr).

Gravitational deformations in the Mesozoic sedimentary rocks utilize pre-existing discontinuities as the main directions in the formation of shear zones and planes. They are usually formed on slopes, where beds are parallel to the slope. Sides of moving blocks are usually controlled by transversal tectonic failures. One of largest deformations of this type occurs on Mt. Ohnište. Extensive deformations are also on the ridge of Ráztocké lazy between Ráztoka and Pohronský Bukovec, in the saddle between Kriváň and Bohúňovo and on the western slope of Mt. Kohút. Different features show on the top of Mt. Salatín, disrupted to parallel crests separated by troughs. Back-walls of cirques are locally also deformed, as observed in the valley heads lying between Mt. Orlová and Mt. Bartková (Škvarček 1986, 1989; Fig. 6.16).

Applying paleoseismic trenching techniques and ^{14}C dating to four trenches dug across three antislope scarps on the south ridge of Mt. Chabenec, one could indicate that the scarps were formed by episodic displacements over the past 9.6 ka (Figs. 6.15 and 6.17). The most recent displacement event (MRE) overlaps in age between 1,410 and 1,860 cal years BP (McCalpin et al. 2019). The longest well-dated record in a single trench (Trench 2) contained four inferred displacement events in the past 6 ka. These four events have created an antislope scarp with $\sim$ 5 m of vertical surface offset. The near-synchronicity of the late Holocene displacements stands in contrast to other dated sackungen sites, where displacements become younger with increasing elevation. Having near-synchronous displacement events at multiple elevations suggests an external trigger, either climatic or seismic (McCalpin et al. 2019).

6.6 Avalanches

The Low Tatra are the largest Slovak mountains in terms of an area above the climatic timberline, i.e. above 1,500 m a.s. l., and are typified by cold mountain climate, the occurrence of snow avalanches and cryogenic and nival processes. The most important part of the Low Tatra Mts. in terms of avalanche occurrence is the Ďumbier part, where as many as 393 source zones is located (Fig. 6.18). There are only 35 avalanche terrains in the Kráľova hoľa part. In the Low Tatra Mts. the starting zones of avalanches cover 1390 ha in total (Žiak 2017). Their minimal extent reaches 0.08 ha and the maximal one is 17.49 ha. The average altitude of these zones is almost 1660 m a.s.l., which is 220 m less than the Slovak average. The average slope is 31°, with the maximum of 57.9°. The modal value of the slope aspect is 104°, i.e. eastern orientation. Track zones in the Low Tatra Mts. exhibit an average length of almost 700 m, which is more than 150 m above the Slovak average. The longest track zone is 2006 m long. The average altitude of the centres of avalanche transport zones is 1491 m, with an average slope of 25°. The predominant orientation of relief in the middle of the avalanche tracks is 284°, i.e. the western aspect. The accumulation zones have 469.5 ha in total, 1.56 ha on

Fig. 6.17 The southern ridge of Mt. Chabenec, where scarps were formed by episodic displacements over the past 9.6 ka. (*photo* P. Ličšák)

average, the minimum area is 0.04 ha and the maximum area covers 17.85 ha. The average altitude of their occurrence is 1386 m a.s.l., the average elevation between the starting and accumulation zones is 280 m and the average slope value is 20.7°. The lowest accumulation zone was found at an altitude of 821 m and the highest located is at 1783 m. The accumulation zones occur mostly at aspect of 125°, i.e. on the south-east slopes (Žiak 2017).

6.7 Human Impact on the Low Tatra Landscape

In the past, mining, charcoal-burning, metallurgy, lumbering and especially shepherding were the main human impacts shaping the Low Tatra Mts. landscape. In several places (e.g. from the mountain saddle Čertovica to Mt. Ďumbier, in the Bocianská Valley, surrounding the settlements Magurka and Dúbrava) one can find traces of mining in the form of small heaps due to the occurrence of iron ore, antimony, gold, silver and copper ores. The highest gold mining tunnels were placed at 1390 m a.s.l. on the northern slope of Mt. Latiborská Hoľa. Gold and silver were mined since the thirteenth century, mining of copper and antimony started from the eighteenth century. In addition, grazing left significant traces not only in the form of vegetation community changes in the area of mountain meadows but also in the timberline lowering by 50–160 m, and thus in the increase of avalanche frequency and soil erosion areas (Midriak 1977). The Low Tatra Mts. have been legally designated as a national park in 1978, as the largest large-scale protected area in Slovakia with 72,842 ha (Štroffek 1982; Klinda 1989). Due to the attractiveness of their landscape and suitable climatic conditions, anthropogenic forms of relief generated by the construction of ski pistes (many slopes, especially in the central southern and northern part of the Low Tatra) and the overall expansion of tourist and recreational centres (Gregorová 2019) are increasing at present. There are 25 ski resorts in the Low Tatra Mts., in which 103 mountain transport facilities and 130 ski pistes have been built.

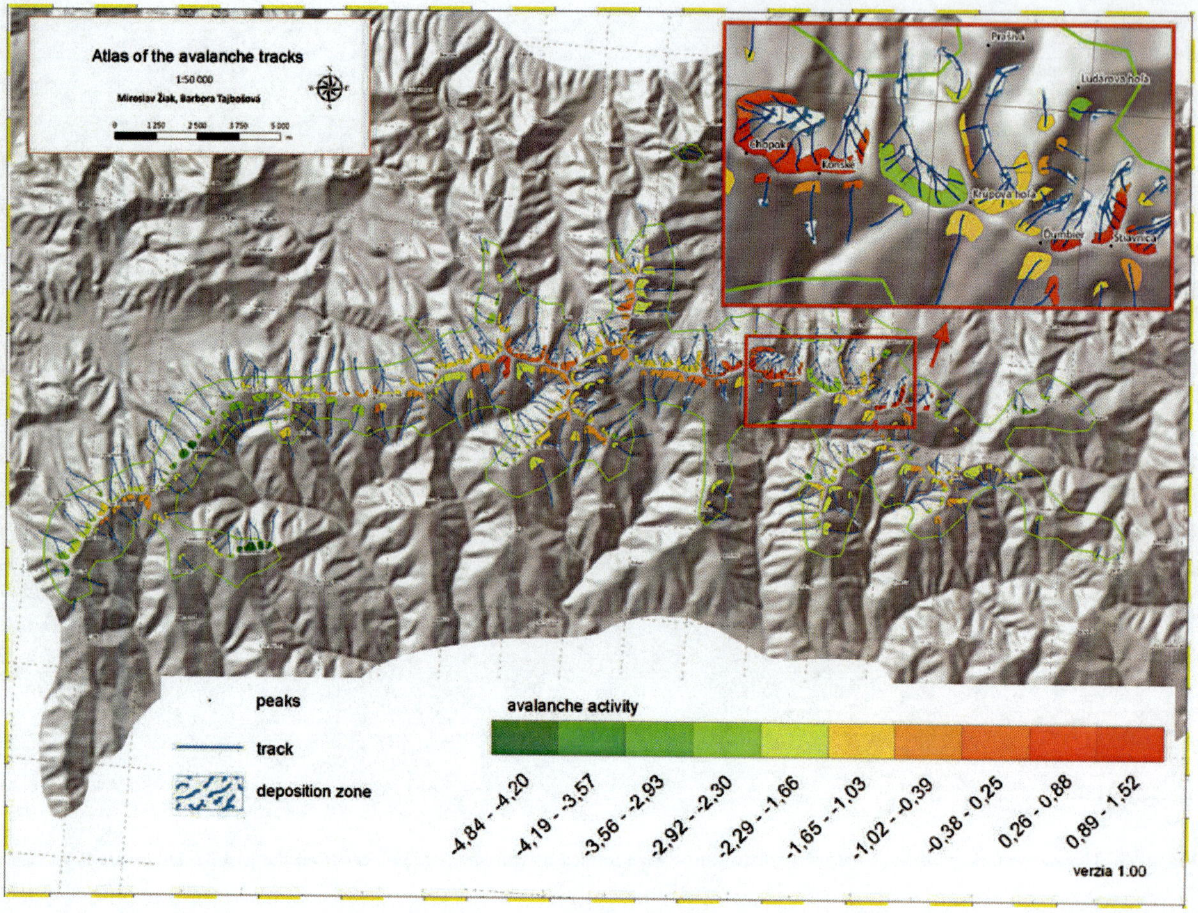

Fig. 6.18 The avalanche tracks and their starting zones in the Low Tatra Mts. (Žiak, 2017)

6.8 Conclusions

The Low Tatra Mts. were uplifted into the form of a mega-anticline in the Miocene. During the Pleistocene, only the uppermost part of the central ridge has been locally shaped by glacial processes. Periglacial climate and processes conditioned, on the one hand, smooth relief of the most elevated sections of the central ridge, as well as of the upper parts of spurs. On the other hand, valley downcutting, mass movement, debris flows and avalanches shaped the spur–ridge system and thus the Low Tatra Mts. roughly got the recent shape. In the northern slopes of the Low Tatra Mts., allogenic karst has developed in the Fatricum and Hronicum nappes. Its fundamental feature is that streams, which start in the crystalline core, incise the limestone–dolomite formations of the nappes and form cave systems. The maximal areal extent of the karst area is represented by the Demänovská and Jánska valleys karst in the central-north part of the Low Tatra Mts., of which the Demänovská Valley karst includes 215 caves of the total length of more than 55 km. Characteristic features of the central ridge of the Nízke Tatry Mts. are numerous, often extensive slope deformations. In the Low Tatra Mts., 393 starting zones of avalanches covering 1390 ha is registered. The mountains have been legally designated as a national park in 1978, as the largest large-scale protected area in Slovakia with 72,842 ha. Due to the attractiveness of their landscape and suitable climatic conditions the mountains are intensively used by tourists throughout the year (Fig. 6.19).

Fig. 6.19 The Vrbické tarn dammed by moraine is the only large lake in the Low Tatra Mts. commemorating the glacial period

Acknowledgements The research was supported by Science Grant Agency (VEGA) of the Ministry of Education of the Slovak Republic and the Slovak Academy of Sciences; 02/0086/21. The authors wish to thank P. Bella, J. Madarás, J. Lacika, P. Hipman, M. Slavický and P. Ličšák for providing photos.

References

Bella P (1993) Poznámky ku genéze Demänovského jaskynného systému. Slovenský Kras 31:43–53

Bella P (2002) K rekonštrukcii planačných povrchov v Demänovských vrchoch na severnej strane Nízkych Tatier. Geographia Slovaca 18:13–20

Bella P (2004) Ďumbiersky kras – kontaktný pruhovitý kras v centrálnej časti Nízkych Tatier. Geomorphologia Slovaca 2:18–29

Bella P, Hercman H, Gradziński M, Pruner P, Kadlec J, Bosák P, Głađzek J, Gąsiorowski M, Nowicki T (2011) Geochronológia vývoja jaskynných úrovní v Demänovskej doline. Nízke Tatry. Aragonit 16(1–2):64–68

Bella P, Haviarová D, Kováč Ľ, Lalkovič M, Sabol M, Soják M, Struhár V, Višňovská Z, Zelinka J (2014) Jaskyne Demänovskej doliny. Štátna ochrana prírody, 191 p.

Biely A, Beňuška P, Bezák V, Bujnovský A, Halouzka R, Hanzel V, Klinec A, Kubeš P, Liščák P, Lukáčik E, Maglay J, Miko O, Molák B, Pulec M, Putiš M, Slavskay M, Vozár J, Vozárová A (1997a) Geologická mapa Nízkych Tatier. 2. vyd. Bratislava: Vydavateľstvo Dionýza Štúra. ISBN 80–969338–7–6.

Biely A, Beňuška P, Bezák V, Bujnovský A, Halouzka R, Hanzel V, Klinec A, Kubeš P, Liščák P, Lukáčik E, Maglay J, Miko O, Molák B, Pulec M, Putiš M, Slavskay M, Vozár J, Vozárová A (1997b) Vysvetlivky ku geologickej mape Nízkych Tatier. 2. vyd. Bratislava: Vydavateľstvo Dionýza Štúra, 1997, 232 p. ISBN 80–969338–7–6.

Biely A, Fusán O (1969) Zum Problem der Wurzelzonen der subtatrischen Decken. Geologicke´ Práce. Správy 42:51–64

Droppa A (1957) Demänovské jaskyne. Krasové zjavy Demänovskej doliny. SAV Bratislava, 289 p.

Droppa A (1972) Geomorfologické Pomery Demänovskej Doliny. Slovenský Kras 10:9–46

Droppa A (1972) Krasové javy Jánskej doliny na severnej strane Nízkych Tatier. Československý Kras 21:73–96

Droppa A (1973) Prehľad preskúmaných jaskýň na Slovensku. Slovenský Kras 11:111–157

Gregorová B (2019) Historickogeografický príspevok k štúdiu vzniku a rozvoja cestovného ruchu južnej strany Ďumbierskych Nízkych Tatier. Studia Historica Nitriensia 23(1):91–108

Gregorová B (2018) Využitie archívnych prameňov a geoinformačných technológií pri výskume dejín ochrany životného prostredia na príklade Bystrej doliny v Nízkych Tatrách. Acta Regionalia: Interdisciplinárne Vedecké Periodikum III(1–2):75–83

Hochmuth Z (1997) Príspevok k chronológii a genéze jaskynných úrovní v Jánskej doline. In: Výskum, využívanie a ochrana jaskýň, zborník referátov z konferencie 8 - 10.10.1997, SSJ, Liptovský Mikuláš, pp. 29 - 35

Hochmuth Z (2008) Krasové územia a jaskyne Slovenska. Geographia Cassoviensis II(2):5–210

Hronček P, Štrba Ľ, Gregorová B (2019) Heritage of the Medieval Human Activity in the Present Landscape of the National Park Low Tatras. In: Public Recreation and Landscape Pretection, 172 – 176

Klinda J (1989) Stanovisko k uvažovanej výstavbe zariadení cestovného ruchu v Národnom parku Nízke Tatry v oblasti Chopku. Poznaj a Chráň 6:10–11

Lauritzen SE (2001) Marble stripe karst of the Scandinavian Caledonides: an end-member in the contact karst spectrum. Acta Carstologica 30(2):47–79

Louček D, Michovská J, Trefná E (1960) Zalednení Nízkych Tater. Sborník Československé Společnosti Zeměpisné 65(4):326–384

Mazúr E, Lukniš M (1978) Regionálne Geomorfologické Členenie SSR. Geografický Časopis 30(2):101–125

McCalpin J, Liščák P, Jelínek R, Zorba M. O, Santacana N (2019) Postglacial deformation history of sackungen on the southern slope of Mount Chabenec, Nízke Tatry Mts., Slovakia. Mineralia Slovaca 51(1):1 – 30

Midriak R (2003) Horské oblasti národných parkov Slovenskej republiky, p 58

Nemčok A (1982) Zosuvy v slovenských Karpatoch. Bratislava, VEDA, p 318

Orvoš P (2005) Kras Krakovej Hole. Spravodaj SSS 36(2):14–20

Orvošová M, Uhlík P, Uher P (2006) Paleokras Ohnišťa – výskum sedimentárnej výplne Veľkého závrtu (Nízke Tatry). Slovenský Kras 44:71–80

Stankoviansky M (1984) Súčasné exogénne reliéfotvorné procesy Ďumbierskych Tatier. Sborník Československé Geografické Společnosti 89(4):285–296

Škvarček A (1978) Glaciation of Mošnica valley in Low Tatras. Acta Facultatis Rerum Naturalium Universitatis Comenianae, Geographica 16:177–190

Škvarček A (1980) Pleistocénne zaľadnenie bazénu Veľkej Oružnej v Nízkych Tatrách. Acta Facultatis Rerum Naturalium Universitatis Comenianae, Geographica 18:13–31

Škvarček A (1983) Niektoré morfologické aspekty doliny Ďurkovej v Nízkych Tatrách. Acta Facultatis Rerum Naturalium Universitatis Comenianae, Geographica 22:97–110

Škvarček A (1986) Niektoré aspekty pleistocénneho zaľadnenia Kráľovohoľských Tatier. Geografický Časopis 38(2–3):236–244

Škvarček A (1989) Zaľadnenie a prejavy iných morfogenetických procesov posledného glaciálu v reliéfe Kráľovohoľských Tatier. Acta Facultatis Rerum Naturalium Universitatis Comenianae, Geographica 28:25–43

Štroffek O (1982) Národný park Nízke Tatry – jeho prírodné hodnoty a poslanie. Spravodajca o Chránených Územiach Slovenska 1:29–32

Turis P, Jasík M (2007) Nízke Tatry, prírodné hodnoty, história a súčasný stav ochrany územia. Správa národného parku Nízke Tatry, 116 p.

Vajs J (1998) Nemodelová. Jelenia Priepasť. Spravodaj SSS 29(4):56–57

Vitásek F (1926) Morfologické studie na jižní straně Nízkých Tater. Sborník Státního Geologického Ústavu Československé Republiky 5:449–467

Žiak M (2017) Lavínová hrozba vo vysokohorskom prostredí Slovenska a Poľska. Geomorphologia Slovaca Et Bohemica 17(2):9–100

Milan Lehotský is a physical geographer and fluvial geomorphologist at the Institute of Geography of the Slovak Academy of Sciences. He was many years head of the Department of Physical Geography, Geomorphology and Natural Hazards. His research topics are responses of fluvial systems to environmental changes, sedimentological connectivity, evolution trajectories, hydromorphology and GIS and remote sensing applications in rivers and landforms research. He is also working as an external lecturer at the Department of Physical Geography and Geoecology, the Faculty of Natural Sciences of the Comenius University in Bratislava.

Bohuslava Gregorová works at the Department of Geography and Geology of Matej Bel University in Banská Bystrica as an assistant professor of human geography. She is specialized in the teaching of Economic Geography, Geography of Tourism, Religious Geography, Behavioural Geography and Geography of Slovakia. Her research interest is focused on spatial models of tourism, theory and methodology of tourism, urban tourism, use and application of GIS in tourism, history of tourism, pilgrimage tourism, regional development in tourism.

Zdenko Hochmuth was an Associate Professor at the Institute of Geography, Faculty of Natural Sciences, Pavol Jozef Šafárik Univeristy in Košice, Slovakia and Dean of the Faculty of Science. His main subjects of research were the development of surface karst landforms and comparative analysis and dating of cave levels. At present, he is interested in paleoclimatology and karst hydrology. He discovered, mapped and described many caves in Slovakia and took part at several speleological expeditions (in Venezuela, Spain, Italy).

7 Horst Structure and Planation Surfaces—Little Carpathians Mts.

Ján Lacika, Ján Urbánek, and Milan Lehotský

Abstract

The Little Carpathians form the most distant, southwestern promontory of the Western Carpathians, wedged between the Záhorská and Podunajská lowlands. They represent mountains individualized and internally differentiated by multidirectional geomorphological lines arranged in a geomorphological grid typical for horst morphostructures. The Little Carpathians are divided into six morphostructural units, which are further differentiated into lower order morphostructural units. In addition to the morphostructural regionalization, the chapter presents the basic scheme of geomorphological development of the Little Carpathians during the neotectonic period. The initial relief of the study area was likely a low ridge individualized in the Badenian period, within roughly similar boundaries as the contemporary mountains. In the changing morphostructural and morphoclimatic conditions, gradual transformation of the relief into a rising horst occurred, which, however, did not behave as an internally homogeneous morphostructure. Its complexity was dictated by different structural-lithological properties of the geological substratum, especially by the diverse developments of its morphostructural units. Therefore, the nature and intensity of transformation processes varied from place to place. In principle, the replacement of old forms by the new ones has progressed from the edge of the mountain range to its interior. So, planation surfaces and subsequent valleys originating in the Neogene can be found in its central part, while the slopes and valleys in the foothills are younger, of Quaternary age. Climatic oscillations of the Quaternary period had a great influence on the transformation of the relief. In the contact area between the mountains and the lowlands, the youngest development of relief is characterized by very intensive anthropogenic impact and its transformation.

Keywords

Morphostructure • Planation surface • Horst • Little Carpathians • Slovakia

7.1 Introduction

The basic geomorphological features of the southwestern periphery of the West Carpathian dome-like mega-morphostructure, including the Little Carpathians, began to form in the lower Badenian (16.5 Ma), when gradual transformation of the paleorelief toward its contemporary state commenced. The composition of the main morphostructural units of the Western Carpathians was developed, and units subject to relative uplift (mountains) were interspersed with subsiding areas (lowlands). The Little Carpathians, the Považský Inovec Mts., and the Tribeč Mts. became the southwestern promontories of the Western Carpathians, spreading into the lowlands of the West Pannonian Basin (Fig. 7.1). Information about the geology of the Little Carpathians can be found in many publications (e.g., Koutek and Zoubek 1936; Buday et al. 1963; Maheľ

J. Lacika (✉)
Department of Geography, Geoinformatics and Regional Development, Faculty of Natural Sciences, Constantine the Philosopher University in Nitra, Nitra, Slovakia

J. Urbánek · M. Lehotský
Department of Physical Geography, Geomorphology and Natural Hazards, Institute of Geography, Slovak Academy of Sciences, Štefánikova 49, 814 73 Bratislava, Slovakia
e-mail: jan.urbanek@zoznam.sk

M. Lehotský
e-mail: geogleho@savba.sk

M. Lehotský and M. Boltižiar (eds.), *Landscapes and Landforms of Slovakia*, World Geomorphological Landscapes,
https://doi.org/10.1007/978-3-030-89293-7_7

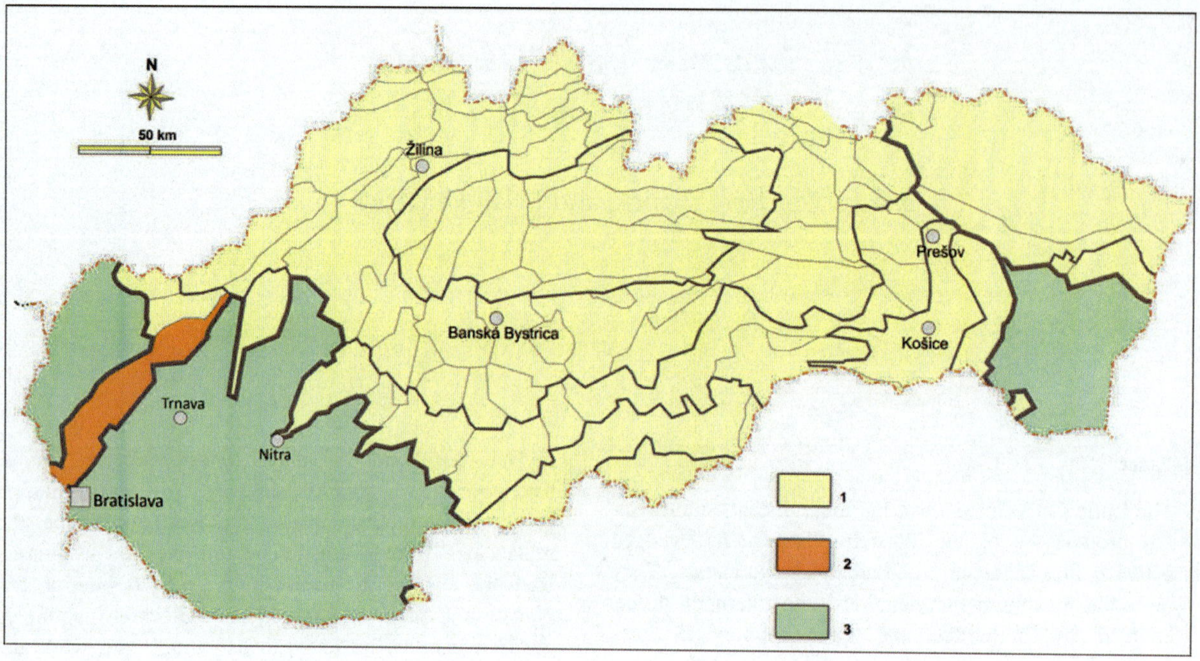

Fig. 7.1 Morphostructural setting of the Little Carpathians (Lacika and Urbánek 2002). 1—the Carpathians, 2—the Little Carpathians, 3—the Pannonian Basin

and Cambel 1972; Began et al. 1984; Salaj 1987; Vaškovský et al. 1988; Polák et al. 2011, 2012; Potfaj et al. 2014, 2015). As far as geomorphological research is concerned, the works of Daneš (1931), Hromádka (1929, 1935), Lukniš (1955, 1972, 1977), Zaťko (1959), Urbánek (1966, 1992), Škvarček (1966), Novodomec (1967), Stankoviansky (1974, 1996), and Lacika (1983) give basic information about spatial variability of landforms and their development in different regions of the Little Carpathians. The synthetic image about geomorphological features of this mountain range has been worked out by Jakál et al. (1990), Lacika and Urbánek (2002) and Urbánek (2014). The Little Carpathians represent the southwestern promontory of the Western Carpathians, individualized into a narrow, long morphology of the horst type. Their length from Devínská brána to Bzince pod Javorinou is 91 km and the width ranges from 2 to 15 km. The Záruby peak (767 m a.s.l.) represents their highest point, the lowest one, at 132 m a.s.l., lies at the point where the Danube River enters into the Danube Plain, and so the local relief is 635 m. According to Lacika and Urbánek (1998), the Little Carpathians belong to the marginal morphostructures of the West Carpathian dome-like mega-morphostructure. Minár et al. (2011) classify them as a part of the Southwest Foreland Region of the Western Carpathians. The Little Carpathians can be classified into six morphostructural units, which are further differentiated into the lower order units.

7.2 Structural Conditions in Relation to Relief

Most of the Little Carpathians belong to the Tatricum unit (Fig. 7.2, Polák et al. 2011), which, unlike in other Western Carpathian core mountains, is divided into a system of partial units including the Prealpine foundation, as well as several, often fundamentally different Mesoisoic successions (Plašienka et al 1991). The area built by the Tatricum is very variable from a structural-lithological and tectonic point of view. The Bratislava and Modra massifs are characterized by monotony due to the significant dominance of granitoid rocks. The diversity of Paleozoic metamorphic rocks conditions high variability of landforms developed on the Pezinok and Pernek tectonic groups. The Pezinské Carpthians north unit is largely built of rocks of the Fatricum and Hronicum tectonic unit. The continuous zone of the Fatricum, extending diagonally across the mountains from Pernek to Smolenice, is associated with relief at lower

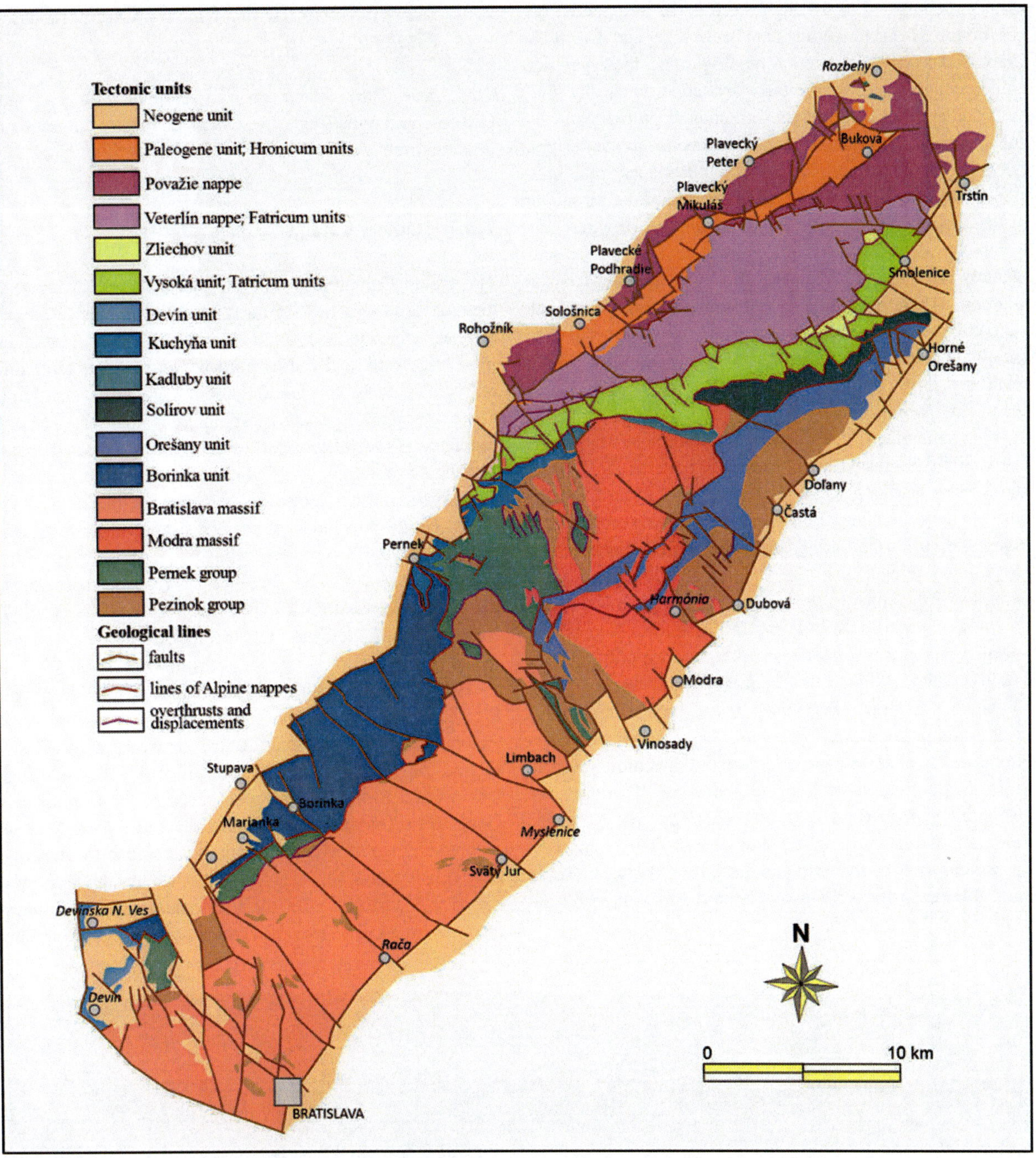

Fig. 7.2 Simplified sketch of tectonic units of the Devínske and Pezinské Carpathians (modified according Polák et al. 2011)

altitudes, compared to the higher setting of monoclinal ridges of the Hronicum unit. The Brezovské and Čachtické Carpathians are quite similar geologically, their geomorphological differences are therefore conditioned by fault tectonics and not by structural-lithological properties of the substratum. From the Upper Cretaceous to the Upper Eocene, conglomerates, sandstones, claystones, and limestones were deposited on the older mantle of the Hronicum in the northern part of the Little Carpathians. However, due to tectonic deformation in the neotectonic period the sedimentary cover was removed, exposing older Mesozoic bedrock. Therefore, in the northern part of the Little Carpathians Mesozoic dolomites and limestones shaped by karstification (e.g., Dobrovodský and Čachtický karst) have a relatively large proportion. The Neogene sediments are sparse and participate very little in the geological variability of the mountains. The Neogene deposits of the Little Carpathians originated in the marginal parts of the Danube and Vienna Basins became elements of the Little Carpathians by their individualization in the new morphological boundaries from the Sarmatian to the Upper Pliocene. It is worth mentioning that in the Devín Carpathians the Sandberg layers (the upper Badenian age) are found almost on the top of the Devínská Kobyla massif. The formation is represented by sands, sandstone, and the presence of fossils (the locality Sandberg, Fig. 7.3). The large area of the Brezovské Carpathians between Jablonice, Trstín, and Dobra Voda settlements is built of Lower Miocene polymictic conglomerates, which are more compact and thus more resistant to denudation than most Neogene sediments. Therefore, we also find them in elevation positions. However, generally they are found in less elevated areas. The transverse depression used by the road and the railway between Trstín and Jablonica settlements is also underlain by these formations (Figs. 7.4 and 7.5).

7.3 Morphostructures of Little Carpathians Mts.

The Little Carpathians can be divided into six morphostructural units (Fig. 7.5), which are further differentiated into sub-units.

7.3.1 Devín Carpathians Unit

The Devín Carpathians unit exhibits the highest degree of tectonic differentiation of the Little Carpathians. This area can be characterized as a Cretaceous block, internally markedly tectonically differentiated morphostructural unit, individualized equally by longitudinal and transverse faults. The expression of passive morphology is not great. The only exception is the western part of the Devínska Kobyla massif and the surroundings of Devín, where Mesozoic limestones with low-resistant Neogene sediments are in contact. The Devín Castle is a block of more resistant, older rocks surrounded by these Neogene deposits, revealed by selective erosion processes. The presence of river terraces in the depression between the Devín castle rock and the Devínská Kobyla massif indicates that the rock represents a former meander core.

The morphostructure of the Devín Carpathians (a—Fig. 7.5) consists of two sub-units: a pair of transverse tectonic-erosion subsidence areas, between which lies a ridge. The southern depression (Fig. 7.5 a.a), called the Devín Gate, forms an antedecent reach of the Danube. The Devín Gate (Fig. 7.6) is a genetically older form than the mountain range of which it is a part. According to Mazúrová (1973), it follows an older tectonic structure, the activity of which was restored after the Pannonian, especially during the Quaternary. The presence of the Danube caused that the

Fig. 7.3 Anthropogenically transformed western slope of the Devínska Kobyla massif (**a**—sand quarry; **b**—quarry). Sand quarry is declared a protected palaeontological and geological locality. (*photo* J. Lacika)

Fig. 7.4 Relief on Neogene sedimentary rocks at the boundary between tectonic units of Pezinské and Brezovské Carpathians (*photo* J. Lacika)

"River Level" (Upper Pliocene–Early Quaternary pediment after Mazúr, 1963) was formed in the vicinity of the contemporary fluvial breach in the Quaternary, conditioning the development of small plateaus on the top of the mountain ridge. Scattered residues of Danube gravels have been preserved on these plateaus, which is why Mazúrová (1973) described them as the level of the plateau Danube terrace. The Quaternary deepening of the Devín Gate was conditioned not only by tectonic uplift of the adjacent mountains, but also by climatic oscillations resulting in the development of a system of Danube river terraces. The total effect of the Quaternary deepening of the Devín Gate is about 100 to 130 m.

To the north of Devínská brána there is a transverse horst (Fig. 7.5 a.b) occupying the Bratislava foothills and the Devínská Kobyla massif. It has many similarities to the Austrian Hainburg Mountains (Hundsmeimer Berge) on the opposite bank of the Danube, which are assigned to the Little Carpathians as the westernmost tip of the Western Carpathians. It does not form a compact whole, but is broken into a system of smaller morphostructural units of lower order with a different geomorphological character. The multidirectional geomorphological lines form a specific geomorphological grid of two elevations and two depressions. The system of plateaus in the upper parts of the foothills is a strongly transformed remnant of the "River

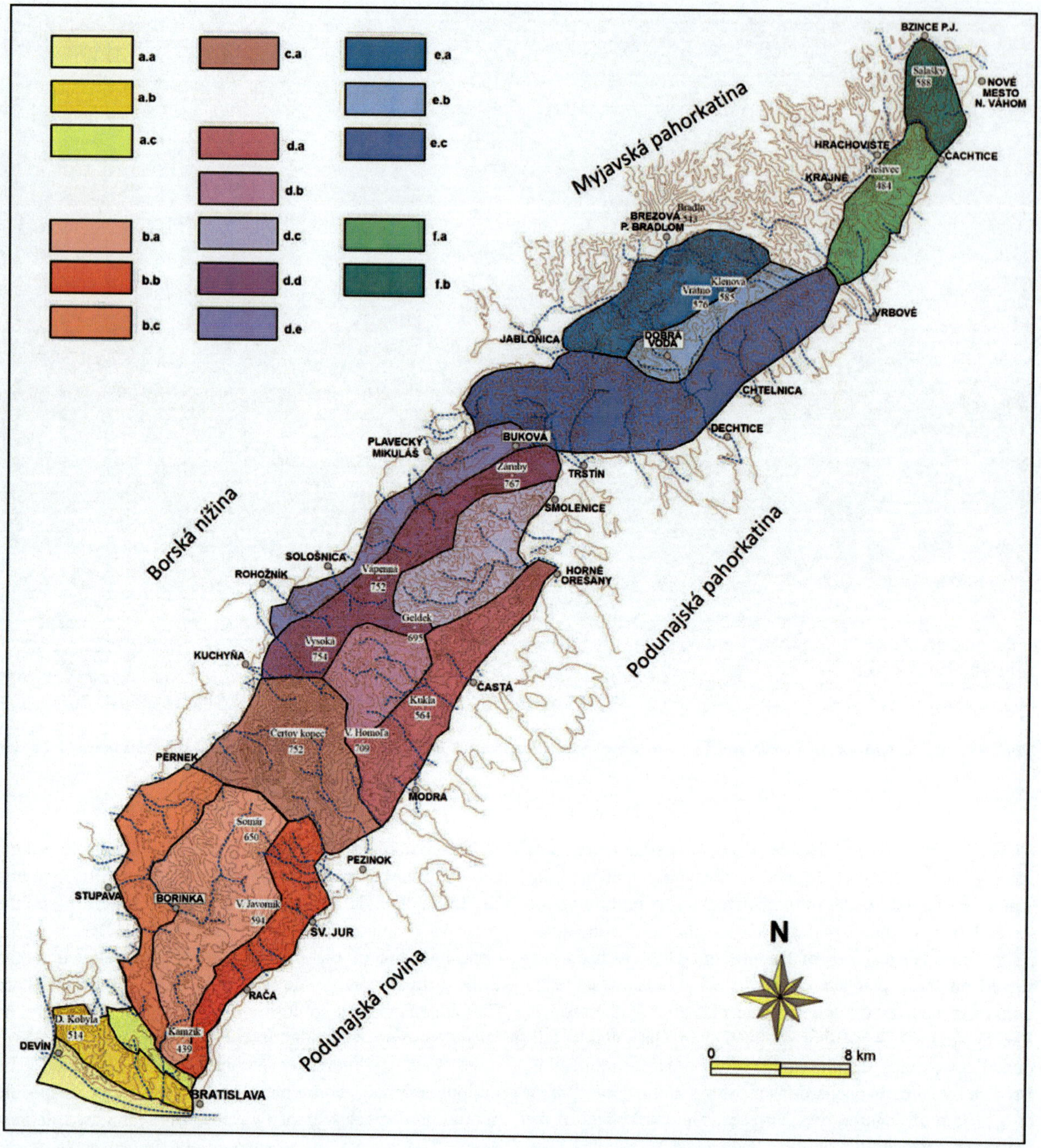

Fig. 7.5 Morphostructural map of the Malé Karpaty Mts. Morphostructural unit of Devín Carpathians: a.a southern tectonic subsidence, a.b transversal horst, a.c northern tectonic subsidence. Morphostructural units of Pezinské Carpathians: Southern unit: b.a central plateau-valley morphostructure, b.b eastern marginal morphostructure, b.c western marginal morphostructure; Core unit: c.a transverse spur ridges morphostructure; Northern unit: d.a eastern marginal elevation, d.b southern central closed depression, d.c northern partially open depression, d.d central longitudinal elevation with monocline ridges, d.e western marginal depression. Morphostructural unit of Brezovské Carpathians: e.a north-western marginal horst, e.b intramountaineous longitudinal depression. Morphostructural unit of Čachtické Carpathians: f.a southern narrow horst, f.b northern massive horst (modified according Jakál et al. 2009)

Fig. 7.6 Devín Gate (Fig. 7.5 a.a., *photo* J. Lacika)

Level" planation surface. The compactness of the horst is significantly disturbed by the water gap of the lower Vydrice brook, called the Mlynská Valley. The process of valley deepening followed the Danube terraces development in the Devín Gate. The water gap has a relatively wide bottom and steep side slopes, rising to a height of approximately 50 m. The isolated Devínská Kobyla massif (514 m a.s.l.) represents the highest part of the Devín Carpathians unit. In the south, the massif borders the Devín Gate, against which it is delimited by steep fault slopes descending in the eastern part of the water gap and in the western part to the river terraces in the Devín urban area. The slope (350 m) on the western side of the massif is one of the highest slopes in the entire southern part of the mountain range. The Devínská Kobyla massif is also bounded by tectonic faults on the northern and eastern sides. Internally, the massif is differentiated into two unequally lifted blocks. The area of the larger south-eastern block—a relatively large plateau—is less elevated, with an altitude of 300–350 m a.s.l. Equivalent flat landforms reduced to small areas are also found on the north-western side of Devínská Kobyla, on the gradually descending ridge of Sandberg. Taking into account their different altitudes, Mičian (2000, 2001) considered them as three generations of planation surfaces, and in addition to so-called "middle mountain planation" and the "River Level" surfaces, so the presence of the sub-middle mountain planation surface known from other mountains of the Western Carpathians is likely (Dzurovčin 1990; Bizubová and Minár 1992; Lacika 1995). Likely, this level was developed during the Pontian-Dacian period. The higher block of Devínská Kobyla appears as an isolated hill rising above the middle mountain planation surface, characterized by dome morphographic features.

The northernmost morphostructural unit of the Devín Carpathians is the northern transverse tectonic subsidence area (Fig. 7.5 a.c) called the Lamačská Gate. It opens to the north-west, narrowing in the opposite direction. In the middle of the gate, there is a saddle and its wide bottom is covered by deluvial deposits and bordered by low steep slopes. Unlike the Devínská Gate, the Lamačšká Gate was not created by any stream, its origin is clearly only tectonic.

7.3.2 Pezinské Carpathians—Southern Unit

The Pezinské Carpathians are the largest geomorphological sub-unit of the Little Carpathians, which can be divided into three parts characterized by different morphological features. The southern part of the Pezinské Carpathians (Fig. 7.7), in comparison to the northern part, appears to be structurally more simple and monotonic. Between the northern and southern parts of the Pezinské Carpathians the third

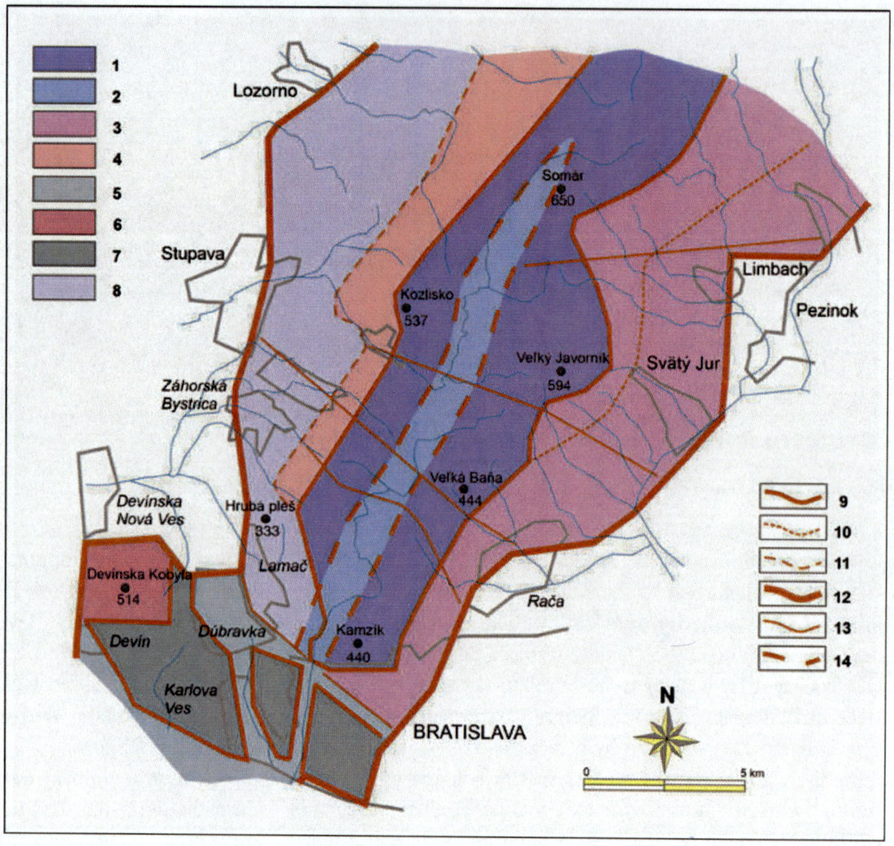

Fig. 7.7 Morphostructures of the southern part of the Little Carpathians. (Urbánek 2014). 1—central part of horst, 2—subsequent valleys, 3—upper eastern foot slopes, 4—lower western foot slopes, 5—gates, 6—Devínska Kobyla block, 7—lower system of blocks with plateaus, 8—piedmont with plateaus, 9—the edge of the high central part of horst, 10—tectonic line dissecting eastern foot slopes, 11—tectonic line on western foot slopes, 12—marginal tectonic line at the foot of horst, 13—transversal tectonic lines disintegrating the central part of horst into the system of step-like slopes, 14—edges of subsequent valleys

morphostructural unit is located and shows higher degree of disaggregation of relief and higher density of transverse geomorphic lines.

In the central plateau-valley morphostructure (Fig. 7.5 b. a) plateaus are the prevailing landform (Fig. 7.8). They are not perfectly flat, but undulating, with local relief up to 50 m. They are covered entirely by the sheet of geest. At several spots, drainless areas occur, in which shallow pools arise during snow melting season and during heavy rainfall. The grouping of landforms such as shallow short valleys, dells, saddles, hummocks, and isolated hills is also typical for this unit and does not interfere with its overall flat character. The individual blocks of the central plateau form a typical horst system, with the altitude from 300 m a.s.l. in the south to more than 600 m a.s.l. in the north and appear to carry fossil planation surfaces of Neogene age, to a certain extent modified by the Quaternary morphognesis. Recognizing similarity with the other mountains of the Western Carpathians, these surfaces can be considered as the middle mountain planation surface developed during the Sarmatian-Pannonian period.

Compactness of the central plateaus is disrupted by the younger systems of the Vydrica and Stpavský brook subsequent valleys. Both valleys are situated on the same geomorphological line of NNE–SSW direction, which is slightly deflected from the longitudinal axis of the mountains.

Fig. 7.8 Southern part of the Pezinské Carpathians unit (photo taken from Pajštún castle, *photo* J. Lacika)

The upper flat and wide segments of the Vydrica Valley with the meandering brook fit in the flat nature of the plateaus. In the middle section, the valley significantly deepens, acquires the "V" cross-section and contains remnants of terraces at the height of 30 m above the valley bottom. The lower, relatively wide section of the valley has undergone significant antropogenic changes in terms of mills, mill races, mill-ponds, residential areas, zoo area, and motorway constructions further downstream.

The upper section of the Stupava Brook Valley is geomorphologically similar to the upper stretch of the Vydrica Valley, although its geological environment is more variable. In the middle section of the valley, a narrow and deep gorge (the right cliff exceeds 200 m) has developed (Fig. 7.9).

The horst of the southern unit of the Pezinské Carpathians is bounded by tectonically influenced marginal morphostructures on both sides (western and eastern ones). The eastern marginal morfostructure (Figs. 7.5 b.b, 7.10) represents a part of the significant morphological and morfostructural line running from the north of Slovakia toward Austria. The morphostructure consists of two basic landforms, i.e., consequent valleys and about 200 m high faceted spurs in between. Due to good accessibility and appropriate soil and climatic conditions the middle and lower parts of spurs are used for vine-growing. The western marginal morphostructure (Fig. 7.5 b.c) differs from the eastern one mainly by the presence of a relatively large foothill originated after the Badenian period, but before the Quaternary and considered as the "River Level" surface (Mazúr 1963).

Fig. 7.9 Gorge in the middle part of the Stupava Brook Valley (*photo* J. Lacika)

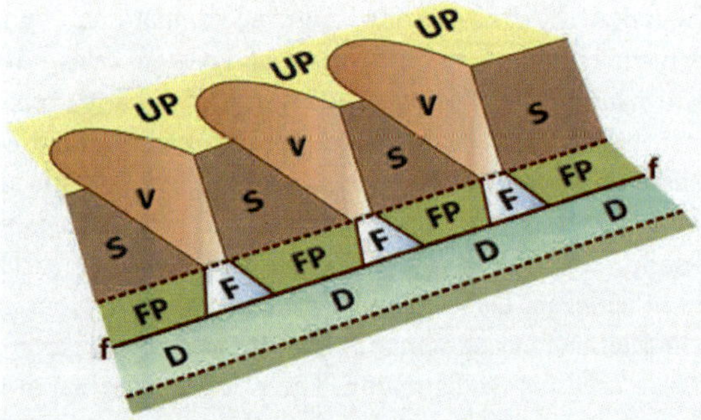

Fig. 7.10 Morphostructural sketch of the marginal slope of the Little Carpathians (Fig. 7.5 b.b, Urbánek 2014). UP—upper plateau, V—valley, S—marginal escarpment, FP—footslope plateau, F—proluvial fan, D—footslope depression, f—foot line

7.3.3 Pezinské Carpathians—Core Unit

The core unit represents the tectonically distorted transverse spur ridges morphostructure (Fig. 7.5 c.a). The unit is characterized by the absence of flat landforms and the occurrence of narrow, deeply incised valleys bordered by spur ridges, of relatively high local relief. Relief disintegration of the unit is conditioned not only by the main tectonic lines constituting the diagonal geomorphologic grid, but also the older fault systems of north–south and east–west directions, which were rejuvenated in certain sections in the neotectonic period. One of these faults predetermined the course of the upper part of the Blatina Valley, which is visible on the map of isobasites (Fig. 7.11), where the tectonic line bends isolines network. The unit is also characterized by a higher degree of anthropogenic transformation

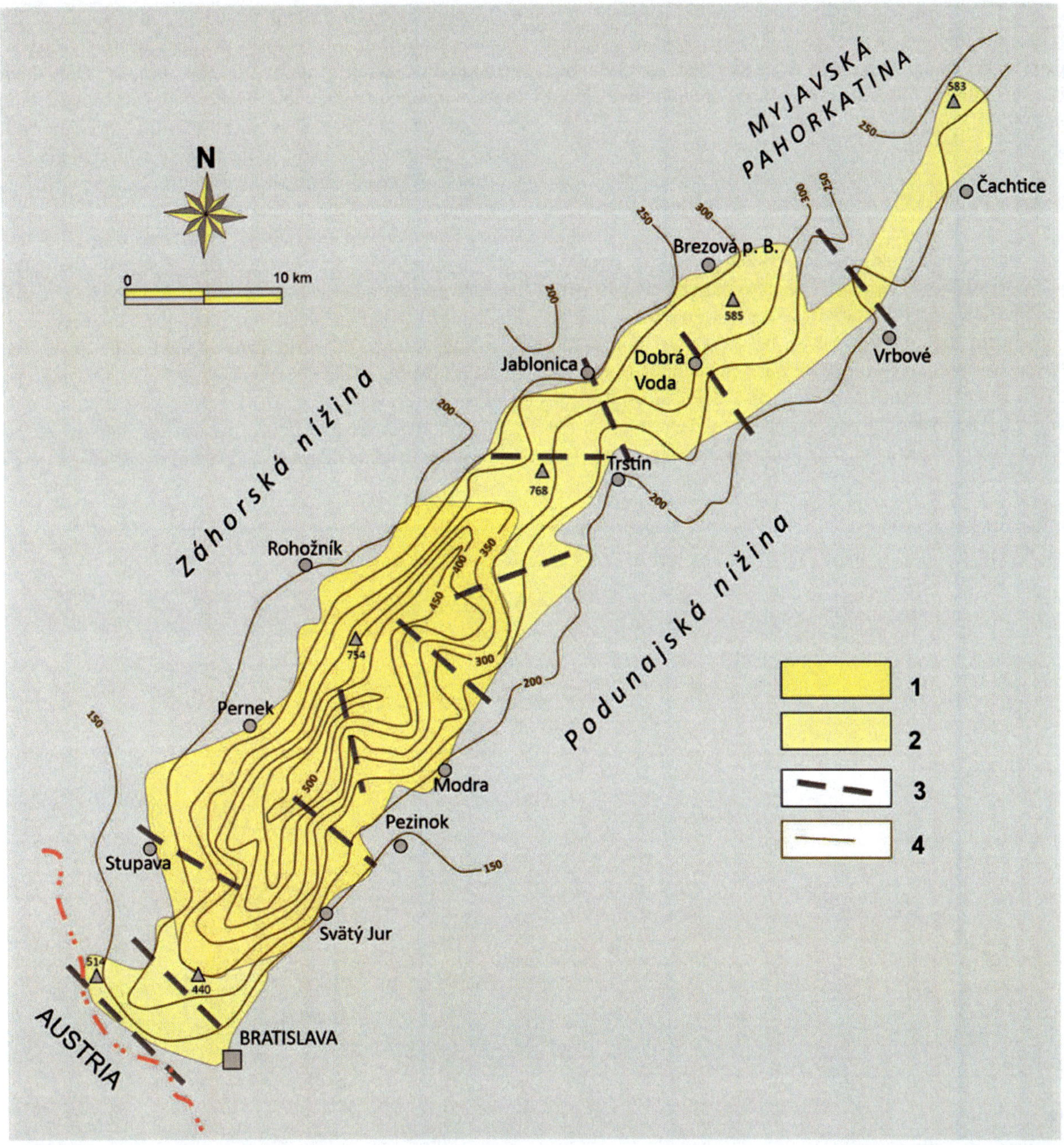

Fig. 7.11 Map of isobasites as isolines connecting the intersections of level contours of the same altitude and valley lines of the third and higher orders. 1—compact part of horst, 2—disintegrated part of horst, 3—morphostructural lines indicated by isobasites, 4—isobasites (interval of 50 m)

of relief due to mining activity. Since the fourteenth-century mining of precious metals occurred, whereas later, since the eighteenth-century local ore deposits have provided primarily antimony ore. The last antimony ore mine was closed in 2006.

7.3.4 Pezinské Carpathians—Northern Unit

Morphological diversity of the northern part of the Pezinské Carpathians is only partially similar to its southern part. One can find some similar landforms, like plateaus, isolated hills, consequent valleys, as well as faceted spurs at margins, however subsequent valleys are missing and new landforms like intramountainous depressions and monocline ridges are present. The overall spatial variability of landforms in the unit is however more complex than in the southern unit. The east-facing marginal elevation stretches on the side of the Danube plain (Fig. 7.5 d.a) and provides relatively distinct geomorphic barrier cut by several brooks that drain intramountainous depressions. Isolated highest hills sticking out of the plateaus, as well as from the foothill level, are built by very resistant quartzites. At the edge of a foothill plateau, the Červený kameň castle was built (Fig. 7.12).

In the middle of the northern unit, there are two depressions. The south-central closed depression (Figs. 7.5 d.b, 7.13) is considered as a result of the large-scale the gravitational nappe collapse, which caused the dividing of the longitudinal foothill depression into two parts in the adjacent area of the Danube Lowland (Feranec and Lacika 1991). The northern, partly open depression (Fig. 7.5 d.c) is facing into the Danube Lowland (Fig. 7.14).

The morphostructural unit of central longitudinal elevation with monocline ridges (Fig. 7.5 d.d) represents not only

Fig. 7.12 Eastern foothills of the Pezinské Carpathians unit (Fig. 7.5 d.a, photo taken from Kukla Hill, *photo* J. Lacika)

Fig. 7.13 Intramountaineous depression in the surroundings of Piesok settlement (Fig. 7.5 d.b, photo taken from Veľká Homoľa Hill, *photo* J. Lacika)

Fig. 7.14 Northern, partially open depression in the surroundings of Lošonc settlement (*photo* J. Lacika)

the highest part of the Pezinské Carpathians northern unit, but also the highest elevation of the Little Carpathians as a whole. However, direct tectonic imprint is indistinct and landform development is mostly considered as a response to selective erosion and non-uniform rock resistance. The resistance of limestones and dolomites conditioned the development of three monoclinal crests (Vysoká 754 m a.s. l., Vápenná 752 m a.s.l. and Záruby 767 m a.s.l., Fig. 7.15), characterized by asymmetric cross-sections. The western part of the northern unit of the Pezinské Carpathians is classified as the western marginal depression (Fig. 7.5 d.e), which opens up into the Borská Lowland (Fig. 7.16). Further to the north, a group of blocks rises and joins the neighboring Brezovské Carpathians morphostructural unit. The longitudinal depression on the north-eastern end of the central longitudinal elevation with monocline ridges unit extends into the Bukovska depression (Fig. 7.17).

7.3.5 Brezovské Carpathians Unit

The Brezovské Carpathians unit is similar to the northern unit of the Pezinské Carpathians in terms of general morphostructural features. However, it has simpler internal structure and the response of landform development to the rock resistance is not so evident. The Brezovské Carpathians unit consists of three sub-units, among which the north-west marginal horst (Fig. 7.5 e.a) looks like a continuation of the northernmost part of the Pezinské Carpathians (Fig. 7.5 d.e). However, the geological substratum is not represented by granitoids but dolomites and limestones, which are subjected to the different degree of karstification. A few isolated hills stick out from the plateau in the central part, including the highest peak of the Brezovské Carpathians, Mt. Klenová (585 m a.s.l.).

The intramontane longitudinal depression bounded by faults lies in the center of the Brezovské Carpathians unit (Fig. 7.5 e.b). The Dobrovodská Basin (Fig. 7.18) is very clear example of this kind of setting. Faults, which condition its formation and bounding units (Fig. 7.5 e.a and e.b), exhibit large seismic activity and belong to an important regional fault system.

The south-eastern horst unit (Fig. 7.5 e.c) is lower and less compact than its north-western counterpart (Fig. 7.5 e. a). Tectonic differentiation is manifested through tilting. Tectonically inclined marginal blocks took shape of toothed asymmetric ridges, whose steeper slopes are oriented toward

Fig. 7.15 Monoclinal ridge of Záruby (photo taken from Smolenice castle tower, *photo* J. Lacika)

Fig. 7.16 Western margin of the Pezinské Carpathians unit (Fig. 7.5 d.e, photo taken from Kršlenica Hill, *photo* J. Lacika)

Fig. 7.17 Bukovská Basin (Fig. 7.5 d.e, photo taken from Ostrý kameň castle, *photo* J. Lacika)

Fig. 7.18 Dobrovodská Basin (Fig. 7.5 e.b, photo taken from Dobrá Voda castle, *photo* J. Lacika)

the mountain interior and the gentle ones face toward the Podunajská nížina Lowland.

7.3.6 Čachtické Carpathians Unit

The Čachtické Carpathians morphostructural unit, representing the axial part of the Little Carpathians horst at its narrowest section, consists of two sub-units. The first one, the southern narrow horst sub-unit (Fig. 7.5 f.a), is very similar to the neighboring Myjavská Upland geologically and partly geomorphologically, and morphostructurally it represents distinctive asymmetrical horst characterized by steeper western slopes and gentle eastern slopes (Fig. 7.19). Despite its width of only 2 km, no transverse valleys have developed. The Jablonka Gorge bounds this sub-unit from the northern massive horst sub-unit (Fig. 7.5 f.b).

Unlike the southern sub-unit (Fig. 7.5 f.a), the northern sub-unit is not only higher and wider, but also morphologically clearly bounded from the side of the Danube Lowland. It is built mainly by limestones subjected to intensive karstification and bearing imprints of a planation surface, from which an isolated ridge of Salašky (588 m a.s.l., Fig. 7.20) sticks out. Along the southern margin of the sub-unit, the "River level" planation surface, attributed to the Váh River, occurs. The cross-section schemes of individual morphostructural units are shown in Fig. 7.21.

7.4 Neotectonic Development of the Little Carpathians

The contemporary geomorphological features of the Little Carpathians are the result of long interaction of the two basic groups of geomorphological processes under changing morphostructural and morphoclimatic conditions. Endogenous processes have created the mountains as a whole, which have given them the basic form of the horst-type positive morphostructure. The uplift history was likely complex, with the rise of the highest central part accomplished in two stages, whereas the lower foothills and transverse depressions were active only in the younger phase. Exogenic geomorphic processes took place in parallel with the endogenous processes, counteracting them and modifying the overall features of the rising horsts. They acted selectively, depending on rock resistance.

The Neogene exogenous morphogenesis took place in the environment of subtropical climate, and was largely effaced by exogenic processes in the younger developmental stages, during the Quaternary. After the climatic

Fig. 7.19 Western margin of the Čachtice Carpathians unit, which is a narrow horst (Fig. 7.5 f.a, photo taken from Čachtice castle, *photo* J. Lacika)

Fig. 7.20 Northern massive part of the Čachtice Carpathians unit (photo taken Čachtice castle, *photo* J. Lacika)

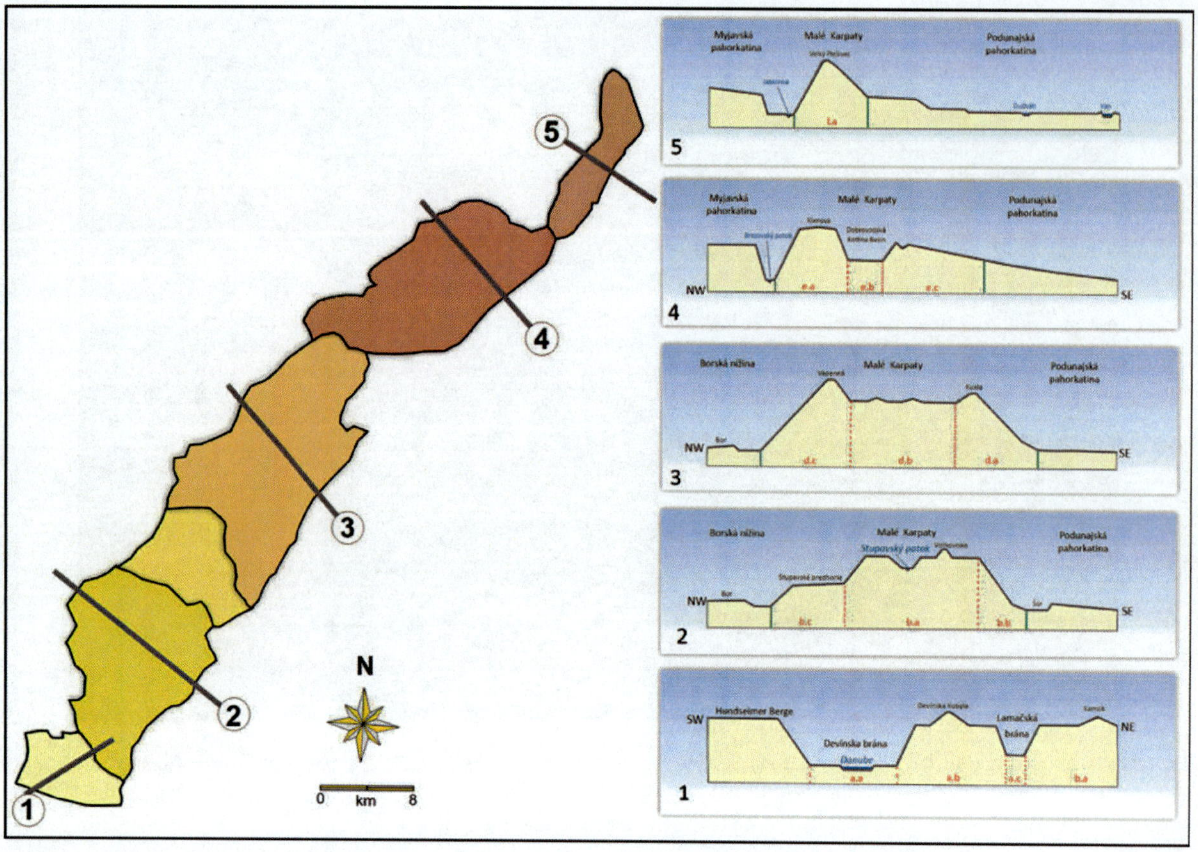

Fig. 7.21 Schematic cross-sections across morphostructural units of the Little Carpathian Mts.

change at the beginning of the Quaternary, thick and uncolidated regolith was exposed to severe destruction and transportation. Gradually, isolated hills and hummocks appeared due to selective weathering and removal of weathering products. Therefore, today we do not deal with the original Neogene surface, but with etchplain as an exposed basal surface that existed under the removed Neogene regolith. Thus, the flat character of the mountains was retained, but at lower altitude. During the Quaternary, several climatic oscillations on the territory of the Little Carpathians left a strong impact on the landscape. During glacial and interglacial periods, the nature and intensity of exogenous processes significantly changed, which led to the evolution of river terraces of the Danube River, as well as terraced fans at foothills. In the Holocene, human activity influenced the Little Carpathians landscape. The specific anthropogenic landforms are found in the wine-growing areas between Bratislava and Horné Orešany Village and throughout the abandoned mining area in the vicinity of Pezinok Town.

7.5 Little Carpathians as a Historical Borderland Area

In conclusion, we would like to draw attention to the remarkable linkages between the nature and man in the Little Carpathians. This elongated, narrow, and relatively low mountain range, having the form of a massive horst, had and still has a significantly dividing effect. Its barrier effect has been translated into the different historical developments of culture as well as of landscapes between two adjacent lowlands. The Little Carpathians are involved in the spatial dualism of the Great Moravian Empire expressed by the existence of the principality of Moravia in the west and the principality of Nitra in the east. Later, they divided the Czech kingdom and the kingdom of Hungary. So, the Little Carpathians horst was a place of long-lasting conflicts and cooperations between the neighboring states. Several castles were built along the mountain edge, whereas the Devín Gate and the historic Czech road following the tectonic line between the Pezinské and Brezovské Carpathians units can serve as very good examples of the particular setting of the Little Carpathians.

Acknowledgements The research was supported by Science Grant Agency (VEGA) of the Ministry of Education of the Slovak Republic and the Slovak Academy of Sciences; 02/0086/21.

References

Began A, Hanáček J, Mello J, Salaj J (1984) Geologická mapa Myjavskej pahorkatiny, Brezovských a Čachtických Karpát, 1:50 000. GÚDŠ, Bratislava

Bizubová M, Minár J (1992) Some new aspect of denudation chronology of the West Carpathians. In: Stankoviansky M (ed) Abstracts of papers. International symposium "Time, frequency and dating in geomorphology", Tatranská Lomnica-Stará Lesná, Bratislava, June 16–21, p 10

Buday T, Cambel B, Kamenický J, Maheľ M (1963) Geologická mapa ČSSR, mapa předčtvrtohorných útvarů 1 : 200 000 M-33-XXXVI Bratislava a M-33-XXXV Wien, Praha, Ústř. Ústav geologický

Daneš J (1931) Ke studiu Malých Karpat po stránce geologické a geomorfologické. Bratslava, Sborník Přír. odb. vlastived. múzea

Dzurovčin L (1990) Geomorfologická analýza strednej časti Slanských vrchov. Kandidátska dizertačná práca, Geografický ústav SAV, Bratislava

Feranec J, Lacika J (1991) Identification and analysis of a "gravity nappe" in the South-Eastern of the Malé Karpaty Mts. by using radar image. In: Proceedings of the eight thematic conference on geologic remote sensing. vol I, Denver, Colorado, pp 663–675

Hromádka J (1929) Průlom Dunajský a půda Bratislavy. Bratislava, Vlastivědný sborník okresu Bratislavského III:161–213

Hromádka J (1935) Zeměpis okresu Bratislavského a Malackého. Bratislava, Vlastivědný sborník okresu Bratislavského II, p 111

Jakál J, Lacika J, Stankoviansky M, Urbánek J (1990) Morfostrukturnyj analiz gornogo massiva Malych Karpat. Geomorfologija 4:37–42

Koutek J, Zoubek V. (1936) Vysvětlivky ku geologické mapě v měřítku 1 : 75 000, list Bratislava 4758. Knih. St. geol. Úst. Čs. Republiky, Praha, 18, 92 pp

Lacika J (1983) Lokalizácia zlomov metódou porovnávania výsledkov morfometrickej a morfogenetickej analýzy reliéfu na príklade južnej časti Malých Karpát. In: Přibyl J, Hrádek M, Kirchner K (eds) Sborník prací 1, Třicet let geomorfologie v ČSAV. Brno, pp 141–148

Lacika J, Urbánek J (1998) New morphostructural division of Slovakia. Geol J 4(1):17–28

Lacika J, Urbánek J (2002) Analógie a odlišnosti morfoštruktúrneho vývoja Malých Karpát a Tríbeča s Pohronským Inovcom. Geomorphologia Slovaca 2(1):44–57

Lukniš M (1955) Správa o geomorfologickom a kvartérno-geologickom výskume Malých Karpát (dolina Vydrice). Geografický Časopis 7(3–4):214–226

Lukniš M (1972) Reliéf. In: Lukniš M (ed) Slovensko 2 - Príroda. Obzor, Bratislava, pp 124–202

Lukniš M (1977) Geografia krajiny Jura pri Bratislave. Univerzita Komenského, Bratislava, p 211

Maheľ M, Cambel B (1972) Geologická mapa Malých Karpát 1: 50 000. GÚDŠ, Bratislava

Mazúr E (1963) Žilinská kotlina a priľahlé pohoria. Vydavateľstvo SAV, Bratislava, 184 pp

Mazúrová V (1973) Príspevok k poznaniu dunajských terás v Devínskej bráne. Geografický Časopis 25(2):112–121

Mičian Ľ (2000) Hypotéza o neogénnych geomorfologických trioláach v Malých Karpatoch na území Bratislavy. In: Lacika J (ed) Zborník referátov z 1. konferencie ASG pri SAV (Liptovský Ján 21–23. 9. 2000), pp 82–85

Mičian Ľ (2001) Príspevok k rozvoju hypotézy o neogénnych geomorfologických trioláach v bratislavskej časti Malých Karpát. Geomorphologia Slovaca 1(1):15–19

Minár J, Bielik M, Kováč M, Plašienka D, Barka I, Stankoviansky M, Zeyen H (2011) New morphostructural subdivision of the Western Carpathians: An approach integrating geodynamics into targeted morphometric analysis. Tectonophysics 502(1–2):158–174

Novodomec R (1967) Geomorfologické pomery povodia Parnej. Geografický časopis 19(3):212–223

Plašienka D, Michalík J, Kováč M, Gross P, Putiš M (1991) Paleotectonic evolution of the Malé Karpaty Mts. – an owerview. Geol Carpath 42(4):195–208

Polák M (ed), Plašienka D, Kohút M, Putiš M, Bezák V, Filo I, Olšavský M, Havrila M, Buček S, Maglay J, Elečko M, Fordinál K, Nagy A, Hraško Ľ, Németh Z, Ivanička J, Broska I (2011) Geologická mapa Malých Karpát 1: 50 000. ŠGÚDŠ, Bratislava

Polák M (ed), Plašienka D, Kohút M, Putiš M, Bezák V, Filo I, Olšavský M, Havrila M, Buček S, Maglay J, Elečko M, Fordinál K, Nagy A, Hraško Ľ, Németh Z, Ivanička J, Broska I (2012) Vysvetlivky ku geologickej mape 1: 50 000. ŠGÚDŠ, Bratislava, 287 pp

Potfaj M, Teťák F (eds), Havrila M, Filo I, Pešková I, Olšavský M, Vlačiky M (2014) Geologická mapa Bielych Karpát (južná časť) a Myjavskej pahorkatiny 1: 50 000. ŠGÚDŠ, Bratislava

Potfaj M, Teťák F (eds), Havrila M, Filo I, Pešková I, Olšavský M, Vlačiky M (2015) Vysvetlivky ku geologickej mape Bielych Karpát (južná časť) a Myjavskej pahorkatiny 1: 50 000. ŠGÚDŠ, Bratislava, 306 pp

Salaj J (1987) Vysvetlivky ku geologickej mape Myjavskej pahorkatiny, Brezovských a Čachtických Karpát 1: 50 000. GÚDŠ, Bratislava, 181 pp

Stankoviansky M (1974) Príspevok k poznaniu krasu Bielych hôr v Malých Karpatoch. Geografický Časopis 26(3):241–257

Stankoviansky M (1996) Reliéf kontaktnej zóny Brezovských Karpát a Myjavskej pahorkatiny so zvláštnym zreteľom na erózne odľahlíky. Geografický Časopis 48(1):47–72

Škvarček A (1966) Geomorfologické pomery strednej časti Malých Karpát. Geografický Časopis 18(2):132–145

Urbánek J (1966) Malé Karpaty a priľahlá časť Podunajskej nížiny v oblasti Jur – Pezinok. Náuka o Zemi, I, SAV Bratislava, 45 pp

Urbánek J (1992) Vývoj dolín južnej časti Malých Karpát. Geografický Časopis 44(2):162–174

Urbánek J (2014) Malé Karpaty – príbeh pohoria. Veda, Bratislava, 143 pp

Vaškovský I (ed), Kohút M, Nagy A, Plašienka D, Putiš M, Vaškovská E, Vozár J (1988) Geologická mapa Bratislavy a okolia 1: 25 000. GÚDŠ, Bratislava

Zaťko M (1959) Geomorfologické pomery povodia Gidra v strednej časti Malých Karpát. Acta Geologica et Geographica Univesitatis Comenianae, 49–84

Ján Lacika is an emeritus Associate Professor of Physical Geography. He was researcher at the Department of Physical Geography, Geomorphology and Natural Hazards of the Institute of Geography SAS in Bratislava (1980–2019) and teacher and research worker of the Department of Geography, Geoinformatics and Regional Development Faculty of Natural Sciences of the Constantine the Philosopher University in Nitra. Originally his main research interest was structural geomorphology (neotectonics, post-volcanic transformation of landforms and development of the valley networks). Later his research was focused on geographical aspects of the natural and cultural landscape development.

Ján Urbánek is a retired geomorphologist. He worked at the Institute of Geography of the Slovak Academy of Science. He also gave lectures of at the Faculty of Nature Science of Comenius University in Bratislava as well as Slovak University of Technology in Bratislava. He was interested in the theoretical geomorphology and geomorphic mapping, slope processes and valley systems, morphostructural geomorpohology and neotectonic development of the Western Carpathians, and geomorphological division of Slovakia. In the years 2000–2008 he was the chairman of the Association of Slovak Geomorphologists.

Milan Lehotský is a physical geographer and fluvial geomorphologist at the Institute of Geography of the Slovak Academy of Sciences. He was many years head of the Department of Physical Geography, Geomorphology and Natural Hazards. His research topics are responses of fluvial systems to environmental changes, sedimentological connectivity, evolution trajectories, hydromorphology and GIS and remote sensing applications in rivers and landforms research. He is also working as an external lecturer at the Department of Physical Geography and Geoecology, the Faculty of Natural Sciences of the Comenius University in Bratislava.

Genesis and Development of the Volcanic Landscape in the Slovenské stredohorie Mts.

8

Ladislav Šimon and Ján Lacika

Abstract

The neovolcanites in the Central Slovakia Volcanic Field originated between 15,5 Ma and 102 ka years ago. Volcanic rocks exist in the mountain ranges of Štiavnické vrchy, Pohronský Inovec, Vtáčnik, Kremnické vrchy, Poľana, Javorie, on the Krupinská planina Plateau and Ostrôžky, in the Zvolenská, Pliešovská and Žiarska kotlina Basins and on the margins of the Podunajská nížina Lowland. Andesite stratovolcanoes prevailed, while the complexes of extrusive domes of acid andesite, dacite and rhyolite rocks are less represented. Volcano-tectonic depressions developed in the stratovolcanoes of Mt. Javorie and the Kremnické vrchy Mts., whereas the 18 km × 22 km caldera with a resurgent horst in its centre is characteristic for the Štiavnica stratovolcano. Relatively young rhyolite volcanism accompanied the subsidence of the Žiarska kotlina Basin. Alkaline basalts represent the final stage of volcanic activity in the territory. Lavas of Ostrá lúka and Devičie, necks near Banská Štiavnica and the volcano of Putikov vŕšok of the middle to late Pleistocene age and the youngest volcano in the Western Carpathians and Central Europe are the relics of their activity. When the volcanic activity ended, the initial volcanic landscape transformed. Fault tectonics and selective erosion-denudation processes in the changing morphoclimatic conditions created a new type of natural landscape, with the mosaics of mountain ranges and basins which form the Slovenské stredohorie. The irregular course of transformation caused the morphostructural variety ensuing of the different types of volcanic-tectonic morphostructures, including conserved segments of the initial volcanic relief.

Keywords

Western Carpathians • Neovolcanites of Central Slovakia Volcanic Field • Stratovolcano • Extrusive dome • Caldera • Monogenetic volcano • Post-volcanic transformation of relief • Morphostructures

8.1 Introduction

Volcanic landscape is formed by volcanic processes and when they cease, the landscape keeps its characteristic appearance for a comparatively long time. However, it gradually disappears and it is replaced by a different landscape type corresponding to the new morphostructural and morphoclimatic conditions of the post-volcanic phase of development. Transformation of volcanic relief, which is the basic type of natural landscape distinctly determining the nature of other components, starts already in the late stage of the origin of a volcano. The origin of a volcano takes place in a certain morphoclimatic environment, under a set of exogenous geomorphological processes. The speed of transformation varies depending on several factors, first of all the nature (type and rate) of exogenous geomorphological processes and structural and lithological properties of the newly created volcanic complex. There are volcanoes of different types and ages in our planet. Very old volcanoes, which were active tens of millions years ago, and the ones that are formed right before our eyes can be studied. It means that there is an ideal study material available for the comprehension of the speed of post-volcanic destruction of volcanoes and proposal of typology of post-volcanic

L. Šimon (✉)
State Geological Institute of Dionýz Štúr, Bratislava, Slovakia
e-mail: ladislav.simon@geology.sk

J. Lacika
Department of Geography, Geoinformatics and Regional Development, Faculty of Natural Sciences, Constantine the Philosopher University in Nitra, Nitra, Slovakia
e-mail: jan.lacika@gmail.com

M. Lehotský and M. Boltižiar (eds.), *Landscapes and Landforms of Slovakia*, World Geomorphological Landscapes,
https://doi.org/10.1007/978-3-030-89293-7_8

geomorphological forms. It also applies to the complicated and very heterogeneous volcanic complex of the Slovenské stredohorie formed during millions of years in changing morphoclimatic and morphostructural conditions. Lacika (2006) addressed this theme in the past, but the new scientific evidence in the sphere of the geological and geomorphological research provides an opportunity to return to the subject.

8.2 Methodology

Three basic data sets are necessary for the recognition and classification of relief transformation:

(a) Data about the initial volcanic relief;
(b) Data about the nature of geomorphologic processes since the origin of the initial volcanic relief until present;
(c) Data about the status of the present-day, transformed volcanic relief.

First of all it is necessary to determine what the initial volcanic relief looked like, i.e. reconstruction. This phase is not the matter of a geomorphologist. The necessary database is obtained from a geologist's work. It is derived from regional geological research focussed on volcanic areas, applying principles of lithofacial analysis and paleovolcanic reconstruction. Geological sources provide information about the age, extent and rate of the volcanic activity, which are valuable entry data for the reconstruction. Reconstruction will be of higher quality if research experience from world regions with active volcanism is simultaneously considered. It is always useful if typologically similar or comparable volcanoes with those that are reconstructed are found. Obviously, the older is the volcanic locality subject to research, the more difficult and hypothetic is the reconstruction of initial volcanic forms.

The second database too is beyond the scope of geomorphological research. Series of interdisciplinary databases such as those related to palynology, paleontology, archaeology, etc. contribute to the recognition of the changing morphoclimatic environment of the given area. Parameters of a morphostructural environment are analysed by a set of methods referred to as morphostructural relief analysis. A number of methods such as mapping of flat relief forms, textures of geomorphological networks and preparation of special morphometric maps (for instance, map of base surfaces) were used in this study.

The present status of the post-volcanic transformation of volcanic relief can be interpreted first of all by means of morphometric analysis and results of geomorphological field research. The resulting product of the geomorphological part of research is formulation of a morphostructural scheme of the given area and its interpretation aimed at the recognition of the final form of the post-volcanic transformation of volcanic relief of the Slovenské stredohorie.

8.3 Geological Structure and Development of Neovolcanites in the Central Slovakia Volcanic Field

8.3.1 Distribution and Age

Volcanism in the territory of central Slovakia created a complicated volcanic structure referred to as the Central Slovakia Volcanic Field (CSVF), active from the Badenian (15.5 Ma) to the Pliestocene. They are represented by the volcanites of the Štiavnické vrchy Mts. including the youngest volcano in Central Europe–Putikov vŕšok near Nová Baňa, volcanites of the Vtáčnik and Kremnické vrchy Mts. and those of the Poľana and Javorie Mts. (Konečný et al. 2001, 2003) (Fig. 8.1). The extensive Štiavnica stratovolcano, occupying an area of more than 2200 km^2 (Konečný et al. 1998a), was formed in the western part CSVF from the Badenian to the Sarmatian. As the stratovolcano developed a large caldera (18 km × 22 km), a subvolcanic intrusive complex of granodiorite and intravolcanic complexes of granodiorite and silicabearing diorite porphyries originated. A number of satellite volcanoes formed in the caldera and on the stratovolcanic slope in the Sarmatian. Eventually, synchronously with the manifestations of rhyolite volcanism, a horst block was lifted up (Konečný et al. 1998b). A complex of superimposed volcanoes of Kremnické vrchy Mts. (Lexa et al. 1998) and Vtáčnik Mts. (Šimon et al. 1997) were formed north of the Štiavnica stratovolcano. The base of the volcanic structure consists of the andesitic Kremnica stratovolcano of Badenian age. The graben of Kremnica dates to the Late Badenian. Subsidence of the graben was compensated by lava effusions and deposition of volcaniclastic rocks more than 1000 m thick (Lexa et al. 1998).

Along the graben boundary faults, a few smaller satellite volcanoes were formed. In the Upper Sarmatian, rhyolite volcanism of the Jastrabá formation was active and accompanied by subsidence of the Žiarska depression. Volcanism of basaltic andesite type was active in the Older Pannonian stage. Its relic is the basaltic volcano of Vlčí vrch (Šimon et al. 2002a). Alkaline basaltic volcanism active in the Pannonian, Pliocene and Pleistocene was the final phase of volcanic activity in the territory. Relics of this volcanism include the lava flows of Ostrá lúka and Devičie, two necks near Banská Štiavnica (Fig. 8.2; Konečný et al. 1998a, b) and the monogenetic volcano of Putikov vŕšok of the Middle to Late Pleistocene age (Šimon et al. 2017, 2019a, b), which

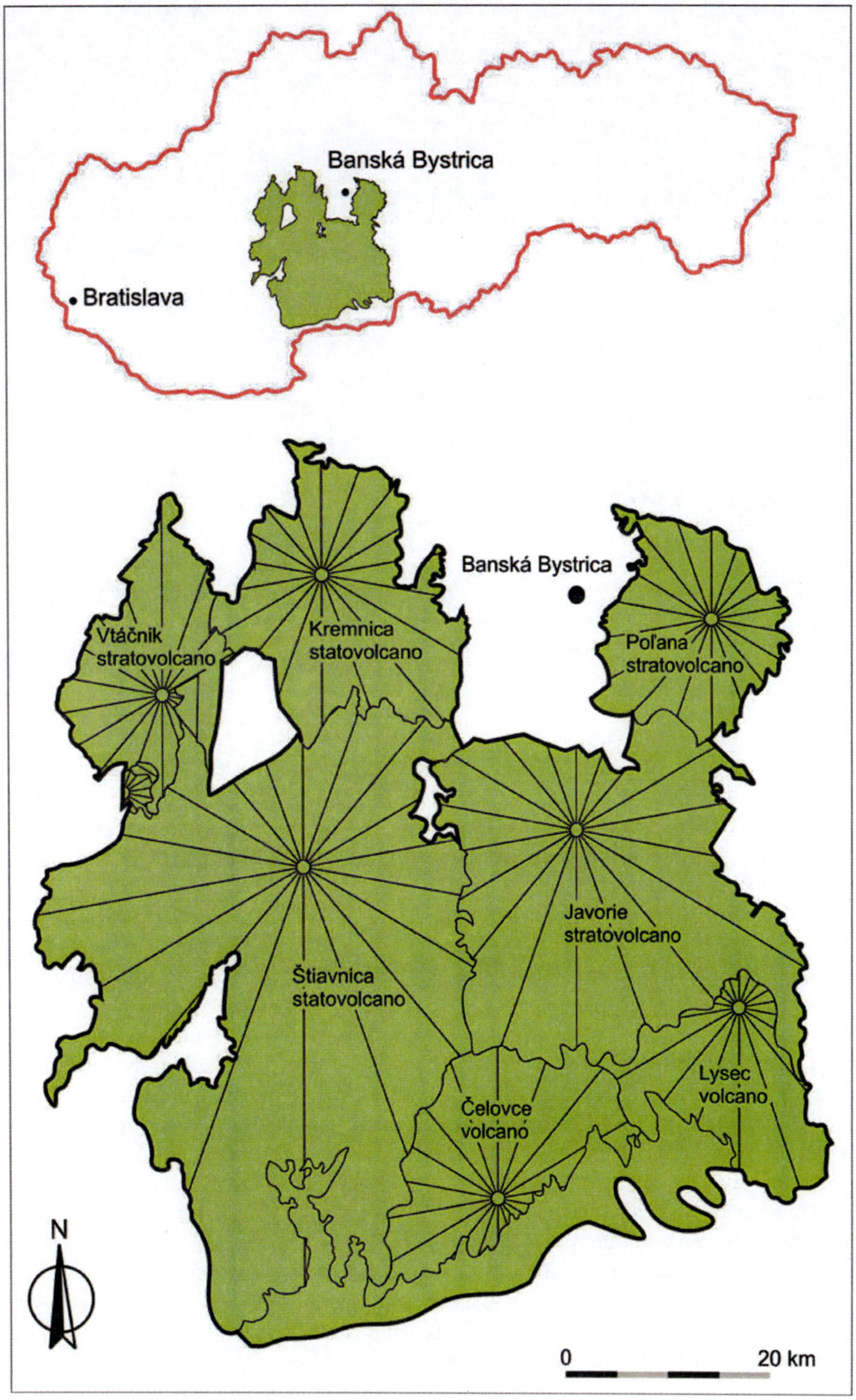

Fig. 8.1 Location of Central Slovakia Volcanic Field and distribution of significant stratovolcanoes and volcanoes in the study area

Fig. 8.2 Relict of a monogenetic volcano of Kalvária (Banská Štiavnica) (*Photo* L. Šimon)

is the youngest volcano of the Western Carpathians (Šimon and Halouzka 1996; Šimon et al. 2002b; Šimon and Maglay 2005). The volcano rests on Quaternary sediments of the river Hron and relics of the volcanic structure of the Štiavnica stratovolcano.

8.3.2 Štiavnica Stratovolcano and Putikov vŕšok Volcano

The Štiavnica stratovolcano, situated in the west of the Central Slovakian neovolcanic region, is the most extensive volcanic structure on the inner side of the Carpathian Arch (Konečný and Lexa 1979, 1984). It continues across the subjacent layer of the stratovolcano in the north-western direction, as far as its northern edge. The northern part of its structure subsided in the area of the Žiarska kotlina Basin. Its eastern edge meets the stratovolcano of Javorie. Its volcano-sedimentary complexes continue to the Ipeľská pahorkatina hilly land in the south and in its south-western part they subsided and are buried under the sediments of the eastern edges of the Podunajská panva Basin. The tallest mountain in the centre of the Štiavnické vrchy Mts. is Mt. Sitno, with elevation of 1009 m, whereas in the western part of the stratovolcano Mt. Veľký Inovec is 901 m high. In the north, the river Hron flows from the east to the west through the Žiarska kotlina Basin; it turns to the south, continues to the southern frontier of the country and joins the river Ipeľ.

The Štiavnica stratovolcano is characterized by a complicated structure, varied volcanic products, a huge caldera and the development of horst structure. Metalogenetic processes that accompanied the development of stratovolcano led to precious metallic and polymetallic mineralization, which was exploited in the mining towns of Banská Štiavnica and Banská Hodruša in the early middle ages. The Štiavnica stratovolcano is built by the following structural units: (a) the bottom structure of stratovolcano, (b) filling of the Štiavnica caldera, (c) upper structure of the stratovolcano and (d) complexes of subvolcanic intrusions. The bottom structure of stratovolcano is from the Badenian stage; filling of the caldera dates from the Upper Badenian to Lower Sarmatian.

The bottom structure of the stratovolcano consists of products of explosive and effusive volcanic activity formed by andesites and flat intrusions of andesite and diorite porphyries. It is divided into a number of volcanic complexes and formations, among which the Sebechleby formation and Žibritov effusive complex stand out (Konečný et al. 1983, 1998a). In the course of the explosive-effusive activity in the Sebechleby formation, andesitic lava flows alternated with eruptions of pyroclastic block and ash flows. Lava flows, which terminated in the littoral zone of the contemporaneous sea, were subjected to destruction and provided material for the origin of thick epiclastic volcanic conglomerates or their blocks. Products of Plinian eruptions in the form of ash-pumice tuffs were deposited in local depressions at the foothills of the stratovolcano. Pyroclastic flows that rolled down the slope crossed the coastline and continued further southward on the sea bottom, changing into submarine mud flows and reaching the distance of even 10–15 km from the coastline.

The second structural unit, which is the filling of the Štiavnica caldera, is formed by the volcano-sedimentary Červená Studňa formation. Its top wall bears 500-m-thick extrusive domes, lava flows and pyroclastic rocks of differentiated andesite of the Studenec formation (Konečný et al. 1998a). Due to denudation volcanic rocks were removed from the central and western part of the horst and extensive exposures of the rocks of the pre-volcanic substratum originated. A denuded subvolcanic intrusive complex formed by the body of granodiorite and diorite, referred to as the Hodruša-Štiavnica intrusive complex, is located in the central part of the horst. A number of stocks and dikes of granodiorite porphyries of the intrusive complexes of Zlatno and Tatiar protrude next to the periphery of the granodiorite intrusion.

The intrusive complex of Banisko is represented by layered intrusions and dikes of silica diorite porphyries or andesite porphyries. The upper structure of the stratovolcano consists of denudation relics of several smaller volcanoes and remnants of effusive and pyroclastic complexes, groups of layers and formations situated in the area of caldera and on the stratovolcanic (Šimon et al. 1995, 1996; Konečný et al. 1998a, b). In the Late Sarmatian, the central part of caldera bulged out and the horst structure developed. The fault system next to the western and north-western edges of the horst, which continued along the eastern margin of the Žiarska kotlina Basin, helped lifting of the rhyolite volcanic rocks of the Jastrabá formation (Konečný et al. 1998b). Volcanic activity ended in the Pannonian and is represented by basaltic andesite or basalt volcanism of the Šibeničný vrch complex in the area of the Žiarska Basin. In its south-eastern part, phreatomagmatic eruptions created the monogenetic volcano of Šibenický vrch with intrusion and effusion in the end of activity (Lexa et al. 1998; Šimon et al. 2002a). During the Pannonian to Pontian, a lava plateau of basaltic lava flows of Ostrá Lúka was formed on the eastern slope of the Štiavnica stratovolcano. Monogenetic volcanoes of Kalvária and Kysihýbel (Fig. 8.2) were formed in the central volcanic zone of the stratovolcano near Banská Štiavnica. They were later removed by denudation.

In the Pliocene to Quaternary, denudation of volcanic structure of the Štiavnica stratovolcano proceeded and led to the origin of erosion valleys. The monogenetic basanite volcano of Putikov vŕšok of the Pleistocene age (102,000 years) indicates the termination of volcanic activity within the Štiavnica stratovolcano (Šimon and Maglay 2005). The volcano rests on Quaternary sediments of the river Hron and the relics of volcanic structure of Štiavnica stratovolcano (Fig. 8.3).

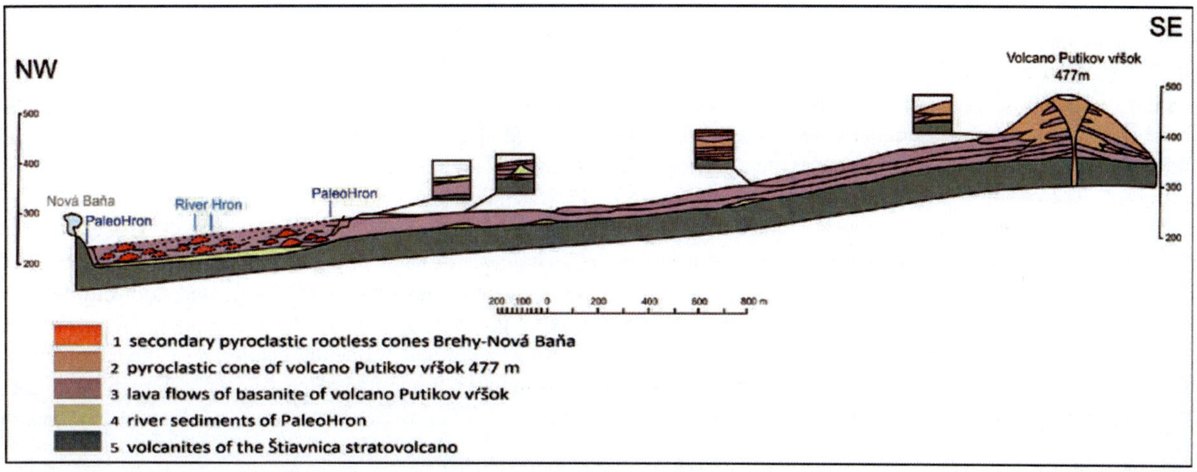

Fig. 8.3 Geological profile of the Putikov vŕšok Volcano (Štiavnické vrchy mts.)

Fig. 8.4 Lava flow from the Putikov vŕšok basanite Volcano (Brehy-Nová Baňa) (*Photo* L. Šimon)

The crater of Putikov vŕšok Volcano (elevation 477 m asl) is located within the ridge of hills above the village of Tekovská Breznica in the Štiavnické vrchy Mts. A pyroclastic cone covered by volcanic bombs was formed next to the crater. Lava flows of nephelinic basanite poured from the crater, creating a lava plateau with lava tongues (Fig. 8.4) next to the river Hron in the top part of a Riss-age terrace.

Prevailingly Strombolian and Hawaiian eruptions but also phreatomagmatic eruptions were responsible for the building of the volcano. Volcanic eruptions were of a low volcanic explosive index (VEI), that is, 1 to 2; in case of hydrovolcanic eruptions the value of VEI increased to 3 or 4. Volcanic explosions produced an ash cloud which rose to the height of 50 m to 1 km. The volcanic activity created a volume of 10,000 m^3 of volcanic materials. The bulk of lava flows poured on a surface of about 4 km^2 and filled a volume of 100 m^3 to 0.5 km^3 as far as the area between the villages of Tekovská Breznicá and Nová Baňa.

Paleovolcanic reconstructions revealed that as the front edges of the lava flows contacted with waters of the river Paleo–Hron, phreatic eruptions gradually gave origin to 16 pseudocraters, which are represented by secondary pyroclastic rootless cones. In the quarry of Brehy–Nová Baňa, relics of four pseudocraters are preserved (Fig. 8.5), deposited on the basanite lava flows (Šimon et al. 2002b, 2019a, b).

8.3.3 Volcanites of the Vtáčnik and Kremnické vrchy Mountain Ranges

Mountain ranges of Vtáčnik and Kremnické vrchy (Fig. 8.6) represent a complicated volcanic form, with the temporal and spatial suite of volcanic structures loaded on one another. The dominant form in this area is the Kremnický graben from the Badenian stage (Konečný et al. 1983; Lexa et al. 1998; Šimon 1991; Šimon et al. 1991, 1997). Its origin was preceded by the activity of an important andesite Kremnica stratovolcano described as the Zlatá Studňa formation (Konečný et al. 1983, Šimon et al. 1996; Lexa et al. 1998) (Fig. 8.7). The formation crops out in the area of uplifted Kremnica horst and was deposited within the Kremnica graben and on its eastern and western sides. The Zlatá Studňa formation in the central volcanic zone of Kremnica stratovolcano is represented by a complex of layered intrusions of metamorphosed rocks, with relics of lava flows, volcanoclastics and andesite dikes.

The basis of the central volcanic zone of the Kremnický stratovolcano consists of subvolcanic bodies of diorites and dioritic porphyries. West of the Kremnica graben the rocks of the Zlatá Studňa formation outcrop near the commune of Remata in the form of an extensive laccolith body of dioritic porphyry. South-east of the town of Banská Bystrica a

Fig. 8.5 Relics of pyroclastics from the pseudocrater Brehy, Putikov vŕšok Volcano (Brehy-Nová Baňa) (*Photo* L. Šimon)

complex of lava flows and volcanoclastics, i.e. the complex of Suchý vrch (Lexa et al. 1998) with traces of development in subaqueous environment, is exposed. East of the Kremnica graben, the Zlatá Studňa formation acquired a typical stratovolcanic structure, with alternating andesitic lava flows with volcanoclastics. Phreatopyroclastic tuff and traces of hyaloclastite brecciation of lava flows are present in the bottom of the formation, suggesting an aqueous environment in the eastern part of the territory in time of the initial developmental stages of the stratovolcanic structure. Lava flows, which transit eastward into a peripheral volcanic zone, alternate with pyroclastic and epiclastic volcanic rocks on the higher levels, closer to the central volcanic zone. The western part of the peripheral volcanic zone of the Kremnica stratovolcano/Zlatá Studňa formation reached as far as the Horná Nitra Basin during the Early to Middle Badenian and so did the bulky Štiavnica stratovolcano with its peripheral volcanic zone (Fig. 8.8).

A thick (200 m) complex of epiclastic and volcanic conglomerates and sandstone rocks referred to as the Kamenec formation (Konečný et al. 1983; Šimon et al. 1997) was deposited in a fluvial-limnic environment in the depression of the Horná Nitra Basin at the foothills of the two stratovolcanoes. Within this formation complex layers of tuffs, pumice tuffs, agglomerates and phreatopyroclastics have been also described (Šimon et al. 1991, 1997). In the subjacent layer, the complex progressively passes into the Handlová formation and Nováky formation complexes, with an increasing presence of the tuffitic coal clays (Šimon et al. 1997).

The Handlová formation (Šimon et al. 1997) is a clay-tuffitic one, with coal beds in the area of the Handlová and Cígeľ coal deposits. The thickness of the complex is between 2 and 50 m. The coal bed crops out near the village of Lehota pod Vtáčnikom, where it was mined in an open-cast mine. The Novácky formation is separated from the Handlová formation primarily or secondarily due to erosion in the area between the villages of Koš and Podhradie. The Koš formation (Šimon et al. 1997), deposited on the Handlová formation and Nováky formation, consists of monotonous dark grey to green-grey clay and calcite clay with fluctuating presence of sand admixture. The abrupt block tectonic and denudation explain why the present thicknesses of strata complex oscillates between several metres and 300 m in subsided blocks.

The Nová Lehota formation (Šimon et al. 1991, 1997) manifests rhyolite volcanism in the form of rhyolite bodies identified by drillings in the Handlová mine and on the surface near the hill Nová Lehota with a height of 575 m (Šimon et al. 1991, 1996, 1997). The first manifestation of the after-coal volcanism in the area of Vtáčnik Mts. was extrusive volcanism of hyperstenic-amphibolic andesites of the Plešina formation (Šimon et al. 1991, 1997). The formation includes groups of extrusive domes accompanied by deposits of volcanoclastics rocks. The Lehota formation (Šimon et al. 1991) rests on the erosion-modelled surface of the Koš formation. Southward, the complex penetrates through the bottom layer of the Kľakovská dolina formation. The thickness of the strata complex varies from several tens of metres to 300 m.

The development of the Kremnica graben that immediately followed provoked dramatic changes in the paleogeography of the territory. The central structural étage of the territory is the filling of the Kremnica graben. The development of the graben in the time interval of the Upper Badenian to Lower Sarmatian took place due to subsidence of an extensive territory, including the central part of the Kremnica stratovolcano, the Žiarska Basin as far as the

Fig. 8.6 View from the south of the Vtáčnik mountain range to the Handlová Basin, in the background are Kremnické vrchy (*Photo* L. Šimon)

northern edge of the Štiavnica stratovolcano and the eastern part of what is today the mountain range of Vtáčnik. Subsidence with an amplitude of more than 1000 m proceeded along the massive fault zones running in the NNE-SSW to N-S direction. Its initial stage was accompanied by volcanism of pyroxenic andesites. Products of this volcanism in the area of the Kremnické vrchy Mts. are included in the Turček formation and in the eastern part of Vtáčnik into the formation of Klakovská dolina, i.e. equivalent lithostratigraphic units. Both formations contain a varied lithologic set of lava flows, hyaloclastic breccias, autochthonous and redeposited pyroclastics and epiclastics of basalts, basaltic andesites, pyroxenic andesites and leucocractic andesites forming the bottom part of the filling of the Kremnica graben (Šimon et al. 1997; Lexa et al. 1998).

Extrusive bodies of the basaltic andesite at Veľký Grič, Kňazov kopec (Figs. 8.9 and 8.10) and Malý Grič (Šimon et al. 1997) crop out outside the principal area of the formation. Dating by the K/Ar method of the rocks of Kľakovská dolina formation yielded an interval 13.2–14.4 Ma (Šimon et al. 1991) and indicates the Upper Badenian. The top part of the filling in the Kremnica graben consists mostly of the products of explosive-effusive volcanism, with occasional extrusions of amphibolic-pyroxenic andesites represented by lava flows and extrusive domes included into the formation of Kremnický štít in the central part of the Kremnické vrchy Mts. and into the Stráň formation in the eastern part of Vtáčnik Mts. (Šimon et al. 1997). Both formations are taken as equivalents, based on the situation and the lithologic/petrographic character of the products.

The Kremnický štít formation is prevailingly an effusive complex comprising of 30-m- to 150-m-thick lava flows. The bottom levels contain first of all pyroxenic andesites with amphibole. The middle levels of the formation contain amphibolic-pyroxenic andesites, whereas lava flows of the biotic-amphibolic-pyroxenic andesite are in the upper levels. The Stráň formation (Šimon et al. 1997) is dominantly built by up to 300-m-thick andesite lava flows, which crop out of the graben blocks in Vtáčnik Mts. to the west of the Žiarska Basin. Based on the positional relationships, both formations, the Kremnický štít and Stráň, are dated to the Upper Badenian to Lower Sarmatian (Šimon et al. 1997; Lexa et al. 1998). Due to the resumed outlet of magmas to the surface reservoirs and processes of their differentiation into andesite magmas, several smaller satellite volcanoes were formed on the edges of the Kremnica graben on the south-eastern slopes of the Kremnické vrchy Mts. and in the western part of the neovolcanic region of Vtáčnik.

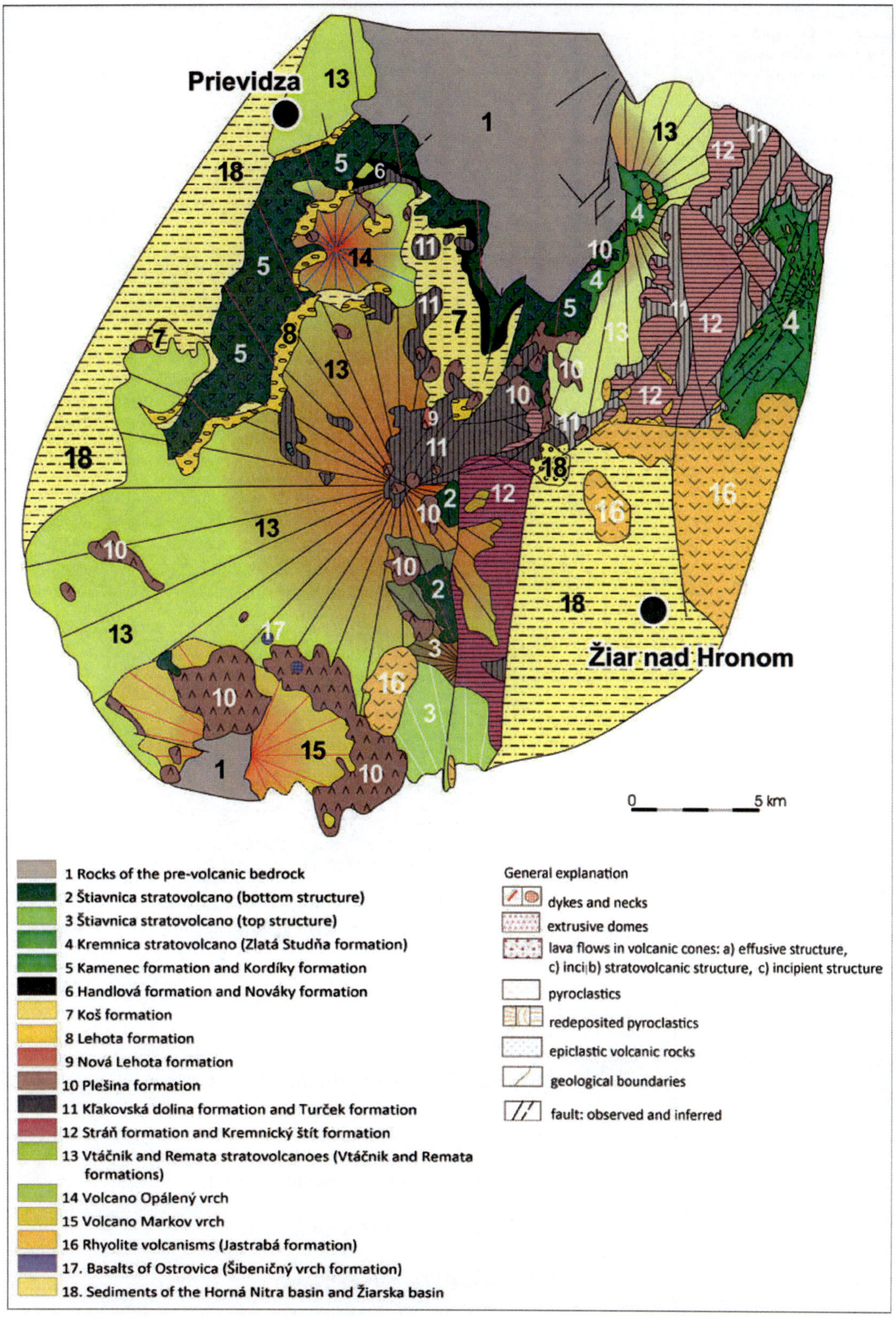

Fig. 8.7 Structural geological scheme of volcanic complexes of the Vtáčnik Mts. and the western part of the Kremnické vrchy Mts.

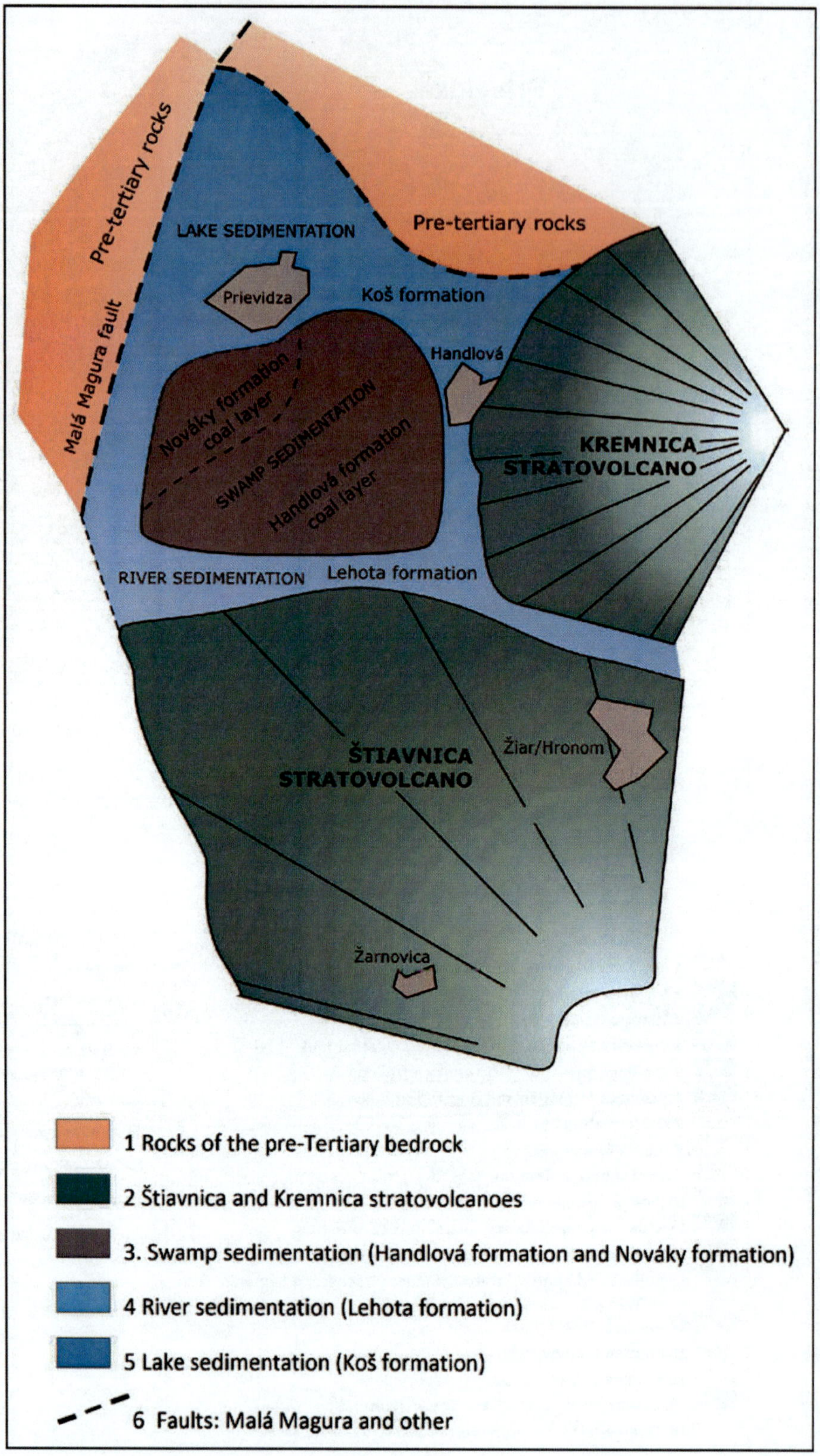

Fig. 8.8 Paleogeographic scheme of the Kremnica and Štiavnica stratovolcanoes in the Badenian, in contact with the Horná Nitra Basin

Fig. 8.9 Extrusion of basaltic andesite—Mt. Kňazov kopec (Vtáčnik) (*Photo* L. Šimon)

Fig. 8.10 Basaltic jointing at Mt. Kňazov kopec (Vtáčnik) (*Photo* L. Šimon)

The Flochová Volcano was formed next to the northern edge of the Kremnica graben. Its relics are represented by the Flochová formation. The Rematský Volcano and its relics, the Remata formation, are near the north-western slope. Sielnica Volcano and its relics in Sielnica formation originated on the south-eastern slopes of the Kremnické vrchy. In its top layer, the Turová Volcano was formed and its relics are part of the Turová formation (Lexa et al. 1998).

In the north-eastern part, the Vtáčnik stratovolcano has arisen and its relics are represented by the Vtáčnik formation, whereas the Markov vrch Volcano is to the south of it and its relics are represented by the Markov vrch formation (Šimon et al. 1997). The dominating volcanic structure in the west of the neovolcanic region was the stratovolcano of Vtáčnik (Šimon et al. 1997). The present-day volcanic structure of stratovolcano's relics is remarkable for its asymmetric shape. Only the western part of the original stratovolcano survived. The eastern part of the stratovolcano was removed by denudation and it advanced to the area of the central volcanic zone. In the central volcanic zone, the stratovolcano is mostly built of thin brecciated lava flows of pyroxenic andesites, sporadically alternating with tuff and agglomerate layers. Lava flows and pyroclastics exhibit traits of periclinal arrangement around the supposed centre at the end of the Kľakovská dolina Valley, where there are several dikes and necks outcropping in the area built principally of older rocks. Autochthonous pyroclasts indicating prevalence of explosive activity in the initial phases of development of the stratovolcano are deposited at the base of the cone. Very thick lava flows alternating with rarer layers of pyroclastic rocks prevail in the transitory volcanic zone, i.e. on slopes of stratovolcanoes. Layers of epiclastic volcanic breccias, conglomerates and sandstones deposited in the lower bottom levels of the volcanic structure accrue with the transition into the peripheral volcanic zone. Parasitic volcanoes used to exist in the northern part of Vtáčnik and the volcanic neck in the area of Opálený vrch and the associated autochthonous pyroclastic flows, agglomerates and tuffs representing the Biela Skala pyroclastics (Šimon et al. 1997) point to their relics. Due to the movement of blocks and rotation of a huge block inclined to the west into the Horná Nitra Basin in the post-volcanic period, the eastern, relatively more uplifted and exposed part of stratovolcano, was subject to severe denudation. The result was the total destruction of the volcanic structure of the eastern part of stratovolcano as far as west as the central zone.

8.3.4 Volcanites of Javorie Mts. and Poľana Mts.

The stratovolcano of Javorie, like the stratovolcano of Poľana, belongs to the group of polygenetic stratovolcanoes, whose development proceeded over several stages in the Neogene, from the Badenian to Sarmatian (Dublan et al. 1997; Konečný et al. 2001, 2003; Šimon et al. 2010, 2012, 2013, 2017, 2019b). The geology of the Javorie and Poľana mountain ranges is of horst-and-graben nature (Konečný et al. 1978; Šimon et al. 2017).

The Javorie stratovolcano in the east of the neovolcanic region is a comparatively extensive volcanic structure, inside which an intrusive complex and volcanotectonic grabens have developed. The structure of stratovolcano consists of (a) bottom stratovolcanic structure, (b) filling of volcanotectonic grabens, (c) intrusive complexes and (d) upper stratovolcanic structure (Konečný et al. 2001, 2003). The bottom of the stratovolcanic structure of Badenian age is mostly buried by the rocks produced by younger volcanic activity of Sarmatian age. The bottom stratovolcanic structure represented by the Stará Huta formation appears on the surface only near the eastern margin of the ranges.

The Stará Huta formation was only confirmed in the central part of the Javorie in a subsided layer in the graben depression under the rocks of the younger, 700-m-thick graben filling. The buried bottom of the stratovolcanic structure is prevailingly built of lava flows of pyroxenic and amphibolic-pyroxenic andesites in the central part. Layered intrusions of andesite to diorite porphyries penetrated its intensively hydrothermally altered bottom part, pierced by younger dikes of andesites. The rocks of the Starohutský complex exposed by denudation next to the eastern edge of the range are built of lava flows of pyroxenic and amphibolic/pyroxenic andesites, alternating with layers of chaotic breccias of pyroclastic flows and epiclastic volcanic breccias in the transitory volcanic zone. The lava flows finish in the southern direction, giving way to the facies of epiclastic volcanic breccias of conglomerates and sandstones.

Brecciation was identified in the eastern part of the graben. It indicates water environment in the initial stage of stratovolcano's development (Konečný et al. 1998). The middle volcanic structure of Javorie consists of the filling of volcanotectonic grabens of Middle Badenian to Lower Sarmatian age, appearing as the lavas of the Blýskavica formation, extrusive domes and protrusions of the Siron formation. The bottom part of the graben consists of lava flows of basic pyroxenic andesite, basaltic andesites and hyaloclastic breccias of the Blýskavica formation. The subsidence of grabens during the effusive activity was compensated by deposition of lavas and 250–300-m-thick hyaloclastic breccias. After the grabens were filled, effusions of lava occurred and transport of fragmented material to the southern slopes of the stratovolcano resulted in the deposition of volcanoclastics rocks. The upper part of the filling in grabens is composed of products of extrusive andesite volcanism of the Siron formation. Petrographic composition of andesite bodies is varied, reflecting the presence of several volcanic forms. Bigger extrusive bodies with fan-like patterns of fluidal textures and belts of autoclastic breccias along the margins testify to the presence of extrusive domes. Protrusions are typically smaller, often with isometric to irregular elliptic profile and steep to vertical patterns of fluidal textures. Laccolite forms are characterized by a range of several square kilometres and distinct auto-metamorphic alterations (Konečný et al. 2003).

Intrusive complexes include those of Lohyňa, Kalinka and Králová. The Lohyňa complex consists of intrusions and extrusions of rhyolites, dacites and andesites. The complex of Kalinka is formed by stock intrusions of dioritic and siliceous diorites. The Králová complex contains stock intrusions of dioritic porphyries. The upper stratovolcanic structure of Sarmatian age includes explosive and effusive andesite volcanism relics, which are represented by the Javorie formation. Rocks of the Javorie formation participate in the structure of the transitory and peripheral volcanic zone. The bottom levels of the stratovolcanic structure are prevailingly made of pyroclastic and epiclastic volcanic rocks, in places alternating with lava flows. The upper levels of the structure are mainly built of lava flows. Relics of lava flows now build the tallest tops and ridges of the Javorie mountain range. During the denudation processes that followed the original stratovolcanic cone was completely removed and was replaced by a depression while denudation revealed intrusions and bodies of the feeding systems. In the consequence of uplift of the eastern part of the region, volcanic complexes of the peripheral volcanic zone were removed and the transitory volcanic zone was considerably reduced (Konečný et al. 2001).

Mt. Poľana is an andesite stratovolcano of medium size, situated next to the north-eastern edge of the Neovolcanic Region. The bottom stratovolcanic structure of Badenian age is represented by the Šutovka formation (Dublan et al. 1997) and mostly buried under the younger products of Sarmatian volcanism. The lower stratovolcanic structure is built of lava flows of pyroxenic andesites and volcanoclastics. Rocks of the bottom structure within the caldera consisting of volcanoclastics and lava flows had been severely hydrothermally changed and penetrated by younger bodies of layered and stock intrusions. The bottom volcanic structure includes products of rhyolite and dacite volcanism of the Strelniky formation of Lower Sarmatian age, associated with the development of the caldera.

During the initial period of volcanic activity, eruption of ash and pumice tuff produced by the Plinian eruptions led to deposits on the northern, western and southern slopes of the stratovolcano and at its foot in the form of ash and pumice lavas, fallen tuff and redeposited tuff. Ash and pumice material was moved from the higher levels of the western slopes of stratovolcano by lahars. Due to the emptying of the upper layers of the magmatic reservoir, the upper parts of the stratovolcanic structure collapsed leading to the origin of a caldera-like structure. The subsiding caldera was initially filled by tuff and pumice material and later by extrusions and effusions of 300-m-thick rhyolite and dacite lavas.

The upper stratovolcanic structure of the Poľana mountain range is composed of products of explosive-effusive volcanism of pyroxenic and amphibolic-pyroxenic andesites of Sarmatian age of Poľana formation. Lower levels of the stratovolcanic structure, with the prevalence of products of explosive activity and lesser representation of lava flows, contain epiclastic volcanic rocks in the bottom and products of explosive activity, that is, pyroclastic rocks alternating with the layers of epiclastics and rare lava flows in its upper level. Epiclastic volcanic rocks gradually take over in the transition area to the peripheral volcanic zone. The upper section of the volcanic structure consists of dominating lava flows of pyroxenic and amphibolic-pyroxenic andesites of the Poľana formation (Šimon et al. 2010, 2013, 2017). In the environment of layered intrusions of andesite porphyries, stock intrusions of andesite and diorite porphyries crop out in the central volcanic zone. Intrusions were accompanied by intensive hydrothermal transformation of the surrounding rocks. Volcanism in the Poľana mountain range produced various genetic types of volcanic rocks of andesite, dacite and rhyolite composition (Figs. 8.11 and 8.12).

8.3.5 Volcanites of the Krupinská planina Plateau

The Krupinská planina plateau is situated in the south-eastern part of the Central Slovakian Volcanic Field. Volcanites of the stratovolcano Javorie reach the Krupinská planina plateau in the north and in the north-west it neighbours with the volcanites of the Štiavnica stratovolcano. The plateau was formed by the volcanic-sedimentary formations and complexes of Badenian age: Príbelce formation, Vinica formation, Čelovce formation and Lysec formation (Konečný et al. 1983).

The Príbelce formation is mostly represented by tuffitic sands with minerals of volcanic origin, which were deposited in a marine environment. Volcanic centres of the Vinica formation originated in the same environment in the form of submarine extrusive domes situated along the Šahy-Lysec volcanic zone in the north-eastern to south-western direction. Extrusive domes in the marine environment were liable to destruction while the volcanic material was redeposited around them or was moved on as lahars. In the central part of the Šahy-Lysec volcanic zone, products of the Čelovce formation were created partly in the marine and prevailingly in the limnic-fluvial environment. Explosive volcanism produced pyroclastic rocks, necks and dikes of pyroxenic andesites, which formed the pyroclastic Čelovce Volcano. Later, products of the Lysec formation were created in a

Fig. 8.11 View of the Poľana Mountains from the village of Očová (*Photo* L. Šimon)

terrestrial or limnic-fluvial environment in the north-eastern part of the Šahy-Lysec volcanic zone. Explosive and intrusive-extrusive andesite volcanism formed the pyroclastic Lysec Volcano.

8.4 Post-Volcanic Transformation of Landform of the Slovenské stredohorie

8.4.1 Morphostructures of the Central Part of the Slovenské stredohorie

8.4.1.1 The Štiavnické vrchy Mts.

The position of the volcanic complex of the Štiavnické vrchy Mts. is central in the geomorphological area of Slovenské stredohorie. Except in the south, where it borders on the Podunajská nížina Lowland, it is surrounded by the neighbouring Neogene volcanic complexes. The geological form, also referred to as the Štiavnica stratovolcano, once outreached the borders of what are the Štiavnické vrchy Mts. today and its products are allochthonous parts of the geological structures of the neighbouring mountain ranges, basins and the Podunajská nížina Lowland. The volcanic paleorelief of the stratovolcano forming the bottom structure of the whole complex (Lower Badenian stage) cannot be considered as the initial relief of the post-volcanic transformation which followed. It practically did not survive in any form; it was either denuded or destroyed by the younger phases of volcanism. In the central part of the Štiavnicka stratovolcano, a large caldera formed in the Upper Badenian stage, which also completely disappeared during the post-volcanic evolution. The Štiavnická brázda furrow identified by Mazúr and Lukniš (1978) in the central part of the mountain range cannot be taken for its transformed form. It is a newly formed erosion-tectonic depression on less resistant volcanic rocks at the intersection of tectonic faults. The extent and rate of the post-volcanic denudation of the Štiavnické vrchy Mts. is attested by the presence of denuded pre-volcanic bedrock (Hodrušsko-štiavnický intrusive complex). The borders of the Vyhnianska brázda furrow, a longitudinal depression in a comparatively dissected Hodrušská hornatina Highland, do not track the structural boundary line between the volcanic and non-volcanic rocks.

The initial volcanic relief for the post-volcanic transformation of the Štiavnické vrchy Mts. appeared only in the third phase of development, when a stratovolcano was formed as an upper structure of the volcanic complex during the Sarmatian stage. After that, volcanism manifested only locally and the final volcanism producing basaltic andesites and basalts was limited to a small area. As Konečný and Lexa (2001) reported, andesitic lava flows filled paleovalleys on the outer flank of the Sarmatian stratovolcano and the post-volcanic relief inversion moved them to elevated

Fig. 8.12 Extrusion of andesite with onion-skin disintegration (Viglaš, Javorie)

positions. Longitudinal elevation of the Kozmálovské vŕšky hills reaching into the Podunajská nížina Lowland as far as the village of Mochovce formed on one of such paleoflows.

The area of the present horst of the Štiavnické vrchy Mts. is substantially smaller than the transformed Neogene volcanic paleo-morphostructure. Peripheral parts of the Sarmatian stratovolcano were considerably denuded or included into the neighbouring horsts (Pohronský Inovec Mts.) or covered by sediments of the subsiding morphostructures (Žiarska kotlina Basin). The greatest reduction took place on the southern side of the stratovolcano, where a great part of the periphery was denuded and partly buried under the sedimentary filling of the West Pannonian Basin. The borders of the Štiavnica morphostructure are prevailingly of tectonic nature. The fault-generated marginal slopes are on the side of the water gaps of the river Hron, of the Žiarska, Zvolenská and Pliešovská Basins. Foothill morphostructures with a similar plateau-like type of relief to those of the Krupinská planina Plateau and the Ostrôžky Plateau are in the south of the Štiavnické vrchy Mts. Presumably, its extent was originally larger, before it was reduced in terms of both the area and volume in favour of the Ipeľská pahorkatina hilly land. In the case of Pohronský Inovec Mts., the periphery of stratovolcano was tectonically uplifted and the result was the formation of a massive marginal horst morphostructures bearing plateau relief (the Lehotská planina Plateau).

The Štiavnica morphostructure itself is internally disintegrated into a system of differently uplifted blocks. Asymmetrical uplift of the western part of the horst gave origin to higher and more dissected Hodrušská hornatina Highland with an isolated block of Mt. Sitno in the central position, which bears erosion-denudation cinders of the andesitic lava flow. The eastern part of the Štiavnica horst consists of Skalka blocks (sub-unit). Distinctly transformed remnants of initial volcanic relief with affinity to two Pliocene volcanoes are in the central part of the Štiavnické vrchy Mts. The pyroclastic cone intersected by a basaltic neck of Kysihýbel is not morphologically distinctive, whereas the volcano of Kalvária is after transformation a quite well-defined morphological form and the landscape dominant of Banská Štiavnica historical town. The basanite volcano of Putikov vŕšok in the Novobanská kotlina Basin is the youngest manifestation of volcanism in Slovakia. As it is only 102,000 years old (Šimon and Maglay 2005), its original volcanic relief has been scarcely transformed.

8.4.1.2 The Pliešovská kotlina Basin

The Pliešovská kotlina Basin is a relatively subsiding morphostructure of specific history; it is a depression lacking the Neogene filling formed in a depression between two Neogene stratovolcanoes and morphologically sharpened fault tectonics. In the northern part of the basin, a complex of basaltic flows outpoured from the Pannonian to Pontian stage. Filling of the basin bottom by volcanic materials resulted in a forced formation of the valley of Neresnica River, which broke over the margin of the Javorie Mts., thus finding a new passage towards the river Hron. The plateau near the village of Ostrá lúka, 185 m above the valley of the river Hron, was formed by relief inversion upon solidified lava flows. A comparatively rare phenomenon occurs in the south, around the village of Pliešovce. The main water divides between the basins of Hron and Ipeľ Rivers which runs along the bottom of the Pliešovská kotlina Basin and not along the mountain ridges. The progressively developing Krupinica (tributary of the river Ipeľ) caused river capture when it shifted to the Pliešovská kotlina Basin by headward erosion of its spring area and tapped the streams in the upper part of the Neresnica Basin (tributary of the river Hron).

8.4.2 Morphostructures of the Western Part of the Slovenské stredohorie

Morphostructural composition of the western part of Slovenské stredohorie spreading west of the middle reach of the Hron River is formed by fault-bound morphostructures: three horsts (Kremnické vrchy Mts., Vtáčnik Mts. and Pohronský Inovec Mts.) and one graben morphostructure (Žiarska kotlina Basin). The Vtáčnik Mts. and Kremnické vrchy Mts. transformed from the original volcanic or volcanic-tectonic type of morphostructures into faulted morphostructures, where the volcanic past is not discernible in relief but it manifests indirectly by the properties of the volcanic rocks. Both horsts are asymmetric with taller partial morphostructral units on the reverse side of the mountain range. The main ridge of the Kremnické vrchy Mts. is situated in the eastern edge of the range, while the similar main ridge of the Vtáčnik Mts. was shifted to the edge on the western side of the mountain range. Parting from these two most elevated marginal blocks second-order morphostructures subside towards each other as far as the depression axis heading from the Handlovská kotlina Basin above the village of Nová Lehota to the Žiarska kotlina Basin. This conspicuous morphostructural element tracks the course of the slowly subsiding zone, in which the volcanic complexes of the western part of Slovenské stredohorie were formed. The southernmost unit, the Pohronský Inovec Mts., is morphologically different from the horsts in the north because the character of its geological structure and relief is more similar to those of the Štiavnické vrchy Mts. east of the water gap of the Hron (Lacika 2005).

8.4.2.1 The Kremnické vrchy Mts.

The origin of the Kremnické vrchy Mts., in contrast to other volcanic mountain ranges, was more complicated and spatially more differentiated (Fig. 8.13). The paleorelief of the lower stratovolcanic structure (Lower Badenian) was subject to complete transformation and not even initial surfaces from the stage, when the Kremnica graben developed in the Upper Badenian and Lower Sarmatian, survived. Thus, volcanic forms from the third and fourth developmental stages in the Sarmatian and the Pannonian might be considered the initial volcanic relief. First of all, several smaller satellite volcanoes became active along the northern and eastern edges of the graben, while the volcano in the south-east of the range neighbouring with the subsiding Žiarska kotlina basin was active in the Late Sarmatian. A specific feature of the Kremnické vrchy Mts. is the prolongation of volcanic activity into the Pannonian, when formation of the volcanic paleorelief was concluded in the origin of a smaller basaltic volcano of Vlčí vrch. During the post-volcanic period, lasting from the Pannonian to the recent age, the complicated volcanic complex of the Kremnické vrchy Mts. transformed into a horst morphostructure with an asymmetrically shifted centre of the tectonic uplift towards the eastern rim. The specific, non-karstic Kunešovská planina Plateau has been preserved in the north-western part of the range (Fig. 8.14), whereas along the south-eastern edge of the range the radial net of valleys suggests preservation of a tiny remnant of the stratovolcano's original periphery (Lacika 1997).

8.4.2.2 Vtáčnik Mts.

The volcanic complex of Vtáčnik is interesting because there are no autochtonous structures of older generations of volcanic activities from the Lower and Middle Badenian stages. Apart from the coal-bearing formations building the neighbouring Hornonitrianska kotlina Basin, the peripheral parts of stratovolcanoes in the Kremnické and Štiavnické vrchy Mts. reach here. Volcanic activity in the space of what is now the eastern part of the Vtáčnik mountain range started only in the Late Badenian, associated with the development of the Kremnica graben. During the Sarmatian stage, three smaller volcanoes arranged in a line, which represent the initial volcanic relief of the Vtáčnik, became active along the western edge of the graben.

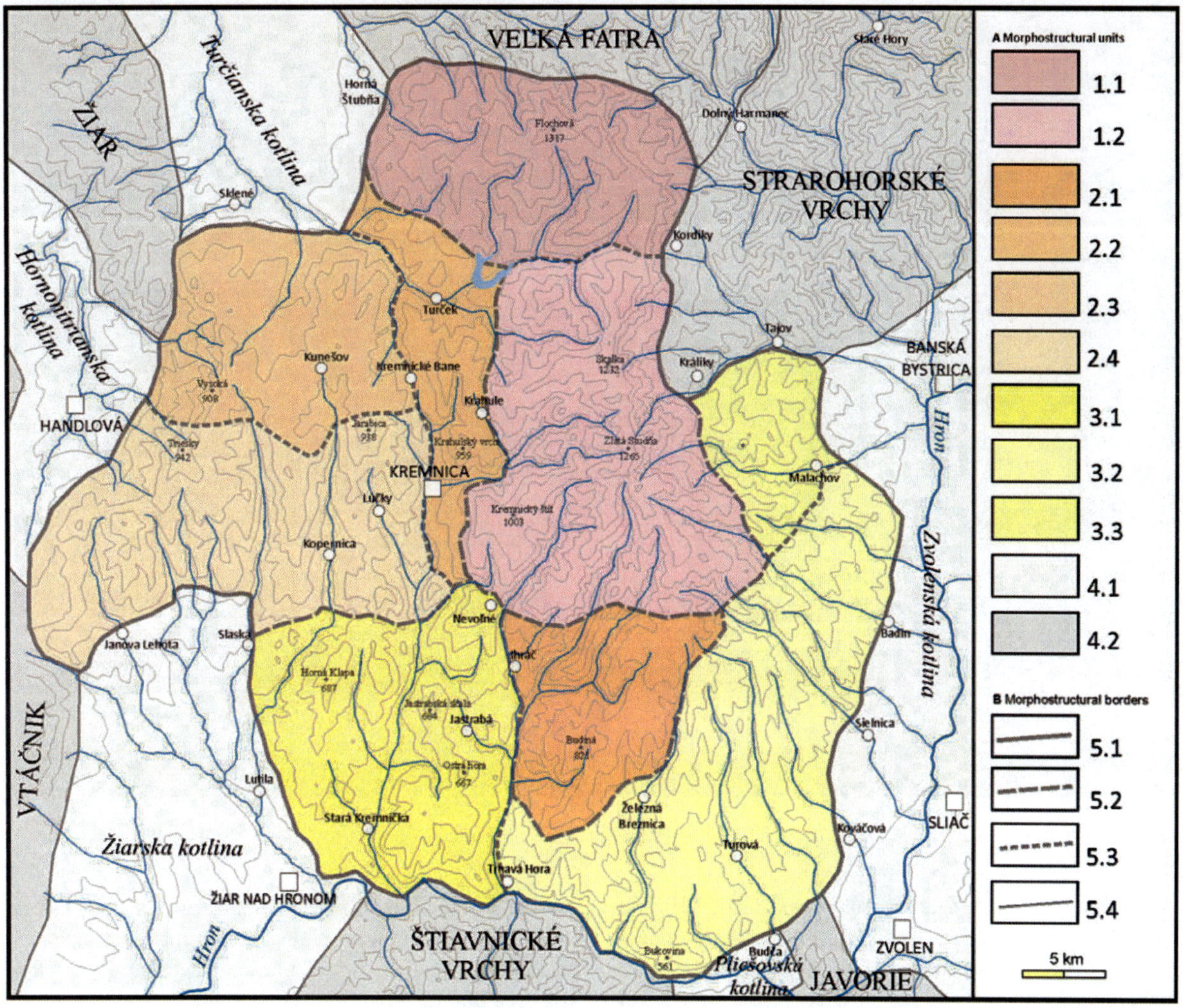

Fig. 8.13 Morphostructural scheme of the Kremnické vrchy Mts. **A Morphostructural unit: 1** Tall partial morphostructures: **1.1** Massive, internally less differentiated, the tallest partial morphostructure of the Flochová, **1.2** Cut, internally more differentiated, very tall partial morphostructure of the Skalka, **2** Moderately tall partial morphostructures: **2.1** cut, internally more differentiated and moderately tall partial morphostructure of Budiná, **2.2** foothill, internally less differentiated and moderately tall partial morphostructures of Krahule, **2.3** plateau-like internally least differentiated and moderately tall partial, morphostructure of Kunešov, **2.4** cut, internally less differentiated and moderately tall morphostructure of Kopernica, **3** Low partial morphostructures: **3.1** foothill internally more differentiated and low partial morphostructures of Jastrabá with parallel allochthonous valley network, **3.2** foothill, internally more differentiated low partial morphostructure of Turová with radially eccentric valley, **3.3** foothill, internally more differentiated low partial morphostructure of Malachov, with treed valley network. **4** Neighbouring morphostructures: **4.1** basin morphostructures: **4.2** mountain morphostructures. **B Morphostructural borders: 5.1** Outer border of morphostructure: **5.2** Inner morphostructural border of the 1st order, **5.3** Inner morphostructural border of the second order, **5.4** Borders of neighbouring morphostructures

The subsequent morphostructural development of the Vtáčnik, regulated by fault tectonics in an unstable morphoclimatic environment, led to a complete destruction of the original volcanic relief forms. Volcanic structure in the present horst appears only as a passive agent. Cascaded relief, crag forms, furrows and small basins are observable in a structurally and lithologically varied environment which supports selective erosion and denudation. An example is the craggy relief of the main ridge of the range (Kláštorská skala, Vtáčnik and Biela skala), linked to the erosion-denudation of lava flow cinders (Fig. 8.15). The horst morphostructure of the Vtáčnik is clearly asymmetric; the main ridge with the tallest mountains is situated in the west and constitutes an area referred to as Vysoký Vtáčnik unit.

This asymmetry in the southern part of the ridge is expressed not only by a greater height but also by the

Fig. 8.14 The Kunešovská planina Plateau in the Kremnické vrchy Mts. (*Photo* J. Lacika)

presence of shorter and steeper valleys and flanks descending to the Hornonitrianska kotlina Basin. The main ridge in the north of Vysoký Vtáčnik unit is shifted to the east, nearer to the Handlovská kotlina Basin. Morphostructural development of the eastern part of the range was greatly influenced by the subsiding Žiarska kotlina Basin, against which the Vtáčnik mountain range is clearly defined by a fault-generated escarpment. The level of the tallest peaks in the Nízky Vtáčnik unit is lower by 200 to 300 m; the territory is dissected by a network of longer valleys and ridges. The dominating landscape elements are the valleys of Kľacká dolina and Prochotská dolina, which widen into small basins in their upper parts.

8.5 Water Gaps of the River Hron, Žiarska kotlina Basin and Pohronský Inovec Mts.

The upper reach of the river Hron, from its spring to the city of Banská Bystrica, is a relic of a very old river network, possibly from before the Neogene stage of the development of the Western Carpathians. The distinct bend of the stream next to Banská Bystrica is a response to the radical change of development when the present morphostructural composition of the south-western wing of the West Carpathian mega-arch was formed. In the Lower Badenian stage, the Paleo-Hron River ran into the volcanoes of Kremnica, which forced the river to seek its passage to the West Panonnian Basin. Inside the system of volcanoes, which later transformed into horst morphostructures, the Hron created a chain of valley gaps connecting basin depressions of graben type. In the middle reach, the Hron mostly flows through the gaps. This occurs in the Žiarska kotlina Basin, where the landscape around the river becomes most open. The basin develops in harmony with the long-term subsidence trend of the volcanotectonic belt in the western part of the Slovenské stredohorie.

According to Konečný et al. (1998), the depression has been subsiding since the Sarmatian, when it was formed between the Neogene volcanoes. Distinct fault delimitation and the great thickness of the sedimentary filling point to tectonic origin (Škvarček 1990). Holec et al. (2015) and Pulišová and Hók (2015) detailed the morphostructural development of the Žiarska kotlina Basin during the youngest stage, considering the effects of fault tectonics on the spatial and elevation differentiation of the Pliocene River Plateau ("roveň" in Slovak) and the Pleistocene terraces. The

Fig. 8.15 Craggy relief on the relics of lava flows (Kláštorná skala, Vtáčnik). (*Photo* J. Lacika)

conspicuous dividing line is the section of the Hron Valley between the Žiarska kotlina Basin and the place where the river leaves the Carpathians and enters the Podunajská nížina Lowland.

A comparatively long gap, which widens into a small basin next to the town of Nová Baňa, originated here. The visual expression of the Novobanská kotlina Basin has been attenuated by the filling consisting of eruption products of the Quaternary volcano of Putikov vŕšok. Southward of the basin the Hron River created a gap following the fault system, which separated its south-western periphery from the Štiavnica stratovolcano (Lacika 2005). It led to the individualization of a separate structure of the Pohronský Inovec Mts. The marginal mountain range of the Slovenské stredohorie with allochthonous volcanic structure is a massive horst with steep peripheral slopes and a flat central part. The presence of the non-karstic Lehotská planina Plateau is the result of the lower degree of internal tectonic disintegration of this morphostructure.

8.6 Morphostructures of the Eastern Part of the Slovenské stredohorie

Two moderately large volcanic complexes of stratovolcano type were formed in the subsiding space of the volcanotectonic belt in time interval from the Early Badenian to the Sarmatian. A more northerly situated stratovolcano of Poľana became active in the transitional morphostructural environment on the dividing line between horst and graben, while the southerly situated stratovolcano of Javorie transformed into horst morphostructure in the north and into transitional morphostructure with the plateau type of relief in the south. Products of an independent volcanic belt, which was active during the Badenian in the Šahansko-lysecký fault belt, complement the volcanic complex forming the geological structure at the edge of the Krupinská planina Plateau and Ostrôžky.

8.6.1 Poľana Mts.

Poľana Mts. seems to be a unique Neogene stratovolcano, which preserved the physiognomy of a volcano. Identification of some features of volcanic paleorelief from the Badenian and Sarmatian stages is surprising. In spite of intensive post-volcanic transformation manifested by denudation and tectonic deformations, the presence of central volcanic depression and outer slopes with radially eccentric valley texture can be still recognized (Figs. 8.16 and 8.17). In the central part of the Poľana massif, with an almost circular ground plan, is a depression interpreted as a collapsed caldera transformed by erosion and tectonic forces. Two different parts are identifiable on the ground plant of the mountain rim. The oval form of the ground plan of the southern part suggests that the caldera was formed by expansion of the central depression through erosion focussed on the fault system. The fault determining the course of the rim from Bukovinka proceeds in the western direction and continues along the valley of Hačava Brook. Lacika (2002) asserted that this gap valley is the locality of the river piracy. The Hačava is the only stream on the outer side of the Poľana Mts. which penetrated inside the massif and tapped the radially concentric net of streams of until then isolated central volcanic depression. The mentioned fault facilitated the process.

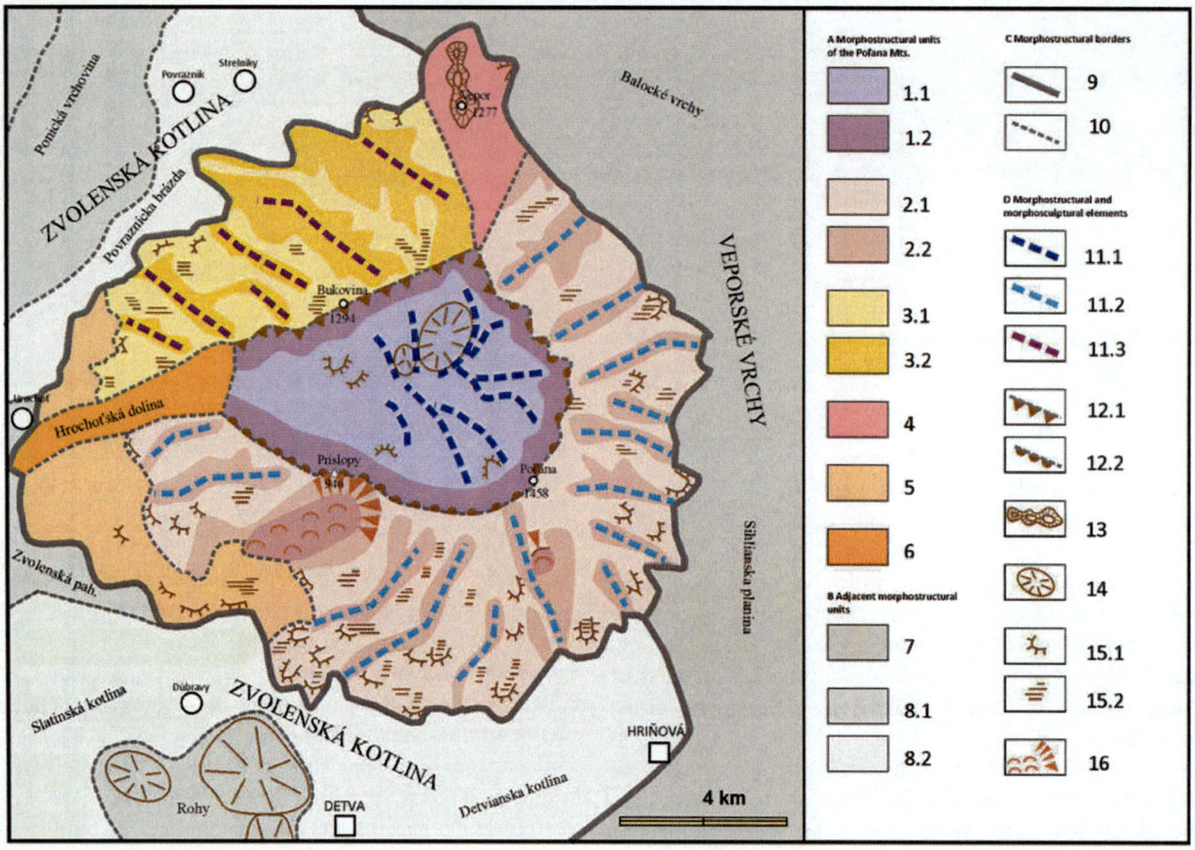

Fig. 8.16 Morphostructural scheme of the Poľana Mts. A. Morphostructural units of the Poľana Mts.: 1—Erosion-tectonic caldera: 1.1 bottom of caldera with radially concentric valley network, 1.2 circumferential slope of caldera ; 2—Outer slopes of stratovolcano with radially eccentric network of ridges and valleys: 2.1 ridges, 2.2 valleys; 3—Outer slopes of stratovolcano with parallel network of ridges and valleys: 3.1 ridges, 3.2 valleys; 4—Morphostructural elevation with erosion cinder of lava flow; 5—Edge of the circumferential slope of stratovolcano reduced by pediplanation ; 6—Gap-formed valley; B. Morphostructural units of neighbouring morphostructures: 7—assive horst morphostructure of the Veporské vrchy Mts. ; 8—Graben morphostructure of the Zvolenská kotlina Basin: 8.1 partial elevations, 8.2 partial depressions; C Morphostructural borders: 9—Outer borders of morphostructures of the first order, 10—Inner borders of morphostructures of the first order; D Morphostructural and morphosculptural elements: 11—Valleys: 11.1 axes of valleys of the radially concentric network, 11.2 axes of valleys of radially eccentric network, 11. 3 axes of valleys of parallel network; 12—slopes of caldera rim: 12.1 erosion slope, 12.2 erosion-fault slope ; 13—erosion cinders of lava flow ; 14—convex elevation on a conserved volcanic body ; 15—structural grade: 15.1 structural slope, 15.2 structural plateau; 16—gravitational deformation

Fig. 8.17 Eroded stratovolcano of Poľana rising above the Zvolenská kotlina Basin (viewed from the south) (*Photo* J. Lacika)

There are two types of relief textures on the outward slopes of the Poľana massif. Radial eccentric texture indicating a better preserved volcanic relief is in the south-west to north quadrant. This texture is impaired and transferred into parallel valley network in the north-western quadrant. Detailed geological research points that probably scarcely a third of the original structure of the Neogene Volcano of Mt. Poľana survived. The question is, therefore, how the traits of the initial volcanic form could have survived on such a considerably reduced initial form? Results of a detailed morphostructural analysis (Lacika 1993, Lacika and Šimon 2004) suggest that the answer should be sought in the specific morphostructural position of Mt. Poľana which is not too exposed in terms of erosion-denudation effects. It represents the case of a volcanic-tectonic morphostructure of transitional type, which was formed on the border of two morphostructures following distinctly different development pathways, i.e. between the Zvolenská kotlina Basin, which is a comparatively complicated morphostructural unit of graben type, with tendency to relative subsidence, and the Veporské vrchy Mts., a massive horst morphostructural unit subject to uplift. The transitory morphostructural position of Mt. Poľana is well visible on the map of isobasites (Lacika 1993). The stratovolcano of Mt. Poľana almost evanesces in the network of isolines derived from contour lines and axes of the third and higher orders. However, their arrangement indicates the presence of a morphostructure hidden under the volcanic complex. Isobasites so to say "denude" the buried edge of the horst morphostructure of the Veporské vrchy Mts. with abruptly dropping slope into the Zvolenská kotlina Basin (Lacika 1993). The eastern part of the volcanic complex of Poľana set down on the high uplifted horst blocks, the central and western parts rest on the tectonically deformed cascade descending to the graben. The mountain ridge of on the northern side of the ranges bearing the erosion cinders of lava flow Vepor-Hrb keeps its elevation with distance from the centre because it has not been deformed by the edge faults. It rests in its entirety on the higher horst blocks. Geomorphological research identified the presence of active faults cutting the comparatively compact mountain range. Tectonic predisposition is probable for the distinct lowering of the caldera rim in the area of the Príslopy saddle and the big circular slides identified by Lacika (1993) at the ends of the Suchohradná and Bystrá dolina Valleys.

8.6.2 The Zvolenská kotlina Basin

The Zvolenská kotlina Basin is a comparatively heterogeneous morphostructural unit. The northern part consisting of the Bystrická vrchovina and Ponická vrchovina Uplands is formed as a special second-order morphostructure of horst type; its nature is not that of a basin and no products of Neogene volcanism exist there. The southern part of the basin is internally differentiated by a second-order morphostructure of graben type. The basins of Sliačska, Slatinská and Detvianska kotlina are the most subsided units with flat relief developed upon the sediments testifying to the Pliocene filling phase. Depressions are separated from each other by relatively less subsided units, with higher and more dissected relief and volcanic structure. Volcanic materials of the lower stratovolcanic structure of Mt. Poľana reach to the territory of the Sliačska pahorkatina hilly land. A group of isolated structural elevations linked with the intrusive bodies of the volcanic depression of Vígľaš that subsided in the Badenian between the stratovolcanoes of Poľana and Javorie is inserted between the Sliačska kotlina and Detvianska kotlina Basins.

8.6.3 The Javorie Mts.

Like the stratovolcano of Štiavnica, the volcanic complex of Javorie stretches beyond the eponymous mountain range. The bottom stratovolcanic structure of Javorie (Badenian age) has been covered by the filling of the volcanic-tectonic graben and a younger stratovolcano of Sarmatian age, which is the initial volcanic relief for the post-volcanic relief transformation. Transformation of the volcanic to a fault-controlled morphostructure still goes on. Tiny segments indicating the presence of the original volcanic relief in a much altered form have been observed. The northern part of the complex was affected by tectonic deformation, which destroyed the central zone of stratovolcano. The central zone shifted as far as the northern edge of the range was broken to such an extent that a part of it was incorporated into the subsiding Zvolenská kotlina Basin and a part of it became a part of a second-order morphostructure of the horst type. In spite of this immense destruction, it is possible to identify a small part of the mountain rim of the central kettle-like depression (caldera?). The transitory and peripheral zone on the southern side of the Javorie stratovolcano was considerably reduced by denudation in the post-volcanic stage of development.

8.6.4 The Krupinská planina and Ostrôžky Plateaus

Two independents transitional morphostructures with a plateau-like relief were individualized from the remnants of the Javorie complex and rocks of the denuded Badenian volcanoes in the Šahy-Lysec volcanic zone. In terms of geomorphology, the Krupinská planina and Ostrôžky Plateaus are considered as two individual units. Nevertheless, from the point of geology they are genetically consistent. Originally, there was a single flat relief of the foreland of Javorie Mts. interpreted by Lukniš (1972) as a foothill volcanic plateau because the layers of tuffites and breccias in the stratovolcano base are almost horizontal. Tectonic uplift of the foreland of the stratovolcano caused the division of the primary volcanic plateau into two parts by a deepening valley system. It gave origin to a non-karstic plateau with two basic subtypes of relief, that is, valleys with a flat bottom and steep circumferential slopes, plateaus and plateau ridges (Fig. 8.18).

Segments of radially eccentric network typical for the volcanic relief on the outer slopes of stratovolcanoes are observable in part of the valley network. The elevation difference between the Krupinská planina Plateau and the

Fig. 8.18 Plateau landscape in the south of the Krupinská planina Plateau (*Photo* J. Lacika)

Ipeľská kotlina Basin is minor and the relief of a more westerly part of the foreland to Javorie Mts. is lower and less dissected. The southern peripheral part of the plateau is slightly higher than the inner part. Supposedly, the edges of the morphostructure were tectonically more uplifted along the reactivated faults of the Šahy-Lysec volcanic zone. Rearranged by faults were probably also the segments of the rectangular valley texture in the basin of the Litava River, with deeply incised meanders. The morphologically distinct border between the Krupinská planina Plateau and the Ipeľská kotlina Basin was indirectly affected by fault tectonics. The increasing amplitude of uplift of the neighbouring blocks led to the retrieval of peripheral slopes of the plateau by pediplanation and progressive enlargement of the Ipeľská kotlina Basin at the cost of the plateau. The Ostrôžky Plateau in the south-eastern edge of the Slovenské stredohorie is a taller, more dissected and tectonically more differentiated morphostructural modification of the Krupinská planina Plateau.

8.7 Conclusion

The Central Slovakian Volcanic Field that coincides with what is now the geomorphological region of Slovenské stredohorie is a product of a multi-phased volcanism in several volcanic centres across a long period between 15,5 Ma and 102,000 years. The dominating feature of the complicated complex of the CSVF was the Štiavnica stratovolcano, spreading over an area of more 2,200 km^2. Within the development of the stratovolcano a huge 18 km × 22 km caldera and a number of satellite volcanoes were created with traits of rhyolite volcanism in the caldera and on the stratovolcano slope. Alkaline basaltic volcanism was the last stage of volcanic activity in the territory. Lavas of Ostrá Lúka and Devičie, necks of Banská Štiavnica and the volcano of Putikov vŕšok of the Quaternary age are relics of this youngest volcanic phase. The monogenetic basanite volcano of Putikov vŕšok of the middle to late Pleistocene age is the youngest volcano in the West Carpathians and Central Europe.

When the volcanic activity ended, initial volcanic landscape distinctly changed and a completely different type of natural landscape was shaped. A new type of natural landscape with the mosaics of mountain ranges and basins of the Slovenské stredohorie originated in the changing climate and under the influence of faulting and selective erosion-denudation processes. Morphostructural research established that the course of transformation of the whole studied region was highly variable and the result is the presence of different types of volcanotectonic morphostructures with preserved segments of the initial volcanic relief (Fig. 8.19). Examples of the transformation process include the denudation remnant of the Neogene Volcano of Mt. Poľana with a central depression and radially eccentric texture of valleys, also present in the morphostructure of the Krupinska planina of a plateau type and in lesser extent in the volcanic-tectonic horst of the Kremnické vrchy mountains. Transformation of the Neogene volcanic landscape in the Štiavnické vrchy, Pohronský Inovec, Ostrôžky a Vtáčnik Mts. reached the stage when the volcanic relief was completely destroyed. The youngest volcano in Slovakia, Putikov vŕšok, is an exception because its volcanic relief has been preserved due to its young age.

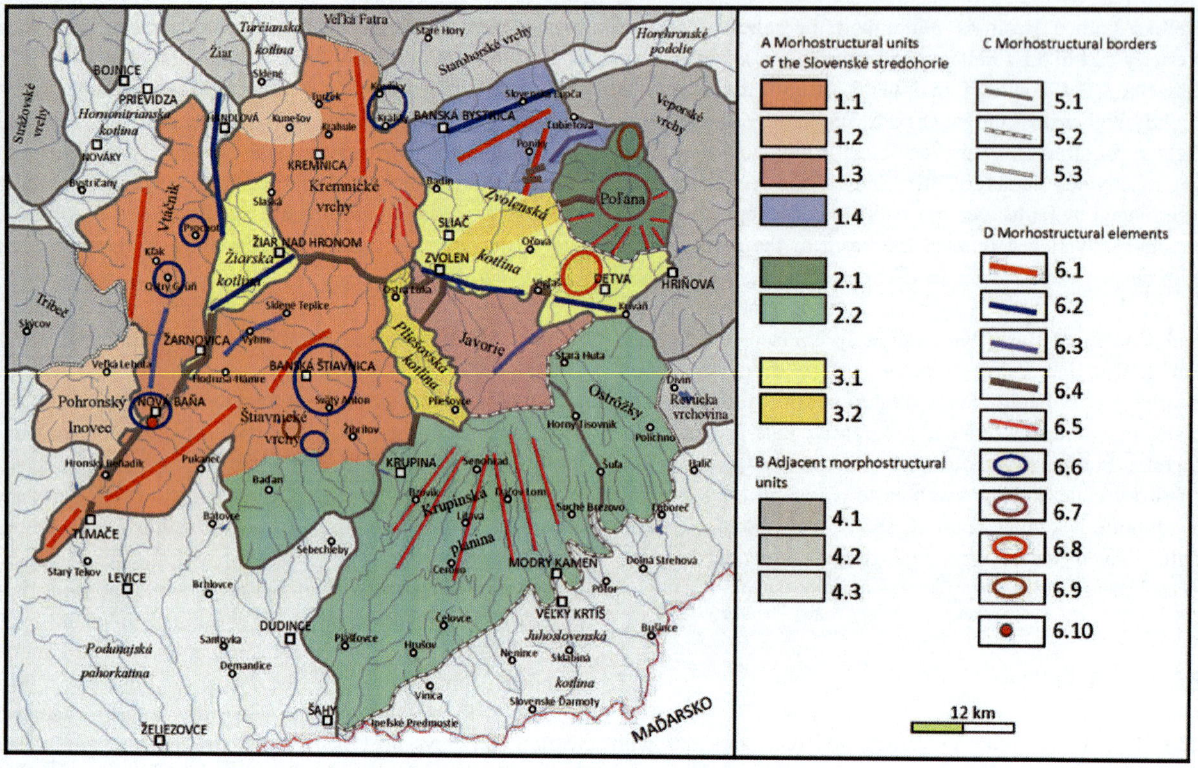

Fig. 8.19 Morphostructural scheme of the Slovenské stredohorie. A Morphostructural units of the Slovenské stredohorie: 1 Horsts morphostructures: 1.1 horst on volcanic structure with total transformed inicial volcanic landform, 1.2 horst on volcanic structure with plateau typee of landform, 1.3 horst on volcanic structure with very strongly transformed inicial volcanic landform, 1.4 horst on not volcanic structure; 2 transitive morphostructures: 2.1 transitive morphostructure with partially transformed initial volcanic landform, 2.2 transitive morphostructure with plateau type of landform; 3 grabens morphostructures: 3.1 graben with the Neogene sediment basement, 3.2 graben with volcanic basement; B Neighbouring morphostructural units: 4.1 high horst morphostructures, 4.2 low horst morphostructures, 4.3 subsided morphostructures (basins and lowlands); C Morphostructural borders: 5.1 border on active tectonic lines, 5.2 geomorphological border with minimal influence of endogenous processes, 5.3 border between passive geological structures; D Morphostructural element: 6.1 axis of horst, 6.2 axis od graben, 6.3 axis of longitudinal structural depression, 6.4 valley gap, 6.5 axis of radial eccentric valley network, 6.6 structural depression, 6.7 erosional caldera, 6.8 structural elevation, 6.9 erosion-denudational elevation (table mountain), 6.10 the Pleistocene volcano with preserved volcanic-type landform

References

Dublan L (ed), Bezák V, Biely A, Bujnovský A, Halouzka R, Hraško Ľ, Köhlerová M, Marcin D, Onačila D, Scherer S, Vozárová A, Vozár J, Žáková E (1997) Vysvetlivky ku geologickej mape Poľany 1: 50 000. ŠGÚDŠ, Bratislava, 238 s. ISBN 80-85314-76-2

Holec J, Medveďová A, Vitovič L, Prokešová R (2015) Neotektonický vývoj Žiarskej kotliny indikovaný geomorfologickou analýzou v prostredí GIS. Geografický Časopis 67(2):181–195

Konečný V, Lexa J, Šefara J (1978) Vzťah vulkanizmu k morfotektonickým štruktúram predvulkanického podložia. Manuskript. archív ŠGÚDŠ, Bratislava

Konečný V, Lexa J, Halouzka R, Dublan L, Šimon L, Stolár M, Polák M, Vozár J, Havrila M, Pristaš J (1998a) Geologická mapa Štavnických vrchov a Pohronského Inovca (Štiavnický stratovulkán). archív ŠGÚDŠ, Bratislava

Konečný V (ed), Lexa J, Halouzka R, Hók J, Vozár J, Dublan L, Nagy A, Šimon L, Havrila M, Ivanička J, Hojstričová V, Miháliková A, Vozárová A, Konečný P, Kováčiková M, Fiľo M, Marcin M, Klukanová A, Liščák P, Žáková E (1998b) Vysvetlivky ku geologickej mape Štavnických vrchov a Pohronského Inovca (Štiavnický stratovulkán I. a II. diel) 1:50 000. ŠGÚDŠ, Bratislava, 473 pp

Konečný V, Lexa J (2001) Stavba a vývoj štiavnického stratovulkánu. Mineralia Slovaca 33:179–196

Konečný V, Lexa J, Šimon L, Dublan L (2001) Neogénny vulkanizmus stredného Slovenska. In: Šimon L, Konečný V, Lexa J (eds) Banská Štiavnica – Mesto na vulkáne. Miner. slov. (Bratislava) 33(3):159–178

Konečný V, Lexa J, Šimon L (2003) Geologic structure and evolution of intravolcanic depressions in the area of Neogene volcanism in Central Slovakia. Mineralia Slovaca 35(3–4):255–290

Konečný V, Lexa J (1979) Štruktúrno-geologická schéma stredoslovenských neovulkanitov v mierke 1: 100 000. Manuskript. archív ŠGÚDŠ, Bratislava

Konečný V, Lexa J (1984) Geologická mapa stredoslovenských neovulkanitov v mierke 1: 100 000. GÚDŠ, Bratislava

Konečný V, Lexa J, Planderová E (1983) Stratigrafické členenie neovulkanitov stredného Slovenska. GÚDŠ, Bratislava, 203 pp

Lacika J (1993) Morfoštruktúrna analýza Poľany. Geografický časopis 45(2–3):233–250

Lacika J (1997) Morfoštruktúry Kremnických vrchov. Geografický časopis 49(1):19–33
Lacika J (2002) Typy riečneho pirátstva vo vulkanických pohoriach slovenských Karpát. Geografický Časopis 54(2):151–164
Lacika J, Šimon L (2004) Postvulkanická transformácia stratovulkánu Poľana (Stredné Slovensko). Geomorphologia Slovaca 4(2):10–17
Lacika J (2005) Transformácia geomorfologických sietí na štiavnickom stratovulkáne (stredné Slovensko). In: Rypl J (ed) Geomorfologický sborník 4. Stav geomorfologických výzkumů v roce 2005: příspěvky z mezinárodního semináře GEOMORFOLOGIE ′05. České Budějovice, Jihočeská universita, 71–74
Lacika J (2006) Postvulkanický vývoj Slovenského stredohoria. In: Smolová J (ed) Geomorfologické výzkumy v roce 2006. Palackého univerzita, Olomouc, pp 129–135
Lexa J, Havrila M, Halouzka R, Hanzel V, Kubeš P, Liščák P, Hojstričová V (1998) Vysvetlivky ku geologickej mape Kremnických vrchov 1: 50 000. ŠGUDŠ, Bratislava, 308 pp
Lukniš M (ed) (1972) Reliéf. In: Slovensko 2 – Príroda. Obzor, Bratislava, pp 124–202
Mazúr E, Lukniš M (1978) Regionálne geomorfologické členenie Slovenskej socialistickej republiky. Geografický Časopis 30 (2):101–125
Pulišová Z, Hók J (2015) Paleonapäťová analýza zlomového porušenia Žiarskej kotliny. Acta Geologica Slovaca 7(2):103–111
Šimon L (1991) Faciálny model formácie Kľakovskej doliny južne od Handlovej. Manuskript. archív ŠGÚDŠt, Bratislava
Šimon L, Lexa J, Halouzka R, Macinská M, Jánová V, Vranovská A, Stolár M, Ďurkovičová J, Vozár J, Novosad P, Sládková M, Wiegerová V (1991) Vysvetlivky ku geologickej mape 1: 25 000 36–311 Janová Lehota. Manuskript. Bratislava, archív ŠGÚDŠ
Šimon L, Konečný V, Halouzka R, Dublan L, Lexa J, Macinská M, Stolár M, Marcin D, Jánová V, Fiľo M, Ardová M (1995) Vysvetlivky ku geologickým mapám 1: 25 000 36–313 (Hliník nad Hronom), 36–331 (Žarnovica). Manuskript. archív ŠGÚDŠ, Bratislava
Šimon L, Konečný V, Dublan L, Lexa J, Kohlerová M, Halouzka R, Pristaš J, Vozár J, Vozárová A (1996) Vysvetlivky ku geologickým mapám 1: 25 000 35–424 (Veľké Pole), 35–442 (Nová Baňa). Manuskript. archív ŠGÚDŠ, Bratislava
Šimon L, Halouzka R (1996) Putikov vŕšok volcano – the youngest volcano in the Western Carpathians. Geol Mag 2:103–123
Šimon L, Elečko M, Lexa J, Kohút M, Halouzka R, Konečný V, Gross P, Pristaš J, Konečný V, Mello J, Polák M, Vozárová A, Vozár J, Havrila M, Kohlerová M, Stolár M, Jánová V, Marcin D, Szalaiová V (1997) Vysvetlivky ku geologickej mape Vtáčnika a Hornonitrianskej kotliny 1: 50 000. ŠGÚDŠ, Bratislav, 282pp
Šimon L, Lexa J, Konečný V (2002a) Pannonian basalt volcano Šibeničný vrch, central Slovakia. In: Michalík J, Šimon L, Vozár J (eds) Proceedings of the XVIIth congress of CBGA, Bratislava, Geologica Carpathica 53, special issue
Šimon L, Pauditš P, Kráľ J (2002b) New data on the Putikov vršok alkali basalt volcano, Central Slovakia. In: Michalík J, Šimon L, Vozár J (eds) Proceedings of the XVIIth congress of CBGA, Bratislava, Geologica Carpathica 53, special issue
Šimon L, Maglay J (2005) Datovanie sedimentov podložia lávového prúdu vulkánu Putikov vŕšok metódou opticky stimulovanej luminiscencie. Mineralia Slovaca 37:279–281
Šimon L, Konečná K, Kováčiková M, Kollárová V, Pauditš P, Maglay J (2019a) Model vývoja monogenetického vulkánu Putikov vŕšok pleistocénneho veku na strednom Slovensku. In: Sborník Otevreného geologického kongresu ČGS a SGS 2019. Praha. Vyd. Česká geologická společnosť
Šimon L, Kollárová V, Kováčiková M (2010) Geologické profilovanie a paleovulkanické rekonštrukcie sarmatsko-bádenskej stavby stratovulkánu Poľana-východ. Manuskript. archív ŠGÚDŠ, Bratislava
Šimon L, Kollárová V, Kováčiková M, Šimonová B (2012) Geologické profilovanie a paleovulkanické rekonštrukcie sarmatsko-bádenskej stavby stratovulkánu Poľana-sever. Manuskript. archív ŠGÚDŠ, Bratislava
Šimon L, Kollárová V, Kováčiková M, Šimonová B (2013) Geologické profilovanie a paleovulkanické rekonštrukcie vulkanickej stavby stratovulkánu Polana-stred. Manuskript. archív ŠGÚDŠ, Bratislava
Šimon L, Kollárová V, Kováčiková M (2017) Paleovulkanická rekonštrukcia vulkanickej stavby v závere Hrochotskej doliny pohoria Poľana. Manuskript. archív ŠGÚDŠ, Bratislava
Šimon L, Kollárová V, Kováčiková M (2019b) Paleovulkanická rekonštrukcia vulkanickej stavby v juhozápadnej časti pohoria Poľana. Manuskript. archív ŠGÚDŠ, Bratislava
Škvarček A (1990) Žiarska kotlina. In: Zaťko M (ed) Analýza vybraných geoekologických komponentov Žiarskej kotliny a okolitých pohorí. Bratislava (Slovenský geologický úrad Bratislava, IGHP Žilina š.p., závod Bratislava, Katedra fyzickej geografie a kartografie PF UK Bratislava)

Ladislav Šimon is a geologist and volcanologist at the State Geological Institute of Dionýz Štúr in Bratislava. His research is focused on geological mapping and paleovolcanic reconstructions of the central Slovak volcanic field of the Tertiary and Quaternary ages. He described and defined the youngest volcano in the Western Carpathians Quaternary age.

Ján Lacika is an emeritus Associate Professor of Physical Geography. He was researcher at the Department of Physical Geography, Geomorphology and Natural Hazards of the Institute of Geography SAS in Bratislava (1980–2019) and teacher and research worker of the Department of Geography, Geoinformatics and Regional Development Faculty of Natural Sciences of the Constantine the Philosopher University in Nitra. Originally his main research interest was structural geomorphology (neotectonics, post-volcanic transformation of landforms and development of the valley networks). Later his research was focused on geographical aspects of the natural and cultural landscape development.

9 Polygenetic Relief in the Foreland of Glacially Sculptured Mountains—Podtatranská kotlina Basin

Ladislav Vitovič, Jozef Minár, Pavel Bella, and Juraj Littva

Abstract

Recent polygenetic morphology of one of the highest-situated basins within the Western Carpathians, in the foreland of the glacially sculptured Tatra Mts., results from the gradual development influenced by tectonic processes as well as climatic conditions since the Miocene. The Podtatranská kotlina Basin, together with the Tatra Mts., belongs to the most neotectonically active morphostructures within the Slovak part of the Western Carpathians, therefore neotectonic processes played a very important role in relief evolution. During the Neogene and partially the Early Pleistocene (?Praetiglian complex), the exhumation of the Tatra Mts. as well as the formation of several planation surfaces took place. Since the Middle Pleistocene (?Elsterian complex), climatic conditions acting in concert with the increasing uplift of the Tatra Mts. caused glaciation and consequently enhanced coarse sediments supply into the glacifluvial as well as fluvial systems. Domination of particular geomorphologic processes changed throughout the Neogene and Quaternary, resulting in various landforms and polygenetic character of the relief.

L. Vitovič (✉)
State Geological Institute of Dionýz Štúr, Bratislava, Slovakia
e-mail: ladislav.vitovic@geology.sk

L. Vitovič · J. Minár
Department of Physical Geography and Geoinformatics, Faculty of Natural Sciences, Comenius University in Bratislava, Bratislava, Slovakia
e-mail: jozef.minar@uniba.sk

P. Bella · J. Littva
Slovak Cave Administration, Liptovský Mikuláš, Slovakia
e-mail: pavel.bella@ku.sk; ; pavel.bella@ssj.sk

J. Littva
e-mail: juraj.littva@ssj.sk

P. Bella
Department of Geography, Faculty of Education, Catholic University in Ružomberok, Ružomberok, Slovakia

Keywords

The Western Carpathians • The Podtatranská kotlina Basin • Neotectonics • Glaciation • River incision • Polygenetic relief • Planation surfaces

9.1 Introduction

The polygenetic morphology of the Podtatranská kotlina Basin (PtK) is a result of complex and comprehensive evolution since the Miocene to the recent. The formation of landforms was an outcome of various geomorphological processes controlled by climate changes but predominantly by changes in tectonic regimes. Geomorphological development of the PtK results from several sedimentary-erosional cycles. The outlines of the PtK and its surroundings developed during the Miocene, while detailed morphological evolution of the PtK took part during the Pliocene and, mainly, during the Quaternary. The occurrence of at least three glaciation stages of the Tatra Mts. in the Middle Pleistocene is responsible for a unique polygenetic relief within the Slovak part of the Western Carpathians. This is caused by the formation of glacial deposits and coarse-grained glacifluvial and fluvial deposits. During the Holocene, acting in concert with ongoing tectonic movements and climate changes, human impact has played an important role in the geomorphological evolution of the PtK (Halouzka 1987; Gross et al. 1990; Halouzka and Raczkowski 1993).

With regard to the traditional old (1.8 Ma) and new world-wide used Pleistocene/Pliocene stratigraphy (Menning and Hendrich 2016; Cohen and Gibbard 2018), adaptation of the older local geochronology to the latter was carried out.

M. Lehotský and M. Boltižiar (eds.), *Landscapes and Landforms of Slovakia*, World Geomorphological Landscapes,
https://doi.org/10.1007/978-3-030-89293-7_9

9.2 Geographical and Geomorphological Settings

The PtK, a vast depressional geomorphic unit, is situated in the northern part of Slovakia. From a historical point of view, the territory consists of the Liptov and Spiš regions, separated by the Štrba dividing block (Vitovič 2020b). From a geomorphological point of view (sensu Mazúr and Lukniš (1978)), the PtK (Fig. 9.1) belongs to the Fatra-Tatra region and consists of three subunits, the Liptovská kotlina Basin (LK, Fig. 9.2), the Popradská kotlina Basin (PpK, Fig. 9.3), and the Tatranské podhorie Piedmont. With its W–E prolonged shape and considerable size (90 × 20–30 km), it is the largest intramountain basin of the Central Western Carpathians. The entire territory covers an area of ~1200 km^2. The bottom of the basin extends from 465 (the Váh River floodplain) to 1600 (the Tatra Mts. foothill) m a.s.l., ~750 m on average, therefore it belongs to the highest-situated basins of the Western Carpathians. The main axis of the PtK is represented by the Váh and the Poprad rivers. Catchments of these rivers belong to the sea-drainage areas of the Black Sea and the Baltic Sea. The dividing line passes in the PtK through the Štrba block. Annual precipitation reaches more than ~1000 mm at the foothill of the Tatra Mts., while in the central part of PpK it falls to ~550–600 mm due to rain shadow of the Tatra Mts. (Faško and Šťastný 2002). As a part of the Central Western Carpathians, it is bounded from the north by the Chočské vrchy Mts., the Tatra Mts., and the Spišská Magura Mts. From the west it is bounded by the Veľká Fatra Mts., while in the south the PtK borders with the Nízke Tatry Mts., the Kozie chrbty Mts., and partially with the Hornádska kotlina Basin. The Levočské vrchy Mts. have a border with the PtK in the east (Fig. 9.1).

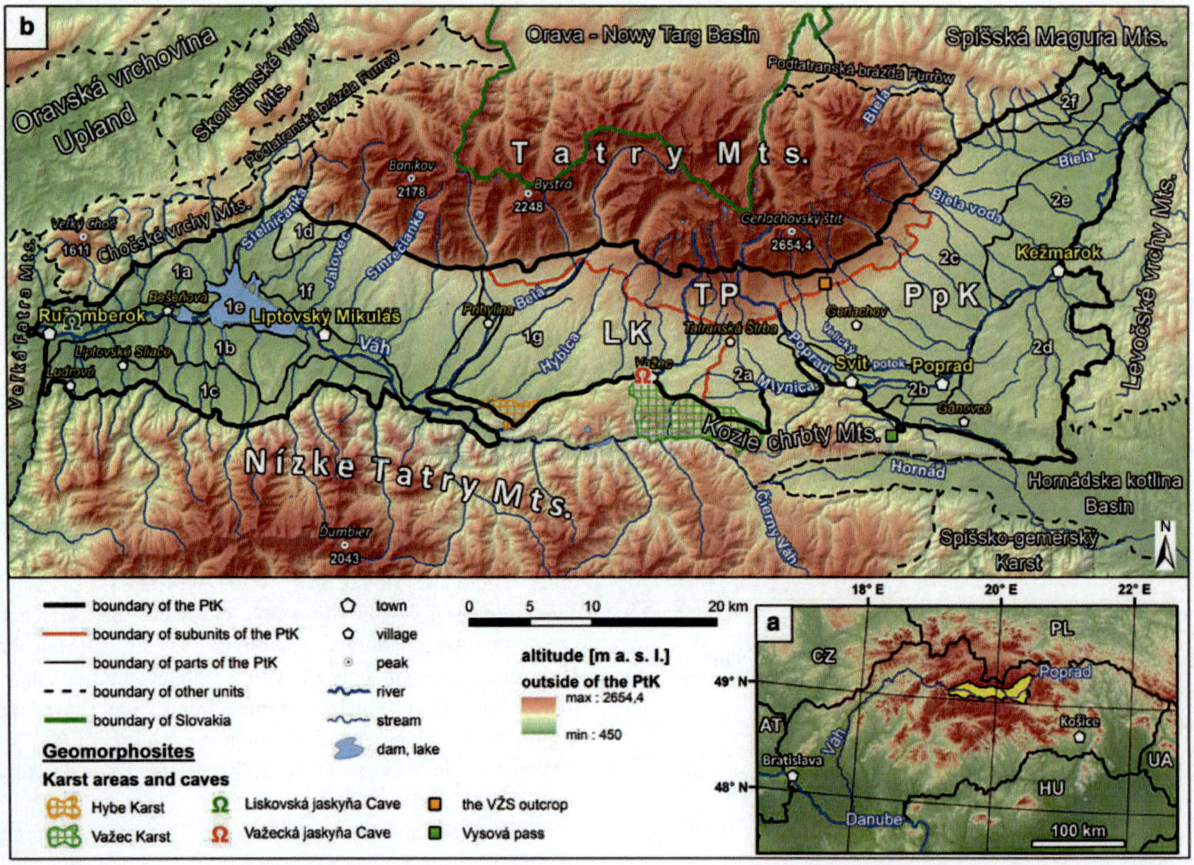

Fig. 9.1 **a** Location of the Podtatranská kotlina Basin within Slovakia. **b** Hypsometry, basic geomorphological division (after Mazúr and Lukniš 1978) and selected geomorphosites of the Podtatranská kotlina Basin and its surroundings. Geomorphologic subunits: LK—Liptovská kotlina (basin), PpK—Popradská kotlina (basin), TP—Tatranské podhorie (piedmont), Numbers refer to geomorphological parts: Liptovská kotlina Basin: 1a: Chočské podhorie Piedmont, 1b: Galovianske háje, 1c: Ľubeľská pahorkatina Hilly land, 1d: Matiašovské háje, 1e: Liptovské nivy, 1f: Smrečianska pahorkatina Hilly land, 1g: Hybianska pahorkatina Hilly land, Popradská kotlina Basin: 2a: Štrbská pahorkatina Hilly land, 2b: Popradská rovina Plain, 2c: Lomnická pahorkatina Hilly land, 2d: Vrbovská pahorkatina Hilly land, 2e: Kežmarská pahorkatina Hilly land, 2f: Vojnianske podhorie Piedmont

Fig. 9.2 The Liptovská kotlina Basin with the Nízke Tatry Mts. in the background and the Western Tatra Mts. on the left side, limited by the Sub-Tatra fault (broken line) (*Photo* L. Vitovič)

Fig. 9.3 Hilly land of the Popradská kotlina Basin, with the High Tatra Mts. (the Sub-Tatra fault marked by broken line) and the Spišská Magura Mts. in the background (*Photo* L. Vitovič)

The PtK constitutes a huge depressional landform, a graben, with hilly topography dominating the basin floor and low-angle slopes, relatively flat ridges, and wide open valleys as the main morphological elements. Along the Váh and Poprad rivers and their longest tributaries, alluvial floodplains occur. Neotectonic processes determined the formation of tectonic landforms and features, mainly along the basin borders, and include faceted slopes together with travertine accumulations and springs. Extensive areas are formed by glacifluvial alluvial fans as well as fluvial terraces and floodplains. Locally, proluvial cones formed too. In the foothills of the Tatra Mts., glacial relief is developed in the form of moraines. Planation surfaces at several altitude levels are developed in the piedmont areas of the PtK as well as in the adjacent parts of surrounding mountains. The prevailing slope inclination within the hilly land is about 3–7°. In the foreland of the Chočské vrchy and Západné Tatry Mts., relief dissection is up to 181 m and slope inclinations are frequently up to 10°. Along the foothills of the High Tatra Mts. (the highest part of the Tatra Mts.), more dissected upland terrain (up to 310 m) exists, with prevailing slope inclination up to 14° (Mazúr 1980).

9.3 Geological Settings and Tectonics

Despite its remarkable size, the PtK represents only a remnant of a vast basin that existed in the Palaeogene. Known as the Central Carpathian Paleogene Basin (CCPB), it originally extended throughout the significant portion of the current northwestern and central Slovakia (see Fig. 3 in Soták et al. 2001; Fig. 1 in Kováč et al. 2016). In the north, the CCPB is bordered by the Pieniny Klippen Belt, which separates the CCPB from the units of the Flysch Belt. Its southern border is difficult to determine, as significant parts of the original basin were eroded (e.g. see Fig. 8 in Vojtko et al. 2015), or covered by younger Neogene sediments (e.g. Kováč et al. 2011) and volcanites (Gross 1978). Besides the PtK, the sedimentary remnants of CCPB are preserved in other areas. They occur chiefly in the basins of Orava, Podhále, Hornád, Turiec, Horná Nitra, and Žilina-Rajec, as well as in the Spišská Magura, Skorušiná, and Levoča Mts., Orava and Šariš Uplands, Pohronské podolie Valley, and in other geomorphological units. Palaeogene sediments are also present in the Malé Karpaty Mts., but their character differs from the rest of the CCPB sediments (Gross 2008). Some authors further distinguish several zones or sub-basins within the CCPB (see Fig. 3.4–1. in Kováčik et al. 2015; Figs. 18 and 19 in Soták 2010; Fig. 1 in Kováč et al. 2016).

The formation of CCPB began after the inversion of earlier Myjava-Hričov wedge-top basins, formed at the top of a consolidated Carpathian orogenic wedge. The ongoing subduction of the Magura Ocean beneath the overriding Inner Carpathian crustal block triggered the collapse of the orogenic wedge. The subsidence of the crust caused the development of a large forearc basin in the northern part of the Inner Carpathian domain. Two mechanisms were proposed by Kováč et al. (1994) as a possible cause of the collapse and subsidence. The first one was subcrustal erosion (an option also proposed by Kázmér et al. 2003), while the second one was subduction retreat of the oceanic lithosphere (later championed by Kováč et al. 2016), potentially enhanced by the extensional collapse of the overthickened continental crust (Plašienka and Soták 2015). Either way, the crustal stretching/thinning and the formation of the forearc basin led to the accumulation of a thick Palaeogene sedimentary succession.

The basin is filled by the sediments of Subtatran Group (after nomenclature of Gross et al. 1984; Gross 2008), which document the basin evolution and are divided into four principal formations (Borové, Huty, Zuberec, and Biely Potok Formation). However, as was noted by Starek and Fuksi (2017), it is important to keep in mind that this division presents a picture that is somewhat simplified, as some sediments were found to have different ages, despite their lithological resemblance. In addition, some of the lithologically identical sediments were deposited in different environments and/or conditions (Soták et al. 2001). Quaternary sediments of various genesis and thickness cover a vast portion of the PtK (Fig. 9.4), from which the glacifluvial and fluvial sediments belong to the most widespread (for more about Quaternary sedimentation see in the following sections).

From the neotectonic point of view, the PtK together with the Tatra Mts. are the most neotectonically active morphostructures of the Western Carpathians (Halouzka 1993). The PtK consists of two basic neotectonic blocks: the LK and the PpK, divided by the Štrba dividing block (Halouzka 1993; Vitovič 2020b). The limitations of the PtK in relation to the adjacent mountains are marked by faults active in the Quaternary (Maglay et al. 1999; Vitovič 2020b), as evidenced by linear faceted slopes, foothill depressions filled with Quaternary deposits, travertine deposits as well as springs of mineral water. The development of several fault systems of different order and age is recorded in the entire area, which results in the formation of partial basin blocks (Franko and Hanzel 1980; Gross 1980; Gross et al. 1990; Halouzka et al. 1999; Jetel 1999; Littva 2017; Vitovič 2018, 2020b).

In the western part of the PpK, three generations of faults were recognized, among which the oldest one is the W–E trending fault system. For instance, the Vikartovce fault, as well as faults bordering the Kozie chrbty Mts. from the north, belong to this category (Fig. 9.5). The Čierny Váh valley is formed on this fault system as well. The course of these faults is not continual, as they are broken (dislocated)

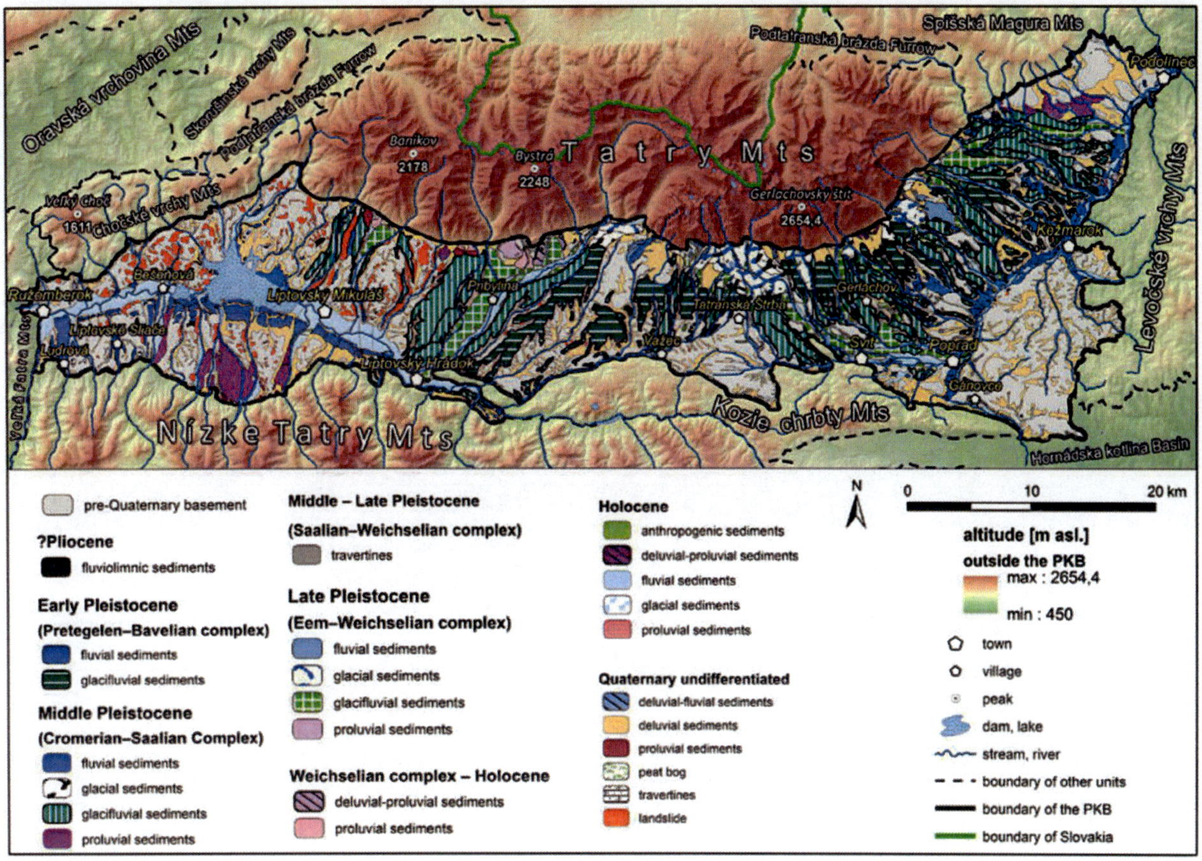

Fig. 9.4 Simplified geological map of the Podtatranská kotlina Basin. Compiled from Gross (1979), Biely et al. (1992), Nemčok et al. (1994), Polák et al. (1997), Gross et al. (1999), modified

by a younger fault system. The activity of the younger fault system with NW–SE trending faults is visible in the tectonic limitations of a partial block between the settlements of Tatranská Štrba and Kežmarok (Fig. 9.5). The youngest faults have N–S direction (Gross et al. 1990). The majority of boundaries of partial blocks of the PpK are limited by NW–SE trending faults (Gross et al. 1999).

The total uplift of the High Tatra Mts. along the Sub-Tatra fault during the Neogene is assumed to be 2500 m (Gross and Köhler 1980). According to Tokarski et al. (2012), the uplift of the whole Tatra Mts. during the Neogene–Quaternary period is assumed to be 4 km, with a considerable part of the total during the Quaternary. Tectonic activity along the Sub-Tatra fault continues until recent times (Gross et al. 1999; Pánek et al. 2020), as also indicated by springs of mineral waters (Jetel 1999).

Vertical dislocation along the Ružbachy fault (E continuation of the Sub-Tatra fault, Fig. 9.5) gradually reduces to several 100 m. Relatively intensive uplift of the PtK during the Neogene to Quaternary period compared to the adjacent morphostructures filled by CCPB sediments (e.g. Orava region) is indicated by the minimum thickness (~1.5 km) of the CCPB strata (Gross et al. 1980; Králiková et al. 2014).

9.4 Selected Geomorphosites—Keys to the Study of the Basin Evolution

Karst, glacifluvial, and fluvial landforms and sediments have the biggest potential to contribute towards knowing and understanding basin development. Till now, this potential has been insufficiently utilized. Presented geomorphosites represent the potential of the recent state of the art and offer an insight into the most valuable geomorphosites of the PtK.

9.4.1 Karst Landforms

Several small carbonate areas, mostly in the eastern and western parts of Liptovská kotlina Basin (Fig. 9.1), are

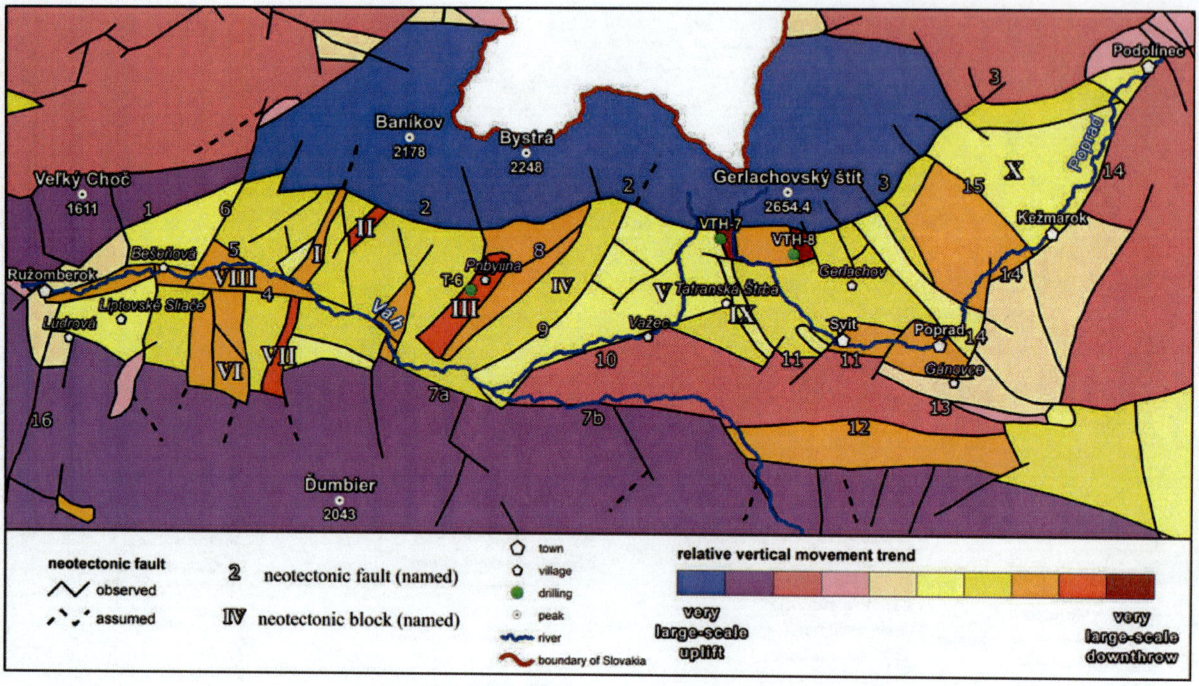

Fig. 9.5 Neotectonic map of the Podtatranská kotlina Basin and its surroundings with names of the main faults and blocks. After Halouzka (1993), Maglay et al. (1999), partially modified. Faults: 1: Choč, 2: Sub-Tatra, 3: Ružbachy, 4: Váh, 5: Trnovec, 6: Kvačany, 7a: Liptovská Porúbka, 7b: Čierny Váh, 8: Horná Belá, 9: Hybica, 10: Važec, 11: Svit, 12: Vikartovce, 13: Gánovce, 14: Poprad, 15: Slavkov, 16: Central Slovak Fault System. Blocks: I: Bobrovec, II: Žiar, III: Vavrišovo, IV: Liptovská Kokava, V: Východná, VI: Liptovský Kríž, VII: Demänová, VIII: Dechtáre, IX: Štrba, X: Spišská Belá

featured by some karst landforms, mainly caves sculpted by allochthonous rivers. Their origin and development are linked with the formation of pediments or river terraces. Within the basin karst, they are classified as the karst of foot plains and terraces (see the typology of karst areas in Slovakia after Jakál 1993).

The Važecká jaskyňa Cave (Fig. 9.6), at the contact of the LK and the Kozie chrbty Mts., represents a 530 m long subhorizontal branchwork cave, partially with a network rhombic pattern guided by several intersected faults. It is located on the left side of the valley of Biely Váh River. The cave is formed in the Middle Triassic Gutenstein Formation of the Hronic Unit, mainly by the inflow of aggressive allochthonous waters and flood-water injections from the Biely Váh River. Its formation was enhanced by waters seeping from precipitation and circulating through the karst aquifer to the valley of Biely Váh River. The cave is situated at the level of the river terrace developed just in front of the cave, 8 m above the recent river bed of the Biely Váh River (Droppa 1962; Bella et al. 2016). Allochthonous gravels and sands with granite pebbles are covered by unsorted sediments that besides prevailing loam contain mainly slightly rounded limestone clasts, as well as fragments of Palaeogene sandstone, weathered granite pebbles, and fossil faunal remains (Volko-Starohorský 1930; Havránek 1935; Droppa 1962). Above the unsorted sediments, fine-grained materials were deposited in several sequences as a result of repeated flood-water injections from the surface river and floods caused by waters circulating through adjacent karstified limestones.

Likewise, the origin of horizontal and subhorizontal segments of the *Liskovská jaskyňa Cave*, the largest cave in the PtK, was linked with an interrupted incision of the river bed and the formation of terraces of the Váh River (Fig. 9.7). This cave is located in the western part of the basin, east of the town of Ružomberok (Droppa 1971). The Liskovská jaskyňa developed inside the eastern edge of the hill Mních (695 m a.s.l.), which is a horst structure composed of Triassic carbonates (a part of the Choč Nappe), uplifted along ENE–WSW-trending faults, active since the Palaeogene (Fig. 9.7). The horst is fractured and bounded also by younger N–S-trending faults (Gross 1971, 1980). The origin of the Liskovská jaskyňa was mostly controlled by these faults, partly by the bedding-planes within the Middle Triassic Gutenstein Formation. It represents a 3D multi-level maze (cave labyrinth), with a length of 4250 m and a vertical span of 72 m. A very high degree of karstification in this area is reflected in a dense network of cave passages, halls, and chambers. More than 4 kms of cave spaces have developed above one another on the ground plan of

Fig. 9.6 Važecká jaskyňa Cave and its nearest surroundings: **a**—the valley of the Biely Váh River with terraces above and in front of the cave, **b**—rich calcite decoration in the cave, **c**—fluvial allochthonous sediments deposited in the cave (*Photo*: P. Bella)

Fig. 9.7 View from the Chočské podhorie Piedmont to the western part of the Liptovská kotlina Basin surrounded by the core mountains, with the expressive horst of Mních hill (*Photo* L. Vitovič)

120 × 100 m (Hochmuth 1983; Jurečka 2002). Lóczy (1878), Janáčik (1968), Droppa (1971, 1972), and Hochmuth (1997) explained the origin of the Liskovská jaskyňa by corrosion and erosion of allochthonous waters sinking from the nearby Váh River, with an additional role of corrosion of seeping meteoric waters. Horizontal and subhorizontal segments in this cave and nearby caves (considered as evolution levels) were correlated with the terraces of Váh River (Janáčik 1968; Droppa 1971; Hochmuth 1977, 1983, 1997).

The most significant cave levels are at an elevation of 38 m, 19 m, and 4 m above the recent river bed of Váh River (Fig. 9.8a). In spite of the fact that the cave is situated on the right bank of the Váh River and near the low-lying accumulation terrace T-Ib (Würm/Weichselian, Fig. 9.8b), typical fluvially modelled morphologies, allochthonous gravels and sand are absent in the cave. Mostly irregular, oval corrosion phreatic morphology (numerous cupolas, smaller spherical and sponge-like hollows, blind chimneys, facets or planes of repose, rock windows and pillars) dominates in the Liskovská jaskyňa (Bella 2005, Fig. 9.8c). Allochthonous fine-grained mud deposited in the cave from the suspension in slow water flow to stagnant water (slackwater facies). Its mineralogical composition indicates the heterogeneous source of material, which has been brought in by the Váh River (Bónová et al. 2014). The piezometric surface of karst groundwater dropped or oscillated following the incision or aggradation of the river bed on the surface. Original phreatic cavities were enlarged, renewed, or remodelled in slowly flowing to stagnant waters or by repeated flood-water injections from the Váh River. Water-table notches on cave walls were sculpted during the phases of the former steady water table (Bella 2005). Upward scalloped channels and vertical conduits indicate that waters ascending along deep faults were probably also involved in the origin and primary development of the cave. Ascending

Fig. 9.8 Liskovská jaskyňa Cave and its surroundings: **a**—the large ? terrace T-V above the cave; **b**—upper and lower cave entrances (arrows), and the accumulation terrace T-Ib in front of the cave; **c**—cave passage with phreatic morphology, partly filled by fluvial allochthonous sediments (*Photo*: P. Bella)

deep waters could have been drained from the cave into the alluvium of Váh River (Bella and Bosák 2012).

9.4.2 Veľká Žltá stena Outcrop

Veľká Žltá stena (VŽS—'Large Yellow Wall') was initially a ~60 m high natural outcrop on the right bank of the Velický Creek in polygenetic (glacifluvial, fluvial, and slope) sediments situated to the west of the village of Tatranská Polianka in the Tatranské podhorie Piedmont. It was revealed as a result of the catastrophic flood in 1813 (Figs. 9.9 and 9.10). The VŽS in its entire vertical profile presents, in normal superposition, a sedimentological record of a complicated evolution not only of the High Tatra Mts. foreland but indirectly of the High Tatra Mts. themselves as well (Lukniš 1973). It is built predominantly by the granodiorite gravel and boulders together with a small portion of flysch, quartzite, and gneiss clasts (Lukniš 1973; Vaškovský et al. 1973; Maglay et al. 2011).

Opinions on the origin of the material were changing. Several authors (e.g. Partsch 1882, and Roth 1885, both in Lukniš 1973) considered the deposits to have a glacial origin. According to Lukniš (1973), these deposits were

Fig. 9.9 The Veľká Žltá stena outcrop from the south. The Weichselian moraine from the Velická valley in the background (*Photo* D. Pošivák)

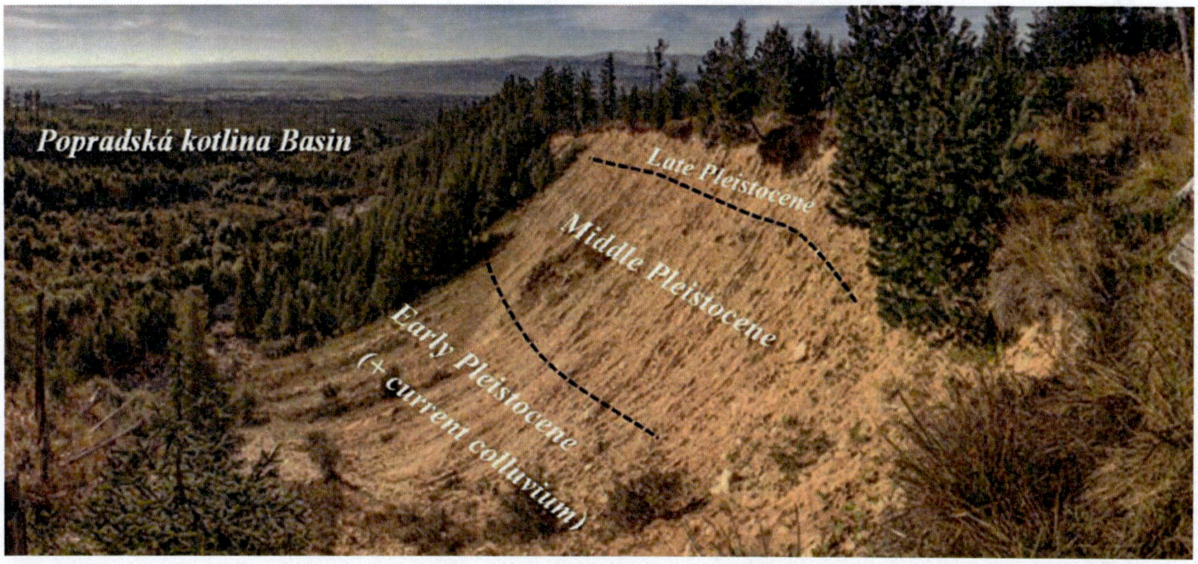

Fig. 9.10 The Veľká Žltá stena outcrop with approximate boundaries of three basic depositional complexes seen from the north, with the Popradská kotlina Basin and the Kozie chrbty Mts. in the background (*Photo* D. Pošivák)

transported predominantly by flowing water, while the upper beds were also affected by solifluction processes. He allowed a possibility of glacifluvial origin (1973, p. 280). In his geomorphological map (Lukniš 1968), the VŽS was identified as a part of glacifluvial fan. In recent studies (Maglay et al. 2011), the material of VŽS is characterized as glacifluvial. According to Pošivák (2019), the state of roundness increases with depth, which could point to fluvial rather than glacifluvial origin of the lower complex.

The entire succession of the outcrop consists of three basic complexes, which were divided into 18 beds (Lukniš 1973, Fig. 9.11). The lower complex, presumably of the Early Pleistocene age (together with the Cromerian complex), is ~40 m thick. It consists of considerably to totally weathered fluvial gravel and boulders disintegrating into the sand. A proportion of large boulders within the complex varies, indicating changes in the transportation energy of Velický Creek during deposition (Lukniš 1973).

The middle complex (?Elsterian), separated from the lower complex by a sharp erosional surface, is ~10 m thick. The deposits are built by a mixture of highly weathered gravel and pebbles, as well as relatively solid gravel and boulders with weathering rinds. The solid gravel is less rounded. The weathered material is considered to be redeposited from the lower parts of accumulation, from the Early Pleistocene deposits (Lukniš 1973).

In the upper complex (?Saalian to ?Weichselian), with a thickness of ~10 m, the proportion of unweathered gravel

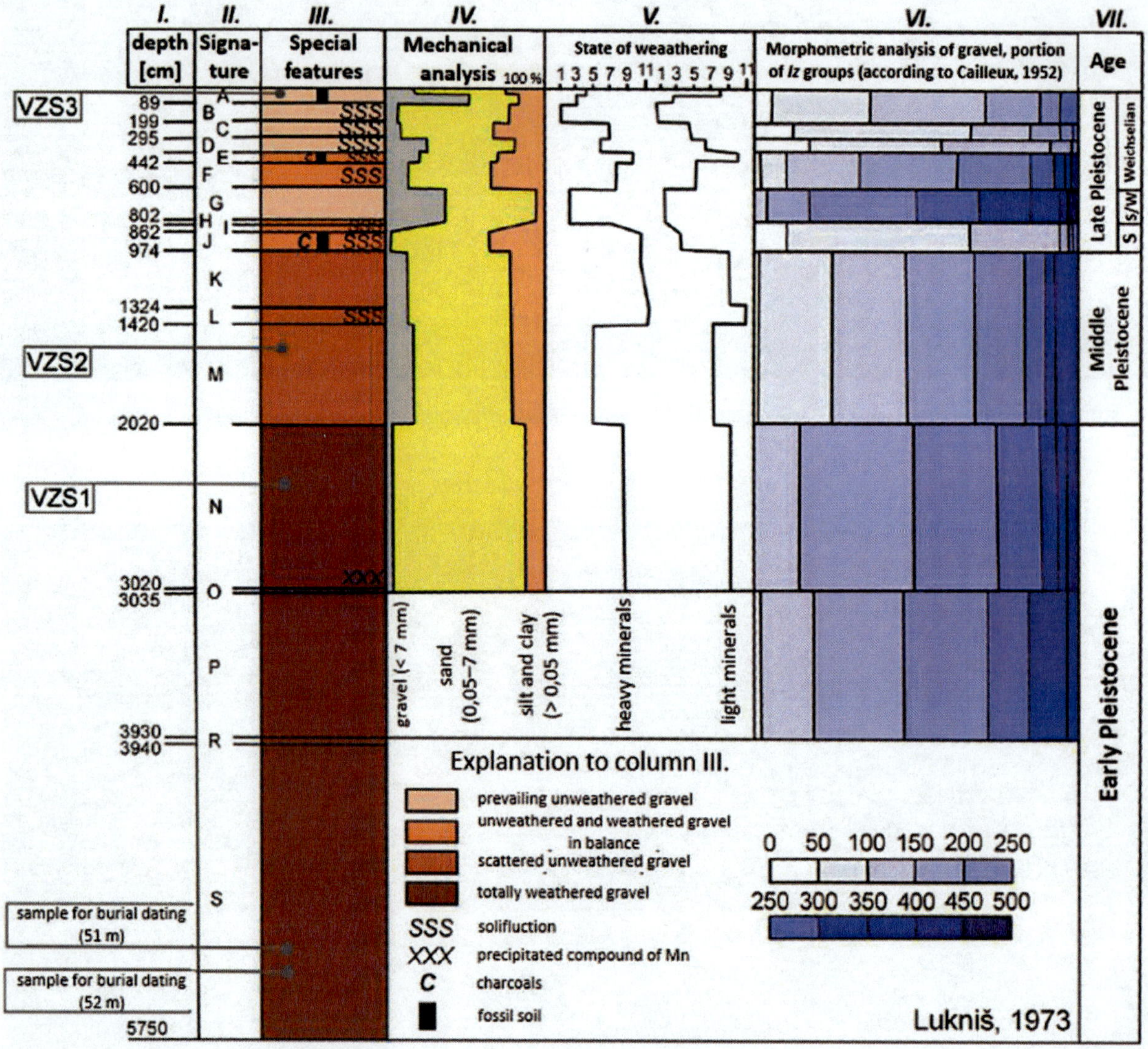

Fig. 9.11 Schematic profile of the Veľká Žltá stena (according to Lukniš 1973, modified by Pošivák 2019). Explanations to column VII.: S: Saalian, S/W: Saalian/Weichselian interglacial (Eemian)

and blocks increases. The matrix is formed by sandy loam containing damaged particles of charcoal. Gravel and cobbles are represented by weathered as well as solid clasts. Solid clasts are less rounded. The percentage of the matrix significantly varies. The beds within this complex are affected by solifluction (Lukniš 1973).

9.4.3 The Dúbrava Crest: Evidence of Neotectonic Activity

A horst morphostructure dividing the Podtatranská kotlina Basin (PtK) from the Hornádska kotlina Basin (HK) is considered as distinct evidence of relatively young neotectonic activity within the Western Carpathians (Figs. 9.1, 9.4, and 9.12). From the PtK side, the Kozie chrbty Mts. are limited by the Svit fault and the Gánovce fault, while from the HK side, the range is limited by a more expressive Vikartovce fault. Fault activity along the northern border can be documented by springs of mineral water, travertine accumulations (in the vicinity of Gánovce village) as well by distinct linearity of the mountain/foothills junction. The evidence of the Vikartovce fault activity is recorded in asymmetric morphology of the crest and the presence of fault scarps and abandoned fluvial deposits, as well in travertine accumulations, perfect linearity, and drainage network changes (Halouzka 1993; Gross et al. 1999; Jetel 1999; Vojtko et al. 2011).

There is sedimentological evidence that streams rising in the Nízke Tatry Mts. (e.g. Bystrá, Vernársky potok) were flowing to the Poprad River (in the territory of the town of Poprad) since the Pliocene/Early Pleistocene transition (? Tiglian/Eburonian complex) (Lukniš 1973). There is geological as well as biological evidence (Madarás et al. 2012) that during the Quaternary, the Poprad River flowed to the HK, receiving the Hornád River as a right-side tributary. But relatively rapid uplift of the Kozie chrbty Mts. forced the Poprad River and the Hornád River to find totally different paths. This resulted in the separation of the PtK from the HK. Therefore, the Poprad River turned its course along the Poprad fault, while the Hornád River shifted along the Vikartovce fault (Halouzka et al. 1999; Vojtko et al. 2011; Madarás et al. 2012). The final redirection of the rivers could have started at the final phase of the Riss (Saalian complex) glaciation (135 ± 14 ka) (Vojtko et al. 2011), however, more probably at the Pleistocene–Holocene transition (Madarás et al. 2012).

Investigation of fluvial deposits in the dry Vysová pass (105–135 m above the Hornád alluvial plain, Fig. 9.1b) was carried out to date fault activity. The luminescence methods (OSL, IRSL) were used together with palynological analyses. The fluvial deposits were dated at 135 ± 14 ka. The average vertical slip rate along the Vikartovce fault (0.0.8–1.0.0 mm/yr) was inferred from the altitude and the age of deposits. According to the results of palynological analyses, a landscape without forest and a rather cool climate were the conditions in which fossil river sediments were deposited (Vojtko et al. 2011).

9.5 Evolution of Polygenetic Relief of the Podtatranská kotlina Basin

9.5.1 Miocene–Pliocene (~23–2.6 Ma)

The post-Palaeogene denudation of a significant portion of Paleogene deposits within the Spišská Belá block (Fig. 9.5) resulted from its uplift along the Poprad fault (Gross et al. 1999). During the Neogene, the PtK is considered to have developed in terrestrial conditions, and the development of

Fig. 9.12 View towards the Vysová pass and the Dúbrava crest (eastern subunit of the Kozie chrbty Mts.). The moraine in the foreland of the Veľká Studená valley dates from the Last Glacial Maximum (see the Sect. 9.5.5 *Late Pleistocene*) (*Photo* L. Vitovič)

an initial river network dates to this time interval (Gross and Köhler 1980; Gross et al. 1990). The primary individualization and relative subsidence of the PtK in relation to the adjacent mountains commenced during the Miocene as evidenced from the fission-track dating of the Tatra Mts., complemented by the sedimentary record, geomorphological evidence, and structural analysis (Králiková et al. 2014). The tectonic evolution of the PtK during the Miocene in relation to the Tatra Mts. was inferred from measured brittle structures in sediments of the Central Carpathian Palaeogene Basin (CCPB). During the Early Miocene, the Tatra Mts. were exhumed in a compressional tectonic regime, which was altered by compressional to the strike-slip regime in the Middle to Late Miocene. Most of the Late Miocene is characterized as a period of tectonic quietness (Králiková et al. 2014). As a consequence of tectonic quiescence, no prominent relief of the Tatra Mts. was formed, as inferred from relatively fine-grained gravel on the Polish foreland of the Tatra Mts. (Tokarski et al. 2012). The Neogene gravel in the Polish foreland does not contain crystalline rocks, which implies that the Tatra Mts. were still covered by Flysch sediments (Tokarski et al. 2012).

This oldest phase of terrestrial development in the southern surroundings of the PtK is represented by the planation surface of the Važec Karst at 1020–1030 m a.s.l., ~ 210–220 m above the recent floodplain of the Biely Váh River. According to Droppa (1962), it is a part of the Miocene peneplain. In the light of new concepts of denudation chronology of the Western Carpathians, it should be the mid-mountain level—an initial planation surface, formed as a tectoplain (Minár et al. 2011), or intramountain level of Zuchiewicz (2011). It was formed during the Late Miocene, consistently with AFT data pointing to decreasing exhumation rates of the Tatra Mts. (Králiková et al. 2014). The planation surface could contain segments of an older peneplain exhumed from below the Palaeogene infill, but it was probably formed also on the Palaeogene sediments within the entire PtK. Remnants of the mid-mountain level on the rim of the surrounding mountains are only little uplifted (Kozie chrbty 1150 m—Droppa 1962, Kobylí vrch in the upland of Belianske Tatra Mts. 1080 m—Bella et al. 2011), but in the central parts of the mountains they are located up to 1500–2000 m a.s.l. This suggests a small altitudinal difference between the basin and the surrounding mountains at the end of Miocene, followed by intensive tectonic uplift of the surrounding mountains during the Pliocene. The late Pontian/Early Pliocene is characterized by the uplift of the Tatra Mts., recorded in ~80 m incision of the Biela River valley (Bella et al. 2011). The tectonic activation could be simultaneous with the Messinian salinity crisis. Relative subsidence of the PtK in relation to the Tatra Mts. is assumed to be 1000–1500 m during the Pontian (or Late Pannonian–Messinian) (Lukniš 1972). During the Pliocene, the relative uplift of the surrounding mountains was probably discontinuous and slowing down, as inferred from a set of subhorizontal epiphreatic segments of the Belianska jaskyňa Cave near the boundary of the PtK (Bella et al. 2011), as well as the highest fluviokarst levels of caves of Demänovské vrchy Mts. (part of the Nízke Tatra Mts.) (Bella 2001). Synchronous formation of two or three pediments is documented in the surrounding of PtK (Michaeli 2001; Zuchiewicz 2011; Bella et al. 2011). One can suppose that flat interflow ridges of relative uplifted blocks in the south of PtK (Vrbovská pahorkatina Hilly land, Štrba dividing block on the boundary of PpK and LK) represent remnants of a final Late Pliocene flat bottom of the PtK.

9.5.2 Pliocene–Pleistocene Transition

Traditionally, the uppermost flat surfaces of intramountain basins of the Western Carpathians (650–950 m a.s.l. in PtK) are considered as remnants of the river-side level that developed as a pediment during the Late Pliocene and/or Early Pleistocene (Lukniš 1972). Droppa (1962, 1967) and Lukniš (1973) termed these uppermost remnants of the flat surfaces of PtK as a 'plateau terrace', considering their Pliocene age. However, with regard to the old (1.8 Ma Quaternary) and new Pleistocene/Pliocene stratigraphy (Menning and Hendrich 2016; Cohen and Gibbard 2018), the Late Pliocene interval in older publications is considered the Early Pleistocene according to the recent stratigraphy. Because of the missing numerical geochronological data, the Pliocene–Pleistocene transition in PtK was rather vague until recently.

The PtK was very probably characterized by lower relief and a larger extent than today during the Late Pliocene. Claystone to clay deposits (thickness ~160 m) at the basis of Quaternary deposits (borehole GH-1 foreland of the High Tatra Mts.) are presumed to be residual deposits of ?limnic origin which were accumulated in the post-Palaeogene depressions. The Pliocene evolution in the western part is represented by remnants of the plateau terrace, which were considered to be formed by fluviolimnic sediments (e.g. Vaškovský 1980; Páleník 1988). The sediments are presumed to have a Pliocene age (e.g. Droppa 1970; Vaškovský 1980; Páleník 1988). While revising the chronology of the PtK development, the age of the plateau terrace in the Western Carpathians could be presumably transferred into the Early Pleistocene age (?Biber/?Praetiglian). This is in accordance with Maglay et al. (2011). Investigation of the absolute age of the accumulation is the scope of recent research. The latest research (Vitovič 2020a) on Bežan hill (669.8 m) located to NE of the village of Liptovské Sliače (Fig. 9.13) revealed that the material has rather only a fluvial origin, being probably covered with proluvial deposits. The

Fig. 9.13 View from the Chočské podhorie Piedmont to the western part of the Liptovská kotlina Basin, with the remains of the Váh River terraces. The highest terrace is situated on the Bežan hill (*Photo* L. Vitovič)

thickness of the terrace accumulation (Bežan hill) was identified to 13.2 m. The surface of the terrace (T-IX) is at ~170 m and the base at 155.5 m above the recent floodplain. Lithological composition (alternating proportions of mainly granites, limestones, quartzites, and palaeovulcanites) reflects the Váh River sediments, together with supply from the Nízke Tatry Mts. Other remains of the terrace are located to the north of the village of Liptovský Trnovec, where only 4 m of fluviolimnic sediments were identified (Páleník 1988). The surface is 682.5 m a.s.l., which is ~140 m (top) and ~135 m (base) above the floodplain of the Váh River. Another residuum of the terrace is preserved to the north of the village of Bobrovec, with the surface at ~790 m a.s.l., underlain by a 16.5 m thick fluvio-limnic accumulation. The base lies ~773.5 m a.s.l. (110.5 m rel. to Jalovec stream) (Páleník 1988). Lithological composition (mainly metamorphic rocks) reflects the dominant sediment supply from the Western Tatra Mts. In the eastern part, the plateau terrace is considered to be located on the western junction of the Levočské vrchy Mts. and the Popradská kotlina Basin. No precise location or elevation was given (Lukniš 1972). After Droppa (1964), up to three individual levels developed as pre-Quaternary (plateau) terraces, however other authors consider their Quaternary age (e.g. Vitovič and Minár 2018). The relative heights of particular remnants of the plateau terrace reaching ~30 m could be linked mainly to younger differential tectonic movements (mentioned later). In general, the plateau terrace is the evidence of the formation of broader and relatively wider Pliocene (?Early Pleistocene) valley bottoms than those of today, with a character of a pediment (Mazúr 1963; Droppa 1962; Lukniš 1972).

The presence of clay deposits in the borehole GH-1, together with the deposits on the Bežan hill, is evidence that the PtK was a landscape covered with lakes during the Late Pliocene (Vaškovský 1980; Gross et al. 1993).

Within the Važec Karst, remnants of the plateau terrace recording the planation of the Basin during the Pliocene are present as well. The erosional karst plateau of Krieslo is situated at ~140 m (~916 m a.s.l.) above the recent floodplain (Droppa 1962). Remnants of the terrace are located as well on the Štrba dividing block, at elevations of ~920 m a.s.l. No accumulations are mentioned as it was an erosional plateau cutting (planating) across Mesozoic and Palaeogene sediments (Droppa 1967). The Lieskovec peak (967 m a.s.l.) in the northernmost part of the Štrba dividing block is considered as a remnant of the ?Pliocene plateau terrace too (Lukniš 1973).

The surface of the Hybe Karst (west of the Važec Karst, Fig. 9.1) represents the river-side level sculpted by pediplanation on subsided carbonate blocks. This planation surface developed on Mesozoic carbonates as well as on adjacent Palaeogene rocks and it was covered by fluvial sediments. Its exhumation started in response to the incision of the valleys of Biely Váh River and Hybica Stream (Droppa 1967, 1968). The recent exhumation and development of surface karst landforms (karren, dolines) have been accelerated due to the intense agricultural land use (Lehotský 2001).

The upper plateau terrace level was also developed in the territory of the Tatra Mts. The rate of incision of the Biela River valley decreased and the epiphreatic subhorizontal part of the Belianska jaskyňa Cave in the Nánosová chodba Passage was formed (Bella et al. 2011). The pediments can be also identified in other marginal parts of the PtK, where they have a relatively large extent; e.g. in the Chočské podhorie Piedmont, Tatranské podhorie Piedmont, Vrbovská pahorkatina Hilly land, as well as in the Kozie vrchy Mts. (that were part of the PtK at that time) (Mazúr et al. 1980).

However, the character and position of river-side level in the most part of the PtK is rather specific in comparison with other Slovak intramountain basins. In the Tatranské podhorie Piedmont it was defined as glacis. This foothill plateau is developed on the polygenetic accumulation, which was later covered by the moraines of the Middle and Late Pleistocene ages. The material of polygenetic accumulation is of various genesis (fluvial, glacifluvial, glacial, and slope deposits), thickness (several metres to tens of metres), fraction (clay to boulders), and age (?Pliocene to Late Pleistocene) (Lukniš 1973, see also Sect. 9.4.2). Accumulation achieved an extremum thickness of >400 m, infilling very small graben structures (Fig. 9.5). Synchronous climatic change and the new stage of tectonic uplift of the Tatra Mts. at the Pliocene–Pleistocene boundary is the most probable reason for the onset of this sedimentation. Scattered remnants of sediments evidence that (?Late Pliocene–) Early Pleistocene deposition originally covered a large part of the Pliocene pediments of the PtK. Pieces of a well-developed pediment immediately above the remnants of accumulation point to a new Early–Middle Pleistocene stage of planation. On the other hand, bedrock revealed below remnants of the accumulation can be interpreted as an exhumed (more dissected) Pliocene pediment. It is in line with the statement of Droppa (1967) that sediments of the oldest Quaternary terrace (T-VIII, sensu Droppa 1967) were accumulated on the erosional surface of the plateau terrace during the Pliocene–Quaternary transition.

During the Late Pliocene, the Liptovská kotlina Basin (LK) was connected with the Popradská kotlina Basin (PpK) and the Hornádska kotlina Basin (HK), without the existence of uplifted elevations of the Štrba dividing block and the Kozie chrbty Mts. As a response to regional extension of NW–SE to NNW–SSE direction during the Pliocene/Early Pleistocene, bounding normal faults as well as longitudinal (axial) faults of the LKB became active. This resulted in the individualization of the basin and its relative subsidence in relation to the adjacent mountains, as evidenced by the faceted slopes limiting the PtK (Gross 1980; Košťálik 1999; Littva 2017). The Sub-Tatra fault has a character of steeply south-dipping parallel planes with the dip of 70–80° (Gross 1980; Gross et al. 1999). The Kozie chrbty Mts. were formed as a separate uplifted morphostructure ('delayed elevations' of Minár et al. 2011), while the HK (part of the Transitional region sensu Minár et al. 2011) relatively subsided compared to the PtK (part of Central region of Minár et al. 2011).

During the ?Pliocene, the Belá River and the Kežmarská Biela Voda River had evidently different courses. The rivers flowed to the south. This is evidenced by quartzite gravel within the oldest accumulations on the foreland of the Tatra Mts. (Lukniš 1973; Pošivák 2019). Subsequently, the streams flowing from the Tatra Mts. into the PpK were deflected to the east, whereas those flowing into the LK were deflected to the west. Both situations reflect the uplift of the dividing block of the Štrba. According to the age of accumulations of the mentioned streams, the migration lasted from the Pliocene–Quaternary transition until the Riss (Saalian complex) glaciation (Lukniš 1973). In terms of revision of the chronology of the PtK development, the migration of the streams could have begun during the Early Pleistocene (?Eburonian complex). The Štrba dividing block is considered to fully act as an elevation since the onset of the Middle Pleistocene (MIS ?19) (Halouzka 1993).

9.5.3 Early Pleistocene (Praetiglian–Bavelian Complex, MIS 100–20)

After the plateau terrace formation, the incision of streams occurred as a response to activation of neotectonic regional uplift. This resulted in the morphological differentiation of the plateau terrace (Lukniš 1972; Baňacký et al. 1993). Accelerated activity along the Sub-Tatra fault, expressed by accelerated exhumation (uplift) rate of the Tatra Mts., occurred at ~6.5–1.0 Ma, as inferred from the AFT dating (Králiková et al. 2014). During the Late Pliocene/Early Pleistocene transition, this resulted in the incision of the Biela River by ~20–25 m (Bella et al. 2011). The drainage divide between the Belá and Biely Váh Rivers developed at the beginning of the Quaternary. The development took place along the border of the Liptovská Kokava and Východná blocks (Fig. 9.5). Subsequently, the Hybica fault formed as a boundary between the blocks (Halouzka 1987). The activity of the Smrečany fault and Liptovský Ondrej fault with N–S direction predetermined the courses of the Smrečianka and Jalovecký streams during the Early Pleistocene (Halouzka 1987). Repeated subsidence of the LK in relation to the adjacent mountains is proved for the Quaternary period. As the bottom of the LK is filled with Quaternary deposits, predominantly developed as terraces, absolute subsidence is not considered to be significant. The amount of subsidence spatially varies, presumed to be around 50 m in relation to the Western Tatra (Halouzka et al. 1999), while in respect to the High Tatra Mts. it is 100–200 m (Hanzel 1979), 400 m (Halouzka 1993), or even 400–600 m (Halouzka et al. 1999) during the Quaternary. The Quaternary tectonic activity in the foreland of the High Tatra Mts. can be proved for several short stages

(Bezák et al. 1993). Since the Pliocene, but mainly during the Quaternary, the PtK has developed into a system of partial tectonic blocks (Fig. 9.5). Vertical tectonic dislocations within the blocks are considered to be ~10–15 m (Halouzka 1987; Halouzka et al. 1999).

The Poprad River valley deflected to the foothills of the Levočské vrchy Mts. by the streams coming from the Tatra Mts. (Lukniš 1973). The course of the river along the foothill was probably driven by the neotectonic activity of the Poprad fault as well (Halouzka et al. 1999). This migration is evidenced by lateral erosion of the foothills, as well as asymmetric preservation of the terraces (Lukniš 1973; Vitovič 2019).

The tectono-sedimentary evolution of the PtK during the Early Pleistocene is recorded predominantly by fluvial (possibly also glacifluvial) and proluvial sediments. Revising the chronology of the PtK development, presumed Neogene well-rounded granodiorite gravel in the Slovak foreland of the High Tatra Mts. identified by Lukniš (1973) could be considered as an Early Pleistocene deposit. But its character (relatively fine-grained compared to ?Quaternary gravel) points to the low dynamics of the High Tatra Mts. streams (Lukniš 1973), which can be consequently related to rather low relief of the terrain presently occupied by the High Tatra Mts. Based on the grain size, these deposits could be related to Neogene gravels identified in the Orava-Nowy Targ Basin (Tokarski et al. 2012). Gravels identified by Lukniš (1973) are significantly kaolinized, highly weathered, and partially modified (transformed) to clayey-sand mass.

Relatively vast alluvial fans were deposited in the foreland of the Tatra Mts. during the Early Pleistocene. The highly weathered material of the fans is traditionally interpreted as mainly glacifluvial, which should indicate glaciation of the Tatra Mts. during the Early Pleistocene (Gross et al. 1990; Halouzka and Raczkowski 1993; Maglay and Halouzka 1999). Glaciation of the Western Tatra Mts. during the Early Pleistocene is assumed by Halouzka (1987). According to Lukniš (1973), moraines and glacifluvial fans of the Early Pleistocene age are highly denuded and covered by solifluction debris within the glacis surface located in the foothills of the Tatra Mts. The Early Pleistocene sediments of the Velický stream deposited on the foreland of the High Tatra Mts. Reach ~40 m thickness. These sediments were analysed in the VŽS outcrop (Fig. 9.1), where Late to Early Pleistocene sediments occur. The formation of the fan of the Mlynica stream took part in the area of the Štrba dividing block (Lukniš 1973).

In general, two generations of fans and terraces developed during the Early Pleistocene: Nová Lesná and Hybe layers (Baňacký et al. 1993; Maglay and Halouzka 1999; Maglay et al. 2011). Predominantly granitoid gravels occur in the accumulations of fans, together with rare quartzites. Within the Hybe layers, the Palaeogene sandstone and metamorphic rock gravel were identified too (Halouzka and Raczkowski 1993). This indicates erosion of the flysch sedimentary cover of the Tatra Mts. which is not preserved anymore (Aubrecht 2014; Králiková et al. 2014). The former extent of large fans can be only reconstructed based on the fans' remnants analyses, as the fans were later eroded and covered with younger deposits. On the foreland of the Kežmarská Biela voda valley, a 15 km broad fan was formed. The fan of the Belá River had probably a similar, or even larger extent, resulting from gradual shifting of the river to the west (Lukniš 1973). The length of the fans is in some cases 4–7 km, the longest remnant of the fan is 14.5 km long (on the left side of the Belá River). The volume of sediments (sediment supply) recorded in the thickness of the deposits can reflect the influence of the uplift of the Tatra Mts. on erosion rates. It ranges from ~10 m at the toe, to ~100 m at the apex of the fans (Hanzel 1979), typically reaching 20–30 m (Maglay et al. 2009). Accumulation does not exceed 20 m thickness (Lukniš 1973) in more distant localities (sites) from the mountain/foreland junction. On the other hand, regarding the old age of the accumulations, the preserved thickness of the fans is considerable (Lukniš 1973). The sediments of alluvial fans are much coarser compared to the residual ?Neogene gravels (Lukniš 1973; Tokarski et al. 2012), which can also point to an increase of the Tatra Mts. uplift. The alluvial fans developed in interconnection with river terraces into which they are interlocked (Droppa 1970; Vaškovský 1980; Maglay and Halouzka 1999). Remnants of the oldest river terraces are preserved mainly along the Váh River. The narrow water gap segment of the Biely Váh valley between the villages of Važec and Uhorská Ves, cut into Mesosoic rocks, started to form at the Pliocene–Pleistocene transition as recorded by the oldest terrace sediments along the valley. The valley segment was considered to have an epigenetic (e.g. Vitásek 1932 in Droppa 1967; Lukniš 1972; Lukniš 1968; Dzurovčin 2000) antecedent (Droppa 1962) or epigenetic-antecedent (Droppa 1967) origin. The epigenetic origin could be evidenced by preserved remnants of the overlying Palaeogene (Palaeocene-Lower Oligocene) Borové formation on both sides of the valley (Gross 2008). The river was shifted to the south, probably due to the aggradation of the Belá River (Lukniš 1972) and other streams rising from the Tatra Mts. (Lukniš 1973). As the faults are present along the valley, the formation of the valley could have been influenced by neotectonic activity as well. The valley of Mlynica stream between the villages of Štrba and Lučivná has a similar origin (Lukniš 1972, 1973). Inferred from borehole analysis, the base of the oldest river terraces in the central part of the LKB is situated at ~124 m (T-VIII) and ~101 m (T-VII) above the floodplain. The base of the oldest terrace of the Poprad River in the eastern part of the PpK is relatively at ~95–96 m (?T-VI–T-V) (Lukniš 1973). The formation of terraces points to regional

Table 9.1 River terraces of the Váh River and Poprad River, their presumed stratigraphy, elevation (of the terrace base), and inferred rate of the incision. The rate of subsidence is calculated for boreholes on the foothill of the Tatra Mts. (Hanzel 1979). Classification of the Váh River terraces compiled from Droppa (1970), Vaškovský (1980), Halouzka (1986), and Maglay et al. (2011). Classification of the Poprad River terraces compiled from Lukniš (1973), Halouzka (1986), and Maglay et al. (2011). The correlation of Alpine with Nordic climatostratigraphy as well as with MIS is based on Menning and Hendrich (2016) and Cohen and Gibbard (2018). The MIS value is assigned to the onset of a particular stage

Terraces		Clinatostratigraphy		MS	Relative elevation [m]		Rate of incision [mm/y]		Rate of subsidence [mm/y]	
		Alpine	North West European		The Váh River	The Poprad River	The Váh River	The Poprad River	VTH-8	VTH-7
IX	Plateau	Biber	Praetiglian–Tiglian complex	100	155.5		0.039–0.041		0.07	
VIII	1st high	Donau	Eburonian–Bavelian complex	63	123.0–125.0	95–96	0.021–0.024			
VII	2nd high	Günz 1	Cromerian complex	19	100.9–102.0		0.09–0.11			
VI	3rd high	Günz 2			79.0–90.6	68–70				
V	1st upper	Mindel 1	Elsterian complex	10	56.4–66.2		0.51–0.68		1.45	
IVb	2nd upper	Mindel 2			41.0–43.0					
IVa	3rd upper	Mindel 3			36.2–37.6	30–36				
III	1st middle	Protoriss	Saalian complex	8	16.8–19.2	20–23	0.12–0.16	0.14–0.16	0.66	0.60
IIb	2nd middle	Riss 1			12.0–15.9	8–12.5				
IIa	3rd middle	Riss 2			−0.4–0.7					
I	Low	Würm	Weicltselian complex	5d	−4.5–10	−5–6			0.24	0.59

uplift in the distal foreland of the Tatra Mts. Nevertheless, the rate of tectonically induced incision of the Váh River is assumed as relatively low (Table 9.1). In contrast, the subsidence trend can be inferred from the depositional settings of sediments in the proximal foreland of the Tatra Mts., where neotectonic depressions are located (Maglay et al. 1999). According to the thickness of deposits (granitoid sand and gravel) close to the village of Vyšné Hágy (borehole VTH-8 (Hanzel 1979)), the displacement along the Sub-Tatra fault during the Early Pleistocene (including Cromerian) is assumed to be ~151 m. Based on the thickness of Early Pleistocene deposits, the average fault slip rate can be calculated to ~0.07 mm/y. Tectonic activity along the sub-Tatra fault varies spatially, as well as chronologically. The relative subsidence of the PtK in relation to the High Tatra Mts. (mentioned above) is much more significant than the relative subsidence of the LK in relation to the Western Tatra Mts. (asumed to be ~50 m) (Halouzka et al. 1999).

The thickness and state of preservation of Quaternary deposits within the PtK is variable. In general, the thickness is changing in relation to the distance from the foothill of the adjacent mountainous area, as well as with the distance along the major rivers. For instance, from the value of more than 400 m at the foothill of the Tatra Mts., it reduces towards the basin axis, where the pre-Quaternary rocks can be exposed (Lukniš 1973; Hanzel, 1979; Maglay et al. 2009). The maximum thickness of identified ?Quaternary sediments (granitoid detritus, debris, sand) is 441.9 m (borehole VTH-7 which did not reach the basement), in the foreland of the High Tatra Mts. close to the settlement of Štrbské Pleso (Hanzel 1979).

9.5.4 Middle Pleistocene (Cromerian–Saalian Complex, MIS 19–6)

Increased neotectonic activity during the Middle Pleistocene resulted in an important change in the morphological evolution of the PtK. Glaciation of the Tatra Mts. resulted in the deposition of glacial sediments in the valleys of the uplifted Tatra Mts., as well as on its foreland. Neotectonic activity is reflected in the increase of slip rate along the Sub-Tatra fault, tectonically induced incision, as well as in accumulation of travertines (Franko 2001; Gradziński et al. 2015; Vitovič 2018, 2019).

The oldest glacial deposits (Smokovec glacial stage, Elsterian) are preserved in the vicinity of mouths of three valleys: Belá, Mengusovská, and Veľká Studená valleys. The height of the degraded moraine rampart is ~4 m (at the stratotype locality of Horný Smokovec) and the length of the remnants is 2–3 km (Hanzel 1979; Halouzka and Raczkowski 1993). Deposition of vast alluvial fans continued in the mountain foreland during the Middle Pleistocene.

The development of the PtK continued through the formation of glacifluvial fans in the foreland of the High Tatra Mts., which are joined with the moraines. In general, the Middle Pleistocene fans are incised into older fans and the Late Pleistocene fans are incised into the Middle Pleistocene fans (Lukniš 1973). According to the new stratigraphy (Menning and Hendrich 2016; Cohen and Gibbard 2018), deposits that were dated at Günz (Cromer complex), are considered the Middle Pleistocene age. Therefore, in the revised chronology of the PtK development, the Gerlachov-Východná layers are interpreted as Middle Pleistocene deposits. Terraced glacifluvial fans of four (Maglay and Halouzka 1999) or five generations developed in the foreland of the High Tatra Mts. during the Middle Pleistocene (Lukniš 1973). The fans in the foreland were deposited on the eroded surface of the Early Pleistocene fans. Therefore, at some sites, the younger material is mixed with highly weathered material of older fans (Lukniš 1973). The fans are built by coarse gravel to boulders along many valleys and have diverse length (Lukniš 1973; Maglay and Halouzka 1999). The width of particular fans can reach 2 km (Hanzel 1979). Their thickness is in general lower, compared to the Early Pleistocene fans (5–15 m, occasionally as much as 30–40 m) (Maglay et al. 2009). Deposits probably related to the Mindel (?Elsterian) and Riss (?Saalian) glaciation are preserved in the VŽS outcrop as well. The Riss (Saalian) glacifluvial fans of the Mlynica stream were deposited to the west of the Štrba dividing block, indicating the stream flow to the Biely Váh River (Lukniš 1973).

During the Mindel glaciation (?Elsterian), the Mengusovce-Lučivná (Rakovec) graben (~2,4 km wide) situated to the west of the village of Važec (Fig. 9.14) subsided, which is indicated by the relatively vast occurrence of the Poprad River deposits. The graben is located between Kolombiarok and Bôrik horsts (Lukniš 1973; Vitovič 2020b). These horsts belong to significantly active Quaternary morphostructures (Gross et al. 1990). The thickness of the graben infill is at least 10 m in its southern part (basement was not reached), while in the northern part it is ~27 m (probably filled with younger deposits too).

The foothill blocks significantly subsided along the Sub-Tatra fault during the Elsterian glaciation as well, which is recorded by a ~106 m thick (borehole VTH-8) sand (mainly) and gravel deposits (?morainic) (Hanzel 1979). Therefore, the tectonic slip rate can be inferred to ~1.5 mm/y, which is the biggest value of the slip rate within the PtK during the Quaternary (VTH–8) (Table 9.1).

Neotectonic development in the western part of the PtK during the Mindel glaciation (Elster) is demonstrated by activation of a relatively younger fault system (at the

Fig. 9.14 The view of the Bôrik horst and Rakovec graben, with the Veľká Žltá stena in the foreground and Kozie chrbty Mts. and Nízke Tatry Mts. in the background (*Photo* J. Holec)

regional scale), with NE–SW and ENE–WSW directions. The fault activity resulted in deflection of Jalovec and Smrečianka streams, caused by the subsidence of Bobrovec and Žiar tectonic blocks (Halouzka et al. 1999). For this territory (foreland of the Western Tatra Mts.), faults striking N–S indicate an older phase of tectonic activity, while faults trending NE–SW (or ENE–WSW) indicate a younger phase (Halouzka 1987). The directions of an older and younger fault system are different in other parts of the PtK. Therefore, this temporal succession of neotectonic activity has rather a local validity (Gross et al. 1999).

Four to five terraces of the Váh River can be recognized within the Günz (Cromerian) and Mindel (Elsterian) glacial periods. The relative elevation of the base of the Váh River terraces is ~80 m (T-VI), ~61 m (T-V), ~42 m (T-IVb), and ~37 m (T-IVa). The upper evolutionary level of the Liskovská jaskyňa Cave is correlated with the terrace T-IV (Mindel 2/Late Elsterian) (Droppa 1971, 1972). The Hybská jaskyňa Cave (Hybe Karst, Fig. 9.1a) was formed by allochthonous waters of the Biely Váh River at the level of bedrock base of the river terrace T-IV (Droppa 1967). Only two terraces of the Poprad River can be identified: ~68 m (T-V) and ~39 m (T-IV). According to Lukniš (1973), the base of the ?Cromerian terrace is ~68–70 m and ?Elsterian terrace occurs at relative elevation of ~30–36 m.

The increased neotectonic activity of the PtK during the Middle Pleistocene can be also proved by significantly increasing rate of tectonically induced incision of the Poprad and Váh Rivers (Table 9.1). During the Elsterian, the rate of incision of the Váh River reached its maximum (~0.6 mm/y). Neotectonic activity along the basin-marginal faults is proved by the oldest travertines in the village of Ludrová. Its age was estimated at 600–750 ka (Franko 2001).

The neotectonic activity continued as a significant control of the development of the PtK during the younger part of the Middle Pleistocene (Riss 1/Early Saalian). It is reflected inter alia in distant glacial deposits in the foreland of the Tatra

Mts., river network changes, spatially differential character of sedimentation within the PtK, as well as in travertine accumulations. The most widespread glacial sediments of the Štôla glacial stage (Early Saalian) reflect an intensive uplift of the Tatra Mts., together with the most extensive glaciation. The sediments (deposited partially over older moraines) are preserved in the vicinity of the mouths of most of the valleys of the High Tatra Mts. and also in the vicinity of the mouths of the easternmost valleys in the Western Tatra Mts. (Bystrá and Kamenistá valleys) (Halouzka and Raczkowski 1993). Glacial sediments of Rakytovec glacial stage (Late Saalian) are deposited partially over older moraines in the foreland of the ~ seven valleys of the High Tatra Mts. and also in the Kamenistá valley (foreland of the Western Tatra Mts.) (Halouzka and Raczkowski 1993). The continuing subsidence of the foothill blocks along the Sub-Tatra fault during the Saalian complex is recorded in thick boulder, gravel, and sand (?morainic) deposits reaching 123 m (borehole VTH-8) and 104 m (borehole VTH-7) (Hanzel 1979). Based on the thickness of ?Saalian deposits, the tectonic slip rate can be calculated to ~0.7 mm/y (VTH-8) or ~0.6 mm/y (VTH-7) (Table 9.1).

Neotectonic development in the SE part of the PtK continued in the tectonic activity of the Poprad block, which began after the formation of terraces T-IV (Late Elsterian) and T-III (?Earliest Saalian) (Halouzka et al. 1999). The wind gap feature (Fig. 9.15) located to the east from the town of Ružomberok in the Váh valley, with a fluvial accumulation of terrace T-V (Early Elsterian), can record neotectonic activity of a W–E trending fault (Littva 2017; Vitovič and Minár 2018). The activity can be proved by travertine deposits (e.g. Vaškovský 1980), asymmetric preservation of terraces (Pešková and Hók 2008), and orientation of the Liskovská cave passages (Littva 2017). The travertines in the village of Bešeňová consist of several generations. Those at the site of Skala were dated at 350–450 ka (Franko 2001) or 350–1200 ka (Gradziński et al. 2015), whereas those in Drienok date at the Riss-Würm transition (Eemian) and Würm (Weichselian, Ložek, and Vaškovský 1972; Franko 2001) as well as to the Holocene to recent (Bešeňová cascade, Vaškovský 1980). Probably due to the activity of a younger fault system of the LK, a relatively huge depression (covering ~1 × 2 km) started to form NW of the Pribylina village. The depression, filled with glacifluvial sediments, was identified through geophysical and drilling survey (Košecký 1968; Janík 1970; Tužinský et al. 1971). The deepest borehole (T–6, 50 m) did not reach the basement (Tužinský et al. 1971). In spite that the area is covered mainly with Würm (Weichselian) deposits, the age of filling material was assessed to Riss (Saalian, Tužinský and Banský 1968).

The development of the PtK during the Saalian complex continued in the formation of next two–three generations of terraced fans in the foreland of the adjacent mountains (Maglay and Halouzka 1999). The width of particular fans reaches up to 800 m (Hanzel 1979). In the foreland of the

Fig. 9.15 Wind gap with the remnant of the terrace T-V formed on the Mních horst, NE of the town of Ružomberok (*Photo* J. Littva, modified)

Western Tatra Mts., they are developed within the Jalovecká and Žiarska valleys, which points to their glacial origin (Halouzka and Raczkowski 1993). Relatively vast alluvial fans developed in the foreland of the Nízke Tatry Mts. too (e.g. Ľupčianska and Křižianka valleys). The largest one, within the Křižianka valley, covers 5 × 3 km, with maximum thickness of the deposits ~20 m. The fan is located in the depression of erosional (Škvarček 1968) or, more probably, neotectonic origin (Vitásek 1923; Vitovič and Minár 2018). In the Váh and Poprad valleys, two–three terraces correlated with alluvial fans can be recognized (Droppa 1970; Vaškovský 1980; Maglay and Halouzka 1999; Vitovič 2018, 2019). The relative elevation of the base of the Váh River terraces is ~18 m (T-III), ~14 m (T-IIb), and ~0 m (T-IIa), while for the Poprad River it is ~31 m (T-III), ~14 m (T-IIb), and ~6 m (T-IIa). Lukniš (1973) identified only two ?Saalian terraces of the Poprad River, with the bases in the relative elevations of 20–23 m and 8–12.5 m. The middle evolutionary level of the Liskovská jaskyňa Cave is correlated with the T-III terrace (Riss 1/Early Saalian). The lower level of the Liskovská jaskyňa Cave, as well as the Važecká jaskyňa Cave, are correlated with the T-II terrace (Riss 2/Late Saalian) (Droppa 1962, 1971, 1972).

Younger neotectonic activity is evidenced along the basin-marginal fault of the Nízke Tatry Mts. by facets and travertine deposits in the village of Liptovský Sliač (Skalica), whose ages are estimated at Mindel-Riss (Holstein) transition (Vaškovský 1980). Neotectonic activity along the transverse fault is evidenced also by travertine of similar age (350–450 ka) (Franko 2001), located in the village of Lúčky (Skalička). Travertines estimated to form in the Middle to Late Pleistocene (Saalian to Weichselian) are situated along the Revúca River, indicating the formation of neotectonic boundaries of the Jazierska depression. Within the depression, the presence of older travertines is presumed too (Pristaš 1997). Neotectonic activity, mainly along the basin-marginal faults and inferred from the travertines of similar age (Maglay and Halouzka 1999), can be identified to the south (SE) of the town of Poprad. Young and significant neotectonic activity in the mentioned territory was also proved by numerical dating of fluvial deposits on the crest of the Kozie vrchy Mts. Dip-slip movement along the Vikartovce fault was calculated to 0.8–1.0 mm/y for the last ~135 ka (Vojtko et al. 2011). Relatively young activity along the Chočské vrchy Mts. boundary fault can be evidenced by faceted slopes as well as travertines in Lúčky (the site Na skale) dated at 139 ka (Gradziński et al. 2015). The asymmetric preservation of terraces in the Kvačianka and Suchý potok valleys, which have a similar orientation as a Choč fault, can indicate neotectonic activity in the western part of the PtK as well (Littva 2017).

9.5.5 Late Pleistocene (Eemian–Weichselian Complex) to Holocene (MIS 5–1)

As a response to change in the regional stress field during the Late Pleistocene to Holocene, extension with general direction parallel with the Western Carpathians orogen (W–E) caused activation of transverse faults stretching generally N–S. Their orientation can vary from NE–SW (in the LK) to NW–SE (in the PpK) direction. The activity of this younger fault system has resulted in the segmentation of the PtK into partial blocks and disintegration of margin-bounding faults (Gross 1980; Littva 2017). In areas, where faults belonging to the younger and older systems cross one another, springs of mineral and thermal waters, as well as travertines occur (Franko and Hanzel 1980; Gross 1980).

Young neotectonic activity is proved by Holocene travertines too, which are ruptured only along NW–SE fractures (Pešková and Hók 2008). The Holocene travertines are located along the axial faults (e.g. Uhorská Ves, Bešeňová), margin-boundary faults (Liptovský Ján, Liptovský Sliač, and Ludrová), and transverse faults (Vitálišovce), indicating their activity (Ivan 1941; Kovanda 1971; Vaškovský 1980; Franko 2001).

The majority of valleys of the Tatra Mts. were shaped by glaciers, resulting in the deposition of moraine sediments on their foreland. The most distal moraines are associated with the Mlynická, Mengusovská, and Studená valleys. During the last glacial stage (Štrbské Pleso stage, Weichselian–Lower Holocene), 3–5 partial stages of glaciation of the Tatra and Western Tatra Mts. were identified (Lukniš 1973; Halouzka 1987; Halouzka and Raczkowski 1993). The Last Glacial Maximum terminal moraine in the foreland of the Veľká Studená Valley (Fig. 9.7) was dated to ~22 ka (MIS 2). The subsequent final retreat of the glacier to its cirque was dated to ~12 ka (Engel et al. 2015). The terminal moraine in the foreland of the Velická Valley was dated to 24.7 ± 1.4 ka (Makos et al. 2014). The thickness of moraine deposits in the foreland of the High Tatra Mts. rises from 28 m (VTH-8), through 67.5 m (VTH-7) to 70–100 m (Halouzka and Raczkowski 1993). Except from few cases, moraine deposition in the Western Tatra Mts. did not reach the foreland and all moraines are confined to the mountainous area. In some places, the thickness of morainic sediments is only 1–5 m. The terminal moraines are washed-out into glacifluvial fans in the valley bottoms. The former glacial streams redeposited the moraine sediments into several branches (segments), which is related to the position of terminal moraines in particular stadial oscillations within the Weichselian glaciation. For instance, there are four branches of the glacifluvial fan emerging from the terminal moraine of the Studená valley. Beyond the Mlynica valley, three branches of the fans trending to both sides of the Štrba dividing block were formed (Lukniš 1973).

The valley floors are filled with young fans, which were incised by post-glacial streams. The fans are accumulated between older fans. In several cases, they are spread into the axial valleys (e.g. Biela, Studený potok valleys). The thickness of the deposits is in some cases very low, resembling only the pavement. In the majority of localities, the thickness reaches 3–4 m (Lukniš 1973). Increased values can be found in the village of Tatranská Lomnica and to the NW of Svit, where the thickness of glacifluvial deposits is 10–15 m (Gross 1979; Hanzel 1979; Halouzka 1987; Halouzka and Raczkowski 1993; Maglay and Halouzka 1999). The highest values were recorded in the Vavrišovo depression, where the infill of the depression reaches 50 m at minimum (borehole T-6, mentioned above). Here, the boundary between the presumed Saalian and Weichselian deposits is unknown. In general, the width of glacifluvial fans in the foreland of the High Tatra Mts. gradually decreases along the streams. In contrast, Holocene floodplains inset into the moraines show downstream increase in their width. The broadest segments of the floodplains are usually located 2–3 km from the moraine fronts. Then, their width reduces again (Lukniš 1973; Hanzel 1979). This phenomenon could reflect the neotectonic activity of transverse faults in the PtK. The length of the fans is variable, with the longest glacifluvial fan located in the Vavrišovo depression, reaching at least 11 km, terminating in the Váh valley.

The bottom deposits are significantly coarse-grained (cobbles, boulders, and blocks) with a low state of roundness, built mainly by granites. In the foothill positions of the forelands, the Weichselian fan deposits cover the older fans, which indicates continuing regional subsidence. In the VŽS outcrop, the Weichselian fan deposits are preserved in the uppermost ~5 m of the accumulation (Lukniš 1973).

Terraced glacifluvial fans along the Poprad River transform from the town of Poprad into fluvial terraces. Within the fluvial and glacifluvial sediments which built the valley bottoms, 2–3 levels with a common rock base can be recognized (Lukniš 1973; Vaškovský 1980). In the LKB, three terraces were recognized, low terrace (T-Ic), upper floodplain (T-Ib), and lower floodplain (T-Ia), respectively. The relative elevation of the surface is usually around 5–6 m (T-Ic), 3–4 m (T-Ib), and 1–2 m (T-Ia) above the contemporary channel (Vaškovský 1980). In the foreland of the High Tatra Mts., the Holocene floodplains are incised 2–7 m into the valley bottom. For the Poprad River valley, the Holocene (floodplain) surface is 2.2 m, and the low terrace is 3.8–4.5 m above the channel. Therefore, only two terraces can be identified, T-Ia (recent floodplain) and T-Ic (low terrace) (Lukniš 1973).

Within the floodplains, coarse bottom sediments are covered with fine-grained, post-glacial Holocene sediments. Floodplain deposits are more coarse-grained than in the foreland of the Western Tatra Mts., reflecting differential tectonic uplift of the mountains (Halouzka and Raczkowski 1993; Maglay and Halouzka 1999).

The thickness of the valley bottom sediments (within the floodplains) is variable, ranging from ~2 m (upper course of the Biely Váh River), through ~10–16 m (confluence of the Váh with the Belá River) to 25 m in the Vavrišovo depression. The floodplain of the Poprad River is underlain by 5–6 m of sediment, while the thickness of the low terrace (T-Ic) is 8–10 m. The young tectonic activity of transverse faults divides the streams into reaches with various thickness of valley bottom deposits (Maglay and Halouzka 1999; Vitovič and Minár 2018). Demonstration of young neotectonic activity in the vicinity of the town of Poprad is expressed by changes in thickness and the width of valley bottom deposits of the Poprad River as well. The thickness of the deposits changes from 4–6.2 m (W of Poprad) to 1.7–2 m (E of Poprad), while the width of the valley bottom ranges from 1300 to 400 m (Lukniš 1973).

The subsidence of the Žiar tectonic block (Fig. 9.5) continued during the Late Pleistocene and Holocene as it is evidenced by young glacifluvial (max ~16 m thick) and proluvial deposits. Neotectonic activity SE of the Važec (crossing of NW–SE fault with NE–SW (Brezovo site)), is evidenced by calcareous tufa deposits of Holocene (Maglay and Halouzka 1999) or recent (Gross 1980; Gross et al. 1990) age, as well as mineral water springs (Gross et al. 1990).

Changes of the river network of the PtK occurred during the Pleistocene–Holocene transition too. The Mlynica stream transformed its course when it shifted from the Biely Váh River to the Poprad River catchment (Lukniš 1973). The Poprad River shifted from the Rakovec depression to the east, where relatively thick Weichselian glacifluvial deposits (reaching 11–14 m) occur (Gross et al. 1990). The neotectonic activity of the Vikartovce fault caused a more significant river network change. According to Gross et al. (1999), redirection of the Poprad River towards NE was caused by the subsidence of the Poprad and Veľká block in the Middle Pleistocene.

Neotectonic activity resulting in the formation of small horst to the south of the town of Poprad can be evidenced by the Small Vikartovce range, parallel with the Vikartovce range. It is a horst formed along the W–E trending faults, along which springs of mineral water together with deposits of calcareous tufas are located (Gross et al. 1999; Jetel 1999). At the crossing with NNE–SSW trending fault, bordering the horst from the west, travertine deposits (Hrádok site) occur (Gross et al. 1999).

In the selected territories of the upper floodplains (e.g. Demänovka, Palúdžanka, etc.) and proluvial cones, formation of peat bogs is taking place (Vaškovský 1980). During the last-glacial stadials, original dissolution morphologies in the Liskovská jaskyňa and Važecká jaskyňa caves were remodelled mainly by slab and block breakdown, forced by frost weathering. The age of cave bear bones (*Ursus speleaus*) from the Važecká jaskyňa Cave was established for over 40 ka by radiocarbon method (Sabol and Višňovská 2007; Laughlan et al. 2012). The fine-grained sediments (slackwater facies) deposited in the Važecká jaskyňa Cave in several sequences as a result of repeated flood-water injections from the surface river and floods have normal magnetic polarity (Brunhes chron, younger than 0.78 Ma). Their uppermost sequences are divided by thin flowstone 10 to 11 ka old (U-series dating). These clastic sediments were deposited in the cave mainly during the Late Pleistocene, while the uppermost sequence was deposited during the Holocene (Bella et al. 2016).

9.6 Conclusions

The polygenetic relief of the Podtatranská kotlina Basin is a result of complicated evolution. Since the Miocene until recent times, evolution has been controlled by a number of processes, related to geodynamic activity (different tectonic regimes) and reflecting climatic changes. As the PtK together with the Tatra Mts. belong to the most neotectonically active morphostructures within the Slovak part of the Western Carpathians, neotectonic processes have significantly controlled the formation of landforms, mainly along the borders with adjacent mountains, resulting in faceted slopes, deposition of travertines, and mineral springs (Vaškovský 1980; Halouzka 1993; Littva 2017).

The evolution of landforms of the PtK was connected with the formation of the adjacent morphostructures too. Uplift associated with the exhumation of the Tatra Mts. very significantly influenced not only the relief of the Tatra Mts., but of the PtK as well. The Tatra Mts. were exhumed in a compressional tectonic regime during the Early Miocene (Králiková et al. 2014), which was followed by a quiet tectonic period (Late Miocene). This resulted in the formation of the mid-mountain level—an initial planation surface, probably within the entire PtK. The surrounding mountains were only little uplifted compared to the PtK at the end of the Miocene, but this was followed by intensive tectonic uplift during the Pliocene. The uplift of the mountains was probably discontinuous and decelerating, resulting in the formation of a flat bottom of the PtK during the Late Pliocene, which is recorded in the particular blocks of the PtK (e.g. the Vrbovská pahorkatina Hilly land, the Štrba dividing block) as well as in the remnants of a plateau terrace (T-IX) in the LK.

As a response to regional extension with NW–SE to NNW–SSE direction during the Pliocene/Early Pleistocene, bounding normal faults, as well as longitudinal (axial) faults of the PtK, were active. This resulted in general shaping of the basin and its relative subsidence in relation to the adjacent morphostructures (Gross 1980; Košťálik 1999; Littva 2017). The increased uplift of the Tatra Mts. during the Early and Middle Pleistocene, connected with climatic changes, resulted in at least three glaciation stages during the Pleistocene. This caused the formation of large glacifluvial fans in its foreland (Halouzka 1987; Halouzka and Raczkowski 1993). Alternation of glacials and interglacials along with neotectonic activity and high sediment supply to streams under periglacial conditions resulted in the formation of flights of river terraces. Along the Váh River, 11 river terraces with individual bedrock levels were formed since the Early Pleistocene (Vitovič 2018), while only six preserved river terraces record the formation of the Poprad River valley (Lukniš 1973; Vitovič 2019).

During the Late Pleistocene to Holocene, the regional stress field changed, resulting in extension with general direction parallel to the strike of the Western Carpathians orogen (W–E). This caused activation of transverse faults trending generally N–S with variations in NW–SE to NE–SW direction. This younger fault system is responsible for the segmentation of the PtK into partial blocks and disintegration of its margin-bounding faults (Gross 1980; Littva 2017).

Acknowledgements This work was supported by the grant of the Scientific Grant Agency of the Ministry of Education, Science, Research and Sport of the Slovak Republic and the Slovak Academy of Sciences (VEGA) No. 1/0146/19 and 1/0602/16 and by the Slovak Research and Development Agency under the contract No. APVV-15-0054.

References

Aubrecht R (2014) Terénne cvičenia zo stratigrafie (Stratigraphic field exercises). Prírodovedecká fakulta Univerzity Komenského, Bratislava

Baňacký V, Halouzka R, Horniš J, Kernáts G, Maglay J, Pristaš J (1993) Geodynamický vývoj územia Slovenska počas kvartéru. In: Geodynamický vývoj a hlbinná stavba Západných Karpár. Zborník referátov zo seminára uskutočneného v dňoch 17. a 18. 12. 1992. GÚDŠ, Bratislava, pp 239–242

Bella P (2001) To the palaeogeomorphological development of fluviokarstic caves in the Demänovské Hills on the northern side of the Nízke Tatry Mts. Geomorphologia Slovaca 1(1):54–63

Bella P (2005) K morfológii a genéze Liskovskej jaskyne. Slovenský Kras 43:37–52

Bella P, Bosák P, Pruner P, Glazek J, Hercman H (2011) The developement of the River Biela Valley in relation to the genesis of the Belianska Cave. Geografický Časopis 63(4):369–387

Bella P, Bosák P (2012) Speleogenesis along deep regional faults by ascending waters: case studies from Slovakia and Czech Republic. Acta Carsologica 41(2–3):169–192

Bella P, Littva J, Pruner P, Bosák P, Šlechta S, Hercman H, Čížková K (2016) Geologická stavba, morfológia a vývoj Važeckej jaskyne. Slovenský Kras 54(1):5–31

Bezák V, Mello J, Nemčok J, Reichwalder P (1993) Tektonický vývoj (Tatier) v terciéri a kvartéri. In: Nemčok J (ed) Vysvetlivky ku geologickej mape Tatier 1 : 50 000. GÚDŠ, Bratislava, pp 113–116

Biely A, Beňuška P, Bezák V, Bujnovský A, Halouzka R, Ivanička J, Kohút M, Klinec A, Lukáčik E, Maglay J, Miko O, Pulec M, Putiš M, Vozár J (1992) Geologická mapa Nízkych Tatier 1: 50 000. GÚDŠ, Bratislava

Bónová K, Bella P, Kováčik M, Bóna J, Petro Ľ, Kollárová V, Kovaničová Ľ (2014) Alochtónne jemnozrnné sedimenty a ich vzťah ku genéze Liskovskej jaskyne (Chočské podhorie, severné Slovensko). Acta Geologica Slovaca 6(2):145–158

Cohen KM, Gibbard PL (2018) Global chronostratigraphical correlation table for the last 2.7 million years, v. 2016a. http://www.stratigraphy.org/upload/QuaternaryChart1.JPG. Accessed 13 June 2019

Droppa A (1962) Speleologický výskum Važeckého krasu. Geografický časopis 14(4):264–293

Droppa A (1964) Výskum terás Váhu v strednej časti Liptovskej kotliny. Geografický Časopis 16(4):313–325

Droppa A (1967) Krasové javy v doline Bieleho Váhu. Geografický Časopis 19(2):141–153

Droppa A (1968) Geomorfologický výskum krasových ostrovov v Liptovskej kotline. Geografický Časopis 20(4):328–342

Droppa A (1970) Výskum riečnych terás v zátopovej oblasti Liptovská Mara. Liptov 1. Vlastivedný zborník. Osveta, Martin, pp 7–34

Droppa A (1971) Geomorfologický výskum Liskovskej jaskyne v Liptovskej kotline. Československý Kras 20:75–84

Droppa A (1972) Výskum riečnych terás v okolí Ružomberka. Liptov 2. Vlastivedný zborník. Osveta, Martin, pp 11–25

Dzurovčin L (2000) Geomorfológia. Vysokoškolské učebné texty. FHPV Prešov 268 pp

Engel Z, Mentlík P, Braucher R, Minár J, Léanni L, Aster Team (2015) Geomorphological evidence and 10Be exposure ages for the Last Glacial Maximum and deglaciation of the Velka and Mala Studena dolina valleys in the High Tatra Mountains, central Europe. Quaternary Sci Rev 124:106–123. https://doi.org/10.1016/j.quascirev.2015.07.015

Faško P, Šťastný P (2002) Priemerné ročné úhrny zrážok 1 : 2 000 000. In Atlas krajiny Slovenskej republiky. Ministerstvo životného prostredia SR, Bratislava; Slovenská agentúra životného prostredia, Banská Bystrica, p 99

Franko O (2001) Pôvod a vývoj minerálnych a termálnych vôd Slovenska v priestore a čase z pohľadu veku travertínov a izotopov O, H a ^{14}C. Podzemná Voda 7(2):26–45

Franko O, Hanzel V (1980) Hydrogeologické pomery Liptovskej kotliny. In: Gross P, Köhler E, Biely A, Franko O, Hanzel V, Hricko J, Kupčo G, Papšová J, Priechodská Z, Szalaiová V, Snopková P, Stránska M, Vaškovský I, Zbořil Ľ (eds) Geológia Liptovskej kotliny. GÚDŠ, Bratislava, p 242

Gradziński M, Wróblewski W, Bella P (2015) Cenozoic freshwater carbonates of the Central Carpathians (Slovakia): facies, environments, hydrological control and depositional history. In: Haczewski G (ed) Guidebook for field trips accompanying 31st IAS meeting of sedimentology held in Kraków on 22nd–25th of June 2015. Polish Geological Society, Kraków, pp 217–245

Gross P (1971) Geológia západnej časti Liptovskej kotliny. Geologické Práce, Správy 56:109–124

Gross P (1978) Paleogén pod stredoslovenskými neovulkanitmi. In: Vozár J (ed) Paleogeografický vývoj Západných Karpát. GÚDŠ, Bratislava, pp 121–145

Gross P (1979) Geologická mapa Liptovskej kotliny 1: 50 000. GÚDŠ, Bratislava

Gross P (2008) Lithostratigraphy of Western Carpathians: Paleogene – Podtatranská group. ŠGÚDŠ, Bratislava, p 78

Gross P, Biely A, Halouzka R, Samuel O, Hanzel V, Šucha P, Kováčik M, Mikuška J, Snopková P, Raková J, Köhler E (1990) Vysvetlivky ku geologickej mape 1: 25 000, list 27–333 Poprad-3. Manuscript. Archív GS SR, Bratislava

Gross P, Köhler E (eds), Mello J, Haško J, Halouzka R, Nagy A, Kováč P, Filo I, Havrila M, Maglay J, Salaj J, Franko O, Zakovič M, Pospíšil L, Bystrická H, Snopková P (1993) Geológia južnej a východnej Oravy, GÚDŠ, Bratislava, p 320

Gross P, Buček S, Ďurkovič T, Filo I, Karoli S, Maglay J, Nagy A, Halouzka R, Spišák Z, Žec B, Vozár J, Borza V, Lukáčik E, Mello J, Polák M, Janočko J (1999) Geologická mapa Popradskej kotliny, Hornádskej kotliny, Levočských vrchov, Spišsko-šarišského medzihoria, Bachurne a Šarišskej vrchoviny 1: 50 000. GS SR, Bratislava

Gross P, Köhler E, Samuel O (1984) Nové litostratigrafické členenie vnútrokarpatského paleogénu. Geologické Práce, Správy 81:103–117

Gross P (1980). Tektonika. In Gross P, Köhler, E, Biely A, Franko O, Hanzel V, Hricko J, Kupčo G, Papšová J, Priechodská Z, Szalaiová V, Snopková P, Stránska M, Vaškovský I, Zbořil Ľ (eds) Geológia Liptovskej kotliny. GÚDŠ, Bratislava, p 242

Gross P et al. (1999) Charakteristika tektonických pomerov a tektonický vývoj územia. In: Gross P, Buček S, Ďurkovič T, Filo I, Maglay J, Halouzka R, Karoli S, Nagy A, Spišák Z, Žec B, Vozár J, Borza V, Lukáčik E, Janočko J, Jetel J, Kubeš P, Kováčik M, Žáková E, Mello J, Polák M, Stráňová Z, Samuel O, Snopková P, Raková J, Zlinská A, Vozárová A, Žecová K (eds) Vysvetlivky ku geologickej mape Popradskej kotliny, Hornádskej kotliny, Levočských vrchov, Spišsko-Šarišského medzihoria, Bachurne a Šarišskej vrchoviny. GÚDŠ, Bratislava, pp 142–157

Gross P, Köhler E, Papšová J, Snopková P (1980) Geológia a stratigrafia sedimentov vnútrokarpatského paleogénu. In: Gross P, Köhler E, Biely A, Franko O, Hanzel V, Hricko J, Kupčo G, Papšová J, Priechodská Z, Szalaiová V, Snopková P, Stránska M, Vaškovský I, Zbořil Ľ (eds) Geológia Liptovskej kotliny. GÚDŠ, Bratislava, p 242

Gross P, Köhler E (1980) Paleogeografia. In: Gross P, Köhler E, Biely A, Franko O, Hanzel V, Hricko J, Kupčo G, Papšová J, Priechodská Z, Szalaiová V, Snopková P, Stránska M, Vaškovský I, Zbořil Ľ (eds) Geológia Liptovskej kotliny. GÚDŠ, p 242

Halouzka R (1986) Z nových poznatkov o stratigrafii kvartéru terasových náplavov riek Západných Karpát (Stredné Pohronie, Orava a Turiec). Spr. Výsk. gúdš 21:167–175

Halouzka R (1987) Stratigrafia a geologicko-paleogeografický vývoj kvartéru v Západných Tatrách a ich predpolí. Čiastková záverečná správa za rok 1987. GÚDŠ, Bratislava (Archive number of report in archive of Geofond: 65146)

Halouzka R (1993) Neotektonické štruktúry a pohyby (vrchný pliocén – kvartér) vybraných kotlín a dolín slovenských Karpát. In: Maglay J, Baňacký V, Halouzka R, Horniš J, Pristaš J (eds). Geodynamický vývoj regiónov Slovenska v období vrcný pliocén – kvartér. GÚDŠ, Bratislava, p 51 (Archive number of final report in archive of Geofond: 78841)

Halouzka R, Maglay J, Baňacký V, Pristaš J, Janočko J, Hók J (1999) Vysvetlivky k neotektonickej mape Slovenska 1: 500 000. GS SR, Bratislava

Halouzka R, Raczkowski W (1993) Kvartér. In: Nemčok J, Bezák V, Janák M, Kahan Š, Ryka W, Kohút M, Lehotský I, Wieczorek J, Zelman J, Mello J, Halouzka R, Raczkowski W, Reichwalder P (1993) Vysvetlivky ku geologickej mape Tatier 1:50 000. Vydavateľstvo Dionýza Štúra, Bratislava, pp 67–96

Hanzel V (1979) Základný hydrogeologický výskum Vysokých Tatier a ich predpolia. Čiastková záverečná správa. GÚDŠ, Bratislava (Archive number of final report in archive of Geofond: 45340)

Havránek F (1935) Važecká jeskyně a její kras (3rd ed). Edice Tatranského a jaskynného múzea, Važec, p 140

Hochmuth Z (1977) Jaskyne v Liskovskom kameňolome. Slovenský kras 15:129–134

Hochmuth Z (1983) Kras Chočských vrchov, Liskovská jaskyňa. In: Hlaváč J, Hochmuth Z, Vozárik P, Valaštiak M (eds) Exkurzný sprievodca krasom Západných Tatier a Chočských vrchov (24. jaskyniarsky týždeň). Príloha Spravodaja Slovenskej speleologickej spoločnosti 14(1), p 23

Hochmuth Z (1997) Príspevok k problematike genézy jaskynných úrovní v Liskovskej jaskyni. Slovenský Kras 35:89–96

Ivan Ľ (1943) Výskyty travertínov na Slovensku. Práce Štátneho geologického ústavu 9:71

Jakál J (1993) Karst Geomorphology of Slovakia. Geographia Slovaca 4, p 38

Janáčik P (1968) Zpráva o výskume a o prieskumných prácach v Liskovskej jaskyni. Slovenský Kras 6:83–86

Janík S (1970) Správa z geofyzikálnych meraní v oblasti Liptovskej kotliny, hydrogeologický prieskum. Ústav užitej geofyziky Brno, závod Bratislava (Archive number of final report in archive of Geofond: 23355)

Jetel J (1999) Hydrogeologické pomerys. In: Gross P, Buček S, Ďurkovič T, Filo I, Maglay J, Halouzka R, Karoli S, Nagy A, Spišák Z, Žec B, Vozár J, Borza V, Lukáčik E, Janočko J, Jetel J, Kubeš P, Kováčik M, Žáková E, Mello J, Polák M, Stráňová Z, Samuel O, Snopková P, Raková J, Zlinská A, Vozárová A, Žecová K (eds) Vysvetlivky ku geologickej mape Popradskej kotliny, Hornádskej kotliny, Levočských vrchov, Spišsko-Šarišského medzihoria, Bachurne a Šarišskej vrchoviny. GÚDŠ, Bratislava, pp 166–193

Jurečka M (2002) Výsledky prieskumnej činnosti v Liskovskej jaskyni (2). Spravodaj Slovenskej Speleologickej Spoločnosti 33(1):30–35

Kázmér M, Dunkl I, Frisch W, Kuhlemann J, Oszvárt P (2003) The Palaeogene forearc basin of the Eastern Alps and Western Carpathians: subduction erosion and basin evolution. J Geol Soc 160(3):413–428. https://doi.org/10.1144/0016-764902-041

Košecký E (1968) Geofyzikálne merania v oblasti Liptovská kotlina. Príloha č. 2 – C. Ústav užitej geofyziky, Brno, závod Bratislava. (Archive number of report in archive of Geofond: 30411)

Košťálik J (1999) Spišská Magura. Geológia, reliéf, geoekológia. Prírodovedecká fakulta UPJŠ, Košice, p 156

Kováč M, Kráľ J, Márton E, Plašienka D, Uher P (1994) Alpine uplift history of the Central Western Carpathians: geochronological, paleomagnetic, sedimentary and structural data. Geologica Carpathica 45(2):83–96

Kováč M, Hók J, Minár J, Vojtko R, Bielik M, Pipík R, Rakús M, Kráľ J, Šujan M, Králiková S (2011) Neogene and Quaternary development of the Turiec Basin and landscape in its catchment: a tentative mass balance model. Geologica Carpathica 62(4):361–379. https://doi.org/10.2478/v10096-011-0027-6

Kováč M, Plašienka D, Soták J, Vojtko R, Oszczypko N, Less G, Ćosović V, Fügenschuh B, Králiková S (2016) Paleogene palaeogeography and basin evolution of the Western Carpathians, Northern Pannonian domain and adjoining areas. Global Planet Change 140:9–27. https://doi.org/10.1016/j.gloplacha.2016.03.007

Kováčik M, Kohút M, Havrila M, Boorová D, Filo I, Sentpetery M, Laurinc D, Olšavský M, Maglay J, Nagy A, Šiomon L, Kollárová V, Kováčiková M, Baráth I, Žecová K, Zlinská A, Černák R, Kordík J, Šoltés S, Kucharič Ľ, Liščák P, Gluch A (2015) Vysvetlivky ku geologickej mape Žiaru 1: 50 000. ŠGÚDŠ, Bratislava, p 420

Kovanda J (1971) Kvartérní vápence Československa. Sborník geologických Věd, Antropozoikum (Praha), Ústřední Ústav geologický (Naklalatelství ČSAV), Praha 7, p 240

Králiková S, Vojtko R, Sliva Ľ, Minár J, Fügenschuh B, Kováč M, Hók J (2014) Cretaceous-Quaternary tectonic evolution of the Tatra Mts. (Western Carpathians): constraints from structural, sedimentary, geomorphological and fission track data. Geol Carpath 65(4):307–326. https://doi.org/10.2478/geoca-2014-0021

Laughlan L, Rabeder G, Sabol M (2012) The fossils and taphonomy of the Važecká Cave – preliminary results. Braunschweiger Naturkundliche Schriften 11:81–86

Lehotský M (2001) Kotlinový kras – geomorfologické pomery a aspekt exhumácie (príklad Hybského krasu). Geografický Časopis 53 (2):111–126

Littva J (2017) Geologický vývoj vnútrohorských kotlín Západných Karpát v období pliocénu a kvartéru. Dissertation thesis, Prírodovedecká fakulta Univerzity Komenského, Bratislava

Lóczy L (1878) Die Liszkovaer Höhle im Baráthegy (Liptauer Comitat). Eine vorgeschichtliche Höhlenwohnung und deren Überreste. Friedrich Killian´s, Budapest, p 55

Lukniš M (1968) Geomorfologická mapa Vysokých Tatier a ich predpolia 1: 50 000. GÚDŠ, Bratislava

Lukniš M (1972) Slovensko 2: Príroda. Obzor, Bratislava, p 917

Lukniš M (1973) Reliéf Vysokých Tatier a ich predpolia. Vydavateľstvo SAV, Bratislava, p 375

Mazúr E (1980) Podtatranská kotlina. In: Encyklopédia Slovenska. Veda, Bratislava, p 575

Madarás J, Marko F, Vojtko R, Šedivá – Kohoutová A, Pekárik L (2012) Zmena rozvodia Popradu a Hornádu v kvartéri: príklad neotektonických pohybov na vikartovskom zlome v pohorí Kozie chrbty (Západné Karpaty). In: 2. Otevřený kongres České geologické společnosti a Slovenskej geologickej spoločnosti, 21–25. 9. 2011 Monínec

Maglay J, Halouzka R (1999) Kvartér. In: Gross P, Buček S, Ďurkovič T, Filo I, Maglay J, Halouzka R, Karoli S, Nagy A, Spišák Z, Žec B, Vozár J, Borza V, Lukáčik E, Janočko J, Jetel J, Kubeš P, Kováčik M, Žáková E, Mello J, Polák M, Stráňová Z, Samuel O, Snopková P, Raková J, Zlinská A, Vozárová A, Žecová K (eds) Vysvetlivky ku geologickej mape Popradskej kotliny, Hornádskej kotliny, Levočských vrchov, Spišsko-Šarišského medzihoria, Bachurne a Šarišskej vrchoviny. GÚDŠ, Bratislava, pp 110–142

Maglay J, Halouzka R, Baňacký V, Pristaš J, Janočko J (1999) Neotektonická mapa Slovenska v mierke 1: 500 000. MŽP SR; GS SR, Bratislava

Maglay J, Moravcová M, Kučera M (2011) Vysvetlivky ku geologickej mape kvartéru Slovenska 1: 500 000. ŠGÚDŠ, Bratislava, p 94

Maglay J, Pristaš J, Kučera M, Ábelová M (2009) Geologická mapa kvartéru – Mapa hrúbky kvartérneho pokryvu [online]. ŠGÚDŠ, Bratislava. http://mserver.geology.sk:8399/arcgis/services/WMS/HKP500/MapServer/WMSServer. Accessed 20 January 2016

Makos M, Dzierzek J, Nitychoruk J, Zreda M (2014) Timing of glacier advances and climate in the High Tatra Mountains (Western Carpathians) during the Last Glacial Maximum. Quat Res 82(1):1–13. https://doi.org/10.1016/j.yqres.2014.04.001

Mazúr E (1963) Žilinská kotlina a priľahlé pohoria (geomorfológia a kvartér). SAV, Bratislava, p 188

Mazúr E, Lukniš M (1978) Regionálne geomorfologické členenie SSR. Geografický časopis 30(2):101–125

Mazúr E, Činčura J, Kvitkovič J (1980) Geomorfológia 1 : 500 000. In: Atlas Slovenskej socialistickej republiky. Slovenská akadémia vied; Slovenský úrad geodézie a kartografie, Bratislava, pp 46–47

Menning M, Hendrich A (2016) Stratigraphic Table of Germany 2016. German Research Centre for Geosciences GFZ, Potsdam

Michaeli E (2001) Georeliéf Hornádskej kotliny. Geografické Prace 9 (2):153

Minár J, Bielik M, Kováč M, Plašienka D, Barka I, Stankoviansky M, Zeyen H (2011) New morphostructural subdivision of the Western Carpathians: An approach integrating geodynamics into targeted morphometric analysis. Tectonophysics 502(1):158–174. https://doi.org/10.1016/j.tecto.2010.04.003

Nemčok J, Bezák V, Biely A, Gorek A, Gross P, Halouzka R, Janák M, Kahan Š, Kotansky Z, Lefeld J, Mello J, Reichwalder P, Raczkowski W, Roniewicz P, Ryka W, Wieczorek J, Zelman J (1994) Geologická mapa Tatier 1: 50 000. GÚDŠ, Bratislava

Páleník M (1988) Sprievodná správa k inžinierskogeologickej mape M 1 : 10 000 Ružomberok. Inžinierskogeologický a hydrogeologický prieskum Žilina. (Archive number of final report in archive of Geofond: 74842)

Pánek T, Minár J, Vitovič L, Břežný M (2020) Post-LGM faulting in Central Europe: LiDAR detection of the >50 km-long Sub-Tatra fault, Western Carpathians. Geomorphology 364:107248. https://doi.org/10.1016/j.geomorph.2020.107248

Pešková I, Hók J (2008) Quaternary dynamics of the Liptovská kotlina Basin inffered from the trabertine deposits disruption and the Váh River terraces assymetry. Mineralia Slovaca- Geovestník 40(3–4):220

Plašienka D, Soták J (2015) Evolution of Late Cretaceous-Palaeogene synorogenic basins in the Pieniny Klippen Belt and adjacent zones (Western Carpathians, Slovakia): tectonic controls over a growing orogenic wedge. Annales Societatis Geologorum Polonieae 85 (1):43–76. https://doi.org/10.14241/asgp.2015.005

Polák M, Bujnovský A, Kohút M, Filo I, Pristaš J, Havrila M, Vozár J, Mello J, Rakús M, Buček S, Lexa J (1997) Geologická mapa Veľkej Fatry 1: 50 000. GS SR, Bratislava

Pošivák D (2019) Geomorfologická lokalita Veľká Žltá stena (Tatranské podhorie) v jej priestorových súvislostiach. Master thesis, Prírodovedecká fakulta Univerzity Komenského, Bratislava

Pristaš J (1997) Kvartér. In: Polák M, Bujnovský A, Kohút M, Pristaš J, Filo I, Havrila M, Vozárová A, Vozár J, Kováč P, Lexa J, Rakús M, Malík P, Liščák P, Hojstričová V, Žáková E, Siráňová Z, Boorová D, Fejdiová O (1997) Vysvetlivky ku geologickej mape Veľkej Fatry 1:50 000. Vydavateľstvo Dionýza Štúra, Bratislava, pp 104–114

Sabol M, Višňovská Z (2007) Cave Bears from the Važecká Cave (Northern Slovakia) science and exhibitions. Scripta Facultatis Scientiarum Naturalium Universitatis Masarykianae Brunensis, Geology 35:145–150

Škvarček A (1968) Niekoľko poznámok ku kvartérno-geomorfologickému vývoju strednej časti Liptovskej kotliny pod Nízkymi Tatrami. Geografický Časopis 20(4):354–358

Soták J (2010) Paleoenvironmental changes across the Eocene-Oligocene boundary: insights from the Central-Carpathian Paleogene Basin. Geol Carpath 61(5):393–418. https://doi.org/10.2478/v10096-010-0024-1

Soták J, Pereszlenyi M, Marschalko R, Milička J, Starek D (2001) Sedimentology and hydrocarbon habitat of the submarine-fan deposits of the Central Carpathian Paleogene Basin (NE Slovakia). Mar Petrol Geol 18:87–114. https://doi.org/10.1016/S0264-8172(00)00047-7

Starek D, Fuksi T (2017) Statistical analysis as a tool for identification of depositional palaeoenvironments in deep-sea fans (Palaeogene formations, Central Western Carpathians, North Slovakia). Acta Geologica Slovaca 9(2):149–162

Tokarski AK, Świerczewska A, Zuchiewicz W, Starek D, Fodor L (2012) Quaternary exhumation of the Carpathians: a record from the Orava-Nowy Targ Intramontane Basin, Western Carpathians (Poland and Slovakia). Geol Carpath 63(4):257–266. https://doi.org/10.2478/v10096-012-0021-7

Tužinský A, Banský V, Gross P, Hanzel V, Hlubocký B, Hornung T, Janík S, Košecký E, Mucha I, Pelikán V, Petráš I (1971) Liptovská kotlina – hydrogeologický prieskum (paleogén, kvartér). Inžinierskogeologický a hydrogeologický prieskum Žilina. (Archive number of final report in archive of Geofond: 30411)

Tužinský Á, Banský V (1968) Ročná správa za rok 1968. Regionálny hydrogeologický prieskum Liptovskej kotliny. Inžinierskogeologický a hydrogeologický prieskum Žilina. (Archive number of final report in archive of Geofond: 21011)

Vaškovský I (1980) Geológia kvartérnych sedimentov. In: Gross P, Köhler E, Biely A, Franko O, Hanzel V, Hricko J, Kupčo G, Papšová J, Priechodská Z, Szalaiová V, Snopková P, StránskaM, Vaškovský I, Zbořil Ľ (eds) Geológia Liptovskej kotliny. ŠGÚDŠ, Bratislava, pp 96–115

Vitásek F (1923) Příspevky k poznání ledové doby v Nižných Tatrách. Sborník Státního Geologického Ústavu Československé Republiky 3:209–224

Vitásek F (1932) Terasy horního Váhu. In Spisy odboru Československé společnosti zeměpisné v Brně, řada A, č. 4. Praha, p 23

Vitovič L (2018) Riečne terasy a intenzita erózie Váhu v Liptovskej kotline. In: Študentská vedecká konferencia PriF UK 2018: zborník recenzovaných príspevkov (25.4.2018 Bratislava). Univerzita Komenského, Bratislava, pp 1345–1350

Vitovič L (2019) Neotectonic evolution of the Poprad River valley in the eastern part of the Popradská kotlina Basin based on the river terraces analysis. In: Študentská vedecká konferencia PriF UK 2019: zborník recenzovaných príspevkov (9.4.2018 Bratislava). Univerzita Komenského, Bratislava, pp 1231–1236

Vitovič L (2020a) Analysis of fluvial systems in the vicinity of Bežan hill (Liptovská kotlina Basin, Western Carpathians). Geografické Informácie 70(3):197–216

Vitovič L (2020b) Odraz morfotektonického vývoja Podtatranskej kotliny v charaktere jej fluviálnych systémov. [Morphotectonic evolution of the Podtatranská kotlina Basin inferred from the character of fluvial sediments]. Dissertation thesis, Univerzita Komenského, Bratislava

Vitovič L, Minár J (2018) Morphotectonic analysis for improvement of neotectonic subdivision of the Liptovská kotlina Basin (Western Carpathians). Geografický Časopis 70(3):197–216. https://doi.org/10.31577/geogrcas.2018.70.3.11

Vojtko R, Marko F, Preusser F, Madarás J, Kováčová M (2011) Late Quaternary fault activity in the Western Carpathians: evidence from the Vikartovce Fault (Slovakia). Geol Carpath 62(6):563–574. https://doi.org/10.2478/v10096-011-0040-9

Vojtko R, Králiková S, Kriváňová K, Vojtková S (2015) Lithostratigraphy and tectonics of the eastern part of Veporské vrchy Mts. (Western Carpathians). Acta Geologica Slovaca 7(2):113–127

Volko-Starohorský J (1930) Važecká jaskyňa (Liptov). Příroda 23:3–8 (zvláštný odtisk)

Zuchiewicz W (2011) Planation surfaces in the polish Carpathians: myth or reality? Part 2. Geographia Pol Spec Iss 84:155–178

Ladislav Vitovič is a geomorphologist at State Geological Institute of Dionýz Štúr in Bratislava. His scientific interest focuses on morphotectonics and Quaternary sediments of the Western Carpathians.

Jozef Minár is Professor of Physical Geography and Geoecology at the Comenius University in Bratislava. His research is focused on the geomorphometry and theoretical geomorphology, tectonic geomorphology of the Western Carpathians and digital geomorphological mapping.

Pavel Bella is Associate Professor of Physical Geography and Geoecology at the Catholic University in Ružomberok and head of the cave protection department at the State Nature Conservancy of the Slovak Republic, Slovak Caves Administration in Liptovský Mikuláš. His research is focused mainly on karst geomorphology and speleology, especially on cave morphology and genesis, hypogene speleogenesis, and the reconstruction of cave development linked with landscape evolution. He also deals with cave geoecology and environmental problems in karst landscape and caves.

Juraj Littva works as a geologist at Slovak Caves Administration at State Nature Conservancy of the Slovak Republic. His work is focused on karst and geology, with a particular interest in the relationship between karst and tectonics as well as the coevolution of caves and landscape.

Limestone Klippen Belt—Atypical Landforms in Flysch Uplands

10

Dušan Plašienka and Ján Novotný

Abstract

The Pieniny Klippen Belt of the Western Carpathians is a peculiar tectonic and geomorphological zone that is only a few km wide, but up to 600 km long. It separates the External Western Carpathians from the Central Western Carpathians. The Pieniny Klippen Belt is exceptional by its picturesque rugged landscape, being formed by numerous, and often isolated cliffy hills called "klippen" surrounded by a smooth relief composing the "klippen mantle". The majority of klippen is composed of comparatively hard, Middle Jurassic to Lower Cretaceous limestones, surrounded by softer in places Lower Jurassic, but mostly Upper Cretaceous to Middle Eocene shales, marls and flysch deposits. In these geological settings, the selective erosion created a very contrasting mosaic of landforms in a relatively small area. In the Pieniny Mts., it has the form of a mountain range with relatively high rock peaks and deep river gorges. The Middle Ages enriched some klippen with castles, which have been preserved to the present, especially in the form of mysterious ruins. All these attributes make the Pieniny Klippen Belt one of the most valuable and interesting landforms in Slovakia, with significant geotouristic potential.

Keywords

Western Carpathians • Pieniny Klippen Belt • Tectonic mélange • Limestone klippen • Rock control • Selective erosion

D. Plašienka (✉)
Department of Geology and Palaeontology, Faculty of Natural Sciences, Comenius University, Mlynská dolina, Ilkovičova 6, 842 15 Bratislava, Slovakia
e-mail: dusan.plasienka@uniba.sk

J. Novotný
Institute of Geography, Slovak Academy of Sciences, Štefánikova 49, 814 73 Bratislava, Slovakia
e-mail: jan.novotny@savba.sk

10.1 Introduction

The Pieniny Klippen Belt (PKB) of the Western Carpathians is a peculiar tectonic and geomorphological zone that is only a few km wide, but up to 600 km long. Forming a backbone of the Western Carpathian orogen (Fig. 10.1), it separates the Cenozoic accretionary prism of the External Western Carpathians (EWC, Flysch Belt) from the Cretaceous basement/cover thrust stack of the Central Western Carpathians (CWC; cf. Froitzheim et al. 2008; Plašienka 2018a). The PKB is exceptional by its picturesque rugged landscape, being formed by numerous, often isolated cliffy hills called "klippen" surrounded by a smooth relief composing the "klippen mantle". The majority of klippen is composed of comparatively hard, Middle Jurassic to Lower Cretaceous limestones, surrounded by softer shales, marls and flysch deposits, which are locally of Lower Jurassic, but mostly represent Upper Cretaceous to Middle Eocene succession.

Besides composition, the extraordinary complex PKB structure resulted from a long-termed deformation in a backstop position between the EWC accretionary wedge and the bulldozing CWC thick-skinned thrust stack (e.g. Plašienka and Soták 2015; Plašienka et al. 2020). As a result, some parts of the PKB show a disorganized, in part nearly chaotic block-in-matrix structure that has been often described as a megabreccia or mélange, interpreted as either tectonic or sedimentary, or both. In general, the PKB was formed by the latest Cretaceous to Middle Eocene thrust stacking of several independent nappe units and by superimposed Late Eocene–Early Miocene out-of-sequence thrusting and wrenching that caused extensive fragmentation of sedimentary successions.

Klippen may have various sizes, from a few metres to several hundred metres long. Uhlig (1904) estimated the number of klippen up to five thousand, Mahel' (1989) to more than two thousand, and Andrusov (1938) counted 125 klippen in the Orava segment of PKB. The term "klippe" has

M. Lehotský and M. Boltižiar (eds.), *Landscapes and Landforms of Slovakia*, World Geomorphological Landscapes,
https://doi.org/10.1007/978-3-030-89293-7_10

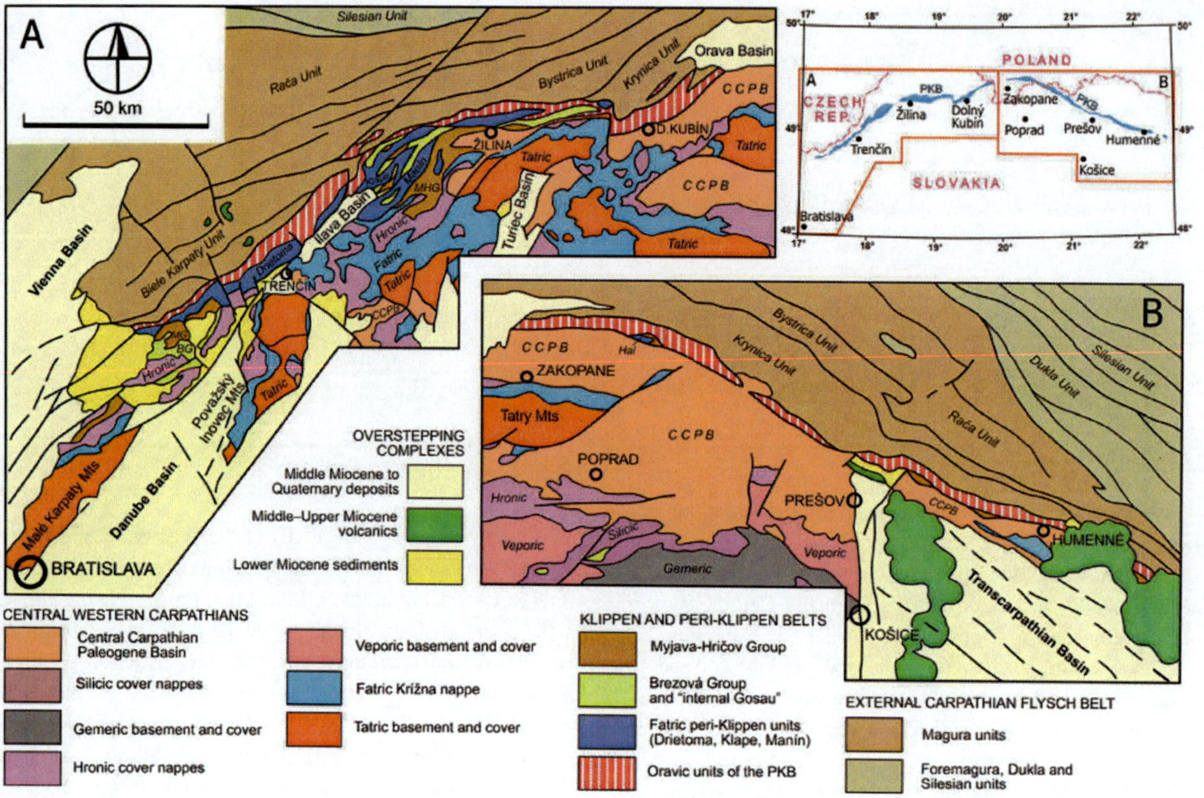

Fig. 10.1 Schematic map showing position of the Pieniny Klippen Belt in the tectonic framework of the Western Carpathians. Abbreviations: CCPB—Central Carpathian Paleogene Basin; BG—Brezová Group (Senonian); MHG—Myjava-Hričov Group (Paleocene–Middle Eocene); Hal—Haligovce Unit

become to be widely used starting from the first half of nineteenth century and the zone of klippen was designated as the "Klippenkalkzug" or "Klippenkalk-Gruppe" (e.g. Stur 1860). Afterwards, Neumayr (1871) introduced the term "Pieninische Klippenzug", i.e. the "Pieniny Klippen Belt" as it is commonly used in the current English-written texts (for the historical overviews see Andrusov 1938, 1964; Scheibner 1968; Birkenmajer 1977, 1986; or Plašienka 2018b).

10.2 Composition and Geological Structure of the Klippen Belt

The PKB as perhaps the most outstanding regional tectonic zone of the Western Carpathians has attracted the attention of geologists for a long time. It was marked like a "tectonic megabreccia", "raisins in cake", "chaotic mélange" or "wonder of nature", which names pertinently express the character of this extremely complex zone. Despite the comparatively small areal extent, a great number of tectonic units, nappes, sedimentary series, successions, developments, formations or morphological klippen types have been distinguished and innumerable opinions about their palaeogeographic settings and tectonic affiliations have been expressed (milestone works of Stur 1860; Neumayr 1871; Uhlig 1890, 1903, 1904, 1907; Andrusov 1931, 1938, 1965, 1968, 1974; Scheibner 1968; Birkenmajer 1977, 1986; Marschalko 1986; Mišík et al. 1996; Mišík 1997). These and numerous other authors provided a range of stratigraphical, lithological and sedimentological observations that are mostly still valid.

The extraordinary tectonic style and rugged surface relief of the PKB are largely conditioned by the material heterogeneity of the various klippen sedimentary successions (Fig. 10.2). Well-bedded basinal Jurassic to Lower Cretaceous successions of the Peri-Klippen Fatric units (Manín, Drietoma) and of the Oravic Pieniny Unit are prone to upright folding, therefore they form "immature" klippen. These are in fact projecting cores of large-scale anticlines—often periclines, particularly in the Manín Unit (Plašienka et al. 2018). In contrast, the thick strong layer of mostly massive Middle Jurassic to Lower Cretaceous limestones of the Oravic Subpieniny Unit (Czorsztyn-type successions;

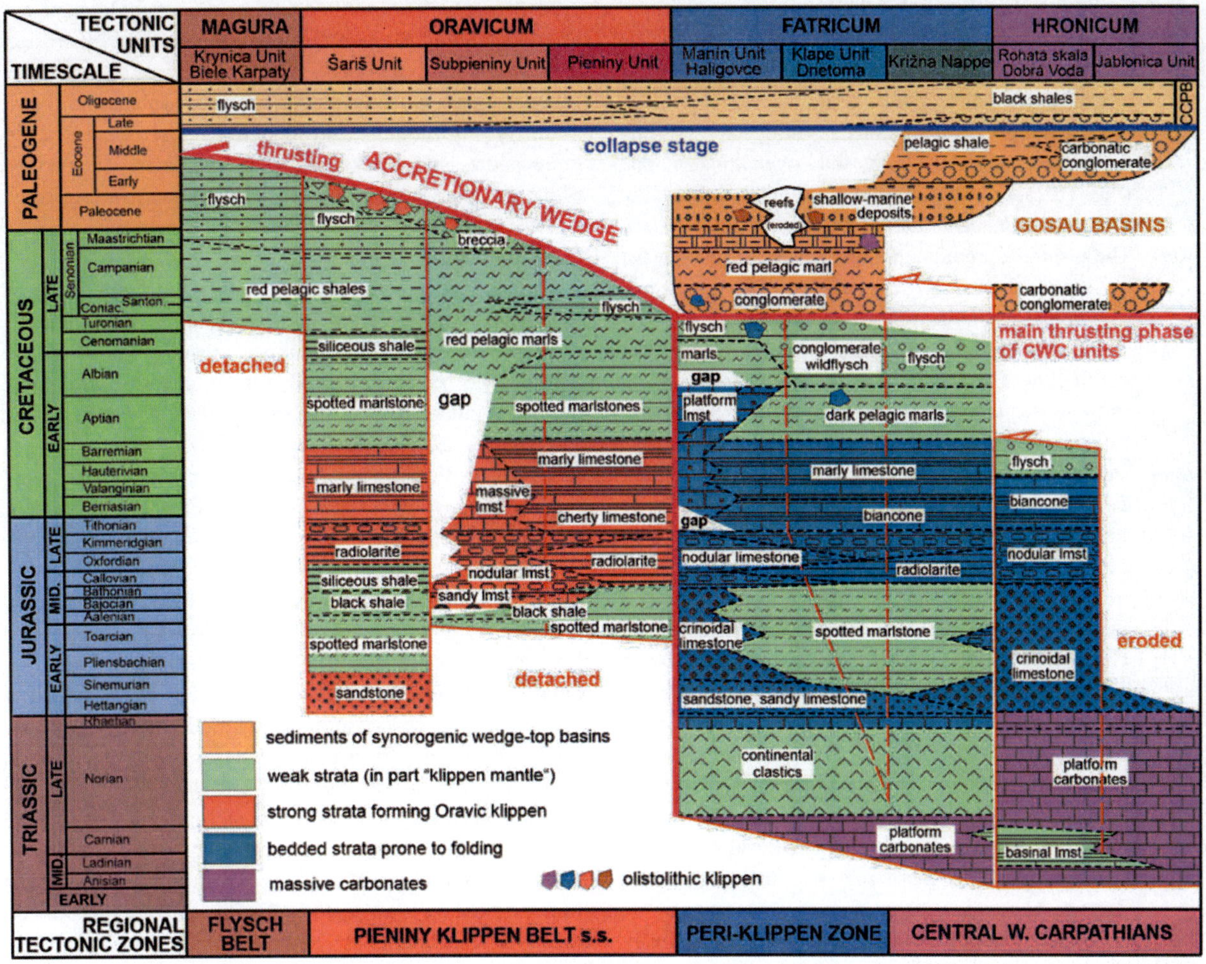

Fig. 10.2 Lithostratigraphic outline of the PKB and contiguous zones showing the material heterogeneity and competence contrasts in principal sedimentary successions and their formations

Fig. 10.2) tends to form nearly isometric blocky klippen—either tectonically separated, or as sedimentary olistoliths.

It is evident that lithology, biostratigraphy and sedimentology of the various klippen formations and successions have been very well studied and are known to great details. However, views on the PKB structure were and remain to be very diverging. In the next text, we shall follow the recent concepts of Plašienka and his co-workers (Plašienka and Mikuš 2010; Plašienka 2012a, b, 2018a, b, 2019; Plašienka and Soták 2015; Plašienka et al. 2012, 2018, 2020), which in part renew the original Uhlig's (1907) ideas and tectonic terminology. Accordingly, PKB in a broader sense (Figs. 10.1 and 10.2) entails units of two groups—Oravic Superunit (PKB s.s.) and Central Carpathian units forming the inner Peri-Klippen zone (Mahel' 1980, 1981). Three Oravic thrust units include the lowermost Šariš Unit (deep-water Jurassic–Middle Eocene succession) overridden by the Subpieniny Nappe (Jurassic–Cretaceous intra-oceanic ridge and slope successions like Czorsztyn, Niedzica, Czertezik, Pruské, e.g. Birkenmajer 1977)) and by the Pieniny Nappe (Jurassic–Cretaceous deep-marine successions of Pieniny, Kysuca, Branisko). The palaeogeographic and tectonic evolution of the PKB progressed in several main stages:

(i) The Oravic sedimentary basin originated by rifting of the European continental lithosphere during the earliest Jurassic, followed by Middle Jurassic breakup of the South Pennine oceanic zone (Piemont–Váh Ocean), with uplift of the Czorsztyn ridge as the rift shoulder. The Pieniny–Kysuca Basin neighboured the ridge from the southern side as a margin of the expanding Váh Ocean.
(ii) Individualization of the Czorsztyn ridge as an intra-oceanic continental ribbon was completed by the Early Cretaceous uplift of the rift shoulder during

oceanic rifting of the North Pennine Valais–Rhenodanubian–Magura Ocean to the north.

(iii) Sequential shortening, detachment and nappe emplacement of Oravic units is recorded by syn-thrusting, coarse-grained wildflysch deposits containing material from the overriding nappe (Fig. 10.2)—latest Cretaceous in the Subpieniny Unit, and Paleocene–Early Eocene in the Šariš Unit.

(iv) The Central Carpathian ("non-Oravic") units consist of the frontal Fatric (Krížna) nappe units—Drietoma, Manín and Klape in western and Haligovce in eastern Slovakia (Fig. 10.1), which were emplaced during the Late Cretaceous over the inner Oravic zones and subsequently incorporated into the PKB structure.

(v) During the early Paleogene (Paleocene–Lutetian), the detached PKB units formed an accretionary wedge growing by frontal accretion and experienced alternating contractional and extensional events that controlled sedimentation in both the foreland trench and wedge-top, Gosau-type piggyback basins.

(vi) Throughout the Paleogene, the western PKB branch (Fig. 10.1) was affected by NW–SE compression and gradual transfer of Oravic units from the front to the rear of the wedge (Fig. 10.2), where their original nappe structure was disintegrated by forward out-of-sequence thrusting, then backtilting and backthrusting, whereas extensive dextral transpression affected the eastern segment during the Late Eocene.

(vii) Further important deformation occurred in the Early Miocene, when the SW–NE compression lead to the narrowing of the eastern segment and sinistral transtension in the western one (Fig. 10.1).

(viii) By the Early–Middle Miocene, the PKB became welded to both the Central and External Carpathian zones, which rotated as a unity by some 50° CCW relative to Europe (Márton et al. 2013). As a result, the PKB occupies a positions of an ancient backstop boundary at the transition from the CWC rigid block to the EWC accretionary wedge (Plašienka et al. 2020).

10.3 Earlier Classifications of Klippen Forms

Several authors attempted categorization of klippen forms, types and subtypes (Birkenmajer 1958; Andrusov and Scheibner 1968; Scheibner 1968; Andrusov 1968, 1974), using both tectonic and morphologic criteria. Although differing in details, these classifications are very similar. In general, they distinguished klippen of non-tectonic and tectonic origin. The non-tectonic sedimentary klippen are either autochthonous, like reef bodies uncovered by erosion of the surrounding sediments, or allochthonous such as older bioherms emplaced in younger deposits by submarine slides, extremely big boulders in conglomerates, olistoliths in slump bodies and blocks in recent landslides, and various "pseudoklippen" like selectively eroded massive sandstone and conglomerate bodies within continuous flysch successions.

Klippen related to tectonic processes can be in situ, tectonically predisposed forms like projecting cores of anticlines with continuous sedimentary successions (the "immature klippen style" of Maheľ 1980), or remnants of cordilleras and horsts inherited from the rifting stage. The proper tectonic klippen are morphologically positive forms, which are tectonically fully individualized from the surrounding sediments (Andrusov 1974). Klippen of the Pieniny type include several subtypes—the Rudín, Vršatec and Pieniny proper, differing in relationships between the klippen and surrounding rocks. In addition, Andrusov (1974) distinguished the diapiric Czerwona skala subtype (relying on the diapiric concept of Birkenmajer 1959) and several types of compound, mainly fold-related klippen.

Based on descriptive morphostructural criteria, Plašienka (2012b) differentiated between the so-called blocky and ribbon klippen in the eastern Slovakian PKB segment. Blocky klippen are more-or-less isometric bodies formed by predominantly Jurassic massive limestones of the Czorsztyn-type successions. They occur in two considerably different settings—as tectonically constricted lenses or boudins within continuous sedimentary successions of the Subpieniny nappe (Fig. 10.3B1), or as olistoliths embedded in mass-flow breccias of the synorogenic flysch of the Šariš Unit (Fig. 10.3B3). In the first case, the internal foliations (bedding and cleavage) show attitudes generally consistent with the surrounding sediments, whereas in the second case they are randomly oriented. The ribbon klippen of the lens-like or lozenge shape are typical for the Pieniny Unit, where they are formed by well-bedded, in places folded Jurassic to Lower Cretaceous limestone and radiolarite formations preserving the structural trends for longer distances. As such, the blocky klippen correspond to the Gruppentypus, and ribbon klippen to the Reihentypus of Uhlig (Uhlig 1890, 1903). The marly-shaly klippen matrix encompassing the blocky or ribbon tectonic klippen has often penetrative scaly fabrics, while the blocky olistoliths are enclosed by more competent and weakly deformed breccias and sandstones.

Despite the current klippen morphology is largely an erosional feature, and the relationships of klippen to the enclosing strata can be rarely examined in outcrops, the main factors controlling the klippen structures have recently been tentatively defined by Plašienka (2018b). The emphasis is

A — separation modes / foliation tilt		foliation missing or unimportant	
		gentle dips <45°	steep dips >45°
		areal klippen groups	linear klippen groups
primary	TECTONIC	isolated blocks in matrix	
		plateaus	rhomboidal blocks
		carapaces	dragon backs
	SEDIMENTARY	olistoliths, slides	
		olistostromes, mass-flows	
secondary	EROSIONAL	periclines and ribs	
		table mts, monadnocks	rock needles and towers
		rugged mosaics	goat backs – cuestas
	GRAVITATIONAL	separate downslope moving blocks	
		chaotic blocks accumulated at landslide toes	
	ANTHROPOGENIC	quarries, carved crags	

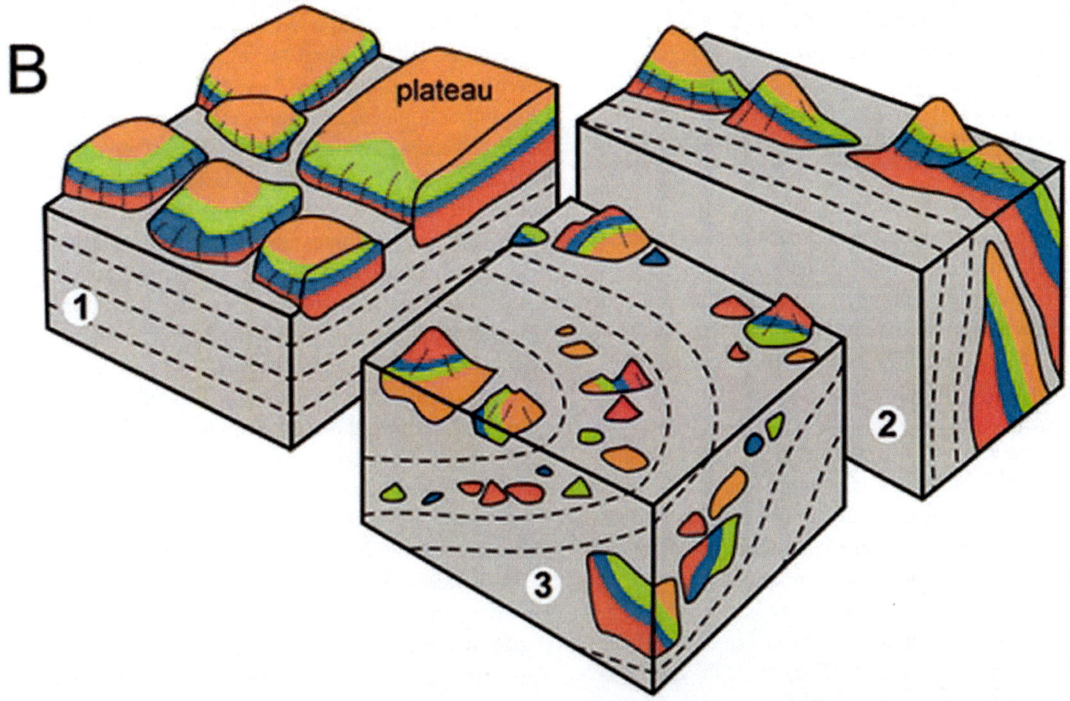

Fig. 10.3 A—Classification of the basic klippen types and forms. Modified after Plašienka (2018b). For details see that work. B—Blockdiagrams illustrating the basic types of primarily separated klippen. Klippen formations are coloured, their matrix is shown grey: 1—plateaus and their carapace-type arrangement in a subhorizontally lying succession; 2—lozenge-shaped blocks and coulisse array (dragon back) in a steeply dipping succession; 3—random distribution of blocky olistoliths and slides embedded in flysch sediments

given on the genetic factors that limited the primary klippen types, i.e. forms that originated before having been affected by surface processes. Nevertheless, the primary characteristics of klippen controlled also the secondary klippen forms shaped by the exogenous modifications, i.e. those that rule the present surface klippen forms and landscape of the PKB.

10.4 Factors Controlling the Klippen Forms

In his classification of the klippen types and forms, Plašienka (2018b) defined the main factors that controlled the klippen origin and shaping. These are the mode of klippen separation (genetic factor), their bedding attitudes and inclination of long klippen axes with respect to horizontal (structural factor), and exogenous processes modifying the klippen forms like erosion, redistribution by gravitational movements and anthropogenic impacts. The influence of these factors on the final klippen forms is schematically classified in Fig. 10.3A. There are also distinguished individual morphological klippen varieties and their mutually related groups, usually various types of short “hogbacks”.

In general, there are two main ways how the klippen could be individualized—the primary mode, whereby klippen originated before being exposed at the Earth’s surface, and secondary when klippen formed or were strongly modified by superimposed surface processes. The primary separation includes either sedimentary (i.e. exogenous) or tectonic (endogenous) modes of separation, while the secondary mode refers to post-origin processes.

10.4.1 Sedimentary Klippen—Olistoliths

There are two types of sedimentary klippen: (i) solitary olistoliths, usually large (many hundreds to several thousand cubic metres), emplaced amidst much younger hemipelagic or flysch deposits; (ii) variously sized olistoliths embedded in younger coarse-grained sandstones or breccia bodies and olistostromes associated with turbiditic flysch sequences.

The occurrences of olistostromes and olistoliths in the PKB were reviewed by Plašienka et al. (2017) in their discussion of the paper by Golonka et al. (2015). They distinguished five basic types, which differ in their tectonic position, genesis, sources and age. From structurally higher to lower and from younger to older, these are: (1) blocks of Thanetian algal-coral reefs embedded in contemporaneous or slightly younger calcareous sandstones and marls associated with tempestites, turbidites and slump bodies (Paleocene–Lutetian Myjava–Hričov Group, cf. Buček and Köhler 2017; Figs. 10.1 and 10.2). They can be regarded as intra-formational olistoliths originated from coeval rim and patch reefs. Type (2) olistoliths are similar to type (1), but older–Upper Cretaceous (Coniacian–Santonian and Maastrichtian). These are formed by re-deposited bioherms associated with conglomerates, allodapic limestones, tempestites and marlstones of nearly the same age (intra-olistoliths), as well as with extra-olistoliths of Lower Cretaceous Urgonian and Middle Triassic Wetterstein platform limestones (Fig. 10.2). Type (1) and (2) olistoliths originated probably in synorogenic wedge-top depressions in a piggyback position, above the deforming accretionary wedge and are affiliated with the Gosau Supergroup (Plašienka and Soták 2015). In the Middle Váh Valley and in the Pieniny Mts, the Gosau sediments form the post-thrusting cover overstepping the Fatric (Klape, Manín, Haligovce and Krížna; Figs. 10.1 and 10.2) and Hronic nappes of the Central Carpathian origin in the so-called Peri-Klippen zone (Mahel’ 1980).

The third olistolith type (3) is represented by the Kostolec group of limestone klippen resting within the Albian–Cenomanian hemipelagic and turbiditic deposits of the southernmost partial subunit of the Manín Unit (Fig. 10.2). The most spectacular olistoliths of type (4) occur in the Klape Unit of the Peri-Klippen zone. These are solitary slide blocks measuring tens to hundreds of metres in diameter, the distinctive Mt. Klapy klippe being the largest one (e.g. Marschalko 1986). The Klape olistoliths are embedded in hemipelagic and distal turbiditic sediments of Aptian–Early Albian age. The provenance of olistoliths in the Manín and Klape units was interpreted by Plašienka (2019).

Genesis and structural settings of type (5) olistostromes and olistoliths in the Oravic units of the PKB that originated in a contractional regime were described by Marschalko et al. (1979), Nemčok et al. (1989), Plašienka and Mikuš (2010), Plašienka et al. (2012, 2017), Plašienka (2012a), Plašienka and Soták (2015). These were formed by fragmentation of the advancing nappe fronts and downslope transportation of discharged debris and olistoliths into the foreland depressions. The Paleocene olistostromes of the Šariš Unit and Maastrichtian olistostromes of the Subpieniny Unit represent syn-thrusting mass-flow deposits composed of material released from destructive frontal edges of overriding Subpieniny and Pieniny thrust sheets, respectively. Olistostromes of the Šariš Unit carry also huge sedimentary klippen of Jurassic–Lower Cretaceous formations derived from the Subpieniny nappe, including the Czorsztyn-type and also Czertezik and/or Niedzica successions (Fig. 10.3B3).

Very rarely, the Paleocene flysch deposits of the Šariš Unit contain olistoliths of massive Lower Cretaceous basalts (Birkenmajer and Lorenc 2008; Spišiak and Sýkora 2009), possibly derived from the Czorsztyn ridge (Spišiak et al. 2011).

10.4.2 Tectonic Klippen

The proper tectonic klippen, which are prevailing in the PKB, originated by tectonic processes in the higher levels of the Earth's crust, as revealed by the essentially brittle deformation mechanisms resulting in individualization of klippen. The klippen are characterized by numerous joints and slickensides, while the enclosing klippen matrix usually shows signs of a semi-ductile scaly fabric. The competence contrast between the klippen and their matrix substantially influenced the shape of the klippen. In the case the contrast was high, e.g. between the massive Jurassic limestones and surrounding shales and marls of the Subpieniny Unit (Fig. 10.2), the klippen form prismatic blocks, often nearly isometric and klippen are always bounded by fault surfaces. Lenticular klippen developed if the competence contrast was low and the klippen strata at least partly retain the stratigraphic continuity with the matrix.

Since bedding is the most common planar anisotropy in the sedimentary rocks, it represents the weakest planar structural grain, along which a klippe can break and separate. Therefore, the surface klippen shapes are largely controlled by their bedding attitude, as the long axes of klippen are usually parallel to bedding. Origin of variously shaped tectonic klippen was characterized in works of Plašienka and Mikuš (2010), Plašienka (2012a, b), and Plašienka et al. (2012).

10.5 Landscape of the Pieniny Klippen Belt

Despite being formed by the same systems of the Oravic tectonic units and sedimentary successions, individual sectors of the PKB differ by the representation of these units and succession and by the internal structure reflecting in part distinct tectonic evolution, for instance, between the western, SW–NE trending, and the eastern, NW–SE striking branches (see Fig. 10.1, e.g. Plašienka et al. 2020). The substantial part of the western branch is characterized by the Peri-Klippen zone composed of the Fatric units that provide a transition to the CWC areas with dominantly mountainous topography. Exceptions are the westernmost PKB sector, where the PKB neighbours the Myjava Uplands composed of the comparatively soft Gosau deposits (in SE), and short sectors in the Váh River Valley adjoining small Neogene pull-apart basins (flatland Trenčín and Ilava depressions; Fig. 10.1). The outer north-western boundary against the Magura units of the Flysch Belt is then usually morphologically distinct.

The eastern branch forms a straight, narrow zone composed exclusively of the Oravic units (except the Haligovce klippe in the Pieniny Mts), which is surrounded from the SW by the flysch deposits of the Central Carpathian Paleogene Basin (CCPB), and from the NE by similar flysch deposits of the Krynica Unit of the Magura nappe system of the EWC Flysch Belt (Fig. 10.1). There, the PKB is morphologically very distinctly expressed by rugged topography, surrounded by a smooth relief of the flysch complexes.

Compared to the geological understanding, the Slovak geomorphological literature devoted to the PKB is relatively poor. Lukniš (1972) in his encyclopaedic work provides the description of the Pieniny Klippen Belt relief of the whole Slovakia. It does not appear as the independent geomorphologic unit, with the exception of the Pieniny Mts; it is mostly delimited as the taxonomic unit of lower hierarchic level with distinctly different relief type within geomorphic units of the Outer Carpathians. Morphologically, it appears as depression in contrast to surrounding morphostructures.

Most of geomorphological contributions are of older date and thus reflect older geological views of the extent and structure of the PKB. Generally, authors treated this landscape as a typical example of the passive morphostructure characterized by selective erosion of the limestone klippen embedded in a weak shaley-marly matrix.

Drdoš (1960), in the study of the relief of the Pieniny Mts., distinguished between two different types of rocks. Sharply cut peaks bind to resistant limestone complexes. Less resistant sandstone and slate formations create a rounded surface or depressions. In addition to differences in rock resistance, Drdoš (1960) emphasized the importance of the system of fissures to which the river network is connected. Directions of streams are thus influenced by both lithology and tectonic structures. In some places, watercourses encircle the resistant rocks of the ridges, while elsewhere they create gorges through them. In addition to epigenetic valleys, Drdoš also admitted signs of antecedence. The relief of the Pieniny Mts. was similarly characterized by Košťálik (1970).

Mazúr (1963) described the relief of the PKB in the Manínská vrchovina Upland, the Kysucká vrchovina Upland and in the Javorníky Mts. According to the resistance of the rocks, he set aside three typical groups of forms—klippen s. s. of hogback type, corresponding to tectonic lens of resistant carbonatic rocks; klippe monadnocks, which are more common than klippen s. s. and built of moderately resistant formations of sandstones and conglomerates; and erosive basins and furrows formed in soft fissile-shale rocks.

Stankoviansky (1983) described the development of the PKB relief in the area of the Myjavská pahorkatina Upland and the Biele Karpaty Mts. according to the morphoclimatic conditions. Older and bigger monadnocks were thought to have formed in a warm humid climate by a combination of fluvial and gravitational processes. Later, in the colder Pleistocene period, smaller klippen formed, whereas the older ones were remodelled. In conditions of periglacial climate, relief was modelled mainly by cryogenic processes.

A decade later, Stankoviansky (1994) followed a newer view of the geological structure of the PKB, which is understood more restrictively than originally. Most sandstone and conglomerate monadnocks (e.g. the Manín Unit) are no longer part of the PKB s. s. Therefore, Stankoviansky (1994) did not classify structural elevations based on rock type, but distinguished monadnocks (largest and most prominent sharp elevations) and periglacial hills (structural elevations of smaller dimensions) according to their size. The third type of landforms is structurally conditioned depressions.

In this way, it is possible to describe the relief of the PKB in most of Slovakia. The boundaries between landforms are significant in the field and they mostly coincide with the boundaries of rock outcrops of different resistance and therefore they are essentially fixed and the significance of current morphogenetic processes is smaller (Urbánek 1986; Stankoviansky 1994).

This description is fully valid mainly in the eastern PKB branch, where the resistant tectonic or sedimentary klippen blocks are distinctly individualized by erosional removal of surrounding sediments and other secondary exogenous processes, such as river abrasion and carving by incised epigenetic valleys (e.g. the Dunajec River Gorge in the Pieniny Mountains). A similar situation occurs in the north-eastern part of the western branch in the Orava region.

However, some parts of the western branch are also influenced by the active neotectonic processes, especially by uplift tendencies of some PKB segments. This was indicated, for example, by the morphostructural analysis of the Kysucké vrchy Mountains in north-western Slovakia (north of Žilina in Fig. 10.1; cf. Mazúr 1963; Novotný 2002, 2003, 2005, 2006).

Massive hogback klippen are cut through by narrow valleys. The higher lifted blocks are more strongly divided and the directions of the valleys are linked to the system of transverse faults. Neotectonic uplift of this area as well as sinking of neighbouring basins (the Žilinská kotlina Basin and the Kysucká kotlina Basin) has been proved (Maglay ed. 1999). This led to intensive headward erosion of streams mouthing in the basins. River piracy (e.g. Lacika 2002) is responsible for the transformation of the valley network (Fig. 10.4).

Secondary factors (Fig. 10.3) that modify the klippen forms and may cause their individualization involve first of all various erosional processes like mechanical abrasion by river flows, chemical dissolution of limestone rocks, exfoliation, etc. Equally important is disintegration and downslope transport of klippen blocks, triggered by gravitational forces. Gravity-driven processes include breakdown of steep cliffs and ridges along vertical fissures, tilting of released blocks, rock falls and avalanches, and sliding, slumping or creeping dependent on the slope inclination and rheology of the underlying or encircling material.

Since typical klippen are built mainly of limestone, the karst phenomenon is also a relatively important element of their modelling. There are mainly tectonic caves, enlarged by corrosion, as well as surface karst phenomena (Hochmuth 2008). Hochmuth (2008) distinguished nine karst areas within the PKB in Slovakia. The most important are the Haligovské skaly in the Pieniny Mts., with the Aksamitka Cave (total length of the corridors 335 m), and the Bošácké bradlá in the Biele Karpaty Mts., with the Landrovská jaskyňa Cave (315 m).

Klippen shapes have also been often modelled by human activities (anthropogenic factor)—from small excavations and local exploitations of raw materials for various purposes to large quarries in which mostly limestones for cement and lime factories are mined (Fig. 10.3). The database of the State Geological Institute of Dionýz Štúr (ŠGÚDŠ 2020) contains more than 20 localities within the PKB registered as mineral deposits.

10.6 Natural and Cultural Value of the PKB

The elevations of the PKB are strongly contrasting forms of relief that contribute to increase the heterogeneity and stability of the landscape. The structure of land cover and land use is very strongly limited by the properties of relief. Although the PKB covers only about 2–3% of Slovakia's territory, it is a very valuable and unique part of the landscape. In addition to the exceptional geological and geomorphological structure, specific calciphile vegetation is also subject to protection.

From the nature protection point of view, the most important area is the Pieninský národný park National Park, which was declared in 1967. It is a continuation of a national park existing in the Polish part of the Pieniny Mts. since 1932. The most valuable part of the park is the Dunajec River Gorge that is 9 km long (Fig. 10.5). The river has created a deep valley across the limestone massif, with steep 200–300 m-high rock walls and meanders. Specific natural conditions are favourable for the occurrence of rare endemic plant species (Lacika and Ondrejka 2009).

The protection of other localities within the PKB is implemented within small-scale protected areas. Most of the major klippen have the status of a Natural monument (a total of 15 objects). Larger groups of klippen form Nature

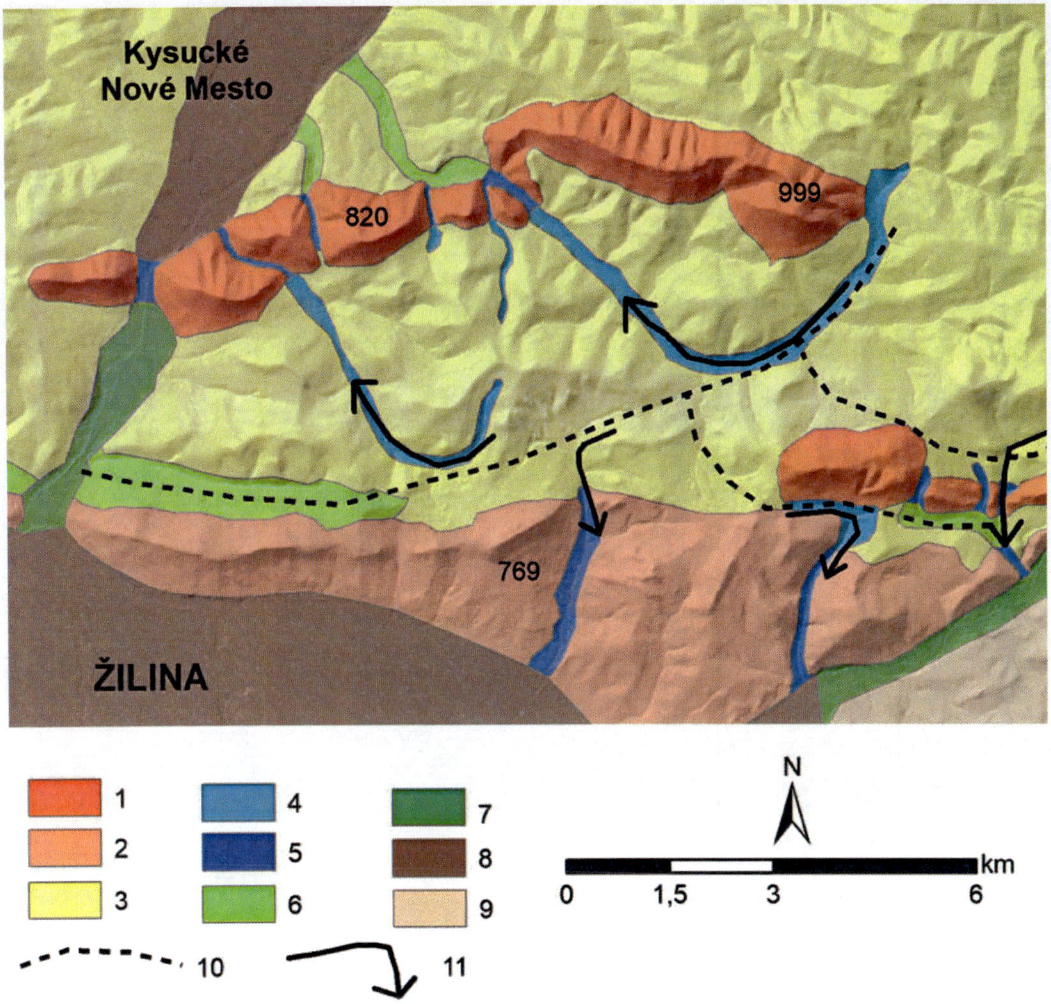

Fig. 10.4 The Kysucké bradlá Klippen, north of the town of Žilina. The case of neotectonic valley network transformation due to river piracy. 1—klippen s.s. (mainly Jurassic to Cretaceous limestones); 2—klippen monadnocks (mostly Cretaceous sandstones); 3—less dissected surface (mostly Cretaceous to Eocene shales, marls and flysch deposits); 4—upper wider parts of valleys of the small streams, locally with the elbow of capture; 5—narrow valley parts cutting through resistant rocks; 6—lower wider valley parts, locally misfit streams; 7—valleys of main rivers; 8—subsiding basins; 9—foothills of the Malá Fatra Mts; 10—presumed previous valley directions; 11—suspected sites of captures

reserves (10). The most important are the Oravské hradné bralo National Natural Monument (Fig. 10.6), the Bielska skala National Nature Reserve and the Manínska tiesňava National Nature Reserve (SAŽP 2007).

The PKB sites are also of undeniable scientific and educational significance. Many geological outcrops could be found which, due to their structure and position, can be used as reference stratigraphic profiles, suitable for calibration of sediment dating within the entire Western Carpathians (Michalík et al. 1997).

Individual klippen also played an important role in the history of human settlement. Significant rocky ridges, sharply contrasting within the surrounding smoothly modelled environment, are often located near the migration routes, which led mainly through the river valleys. They were thus more accessible than similarly formed rock formations located in remote parts of the mountains. Many of klippen have therefore become natural refuges and cult places, provided a strategic location on elevated and more easily defended places, as well as building material for the

Fig. 10.5 The Dunajec River Gorge in the Pieninský National Park (*Photo* J. Lacika)

construction of fortifications. Archaeological findings document more than 20 Slavic hillforts and fortified settlements of ancient age (Slovanské hradiská 2020) within the PKB, especially in the Váh River and the Orava River Valleys.

In the Middle Ages, castles were built on several klippen, which were the centres of administrative and military power at that time. Lacika (2016) registered 11 castles within the PKB. Only the Oravský hrad Castle has been preserved, the others are currently in ruins (Fig. 10.7).

10.7 Touristic Promotion

Klippen significantly increase the natural and aesthetic value of the landscape. From the point of view of international tourism, the already mentioned Pieninský NP has the highest potential. In addition to the exceptional natural environment, it also offers elements of traditional Goral culture, which are reflected in the local cuisine, architecture and folk art. One of the biggest attractions of the Pieniny Mts. is a rafting trip

Fig. 10.6 The Oravské hradné bralo National Natural Monument with the Oravský hrad Castle (*Photo* J. Lacika)

across the Dunajec Gorge (Fig. 10.8), during which visitors can admire the beauty of Pieniny's ridges from a special perspective. Every year, more than 80,000 visitors (PIENAP 2020) take part in these cruises on the Slovak side.

The Oravský hrad Castle is also an important destination. It rises on a massive limestone ridge above the Orava River. The castle dates from the thirteenth century and is visited annually by about 200,000 visitors (Oravské múzeum 2020). Several well-known films were made in the castle, the most famous of which is probably the silent German Expressionist horror film "Nosferatu: A Symphony of Horror" from 1922, inspired by Bram Stoker's Novel Dracula (Oravské múzeum 2020).

Other localities within the PKB have regional or local tourist potential. The most attractive are the Vršatské bradlá Klippen, the Manínska tiesňava Gorge and the ruins of the Vršatec Castle, the Lednica Castle (Fig. 10.9) or the Kamenica Castle. Most sites are accessible by educational hiking trails.

10.8 Conclusion

The Pieniny Klippen Belt is a testament to the complicated development of the Western Carpathians. Tectonic processes dispelled, compressed and tore the sedimentary rock formations deposited in the Mesozoic seas with fascinating force. This created a narrow zone of limestone cliffs, which, as a wall, separate the Inner and Outer Carpathians.

The unique character of the relief of the PKB results from its overall spatial composition. The complicated geological structure caused that selective erosion created a very contrasting mosaic of landforms in a small area. Meanwhile, the mosaic in individual localities is always somewhat different. It is determined by the facts that the geological structure of the PKB is not homogeneous and that the position of the PKB is always different regarding the neighbouring morphostructures (Novotný 2006).

Fig. 10.7 The ruins of the Vršatec Castle (*Photo* J. Lacika)

Fig. 10.8 Rafting in the Dunajec River Gorge in the Pieninský National Park (*Photo* J. Lacika)

Fig. 10.9 The Lednica Castle ruins, standing on the "dragon back" klippe in the Vršatské bradlá Klippen (*Photo* J. Lacika)

Fig. 10.10 Panoramic view of the Pieniny Mts. (*Photo* J. Lacika)

In the Pieniny Mts., it has the form of a mountain range with relatively high rock peaks (Fig. 10.10) and deep river gorges (Fig. 10.11). In the valley of the Orava River, it creates steep rock towers, reflected on the water surface. In the Biele Karpaty Mts., the Vršatské bradlá Klippen form an impressive rock town (Fig. 10.12). In the vicinity of the town of Žilina, it has the character of wooded hills. Elsewhere, they form small ridges, isolated or clustered, scattered across the surrounding country (Fig. 10.13). The Middle Ages enriched some klippen with castles, which have been preserved to the present, especially in the form of mysterious ruins. Others were used for the extraction of mineral resources, mainly limestone for cement production and construction purposes. Thanks to this, it is possible to find rock outcrops, allowing to study the geological history of this area.

All these attributes, in conjunction with the relatively easy access resulting from its location, confirm that the Pieniny Klippen Belt is one of the most valuable and

Fig. 10.11 The Dunajec River Gorge in the Pieniny Mts. (*Photo* J. Lacika)

Fig. 10.12 The Vršatské bradlá Klippen (*Photo* J. Lacika)

Fig. 10.13 The Kyjovské bradlá Klippen in the eastern part of the Pieniny Klippen Belt (*Photo* J. Lacika)

interesting landforms in Slovakia, with significant geo-touristic potential.

Acknowledgements The authors are thankful for the financial support of the PKB research from the Slovak Research and Development Agency (project APVV-17-0170) and Slovak VEGA Grant Agency (project no. 2/0052/21). We also wish to thank Dr. Ján Lacika for kindly providing the photos and to an anonymous reviewer for the thorough and constructive review.

References

Andrusov D (1931) Étude géologique de la zone des Klippes internes des Carpathes Occidentales. I-iére partie: Introduction, II-ième partie: Stratigraphie (Trias et Lias). Rozpr Státního Geol Ústavu ČSR 6:1–167

Andrusov D (1938) Étude géologique de la zone des Klippes internes des Carpathes Occidentales, IIIe partie: Tectonique. Rozpr Státního Geol Ústavu ČSR 9:1–135

Andrusov D (1964) Geologie der tschechoslowakischen Karpaten I. Akademie Verlag, Berlin, Verlag der Slowakischen Akademie der Wissenschaften, Bratislava

Andrusov D (1965) Geologie der tschechoslowakischen Karpaten II. Akademie Verlag, Berlin,Verlag der Slowakischen Akademie der Wissenschaften, Bratislava

Andrusov D (1968) Grundriss der Tektonik der nördlichen Karpaten. Verlag der Slowakischen Akademie der Wissenschaften, Bratislava

Andrusov D (1974) The Pieniny Klippen Belt. In: Maheľ M (ed) Tectonics of the Carpathian-Balkan regions. Dionýz Štúr Geological Institute, Bratislava, pp 145–158

Andrusov D, Scheibner E (1968) Classification of "Klippes" or "Klippen". In: Malkovský M (ed) Report of the twenty-third international geological congress, proceedings of section 3 – Orogenic belts. Academia, Prague, 19–23 August 1968, pp 93–102

Birkenmajer K (1958) Przewodnik geologiczny po pienińskim pasie skałkowym. Część I – szkic geologiczny pasa skałkowego (in Polish: Geological guidebook of the Pieniny Klippen Belt. Part I – geological sketch of the Klippen Belt). Wydawnictwa geologiczne, Warszawa

Birkenmajer K (1959) Diapiric tectonics in the Pieniny Klippen Belt (Carpathians). Bulletin de l'Académie Polonaise des Sciences, sér. sci. chim., géol., géogr. 7:123–128

Birkenmajer K (1977) Jurassic and Cretaceous lithostratigraphic units of the Pieniny Klippen Belt, Carpathians, Poland. Stud Geol Pol 45:1–158

Birkenmajer K (1986) Stages of structural evolution of the Pieniny Klippen Belt, Carpathians. Stud Geol Pol 88:7–32

Birkenmajer K, Lorenc MW (2008) Lower Cretaceous exotic intraplate basaltoid olistolith from Biała Woda, Pieniny Klippen Belt, Poland: geochemistry and provenance. Stud Geol Pol 131:237–246

Buček S, Köhler E (2017) Palaeocene reef complex of the Western Carpathians. Slovak Geol Mag 17(1):3–163

Drdoš J (1960) Príspevok k morfológii Pienin. Geogr časopis 12:38–61

Froitzheim N, Plašienka D, Schuster R (2008) Alpine tectonics of the Alps and Western Carpathians. In: McCann T (ed) The Geology of Central Europe, vol 2. Mesozoic and Cenozoic. Geological Society Publishing House, London, pp 1141–1232

Golonka J, Krobicki M, Waśkowska A, Cieszkowski M, Ślączka A (2015) Olistostromes of the Pieniny Klippen Belt, Northern Carpathians. Geol Mag 152(2):269–286

Hochmuth Z (2008) Krasové územia a jaskyne Slovenska. Geogr Cassoviensis 2(2):1–210

Košťálik J (1970) Geomorfologické pomery Penín. In: Karniš J et al. Peniny. Fyzickogeografická charakteristika. Geografické práce, Prešov 1(2)

Lacika J (2002) Typy riečneho pirátstva vo vulkanických pohoriach slovenských Karpát. Geogr Časopis 54(2):151–164

Lacika J (2016) Vybrané geografické aspekty rozšírenia hradov na Slovensku. In: Duchoňová D, Fundárková A (eds) Hrady a hradné panstvá na Slovensku: dejiny, majitelia, prostredie. – Historický ústav SAV – VEDA, Bratislava, pp 179–205

Lacika J, Ondrejka, K (2009) Národné parky. Dajama, Bratislava

Lukniš M (1972) Reliéf. In: Lukniš M (ed). Slovensko 2 – Príroda. Obzor, Bratislava, pp 124–202

Maglay J (ed) (1999) Neotektonická mapa Slovenska (1:500 000). GSSR, Bratislava

Maheľ M (1980) Pribradlové pásmo, charakteristika a význam (English summary: the Peri-klippen zone: nearer characterization and significance). Mineralia Slovaca 12(3):193–207

Maheľ M (1981) Island character of the Klippen Belt; Vahicum – continuation of Southern Penninicum in West Carpathians. Geol zborník – Geol Carpathica 32(3):293–305

Maheľ M (1989) Bradlové pásmo z aspektu geodynamického modelu (English summary: the Klippen Belt from the aspect of the geodynamic model). Mineralia Slovaca 21:99–108

Marschalko R (1986) Vývoj a geotektonický význam kriedového flyšu bradlového pásma (English summary: evolution and geotectonic significance of the Klippen Belt Cretaceous flysch in the Carpathian megastructure). Veda, vyd. SAV, Bratislava

Marschalko R, Haško J, Samuel O (1979) Záskalské brekcie a proces vzniku olistostrómov (English summary: Záskalie breccias and genesis of olistostromes). Geol Práce Správy 73:75–88

Márton E, Grabowski J, Plašienka D, Túnyi I, Krobicki M, Haas J, Pethe M (2013) New paleomagnetic results from the Upper Cretaceous red marls of the Pieniny Klippen Belt, Western Carpathians: evidence for general CCW rotation and implications for the origin of the structural arc formation. Tectonophysics 592:1–13. https://doi.org/10.1016/j.tecto.2013.01.027

Mazúr E (1963) Žilinská kotlina a priľahlé pohoria (Geomorfológia a kvartér). Vydavateľstvo SAV, Bratislava

Michalík J, Baráth I, Boorová D, Halásová E, Mišík M, Ožvoldová L, Rakús M, Reháková D, Salaj J, Soták J (1997) Stratigrafické národné referenčné profily na Slovensku. Mineralia Slovaca 29 (2):159–163

Mišík M (1997) The Slovak part of the Pieniny Klippen Belt after the pioneering works of D. Andrusov. Geol Carpathica 48:209–220

Mišík M, Aubrecht R, Sýkora M, Ožvoldová L (1996) New lithostratigraphic units in the Klippen Belt. Slovak Geol Mag 1:17–19

Nemčok J, Kullmanová A, Ďurkovič T (1989) Vývoj a stratigrafické postavenie gregorianskych brekcií bradlového pásma na východnom Slovensku (English summary: facies- and stratigraphical analyses of Gregorianka breccias in Klippen Belt of East Slovakia). Geol Práce Správy 89:11–37

Neumayr M (1871) Jurastudien. 5. Das Pieninische Klippenzug. Jahrbuch der kaiserlich-königlichen geologischen Reichsanstalt 21 (4):451–536

Novotný J (2002) Reliéf bradlového pásma Kysuckej vrchoviny. Geomorphol Slovaca 2(1):66–78

Novotný J (2003) Influence of neotectonics on the relief development of the Pieniny Klippen Belt (example from the Kysucká vrchovina Upland). Geomorphol Slovaca 3(2):30–37

Novotný J (2005) Pasívne morfoštruktúry Kysuckých bradiel a ich neotektonická transformácia. Geomorphol Slovaca 5(2):49–62

Novotný J (2006) Geomorfologická analýza Kysuckých bradiel (Geomorphologic Analysis of the Kysuce Klippen). Geogr Slovaca 22:1–158

Oravské múzeum (2020) Oravský hrad. https://www.oravskemuzeum.sk/expozicie/oravsky-hrad/. Accessed 20 Aug 2020

PIENAP (2020) Pieninský národný park. http://pienap.sopsr.sk/. Accessed 20 Aug 2020

Plašienka D (2012) Jurassic syn-rift and Cretaceous syn-orogenic, coarse-grained deposits related to opening and closure of the Vahic (South Penninic) Ocean in the Western Carpathians – an overview. Geol Q 56:601–628. https://doi.org/10.7306/gq.1044

Plašienka D (2012a) Early stages of structural evolution of the Carpathian Klippen Belt (Slovakian Pieniny sector). Mineralia Slovaca 44:1–16

Plašienka D (2018a) Continuity and episodicity in the early Alpine tectonic evolution of the Western Carpathians: how large-scale processes are expressed by the orogenic architecture and rock record data. Tectonics 37(7):2029–2079. https://doi.org/10.1029/2017TC004779

Plašienka D (2018b) The Carpathian Klippen Belt and types of its klippen – an attempt at a genetic classification. Mineralia Slovaca 49(1):1–24

Plašienka D (2019) Linkage of the Manín and Klape units with the Pieniny Klippen Belt and Central Western Carpathians: balancing the ambiguity. Geol Carpath 70(1):35–61. https://doi.org/10.2478/geoca-2019-0003

Plašienka D, Mikuš M (2010) Geologická stavba pieninského a šarišského úseku bradlového pásma medzi Litmanovou a Drienicou na východnom Slovensku (English summary: Geological structure of the Pieniny and Šariš sectors of the Klippen Belt between Litmanová and Drienica villages in eastern Slovakia). Mineralia Slovaca 42:155–178

Plašienka D, Soták J (2015) Evolution of Late Cretaceous-Palaeogene synorogenic basins in the Pieniny Klippen Belt and adjacent zones (Western Carpathians, Slovakia): tectonic controls over a growing orogenic wedge. Ann Soc Geol Pol 85(1):43–76. https://doi.org/10.14241/asgp.2015.005

Plašienka D, Soták J, Jamrichová M, Halásová E, Pivko D, Józsa Š, Madzin J, Mikuš V (2012) Structure and evolution of the Pieniny Klippen Belt demonstrated along a section between Jarabina and Litmanová villages in Eastern Slovakia. Mineralia Slovaca 44 (1):17–38

Plašienka D, Michalík J, Soták J, Aubrecht R (2017) Discussion of "Olistostromes of the Pieniny Klippen Belt, Northern Carpathians." Geol Mag 154(1):187–192

Plašienka D, Šimonová V, Bučová J (2018) Nucleation and amplification of doubly-plunging anticlines – the Butkov pericline case study (Manín Unit, Western Carpathians). Geol Carpath 69(4):165–181. https://doi.org/10.1515/geoca-2018-0022

Plašienka D, Bučová J, Šimonová V (2020) Variable structural styles and tectonic evolution of an ancient backstop boundary – the

Pieniny Klippen Belt of the Western Carpathians. Int J Earth Sci 109:1355–1376. https://doi.org/10.1007/s00531-019-01789-5
SAŽP (2007) Štátny zoznam osobitne chránených častí prírody SR. http://uzemia.enviroportal.sk. Accessed 20 Aug 2020
Scheibner E (1968) The Klippen Belt of the Carpathians. In: Maheľ M, Buday T (eds) Regional geology of Czechoslovakia. Part II – Western Carpathians. Academia, Praha, pp 304–371
ŠGÚDŠ (2020) Ložiská. https://gis.geology.sk/arcgis/services, Name: wgs_registreGeofondu/loziskaSR_wgs. Accessed 20 Aug 2020
Slovanské hradiská (2020) http://www.hradiska.sk/. Accessed 20 Aug 2020
Spišiak J, Sýkora M (2009) Geochémia a mineralógia bazaltov z Hanigovce – pročské vrstvy (in Slovak: Geochemistry and mineralogy of the Hanigovce basalts – the Proč beds). In: Jurkovič Ľ, Slaninka I, Ďurža, O (eds) Geochémia 2009. Konf., Symp., Sem. State Geol. Inst. D. Stur, Bratislava, pp 106–109
Spišiak J, Plašienka D, Bučová J, Mikuš T, Uher P (2011) Petrology and palaeotectonic setting of Cretaceous alkaline basaltic volcanism in the Pieniny Klippen Belt (Western Carpathians, Slovakia). Geol Q 55(1):27–48
Stankoviansky M (1983) Vplyv litologicko-štruktúrnych vlastností hornín na geomorfologické pomery bradlového pásma na príklade jeho západného úseku medzi Podbrančom a Moravským Lieskovým. In: Přibyl J, Hrádek M, Kirchner K (eds) Třicet let geomorfologie v ČSAV: Sborník referátů z geomorfologické konference, Lipovec 4. – 6. 10. 1982. Sborník prací 1. ČSAV, Brno, pp 133–140
Stankoviansky M (1994) Hodnotenie reliéfu povodia Vrzávky so zvláštnym zreteľom na jeho súčasnú modeláciu. Geogr Časopis 46:267–282
Stur D (1860) Bericht über die geologische Uebersichts-Aufnahme des Wassergebietes der Waag und Neutra. Jahrbuch der k.k. geologischen Reichsanstalt 11:17–151
Uhlig V (1890) Ergebnisse geologischer Aufnahmen in den westgalizischen Karpathen. II. Der pieninische Klippenzug. Jahrbuch der k. k. geologischen Reichsanstalt 40(3–4):559–824
Uhlig V (1903) Bau und Bild der Karpathen. In: Diener C, Hoernes R, Suess FE, Uhlig V (eds) Bau und Bild Österreichs. F. Tempsky, Wien und G. Freytag, Leipzig, pp 649–911
Uhlig V (1904) Über die Klippen der Karpathen. Congrès Géologique International Compte Rendu de la IX. Session, Vienne 1903, Premier Fascicule. Imprimerie Hollinek Fréres, Vienne, pp 427–453
Uhlig V (1907) Über die Tektonik der Karpathen. Sitzungsberichte der Kaiserischen Akademie der Wissenschaften, matematisch-naturwissenschaftliche Klasse 116(part I):871–982
Urbánek J (1986) Geomorfologické pomery Bestín a priľahlej časti Bošáckych bradiel. Geogr Časopis 38:300–321

Dušan Plašienka is a Professor of Geology at the Department of Geology and Palaeontology, Comenius University in Bratislava. He is engaged in research and teaching of the regional geology and early Alpine tectonic evolution of the Western Carpathians, particularly in the Vepor-Gemer region, Malé Karpaty and Považský Inovec Mts, and in the Pieniny Klippen Belt.

Ján Novotný is a geomorphologist at Institute of Geography, Slovak Academy of Sciences,Bratislava. His research interests focus especially on morphostructures, fluvial geomorphology, geomorphic division of the Western Carpathians and geodiversity.

Results of the Morphotectonics and Fluvial Activity of Intramountain Basins: The Turčianska Kotlina and Žiarska Kotlina Basins

11

Ján Sládek, Ladislav Vitovič, Juraj Holec, and Jozef Hók

Abstract

This contribution presents the results from two typical intramountain basins of the Western Carpathians in light of past geomorphological research. Intramountain basins are characteristic negative morphostructures of the Western Carpathians. The basins are tectonically separated from their surrounded mountains. The paper is divided into two parts separately dealing with the morphostructure and landform evolution of these basins, supported by our own research. In the part of the paper that reports on the Žiarska kotlina basin we present some basic knowledge about initial georelief formation followed by fluvial activity resulting in the creation of different landforms during thc Quaternary period (e.g., fluvial terraces, gullies etc.). The Turčianska kotlina basin is presented in light of its morphostructural evolution in connection with its surrounding mountain morphostructures. The morphostructure evolution is presented based on uncommon morphostructural analyses (isobase surfaces, morphotectonic network analysis).

J. Sládek (✉)
Department of Physical Geography, Geomorphology and Natural Hazards, Institute of Geography, Slovak Academy of Sciences, Štefánikova 49, 814 73 Bratislava, Slovakia
e-mail: geogslad@savba.sk

L. Vitovič
State Geological Institute of Dionýz Štúr, Bratislava, Slovakia
e-mail: ladislav.vitovic@geology.sk

J. Holec
Slovak Hydrometeorological Institute, Bratislava, Slovakia
e-mail: juraj.holec@shmu.sk

J. Hók
Department of Geology and Palaeontology, Faculty of Natural Sciences, Comenius University, Bratislava, Bratislava, Slovakia
e-mail: jozef.hok@uniba.sk

Keywords

The Western Carpathians • Morphotectonics • Morpholineaments • River terraces • Neotectonics

11.1 Introduction

Intramountain basins are characteristic landform features of the Inner Western Carpathians (Mazúr 1964). Their evolution is connected to the neotectonic evolution of the Western Carpathians. The formation of the Carpatho-Pannonian system began before the neotectonic period at the end of the Miocene, when the tectonic micro plates of ALCAPA and Tisza-Dacia started to create the Carpatho-Pannonian area (Minár et al. 2011).

The Turčianska kotlina basin (TKB) and the Žiarska kotlina basin (ŽKB) belong to the Central Carpathians Fault System (CCFS) basins (Nemčok and Lexa 1990; Kováč and Hók 1993). Their origin relates to the dextral transtensional tectonic activity of the CCFS during the late Miocene (Hók et al. 1998; Kováč et al. 2011; Pulišová and Hók 2015). The present-day orientation of the stress field is characterised by the orogen-parallel extension oriented in NE–SW to E–W directions (Hók et al. 2016). One of its most typical tectonic features is the long-lasting activity of master faults present along the western margin of the basins, resulting in typical half-graben structures involving vertical displacements of 1500–2000 m (Konečný et al. 2003). The Miocene–Pliocene infill of the basins contains mainly clays and sandstones, with an admixture of conglomerates (TKB) and volcanoclastics (ŽKB).

The aim of this chapter is to present the landscape evolution of the Turčianska kotlina and Žiarska kotlina intramountain basins and their surroundings from a morphostructural point of view.

M. Lehotský and M. Boltižiar (eds.), *Landscapes and Landforms of Slovakia*, World Geomorphological Landscapes,
https://doi.org/10.1007/978-3-030-89293-7_11

11.2 Physical Geographic Settings

11.2.1 The Žiarska Kotlina Basin

The Žiarska kotlina Basin, a low-lying geomorphic unit, is situated in the central part of Slovakia (Fig. 11.1). From a historical point of view, the territory has been part of the Tekov region. According to Mazúr and Lukniš (1978), the ŽKB as a geomorphological unit also includes the Žarnovické podolie valley sub-unit at its southern margin. The ŽKB is bounded by three volcanic mountain ranges, the Vtáčnik Mts. at the west, the Kremnické vrchy Mts. at the north and east, and the Štiavnické vrchy Mts. in the southeast. It is a relatively small intermountain basin ($\sim 17 \times 11$ km) with the area of $\sim$ 120 km^2. The surface of the basin extends from $\sim$ 225 (the Hron River floodplain in the south) to 470 (the Kremnické vrchy Mts. foothill) m a. s.l., thus it belongs to the moderately elevated basins of the Western Carpathians. The highest parts of the basin are the remnants of the oldest relief from the Late Pliocene. The Hron River represents the main axis of the ŽKB. Its catchment belongs to the drainage area of the Black Sea. Annual precipitation reaches $\sim$ 700–900 mm (Mazúr 1982; Bizubová et al. 1990; Faško and Šťastný 2002).

From a geomorphological point of view, the ŽKB as a geomorphological unit represents a subsided landform, a graben, with morphologically distinct boundary slopes formed along normal faults (Halouzka et al. 1999a, b). The morphology of this tectonic depression can be characterised by smooth low relief, where the basin's hilly land predominates. Two basic morphographic levels can be identified: the lower hilly land (in most of the area) and upper hilly land (in the northern and northeastern parts). The planar character of relief can be identified along the broad valley bottoms (e.g., the Hron River), mainly in the southern part of the ŽKB. The smooth basin floor morphology is reflected in prevailing slope inclinations. Within the central and southern parts, slopes are about 2–6°, while in the northeastern part they can be up to 10° (Mazúr 1982).

The most extensive fluvial landforms are represented by floodplains and river terraces. The width of the floodplain of the Hron River varies from 1 to 2.5 km. Its broadest segment is in the vicinity of the village of Dolná Trnávka, while its narrowest segment is formed along the Žarnovické podolie valley. Locally, two levels of the floodplain (lower and upper) are developed (Mazúr 1980; Halouzka 1998a; Konečný et al. 1998). The width and depth of floodplain accumulation can record neotectonic activity along the Hron River and the most intensive subsidence is assumed to occur in the southeastern part of the ŽKB (Halouzka et al. 1999a, b). River terraces are developed on several levels, mainly along the Hron River and some of its tributaries (more in Sect. 11.3.2). Proluvial cones and alluvial fans were formed by streams flowing from the adjacent mountains. The largest fans were formed by the Lutila and Prochoť streams. River terraces, cones, and fans are partially covered by loess loam (Mazúr 1982; Bizubová et al. 1990; Halouzka 1998a; Halouzka et al. 1999a, b).

The direction of the main right-side tributaries of the Hron River is determined by the NW–SE trending faults (the

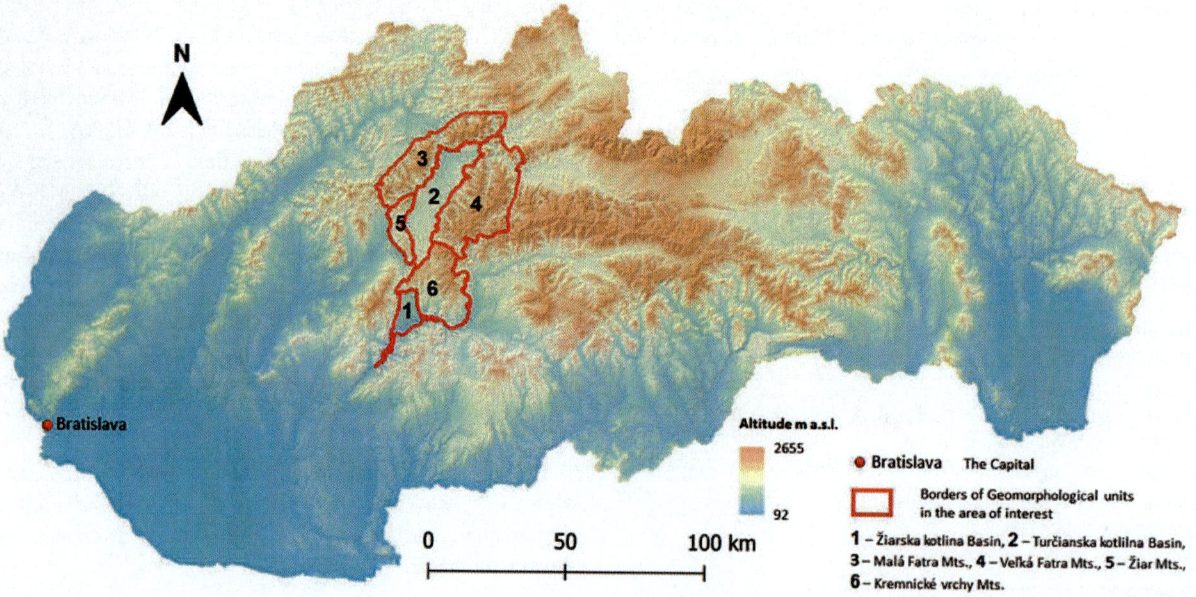

Fig. 11.1 Location of study areas in Slovakia. Source of background digital elevation model—DMR 3.5—Geodetic and Cartographic Institute

Lutila and the Prochoť faults). Neotectonic activity of the basin-margin faults is recorded in fluvial erosion in junction areas, the presence of bedrock exposures in river channels, and the non-graded longitudinal profiles of the streams (mainly Lutila and Slaský streams) (Bizubová et al. 1990).

11.2.2 The Turčianska Kotlina Basin

The Turčianska kotlina basin, another low-lying geomorphic unit, is situated in the western part of central Slovakia (Fig. 11.1). From a historical point of view, the territory has been part of the Turiec region. The TKB as a geomorphological unit belongs to the Fatra-Tatra region and consists of six subunits (Mazúr and Lukniš 1978). The Šútovské podhorie Piedmont is in contact with the Malá Fatra Mts. in the north, while the Sklabinské podhorie Piedmont and the Mošovská pahorkatina hilly land are situated in the east, along the contact with the Veľká Fatra Mts. The Diviacka pahorkatina hilly land is situated in the south and limited by the Kremnické vrchy Mts. and the Žiar Mts. The Valčianska pahorkatina hilly land in the west borders the Malá Fatra Mts., while the Turčianske nivy part is situated in the central part. The TKB with its prolonged shape ($\sim$40 $\times$ 15 km) in a NNE–SSW direction has an area of $\sim$ 470 km^2. The elevation of the TKB is between 400 and 700 m a.s.l. and it belongs to the most elevated basins of the Western Carpathians. Two basic morphographic levels are predominant in the basin's morphology. The plain character of relief occurs along the floodplains and river terraces of the Váh and the Turiec rivers, while the basin hilly lands predominate in the marginal parts of the basin. The TKB is a tectonic depression (graben), which formed originally from two individual subsided blocks (southern and northern one), separated by the partially uplifted block (Mazúr 1982; Halouzka et al. 1999a, b). Hydrologically, the area belongs to the catchment of the Váh River, which flows through the northern part of the TKB. However, the Turiec River is the main river of the TKB flowing from the Kremnické Vrchy Mts. The spring of the river is approx. 1090 m a.s.l. and its flow is diverted to the north. The Turiec River course is pushed towards the western part of the basin by alluvial fans from the Veľká Fatra Mts., but it also reflects the westward tilting of the basin floor. Annual precipitation reaches $\sim$ 800–900 mm (Faško and Šťastný 2002).

11.3 The Žiarska Kotlina Basin

11.3.1 The Formation of Initial Relief (Pliocene)

The evolution (ŽKB) during the Pliocene is connected with the differential synsedimentary subsidence of blocks, resulting in lake formation and their consequent regression, forming a riverside planation level. The system of interconnected lakes within the basins along the Hron River (the Pohronie region) is recorded in the youngest subaquatic fluviolacustrine deposits of the Lower Pliocene. They are represented by the infill of the former basin surface (fluviolacustrine sand, silt, sandy clay, and clay) as well as by boulder-block deposits of a river delta (west of the village of Jánova Lehota, situated in the northern part of the ŽKB) that entered the receding freshwater lake. The initial river network is considered to have developed during the Late Pliocene, which has been documented in fluvial deposits (mostly gravel and sandy gravel interbedded with clay, called the Hron Gravel formation). This represents the oldest phase of terrestrial development connected to the initiation of neotectonic activity, resulting in the formation of the oldest river terrace (Lukniš 1972, Halouzka 1998b); this can be referred to as a plateau terrace (sensu Maglay et al. 2011) which is thought to have an Early Pleistocene age (?Biber/?Pretegelen complex). (Halouzka 1986b).

The initial morphology of the ŽKB of purported Late Pliocene age is considered to have a plain character. The initial plain was subsequently dissected and denuded, resulting in its recent morphology. The former relief has been documented in the most elevated parts of ridges of the basin's hilly land (situated 333–336 m asl.) and on isolated (scattered) elevations in the foothills of the Kremnické vrchy Mts. (situated 464–469 m asl.). Relicts of the initial relief are also preserved in the marginal parts of the adjacent mountains. The remnants of the initial relief are generally tilted towards to the Hron River (Bizubová et al. 1990).

The state of preservation of the Late Pliocene plateau terrace considerably varies in space (e.g., Škvarček 1973; Halouzka 1998b). According to Ispaits (1943 in Škvarček 1973), the terrace levels situated 43 m above the floodplain, are considered to have a Pliocene age. The relative altitude of the riverside level in Žarnovica (Žarnovické podolie basin) is 170–180 m, while on the southern border (in the area occupied by tectonically deformed volcanic rocks) it is up to 150 m (Škvarček 1973). Its vertical position varies between the basin and the water gaps of the Hron River valley, resulting from differentiated neotectonic activity (Škvarček 1973; Mazúrová 1978). The plateau terrace is located at the relative height of 90–120 m within the basin, while it is 200–250 m above the river in water gap reaches (Mazúrová 1978). The largest remnants of the terrace are situated to the S and SE of the town of Žiar nad Hronom, at the relative height of 150–170 m. The thickness of this sedimentary body is quite substantial (up to 20 m) (Halouzka 1986a, b; Holec et al. 2015). In other locations, it is preserved only residually (e.g., west of the Jánova Lehota village and in the vicinity of Žarnovica).

11.3.2 Fluvial Activity During the Quaternary

The development of relief during the Quaternary was influenced mainly by neotectonic activity and climatic change, which resulted mainly in the formation of terraces and alluvial fans.

River terraces in the ŽKB, predominantly situated below the plateau terrace (riverside level), have been investigated by many geologists and geomorphologists who focussed mainly on the Hron River terraces. Four Pleistocene terraces of the Hron River were identified by Hromádka (1935 in Škvarček 1973), while Kettner (1928 in Škvarček 1973), together with Ispaits (1943 in Škvarček 1973) mentioned only three terraces. There are indeed three terraces, occurring in sequence with alluvial fans of the right-side tributaries, with the relative altitudes of the terrace bases at 5–10, 20–25, and 50–60 m (Lukniš 1972). More detailed research was carried out by Škvarček (1973), who identified six terraces and a recent floodplain, locally developed on two levels. The terraces were divided into high (with altitudes of the bases at 80–110, 50–65, and 27–33 m above the river), middle (16–20 and 10–14 m), and low (connected with the floodplain's base level). The thickness of the valley bottom infill is ~ 12 m (in the village of Dolná Ždaňa), which is considered to be a high value regarding the entire valley length. The high and middle terraces are often covered with loess loam (5–10 m thick). The terraces are preserved mainly in the vicinity of tributary junctions (Škvarček 1973). The total Quaternary erosional effect of the Hron River is estimated to be 85–115 m (Škvarček 1973). Similar results were reported by Mazúrová (1978), who divided the terraces into high (80–110, 62–70, 36–45 m), middle (20–26, 10–14 m), and low (-4–8 m) levels, also reflecting two levels of floodplain. No precise spatial distribution and age of the terraces has been given in the previous investigations (Škvarček 1973; Mazúrová 1978). Mazúr (1980) identified five terraces, with the relative altitudes of the bases at 5–10, 16–25, 28–35, 50–60, and 90–100 m. An investigation of the terraces in the southeastern part of the ŽKB was conducted by Halouzka (1986a), with the results being later incorporated into the geological map of the whole ŽKB (Konečný et al. 1998).

According to Halouzka (1998a, b), the following terrace levels were distinguished within the ŽKB. Among the high terraces, the remnant of the uppermost level is mapped on the border with the Kremnické vrchy Mts. The second and third high terrace levels remain undistinguished and are mapped as one unit. The heights of terrace bases are marked in Table 11.1. Among the upper terraces, two levels with a

Table 11.1 River terraces of the Hron river within the Žiarska kotlina basin, their presumed stratigraphy (according to Halouzka 1998a, b) and relative elevation of the base (according to Halouzka 1998a, b and Holec et al. 2015). Area I: the town of Žiar nad Hronom and its nearest surroundings, area II: between the villages of Dolná Trnávka and Dolná Ždaňa, area III: between the settlements of Žarnovica and Voznica. Where two values of elevation are present, numbers in brackets refer to values according to Halouzka (1998a, b)

Terraces		Climato-stratigraphy	Numerical age (based on OSL dating)	Area I		Area II		Area III	
				Left bank	Right bank	Left bank	Right bank	Left bank	Right bank
				Relative height above the Hron River [m]					
VII	1th high	Donau		–	130	–	–	–	–
VIb	2nd high	Older Giinz		–	–	–	–	–	–
Via	3rd high	Younger Giinz		–	90	–	87	–	–
V	1th upper	Mindel 1		–	83–88	–	–	–	–
IVb	2nd upper	Mindel 2		–	60–65	–	52–55	–	–
IVa	3 upper	Mindel 3		–	–	–	43–48	45–50	–
III	1st middle	Preriss		1	24–26 (21–28)	–	35–38	–	25–35
lib	2nd middle	Older Riss	> 211 + 32 ka	3–12	15–18 (8–13)	–	25–27	–	–
Ha	3rd middle	Younger Riss		1	–	0	6–8 (20–23)	3–8	4–9
Ic	Low	Wiirm		−8	–	−6	–	−5	–

sparse occurence are mapped within the basin, mostly between the settlements of Lovča and Dolná Trnávka. The middle terraces have more frequent occurrence, and are divided into three levels—the first, second, and third middle terrace, whose ages are assigned to pre-Riss and Riss periods. Halouzka (1998a, b) mentioned the postgenetic differentiation of middle terraces in the longitudinal profile of the river, pointing out the neotectonic activity within the region. Low terraces are present along the river in the neighbourhood of the present-day floodplain.

The numerical age of the second middle terrace accumulation (Table 11.1) was obtained by OSL dating. However, only a minimum age (>211 ± 32 ka) was estimated (Holec 2015). The river terrace distribution is not symmetrical within the basin, which corresponds to the position of the Hron River near the southern basin boundary (Fig. 11.2). Within the water gaps, no asymmetry has been recorded (Škvarček 1973). The asymmetry of terrace heights can be observed as well: the terraces on the right bank of the Hron River are more elevated, which can be ascribed to the non-uniform tectonic uplift of the area. This asymmetry is present within the broader part of the basin, although it is not observed in the Žarnovické podolie basin. The cross-sections of the Hron River valley reflecting this asymmetry are shown in Fig. 11.3. The results of the granulometric analyses of the middle terraces' gravel on the right bank of the Hron River imply proluvial origin, which is supported by the morphoposition as well (the former confluences of the Hron River and its tributaries) (Holec et al. 2015).

11.3.3 Thickness of Quaternary Deposits

Three areas with significant thicknesses of Pliocene to Quaternary sediments can be found within the ŽKB (Fig. 11.4). The biggest thickness can be found on the left bank of the Hron river, southward from the town of Žiar nad Hronom. This area corresponds to the Pliocene surface at the boundary of the Štiavnické vrchy Mts. and the Žiarska kotlina basin. The second area is located at the northern boundary of the basin, whereas the third area with a significant thickness coincides with the belt of fluvial sediments along the Hron river valley, comprising floodplain and terrace sediments together with proluvial sediments of the tributaries. On the other hand, the central part of the basin has only a thin cover of Quaternary sediments, with frequent outcrops of Neogene sediments. The character of the Quaternary and Neogene sediments is documented in Fig. 11.5.

11.3.4 Gully Network and Slope Deformations

Gullies and slope deformations, especially landslides, are landforms which are present in the territory of the Žiarska kotlina basin (Fig. 11.6) and indirectly point out to tectonic activity in the area. The higher density of these phenomena is linked to slopes created by fault activity. A higher density of gully networks is present along the uplifted area on the right bank of the Hron river. The gullies in the Žiarska kotlina basin were formed in three settings: (a) generally in a direction perpendicular to the Hron valley, (b) perpendicular to those representing type (a), and (c) perpendicular to the right-side tributaries of the Hron river (i.e., generally parallel to the Hron river). The highest gully density can be observed in the area west from the town of Žiar nad Hronom and to the north of the village of Lovča, with a density of more than 2.2 km/km^2.

The highest density of landslides can be observed along the western and southwestern borders of the basin, consisting of the steep slopes of the Vtáčnik mountains. The southwestern part of the basin is lined with the group of springs, which together with a higher landslide density point out to neotectonic activity in the area. The second important area with a high landslide density is located along the right tributaries of the Hron river, where they are triggered mainly by lateral erosion and favoured by local lithology, with Quaternary sediments situated on Neogene sands and clays.

11.4 The Turčianska Kotlina Basin and Its Surroundings

11.4.1 Overwiev of Geomorphological Research

The intensification of geomorphological research of the Turčianska kotlina Basin dates back to the 1960s. One of the first modern geomorphological studies in the area of the TKB was conducted by Mazúr (1963). This publication contains a short part dedicated to the Malá Fatra Mts., which are the western part of our area of interest. In his subsequent papers (Mazúr 1964, 1976, 1979) dedicated to the morphostructural analysis of the Western Carpathians, the evolution of the TKB and its surroundings is only briefly mentioned. During that time Činčura (1967, 1969) wrote about the landform evolution of the southern part of the TKB. More systematic research of the georelief of the TKB began in the 1990s, focussing on lithology, georelief settings, and relative vertical movements (e.g., Bizubová and

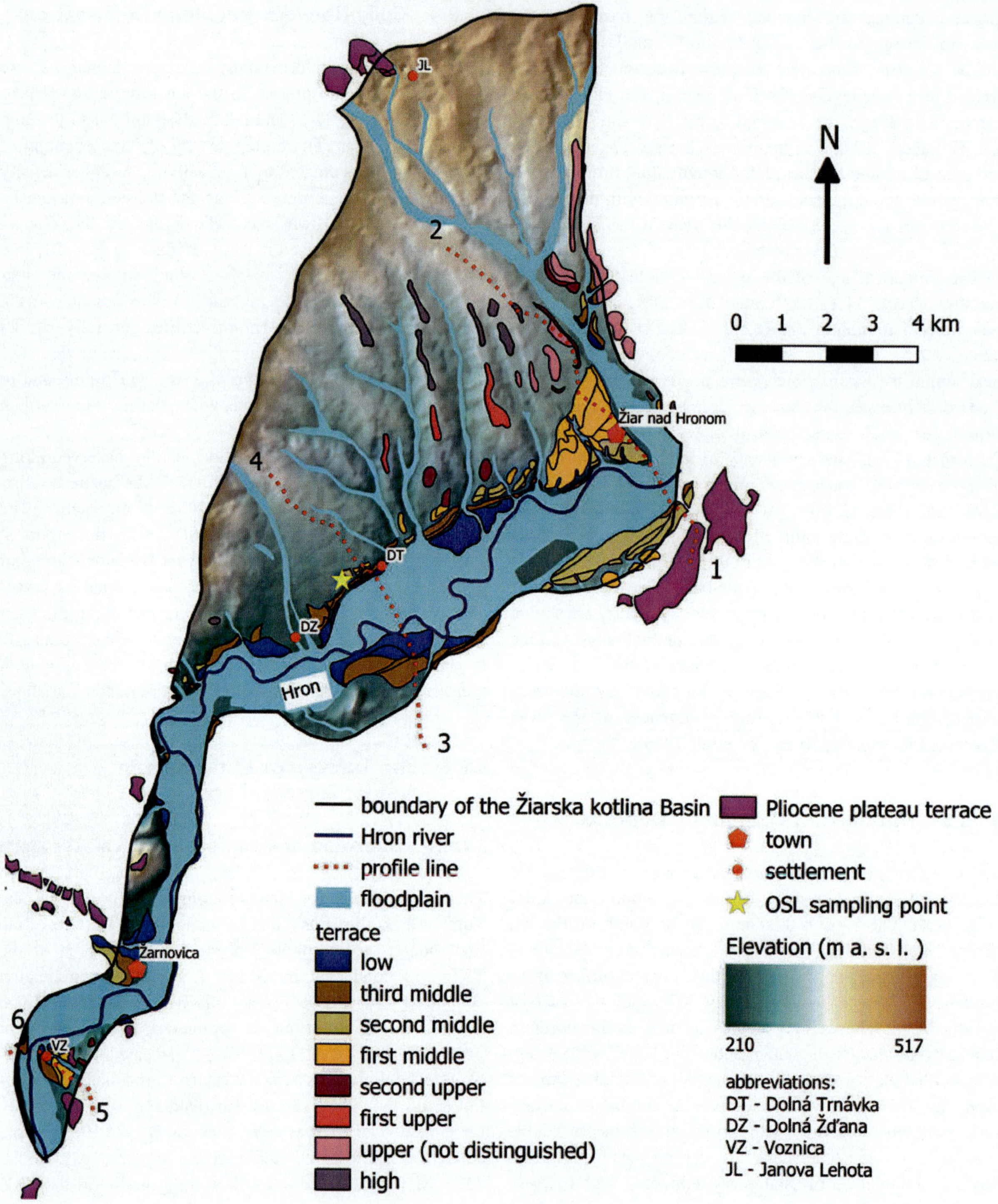

Fig. 11.2 Hypsometry of the Žiarska kotlina basin and river terraces of the Hron river and its tributaries. The delimitation of the basin is in line with Mazúr and Lukniš (1978). Profile lines refer to Fig. 11.3

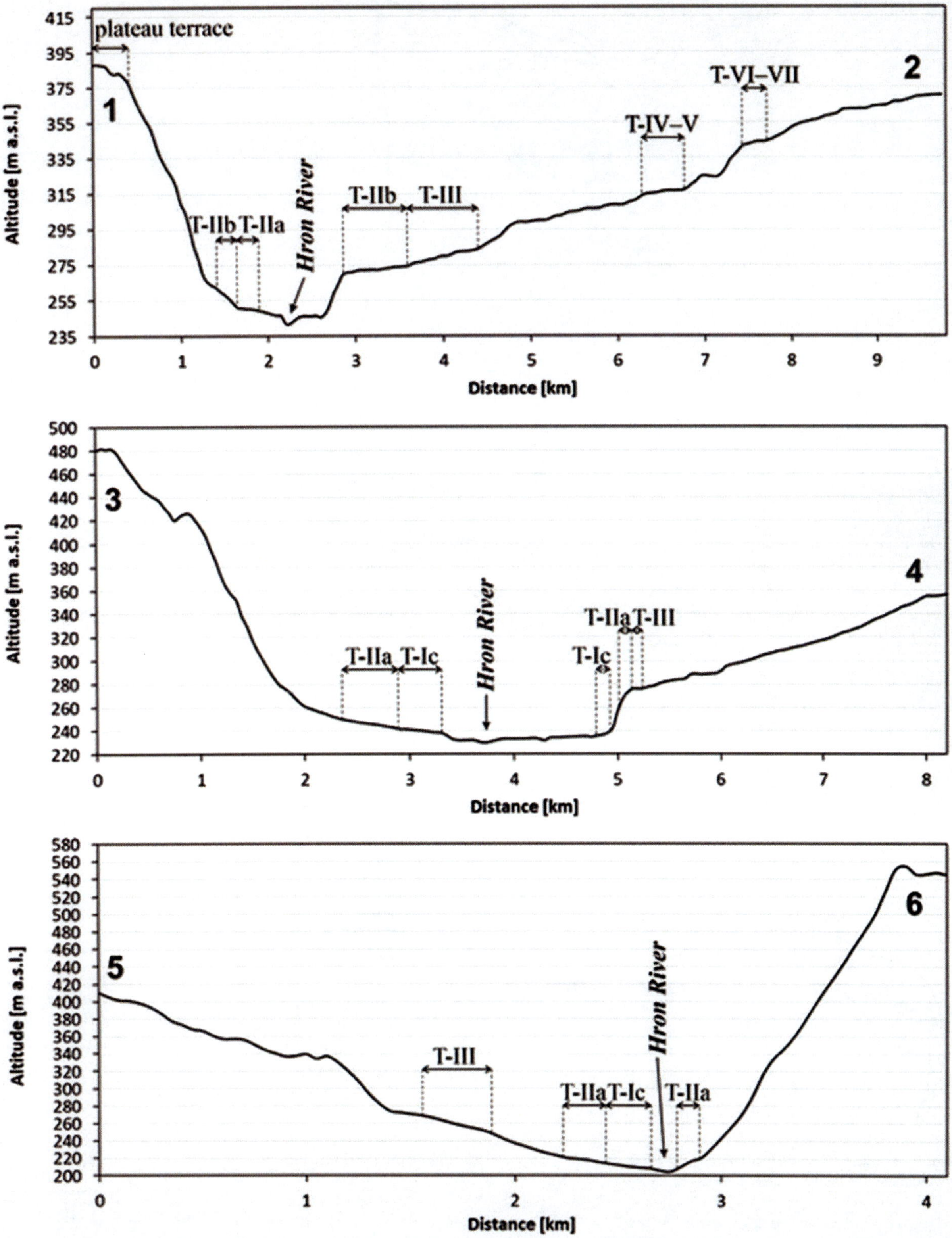

Fig. 11.3 Transverse profiles (1–2, 3–4, 5–6) of the Hron valley to show the position of river terraces (terraces according to Halouzka 1998a)

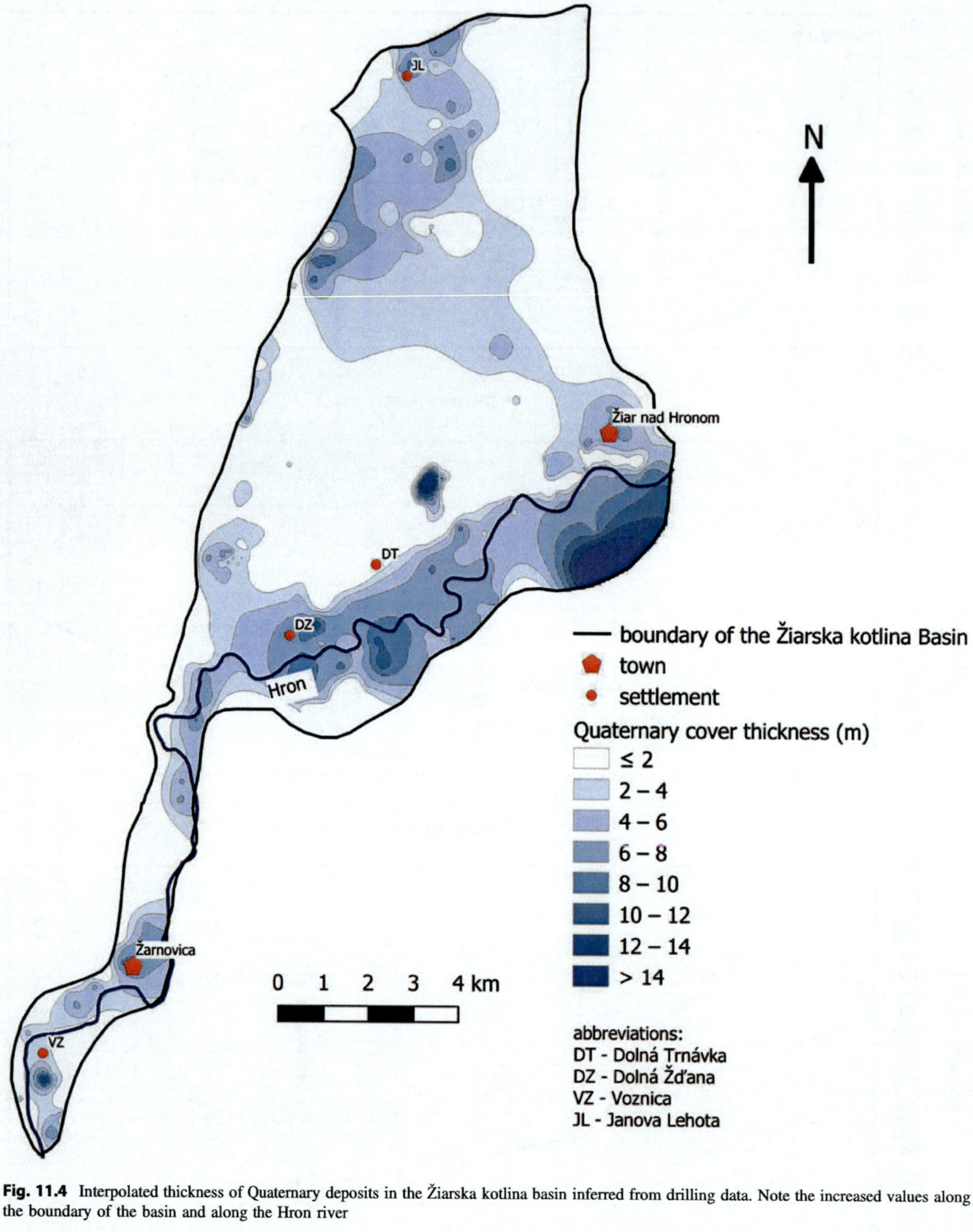

Fig. 11.4 Interpolated thickness of Quaternary deposits in the Žiarska kotlina basin inferred from drilling data. Note the increased values along the boundary of the basin and along the Hron river

Fig. 11.5 The boundary between Neogene and Quaternary sediments in the Žiarska kotlina basin revealed in an excavation for road construction. The boundary is marked by the hammer. (*Photo* A. Medveďová)

Machová 1994; Minár and Bizubová 1994; Minár and Tremboš 1994a, b; Minár 1996), mainly in the southern part of the TKB.

The contact zones of between the TKB and the Malá Fatra Mts., TKB, and the Veľká Fatra Mts., and TKB and the Žiar Mts. have been described by Bizubová (1999, 2002; Bizubová and Barka 2002). Following these papers (Bizubová et al. 2005; Sládek 2006, 2007; Sládek and Bizubová 2007, 2008; Minár and Sládek 2009), we present some uncommon methods of morphostuructural analysis that involve isobase surface reconstruction, and geomorphological and morphotectonic network analyses.

11.4.2 Geological and Geomorphological Settings

The TKB and surrounding mountain ranges make up a morphostructural half-graben complex (Kováč et al. 2011), with a wide spectrum of geomorphic shapes and landform types. There are wide and flat river floodplains and terraces, as well as narrow rock ridges and canyons. The vertical disection is also considerable. The valley bottoms of the major rivers of Váh and Turiec are located at about 300 m a.s.l. The highest point is in the Malá Fatra Mts.—Veľký Kriváň (1709 m a.s.l.). The altitude of the surrounding

Fig. 11.6 Gully with outcrops of Neogene sediments. (*Photo* A. Medveďová)

mountain ranges varies around 1000—1500 m a.s.l., except for the Žiar Mts., where most of the main ridge is below 1000 m a.s.l. The borders of major morphostructures, including the TKB, follow Cenozoic faults which are also visible in the landscape, e.g., as facets (Fig. 11.7).

The geomorphological settings and lithology of the surrounding morphostructures must be known for morphostructural analyses, so in the next part we briefly introduce the mountain ranges around the TKB.

11.4.2.1 The Core Mountains

The Malá Fatra Mts., Veľká Fatra Mts., and Žiar Mts. belong to the core mountains of the Internal Western Carpathians. A general scheme of a core mountain range is that it consists of a crystalline core, Mezozoic sedimentary cover, and younger cover nappes. These mountain ranges create horst morphostructures surrounding the TKB. The core contains Late Paleozoic granitoid and metamorphic rocks (various type of Palaeozoic granites, granodiorides, and mylonites along fault lines). Apatite fission tracks analyses yielded ages between 9 and 20 Ma (Danišík et al. 2010; Králiková et al. 2014). The Mezozoic cover is represented by carbonate sediments of Triassic to Cretaceous ages and includes various limestone types, dolomites, sandstones, quartz rocks, claystones, sandstones, conglomerates, and slates.

The lithological settings determined various morphostructures and landform types. A relatively sharp modelled georelief is associated with the Hronicum cover nappe (limestones, massive dolomites), where canyons, gorges, and crags are common, such as at Diery, Tiesňavy, Mt. Veľký and Malý Rozsutec, Mt. Tlstá, Mt. Ostrá, and others. These structures create a morphological contrast to the otherwise smooth georelief of the Fatricum nappe (various types of limestones, dolomites, marly limestones, marls) present at Mt. Ostredok, Mt. Krížna, Mt. Borišov, etc. (Fig. 11.8). In both Fatras and the Žiar Mts. numerous karstic landforms such as caves, lapies, sinkholes, etc. can be found. Remains of nival processes and related landforms such as nivation

Fig. 11.7 The faceted slopes of the northern part of the Malá Fatra Mts. are the morphological evidence of young tectonics. (*Photo* J. Sládek)

Fig. 11.8 Morphological contrast between peak shapes in the Fatricum nappe (Mt. Stoh—1607 m a.s.l., first from the right) and the Hronicum nappe (Mt. Veľký Rozsutec—1610 m a.s.l., background) in the northern part of the Malá Fatra Mts. (*Photo* J. Sládek)

Fig. 11.9 Nivation hollow with an avalanche source zone in the top part of the Veľká Fatra Mts. (Mt. Frčkov—1586 m a.s.l.). (*Photo* J. Sládek)

hollows, avalanche tracks, and cones occur in deforested areas of the Veľká and Malá Fatra Mts. (Fig. 11.9). Several travertine mounds are present along the eastern border of the Veľká Fatra Mts.

11.4.2.2 The Neovolcanic Kremnické Vrchy Mts.

The Kremnické vrchy Mts. belong to the neovolcanic region of Slovakia. The mountains enclose the TKB on its south-eastern side and their morphological pattern reflects several stages of volcanic evolution. The highest peak is Mt. Flochová (1317 m a.s.l.), located on a wide, flat remnant of a lava flow. The northern part of the mountains is characterised by smooth georelief, with different vertically spaced wide plateaus which are separated by steep slopes. The characteristic landform features of the plateau edges are sharply cut rock towers. Recent georelief is the result of the exogenic transformation of former volcanic georelief and the vertical tectonic movements of plateaus (Lacika 1997). Differences in rock resistance have resulted in numerous landslides, mainly on the border of the mountains. The volcanic rock formations of the Kremnické vrchy Mts. Comprise a complex of volcanic landform shapes from basalts, volcanoclastic rocks, andesites, and rhyolites dated as having formed during the Badenian to Pannonnian stage (Lexa et al. 1998).

The bedrock under the neovolcanics of the northern part of the Kremnické vrchy Mts. (area of interest) formed before the Cenozoic period is similar to the surrounding core mountains (Veľká Fatra Mts., Žiar Mts., Starohorské vrchy Mts.). This bedrock consists of crystalline and Mesozoic rocks of the core mountains.

The overburden of the Kremnické vrchy Mts. has been created by andesites and its epiclastic breccias, and rhyolites. The remains of lava flows from these rocks are morphologically visible in the landscape (Fig. 11.10). Except for the remains of rhyolite, which create massive but relatively short remains of lava flows.

Fig. 11.10 The Bralova skala (826 m a.s.l.)—remains of a lava flow. (*Photo* J. Sládek)

11.4.2.3 The Turčianska Kotlina Basin

The TKB creates a half-graben depression unit (Rakús and Hók 2002; Kováč et al. 2011). The landscape is smooth and has been remodelled by periglacial and fluvial processes. The river terrace systems are well-preserved near the junction of the Váh and Turiec rivers. The remains of the oldest terraces are 45–90 m above the Turiec and Váh rivers, with their relative heights decreasing to 15–30 m toward the south. The thickness of gravel accumulation reaches 2–3 m. The most common terraces are called „middle terraces " (Riss), occurring 10–30 m above the river. The number of these terraces differs in the northern (3) and southern (2) parts of the basin. The northern terraces have an age associated with the Riss glaciation while the southern terraces were partly formed during Mindel glacial stage. The thickness of gravel accumulation of these terraces exceeds 20 m. The „low terraces " of late Pleistocene age occur 3–8 m above the river. They are located only in northern part of the TKB. Their thickness is about 15 m (Kováč et al. 2011).

The TKB subsided along the faults delimited its basement. The fault tectonics are also visible inside the basin manifested by several mineral water springs, travertine heaps, tectonically based delens, tectonic crushed rocks, and so on. The pre-Cenozoic bedrock of the TKB has been vertically shifted by tectonic movements to many tectonic blocks.

The oldest rock formations are represented by crystalline and metamorphic rocks sourced from the surrounded mountains. These rocks outcrop on surface just on the border of the TKB due to more than 1200 m of sedimentary infill on its western border (Kováč et al. 2011).

The Paleogene infill (breccias, conglomerates, claystones) is located in the north-east part of the basin. The Neogene sediments originated from the surrounding mountains.The TKB was a predominantly isolated lake during the middle Miocene and therefore, sedimentary formations are often dated only on the basis of their superposition. Fluvial activity had a great impact during the Quaternarry period. The remains of alluvial cones and different terrace systems are displaced at the bottom of the whole basin. Several locations of the travertines, peat moores, and highmoores are located on the bottom of the basin.

11.4.3 The Morphostructural Analysis

This part of the contribution deals with the non-traditional methods of morphostructural analysis—isobase surfaces, morpholineaments, and morphotectonic network. These analyses should support the transtension regime of paleotension field theory as described in Kováč et al. (2011).

Isobase surfaces

An isobase surface is a theoretical surface which is cut to thalwegs of certain order. Isobase surface creation consist of several steps. The 1st step is thalweg (stream) delimitation and its ordering (Fig. 11.11a). We used stream ordering according to Strahler (1952). The 2nd step involves connecting points with equal altitude on thalwegs of the same order. The lines are named as isobase lines. The 3rd step is to generate the isobase surface of a certain order. For isobase surfaces of 2nd and higher orders we use isobase lines of 2nd, 3rd, 4th … orders. For isobase surfaces of 3rd and higher order we use isobase lines of 3rd, 4th, 5th… order (see Fig. 11.11b). For visualisation, we can created a digital elevation model (DEM) (Fig. 11.12) from isobase lines of certain orders. The isobase surfaces represent general landform features—morphostructures.

An isobase surface of a certain order can simulate a theoretical state of georelief in conditions when all erosion and denudation processes had been stopped. Simply said—we removed the georelief above the local erosion base of a certain order. Such a surface was smoother and more uniform, but older and wider morphostructures were more visible. According to Spiridonov (1975) the isobase surfaces defined by thalwegs of a higher order represent older morphostructures which mean that isobase surfaces of the 1st and higher orders represent younger morphostructures than isobase surfaces of 2nd, 3rd, 4th, and higher orders. The isobase surfaces can be constructed in areas with well developed river (valley) networks.

In the area of interest (Fig. 11.1—geomorphological units 1, 2, 3, 4) we used a semiautomatic construction of isobase surfaces according to Jedlička et al. (2015). Above these surfaces we delimited some linear features which may have a connection with (morpho)tectonics. A sketch of the possible morphostructure evolution based on isobase surface reconstruction is displayed in Fig. 11.12. In the series of images of isobase surfaces of 1st—5th orders we can see a simplification of the larger morphostructure units of the Malá and Veľká Fatra Mts. Žiar Mts. and Kremnické vrchy Mts. and simultaneously the TKB as an intramountain depression.

Linearly Arranged Georelief Elements-Morpho lineaments.

Linear features—lineaments—can be identified and delimited in recent georelief, in an isobase surface or a geological environment. These linear features can be arranged into perpendicular systems—networks. If linear elements are derived from contours or a 3D model, information about the geomorphological network can be obtained. This network consists of morpholineaments (linear features of georelief). Linearly arranged lineaments related to tectonic features like faults (tectolineaments) create tectonic networks. Figure 11.13 displays an example of morpholineaments derived from contours and a DEM.

According to Urbánek (1993), the geomorphological network is a continuously coherent unit consisting of several parallel running groups of geomorphologic lines which are mutually crossed. These lines are visible in georelief as lined up geomorphologic landforms. Such a geomorphological line can consist of several landform types such as valleys, saddles, ridges, foothills, etc.—morpholineaments. Their mutually perpendicular arrangement has been called a

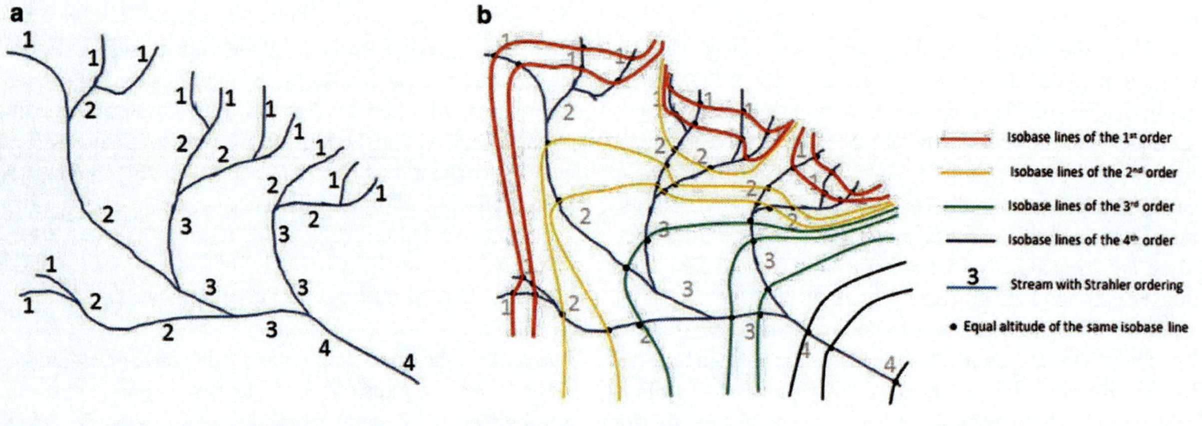

Fig. 11.11 Basic principles of isobase line construction based on Strahler´s stream ordering

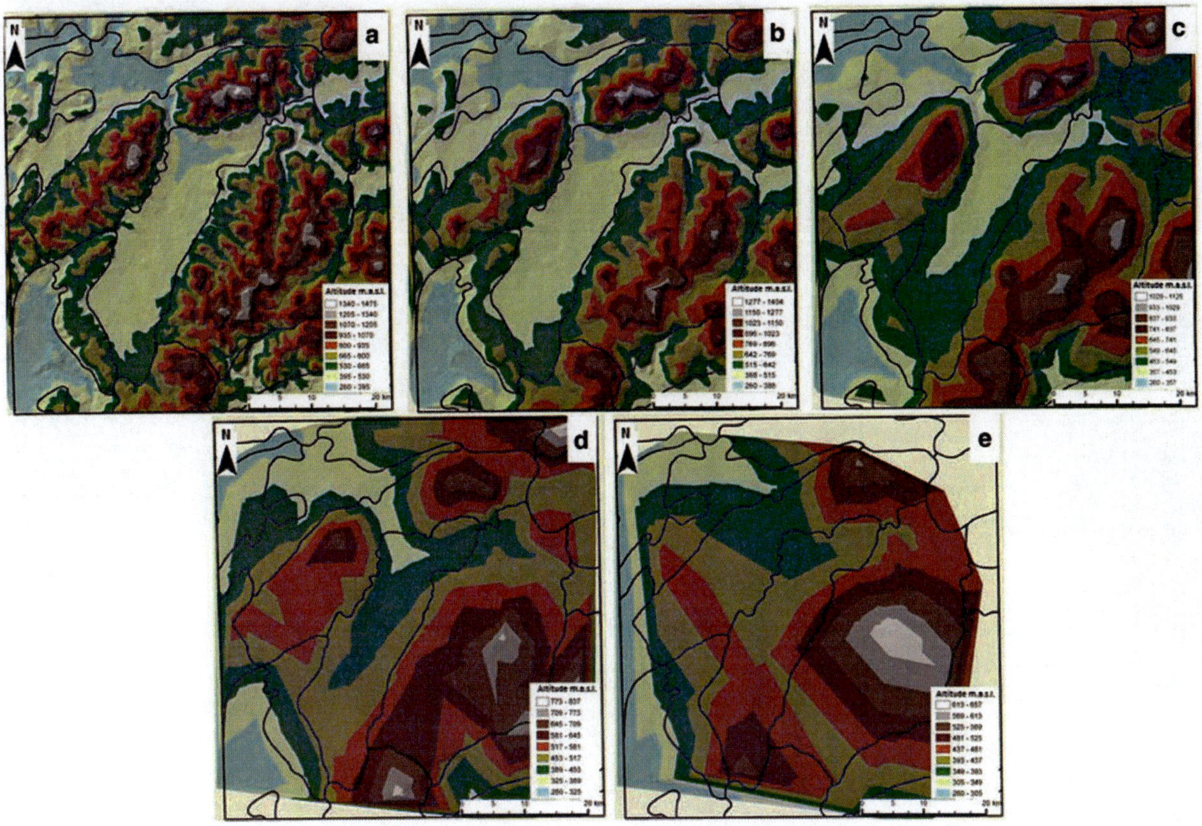

Fig. 11.12 The possible stages of morphostructural evolution in the surroundings of the Turčianska kotlina Basin, based on isobase surface reconstruction. **a**—isobase surfaces of 1st and higher order, **b**—isobase surfaces of 2nd and higher order, **c**—isobase surfaces of 3rd and higher order, **d**—isobase surfaces of 4th and higher order, **e**—isobase surfaces of 5th and higher order

geomorphological network. Similarly, we can apply such a method to tectonic linear elements—tectolineaments. In this case, information can be obtained about tectonic networks.

From a morphostructural point of view, it is important to see and analyse neotectonic features in recent georelief. For this reason, we tried to find coincidences between geomorphological and tectonic networks. These networks consist of lineaments derived from DEM and contours of a 1:50 000 scale (morpholineaments) and faults (tectolineaments) taken from geological maps of a 1:50 000 scale. If there is a coincidence between the networks created we can infer that the geomorphological network originated in the tectonic network. In such a case, a morphotectonic network can be referred to. More about these analyses in the TKB area has been reported in Sládek (2010).

Geomorphological Network Constructed Above Isobase Surfaces

Geomorphological network reconstruction based on isobase surfaces of several orders can provide a helpful visualisation tool for identifying horizontal movements of morphotectonic fields in time (Minár and Sládek 2009). For this purpose, we used isobase surfaces of 1st–5th orders in the area of interest.

The spatial change of morpholineaments is sketched in Fig. 11.14a–e. The rose diagrams (Fig. 11.15) of these networks show an approximately orthogonal system of lineaments of each isobase surface stage and also different azimuths of the main network lines. The superposition of these lineaments (Fig. 11.16) shows the possible rotation of

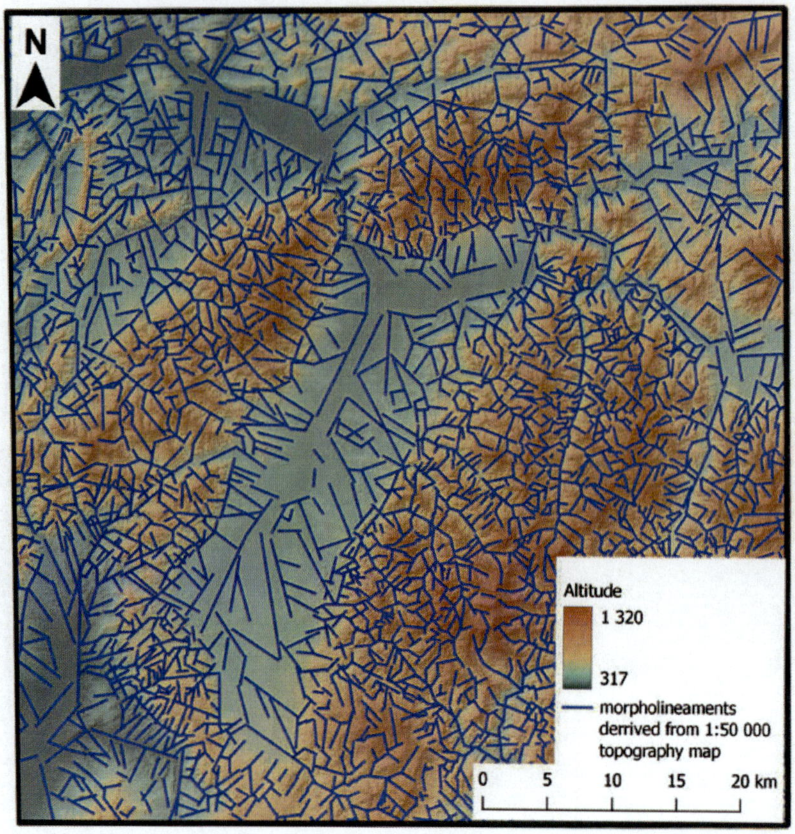

Fig. 11.13 Morphotectonic network of recent georelief

network generations in time and space. The outlined orthogonal system rotation supports some early research results from this area, e.g., Sládek (2008), Minár and Sládek (2009), and Kováč et al. (2009). The network rotations occur both clockwise and counterclockwise. This rotation can be explained as a response to paleotension field changes. The paleotension field might react to compression stress changes from the NW–SE during the Lower Miocene through to the NE-SW to the recent NW–SE (NNW-SSE, respectively) direction (Hók et al. 2000). If we synchronise the results from the abovementioned papers with our results in Fig. 11.14, 11.15, 11.16, the network constructed above isobase surfaces of the 5th order (Fig. 11.14e) might be representative of the Lower Miocene. The network constructed above isobase surfaces of 2nd and higher order (Fig. 11.14b) might be representative of the Upper Miocene. The N-S direction on the isobase surfaces of 5th and higher order is not visible. This can be explained as a result of young tectonic activity. This direction is more visible in recent georelief, eventually in rose diagrams of isobase surfaces of the 2nd and 3rd orders (Fig. 11.16) represented by short lineaments with N-S and its perpendicular E-W oriented lines.

Based on the above, we support the evolution of the TKB as a rift and later post-rift system as presented by Kováč et al 2011. Later, the two different depocentres were created (northern and southern parts of the basin) together with the half-graben morphostructure evolution.

11.4.4 Morphotectonics and Human Activities

Morphotectonically conditioned landforms in the area of interest are used for numerous human activities (e.g., ski slopes, mineral water sources, etc.). A large part of the Veľká Fatra Mts. and the northern part of the Malá Fatra Mts. consist of national parks. The fauna and flora as well as morphologically dominant features based on tectonic

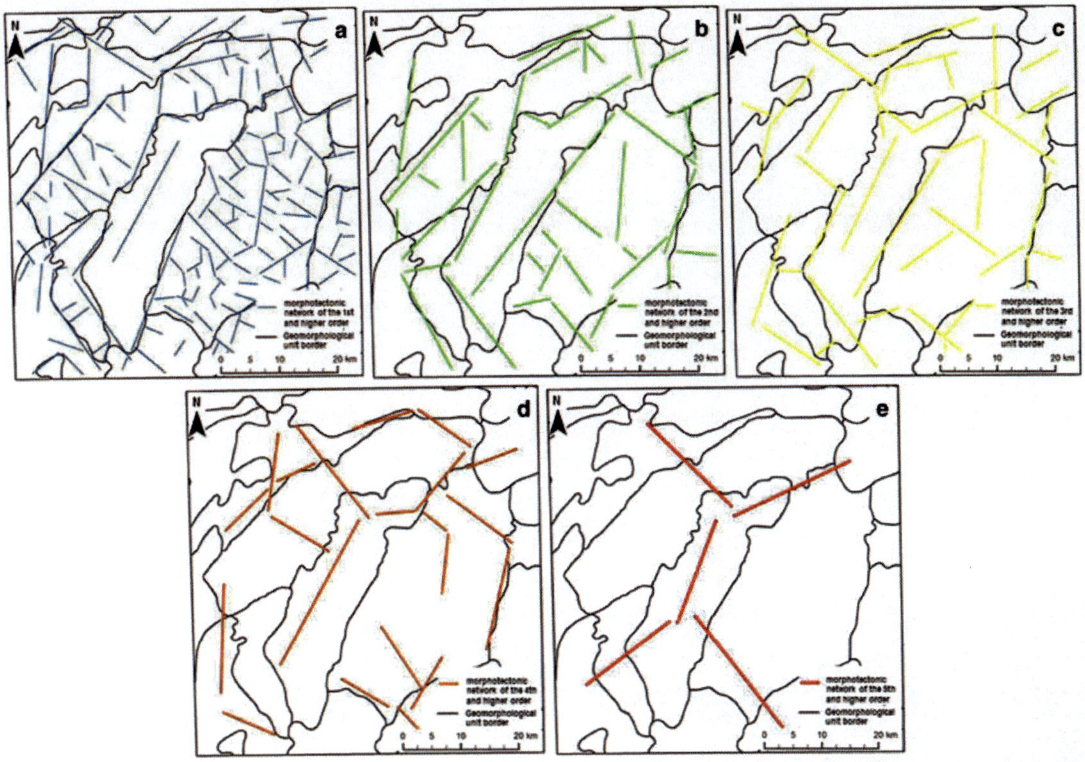

Fig. 11.14 Spatial change of morpholineaments constructed above isobase surfaces of: **a**—1st order, **b**—2nd order, **c**—3rd order, **d**—4th order, **e**—5th order

predisposition are protected as national heritage. The numerous natural reserves binding to the morphotectonic features with the highest degree of protection occur in the area of interest (e.g., nappe remains of Rozsutec, Kľak, Tiesňavy, and Diery canyons, etc.) (Fig. 11.17). Based on the lithology there are karstic landforms—scrapes, sinkholes, and caves—the Mažarná cave (Fig. 11.18), Harmanecká jaskyňa cave, and numerous publicly inaccessible caves.

Travertines (Rojkov, Jazierce) provide evidence of Quaternary neotectonics, as well as mineral water sources (Fig. 11.19) on the bottom of the TKB (Fatra, Budiš, Kláštorná) which are commercially used and bottled, or used as in Spas—Turčianske Teplice for treatment of musculoskeletar and other disorders.

Furthermore, the shape of a basin and surrounding morphostructures are attractive for all season hiking.

The foothills are known as ski resorts (Valčianska dolina—Fig. 11.20, Jasenská dolina). The mountain ski resorts of the Malá and Veľká Fatra Mts. (Martinské hole—Fig. 11.21, Malinô Brdo, Donovaly) reach higher altitudes and thanks to their disposition to tectonic as well as erosion and denudation or landslide modelled slopes, they are suitable for more demanding skiers.

The direct contact with rocks and morphotectonics is also connected to rock climbing. For this activity, the tectonic mirros, faults, and exposed rock cliffs are used. Although large parts of the area of interest are protected, exceptions for climbing in some areas in the Blatnická dolina valley, Sučany, Malinné, Višňové, Bystrička valley, etc.)

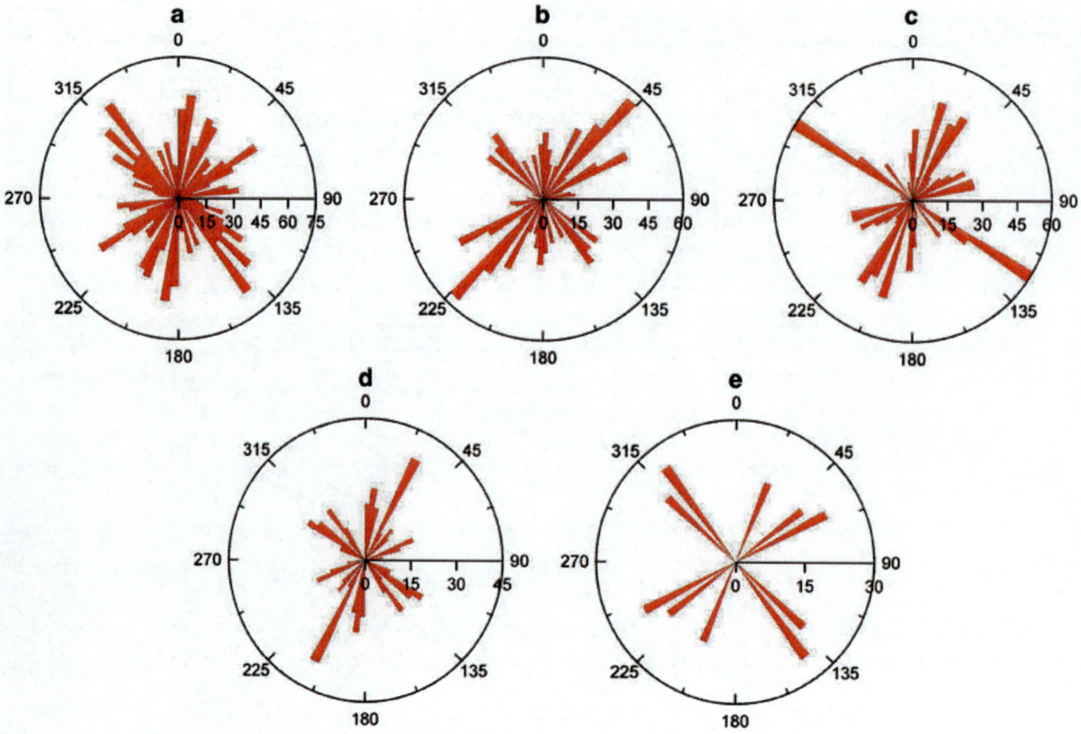

Fig. 11.15 Rose diagrams of main morpholineament directions constructed above isobase surfaces: **a**—1st and higher order, **b**—2nd and higher order, **c**—3rd and higher order, **d**—4th and higher order, **e**—5th and higher order

(Fig. 11.22) are granted. In 2013 for hiking community the Ferrata of Mountain Rescue Service (Fig. 11.23) in Pivovarský potok brook (Malá Fatra Mts.) was built. The path is built along a tectonic fault inside a granitoid and paragneiss crystalinic core. The height elevation of the path is about 785 m and it has a total length of 12 km.

11.5 Conclusions

Intramountain basins are typical morphotectonic features of Slovakia (Western Carpathians). The ŽKB and TKB belong to this mosaic of intramountain basins. The basin borders follow tectonic faults. Some intrabasin faults also have a significant influence on basin morphology and landscape shapes. The ŽKB is characterised by its position between neovolcanic mountains with an impact on its evolution and sedimentary infill. One of the famous features of the basin are gullies. Their density reaches more than 2.2 km/km^2.

The second intramountain basin—the TKB has been defined as a half-graben deppression between core mountain ranges. The sedimentary infill of the TKB is irregulary distributed at the bottom of the basin. Most of the sedimentary infill of the TKB is sourced from the surrounding mountains. The tilting of the half-graben morphostructure of the basin has caused an extreme depth (more than 1200 m) of infill at the western border of the basin. This part of the contribution was aimed at conducting a morphostructural analysis which is not used as a common tool of morphotectonic research (morphotectonic network reconstruction, isobase surfaces reconstruction). The presented methods support the results of earlier researches of the TKB area and outlined some new possibilities of morphostructure evolution.

Acknowledgements This paper was supported by the Scientific Agency of the Ministry of Education, Science, Research and Sport of the Slovak Republic and the Slovak Academy of Sciences under the contract VEGA 2/0052/21.

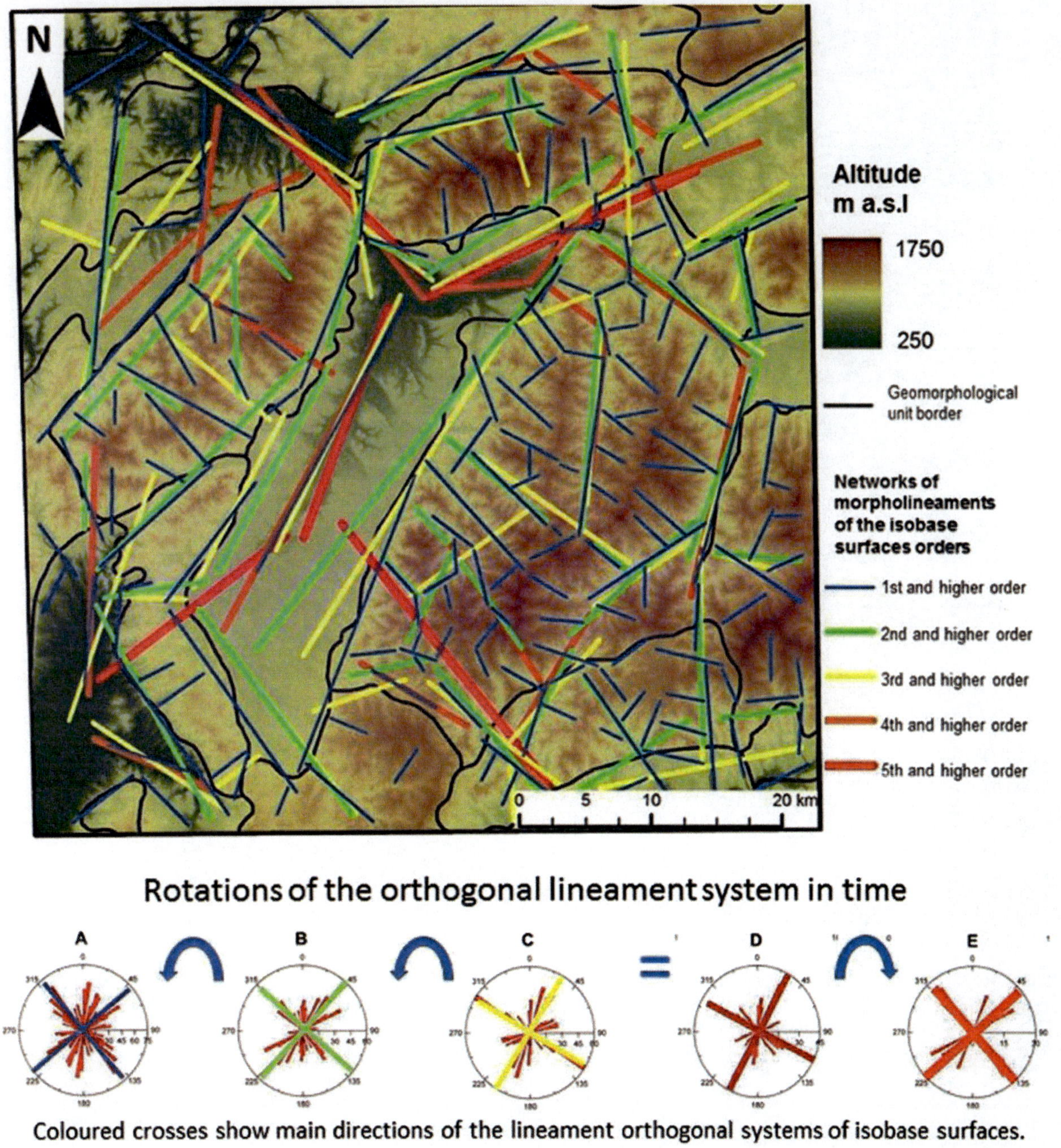

Fig. 11.16 Superposition of morpholineaments constructed above the 1st—5th isobase orders. See the clockwise/counterclockwise rotations of the main lineaments

Fig. 11.17 The Veľký Rozsutec (1610 m a.s.l.)—thrust outlier of the Hronicum nappe overlying the Fatricum nappe. (*Photo* J. Sládek)

Fig. 11.18 The Mažarná cave. (*Photo* J. Sládek)

Fig. 11.19 **a** The natural source of mineral water near the Budiš village. **b** The Kláštorná mineral water bottler. (*Photo* J. Sládek)

Fig. 11.20 The Valčianska dolina ski resort: a small resort on the border between the Turčianska kotlina basin and slopes of the Malá Fatra Mts. (*Photo* J. Sládek)

Fig. 11.21 Martinské Hole ski resort in the top part of Malá Fatra Mts. (*Photo* J. Sládek)

Fig. 11.22 One of the official rock climbing areas in the mouth of the Gaderská dolina valley—Veľká Fatra Mts. (*Photo* J. Sládek)

Fig. 11.23 The part of "via ferrata" in the Malá Fatra Mts. (*Photo* Peter Bebjak)

References

Bizubová M (1999) Niektoré morfotektonické črty kontaktných zón Turčianskej kotliny s okolitými pohoriami. In: Geografia XI. Časť B: Sborník prací Pedagogické fakulty Masarykovy Univerzity Sv. 145, Řada přírodních věd, č. 22. Brno: Masarykova univerzita, 300–307pp

Bizubová M (2002) Príspevok k poznaniu zarovnaných povrchov pohoria Žiar. Geomorphologia Slovaca. 2(1):58–65

Bizubová M, Barka I (2002) Príspevok k poznaniu geomorfologickej situácie v okrajových častiach pohoria Žiar. In: Geomorfologický sborník 1: Stav geomorfologických výzkumů v roce 2002 – příspěvky k mezinárod. semináru. Brno, Masarykova univerzita v Brně, 19–21p

Bizubová M, Machová Z (1994) Náčrt geologických pomerov Diviackej pahorkatiny pre potreby fyzickogeografického výskumu. Acta Facultatis Rerum Naturalium Universitatis Comenianae, Geographica 35:9–16

Bizubová M, Machová Z, Mičian Ľ, Minár J, Škvarček A, Zaťko M (1990) Geomorfologické pomery. In Analýza vybraných geoekologických komponentov Žiarskej kotliny a okolitých pohorí. Záverečná správa úlohy HZ 11/90 – VČ: 122

Bizubová M, Minár J, Sládek J (2005) Vývoj pohoria Žiar v kontexte denudačnej chronológie Západných Karpát. Geomorphologia Slovaca 5(2):5–15

Činčura J (1967) Príspevok k veku poriečnej rovne v Západných Karpatoch (na príklade južnej časti Turčianskej kotliny). Geografický Časopis 19(4):316–326

Činčura J (1969) Morfogenéza južnej časti Turčianskej kotliny a severnej časti Kremnických vrchov. Náuka o Zemi, Geographica 4:5–67

Danišík M, Kohút M, Broska I, Frisch W (2010) Thermal evolution of the Malá Fatra Mountains (Central Western Carpathians): insights from zircon and apatite fission track thermochronology. Geol Carpath 61(1):19–27. https://doi.org/10.2478/v10096-009-0041-0

Faško P, Šťastný P (2002) Priemerné ročné úhrny zrážok. 1: 2 000 000. In: *Atlas krajiny Slovenskej republiky*. Bratislava: Ministerstvo životného prostredia SR a Basnká Bystrica: Slovenská agentúra životného prostredia, p 99

Halouzka J (1986a) Základná geologická mapa ČSSR 1:25 000, list 36–314 Žiar nad Hronom. Bratislava: GÚDŠ

Halouzka R (1986b) Z nových poznatkov o stratigrafii kvartéru terasových náplavov riek Západných Karpát (Stredné Pohronie, Orava a Turiec). Spr. Výsk. gúdš 21:167–175

Halouzka V (1998a) Kvartér. In Vysvetlivky ku geologickej mape Štiavnických vrchov a Pohronského Inovca (štiavnický stratovulkán), 2. diel. Bratislava: Vydavateľstvo Dionýza Štúra, 1998. 466 s

Halouzka V (1998b) Pliocén. In Vysvetlivky ku geologickej mape Štiavnických vrchov a Pohronského Inovca (štiavnický stratovulkán), 2. diel. Bratislava: Vydavateľstvo Dionýza Štúra, 1998. 466 p

Halouzka R, Maglay J, Baňacký V, Pristaš J, Janočko J, Hók J (1999a) Vysvetlivky k neotektonickej mape Slovenska 1:500 000. Bratislava (GS SR)

Halouzka R, Maglay J, Baňacký V, Pristaš J, Janočko J, Hók J (1999b) Vysvetlivky k neotektonickej mape Slovenska 1:500 000. Bratislava, GS SR, 48 p

Hók J, Kováč M, Rakús M, Kováč P, Nagy A, Kováčová - Slamková M, Sitár V, Šujan M, (1998) Geologic and tectonic evolution of the Turiec depression in the Neogene. Slovac Geol. Mag. 4(3):165–176

Hók J, Bielik M, Kováč P, Šujan M (2000) Neotektonický Charakter Územia Slovenska. Mineralia Slovaca 32:459–470

Hók J, Kysel R, Kováč M, Moczo P, Kristek J, Kristeková M, Šujan M (2016) A seismic source zone model for the seismic hazard assessment of Slovakia. Geol Carpath 67(3):273–288. https://doi.org/10.1515/geoca-2016-0018

Holec J (2015) Systémy riečnych terás vo vybraných morfoštruktúrnych regiónoch Západných Karpát. Prírodovedecká fakulta Univerzity Komenského, Bratislava, Dizertačná práca

Holec J, Medveďová A, Vitovič L, Prokešová R (2015) Neotektonický vývoj Žiarskej kotliny indikovaný geomorfologickou analýzou v prostredí GIS. Geografický Časopis 67:181–195

Jedlička K, Sládek J, Šilhavý J (2015) Semiautomatic construction of isobase surfaces: A case study from the central Western Carpathians. Comput Geosci 78:73–80. https://doi.org/10.1016/j.cageo.2015.02.012

Konečný V, Lexa J, Halouzka R, Dublan L, Šimon L, Stolár M, Nagy A, Polák M, Vozár J, Havrila M, Pristaš J (1998) Geologická mapa regiónu Štiavnických vrchov a Pohronského Inovca. Bratislava, Ministerstvo životného prostredia SR

Konečný V, Lexa J, Šimon L (2003) Geologic structure and evolution of intravolcanic depressions in the area of Neogene volcanism in central Slovakia. Mineralia Slovaca 35:255–290

Kováč M, Minár J, Hók J, Vojtko R, Bielik M, Šujan M, Králiková S, Simovičová Z, Sládek J (2009) From the Miocene subsidence to Pliocene-Pleistocene tectonic inversion: a case study from the Turiec Basin (Western Carpathians). In: Annual SourceSink-Meeting 2009 (Salzburg, 26–27. 11. 2009). Salzburg: Universität Salzburg. 14pp

Kováč M, Hók J, Minár J, Vojtko R, Bielik M, Pipík R, Rakús M, Kráľ J, Šujan M, Králiková S (2011) Neogene and Quaternary development of the Turiec Basin and landscape in its catchment: a tentative mass balance model. Geol Carpath 62(4):361–379. https://doi.org/10.2478/v10096-011-0027-6

Kováč P, Hók J (1993) Central Slovakia fault system – field evidence of strike slip. Geol Carpath 44(3):155–159

Králiková S, Vojtko R, Andriesen P, Kováč M, Fügenschuh B, Hók J, Minár J (2014) Late Cretaceous-Cenozoic thermal evolution of the northern part of the Central Western Carpathians (Slovakia): revealed by zircon and apatite fission track thermochronology. Tectonophysics 615–616:142–153. https://doi.org/10.1016/j.tecto.2014.01.002

Lacika J (1997) Morfoštruktúry Kremnických Vrchov. Geografický Časopis 49(1):19–33

Lexa J, Halouzka R, Havrila M, Hanzel V, Kubeš P, Liščák P, Hojstričová V (1998) Vysvetlivky ku geologickej mape Kremnických vrchov 1:50 000. Geologická služba Slovenskej republiky, Bratislava

Lukniš M (1972) Slovensko 2: Príroda. Bratislava, Obzor, 917p

Maglay J, Moravcová M, Kučera M (2011) Vysvetlivky ku geologickej mape kvartéru Slovenska 1 : 500 000. ŠGÚDŠ. Bratislava, p. 94.

Mazúr E (1963) Žilinská kotlina a priľahlé pohoria. Bratislava, Vydavateľstvo Slovenskej akadémie vied

Mazúr E (1964) Intermountain basins – a characteristic element in the relief of Slovakia. Geografický Časopis 16(2):105–126

Mazúr E (1976) Morphostructural features of the West Carpathians. Geografický Časopis 28(2):101–111

Mazúr E (1979) Morfoštruktúry Západných Karpát a ich vývoj. Acta Facultatis Rerum Naturalium Universitatis Comenianae, Geographica 17:12–34

Mazúr E (1980) Žiarska kotlina. Encyklopédia Slovenska. Bratislava, Veda, 575p

Mazúr E (1982) Encyklopédia Slovenska, 6 zväzok. Bratislava, Veda, 776pp

Mazúr E, Lukniš M (1978) Regionálne Geomorfologické Členenie SSR. Geografický Časopis 30(2):101–125

Mazúrová V (1978) Terasy riek čs. Karpát a ich vzťah k terasám Dunaja. Geografický Časopis 30:281–296
Minár J (1996) Geomorfologická regionalizácia Diviackej pahorkatiny. Acta Facultatis Rerum Naturalium Universitatis Comenianae, Geographica 39:37–48
Minár J, Bizubová M (1994) Vývoj reliéfu južnej časti Turčianskej kotliny. Acta Facultatis Rerum Naturalium Universitatis Comenianae, Geographica 35:25–34
Minár J, Bielik M, Kováč M, Plašienka D, Barka I, Stankoviansky M, Zeyen H (2011) New morphostructural subdivision of the Western Carpathians: an approach integrating geodynamics into targeted morphometric analysis. Tectonophysics 502:158–174. https://doi.org/10.1016/j.tecto.2010.04.003
Minár J, Sládek J (2009) Morphological network as an indication of a morphotectonic field in the central Western Carpathians (Slovakia). Z Geomorphol Supp 53(2):23–29. https://doi.org/10.1127/0372-8854/2009/0053S3-0023
Minár J, Tremboš P (1994a) Analýza georeliéfu ako podklad pre komplexný krajinnoekologický výskum (modelové územie „Rudno“). Acta Facultatis Rerum Naturalium Universitatis Comenianae, Geographica 35:35–48
Minár J, Tremboš P (1994b) Prírodné hazardy – hrozby, niektoré postupy ich hodnotenia. Acta Facultatis Rerum Naturalium Universitatis Comenianae, Geographica 35:173–194
Nemčok M, Lexa J (1990) Evolution of the basin and range structure around the Žiar Mountains range. Geol Zbor Geol Carpath 41 (3):229–258
Pulišová Z, Hók J (2015) The palaeostress analysis of the fault rupture of the Žiarska kotlina Basin. Acta Geologica Slovaca 7(2):103–111
Rakús M, Hók J (2002) Geológia Turčianskej kotliny. manuskript, Archív ŠGÚDŠ Bratislava
Sládek J (2006) Geomorfologická analýza pohoria Žiar s dôrazom na detailný geomorfologický výskum Vyšehradského sedla. Prírodovedecká fakulta Univerzity Komenského, Katedra fyzickej geografie a geoekológie, Diplomová práca. Bratislava
Sládek J (2007) Rekonštrukcia paleoreliéfu v oblasti Vyšehradského sedla. In: Zborník recenzovaných príspevkov, Študentskej vedeckej konferencie, 2. zväzok. Bratislava. Kartprint, 253–255 pp
Sládek J (2008) Nové poznatky o vývoji georeliéfu pohoria Žiar. Acta Geographica Universitatis Comenianae. 51:155–169
Sládek J (2010) Odraz morfotektonického poľa v geomorfologickej mriežke oblasti Turčianskej kotliny. Prírodovedecká fakulta Univerzity Komenského v Bratislave, Dizertačná práca. Bratislava, 126 p
Sládek J, Bizubová M (2007) Niektoré prístupy k poznaniu georeliéfu pohoria Žiar. In: Geomorfologický sborník 6: Sborník abstraktů a exkurzní průvodce, Stav geomorfologických výzkumů v roce 2007. Ostrava. Ostravská Univerzita, 51 p
Sládek J, Bizubová M (2008) Vyšehradské sedlo - kľúč k poznaniu vývoja pohoria Žiar? Acta Geographica Universitatis Comenianae, Geographica 50:195–203
Spiridonov AI (1975) Geomorfologičeskoje kartografirovanije. Nedra, Moskva
Strahler AN (1952) Hypsometric (area-altitude) analysis of erosional topology. Geol Soc Am Bull 63(11):1117–1142. https://doi.org/10.1130/0016-7606(1952)63[1117:HAAOET]2.0.CO;2
Škvarček A (1973) Náčrt kvartérneho vývoja horského úseku doliny Hrona. Geografický Časopis 25(2):136–145
Urbánek J (1993) Geomorfologické formy tektonického pôvodu (identifikácia a mapovanie). Mineralia Slovaca 25:131–137

Ján Sládek is a geomorphologist at the Institute of Geography of Slovak Academy of Sciences in Bratislava as well as he works as a professional UAV operator and beta tester of laser scanner products in private sector. His research is focused on morphotectonics, morphostructure analysis and fluvial landscape monitoring by using of UAVs. The region of his interest covers Slovak part of Carpathian Mountain Range mainly surroundings of Turčianska kotlina Basin and other core mountains.

Ladislav Vitovič is a geomorphologist at State Geological Institute of Dionýz Štúr in Bratislava. His scientific interest focuses on morphotectonics and Quaternary sediments of the Western Carpathians.

Juraj Holec is a researcher at Slovak Hydrometeorological Institute, Department of Climatological Service. He has worked in fields of fluvial geomorphology, research of Quaternary, Natural Hazards, Spatial Climatology, Geographical Information Systems and Geostatistics.

Jozef Hók is Associate Professor of Geology at the Department of Geology and Paleontology, Comenius University in Bratislava. The focus of the work is the structural geology, tectonics and regional geology of the Western Carpathians.

12 Inland Delta and Its Two Large Rivers: Danube Plain, the Danube and Váh Rivers

Milan Lehotský, Juraj Maglay, Juraj Procházdka, and Miloš Rusnák

Abstract

The Danube Plain (Podunajská rovina) hosts the largest fluvial system in Slovakia. In the past the territory of the Danube Plain has operated as a dynamic fluvial system (inland delta) with its own anastomosing, migrating, meandering and braided river channel patterns and the development of several fluvial terraces, levees, abandoned channels and aeolian landforms. The Danube and Váh rivers as watercourses with a predominantly gravel channel bottom and laterally active from the transitional period of the Pleistocene/Holocene to the Medieval Period conditioned the development of the Danube Plain. Due to the abundance of gravels, groundwater, fertile soils, the presence of relatively large rivers and high density of settlements, the human impact on the Danube Plain resulted in a dense network of routes of different categories, drainage and irrigation canals usually equipped by pumping station, artificial levees and embankments as well as artificial lakes and two dams (Gabčíkovo waterworks on the Danube and the Kráľová dam on the Váh). The Danube river and its head-race canal serve as one of the main navigation routes in Europe. The Danube inter-dike area has become the Danube Meadows Protected Landscape Area since 1998.

Keywords

Danube Plain • Fluvial landforms • Aeolian landforms • Danube • Váh • Recent development • Slovakia

12.1 Introduction

The Danube Plain hosts the largest fluvial system in Slovakia, which is characterized by the most intense morphodynamics in both the transitional period of the Late Pleistocene/Holocene and subsequently in the Holocene when this area was morphologically transformed to its current form. In the past, the territory of the Danube Plain has operated as a dynamic fluvial system (inland delta) with its own anastomosing, migrating, meandering and braided river channel patterns and the development of several spatio-temporal groups of fluvial landforms. This is manifested also by the complexity and variability in facial structure and grain size of their sediments. The Danube and Váh rivers as watercourses with a predominantly gravel channel bottom and laterally active from the transitional period of the Pleistocene/Holocene to the Medieval Period mainly conditioned the development of the Plain.

The Danube Plain is characterized by very low local relief, which rarely exceeds 3–4 m, apart from the zones of occurrence of aeolian landforms where it reaches up to 15 m. The absolute height ranges from 107 m a. s. l. in Komárno town to 130 m a. s. l. in Bratislava. Another characteristic feature of the Plain is its aridity, with the average annual precipitation total ranging between 530 and 700 mm/year, which is the lowest in Slovakia. The Danube Plain is characterized by morphotectonically undifferentiated, planar to slightly undulating accumulation-erosion to

M. Lehotský (✉) · M. Rusnák
Department of Physical Geography, Geomorphology and Natural Hazards, Institute of Geography, Slovak Academy of Sciences, Štefánikova 49, 814 73 Bratislava, Slovakia
e-mail: geogleho@savba.sk

M. Rusnák
e-mail: geogmilo@savba.sk

J. Maglay
State Geological Institute of Dionyz Stur, Mlynská dolina 3962/1, 817 04 Bratislava, Slovakia
e-mail: juraj.maglay@geology.sk

J. Procházdka
Department of Physical Geography and Geoinformatics, Faculty of Natural Sciences, Comenius University, Mlynská dolina, 6842 15 Bratislava, Slovakia
e-mail: juraj.prochazka@uniba.sk

M. Lehotský and M. Boltižiar (eds.), *Landscapes and Landforms of Slovakia*, World Geomorphological Landscapes,
https://doi.org/10.1007/978-3-030-89293-7_12

accumulation fluvial relief, with an average slope of 1.5°–2°. The recent evolution of the Danube and Váh river channel patterns, which border the Plain from the south to the east, gives the added value in understanding the last phase of the development of the Plain that involved human interventions in the riverine landscape. Besides, the Plain is rich in underground waters (the richest area in Central Europe), thermal underground waters, gravels and fertile soils, which are used for relevant purposes such as irrigation, drinking water supply, recreation and construction as well as agriculture.

12.2 Geology and Tectonics of the Danube Plain

In terms of regional geological division (Vass et al. 1988a, b), the Danube Plain is part of the promontory of the Danube Basin (now referred to as the Slovak part of the Danube basin). Most of the region is formed by a unit distinguished within the Danube Basin–Gabčíkovská Basin. Approximately in its centre lies a young neotectonic subsidence structure—the Gabčíkovo Depression. The bulk of its basement consists of Neogene sediments of Lower Badenian to Pliocene age. The pre-Cenozoic fundament consists mostly of the rocks belonging to the Tatricum crystalline formations, the northern and southern Veporicum, and the sediments of the Paleozoic and the Mesozoic of the Silicicum unit. The fundament is discordantly overlain by the transgressional Cenozoic sedimentary fill. At the intersection of the Neogene sedimentary fill with the rocks of tectonic units of the fundament, there are several buried isolated stratovolcanic centres (Vass 2002). The Neogene, and locally Palaeogene, sedimentary filling of the basin reaches a mean thickness of about 5500 m (Fusán et al. 1987), increasing in the Gabčíkovo Depression area up to 8500–9000 m (Hrušecký 1999). The Tertiary fill is covered by a coarse transgressive and discordant formation consisting of Quaternary fluvial deposits of a unique inland Danube delta, laterally limited by a smaller delta of the Váh, Nitra and Žitava rivers. Their deposits occur in superposition, and only along the edges of the Plain and at the mountain feet, there are accumulation terrace staircases. Holocene fluvial deposition, aeolian and organogenic sediments are common on Pleistocene fluvial sediments. In the marginal NW part of the Little Carpathians there are numerous occurrences of deluvial, deluvio-proluvial and proluvio-alluvial sediments (Bezák et al. 2008).

The current shape of the Danube Basin is the result of a complex geological development in time and space (Adam and Dlabač 1961, 1969). One of the manifestations of this development is the occurrence of several tectonically conditioned depocenters (depressions) within its territory, which recorded maxima of subsidence activity in different time periods and different intensities. One such depression is also the Gabčíkovo Basin (Vass et al. 1988b; Kováč et al. 2001). In the early to middle Pannonian, sedimentation took place in the Gabčíkovo Basin in the brackish environment of a shallow lake, which was filled mainly with delta sediments transported by rivers from the emerging Alpine-Carpathian orogen. The oldest Quaternary sediments in the region are considered to be layers of fluvial to fluvial-limnic sediments, which form the basal part of the Quaternary filling of the Gabčíkovo Depression centre and are located at a depth of about 450–500 m (Scharek et al. 1998). Fluvial sediments of the Early Middle Pleistocene (Mindel) participate in the construction of the basal part of the Quaternary sedimentary filling (inland delta) of the Gabčíkovo Basin and were probably laid down by the already existing Danube.

The overburden of this fluvial accumulation is represented by fluvial sediments of the middle complex of the "Danube gravel series" (Janáček 1967) of Middle Pleistocene (Riss) age (Vaškovský and Vaškovská 1977). The Upper Pleistocene fluvial accumulation (Würm) forms the uppermost part of the middle complex of the Danube gravel series and its occurrence area within the Danube Plain is identical with the extent of the overlying Holocene accumulation. Holocene accumulation is represented by sandy gravels, calcareous sands and silts, fluvial-organic, organogenic and palustrine sediments. The relief of the Plain is typified by the occurrence of a dense network of abandoned channels, mostly completely filled by sediment, and branch channels in the aquatic stage within the recent Danube river floodplain, in many places anthropogenically modified. As far as tectonics is concerned, the area of the Slovak part of the Danube Basin and marginal structures are characterized by fracture tectonics and divided into a number of structural tectonic units characterized mainly by vertical movement tendencies. The Danube Plain as a geomorphological phenomenon represents a subsiding neotectonic unit of the Gabčíkovo Depression. Its edge, on the other hand, belongs to a series of units showing positive (rising) trends of variable intensity, geomorphologically defined as the Danube Uplands. This two-part aspect reflects the basic Quaternary tectogenesis of the area. To a significant extent, it directly bears on the Quaternary sedimentation space and, at the same time, on the pattern of erosion and accumulation in the geomorphological development, and thus the evolution of relief is also guided by it.

12.3 Plain Landforms

In terms of the geomorphological division by Mazúr and Lukinš (1978), the Danube Plain (Podunajská rovina-3.2, Fig. 1.1) is part of the Western Pannonian Province, the Lower Danube Basin Sub-Province and Podunajská Lowland (Podunajská nížina) unit (Figs. 1.1 and 1.2). The spatial pattern and classification of landforms of the Danube Plain and the courses of the Danube and Váh rivers are shown in Fig. 12.1.

12.3.1 Würm Terraces and Levees

The most significant morphologically is the occurrence of the so-called "the Würm Prater terrace" (Fink 1960), which extends into the territory from Austria and its edge extends along the Slovak-Austrian state border from the village of Kittsse to the junction of the Slovak-Austrian-Hungarian border, from where it continues to the Hungarian territory (Fig. 12.2). The surface of the terrace rises 4–6 m above the Danube floodplain and 6–8 m above the contemporary water level in the Danube channel itself. The terrace is formed by sandy gravels with an occasional transition to gravelly sands, which are waterlogged at the depth of 7–9 m.

To the NE of Bratislava, the terrace occupies a relatively large area and, compared to the above-mentioned right-bank

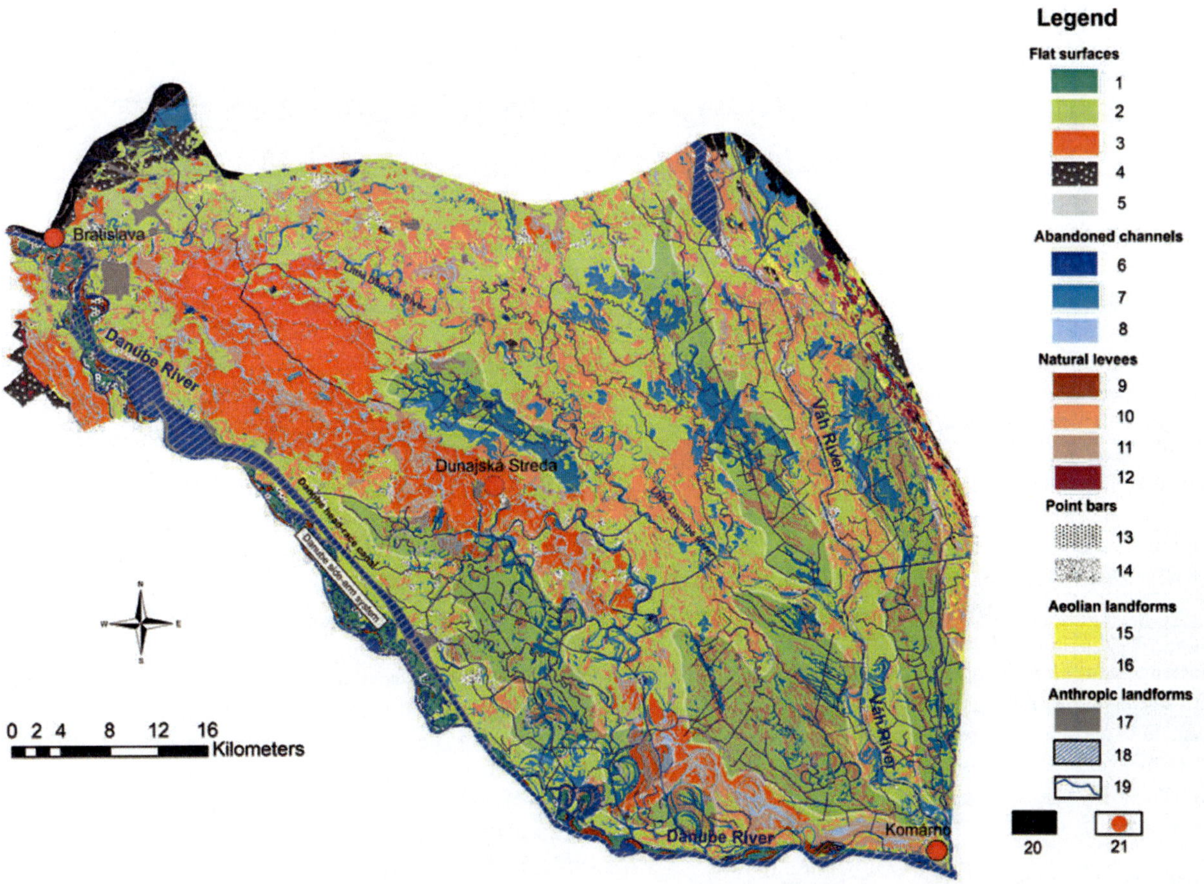

Fig. 12.1 Landforms of the Danube Plain. Flat surfaces: 1. Anthropocene floodplain; 2. Holocene floodplain; 3. Late Würm/Holocene floodplain ("core" area); 4. Würm terraces; 5. Backswamp depressions: Abandoned channels: 6. Anthropocene-abandoned channels; 7. Holocene relicts of the abandoned channel and oxbow lakes; 8. Relicts of the abandoned channel and oxbow lakes in the "core" area natural levees; 9. Anthropocene natural levees; 10. Holocene natural levees; 11. Late Würm/Holocene natural levees; 12. Würm natural levees; Point bars: 13. Anthropocene point bars; 14. Holocene point bars; Aeolian landforms: 15. Holocene aeolian landforms; 16. Late Würm/Holocene aeolian landforms; Anthropic landforms: 17. Large superficial building phenomena; 18. Dams and inter-dike floodplains; 19. Irrigation and drainage channel; Other symbols: 20. Mountain and hilly land margins; 21. Settlements

Fig. 12.2 Right-bank low terrace (the "Prater" terrace) of the Danube near the Slovak-Austrian state border. The surface of the terrace morphologically noticeably rises about 4.5 m above the surface of the Late Würm/Holocene floodplain ("the above floodplain terrace") (*Photo* J. Maglay)

terrace, has a greater slope and is dissected by streams flowing from the Little Carpathians. The surface of the terrace is relatively subtle, rising only 1–3 m above the Holocene floodplain of the Danube and often merging with the recent Holocene floodplain. Another occurrence of the terrace is located in the NE, on the edge of the Plain, where it rises 2–6 m, and locally up to 8 m above the contemporary water level in the Váh channel (Pristaš et al. 2000). In several places, the terrace is covered by a 1–3 m thick loess cover and aeolian sands up to 10 m thick. The surface of the terrace is, therefore, due to aeolian activity, slightly, sometimes even significantly undulating. Another factor behind the unevenness of the terrace surface is the lateral displacement of the Váh channel and its arms, resulting in its arrangement into the form of disturbed longitudinal mounds of the NW–SE direction. Due to lateral erosion of streams, isolated, elevated (approx. 1–3 m high) levees appear within the low terraces and rise above its surface (Maglay et al. 2017, 2018) (Fig. 12.3).

12.3.2 Late Würm/Holocene Landforms

The Late Würm/Holocene floodplain represents the main body of the Danube alluvial fan (inland delta) disturbed by lateral erosion. Its highest axial part, running from Bratislava to Komárno, is referred to as the core area of the Plain (Lukniš and Mazúr 1959; Fig. 12.4). Its part running along the right-bank low "Prater" terrace is morphologically defined as "the above floodplain terrace". According to the current findings based on ^{14}C dating (Maglay et al. 2017, 2018), this transitional fluvial formation corresponds to the Late Glacial (11,703 cal. BP) with a transition to the Lower Holocene, to the Pre-Boreal (11,734–10,203 cal. BP) and the Boreal (10,203–8,900 cal. BP).

The redeposited subsoil of sandy gravels and gravelly sands indicates short-term lateral displacements of channels due to sudden floods, accompanied by channel lateral migration, their branching and locally braiding.

Late Würm/Holocene levees represent low, more or less isolated sandy elevations on the Late Würm/Holocene floodplain. Unlike aeolian landforms, levees rise above the floodplain surface by only 2–3 m on average.

Relicts of Late Würm/Holocene abandoned channels have survived as shallow linear depressions within the Late Würm/Holocene floodplain and most of them have been levelled by ploughing. Based on numerical dating by the AMS method, the age of sediments was determined to be 9,700 y. BP (Maglay et al. 2017, 2018).

The morphological manifestations of the Upper Pleistocene/Holocene aeolian sedimentation can be found in the form of isolated small barchans, through compound dunes, to spatially more pronounced dunes of various shapes (Fig. 12.5). They are formed by Upper Pleistocene sands

Fig. 12.3 Outcrop within a natural levee of Würm age in Štrkovec village near the Váh river dated to 13,370 ± 90 BP–13,410 ± 40 BP based on AMS (Maglay et al. 2017, 2018) (*Photo* M. Vlačiky)

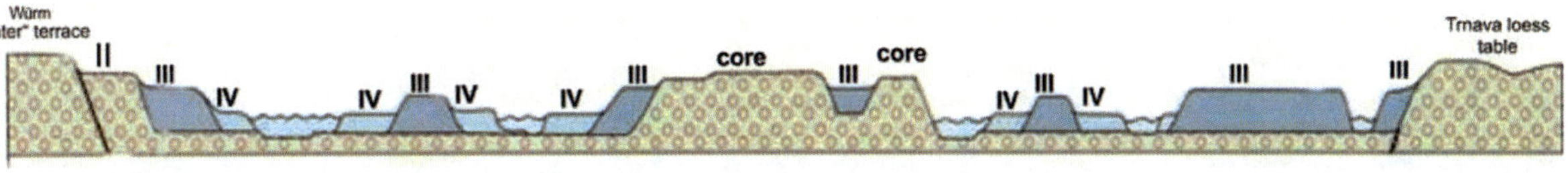

Fig. 12.4 Morphological position of the "core area" within the territory of the Danube Plain and adjacent hilly land. Note the same position of the "core" and the Late Würm/Holocene—"the above floodplain terrace" (II) and the lower position of the Holocene (III) and Anthropocene (IV) floodplains (source Maglay et al. 2017, 2018)

originated from levees and sandy sheets, and drifted during the postglacial and older Holocene, mostly for a short to medium distance.

12.3.3 Holocene Floodplain

The flat relief of the younger floodplain occupies the largest area of the Danube Plain along the edges of the Late Würm/Holocene floodplain as a lower level of the Plain surface. Point bars occur as slight elevations, tied to the convex part of the channel bend. They occur exclusively outside the Late Würm/Holocene floodplain, near the current watercourses in the zones of levees, but also at the bottom of abandoned channels. These landforms bypass "the core area" from both sides and border the contemporary channels of the Danube, Čierna voda, Little Dunaj and Váh rivers. In contrast to the older, above-described levees, these morphologically less pronounced forms are found in large areas of the Holocene floodplain. Their occurrence is tied either to the wider surroundings of the Little Danube and Váh river channels or to slightly elevated zones located along the abandoned sediment-filled branches.

The development of relict abandoned channels represents a response to the change in the hydrodynamic regime of watercourses in the Danube Plain, from a multi-channel

Fig. 12.5 Remnant of a dune after removal of most of the sand near Kameničná village. Originally, the dune reached a height of about 14 m. Its upper part was dated to 9,610 BP, and the lower one to 10,320 BP (Maglay et al 2017, 2018) (*Photo* J. Maglay)

anastomosing river system to a single-channel meandering one. The Holocene abandoned arms survived as linear concave landforms and are in marked contrast to the flat surface of the Holocene floodplain. Channel migrations and avulsions are manifested either as individual linear forms or as a dense network of bends over a relatively large area.

Like the older aeolian forms, the Holocene ones represent an important landscape-shaping element and complete the overall planar landscape character of the Plain. The young aeolian landforms form various irregular landforms, manifesting as small isolated islet-like ridges to small hills (Fig. 12.6). To a lesser extent, they appear as barchans or as more continuous linear structures of connected and parallel dunes, occasionally passing into more pronounced barchans and loaf-shaped dunes. The dunes originated by sand drift from older sandy fluvial sediments over short distances in dry climatic periods.

In addition, for landscapes characterized by the presence of a complicated anastomosing system, as well as by the development of levees along main channels, the occurrence

Fig. 12.6 The barchan dune in the Nesvady village (*Photo* J. Maglay)

of large depressions—backswamps is typical. They are disconnected from surface waters but show connectivity with underground waters. This kind of landform can also be found in the Podunajská Plain along the main river channels. Their periodical inundation negatively affected crop production, so many drainage canals and pumping stations have been constructed in the past and are maintained up to the present.

12.3.4 Anthropocene Floodplain and Recent Development of the Danube Riverine Landscape

The floodplains of the Danube river, which have been delimited by artificial levees since the sixteenth century, represent the youngest landforms of the Plain. So, their hydrological and sedimentological connectivity with the adjacent area of Holocene floodplains has been totally limited.

With a total length of 2850 km and a drainage basin of 801,463 km^2, the Danube river is the second-largest fluvial system in Europe. Natural conditions in the Late Glacial and mainly postglacial period and their stabilization under anthropogenic influence resulted in the present state of the Slovak reach of the Danube, which bifurcates into the main channel (The Danube), the Little Danube and the Mosoni Danube. According to the Danube River classification (Sommerhäuser et al. 2003), the Danube in Slovakia (length 172 km) belongs to the section of "Lower Alpine foothills Danube" and its sub-section of "the inland delta", whereas further downstream it belongs to the section of "Hungarian Danube Bend" and its first sub-section composed of an anabranching system.

The late Neolithic period and the Bronze Age (ca. 4500–2500 BC) were periods of relatively low discharges and low flood activity in the basins of the Upper Danube (Spurk et al. 2002). Archaeological investigations in the historical centre of Bratislava confirm that at that time there were particularly favourable conditions for human settlements directly on the Danube floodplain (Buch and Heine 1995). On the other hand, during more humid climatic micro-periods in part of the Eneolithic and later in the earlier Iron Age, it was only possible to build settlements in areas at a minimum altitude of 136–137 m a.s.l. (Baxa 1990). The Roman period also represented relative sedimentation calmness and building the military fortress along the Danube (Limes Romanus) (Fig. 12.7).

Fig. 12.7 Remnants of the Gerulata fortress from the Roman period built near Bratislava as one of the military points of "Limes Romanus" (*Photo* J. Lacika)

The phase of increased flood and river activity lasted only shortly and began in the early Middle Ages, about 580–770 AD (Buch and Heine 1995). During Medieval Age, the most significant anabranch of the Danube in the Danube Plain was the Čalov, considered as the medieval Little Danube because the Čalov arm did not enter into the Váh river at Kolárovo, but into the Danube at the village of Čičov (Fig. 12.8). Thus, the Žitný Ostrov Island, in its present form, did not exist in the thirteenth and fourteenth centuries. In the subsystem of southern arms, the longest (at least 50 km) and geographically most significant was the "Čelic" arm, separated from the Danube at the village of Čilistov. A note from 1272 (Ortvay 1892) reveals that even in the thirteenth century the Čelic did not carry much water and was located in the same place as the present Danube main channel near the village of Bodíky. The main channel must have been situated more to the southwest at that time. Written records show that the Danube in the stretch Bratislava–Rusovce had a similar character as it did later, in the eighteenth and nineteenth centuries.

In connection with hydro-climatic manifestations of the Little Ice Age, the frequency and magnitude of floods increased in many European rivers in the second half of the sixteenth century (Brázdil et al. 1999). Also, at the Danube, large and numerous floods affected Bratislava and Žitný Ostrov Island. At that time, the Čalov (=Little Danube) reached a width of 150–200 m already at Bratislava in the present-day settlement of Ružinov (Pišút 2005). Since the late seventeenth century the discharge in the Little Danube has definitely decreased, probably as a result of terrain changes in Bratislava and gradual shifting of its upper entrance farther away from the city (Baxa 1990). According to the Mikovíni map (1735), the main Danube channel formed six large regular meanders between Šamorín and Sap in the first third of the eighteenth century. Since the 1720s, an increasing activity of the main channel and its side arms was demonstrated not only by frequent dike breaches but particularly by undermining and destruction of protective dikes, which had been built all along the Danube at that time. In the eighteenth century, the surface runoff increased as a result of introducing new agricultural methods and crop

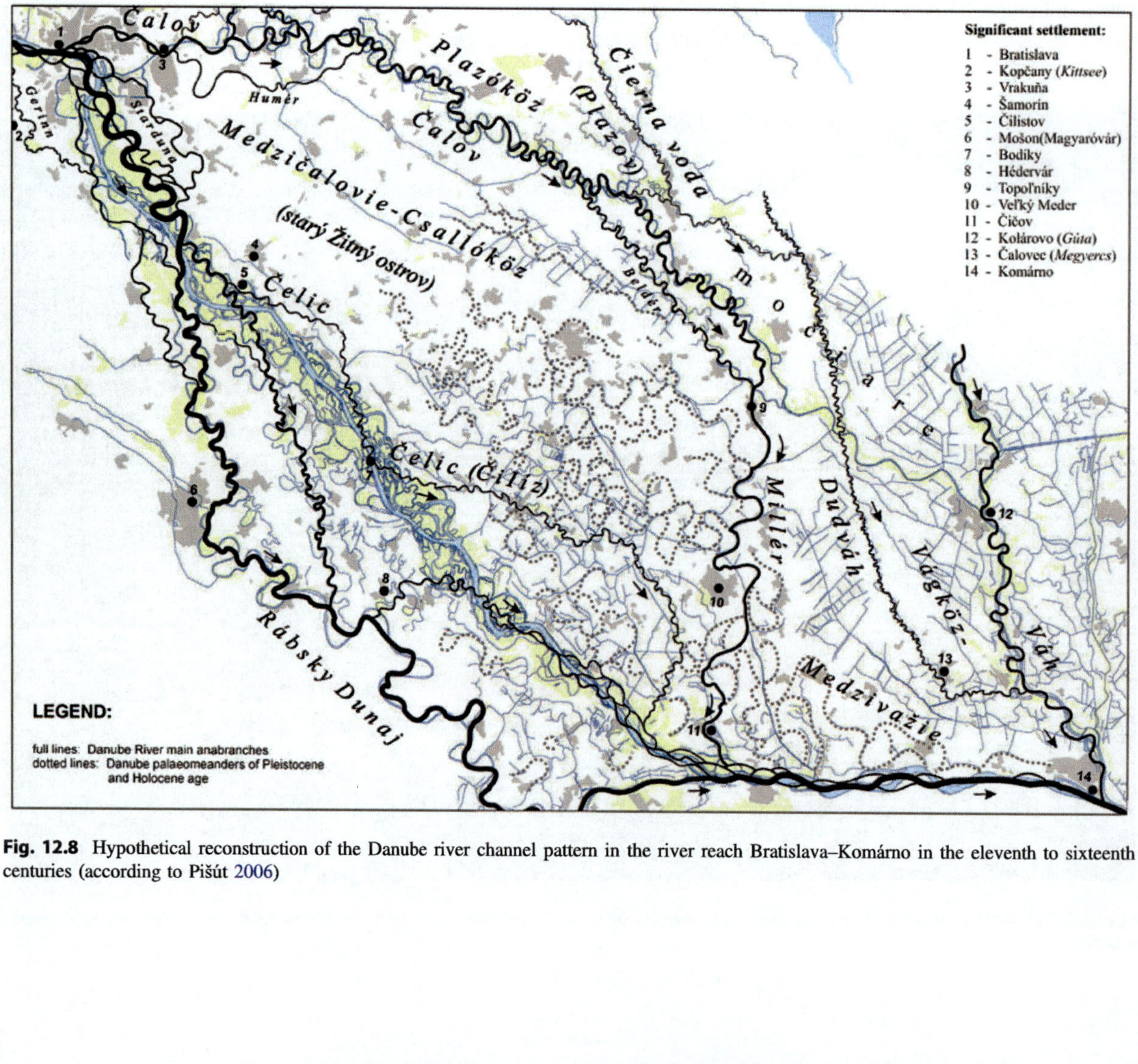

Fig. 12.8 Hypothetical reconstruction of the Danube river channel pattern in the river reach Bratislava–Komárno in the eleventh to sixteenth centuries (according to Pišút 2006)

plants. The general increase of sediment supply resulted in the increased lateral activity of the rivers. In the Danube, it was manifested by accelerated development and change of its meander size. The sediment migration was indicated by a considerable increase in the number and total area of gravel bars. Bars and new islands pushed the stream laterally and caused bilateral widening of the riverbed (Pišút 2002, 2005).

Already in the distant past, the Danube waters were manipulated and smaller, periodically active arms were partitioned and dammed for the purposes of fishing (Füry et al. 1986). Numerous pile and wicker constructions, especially stable beluga weirs, dammed the side arms or in the case of narrower arms dammed the whole stream, and these could have had a considerable impact on the deposition in the riverbeds (Alapy 1933). The most typical technical appliances of the feudal period—boat and pile mills also could have influenced the erosion–sedimentation processes by enhancing lateral erosion with surf, the transformation of water streams into mill-races, the construction of submerged dams or by the sedimentation downstream from the mills (Martinka 1956). In the fifteenth to seventeenth centuries, lateral erosion was probably the immediate cause of doom for some medieval villages situated near the river. Regarding expenses and organization, the most prodigious works attempted in the eighteenth and in the first half of the nineteenth century to divert the stream in a secure direction were artificial cut-offs. In the 1830s, water-concentrating structures started to be gradually built to ensure sufficient navigation depth in the Danube riverbed in order to develop steam navigation. In 1829, the First Danube Steam Navigation Company was founded. In 1831, regular traffic started on the Vienna line to Budapest (Pišút 2006). At the end of the nineteenth century and in the early years of the twentieth century the main meandering channel was converted for navigation purposes into a 300 m wide (average annual discharge 2025 m^3 s^{-1}), single-thread straight channel reinforced with rip-raps on banks and artificial levees. Such a condition, involving channel maintenance works (gravel mining, groyne maintenance) and relatively well-functioning hydrological and sedimentological connectivity, mainly in the side-arm system area near the town of Gabčíkovo as well as upstream and downstream, lasted till 1992. Then, after the Gabčíkovo waterworks, consisting of the Čunovo diversion weir, the 29 km-long head-race canal ending by the Gabčíkovo weir and the 10 km-long trail-race canal were launched into service, and the original Danube channel became replaced by the 40 km-long by-passed channel (*"The Old Danube"*). The position and structure of the Gabčíkovo waterworks as well as the successive simplification of the side-channel pattern of the Danube River is depicted in Fig. 12.9.

The Gabčíkovo waterworks was put into operation in October 1992 (Fig. 12.10). Electricity production, flood protection, improvement of navigation, recreation and groundwater management have been the main reason for its construction. As far as electricity production is concerned, there is an installed capacity of 720 MW in eight turbine/generator sets. The take-off structure from the head-race canal at Dobrohošť provides permanent watering, or even an artificial simulation of floods on the side-arm system (Kocinger 2001, Fig. 12.11). Apart from these economic purposes, a white-water sports facility at the Čunovo weir is used for amateur as well as professional meetings and competitions and bicycle paths on both sides of the canal from Bratislava to Gabčíkovo as well as along the whole length of the left bank downstream of Gabčikovo have been constructed to operate as a section of the Danube cycle path.

12.3.5 Anthropocene Floodplain and Recent Development of the Váh Riverine Landscape

The Váh river is the longest river of Slovakia (403 km) and about a half of its length is on the Podunajská nížina lowland. Prior to channel regulation, the Váh had three segments with different channel patterns on the Podunajská nížina lowland. The upstream segment (from Beckovská brána port to the town of Piešťany) exhibits a wandering pattern, whereas the downstream one in the Danube Plain (between Hlohovec and Komárno) shows a meandering pattern (Fig. 12.12).

An analysis of sinuosity according to the maps of the I. military survey of Habsburg Empire (Arcanum 2004) confirmed the relation between the character of river channel pattern and active faults (Maglay et al. 1999). The highest values of sinuosity index are associated with sections of maximal differences in vertical movement tendency between two neighbouring tectonic blocks, mainly in the transitional area (Sect. 5) from the Podunajská pahorkatina hilly land to the Podunajská rovina (Fig. 12.13). Neotectonic movements conditioned the lateral development of the lower Váh river system in historic and present times (Procházka 2017). There is an abrupt change of the Váh channel direction to the south-east as it passes from the hilly land to the Podunajská rovina plain. This change of direction is according to Lukniš (1972) attributed to the Danube river, which deposited a large alluvial fan of Žitný Ostrov Island and so forced the Váh river channel to the south-east. At present, the reason behind the shifting of the Váh river channel to the east is seen in the tilting of the Podunajská nížina lowland (mainly Trnavská pahorkatina hilly land) to the south-east (Dlabač

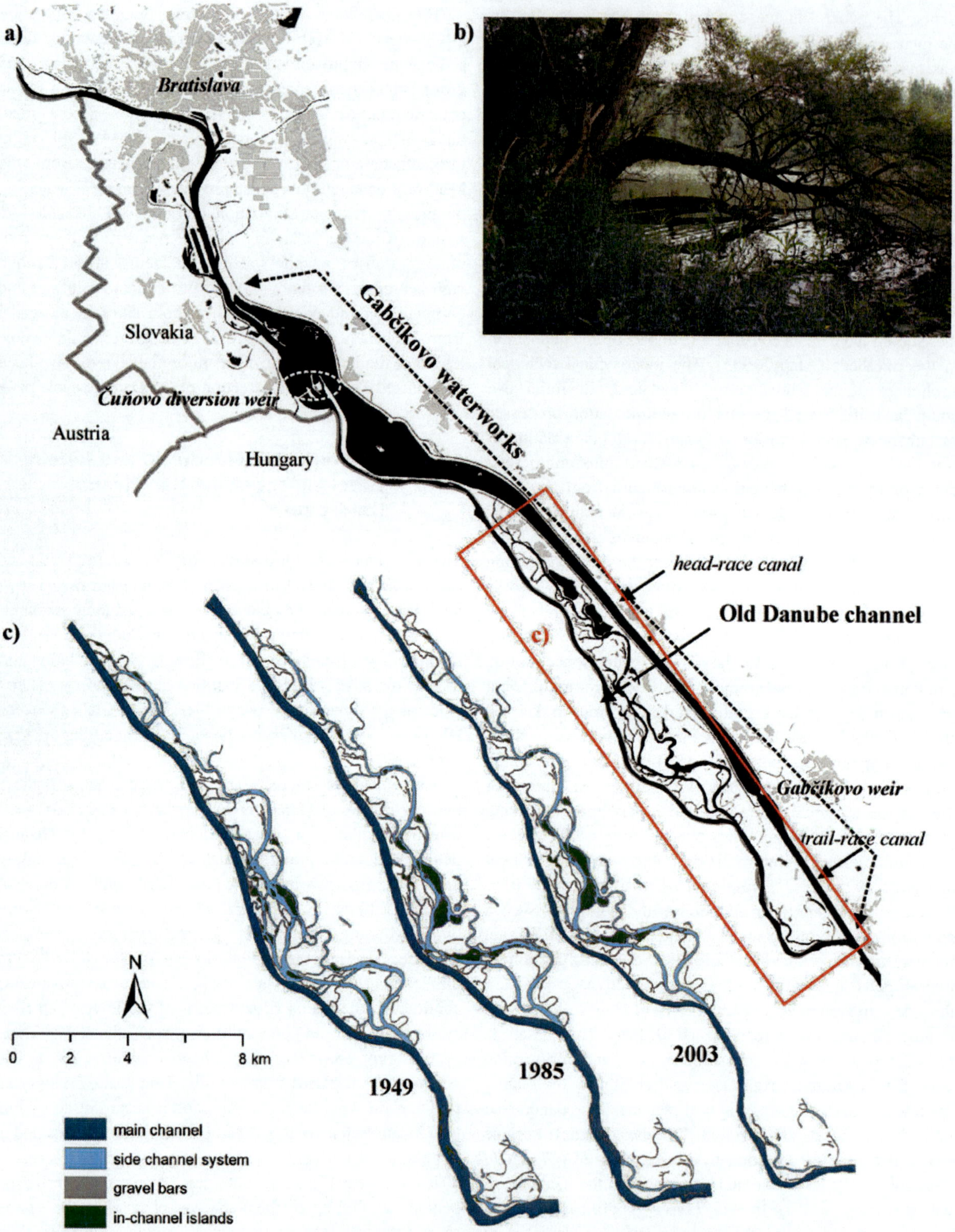

Fig. 12.9 The Gabčikovo waterworks and the side-channels pattern development from 1949. **a** localization of the Gabčikovo waterworks, the Old Danube channel and the branch channels system; **b** an example of the contemporary character of the branch channel system; **c** gradual degradation of the branch channel system, disconnection from old Danube channel and the Old Danube narrowing after 1992 (onset of operation of the Gabčíkovo waterworks)

Fig. 12.10 Oblique view of the Gabčikovo weir and upstream head-race canal (*Photo* https://www.24hod.sk/files/gab.jpg)

1960; Maglay 2010). An earlier shifting of the Váh river channel to the east during the Pleistocene is documented also by other lines of evidence. One of them is the consecutive migration of human settlements to the east that followed lateral shifts of the river channel in the past (Ištok and Ižof 1990). Another one consists in the Middle Pleistocene gravel deposition of the Váh river, found near the villages of Opoj and Majcichov, where they rest on the Neogene clays and are covered by loess. The tendency of the Váh river channel to shift to the east during the Quaternary persists in recent times. Ištok (1978) assessed it at 0.5–1 m per year during the 1970s, 0.9–1.2 km between the villages of Posádka and Horná Čepeň and 1.5–1.9 km on the meander near the town Sereď for the Holocene period.

Heavy minerals from the Váh river basin, found in the lower parts of the Middle Pleistocene deposits in the core area of the Danube Plain document shifting of the channel in this direction (Horniš and Priechodská 1979). Lateral shift of the lower Váh river channel during the Quaternary is also conditioned by rapid tectonic subsidence of the area around Komárno and Kolárovo (Vaškovský 1977; Dlabač 1960). Later, another change of direction occurred due to tectonic movements; the Váh between Šoporňa and Komoča has a tendency to shift to the south-west (Lukniš 1969). Subsiding character of the Váh river segment around Zemné, Kolárovo and Kameničná is reflected by a high index of channel sinuosity (Procházka and Pišút 2015; Fig. 12.14).

In the lower part of the Žitný Ostrov Island, we can identify a 20 km-long river channel—paleo-Dudváh into which the Little Danube ran in the past (Hladký and Závodný 2015; Fig. 12.14). Its abandonment was conditioned by a 16 km-long avulsion channel of the Little

Fig. 12.11 Weir and one of the branches in the protected branch channel area of the Danube. There are nine steps of weirs (A-J) maintaining the quasi-original character of the arm system (*Photo* M. Lehotský)

Danube headed towards the east and joined the Váh river near Kolárovo town, probably during the millennial flood in 1501 (Pišút et al. 2010). Another example of avulsion of the Váh channel was found near the village of Selice, where the local Hungarian toponym *Hajó út*—"Ship Route" of the abandoned channel indicates that this phenomenon occurred. The "Ship Route" name surely indicates the Váh main channel as the main route for sailing. (Medňanský 2007). Furthermore, some parameters of this channel and its levée correspond to the main channel (Fig. 12.14). According to historic records, we can date this avulsion to the beginning of the fifteenth century. The elevated levée is also indicated by land use. While the lower water-table and lower flood-risk on the levée enabled to use the soil as arable land, wet meadows prevailed in the neighbourhood of the levée (Fig. 12.15). The abandonment of *Ship Route* channel and the avulsion of the Little Danube towards Kolárovo are tectonically induced changes of the lower Váh river channel position. Construction of embankments was realized mostly on the Podunajská Plain, where flood-protection levees were built almost continually in the eighteenth century, while further upstream they were built only around residential areas (Procházka 2017). The oldest information about channel regulation works is from the surrounding of Leopoldov, dating from the seventeenth century.

Thanks to historic maps, we can find that one of the routine regulation works was the isolation of side channels (that were cut off from the main channel) and consequently, the decrease in the number of bars, their area and average size (Procházka 2020). Thanks to these works, the floodplain

Fig. 12.12 The Váh river segment, primarily with simple meandering channel pattern on the Podunajská rovina plain (on the maps of the II. military survey of Habsburg Empire; from Jankó et al. 2005)

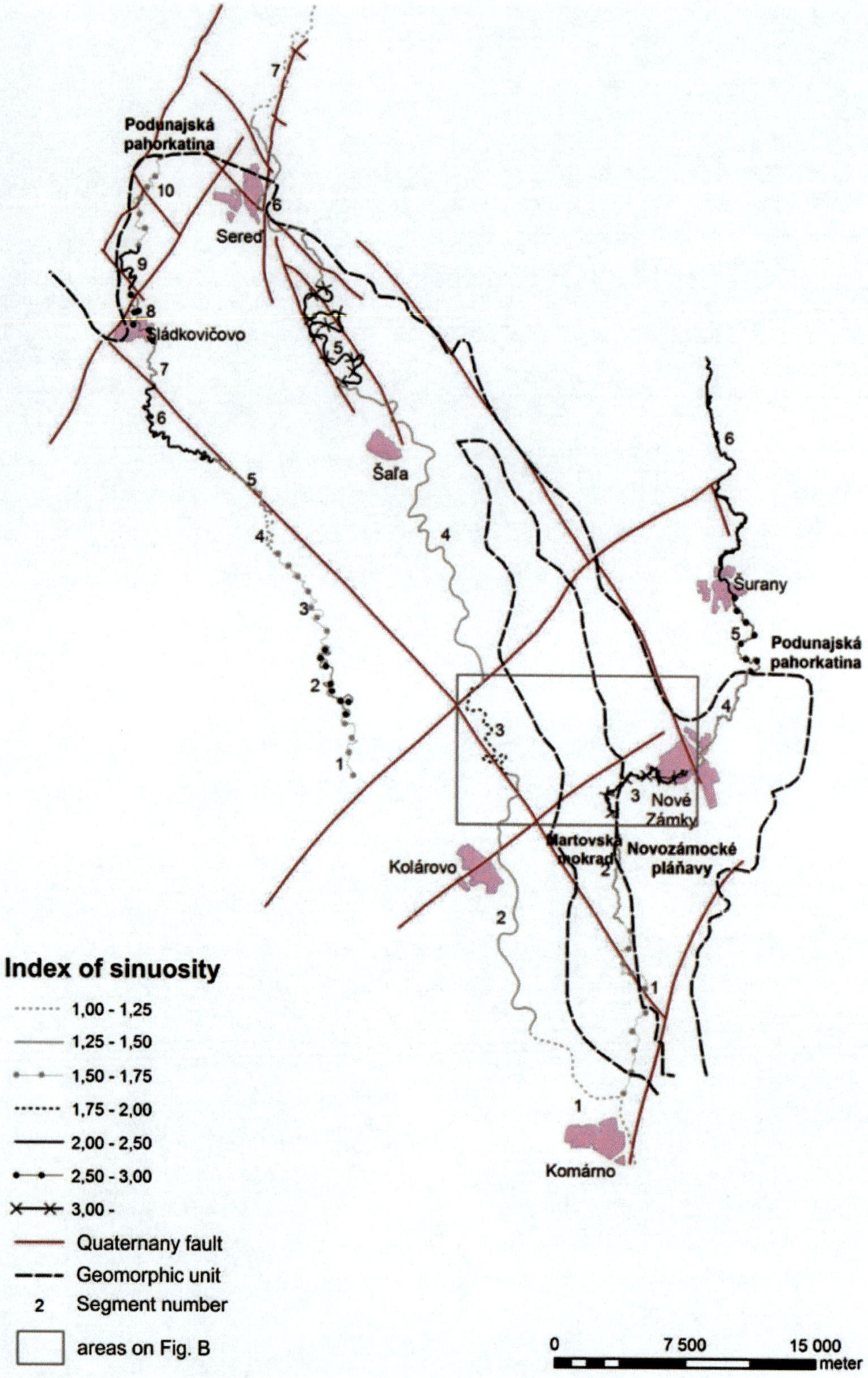

Fig. 12.13 Segments of the lower Váh river and their indices of sinuosity (Procházka 2013)

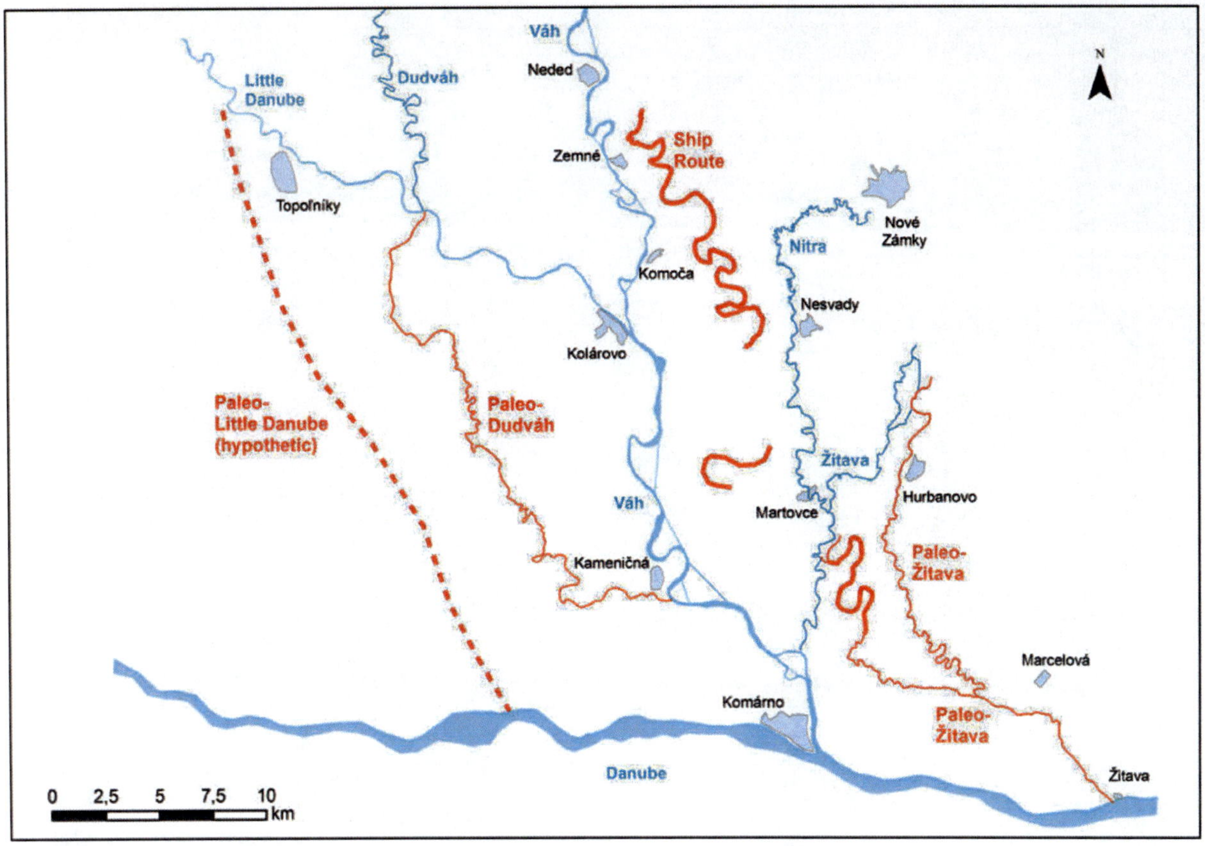

Fig. 12.14 The most significant changes of the river network of lower Váh (segment Šaľa-Komárno) and its surroundings in historic times. Red lines—original reaches of river channels (Procházka and Pišút 2015)

was partially drained off and agriculture conditions improved (Pišút et al. 2016). The length of the Váh river channel segment on the Podunajská rovina plain was shortened in 1782–1900 by 18%. Regulation works were not always effective and, on many sections, they had to be repeated. The number of bars along the whole lower Váh decreased by half (from 240 to 117), mainly on the Podunajská rovina. The total area of bars decreased from 866 to 285 ha (by 80% on the Podunajská rovina and by 50% on the Podunajská pahorkatina). The biggest area and number were indicated in 1840–1882 when their number and area increased. After that time, their area significantly decreased mainly due to channel training. The most important period of river regulation of the Váh channel occurred in the twentieth century. The works, including building of the drainage canal system, conditioned stabilization of the channel, disappearance of wet meadows and partially also riparian forests (nowadays in these areas there is arable land) (Jančovič and Petrovič 2012). Due to large-scale construction works of housing units during the period of socialism, many gravel pits were built. The most important anthropogenic intervention to the Váh river channel was the construction of two big dams—Sĺňava next to Piešťany (1959) and Kráľová between Sereď and Šaľa (1985). Regulated discharge in the main channel and the effect of "hungry water" below dams caused bed incision and channel narrowing.

12.4 "New" Danube Plain

Due to the richness of gravel, groundwater, fertile soils and the presence of relatively large rivers as well as the high density of settlements, the human impact on the Danube Plain resulted in a dense network of linear anthropic landforms like routes of different categories, drainage and irrigation canals usually equipped by pumping station, artificial levees and embankments as well as localized impacts like artificial lakes (Fig. 12.16), dumps, mounds of gravel and areas of building construction (Gabčíkovo waterworks on the Danube and the Kráľová dam on the Váh rivers, Bratislava airport, large factory zones). The Danube river and its head-race canal serve as one of the main navigation routes in Europe. The gravel sedimentary fill of the Danube Plain as

Fig. 12.15 "Ship Route" channel in the half of nineteenth century (in the middle). On the left is the active Váh main channel (map of II. military survey of Habsburg Empire, from Jankó et al. 2005).

well as the presence of large rivers determine its richness in groundwater. In the central part of the Danube Plain, in the area of the Žitný Ostrov island (the area between the Danube and Little Danube), groundwater reserves, expressed in terms of usable amount, reach approximately 20,400 $l\ s^{-1}$, which represents the largest reserves of underground waters in Slovakia and Central Europe. So, this part of the Danube Plain represents the largest protected water management area in Slovakia. Favourable hydrogeological conditions of the plain were also reflected by drilling six large wells, which supply the city of Bratislava and its surroundings drinking water. Another deeper wells provided thermal waters which are used for recreation purposes in aquaparks (Dunajská Streda, Veľký Meder). The Danube inter-dike area of the plain has become the Danube Meadows Protected Landscape Area since 1998. It consists of five separate parts, where the anastomosing fluvial system with the network of side channels and wetlands, as the largest protected area, is found in the south-central part of the Danube Plain along the Old Danube channel.

Fig. 12.16 Artificial lake created by gravel mining and filled by groundwater. Lake banks are used for house construction as a new suburban residential area as well as for public recreation (*Photo* J. Lacika)

Acknowledgements This work was supported by the grant of the Scientific Grant Agency of the Ministry of Education, Science, Research and Sport of the Slovak Republic and the Slovak Academy of Sciences (VEGA) No. 1/0781/17 and No. 2/0086/21 and by Slovak Research and Development Agency under the contract No. APVV-15-0054. The authors wish to thank J. Lacika and M. Vlačiky for providing photos.

References

Adam Z, Dlabač M (1961) Nové poznatky o tektonice Čs. části Malé dunajské nížiny. Věstník ÚÚG 36(3):188–198

Adam Z, Dlabač M (1969) Vysvětlivky k mapám mocností a litofaciálního vývoje Podunajské nížiny. Zbor. geol. Vied, Západ. Karpaty, rad ZK 11:156-172

Alapy G (1933): A csallóközi halászat története. Komárno

Arcanum (2004) Az első katonai felmérés. (DVD). Budapest, HM HadtörténetiIntézetés Múzeum Térképtára; Arcanum Adatbázis Kft

Baxa P (1990) Vplyv Dunaja na osídlenie historického jadra Bratislavy v praveku. Študijné Zvesti Archeologického Ústavu SAV 26:129–143

Bezák V (ed.), Elečko M, Fordinál K, Ivanička J, Kaličiak M, Konečný V, Kováčik M, Maglay J, Mello J, Nagy A, Polák M, Potfaj M, Biely A, Bona J, Broska I, Buček S, Filo I, Gazdačko Ľ, Grecula P, Gross P, Havrila M, Hók J, Hraško Ľ, Jacko (ml.) S, Jacko (st.) S, Janočko J, Kobulský J, Kohút M, Kováčik M, Lexa J, Madarás J, Németh Z, Olšavský M, Plašienka D, Pristaš J, Rakús M, Salaj J, Siman P, Šimon L, Teťák F, Vass D, Vozár J, Vozárová A, Žec B (2008): Prehľadná geologická mapa Slovenskej republiky 1: 200 000. Ministerstvo životného prostredia SR, ŠGÚDŠ, Bratislava. ISBN 978-80-89343-21-8

Brázdil R, Glaser R, Pfister Ch, Dobrovolný P, Antoine JM, Barriendos M, Camuffo D, Deutsch D, Enri S, Guidoboni E, Kotyza O, Sanchez-Rodrigo FL (1999) Flood events of selected European rivers in the sixteenth century. Climatic change 43: 239-285. doi: https://doi.org/10.1023/A:1005550401857

Buch MW, Heine K (1995) Fluvial geomorphodynamics in the Danube River valley and tributary river systems near Regensburg during the Upper Quaternary – theses, questions and conclusions. Z Geomorphol Supp 100:53–64

Dlabač M (1960) Poznámky kevztahumezitvarem povrchu a geologickou stavbou Podunajské nížiny. Geologické Práce 59:69–100

Fink J (1960) Leitlinien einer österreichischen Quartärstratigraphie. Mitt Geol Ges 53:249–266

Füry J, Décsi L, Stanovský J (1986) Povodie Dunaja včera a dnes. Povodie Dunaja 1:15–35

Fusán O, Biely A, Ibrmajer J, Plančár J, Rozložník L (1987) Podložie terciéru vnútorných Západných Karpát. GÚDŠ, Bratislava, 123 p.

Hladký J, Závodný A (2015) Hydronymia Žitného ostrova. Trnavská Univerzita v Trnave, Pedagogická fakulta, p 319

Horniš J, Priechodská Z (1997) Použitie ťažkých minerálov pri štúdiu sedimentárneho komplexu Žitného ostrova. Mineralia Slovaca 11 (6):529–536

Hrušecký I (1999) Central part of the Danube Basin in Slovakia: geophysical and geological model in regard to hydrocarbon prospection. Explor Geophys Remote Sens Environ 6(1):2–55

Ištok P (1978) Poznámky k podielu tektoniky na vývoji reliéfu Podunajskej nížiny v okolí Serede. Geogr Časopis 30(1):75–82
Ištok P, Ižof J (1990) Podmienky vzniku a vývoja osídlenia krajiny dolného toku Váhu vo svetle geografických a archeologických prieskumov. Študijné Zvesti Archeologického Ústavu SAV 26:145–168
Janáček J (1967) Stratigrafické poznatky v mladých sedimentoch centrálnej časti Podunajskej panvy. In: Dielčia ZS za r. 1965 – 1966: Nové poznatky o tektonike centrálnej časti Podunajskej panvy. Archív ŠGÚDŠ, 5–11
Jančovič P, Petrovič F (2012) Trendy vývoja kultúrnej krajiny medzi mestami Piešťany a Hlohovec. Životné Prostredie 46(1):34–37
Jankó A, Oross A, Timár G (2005) A másodikkatonaifelmérés 1819–1869. (DVD). Budapest: HM HadtörténetiIntézetés Múzeum Térképtára; Arcanum Adatbázis Kft
Kováč M, Baráth I, Halouzka R, Joniak P, Sliva Ľ, Vojtko R (2001): Neotektonický vývoj hrastu Považského Inovca a blatnianskej priehlbiny Dunajskej panvy „Slip rate analysis". Ms., Archív PriF UK, Bratislava
Lukniš M (1969) Poznámky k vývinu reliéfu Podunajskej nížiny v okolí Nových Zámkov. Stud Geogr 1:45–51
Lukniš M (1972) Reliéf. Slovensko II. Príroda
Lukniš M, Mazúr E (1959) Geomorfologické regióny Žitného ostrova. Geografický časopis 11(3):161–206
Martinka M (1956) Historickogeografické Črty Žitného Ostrova. Geografický Časopis 7:134–139
Maglay J (2010) Geomorfológia Trnavskej pahorkatiny v závislosti na štruktúrnom vývoji a charaktere usadenín vrchného pliocénu a kvartéru. Dizertačná práca. Prírodovedecká fakulta Univerzity Komenského, Bratislava
Maglay J, Halouzka R, Baňacký V, Pristaš J, Janočko J (1999) Neotektonická mapa Slovenska 1 : 500 000. Bratislava: Ministerstvo životného prostredia Slovenskej republiky
Maglay J (ed), Fordinál K, Nagy A, Kováčik M, Šefčík P, Vlačiky M, Šimon L, Moravcová M, Zlocha M, Fričovská J, Zlinská A, Žecová K, Baráth I, Liščák P, Ondrášiková B, Gluch A, Kucharič Ľ, Zeman I, Kubeš P, Benková K, Bottlík F, Marcin D, Michalko J, Baláž P, Stupák J, Tuček Ľ (2017) Vysvetlivky ku geologickej mape regiónu Podunajská nížina – Podunajská rovina v mierke 1 : 50 000. ŠGÚDŠ, 379 p
Maglay J, Fordinál K, Nagy A, Vlačiky M, Šefčík P, Fričovská J, Moravcová M, Kováčik M, Baráth I, Zlocha M (ed) (2018) Geologická mapa Podunajskej nížiny – Podunajskej roviny. Regionálne geologické mapy Slovenska 1 : 50 000. ŠGÚDŠ
Kocinger D (2001) Gabčíkovo part of hydroelectric power project and Joint slovak-hungarian monitoring of environmental impact. http://www.vvb.sk/gabcikovo.gov.sk/index.php?page=english-section. Accessed 10 Sept 2020
Mazúr E, Lukniš M (1978) Regionálne geomorfologické členenie SSR. Geografický časopis 30(2):101–125
Medňanský A (2007) Malebná cesta dolu Váhom. Vydavateľstvo Spolku Slovenských spisovateľov, Bratislava, 184 p.
Ortvay T (1892) Pozsony város története. I. zv., Pozsony (Prešporok)
Pišút P (2002) Channel evolution of the pre-channelized Danube river in Bratislava, Slovakia (1712–1886). Earth Surf Proc Land 27:369–390. https://doi.org/10.1002/esp.333
Pišút P (2005) Príspevok historických máp k rekonštrukcii vývoja koryta Dunaja v oblasti uhorsko – rakúskej hranice (16–19. storočie). In Pravda J (ed) Historické mapy, Kartografická spoločnosť SR a Geografický ústav SAV, pp 167–181
Pišút P (2006) Changes in the Danube riverbed from Bratislava to Komárno in the period prior to its regulation for medium water (1886–1896). Slovak - Hungarian Environmental Monitoring on the Danube. www.vvb.sk/old.gabcikovo.gov.sk/doc/moson/index.htm
Pišút P, Břízová E, Čejka T, Pipík R (2010) Paleofloristic and paleofaunistic analysis of Dudváh River oxbow and implication for Late Holocene paleoenvironmental development of the Žitný ostrov Island (SW Slovakia). Geol Carpath 61(6):13–533. https://doi.org/10.2478/v10096-010-0032-1
Pišút P, Procházka J, Matečný I, Bandura P (2016) Vývoj koryta Váhu pri Leopoldove v 17.-20. storočí a odozva rieky na zásahy človeka. Univerzita Komenského v Bratislave, 272 p
Pristaš J, Elečko M, Maglay J, Fordinál K, Šimon L, Gross P, Polák M, Havrila M, Ivanička J, Határ J, Vozár J, Tkáčová H, Tkáč J, Liščák P, Jánová V, Švasta J, Remšík A, Žáková E, Töröková I (2000) Vysvetlivky ku geologickej mape Podunajskej nížiny - Nitrianskej pahorkatiny 1: 50 000. ŠGÚDŠ, Bratislava, 250 p. ISBN 80-88974-26-7
Procházka J (2013) Pôdorysná vzorka riečneho koryta dolného Váhu, Nitry a Dudváhu s osobitným zreteľom na vplyv faktora tektoniky. Geomorphol Slovaca Bohem 13(2):19–30
Procházka J (2017) Geomorfologická odozva environmentálnych zmien riečneho koryta dolného Váhu. Dizertačná práca, Prif UK. Bratislava, 159 p
Procházka J (2020) Významné historické zmeny fluviálneho systému dolného Váhu a význam mechanizmu avulzií v jeho vývoji. Geogr Cassoviensis 14(1):92–108. https://doi.org/10.33542/GC2020-1-06
Procházka J, Pišút P (2015) Regulácie koryta nížinného meandrujúceho vodného toku v období r. 1782–1900 (na príklade rieky Váh v úseku Sereď-Komárno). Geographia Cassoviensis 9(1):44–55
Scharek P, Herrmann P, Kaiser M, Pristaš J, Tkáčová H (ed) (1998) Danube region Vienna – Bratislava – Budapest. Map of genetic types and thickness of Quarternary sedimensts 1 : 200 000. DANREG (Danube region Environmental Geology Programme). Magyar Allami Földtani Intezet (Geological Institute of Hungary), Budapest
Sommerhäuser M, Robert S, Birk S, Hering D, Moog O, Stubauer I, Ofenböck T (2003) Final report, Activity 1.1.6 "Developing the typology of surface waters and defining the relevant reference conditions", UNDP/GEF DANUBE REGIONAL PROJECT (DRP) "Strengthening the Implementation Capacities for Nutrient Reduction and Transboundary Cooperation in the Danube River Basin"
Spurk M, Leuschner HH, Baillie MGL, Briffa KR, Friedrich M (2002) Depositional frequency of German subfossil oaks: climatically and non-climatically induced fluctuations in the Holocene. Holocene 12 (6):707–715. https://doi.org/10.1191/0959683602hl583rp
Vass D (2002) Litostratigrafia Západných Karpát: neogén a budínsky paleogén. ŠGÚDŠ, Bratislava, 202 p
Vass D, Began A, Gross P, Kahan Š, Krystek I, Köhler E, Lexa J, Nemčok J, Růžička M, Vaškovský I (1988a) Mapa regionálneho geologického členenia Západných Karpát a severných výbežkov Panónskej panvy na území ČSSR 1:500 000, GUDŠ, Bratislava
Vass D, Began A, Gross P, Kahan Š, Krystek I, Köhler E, Lexa J, Nemčok J, Růžička M, Vaškovský I (1988b) Vysvetlivky k mape regionálneho geologického členenia Západných Karpát a severných výbežkov Panónskej panvy na území ČSSR 1:500 000. ŠGUDŠ, Bratislava, 65 p
Vaškovský I (1977) Kvartér Slovenska. Bratislava: ŠGUDŠ, Bratislava. 247 p
Vaškovský I, Vaškovská E (1977) Regionálny kvartérno-geologický výsrrkum Žitného ostrova. Bratislava, ČZS za r. 1974–1977, MS, archív ŠGÚDŠ, 109 p

Milan Lehotský is a physical geographer and fluvial geomorphologist at the Institute of Geography of the Slovak Academy of Sciences. He was many years head of the Department of Physical Geography, Geomorphology and Natural Hazards. His research topics are responses of fluvial systems to environmental changes, sedimentological connectivity, evolution trajectories, hydromorphology and GIS and remote sensing applications in rivers and landforms research. He is also working as an external lecturer at the Department of Physical Geography and Geoecology, the Faculty of Natural Sciences of the Comenius University in Bratislava.

Juraj Maglay is a geologist at the State Geological Institute of Dionýz Štúr in Bratislava. He coordinates and implements geological research and mapping of Quaternary deposits within the territory of Slovakia. Based on the results of sedimentological, stratigraphic, neotectonic and geomorphological research, supported by mineralogical, geochemical and paleontological data, he cooperates in compiling basic and regional geological maps (1: 25 000, 1: 50 000) and purpose and thematic maps of the Slovak Republic, such as Map of genetic types of Quaternary deposits 1: 500 000, Map of Quaternary thickness 1: 500 000 and Map of Quaternary deposits of the Slovak Republic 1: 200 000.

Juraj Procházka is Assistant Professor of Geomorphology at the Comenius University in Bratislava. In the long term his work is focused on river behaviour of alluvial rivers during historic times (prior tu anthropic regulations). Nowadays his main scientific interest is neotectonic evolution of Zvolenská kotlina basin as reflected in Pliocene and Quaternary fluvial deposits.

Miloš Rusnák is a fluvial geomorphologist at the Institute of Geography of the Slovak Academy of Sciences (Department of Physical Geography, Geomorphology and Natural Hazards). His research topics are fluvial geomorphology, spatial data processing in GIS, UAV data acquisition and processing, fluvial processes and sediment connections in gravel-bed rivers and remote sensing applications in rivers and landforms research. He is the author and co-author of several papers dealing with fluvial system evolution in the Outer Western Carpathians.

13 Unique Floodplain and Aeolian Landforms: Záhorská Nížina Lowland

Marián Jenčo and Juraj Maglay

Abstract

Extensive braiding and voluminous deposition of gravelly sands and sands of the Lower Morava river was due to significant loss of gradient and rock composition of the drainage basin. Open position in respect to westerly winds created optimal conditions for aeolian redeposition of sands from the alluvium of the Morava river onto the plains between the river and the Little Carpathians. As a result, coversands or dunes occur over large areas of the Záhorská nížina Lowland. Wind-blown sands are quartz-dominated and dunes are fixed mostly by pine forests. As a result of military training activity and related closure of the territory, several large areas of open dunes with rare habitats were preserved. The Morava river formed a wide floodplain in the western part of the Záhorská nížina Lowland. Due to the floods, there were frequent changes in the hydrographic network during the Holocene, reflected in the rapidly changing microrelief of the floodplain. At present, the floodplain of the Lower Morava river became a great non-regulated flooded area with oxbow lakes and other morphological depressions.

Keywords

West Carpathians • Borská nížina Lowland • Quaternary • Floodplain • Sand dunes • River terraces

M. Jenčo (✉)
Department of Physical Geography and Geoinformatics, Faculty of Natural Sciences, Comenius University in Bratislava, Mlynská dolina, Ilkovičova 6, 842 15 Bratislava, Slovakia
e-mail: marian.jenco@uniba.sk

J. Maglay
State Geological Institute of Dionýz Štúr, Mlynská dolina 1, 817 04 Bratislava, Slovakia
e-mail: juraj.maglay@geology.sk

13.1 Introduction

Záhorská nížina Lowland (ZNL) is the westernmost plain area of Slovakia. It represents the northeastern marginal part of the Vienna Basin and is separated from the Danube Lowland by the Little Carpathians. The ZNL connects with the Lower Morava Valley of the South Moravian Basin on the northwestern side. The lower reach of the Morava river creates a sharp western border of ZNL.

ZNL is divided into two very different geomorphological units: Chvojnická pahorkatina Upland in the north and Borská nížina Lowland (BNL) in the south. Chvojnická pahorkatina Upland has a slightly undulating terrain. It is arable land with small forest areas on the top of the hills and in the valleys and dells, which need protection against gully erosion. BNL is forested flat land, even though one of the characteristics, in spite of its relatively small size, is geographical diversity. Pine monocultures of sand plains are in the contrast to riparian forests in the Morava river floodplain. Both fluvial and aeolian forms are distinctive features of the BNL. In the floodplain, natural levees alternate with oxbow lakes and backswamps, whereas in the sand plains, dry dunes alternate with wet interdune depressions. Various hydrological conditions together with rock composition are reflected in the contrast of the soil cover. Fluvisols, Gleysols, Regosols, Cambisols, Podzols and also Umbrisols occur in the area, whereby Umbrisols, Cambisols and Podzols are relatively rare soils at low altitudes in Slovakia.

The diversity of fluvial and aeolian landforms and the presence of extensive riparian forests along the Morava river and large pine forests makes BNL a very attractive land from the human perspective. For this reason, the followed text will focus on this part of ZNL.

M. Lehotský and M. Boltižiar (eds.), *Landscapes and Landforms of Slovakia*, World Geomorphological Landscapes,
https://doi.org/10.1007/978-3-030-89293-7_13

13.2 Geographical and Geomorphological Setting

Borders of geomorphological subunits of BNL are shown in Fig. 13.1. Záhorské pláňavy Plains, which emerge from the Dolnomoravská niva Floodplain in the western part of BNL, is an area with remnants of Pleistocene terraces. The spatial distribution of river terraces is a result of the migration of the Morava river in the westerly direction. Bor is an elevated area in the central part of BNL, with unique aeolian landforms. Marsh land with peat bogs in the subsided Podmalokarpatská zníženina Depression separates them from the Little Carpathians.

The structure of natural territorial units reflects the prevailing direction of main Quaternary faults. The system of the low neotectonic blocks of Dolnomoravská niva Floodplain crosses with the system of low neotectonic blocks of Podmalokarpatská zníženina Depression, which continues to Austria. The intersection area is created by a strongly subsided block of Zohor depression. Quarternary filling of Zohor depression is up to 86 m thick (Maglay et al. 2012). The Zohor depression separates the block of the Stupava-Nováves elevation from the Malacky block. The only one of the morphologically more significant faults that do not take into account the structure of natural territorial units is Rudava fault. The Rudava fault (E–W direction), which passes through the central part of Bor and Záhorské pláňavy Plains, separates the Malacky block from the internally more segmented Lakšár block (Fig. 13.1).

Clay sediments of the Pannonian Sea create the key ingredient of the impermeable Neogene basement of BNL. Afterwards, the regression of the Pannonian Lake at the end of the Pliocene exposed clay sediments, which were subsequently covered by Quaternary deposits. Most common are Pleistocene sands and sandy gravels of river terraces. Sandy loams on floodplains and wetland sediments are younger deposits from the Holocene. Material of aeolian sands of the Bor has been transported from deflation zones in the Morava river floodplain and its terraces. Outcrops of Neogene bedrock often occur on the Lakšárska pahorkatina Upland. It is part of Bor, with the highest hill of BNL (Mária Magdaléna, 297 m a.s.l.). Outcrops are covered by weathered Middle Badenian and Late Carpathian clays, claystones and marls and rise above loamy to sandy-loamy deluvial deposits. Proluvial sediments are present in the foothill of the Little Carpathians (Fig. 13.2).

Sorting of minerals of aeolian sands rises from west to east. In this direction, the content of resistant minerals (garnet, staurolite) in the heaviest fraction increases and the percentage of less resistant minerals, such as amphibole, decreases (Minaříková 1965; Vaškovská 1963, 1971). It is a result of the transport length. The prevailing aspect of linear dunes and lee sides of dunes with steeper slopes indicate transportation of sands by western and northwestern winds, particularly intensive during glacial periods. Soil has been stripped of its protective vegetation cover as a result of low temperatures. The last mass deposition of aeolian sands took place in the climatic phase of the last glacial of Late Pleistocene. The onset period as well as the subsequent colder phases of the Last Glacial Period climate cycle in the Vienna Basin area was still characterized by a cooler and dry climate (Moravcová 2010).

Forestation of Central Europe in the first half of the Holocene created the conditions for the stabilization of drift sands by forest crop. After centuries of human intervention protected forests have been destroyed. Therefore, it was necessary to fix sands by planting new trees. Artificial afforestation began on a larger scale in the mid-seventeenth century (Konôpka et al. 2012). Afforestation was massive in Maria Theresia's times in the eighteenth century and probably gave the core part of the lowlands its geographic name. Borina, bor or bôr are the names of pine stands in the Slovak language. Production pine and oak-pine forests are growing at present. Scots pine (*Pinus sylvestris*) with high ecological valency dominates; oak is represented to a lesser extent, especially *Quercus petraea*.

Pine and oak-pine forests on the Bor and Záhorské pláňavy Plains are closely connected with the soil cover. The content of silica in aeolian sands is up to 90%. Soil matter is therefore poor in nutrients and soils are acidic. The waste of pine needles further increases the acidity of the environment. Extreme acidity creates conditions for podzolization (Bublinec 1974). The most common soils are Arenosols, but Cambisols and Podzols occur here too. Humic soils, including Umbrisols, have developed in places with sufficient moisture to grow the herbaceous undergrowth. The soil cover is complicated by local hydrological conditions, which in deeper depressions are suitable for the formation of hydromorphic soils, overgrown by birch-alder forests. On the Bor and Záhorské pláňavy Plains, birch-oak and oak-hornbeam forests are also preserved.

Haplic, Gleyic and Mollic Fluvisols, associated with Gleysols and Eutric Histosols are dominant soils on the alluvial plain of Morava river. Softwood alluvial forests and hardwood alluvial mixed forest with dominant *Quercus robur* and *Fraxinus angustifolia* subsp. *danubialis* alternate with wet meadows.

To understand the recent climate evolution in the region, it is necessary to mention mesotrophic peat bogs on the alluvial plain of Morava river and in particular, unique, old peat bogs on the Podmalokarpatská zníženina Depression. Peat layers are a very valuable source of information about the changes in vegetation cover and climate development in the Holocene (Krippel 1986).

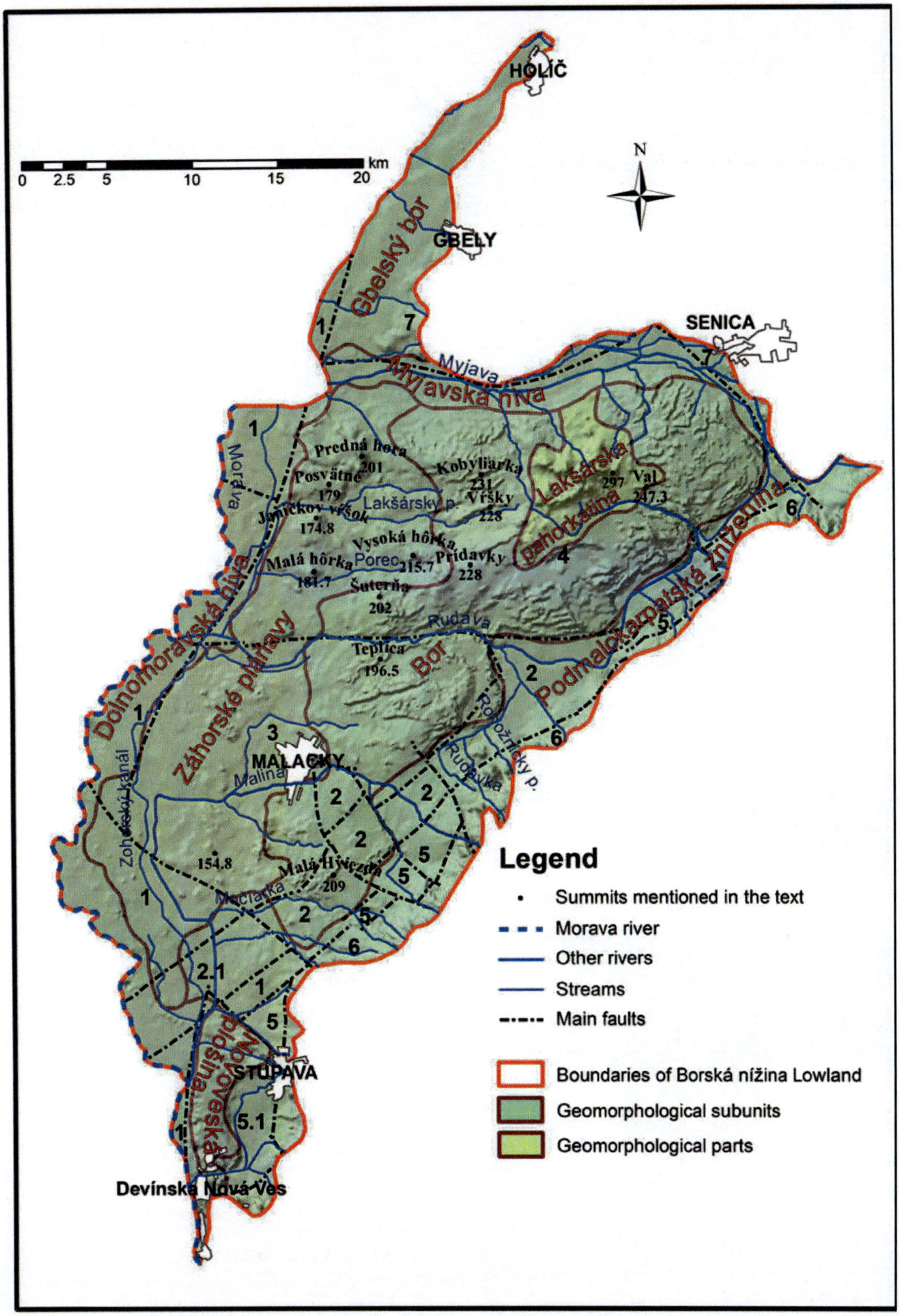

Fig. 13.1 Geomorphological subunits and structural-tectonic scheme of the Borská nížina Lowland—numbers represent the systems of neotectonic blocks or their individual blocks: 1 blocks of Dolnomoravská niva Floodplain; 2 blocks of Podmalokarpatská zníženina Depression; 2.1 block of Zohor depression; 3 Malacky block; 4 Lakšár block; 5 Sub-Little Carpathian blocks; 5.1 block of Stupava-Nováves elevation; 6 horst structures at the edge of Little Carpathians; 7 uplifted blocks at the edge of Chvojnická pahorkatina Upland. Data source Mazúr and Lukniš (1986), Fordinál et al. (2012)

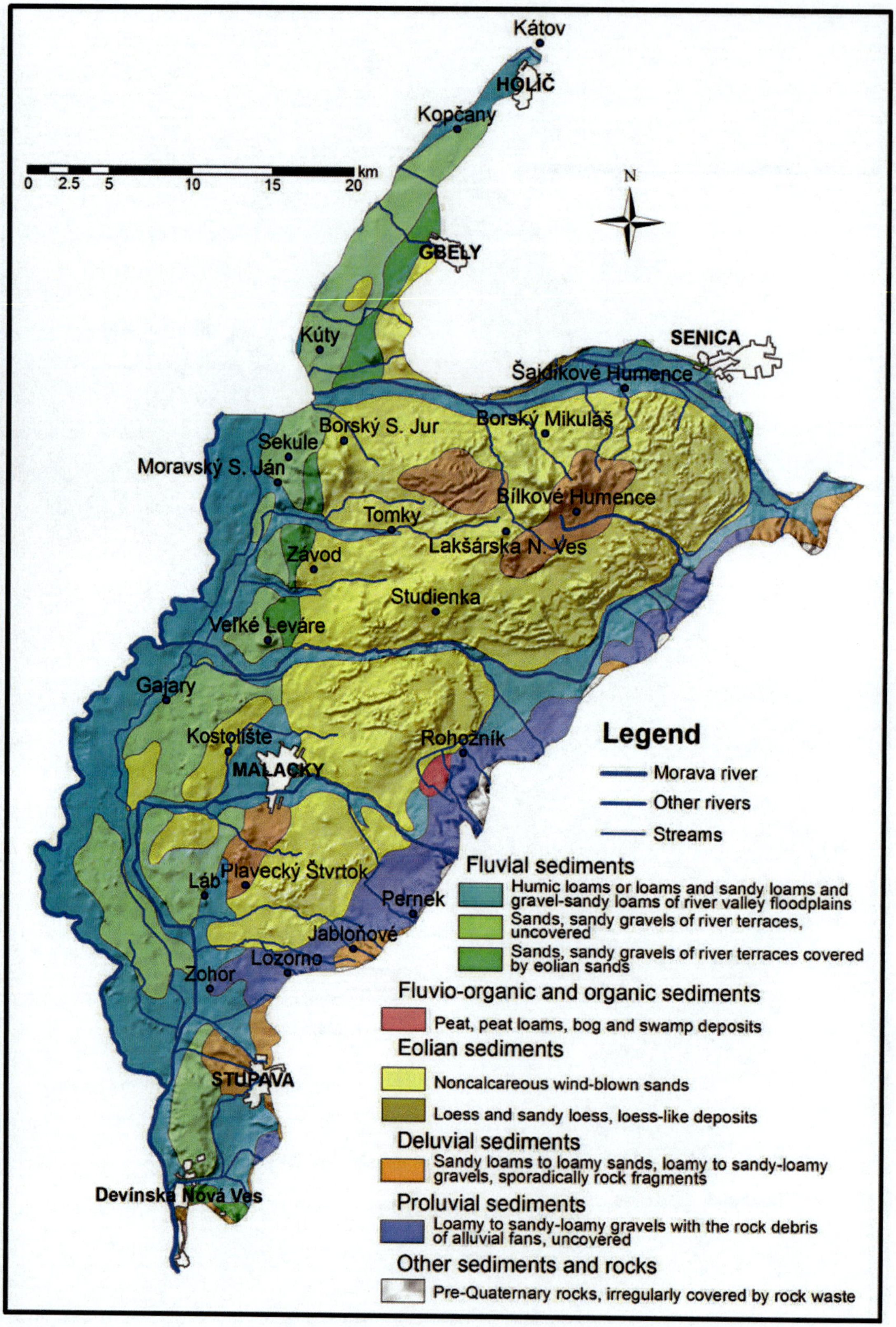

Fig. 13.2 Quaternary deposits of the Borská nížina Lowland. Data source Maglay and Pristaš (2002)

13.3 Fluvial and Proluvial Sediments and Landforms

The activity of the Morava river and its tributaries was fundamental in forming large parts of the ZNL. In its southern part, it was even the decisive factor which, either directly or together with wind activity, formed the character of the landscape.

Early Pleistocene (the Calabrian Stage, specifically from the Menapian glaciation) fluvial gravel deposits are the earliest terrace accumulations. Small deposits of gravel have been preserved in the vicinity of the village of Borský Svätý Jur within the Predná hora elevation at 201 m a.s.l. Slightly larger accumulations are located to the north from the village of Studienka, at even higher altitudes (from the West to the East: Vysoká hôrka 215.7 m a.s.l., Dúbrava 222 m a.s.l., Prídavky 228 m a.s.l.).

Fluvial accumulation of the fine-grained gravels to sands and sandy gravels from the older part of the Middle Pleistocene (Chibanian Stage, specifically from the Elsterian glaciation) form the highest acknowledged level of the Morava river terraces. The best-preserved terrace from this level is the Devínska Nová Ves terrace in the southern geomorphological subunits of BNL–Novoveská plošina terrace. The surface of the sandy gravel of the exposed high terrace is located up to 30 m above the Morava river floodplain. An interesting petrographic fact is that rocks from the Dyje river basin (a more northerly right tributary of the Morava river), to a large extent contribute to the composition of older, non-Little Carpathian fluvial sediments of the BNL (Minaříková 1973; Musil 1993). Rocks from the Dyje river basin are mainly represented in older sandy gravels of this terrace.

The sandy gravels of other high terraces cover some elevations in the northern part of Záhorské pláňavy Plains and are located relatively close to the current bed of Morava river. Sandy gravels deposited in this zone rise to the surface on flat ridges (Posvätné 179 m a.s.l., Janíčkov vŕšok 174.8 m a.s.l., Malá hôrka 181.7 m a.s.l., elevation (172 m a.s.l.) on which the village of Veľké Leváre was founded). Sandy gravels of high terraces are located also further from the current bed of the Morava river (Šuterňa 202 m a.s.l.).

Accumulation of sands during the older part of the Middle Pleistocene created conditions for the formation of high terraces of finer material. The zone of sands and fine-grained gravels extends with small interruptions east of the main railway between the Rudava river floodplain and floodplain of Močiarka stream. This zone with very well-preserved remnants of the high terrace is 15 km long and up to 4 km wide. In the west, it is immediately followed by the zone of the younger middle but also low terraces.

The zone of younger terraces from the Riss period (Chibanian Stage, specifically from the Saalian glaciations) continues in the north to south direction. Sands and fine-grained gravels of higher middle terrace occur between the village of Závod and settlement Tomky. Terrace deposits between the villages of Sekule and Borský Svätý Jur and between Moravský Svätý Ján and Závod are composed of coarser material. In the second case, these sediments are largely covered with aeolian sands. The higher middle terraces are also associated with large areas of sandy gravels to the north from the village of Kúty in Gbelský bor. In the surroundings of the town of Senica, on the Chvojnická pahorkatina Upland, the sandy gravels and gravels of higher middle terraces are buried under a thick layer of loess.

The largest compact area of the lower middle terrace from the younger part of Saalian glaciation forms a right triangle with a 5 km-long base. Vertex of the right angle is located on the eastern edge of the village of Gajary, and the village of Kostolište is located approximately halfway through the hypotenuse. On the eastern side, there is a flat surface of sandy gravels covered with aeolian sands. On the northwestern side, there is a smaller part of the terrace built of finer-grained material.

Fluvial sediments of the Morava river and its tributaries, the so-called bottom accumulation from the Late Pleistocene period at BNL occupy vast areas of the Quaternary depositional area. However, at the level of current floodplains they are mostly covered by Holocene fluvial sediments and on adjacent low terraces by aeolian sands from the younger phase of Late Pleistocene (Upper Pleniglacial).

The youngest fluvial accumulations of the Morava river and its larger tributaries, Myjava river and Rudava river, form a narrow flat area, especially in the western and northern part of the BNL. The Holocene silty sands to sands of these rivers are deposited above the sediments of gravelly sand to sandy gravel bottom accumulation. The boundary between bottom accumulation sediments and overlying floodplain sediments is erosive and shows a varying degree of discordance. The bottom accumulation deposits crop out from beneath the alluvial sediments on the edge of the floodplain in the form of uncovered low terraces, locally within the floodplain in the form of channels and nearby-channels natural levees, or artificially exposed areas for gravel exploitation (e.g. villages of Veľké Leváre and Zohor).

The floodplains have a morphotectonically undifferentiated, flat, in places slightly wavy, accumulation-erosional to accumulation fluvial relief, with an average slope of 1.5° (max. 2°). Morphologically, it is a fluvial flat or fluvial system in places with a transition to fluvial and peat wetlands. The lithological resistance of river sediments to subsequent erosion-denudation processes is negligible.

The floodplain accumulation of the Morava river occurs due to sudden changes in the hydrodynamic regime of the flow and hence show a variety of lithofacies and laterally and horizontally changing formation. Due to the floods, there were also frequent changes in the hydrographic network of the Morava river during the Holocene, reflected in the rapidly changing microrelief of the floodplain. Sub-recent accumulation erosional or local erosional processes, mostly acting laterally, caused the floodplain of the Morava river to be made of two morphologically differentiated sub-levels. They are referred to as higher and lower floodplains. Both sub-levels have a slightly different, climatically conditioned lithological and sedimentological character of the accumulation.

The higher (older) sub-level, with a height of 1.5–2 m above the lower sub-level, occupies the majority of the Morava river floodplain. As part of the accumulation, clayey loams to clayey fine-grained sands are present at the base. Redeposited gravels and sands of bottom accumulation occur in the direction of active flow. On the basal sediments there is a layer of fossil alluvial soil (Preboreal-Atlantic), up to 0.5 m thick. In its overburden, loamy, silty and clayey fine-sandy humus-rich sediments of the floodplain facies are widespread. Depending on the location, the presence and character of the sandy and fine gravel component in the sediment varies, as does the content of organic sediments. The total thickness of the alluvial sediments of the higher stage of the Morava river ranges from 1.5 to 2.5 m (Baňacký and Sabol 1969). Greater thickness was found only in local neotectonic depressions to the west of Kúty and to the south of Zohor, where it may reach 3–3.5 m.

The lower (younger) alluvial sub-level forms the nearby channel facies of the mainstream. Its surface is in the range of 0.5–1 m above the mean water level of the Morava river. The level results from a younger Holocene lateral erosion caused mainly by changes in the Quaternary climate. The sediments of this sub-level also overlie, partly discordantly, sandy gravels and sands of the bottom accumulation and the redeposited gravels of the nearby channel facies. The base contains clayey loams to clayey fine-grained sands or locally only clays. Towards the active flow of the Morava river, redeposited subsoil gravels and sands are also present. Above the basal horizon, a layer of silty fine-grained sand to silt is found. The whole horizon reaches a thickness of 0.5–1.5 m. The humus-rich horizon of the fossil floodplain soil, typical for the older sub-level, is not found in the accumulation of the younger Holocene.

The Morava river floodplain represents a fluvial system of accumulation erosion processes of an alluvial river (a river flowing over its own sediments and partly also over the sediments of an adjacent stream). It is characterized by high morphodynamics from the transitional period of the Late Pleistocene/Holocene through the Holocene to the Subrecent when this area was permanently morphologically transformed. The remodelling intensity has slowed down only due to human interventions.

The variously elevated floodplain surface is differentiated by branching channels and interconnected channels and a network of oxbows. The oxbows are in various stages of development, from buried to anthropogenically modified.

The plateau of the higher floodplain sub-level is often variegated, with unique longitudinal ridges formed by sands of natural levees and young aeolian sandy accumulations. Fluvial-organic, organogenic and palustrine sediments, fen peats and humoliths have developed as the fillings of oxbows and other depressions of the floodplain.

The alluvial cones of the Little Carpathian rivers and streams fill up a large part of the Podmalokarpatská znížená Depression and extend to the southern Bor. Proluvial sediments of southern Bor are covered with aeolian sands, but to the southeast from the town of Malacky, they are exposed. To the north of the village of Rohožník, the front of the middle alluvial cone of the Rohožnícky potok stream is buried under the aeolian sands. Downstream of the confluence with Rudavka stream, the Rohožnícky potok is now diverted to the northeast as a result of fault activity in the central part of the Podmalokarpatská znížená Depression.

13.4 Aeolian Landforms

The quantity of aeolian sands and dune forms on the ZNL has long attracted the attention of geoscientists. The first extensive research, focused on the origin, evolution and characteristics of sands in this area, was carried out by Hromádka (1935). Aeolian sands of ZNL did not escape the attention of scientists who explored the sands of the Lower Morava Valley (Vitásek 1942; Pelíšek 1963). Janšák's work (1950) is also interesting. Lithological characterization of aeolian sands was presented by Minaříková (1965, 1973) and Vaškovská (1963, 1971). Complex results of geological mapping of BNL were published for the first time in the form of geological maps at a scale of 1:50,000 (Baňacký and Sabol 1973). The most recent results of research and mapping have been summarized in the Geological Map of the Záhorská nížina Lowland (Fordinál et al. 2012). Within the legend of this map, really two categories of surface aeolian sands are selected: (a) aeolian fine-grained non-calcareous sands from the Late Würm (Late Weichselian, ~15–12 ka BP)–Holocene period and aeolian sands which covered Late Pleistocene terraces and alluvial cones and (b) aeolian fine-grained non-calcareous sands of dunes and moving dunes and aeolian sands from the Würm (Weichselian, ~70–15 ka BP), which covered older terraces and alluvial cones.

Numerous authors (Hromádka 1935; Vaškovská 1963; Minaříková 1973; Maglay et al. 2012) have defined several zones of aeolian sands of ZNL. The largest zone is the central zone of aeolian sands, which occupies almost the entire Bor and northeastern part of the Záhorské pláňavy Plains. This zone is approximately triangular in shape. Nearly continuous zone of sand extends to the east from the line connecting Borský Svätý Jur, Malacky and Lozorno (Fig. 13.2). The northern border of the central zone is formed by the southern margin of the Myjavská niva Floodplain and in the southeast, it is bounded by the Podmalokarpatská zníženina Depression. The waterlogged parts of the Podmalokarpatská zníženina Depression represent a significant barrier to the spatial distribution of aeolian sands to the east. Therefore, the boundary between aeolian sands and alluvial sediments, although not regular, is very sharp. Within the zone, the continuity of the aeolian sand cover is interrupted by narrow strips of alluvial sediments along the rivers and streams flowing from the Little Carpathians or springing on the Lakšárska pahorkatina Upland. Neogene sediments also reach the surface on the Lakšárska pahorkatina Upland and in the west, there are also sands and gravels of the river terraces. Deluvial-fluvial sediments occur in inter-dune depressions and dells. The most common are outwash loams to loamy sands.

In the central zone, most aeolian sand forms are present. Flat or undulating sand sheets, dunes of various sizes, their denudation remnants, and also extensive composite dunes and dune complexes have been preserved. These forms create different types of aeolian relief. The sand sheet has often a table land character. The surface is flat or slightly wavy. Elsewhere, the surface may be regularly wavy, although it remains flat in other places and occasionally single dunes may occur (Maglay et al. 2012). Such a land surface is typical of the aeolian sand cover of the Late Pleistocene terraces between Malacky and Veľké Leváre. Similar is the aeolian sand cover of low alluvial cones between Malacky and village of Jabloňové.

Large areas in the northeastern part of the central zone are more dissected and depressions are often waterlogged. Landscape has the character of moderately dissected hilly land. Due to multi-cycle development of sand deposition, the waviness is often irregular and chaotic (Fig. 13.3 and 13.4). The aeolian sand cover in the south part of the central zone is similar at different sites. These are isolated localities of dune complexes to the southeast of Malacky (Vašková, Hrubá Hviezda, Malá Hviezda, Pernecké vŕšky).

The largest dunes occur in the western and more northerly part of the zone, where they rise above less dissected plains. These sand elevations also form several kilometres long ridges up to 30 m high. In the south of the central zone, around the villages of Zohor and Lozorno, aeolian sands wedge out or they only form very thin covers. The maximum thickness of aeolian sands (50 m) is recorded in the

Fig. 13.3 Middle wavy surface in the central part of the Bor

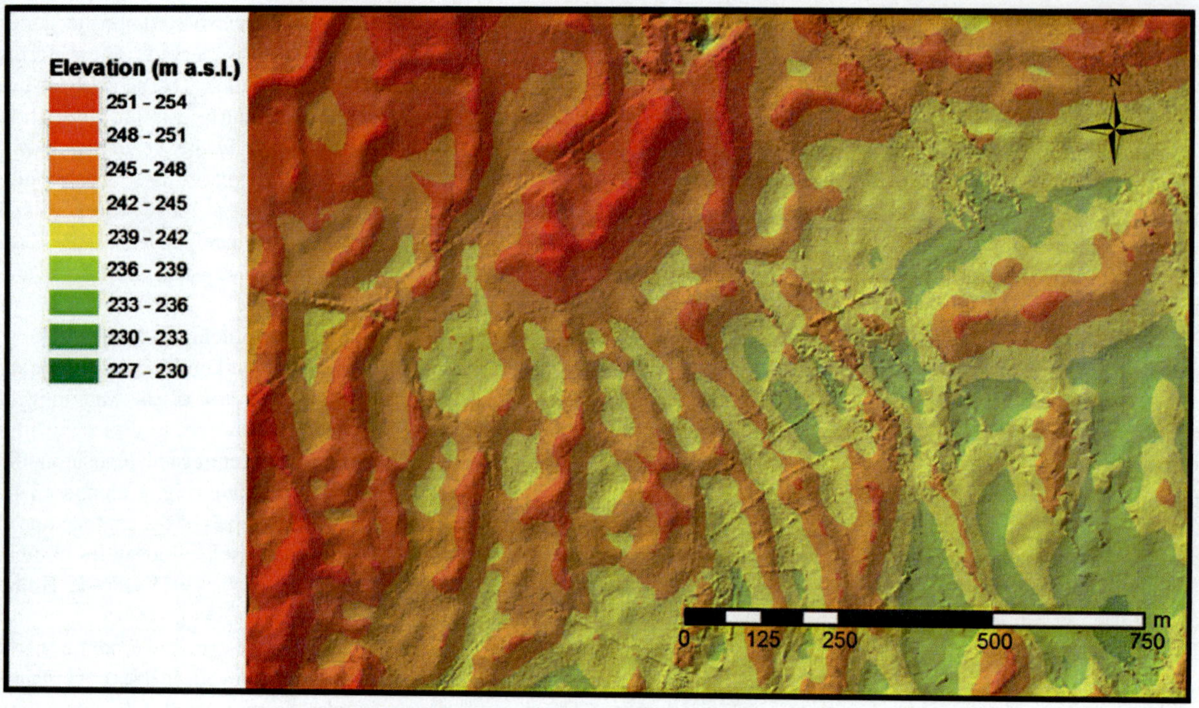

Fig. 13.4 Irregularly undulated sand sheet (elevation model from LiDAR data). Data source NFC-IFRI Zvolen 2019

neotectonic depression, to the southwest of Malá Hviezda, where aeolian sands from the Middle Pleistocene are buried under younger aeolian sands.

Another greater concentration of aeolian sands is formed by the Láb-Gajary zone on the terraces of the Záhorské pláňavy Plains, between the villages of Láb and Malé Leváre. Rudava river floodplain forms the border of more compact aeolian deposits in the north. The Láb-Gajary zone is gripped by an atectonic Malacky-Láb depression and the Morava river floodplain in the east to west direction. The bottom accumulation in low terraces is covered with younger aeolian sands. In the northern half of the zone, aeolian sands also cover the lower middle terraces. These depositions were found to be the cause of stream diversions to the south in the Malacky-Láb depression. On the edge of the zone, e.g., northeast from the village of Gajary and around the villages of Kostolište and Láb or Plavecký Štvrtok (Piesočné, Beničovské), the dune complexes or rounded isolated dunes were formed (Fig. 13.5). Aeolian sand cover reaches thickness up to 8–13 m. Mostly, however, sand cover layers on the low river terraces further west are not as thick (2–3 m) and the sand surface is only slightly wavy. The same is true for the Devínska Nová Ves terrace, which in the northern part is covered by up to 3 m thick layer of young (Late Pleistocene/Holocene) aeolian sands. These sand deposits and sand deposits in the Little Carpathians foothills between Stupava and Lozorno are part of the Stupava-Nováves zone of aeolian sands.

The aeolian sands, to a lesser extend are preserved in the elevated places of the Morava river floodplain. These sand islands create the Morava river floodplain zone. The small sand sheets and initial dunes are situated also on the northeast of BNL, along the southern margin of the Chvojnická pahorkatina Upland and on the foothills of Little Carpathians.

Geomorphological subunit of BNL called Gbelský bor (Fig. 13.1) is separated from the Záhorské pláňavy Plains by the Myjavská niva Floodplain. Aeolian sands were deposited here in several places around the village of Kopčany. To the south, there is a larger area of aeolian sands. These sands lie on the middle terrace of the Morava river and are a loose continuation of the Láb-Gajary zone. The northernmost occurrence of aeolian sands in the Záhorie region is recorded between the villages of Kátov and Vrádište. The sands cover the fluvial-prolluvial cone of the Chvojnica river on the border of the Lower Morava Valley and Chvojnická pahorkatina Upland. These sands belong to the Morava river floodplain sands group (Baňacký et al. 1996).

As a result of several periods of increased aeolian activity older aeolian forms were destructed and new forms were created. Optically stimulated luminescence (OSL) dating of aeolian sands confirmed that deposition of sands, which

Fig. 13.5 Rounded dune. Cote 154,8 m a.s.l. in Beničovské locality (Plavecký Štvrtok cadastral area)

covered most of the territory, are from the Last Pleistocene period (Borský Mikuláš—12.150 ± 600 years BP, Plavecký Štvrtok—14.710 ± 710 years BP, Šajdíkové Humence—16.130 ± 765 years BP) (Fordinál et al. 2013). The question remains whether there was a sandy deposition already before the last glaciation. The sedimentary record of aeolian activity before the Late Pleistocene (pre-Weichselian) is represented by aeolian sands preserved in normal sequence within the zones of subsided tectonic blocks.

Saltating sand grains get trapped by a surface barrier and begin to accumulate. The wind pushes sand grains to the top of the pile. Sand grains on the lee side fall and create the steepest slope. The slope angle of the slip face can be close to the angle of repose for sand and is always larger than the slope angle on the windward side. As dunes migrate, they can collide and stack on top of one another or connect. Thus, composite dunes are formed. Prolonged ranges of transverse dunes are frequent in BNL. Elevations can be several kilometres long. Transverse dunes lie perpendicularly to the wind. Contours of the composite transverse dune on the map are usually wavy (Fig. 13.6). It is the result of an arc shape of dunes before the connection. Of course, the development of dunes is in most cases much more complicated and, therefore, the formation of dune complexes is not rare.

One of the barriers to the distribution of sand is water. If a migrating dune reaches a swamp or river flow and is unable to cross it, it copies the bank line. The shape of the dune loses a visible dependence on the wind direction (Janšák 1950). If the sand hits water it becomes wet and the individual grains are subject to adhesion forces. The sand acquires the properties of a more compact material and the sand stops moving. Neogene bedrock of BNL causes the groundwater level to be often close to the surface. In such cases, capillarity can cause higher sand moisture, even in non-swampy areas (Jenčo et al. 2018). Wetlands in the higher parts of BNL are not rare. The original name of Bor was 'Búr'. According to some authors, geographical names associated with this word are related to wetlands (Kmeť 1901; Janšák 1950). The word 'Búr' most often occurs in the original geographical names in the central and especially the northern parts of Bor and Záhorské pláňavy Plains. A barrier similar to water is the vegetation cover. The result of these factors, which not only affected but also changed in the time before stabilization of the dunes, is a very complicated relief. However, the direction of the major axes of the individual dunes is not always dependent on the wind direction. Very often it is identical with the direction of streams or wetland margins.

Morphologically representative transverse dunes were created in an area to the northwest of the village of Lakšárska Nová Ves. These are three parallel transverse dunes oriented in the southwest to northeast direction. The name of the northernmost dune is Kobyliarka (231 m a.s.l.). The southernmost dune with the name of Vŕšky (228 m a.s.l.) follows the flow of the Lakšársky potok. The name of the middle dune is Košariská (222.9 m a.s.l.). The average length of the elevations is 5 km and the relative height does not exceed 25 m. The width of the valley between the dunes

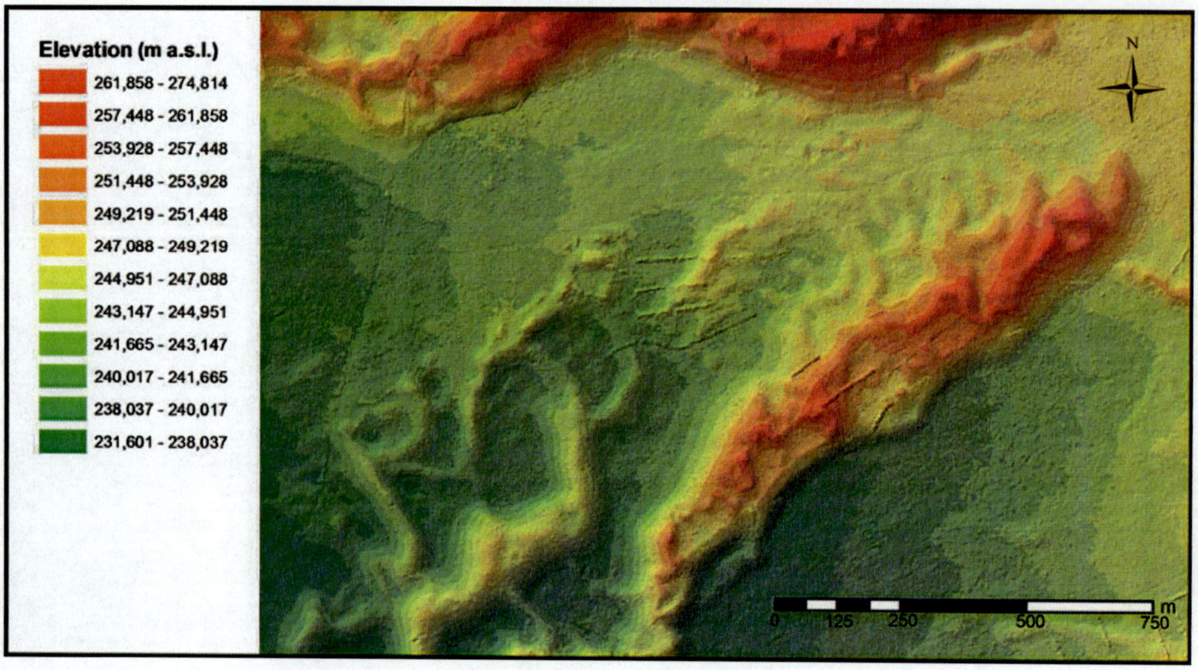

Fig. 13.6 Serrated forms of waviness. Part of the dune of Košariská above the locality V bahne (elevation model from LiDAR data). Data source NFC-IFRI Zvolen 2019

does not exceed 1 km. High isolines on the northwestern windward sides of these dunes are markedly wavy (Fig. 13.6). The thickness of the aeolian sand cover on the nearby isolated dune (Lakšárska duna), which is located on the northwestern periphery of Lakšárska Nová Ves, reaches 40 m.

Mešterova lúka (190 m a.s.l.) and Teplica (196.5 m a.s.l.) is a 4 km-long composite transverse dune to the north of the road between Malacky and Studienka. It is bordered by a significant depression from the south, which is still sloppy. In the northeast, the dune merges with the sand cover on the left bank of Rudava river, relative heights rarely exceed 15 m and the slope of the leeward slope does not exceed 10° (Fig. 13.7). The windward side is an example of the waving shape of the transverse dune. The teeth-like spurs are 100–200 m long. The southern terrain level of the Orlovské vŕšky rises over the depression of Bahno ('mud' in English) and in the east it merges with the chaotic relief of open dunes of the locality Krížnica intensively used by the military.

In BNL, except for isolated sand ridges maintaining southwest to northeast direction, there are also terrain levels following other directions. Several sand elevations in Bor border the inner side of the arc of the upper Rudava river. Some authors consider these elevations to be part of the longest sand level of BNL (Janšák 1950; Škvarček 1981). A significant elevation of the north to south direction was formed along the contact of Bor with the Podmalokarpatská znížena Depression between the village of Pernek and the Rudava river floodplain. This elevation is called Čertova hrádza. In the southern part the slopes of the leeward side are among the steepest in BNL (nearly 30°).

Longitudinal dunes differ from transverse dunes by the axial symmetry. Such is the sand ridge called Val (247.3 m a.s.l.) to the northeast from the village of Bílkové Humence. The dune is S-curved, follows east to west direction, is 2.3 km long, up to 200 m wide and 15 m high.

Aeolian processes were changing the surface of ZNL until the recent past. Redeposition of sands can occur at any time if the following three conditions are met: insufficient substrate stabilization, sufficient wind energy and adequate sand fraction (Nehyba and Havlíček 2001). Such conditions have been created even in the Holocene. The age of some samples of aeolian sands from Borský Mikuláš was determined by OSL at 1,215 ± 15 years to 460 ± 10 years BP (Fordinál et al. 2013). The renewed redeposition of sands in medieval times was probably related to deforestation.

13.5 Protection of Landscapes

Redeposition of aeolian sands occurs in the present times. It results from the way of land use. Potential for deflation increases in the event of disruption or removal of topsoil. This occurs by military activity and after timber harvesting

Fig. 13.7 The leeward (steeper) slope of the Mešterova lúka

when invasive techniques of soil preparation for afforestation are used. Figures 13.4 and 13.6 show short linear features which are anthropogenic ramparts formed by the accumulation of tree stools and root systems with surface soil matter by the bulldozers.

The exposure of the sand cover caused by forest techniques is only temporary and does not create conditions for the formation of specific complexes of open sands that can become hosts of rare plants and animals. On the other hand, a number of open dunes, and thus habitats of rare flora and fauna, are threatened by overgrowing trees in the Záhorie Military District. Záhorie military training area was founded in 1923. Today, with an area of 27,650 ha it occupies, except the northwestern part, almost the rest of the central zone of aeolian sands. The disruption of soil cover by military activity and the special entry regime were the reasons why many rare habitats have been preserved here. Large artillery impact areas and other military areas (Kotlina, Šranecké piesky, Bežnisko, and Široké) are now part of the European network of protected areas NATURA 2000. Dry open dunes and sand sheets often alternate with wetter depressions. Natural communities that occupied dry habitats represent a rare non-forest habitat of Pannonic inland dunes and a habitat of European dry lowland heath.

The western part of the Dolnomoravská niva Floodplain, with its diverse wetlands, is a Ramsar Site. Most of the Morava river floodplain Ramsar site is included in the Chránená krajinná oblasť Záhorie protected landscape area and a few national nature reserves. Periodically flooded areas and areas impacted by groundwater represent slightly altered alluvial forests and fen meadows. The largest complex of fen meadows with purple moor-grass (*Molinia caerulea*) is protected in Abrod nature reserve along the Porec stream.

13.6 Conclusion

Today's face of ZNL and especially its southern part (BNL) is the result of the location of the region and neotectonic movements. The region had the potential for the location of base level for contact structure between the West Carpathians and the Bohemian Massif. The system of neotectonic blocks and faults in the region underpinned the current river network. The Morava river floodplain today geographically coincides with the system of the low blocks of the Dolnomoravská niva Floodplain. The system of Morava river terraces is the component of the western parts of the higher Malacky block and Lakšár block. Fluvial sediments of terraces are composed of sandy gravels, in particular the gravel sands to sands. Along with the alluvial sand sediments of the Morava river floodplain enough sand fraction was available for redeposition.

The land openness to the northwest and west winds exposed soil surface and poor vegetation cover in both Late Pleistocene and Pleistocene/Holocene periods were the causes of massive aeolian transport of sand in the easterly direction. This transport has continued during several cycles, almost to the present day. The transport of sand was so intense that greater part of Bor and Záhorské pláňavy Plains is now covered with aeolian sands. Massive transport of sands proved able to bury the Neogene deposits and proluvial fans of the Little Carpathians rivers and streams in the east when the spread of fans to the west was limited by the neotectonic movements in the Podmalokarpatská zníženina Depression. Plains of BNL in the warmer cycles of Holocene were overgrown with forest. The sand was fixed and redeposition ceased. It was resumed when soil surface was exposed, for instance, as a result of human activity.

Acknowledgement This contribution was prepared with the support of the Operational Programme Integrated Infrastructure within the project FOMON - ITMS code 313011V465, funded by the ERDF.

References

Baňacký V, Sabol A (1969) Základný geologický výskum kvartéru Záhorskej nížiny. Záverečná správa, GÚDŠ, Bratislava, 149 pp

Baňacký V, Sabol A (1973) Geologická mapa Záhorskej nížiny. GÚDŠ, Bratislava

Baňacký V, Elečko M, Vass D, Potfaj M, Slavkay M, Iglárová Ľ, Čechová A (1996) Vysvetlivky ku geologickej mape Chvojnickej pahorkatiny a severnej časti Borskej nížiny 1:50000. Vydavateľstvo Dionýza Štúra, Bratislava

Bublinec E (1974) Podzolový pôdotvorný process pod borovicovými porastmi Záhoria. Veda, vyd. SAV, Bratislava, 119 pp

Fordinál K, Maglay J, Elečko M, Nagy A, Moravcová M, Vlačiky M, Kučera M, Polák M, Plašienka D, Filo I, Olšavský M, Buček S, Havrila M, Kohút M, Bezák V, Németh Z (2012) Geologická mapa Záhorskej nížiny (Geological map of the Záhorská nížina Lowland) 1: 50000. ŠGÚDŠ, Bratislava

Fordinál K, Maglay J, Nagy A, Elečko M, Vlačiky M, Moravcová M, Zlinská A, Baráth I, Boorová D, Žecová K, Šimon L (2013) Nové poznatky o stratigrafii a litologickom zložení neogénnych a kvartérnych sedimentov regionu Záhorská nížina. Geologické práce, Správy 121:47–87

Hromádka J (1935) Zemepis okresu bratislavského a malackého, 2: Malé Karpaty, Záhorská nížina, Podunajská nížina pri Bratislave. Vlastivedný zborník okresu bratislavského a malackého, 2. Bratislava (Nákladom učiteľstva)

Janšák Š (1950) Eolické formácie na Slovensku. Sborník SAV 2 (1–2):37–41

Jenčo M, Matečný I, Putiška R, Burian L, Tančárová K, Kušnirák D (2018) Umbrisols at lower altitudes, case study from Borská lowland (Slovakia). Open Geosci 10(1):121–136

Kmeť A (1901) Sitno. Tlačou Karla Salvu, Ružomberok

Konôpka J, Greppel E, Lipták J (2012) Prirodzená alebo umelá obnova porastov na Záhorí. Lesnícky Časopis - Forestry Journal 58:100–110

Krippel E (1986) Postglaciálny vývoj vegetácie Slovenska. Veda, vyd. SAV, Bratislava. 307 pp

Maglay J, Pristaš J (2002) Kvartérny pokryv 1: 1 000 000. In Atlas krajiny Slovenskej republiky, Ministerstvo životného prostredia SR, Bratislava

Maglay J (2012) Neotektonika. In: Fordinál K. (eds), Vysvetlivky ku geologickej mape Záhorskej nížiny 1: 50000. ŠGÚDŠ, Bratislava, pp 146–149

Maglay J, Moravcová M, Vlačiky M (2012) Kvartér. In: Fordinál K. (eds), Vysvetlivky ku geologickej mape Záhorskej nížiny 1:50000. ŠGÚDŠ, Bratislava, pp 83–143

Mazúr E, Lukniš M (1986) Geomorfologické členenie SSR a ČSSR. Časť Slovensko. Slovenská kartografia, Bratislava

Minaříková D (1965) Mineralogický výzkum eolických písků Záhorské nížiny s použitím matematické statistiky. Geologické práce, Správy 35, GÚDŠ, Bratislava

Minaříková D (1973) Petrografie kvartérních sedimentů Záhorské nížiny. Sborník geologických věd, řada A, antropozoikum, svazek 9, Ústřední ústav geologický, Praha

Moravcová M (2010) Zmeny prírodného prostredia Slovenska a Moravy na hranici pleistocén holocén (prvá polovica OIS 3 – začiatok OIS1). Geologické Práce, Správy 116:9–72

Musil R (1993) Geologický vývoj Moravy a Slezka v kvartéru. In: Přichystal A, Ostová V, Suk M: Geologie Moravy a Slezska. Moravské zemské muzeum a sekce geologických věd PřF MU, Brno

Nehyba S, Havlíček P (2001) Granulometrie kvartérních sedimentů v soutokové oblasti Moravy s Dyjí. Granulometry of Quarternary sediments in the influence area of Morajva and Dyje Rivers. Zprávy o Geologických Výzkumech 34:84–87

Pelíšek J (1963) Charakteristika vátych písku Slovenska. Geologické Práce, Zošit 64:103–117

Škvarček A (1981) Geomorfologické pomery Borskej nížiny. AFR-NUC, Geographica 19:165–183

Vaškovská E (1963) Niektoré nové poznatky o eolických pieskoch Záhorskej nížiny. Geologické Práce, Zošit 64:121–140

Vaškovská E (1971) Litologicko-faciálna charakteristika genetických typov kvartérnych sedimentov Záhorskej nížiny. Geologické práce, Správy 55, GÚDŠ, Bratislava

Vitásek F (1942) Dolnomoravské přesypy. Práce Moravské přírodovědecké společnosti, 14, 9, Brno

Marián Jenčo is an Assistant Professor at the Department of Physical Geography and Geoinformatics, Faculty of Natural Sciences of Comenius University in Bratislava. His professional interest includes geomorphometric modelling, solar radiation modelling, landscape ecology and soil conservation.

Juraj Maglay is a geologist at the State Geological Institute of Dionýz Štúr in Bratislava. He coordinates and implements geological research and mapping of Quaternary deposits within the territory of Slovakia. Based on the results of sedimentological, stratigraphic, neotectonic and geomorphological research, supported by mineralogical, geochemical and paleontological data, he cooperates in compiling basic and regional geological maps (1:25000, 1:50000) and purpose and thematic maps of the Slovak Republic, such as Map of genetic types of Quaternary deposits 1:500000, Map of Quaternary thickness 1:500000 and Map of Quaternary deposits of the Slovak Republic 1:200000.

14 Fan-Shaped Drainage Network, Glacis and Loess Tables: Východoslovenská Nížina Lowland

Dušan Barabas and Ján Bóna

Abstract

The relief of the Eastern Slovak Lowland is the result of Quaternary geological processes and fluvial modelling of the Eastern Slovak rivers. These processes are related to changes in the geological structure within the Great Danube Basin (Alföld). In addition to fluvial and proluvial deposition, processes of weathering and slope modelling under periglacial conditions (formation of glacis or cryoglacis) and intensive aeolian activity (loess plateau and wind-blown sands) occurred. Crustal movements in the Eastern Slovak Lowland, active since the end of the Pliocene and during the Quaternary, influenced the morphological structure, giving impetus to the transformation of the river network. Local subsidence centres with intensive aggradation, reaching 100 m or more in specific areas, became zones of concentration of surface water, reflected in confluences of several main streams (e.g., Ondava and Latorica rivers). Concentration of streams in a narrow space allowed for the formation and development of a fan-shaped river network (the so-called complex fan), typical for the region. The nature of precipitation, runoff conditions and sediment properties have become an ideal environment for the formation of heavy (clayey) soils, the proportion of which is high in the total lowland area. Over the last centuries, extreme natural hydrological-pedological conditions necessitated considerable effort from people to make the area suitable for agriculture and other land use. The water management schemes allowed for more intensive economic use of the country in the Eastern Slovak Lowland.

D. Barabas (✉) · J. Bóna
Institute of Geography, Faculty of Natural Sciences, Pavol Jozef Šafárik University in Košice, Jesenná 5, 040 01 Košice, Slovak Republic
e-mail: dusan.barabas@upjs.sk

J. Bóna
e-mail: jan.bona@upjs.sk

Keywords

Carpathians • Eastern Slovak Lowland • Neotectonics • Drainage network • Water management

14.1 Introduction

The specific geological, morphostructural and hydrological conditions in which the Východoslovenská nížina (Eastern Slovak) Lowland was established heralded the possibilities of its land use. During the Quaternary, the river network underwent a major transformation. This transformation occurred mainly in the plains, where the low-lying areas were filled with fluvial sediments, and to a lesser extent affected the marginal uplands, where the proluvial deposits formed. Periglacial conditions with intensive solifluction allowed for the formation of glacis surfaces, which, despite their foothill position, were often buried under proluvial and, less often, fluvial sediments. They have been preserved on the water-dividing ridge at the foothills lining the lowland. Aeolian sediments—loess and wind-blown sands—formed vast areas in neotectonically uplifted areas or in the areas subject to small-scale downthrow (Baňacký 1988; Baňacký et al. 1987, 1988, 1989). In terms of geomorphological division (Mazúr and Lukniš 1980), the loess plateaus of Trebišov, Malčice, Závadka (Trebišovská, Malčická, Závadská tabuľa) and the Medzibodrožie area (Medzibodrocké pláňavy) consist of wind-blown sands. Significant subsidiary structures and differentiated vertical movements formed a fan-shaped texture of the river network. In addition to the main fan-shaped river network of Bodrog, Laborec and Ondava rivers created their own fans, too (Lukniš 1972). During the Holocene, in the dominant part of the territory river (alluvial) sediments were accumulated—mainly clayey and sandy loams. Contrary to the other lowland areas of Slovakia, such as the Danube Lowland (Maglay et al. 2006) and the Záhorská nížina Lowland (Fordinál et al. 2012), the

M. Lehotský and M. Boltižiar (eds.), *Landscapes and Landforms of Slovakia*, World Geomorphological Landscapes,
https://doi.org/10.1007/978-3-030-89293-7_14

fluvial sediments in the Východoslovenská nížina (Eastern Slovak) Lowland were derived and transported to the sedimentation area from mountainous areas, which consist mainly of flysch rock complexes.

Anthropogenic transformation formed the current scenery of the Eastern Slovak Lowland. Firstly, by reducing the extent of swampy areas and regulating streams, one enabled full use of the terrain, without the relatively frequent floods. This was associated with the extensive development of agriculture and the related industry. On the other side, anthropogenic transformation reduced the rate of surface water infiltration to the groundwater level and increased outflow intensity (Kupčo 1995).

The chapter presents the complexity of relations within the Eastern Slovak Lowland, which played an important role in the development of its proper use. The contribution documents the development of the territory as well as the reasons for its crucial changes. It outlines the transformation in the context of anthropogenic activities and evaluates their impact on the Eastern Slovak Lowland.

14.2 Geomorphology, Geology and Tectonic Development

The Eastern Slovak Lowland represents the north-eastern extremity of the Great Danubian Plain (Mazúr and Lukniš 1980). It is situated between the Slanské vrchy and Vihorlatské vrchy Mts. Geomorphologically, the lowland includes two areas: (1) the Eastern Slovak Upland, which represents a peripheral part of the mountains and forms a transitional area between the surrounding mountains and (2) the Eastern Slovak Plain (cf. Fig. 14.1). From the geological–geomorphological point of view, the Eastern Slovak Lowland represents a relatively young structural plain with geomorphic features created by accumulation and simultaneous block faulting (Kvitkovič 1961).

Its western boundary, generally of N–S direction, is formed by a distinct morphostructure of the Slanské vrchy Mts. neovolcanites (Fig. 14.2a). In the south-western part, the lowland is bordered by the elevation of the Zemplínske vrchy Mts. (Fig. 14.2b). This horst morphostructure of NW–SE direction is composed of Palaeozoic-Triassic rock complexes belonging to the Zemplinicum Unit. The south-eastern border is formed by the buried Seredne elevation (Seredne horst sensu Rudinec 1989) of NE–SW direction. The northern boundary of the Eastern Slovak Lowland is shaped by the Nízke Beskydy Mts. consisting of sedimentary rocks, which are members of the Central-Carpathian Paleogene Basin sedimentary sequences. The morphostructure of the Vihorlat and Popriečny neovolcanites (Fig. 14.2c) is situated in the NE part of the Eastern Slovak Lowland, and towards the south-west there is a markedly sinking morphostructure of the Humenské vrchy Mts., built by the Mesozoic rocks of the Krížna Nappe.

Geologically, the Eastern Slovak Lowland represents a unit occurring in the central and eastern parts of the East Slovak Basin in the Western Carpathians. On the Slovak territory, it has a significantly autonomous position within the north-western part of the large Transcarpathian Depression, spreading mainly in Ukraine and subsequently in Romania (Kováč 2000; Vass 2002).

The East Slovak Basin (Kováč et al. 1995; Kováč 2000) developed firstly as a fore-arc ("pull-apart" type) basin, later (at the end of the Badenian) as an extensive back-arc (intra-arc) basin, and finally as an extensive intra-mountain basin of the Pannonian system during the Late Miocene. The sedimentation in this basin proceeded from the Lower Miocene with the presumed hiatus in the Ottnangian, then without any interruption from the Karpatian to the Recent. Given that the East Slovak Basin developed in several stages under specific geotectonic conditions, the succession of Neogene sediments reaches a cumulative thickness of 8–9 km (Vass et al. 2000).

The basin was part of a large epicontinental sea during the middle and partly the late Miocene. In the Pannonian period, it became an integral part of a large Pannonian lake, from which it was gradually separated in the Pliocene (Kováč 2000). After that, the sedimentary environment changed, producing fluvial, proluvial and aeolian features (Baňacký 1980).

The development of the East Slovak Basin was accompanied by extensive volcanism—rhyolitic (during early to middle Miocene) and andesitic (during the middle to late Miocene). In that time the stratovolcanic structures of the Slanské vrchy Mts., Vihorlat and Popriečny Mts., formed by alternating lava flows and pyroclastic series, were transformed into neovolcanic mountains (Kaličiak and Žec 1995; Lexa et al. 2010).

14.3 Tectono-Sedimentary and Geomorphic Evolution During the Quaternary Period

The late Pliocene period can be considered as an initial stage of the Eastern Slovak Lowland tectono-sedimentary development during the Quaternary. Late Pliocene fluvial-limnic sands and clays were continuously deposited until the Lower Pleistocene (Baňacký et al. 1987, 1989). They occur at the depth of 70 m in the Strážne-Trakany depression. Limnic sediments of the same age are deposited in both the Michalovce-Sliepkovce depression at the depth of 60–65 m and in the Hraň depression at a depth of 60 m. This discrepancy ultimately points to different intensities of basin sinking (Baňacký et al. 1993).

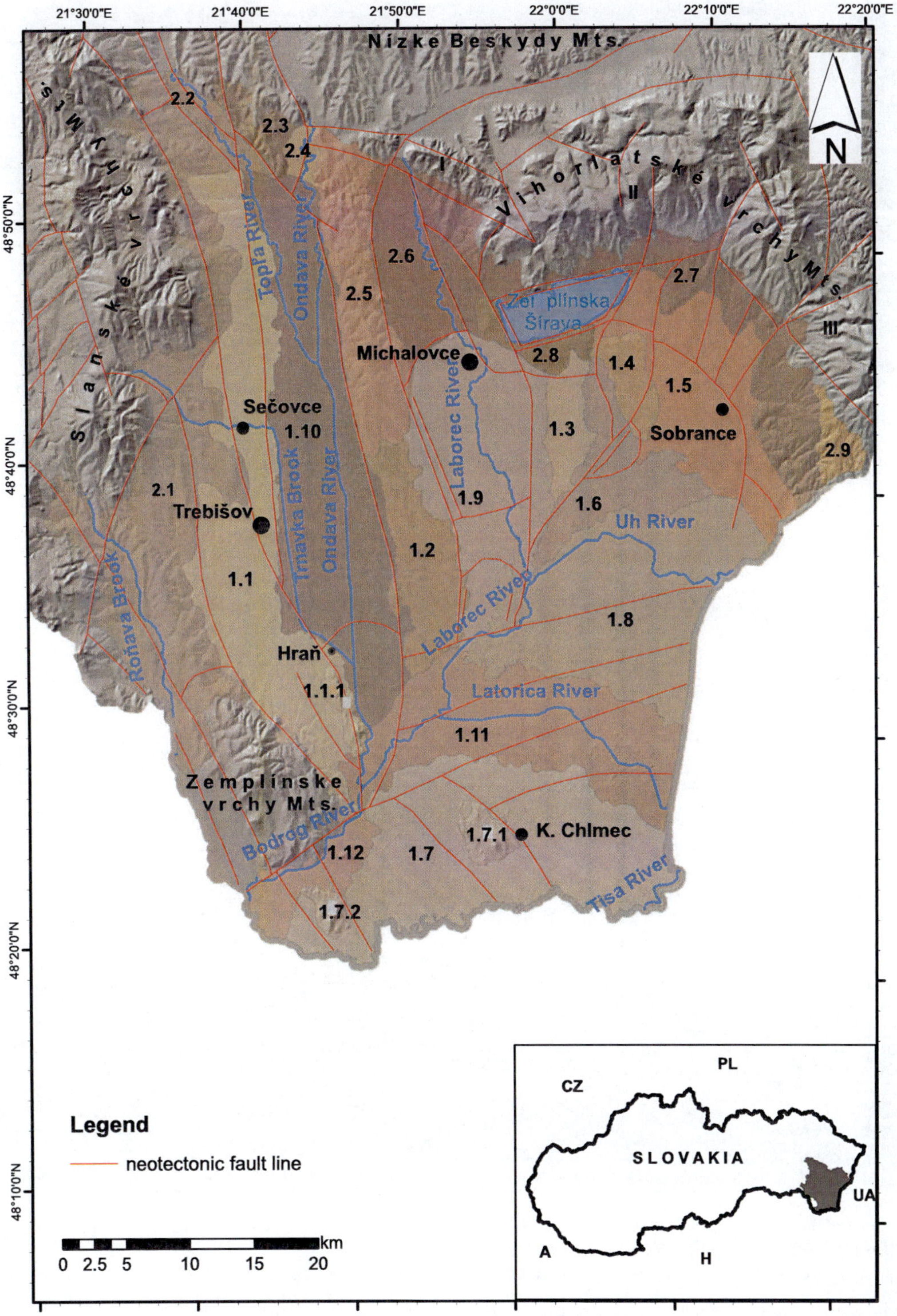

Fig. 14.1 Geomorphological division (sensu Mazúr and Lukniš 1980, imaged in different contrasting colours) of the Východoslovenská nížina (Eastern Slovak) Lowland and direction of neotectonic faults (Maglay et al. 1999). *Explanations:* 1 Eastern Slovak Flat: 1.1 Trebišovská tabuľa, 1.1.1 Veľký vrch, 1.2 Malčická tabuľa, 1.3 Iňačovská tabuľa, 1.4 Závadská tabuľa, 1.5 Sobranecká rovina, 1.6 Senianská mokraď, 1.7 Medzibodrocké pláňavy, 1.7.1 Chlmecké pahorky, 1.7.2 Tarbucka, 1.8 Kapušianske pláňavy, 1.9 Laborecká rovina, 1.10 Ondavská rovina, 1.11 Latorická rovina, 1.12 Bodrocká rovina; 2 Eastern Slovak Upland: 2.1 Podslanská pahorkatina, 2.2 Toplianská niva, 2.3 Vranovská pahorkatina, 2.4 Ondavská niva, 2.5 Pozdišovský chrbát, 2.6 Laborecká niva, 2.7 Podvihorlatská pahorkatina, 2.8 Zalužická pahorkatina, 2.9 Petrovské podhorie; Vihorlat Mts.: I Humenské vrchy, II Vihorlat, III Poprieční. DTM visualization. *Source* Národný geoportál; *Performed by:* D. Barabas

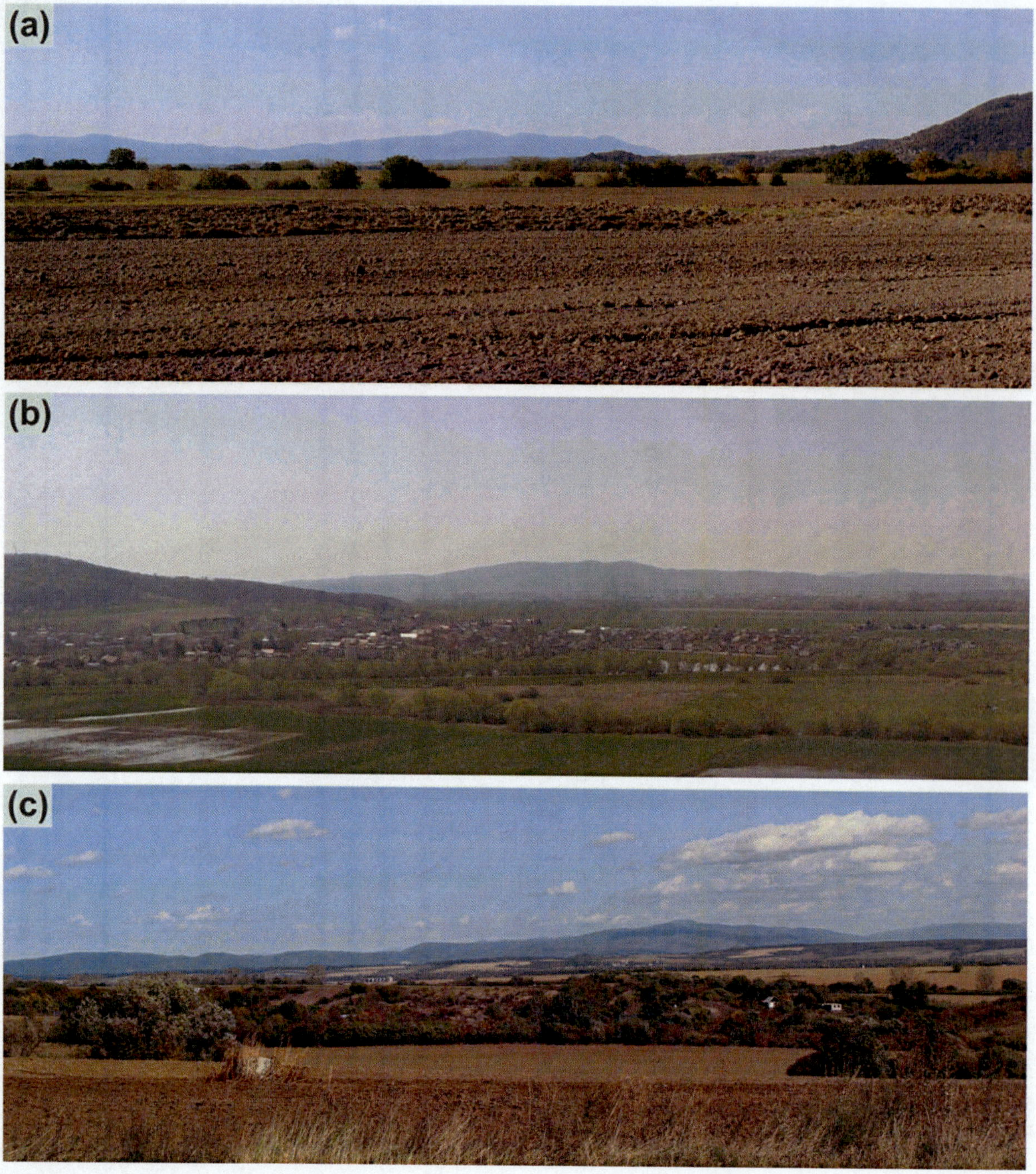

Fig. 14.2 **a** Relief of the Zalužická pahorkatina hilly land (2.8), with the volcanic massif of Medvedia hora Hill (Vihorlat area) on the right and the Slanské vrchy Mts. on the horizon (*photo* J. Bóna); **b** relief of the Ondavská rovina (1.10) and Trebišovská tabuľa (1.1, in the background). On the left, there is Avaš Hill—a volcanic andesite massif of 237 m a.s.l., aeolian sands are developed at the footslope of the Zemplínske vrchy Mts., visible on the horizon (*Aerial oblique photo* M. Lacko); **c** fluvial relief of the Ondavská rovina plain (1.10) with the elevated structure of the Pozdišovce horst (2.5) in the background; volcanic massifs of Vihorlat (II) and Popriečny (III) are visible on the horizon. The symbols in brackets correspond to Fig. 14.1. Photo J. Bóna

The formation of the valleys was initialized within the "river-side" (submountainous) level, with remnants of this initial surface preserved in the peripheral hilly lands of the lowland of Eastern Slovak Lowland (Baňacký et al. 1993). Relative elevations of this level range from 80 to 100 m (Kvitkovič 1961; Baňacký 1968).

The Quaternary geological–geomorphological processes in this area led to the deposition of various genetic groups of sediments (deluvial, proluvial, fluvial and aeolian) (Fig. 14.3), whose genesis and distribution were significantly influenced by neotectonics and exogenous factors (Baňacký 1980). In this period, the fluvial activity of the Eastern Slovakian rivers, weathering processes and slope modelling under periglacial conditions, and intensive aeolian activity occurred, leaving an evident geomorphic legacy.

The Eastern Slovak Lowland territory shows complicated tectonic and morphostructural composition. It is characterized by a relatively higher number of tectonic blocks regarding its total area with vertical downthrow movements during the Quaternary. Faults following general N–S and ENE–WSW directions, subordinately trending NW–SE, were active in the area (cf. Baňacký 1986; Maglay et al. 1999; Fig. 14.4).

The Laborec River fluvial sediments (sand and gravel) filled the Michalovce-Sliepkovce depression during the Early Pleistocene. The depression subsided during this time by about 10–12 m. The Pozdišovce horst—a longitudinal neotectonic elevation of N–S direction (Baňacký 1968, 1980), extending from the Upland (Hilly land) to the East Slovakian Flat (Mazúr and Lukniš 1980)—has begun to form between the Ondava and Laborec river valleys (Figs. 14.5 and 14.6a). The Hraň depression that occurs to the north of the Ondava and Latorica rivers confluence is filled by accumulations of fluvial sands and gravels, and the total subsidence of this structure reached 10 m during the Early Pleistocene. A slightly different situation is recorded in the Strážne-Trakany depression, in which the sedimentary fill has a fluvial clayey-sandy character and the subsidence in the mentioned period reached 5–15 m.

During this period, the areas beyond the reach of fluvial erosion were shaped by slope processes, mainly solifluction (Dzurovčin 1998, 2001, 2010). Such areas comprise mainly foot slopes of the volcanic mountains, inclined towards the adjacent lowland. The upper parts of the surfaces perfectly follow the circular boundary line of their mountain foothills, linking to the "river-side" denudation surface level. In the Slanské vrchy and Vihorlatské vrchy Mts., the base of the volcanoes has been re-modelled by sedimentation processes, resulting in erosive glacis surfaces, inclined at an angle of 2–5°. These erosional piedmont plains were formed on geomorphologically poorly resistant sedimentary and volcanic-sedimentary rocks, and they are often overlapped by several-metres-thick accumulations of younger proluvial (Fig. 14.6b) and deluvial sediments. Considering that they were formed under periglacial conditions during the Early Pleistocene, they can be characterized as cryoglacis (Dzurovčin 1990, 2010).

The geomorphic development in the Middle Pleistocene was also very intensive. In the more elevated part of the lowland, there are distinctive proluvial fans consisting of loamy gravels (Fig. 14.6b), which were washed out from the surrounding volcanic mountains (Baňacký 1988). Accumulations of fluvial sandy gravel sediments indicate subsidence of about 38–40 m in the Michalovce-Sliepkovce depression and approximately 18 m in the Hraň depression. The deposits show sandy character in the Strážne-Trakany depression, and downthrows reach approximately 30 m (Baňacký et al. 1993).

During the Late Pleistocene there was significant proluvial sedimentation, and at the same time, loess accumulation progressed. The formation of featureless loess blankets continued on the plain (Baňacký 1988; Baňacký et al. 1987, 1988, 1989; Košťálik 1999). They were subsequently differentiated by the development of the drainage pattern, and currently represent morphologically striking sub-areas in the southern part of the Trebišovská tabuľa (plateau), reaching down to the foothills of the Zemplínske vrchy Mts., and also the Malčická, Iňačovská and Závadská tabuľa plateaus (sensu Mazúr and Lukniš 1980).

Sedimentation of wind-blown sands is a significant phenomenon for the end of the Pleistocene and the Holocene (Fig. 14.6c, d). These sands are dominant in the Medzibodrožie area (Medzibodrocké pláňavy) and in the Latorica river valley, where they appear as sand dunes (Fig. 14.6d) and ridges, and continuous blankets of wind-blown sands in the flat part of the lowland. Morphologically, sand accumulation areas from elevations of various shapes, which rise above the surrounding fluvial relief of the Latorica, Bodrog, Tica, Malá Krčava and Tisza rivers, which, along with Ondava and Laborec rivers, were the source of wind-blown sands. They were blown to the swamplands affected by frequent floods, therefore they are preserved only sporadically (Baňacký et al. 1988, 1989; Košťálik 2009).

The above-mentioned depressions, in which the downward trend resulted in subsidence values ranging from 15 to 40 m, have continually evolved. During this period, new subsiding structures were formed, mostly filled by flood-derived clayey loams, organic muds and peats. In the Drahňov, Senné (Fig. 14.6e) and Šírava (Sub-Vihorlat, Fig. 14.6f) partial depressions (sensu Baňacký et al. 1986) the downthrows reached 5–20 m and persisted during the Holocene up to Recent (Baňacký et al. 1993). Holocene downthrows in selected depressions of the Eastern Slovak Lowland recorded significant intensity (−0.72 to −1.06 mm/year; Kvitkovič 1993). Data of recent accurate

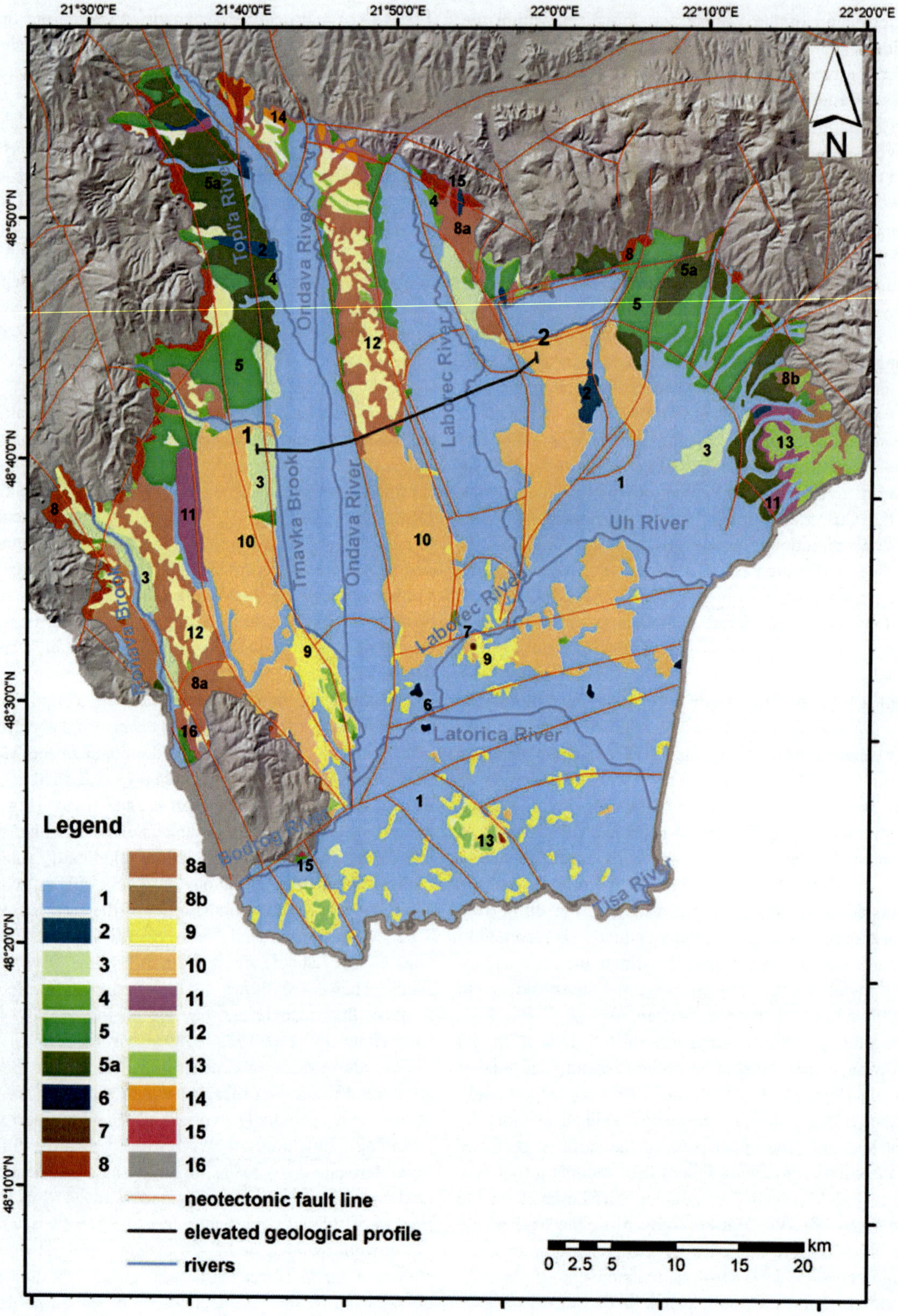
21°30'0"E
21°40'0"E
21°50'0"E
22°0'0"E
22°10'0"E
22°20'0"E
48°50'0"N
48°40'0"N
48°30'0"N
48°20'0"N
48°10'0"N
N
Topľa River
Ondava River
Trnavka Brook
Laborec River
Uh River
Latorica River
Bodrog River
Tisa River
Roňava Brook
Legend
1
2
3
4
5
5a
6
7
8
8a
8b
9
10
11
12
13
14
15
16
neotectonic fault line
elevated geological profile
rivers
km
0 2.5 5 10 15 20

◀ **Fig. 14.3** Genetic types of Quaternary deposits of the Východoslovenská nížina (Eastern Slovak) Lowland (Maglay et al. 2008b) and direction of neotectonic faults (Maglay et al. 1999). *Explanations:* Genetic types of Quaternary deposits: 1 fluvial alluvial (floodplain) sediments, 2 fluvial terraces sediments, 3 fluvial terraces sediments with the cover of loess, aeolic-deluvial loess-like loams, 4 proluvial sediments in flood-plain alluvial cones, 5 proluvial sediments in terraced alluvial cones, 5a proluvial, terraced sediments, with the cover of loess, aeolic-deluvial loess-like loams, 6 fluvial-organic to paludinal (palustrial) sediments, 7 organogenic sediments, 8 deluvial sediments (lithofacially undistinguished), 8a deluvial loams, 8b deluvial sediments (coarse, blocks), 9 aeolian sands, 10 loess, 11 aeolian-deluvial sediments (loess-like loams); discontinuous covers of unspecified slope debris and screes on: 12 sedimentary rocks of the Neogene, 13 Neogene volcanic rocks, 14 sedimentary rocks of the intra-Carpathian Paleogene, 15 carbonate rocks of the Mesozoic, 16 mostly metamorphic and sedimentary rocks of the Paleozoic. DTM visualization. *Source* Národný geoportál; *Performed by:* D. Barabas

levelling measurements confirm this trend (e.g., Kvitkovič and Vanko 1972; Vanko 1988).

The territory of the Eastern Slovak Lowland and the adjacent geomorphological units is drained by the Bodrog river network (Latorica, Uh, Laborec, Ondava and Topľa rivers). The type of the Eastern Slovak Lowland river network in a rough outline reflects the underlying geological structure. Its texture was modified by morphotectonic development of the area during the Pliocene and Quaternary. However, neotectonics performed a dominant role in the development of the Eastern Slovak Lowland river network (Baňacký 1965, 1968; Baňacký et al. 1965, 1987, 1989). This was also reflected in the wider context of the Eastern Slovak Lowland (Borsy and Félegyházi 1983) as a part of the Great Danube Basin. The distribution and formation of the river network were predetermined by the main as well as local subsidence centres (Fig. 14.7). The crucial recipient for the whole area is Tisza river, which reaches the territory of Slovakia only as a border stream of 5 km long. Nevertheless, the development of its river bed influenced the development of the river network of the whole East Slovak Lowland.

The development of Tisza river network (Dzurovčin 2011), which drains the territory of the Eastern Slovak Lowland through its various tributaries, is related to the main subsidence centre located in the Great Hungarian Plain (Alföld). The total effect of Quaternary subsidence reaches 650–700 m and more, as indicated by the positions of Quaternary sediments in southeast Hungary, in the territory between Szolnok and Szeged towns (Franyó 1992; Gábris and Nádor 2007). Gábris and Nádor (2007) pointed out that three river systems, namely Danube, Tisza and Bodrog rivers, probably aggraded their material in Alföld during the Middle Pleistocene. The relative uplift of the area as a result of neotectonic movements to the south of the recent floodplain of Tisza river created a prerequisite for the position change of its channel. In the Late Glacial, the Tisza river formed a great loop of NE direction towards the current border of Slovakia and Ukraine, where its channel followed a local subsidence centre (Kvitkovič and Plančár 1975; Fig. 14.7). After reaching the maximum curvature towards the N (NW), the Tisza river returned to the original southwest direction, creating a distinctive loop near the villages of Veľké Trakany and Malé Trakany. These changes in the course of the river resulted in the capture of the waters of the Eastern Slovak Lowland by the Tisza River (i.e., river piracy of the Tisza river in the Hungarian territory) and a change in the texture of the river network, according to Borsy and Felegyházy (1983). During the Holocene there was a very intensive lateral erosion of the Tisza river in the southern part of the lowland. Tisza and Latorica rivers have closed a vast area in the southern part of the lowland, the relief of which is the result of the fluvial and eolian activity. An extensive network of abandoned arms of Tisza and Latorica rivers is evidence of intense fluvial activity (Fig. 14.8), which could have developed only under conditions of subsidence. The preservation of the Latorica river system is a consequence of intensive aggradation of fluvial material, which has also become a barrier to the drainage of the Latorica waters by the Tisza river system (cf. Lukniš 1973) in Ukraine. This zone, situated between Latorica and Tisza rivers, has a maximum width of 3 km at its narrowest point.

Local subsidence centres (Figs. 14.4 and 14.7) of the Eastern Slovak Lowland, located next to its southern and southeastern borders—the Hraň, Bežovce and Strážne--Trakany depressions (Baňacký 1980, 1986)—became the basis for the development of the Eastern Slovak Lowland river network subsystem texture.

The thickness of Quaternary sediments reaches a maximum in the above-mentioned depressions, i.e., 70–80 m, locally as much as 116 m (Baňacký 1986; Kvitkovič 1993; Maglay et al. 2008a). In these places, the Ondava, Laborec and Latorice river networks were connected. The Tisza river partially used the valley modelled by the Bodrog river in Hungary, and also the texture of the Bodrog river system was shortened. It changed from the texture of parallel streams to the complex (compound and multilevel) fan texture—complex shapes of fan drainage patterns (sensu Dub 1957; Lukniš 1972) during the Late Pleistocene and Holocene. The reason for this texture change was the different intensity of neotectonic vertical movements (cf. Fig. 14.4).

The subsidence of the Eastern Slovak Lowland caused activation of erosion processes, which ensured downcutting of folded morphostructures of the flysch Carpathians to the

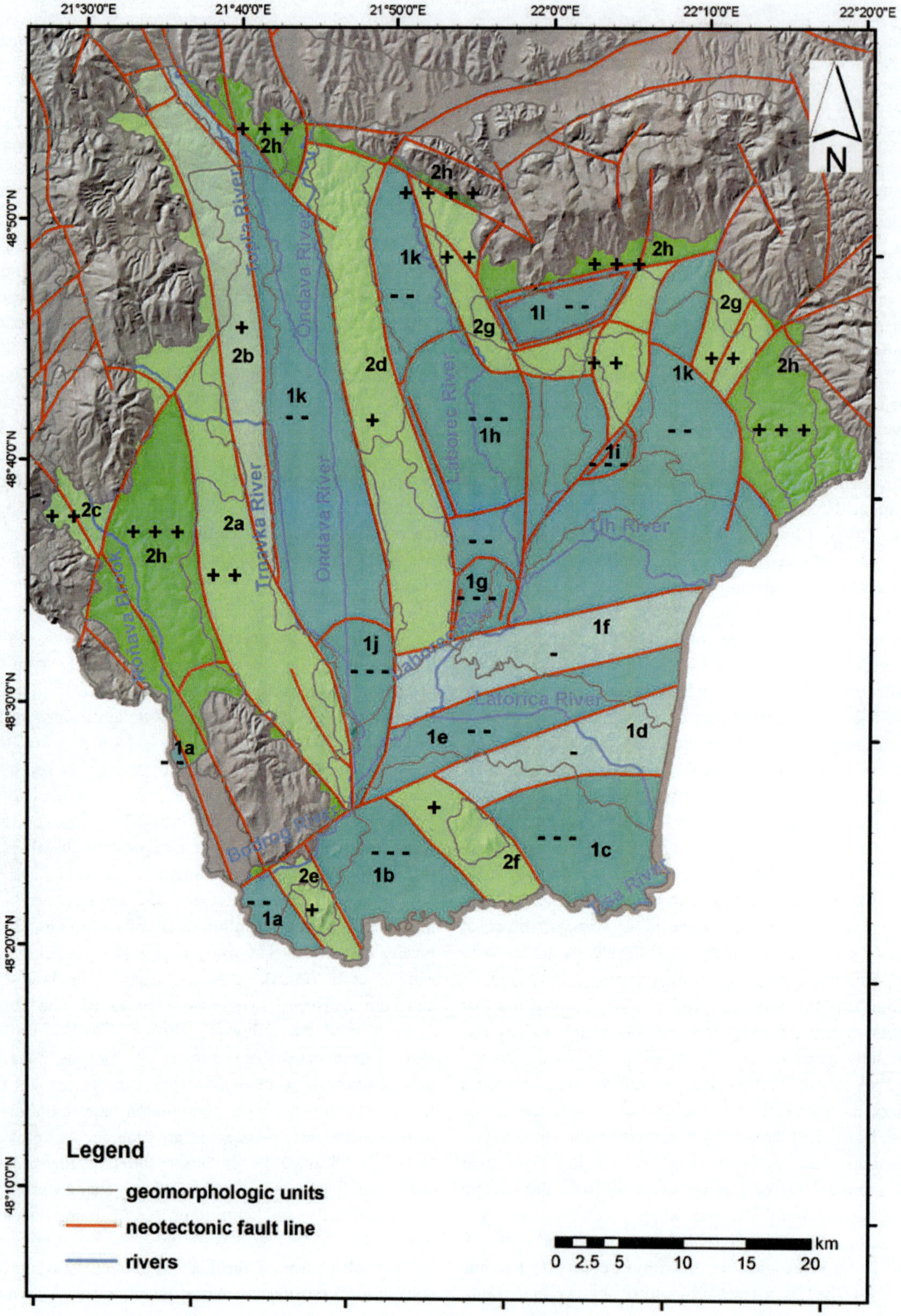

21°30'0"E
21°40'0"E
21°50'0"E
22°0'0"E
22°10'0"E
22°20'0"E
48°50'0"N
48°40'0"N
48°30'0"N
48°20'0"N
48°10'0"N
N
Topla River
Ondava River
Trnavka River
Laborec River
Uh River
Latorica River
Bodrog River
Tisa River
Ondava Brook
1a
1b
1c
1d
1e
1f
1g
1h
1i
1j
1k
1l
2a
2b
2c
2d
2e
2f
2g
2h
Legend
geomorphologic units
neotectonic fault line
rivers
0 2.5 5 10 15 20 km

◂ **Fig. 14.4** Neotectonic pattern of the Východoslovenská nížina (Eastern Slovak) Lowland (Maglay et al. 1999; Baňacký 1986), with indications of the intensity of relative vertical movements of tectonic blocks (downthrow [–] or uplift [+]), the boundaries of geomorphological units after Mazúr and Lukniš (1980). *Explanations:* subsiding morphostructures*:* 1a Roňava depression, 1b Strážne depression, 1c Trakany depression, 1d Poľany horst, 1e Bežovce depression, 1f Maťovce horst, 1g Drahňov (Dolná Duša) depression, 1h Michalovce-Sliepkovce depression, 1i Senne depression, 1j Hraň (Dolná Ondava) depression, 1k subsidence morphostructures of the alluvial plain, 11 Šírava (Sub-Vihorlat) depression; uplifted morphostructures: 2a Sečovce horst of the foothill morphostructure of Slanské vrchy Mts., 2b slightly depressed part of the foothill morphostructure of the Slanské vrchy Mts., 2c horst morphostructure of the margins of the Slanské vrchy Mts., 2d Pozdišovce horst, 2e Tarbucka horst, 2f Chlmec horst, 2g Zalužice horst, 2h foothill morphostructures of marginal mountain range (piedmont terrace). DTM visualization. *Source* Národný geoportál; *Performed by:* D. Barabas

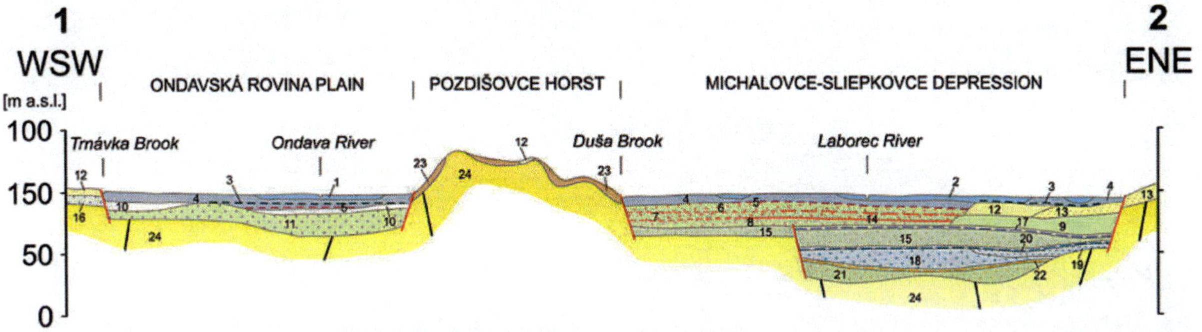

Fig. 14.5 Geological profile (Baňacký 1988; modified) across the Ondavská rovina plain (1.10; 1k), elevated structure of the Pozdišovce horst (2.5; 2d) and the Michalovce-Sliepkovce depression (1.9; 1h). The symbols in brackets correspond to Figs. 14.1 and 14.4, respectively. The profile is located in Fig. 14.3. Explanations*:* Late Holocene—fluvial sediments: 1 fine-grained sands (Subatlantic—Subrecent), 2 loams (Subboreal—Subatlantic), fluvial-organic sediments: 3 loams; Holocene unspecified—fluvial sediments: 4 predominantly loams, Upper Pleistocene—fluvial-organic sediments (Late Würm): 5 sandy loams with organic admixture, fluvial sediments (Würm–Late Würm): 6 sandy gravels (W^3), 7 sandy gravels (W^2), 8 sandy gravels (W^1), 9 loams, 10 sands, 11 sandy gravel, aeolian sediments (Würm): 12 loess, loess loams, aeolian-deluvial sediments (Würm): 13 non-calcareous loess loams, fluvial-organic sediments (interstadial $W^{1/2}$): 14 loams; Middle Pleistocene—fluvial sediments (Riss): 15 sandy gravels, deluvial-fluvial sediments (Riss): 16 sandy gravels; fluvial-organic sediments (interglacial R/W): 17 loams, fluvial sediments (Mindel): 18 sandy gravels, fluvial-deluvial sediments (Mindel): 19 loams, deluvial-fluvial sediments (interglacial M/R): 20 mostly loams; Lower Pleistocene—fluvial sediments (Biber—Günz): 21 sandy gravel, fluvial-organic sediments (interglacial G/M): 22 loams; Quaternary unspecified—deluvial sediments: 23 mostly loams; Neogene unspecified—clastic sediments: mostly clays and gravel; black lines—faults observed; red lines—neotectonic (Quaternary) faults

north of the lowland, in the Nízke Beskydy Mts. Headward erosion processes transformed the river network in the headwater parts of the drainage basins (the main Carpathian ridge), near the main European divide (Leško 1952a,b; Vojtko et al. 2012; Lacika and Lehotský 2013). Transformation of the river network also occurred in the southern part of the East Slovak Lowland. Changes in the structure of the river network subsequently created conditions for the formation of the Bodrog River complex fan. In addition to the Bodrog River, a tributary of the Uh River–Okna Brook created its lower-order fan (NE edge of the Eastern Slovak Lowland).

The natural transformation of the river network was largely related to the neotectonic movements, where the rates of relative (positive and negative) vertical movement tendencies were spatially differentiated. The emerging depressions and/or elevations have become crucial for the formation of river systems.

14.4 Spatial Structure of the Soil Cover of the Eastern Slovak Lowland

The soil cover (Džatko 2002; Vilček 2004) of the Eastern Slovak Lowland was created as a result of the soil-forming substrate (substratum) characteristics, hydrological processes and also morphological and climatic conditions. The crucial process which influenced the character of the Eastern Slovak Lowland soil cover was fluvial modelling in periglacial conditions. Formation and spatial structure of aeolian

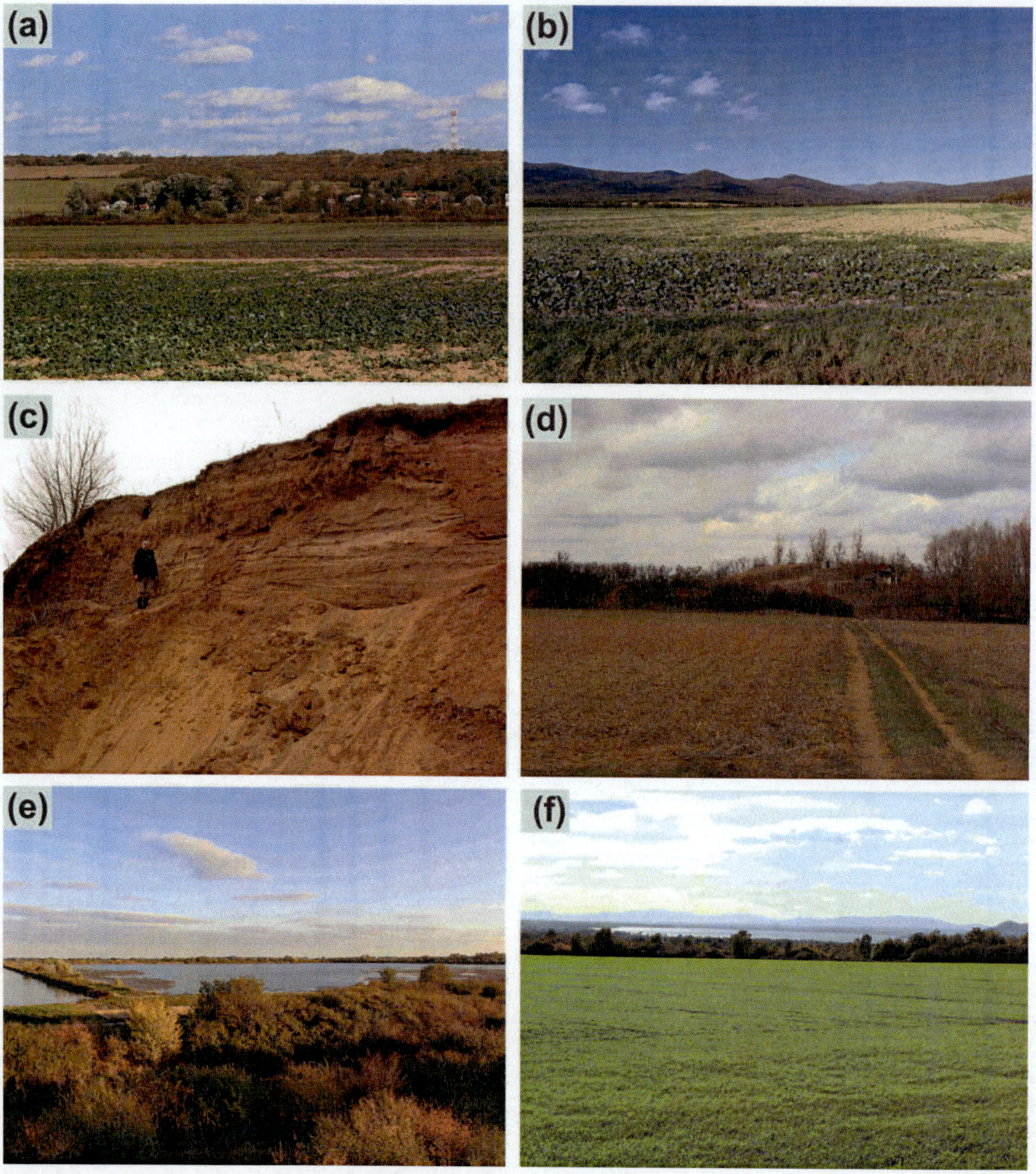

Fig. 14.6 **a** Elevated structure of the Pozdišovce horst (2.5, elevation 58 m a.s.l.) and fluvial relief of the Ondavská rovina plain (1.10, near Trhovište village); **b** middle Pleistocene (Riss) loamy gravels of the proluvial fan formed by the Okna Brook, the mountains of Vihorlat (II) visible on the horizon; **c** cover of aeolian (wind-blown) sands with typical cross-bedding on the slopes of the Moľva andesite volcanic body of 124 m a.s.l. (abandoned sandpit to the east of Sirník village); **d** dome dune formed by aeolian (wind-blown) sands situated westward of Svätuše village (Medzibodrožské pláňavy 1.7); **e** national natural reserve of the Senné rybníky fishponds, built in the wetland area of the Senné depression (1i); **f** the Zemplínska Šírava water reservoir (artificial lake) was built in the years 1961–1965 in a significant wetland area of the Šírava (Sub-Vihorlat) depression (1l). On the horizon, the Slanské vrchy Mts., and extrusive andesite body of the Medvedia hora Hill belonging to the Vihorlat unit (II) on the right. The symbols in brackets correspond to Figs. 14.1 and 14.4. Photo J. Bóna

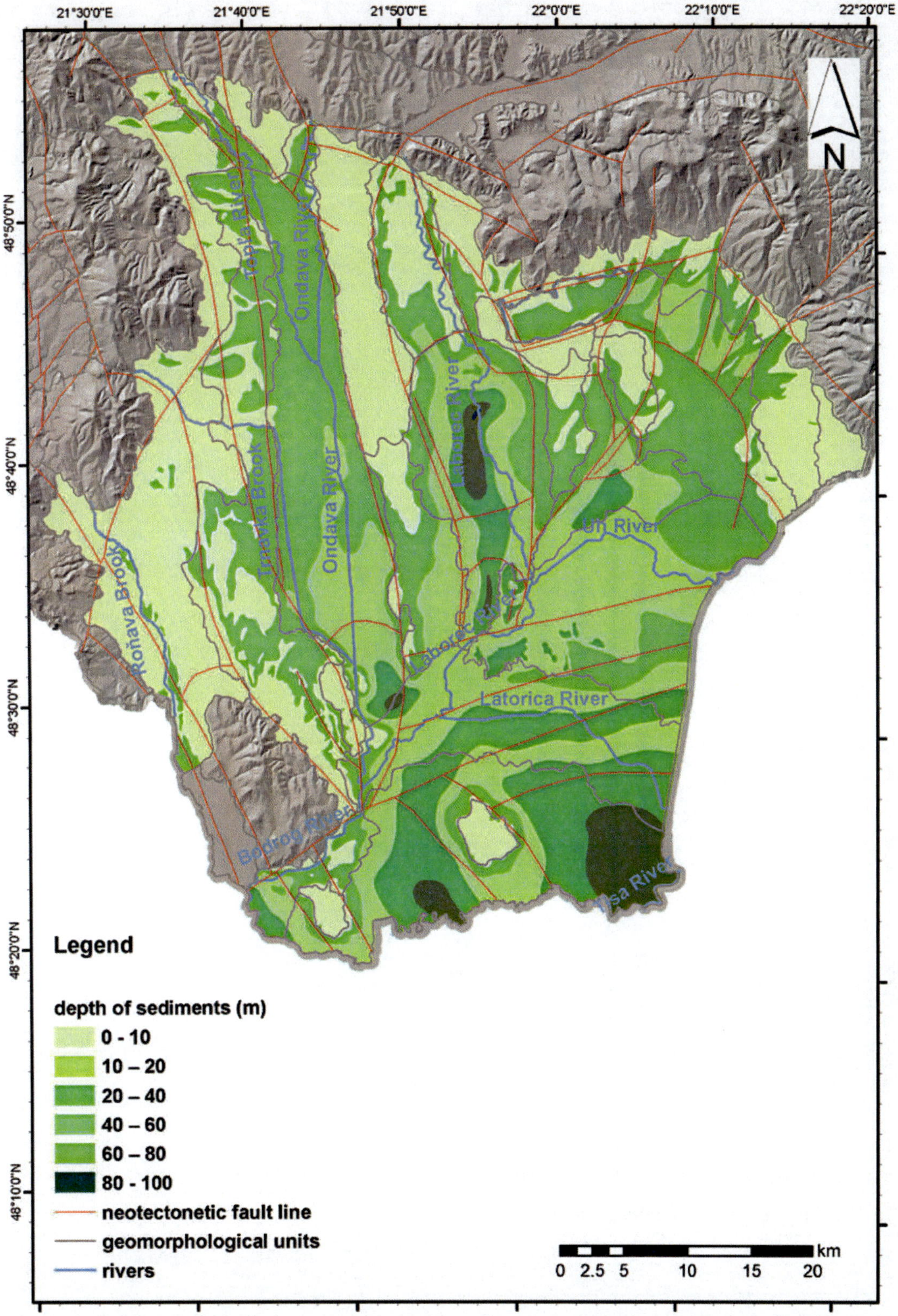

Fig. 14.7 Subsidence centres inferred from the thickness of the Quaternary deposits (Maglay et al. 2008a), with neotectonic faults (Baňacký 1986; Maglay et al. 1999) and boundaries of geomorphological units (Mazúr and Lukniš 1980); DTM visualization. *Source* Národný geoportál; *performed by:* D. Barabas

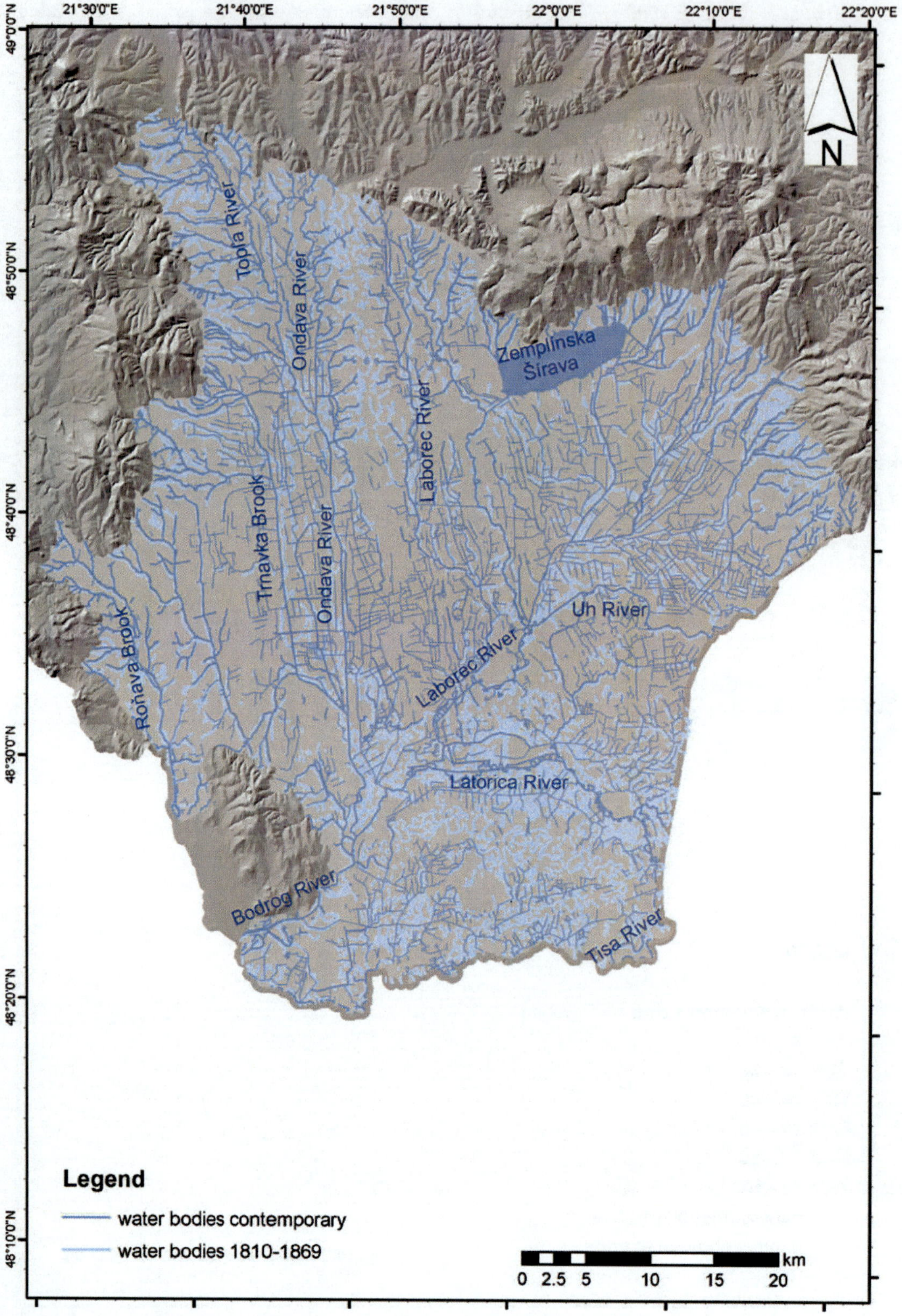

Fig. 14.8 Drainage pattern changes in the Východoslovenská nížina (Eastern Slovak) Lowland are evident from the comparison of the present situation with that from the II. Military mapping (1810–1869). *Source* Národný geoportál; *Performed by:* D. Barabas

sediments and dominant presence of fluvial–alluvial (floodplain) sediments (Fig. 14.3), which are the most widespread soil-forming substrate on Eastern Slovak Lowland, was influenced by neotectonics and the related morphological structure of the area. The presence of clayey detritus (fluvial loams) is important for understanding the hydro-pedological processes that take place in Eastern Slovak Lowland. Combination of the existence of complex (multilevel) fan drainage patterns with a subsidence centre and a significant thickness of fluvial sediment accumulation created ideal conditions for the formation of hydromorphic soils which belong to the very heavy to heavy soils category (Fluvisoil, Gleysoil and Planosols). These soils (47.2%) are the most widespread in the studied area. Regular supply of the fluvial clayey (locally fine-grained) detritus originated from the flysch mountains source areas created conditions that hindered intensive seepage of surface water to the ground level water. Therefore, it formed surface runoff and the accumulation of water in the depressions (Fig. 14.10a, b). These conditions allowed the formation of extremely heavy soils—gley soils (11.8%), which are widespread mainly in subsidence centres and at the river confluence forming river complex fan (Fig. 14.7, Table 14.1). The upland parts with the accumulation of proluvial sediments or elevation (horst) structures with the presence of loess sediments enabled the formatting of higher quality soil covers (Figs. 14.4 and 14.9). The percentage of the high-quality soils in Eastern Slovak Lowland is almost 24.8% (Chernozem, Haplic Luvisols, Luvisols; Table 14.1). Apart from Cambisoils, with 5.5% proportion in the marginal parts of the lowland which are attached to the mountains (upland), the percentage of other types of soils is almost negligible. Out of non-agricultural areas, there are mainly the settlements and forests (Table 14.1).

14.5 Landscape, Man and Water Management

In the lowlands, the close relationships between the river network and human activities are distinctive from other areas (Hanušin 1949; Hreško et al. 2014). The threats and risks of floods, inseparably linked with the human use of the environment, dictated anthropogenically conditioned transformation of the river network. This close relationship resulted in significant human intervention in the river system, which deformed not only the river network but also the drainage conditions in the surrounding areas. The development of the Eastern Slovak Lowland river network is very closely related to the historical development of the settlement pattern. The earliest settlements identified so far were dated for 300 years B.C. and are considered to be Celtic. The occurrence of archaeological sites is usually tied to the elevated sites of natural levees, upland areas and elevations formed by aeolian sediments (sand dunes). Some localities are characterized by the existence of several cultural layers. This phenomenon points to a permanent settlement (Miroššayová 2019). Until the first half of the nineteenth century, we can talk about an important impact of the drainage pattern on the spatial structure of settlements. Since the mid-nineteenth century, however, there has been a significant change in the behaviour of the population. The progress and qualitative

Table 14.1 Proportions of the soil types areas in km^2 and % of the Eastern Slovak Lowland (Šály et al. 2000)

Soils	Area (km^2)	Area (%)
Fluvisols	817.30	33.55
Chernozem	98.02	4.02
Mollic Fluvisols	50.50	2.07
Haplic Luvisols	113.90	4.68
Luvisols	392.70	16.12
Regosols	54.89	2.25
Cambisols	133.20	5.47
Planosols	43.40	1.78
Rendzic Leptosols	0.26	0.01
Gleysols	287.68	11.81
Solonchaks and Solonetz	4.70	0.19
Leptosols	3.78	0.16
Headquarters	171.90	7.06
Forests	155.20	6.37
Water bodies	62.50	2.57
Other areas	47.00	1.93

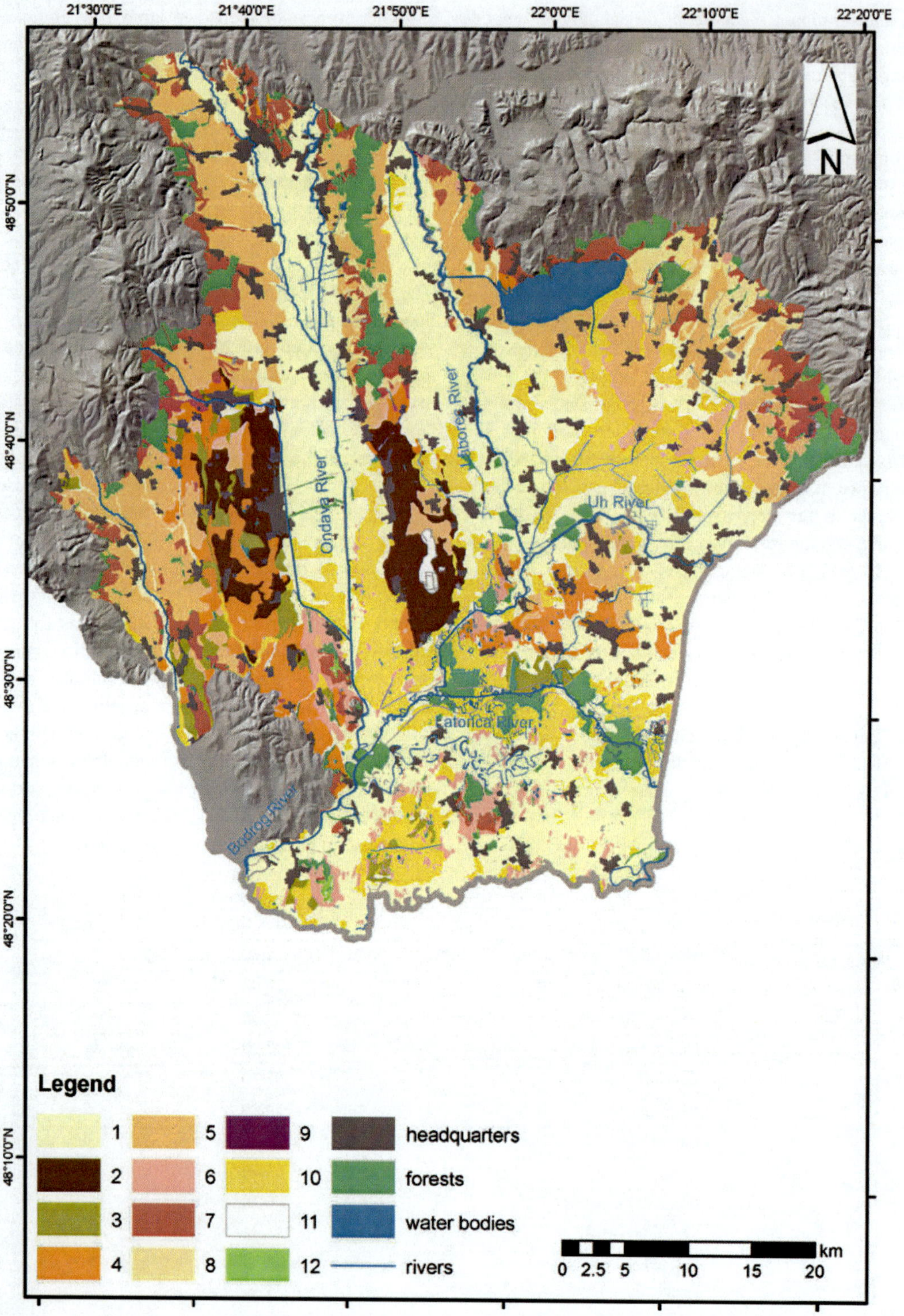

Fig. 14.9 Spatial distribution of soil types on the Východoslovenská nížina (Eastern Slovak) Lowland (Šály et al. 2000). Explanations: 1 Fluvisols, 2 Chernozem, 3 Mollic Fluvisols, 4 Haplic Fluvisols, 5 Luvisols, 6 Regosols, 7 Cambisols, 8 Planosols, 9 Rendzic Leptosols, 10 Gleysols, 11 Solonchaks and Solonetz, 12 Leptosols. *Source* Portal. vupop.sk; *Performed by:* D. Barabas

change in management meant increasing demand for the quantity and quality of the agricultural fund. Social changes were associated with the Enlightenment reforms of Empress Maria Theresa and her son Joseph II. provoked increased spending for the state, which could be only secured by raising taxes (Hanušin 1949). The increase in taxes depended on intensified land use. This was possible only through water management. According to Hanušin (1949), the inundation area made up 17% of the area within the Trebišov town district, in the Michalovce town district it was 12%, and in the Vranov nad Topľou town district it was 6.7%. This potential, together with the intensification of agriculture, ensured an increase in production and hence, higher taxes. At the instigation of Maria Theresa, Dutch technicians were called to draw up a water management plan. However, this was not implemented. Continuous damage to property as a result of floods was the reason for future water management. After catastrophic floods in the early nineteenth century, in 1827 the Royal Commission introduced the first concrete measures to protect against flooding. In the years 1845–1846, there was an organized flood control campaign of the watercourses (channel straightening, construction of storm-surge barriers).

The construction of the first canals occurred in the most flooded parts of the lowland—the Ondavská rovina plain (Fig. 14.10). It is an area with a small to medium relative vertical downthrow (Maglay et al. 1999), the lowest slope and the highest density of the river network. The artificial river mouth of Ondava river to Latorica river, as well as the overall modification of the Ondava river channel partially helped but did not solve the basic problem of drainage of land usable for agriculture. An expectation to systematically address the drainage of the lowland was the reason for the invitation of an Italian scientist and engineer Pietro Paleòcapa, who was asked to solve the problem for the Tisza river and its tributaries. However, the bulk of drafted proposals were not implemented. Later in 1852–1854, a map of the inundation area was drawn up as a basis for the implementation of water management proposals. After a number of local modifications, systematic water management was undertaken in the middle of the nineteenth century. Channel straightening was introduced and canal systems were built. During the building of the canals and straightening of the river network, fluvial sedimentary material was spread along the canal banks or used to build the storm-surge barriers. This resulted in the creation of barriers through which water cannot flow over the surface (Fig. 14.10b). This stage ended in 1888.

The second stage of work focused on the solution of water management problem in the Eastern Slovak Lowland took place in the late nineteenth and early twentieth centuries. The aim of this stage was to divert internal waters to the recipients. For this purpose, pumping stations were built (Fig. 14.10e, f). In total, 16 dewatering pump stations with a capacity of 112.8 m^3 s^{-1} were built in the Eastern Slovak Lowland. The last stage of water management was carried out in the second half of the twentieth century. In order to drain the agricultural area and adjust the soil water regime, systematic drainage was carried out. In total, 112,565 ha of agricultural land were drained. The largest areas drained by systematic drainage are located in the district of Michalovce town (Ivančo and Kupčo 1995).

The most significant transformation of the river network texture occurred in the Eastern Slovak Flat, marginally the transformation also affected the streams in the upland. The streams were shortened due to human efforts to intensify the land use of the Eastern Slovak Lowland. The shortening of streams as a result of the water management is also related to tributaries from the Eastern Slovak Upland. The shortening of streams flowing from the upland was a response to the fact that these tributaries formed extensive wetlands on the plain, caused by natural levees along their recipient rivers. The streams flowing from the Slanské vrchy Mts. to the east or southeast turn towards the south where they cross the border between the upland and the flat parts of the area.

The change of direction was caused by the presence of neotectonic faults (Baňacký et al. 1987), which follow the general N–S trend, and in these places, they almost perfectly coincide with the border between the upland and the plain. Streams that flowed towards the flat area reduced their kinetic energy, and deposition of fluvial materials started between the natural levees created by the recipient river and the more elevated terrain. The streams flowing from the WSW direction changed to the SSE direction, and this was evidently related to neotectonic movements. In the past, these streams fed an extensive network of useless swamps, while flowing towards the recipients. At present, the original drainage network is completely changed. Streams were shortened to increase the slope and rate of water runoff. At the same time, it prevents water from spilling out of the river channel, which allowed for agricultural land use of previously marshy areas. These interventions in the river network can also be seen on maps of II. Military mapping (1810–1869), where a sudden change of channel direction related to, for example, the Bačkovský potok Brook (Fig. 14.11), Trnávka and Chlmec brooks can be observed. The Bačkovský potok Brook is shortened by 4 km. Other streams (e.g., Trnávka and Chlmec brooks) have also been shortened.

A total length of 316 km of watercourses was regulated and 453 km of storm-surge barriers were built (Ivančo and Kupčo 1995). Besides, streams were shortened in the total length of 110 km. Water management meant straightening the streams and also their shortening.

There was also an ongoing process of building drainage canal systems and reducing abandoned arms by isolating

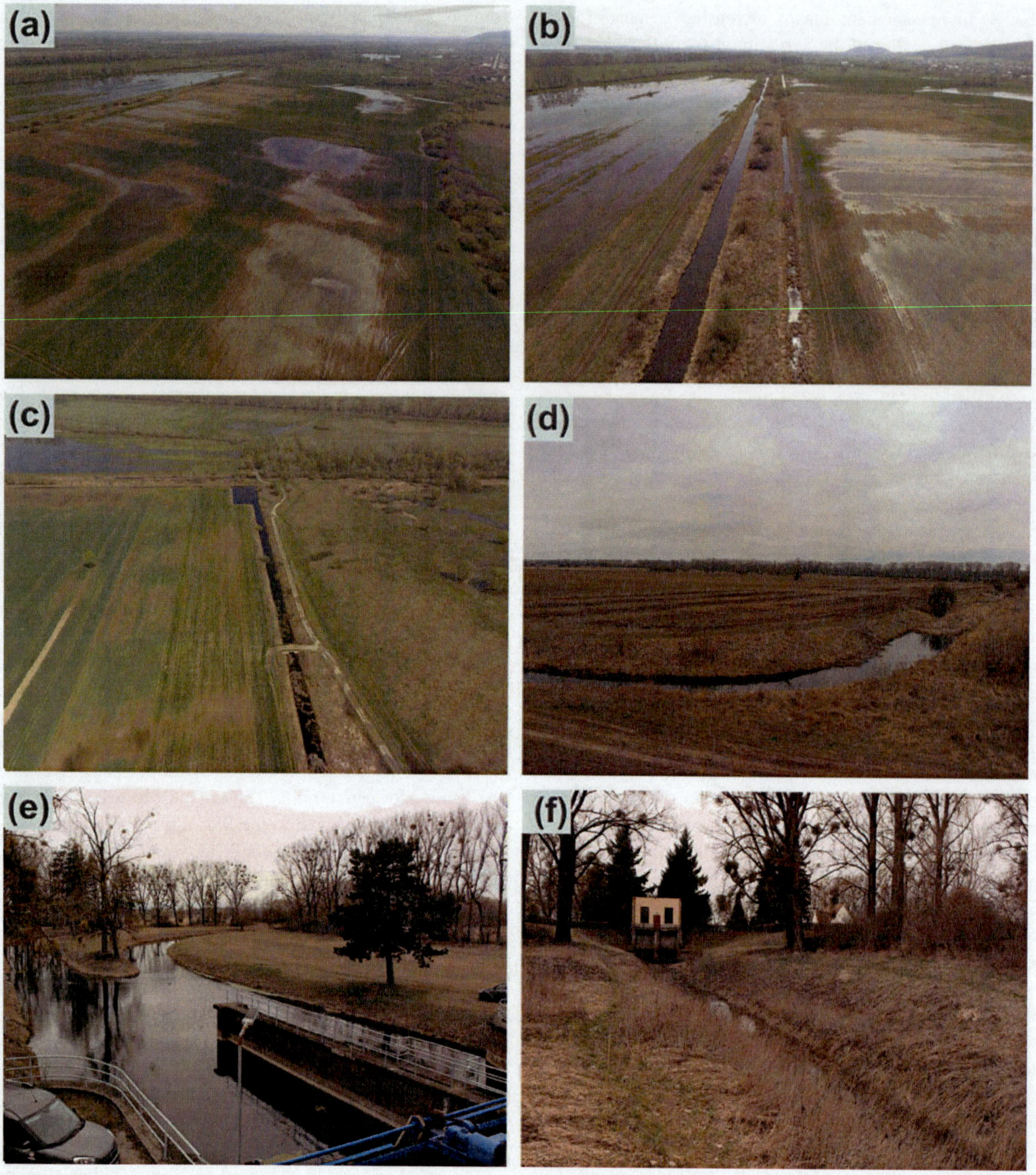

Fig. 14.10 **a** Waterlogged areas (fields) on the Ondavská rovina plain (1.10) near the village of Hraň between Ondava river and Trnávka brook, the symbols in brackets correspond to Fig. 14.1; **b** waterlogged agricultural parcels and Kopaný jarok drainage canal with a parallel drainage canal with distinctive features of eutrophication; **c** Neľov drainage canal and perpendicularly oriented Kopaný jarok drainage canal near the village of Hraň in the background; **d** Neľov drainage canal loop—view from the storm-surge barrier of the Trnávka Brook; **e** mouth of the Kopaný jarok drainage canal (left) and the Hraň drainage canal into the Hraň dewatering pump station, was built in 1938–1948; **f** canal outflow leading from the Hraň dewatering pump station to the Ondava River artificial river bed. *Aerial oblique photo* M. Lacko a, b, c—2018; *photo* J. Bóna d, e, f

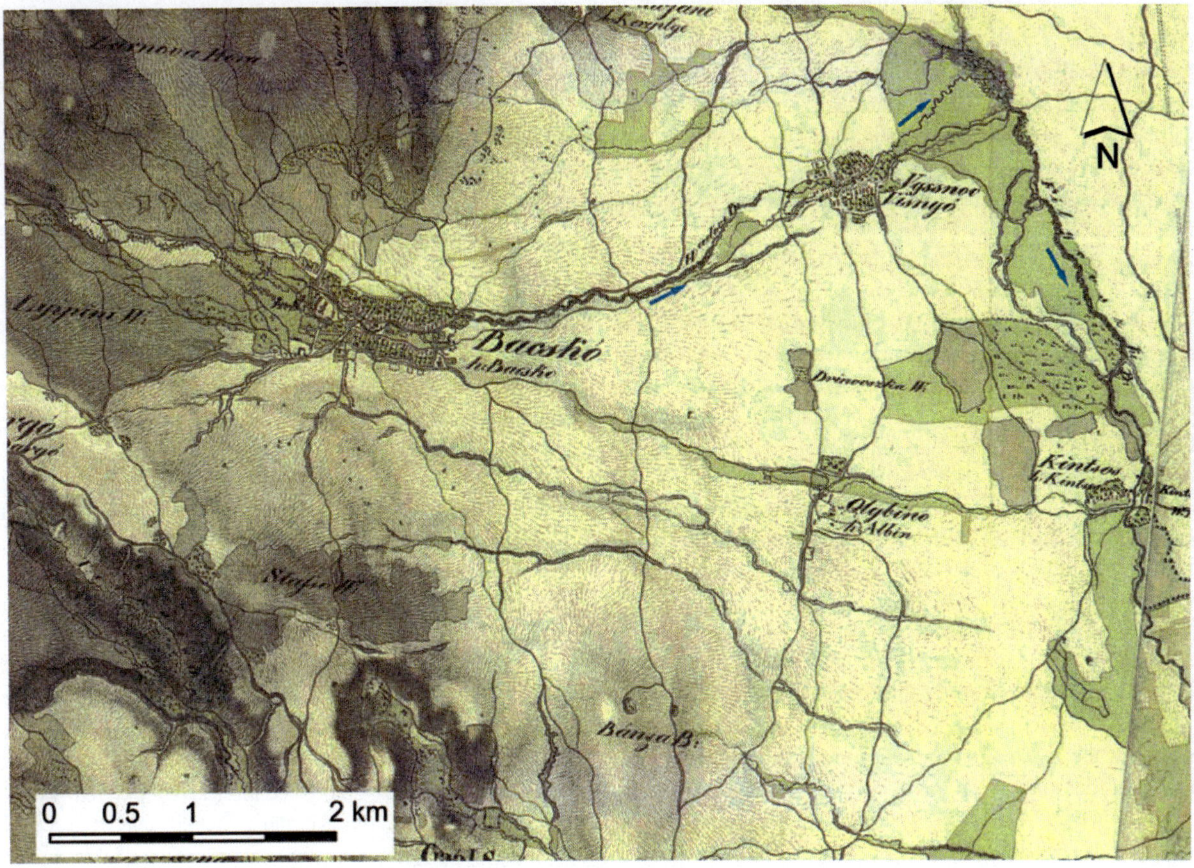

Fig. 14.11 Sharp, tectonically controlled direction change of the channel of the Bačkovský potok Brook from ENE–WSW to NNW–SSE near the village of Višňov, shown on the topographic map of the II. Military mapping from 1810–1869. *Source* Národný geoportál; *Performed by:* D. Barabas

them from the main stream through flood defences and drying up. All that led to changes in particular water bodies (Table 14.2).

Despite significant water management modifications of the Eastern Slovak Lowland river network, there are still areas with the highest subsidence (medium-scale downthrows) that remain affected by waterlogging or have lower utilization rates for agricultural purposes. The reason is not only the subsidence but also the presence of very heavy soils (clayey soils) that are typical of these localities (Ivančo et al.

Table 14.2 Changes in the length and slope of selected recipients, when comparing maps of the II. Military mapping (1810–1869) and the current state

Rivers name	Elevation begin (m a. s.l.)	Elevation mouth (m a.s.l.)	Length stream (km) (II. military mapping)	Length stream (km) (current state)	Difference (km)	Slope (m/km) (II. military mapping)	Slope (m/km) (current state)
Latorica River	99.20	95.30	67.44	31.48	35.96	0.0578	0.1240
Laborec River	131.00	95.00	87.56	56.66	30.90	0.3529	0.5454
Ondava River	120.00	95.30	71.36	57.88	13.48	0.3461	0.4267
Topľa River	133.00	103.10	37.78	31.80	5.98	0.7914	0.9403
Uh River	101.00	97.00	29.05	21.26	7.79	0.1377	0.1881
Bodrog River	95.30	94.30	30.79	14.88	15.91	0.0325	0.0672
Σ	–	–	323.98	213.96	110.02	0.2864	0.3820

2003). The Senianska mokraď (Fig. 14.6e), Ondavská (Fig. 14.10a–c) and Latorická rovina (Fig. 14.13a) areas (sensu Mazúr and Lukniš 1980) represent extreme examples, where there was significant meandering of the streams and formation of abandoned arms (Fig. 14.12a), which are partially preserved to the present (Latorica Protected Landscape Area in the Latorická rovina and Medzibodrocké pláňavy units, Fig. 14.13b–d). Despite water management, these areas are still very waterlogged, especially in spring.

The most significant transformation of the river network was due to anthropogenic activity. During 150–180 years the character of the river network of the whole lowland has completely changed. It was a combination of qualitative and quantitative changes. The most important factor was the effort to effectively use the maximum possible area for agricultural purposes (Fig. 14.12a, b). In terms of quality, naturally meandering streams (rivers) were changed to straightened channels with revetment (paved) banks. The texture and ratio between streams, channels and abandoned arms has also changed (Table 14.3).

Very often, the transformation of the river network resulted in the conversion of smaller streams into canals, which led to a decrease in their length. There was also a significant reduction in the number of abandoned arms, which in the past represented an obstacle to the use of the lowland for agricultural purposes. The conditions of the lowland became better for water management because the total length of water bodies has increased (Table 14.3). This situation should help to increase the overall period of water retention in the country and contributes to more efficient water management.

The recognition of a long-term declining trend in groundwater levels in the Eastern Slovak Lowland (Šťastný et al. 1998; Šútor et al. 1995) contradicts the idea of improving water budget (increasing the amount of water) by extending water bodies. The main reason for this mismatch is a significant change in the slope of the watercourse (Table 14.2). Increasing the slope, in extreme cases as much as twice due to shortening the stream, e.g., in the Latorica river and Bodrog river, in half of their original length, resulted in an increase of the flow rate by almost 50%. This situation results in a faster outflow of water from the territory, shortening the residence time of water in the territory and reducing the area of infiltration. The second reason is the significant drainage of the abandoned arms (systematic drainage), which have significantly prolonged the period of water retention in the country. Surface water of depressed areas with very heavy soils infiltrates and recharges the groundwater very slowly. In the case of heavy and very heavy soils (clayey soils), the rate of infiltration is in the range of 1–20 cm/day (Šútor 1985; Novák et al. 1986; Šútor et al. 1995). The acceleration of infiltration was ensured by systematic drainage, supported by deep ploughing (Kravčík and Barabas 1988; Kravčík et al. 1990).

14.6 Conclusions

The Eastern Slovak Lowland is a country in which water plays a crucial role. The formation of fluvial relief, which dominates in the Eastern Slovak Lowland, influenced the character of the soil cover. The accumulation of fine clay sediments transported from the sedimentary flysch source areas of the main recipients was the main reason for the formation of heavy (clayey) soils. In the Eastern Slovak Lowland area, they take up almost 20% in the category of very heavy soils and 22% in the category of heavy soils (Lorenčík et al. 1979). Hence, 42% of total area of heavy soils are located mainly in depressions, i.e., in neotectonically downthrown areas. These heavy soils reduce the rate and intensity of water infiltration. The combination of neotectonic vertical movements and soil cover texture crucially influences the development of the Eastern Slovak Lowland. However, ignorance of these conditions by people and technical interventions into the country have contributed to changes that may cause serious problems in the use of this area in the future. Gomboš et al. (2018) pointed to an increase in the evapotranspiratory deficit over the years 1970–2015. The trendline shows an increase of this deficit for more than 50% over the reported period. The evidence for that is a drop in groundwater levels (Mati et al. 1995; Šťastný et al. 1998). If there is an ongoing ambition for agricultural land use of the Eastern Slovak Lowland in the future, we will not avoid, in extreme cases, intensive irrigation. On the other hand, there may be a possibility to use sub-territories that are currently not being used due to the high groundwater level. Certainly, the use of the Eastern Slovak Lowland area has never been and will not be as efficient as other lowland areas of Slovakia. This is due to non-uniform recent vertical (downthrow) movements trends of tectonic blocks, and the related hydrological conditions in connection with the character of fluvial sediments and the development of specific soil types.

The potential of the Eastern Slovak Lowland soil cover for agricultural production is limited. This is mainly determined by the intensity of anthropogenic interventions. Man-made interventions enabled the usage of the previously unused land. Despite extensive investments (storm-surge barrier, hydromelioration), the use of the production potential of hydromorphic soils is problematic due to the high groundwater level and hydro-physical characteristics of these soils (Šútor et al. 1995).

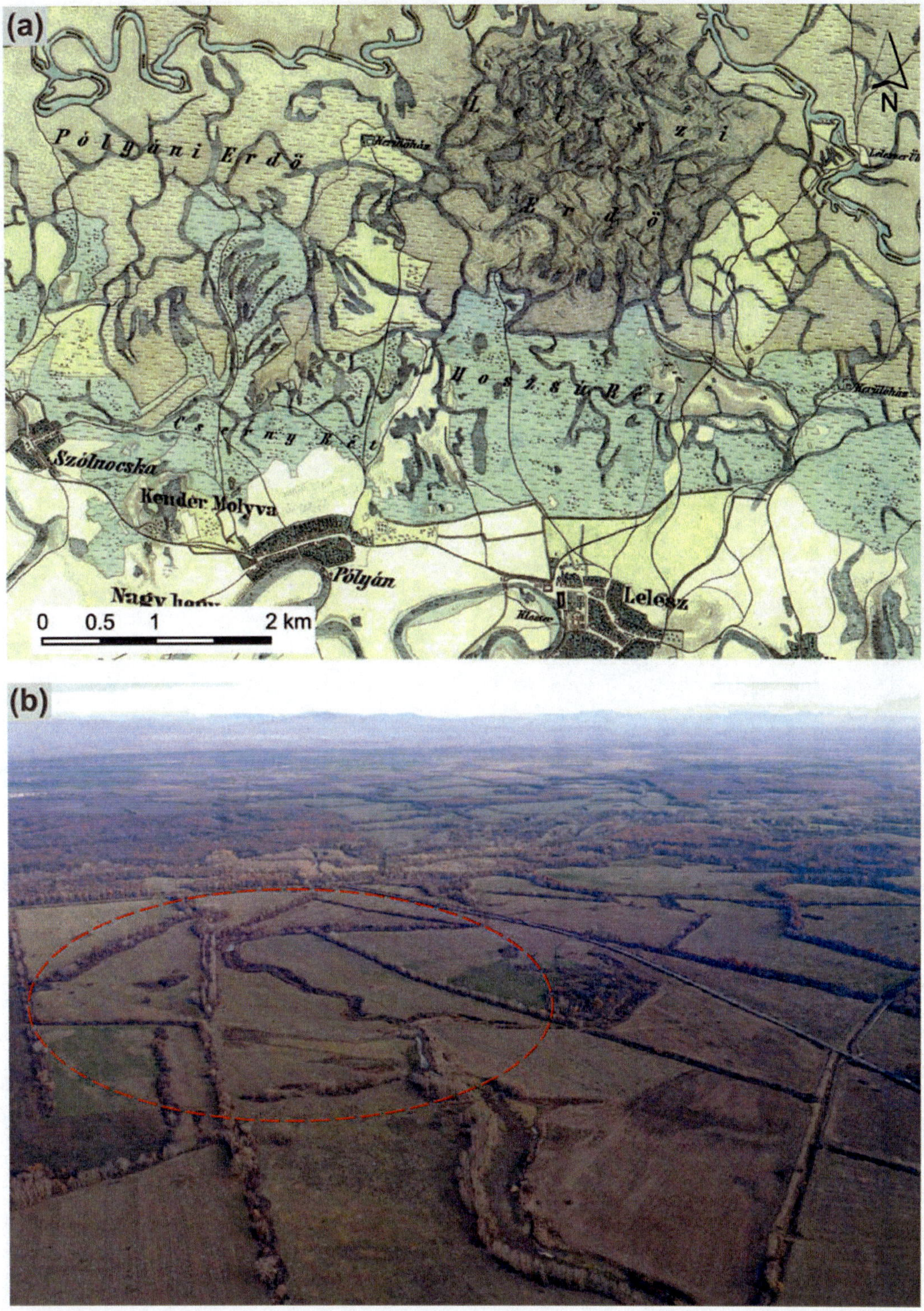

Fig. 14.12 **a** An example of the Leleský les Forest (on the map in Hungarian—Leleszi Erdö) swampy area and the dense network of abandoned (meander) arms in Medzibodrožie area, south of the Latorica River. It is shown on the topographic map of the II. Military mapping from 1810 to 1869. **b** Floodplain forests of Latorica Protected Landscape Area, the Slanské vrchy Mts. visible on the horizon. Leleský les Forest (red ellipse) is situated to the north of the village of Leles; the transformation of a swampy country with abandoned (meander) arms into an agricultural country was caused by anthropogenic activity (hydromelioration). *Source* Národný geoportál, *Performed by:* D. Barabas **a**; *aerial oblique photo* J. Kaňuk **b**

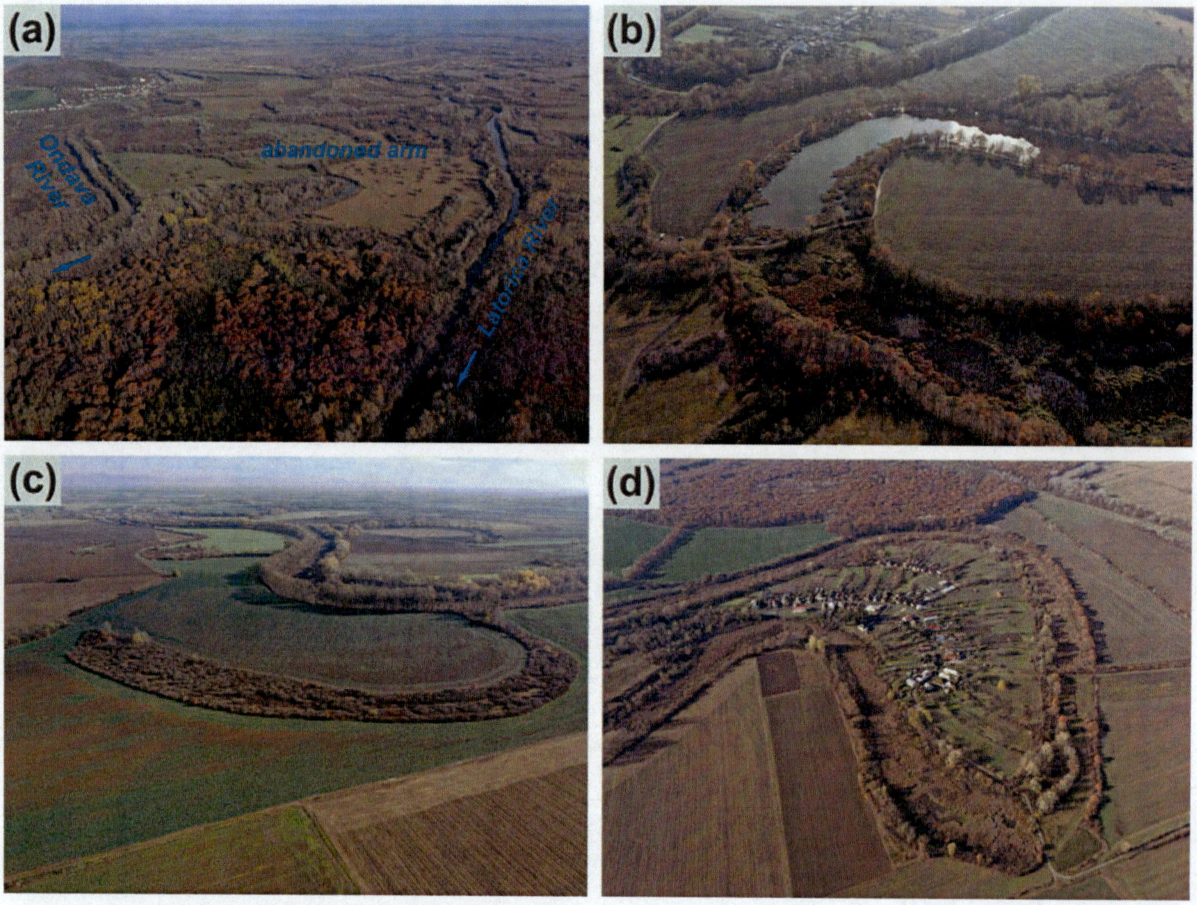

Fig. 14.13 a Area ca. 500 m NE from the confluence of the Ondava and Latorica rivers with the occurrence of floodplain forests—an abandoned arm between the rivers represents a meander remnant of the Latorica river (view from SW). The Veľký vrch Hill (272 m a.s.l.), which is an andesite extrusive body, with the village of Brehov on its southeast slope is on the left; **b** abandoned arm of the Tica river near the village of Svätá Mária-Pavlovo (view from NW); **c** abandoned arm (meander scar) of the Tica River between the villages of Rad and Svinica (view from NW); **d** abandoned arm of the Tica river surrounds the village of Rad (view from E). The abandoned arms show evidence of eutrophication. *Aerial oblique photo* J. Kaňuk

Table 14.3 Total length of water streams, canals and abandoned arms. The ± sign indicates an increase or decrease in the length of the streams (current state) in proportion to the II. Military Surveying (1810–1869)

Water bodies	Length stream (km) (II. military mapping)	Length stream (km) (current state)	Difference (km)
Stream	2043.6	1463.3	–580.3
Canal	37.2	2547.8	+2509.8
Abandoned arm	1609.4	122.9	–1486.5
Σ	3690.2	4134.0	+443.8

Acknowledgements This study was supported by Grant VEGA 1/0798/20 and Intereg HUSKROUA/1702/8.1/0065/GeoSES. For constructive comments on the manuscript, the authors thank Prof. Dr. hab. P. Migoń, Dr. M. Lehotský and Dr. J. Maglay.

References

Baňacký V, Harčár J, Sabol A (1965) Neue Kentnisse über den einfluss der Tektonischen bewegungen auf die Quartare entwicklung der Slowakischen niederungen. Geologické Práce, Zprávy 36:63–73 (in German)

Baňacký V (1965) Outline of the development of the quaternary in the middle stream of the river-Basin Laborec in Relation to the Young Tectonic Movements. Sborník Východoslovenského múzea v Košiciach, Séria A – Geologické vedy VI(A):95–109 (in Slovak, English Summary)

Baňacký V (1968) Geological History of the northern part of the East Slovakian Lowland in the quaternary. Geologické Práce, Zprávy 44–45:241–262

Baňacký V (1980) New data on young (Quaternary) tectonics in East Slovakian Lowland. Geologický Zborník – Geol Carpath 31:603–609

Baňacký V (1986) Mapa kvartérnych neotektonických štruktúr Východoslovenskej nížiny. SGÚ – GÚDŠ, Bratislava (in Slovak)

Baňacký V (1988) Geological map of the northern part of the East Slovakian Lowland (1:50 000). GÚDŠ, Bratislava (in Slovak)

Baňacký V (ed), Vass D, Kaličiak M, Remšík A, Pospíšil Ľ (1987) Explanations to the geological map of the northern part of the East Slovakian Lowland (1:50.000). GÚDŠ, Bratislava 117 pp (in Slovak, English Summary)

Baňacký V (ed), Elečko M, Kaličiak M, Lexa J, Straka P, Vass D, Vozár J, Vozárová A (1988) Geological map of the southern part of the East Slovakian Lowland and Zemplín Hills (1:50.000). GÚDŠ, Bratislava (in Slovak)

Baňacký V (ed), Elečko M, Kaličiak M, Straka P, Škvarka L, Šucha P, Vass D, Vozárová A,Vozár J (1989) Explanations to the geological map of the southern part of the East Slovakian Lowland and Zemplín Hills (1:50.000). GÚDŠ, Bratislava 143 pp (in Slovak, English Summary)

Baňacký V, Halouzka R, Horniš J, Kernats G, Maglay J, Pristaš J (1993) Geodynamicky vyvoj uzemia Slovenska počas kvarteru. In: Rakús M, Vozár J (eds) Geodynamický model a hlbinná stavba Západných Karpát. Konf., Symp., Sem. GÚDŠ, Bratislava, pp 239–242 (in Slovak)

Borsy Z, Félegyházi (1983) Evolution of the network of water courses in the northeastern part of the great Hungarian plain from the end of the Pleistocene to our days. Q Stud Poland 4:115–124

Dub O (1957) Hydrológia, hydrografia, hydrometria. Slovenské vydavateľstvo technickej literatúry, Bratislava 485 pp (in Slovak)

Dzurovčin L (1990) Geomorfologická analýza strednej časti Slánskych vrchov. Ph.D. thesis, Institute of Geography, SAS, Bratislava 161 pp (in Slovak)

Dzurovčin L (1998) The morphosculpture of East Slovakia neovolcanic Mountains. Acta facultatis studiorum humanitatis et naturae Universitatis Prešoviensis, Prírodné vedy XXIX – Folia geographica (1):77–112 (in Slovak, English Summary)

Dzurovčin L (2001) Stages of Relief development of Stratovolcanoes in the East Slovakia. Acta facultatis studiorum humanitatis et naturae Universitatis Prešoviensis, Prírodné vedy XXXV – Folia geographica (4):41–56 (in Slovak, English Summary)

Dzurovčin L (2010) Reliéf neovulkanických pohorí Východného Slovenska. In: Mezei I, Barabas D (eds) Geografické poznatky bez hraníc – výber z maďarských a slovenských príspevkov z fyzickej a humánnej geografie. UPJŠ v Košiciach, Košice, pp 142–146 (in Slovak)

Dzurovčin L (2011) Types of river networks in the East Slovakia. Geographia Cassoviensis V(2):31–34 (in Slovak, English Summary)

Džatko M (2002) Evaluation of the productive potential of agricultural soils and soil-ecological regions in Slovakia. VÚPOP, Bratislava 88 pp. (in Slovak)

Fordinál K (ed), Maglay J, Elečko M, Nagy A, Moravcová M, Vlačiky M, Kučera M, Polák M, Plašienka D, Filo I, Olšavský M, Buček S, Havrila M, Kohút M, Bezák V, Németh Z. (2012) Geological map of the Záhorská nížina Lowland (1:50.000). ME SR and ŠGÚDŠ, Bratislava (in Slovak and English)

Franyó F (1992) The thickness of the quaternary sediments in Hungary – Map (scale 1:500.000). Geological Institute of Hungary, Budapest

Gábris G, Nádor A (2007) Long-term fluvial archives in hungary: response of the Danube and Tisza rivers to tectonic movements and climatic changes during the Quaternary: a review and new synthesis. Q Sci Rev 26:2758–2782. https://doi.org/10.1016/j.quascirev.2007.06.030

Gomboš M, Kandra B, Balejčíková L (2018) Impact of drought on the soil water regime elemements under the East slovakian lowland conditions. In: Rožnovský J, Litschmann T (eds) Hospodaření s vodou v krajině (21– 22. 6. 2018), Třeboň, pp 1–12 (in Slovak, English Summary)

Hanušin J (1949) Vodné družstvo na Ondave – storočnica vodohospodárskej činnosti. Roľnícka osvetová služba číslo 26, Bratislava 155 pp (in Slovak)

Hreško J, Petrovič F, Hrozenská S, Mišovičová R, Rybanský Ľ, Petluš P (2014) Archetypes of Lowland Fluvial Landscape of Slovakia. Životné prostredie 48(1):33–37 (in Slovak, English abstract)

Ivančo J, Kupčo M (1995) Vývoj a regulácia vodných pomerov na Východoslovenskej nížine. In: Šútor J (ed) Hydrológia Východoslovenskej nížiny, 1. vyd., Media Group, Michalovce, pp 116–153 (in Slovak, English Summary)

Ivančo J, Pavelková D, Mati R (2003) Dynamika zásob vody v pôdnom profile ťažkých pôd na Východoslovenskej nížine. Acta Hydrol Slovaca 4(1):74–85 (in Slovak, English abstract)

Kaličiak M, Žec B (1995) Review of Neogene volcanism of Eastern Slovakia. Acta Volcanol 7:87–96

Košťálik J (1999) Loess and fossil soils of East Slovakia and Rimava Basin – their genesis and chronostratigraphy. Faculty of Science, Pavol Jozef Šafárik University in Košice, Košice 191 pp (in Slovak, English Summary)

Košťálik J (2009) A Contribution to the Study of Genesis of Eolic Sands in the Southern Part of Východoslovenská nížina Lowland. Geographia Cassoviensis III(2):81–91 (in Slovak, English Summary)

Kováč M (2000) Geodynamic, paleogeographic and structural evolution of the Carpathian-Pannonian region in Miocene—A new view on the Neogene basins of slovakia. Veda, Bratislava 204 pp (in Slovak)

Kováč M, Kováč P, Marko F, Karoli S, Janočko J (1995) The East Slovakian Basin—A complex back-arc basin. Tectonophysics 252 (1–4):453–466. https://doi.org/10.1016/0040-1951(95)00183-2

Kravčík M, Barabas D (1988) Vplyv úrovne hladín podzemných vôd na odtokové pomery Východoslovenskej nížiny. Vodní hospodářství 1988a(12):309–315

Kravčík M, Barabas D, Kochaník J (1990) Dynamika hydrologických procesov na ťažkých pôdach Východoslovenskej nížiny. Vodní hospodářství 1990(10):401–408

Kupčo M (1995) Povrchové vody. In: Šútor J (ed) Hydrológia Východoslovenskej nížiny, 1. vyd. Media Group, Michalovce, pp 154–200 (in Slovak, English Summary)

Kvitkovič J (1961) Príspevok k poznaniu neotektonických pohybov vo Východoslovenskej nížine a priľahlých oblastiach. Geografický časopis 13(3):176–194 (in Slovak, English Summary)

Kvitkovič J (1993) Intensity of vertical tectonic movements of the Earth's crust in thelowlands of Slovakia in the Holocene. Geografický časopis 45(2–3):213–232 (in Slovak, English Summary)

Kvitkovič J, Plančár J (1975) Analyse der Morphostrukturen aus dem Aspekt gegenwärtiger bewegungstendenzen in beziehung zum geologischen Tiefbau der Westkarpaten. Geografický časopis 27 (4):309–325 (in Slovak, German Summary)

Kvitkovič J, Vanko J (1972) Recent crustal movements in the region of eastern Slovakia. Geografický Časopis 24(2):151–163

Lacika J, Lehotský M (2013) Morphostructural relief analysis as a tool for comprehension of the development and changes to river networks—The example of north-eastern Slovakia. Geografický Časopis 65(3):251–268

Leško B (1952a) Vplyv geologickej stavby na vodnú sieť Laborca. Geologický Sborník SAV 3(1–2):45–58 (in Slovak, German Summary)

Leško B (1952b) Pirátstvo rieky Udavy. Zemepisný Sborník SAV a U 4(3–4):89–106 (in Slovak, German Summary)

Lorenčík L et al (1979) Koncepcia ochrany zúrodňovania a využívania pôdneho fondu na východoslovenskej nížine. Technical reeport, KPVS, Michalovce 185 pp (in Slovak)

Lukniš M (1972) Reliéf. In: Lukniš M (ed): Slovensko 2, Príroda. Obzor, Bratislava, pp 124–202

Lexa J, Seghedi I, Németh K, Szakács A, Konečný V, Pécskay Z, Fülöp A, Kovacs M (2010) Neogene-quaternary Volcanic forms in the Carpathian-Pannonian region: a review. Cent Eur J Geosci 2 (3):207–270. https://doi.org/10.2478/v10085-010-0024-5

Maglay J (ed), Halouzka R, Baňacký V, Pristaš J, Janočko J (1999) Neotectonic Map of Slovakia (1:500.000). Publ. Ministry of the Environment of the Slovak Republic and Geological Survey of the Slovak Republic, Bratislava (in Slovak and English)

Maglay J (ed), Pristaš J, Nagy A, Fordinál K, Buček S, Havrila M, Kováčik M, Elečko M, Baráth I (2006) Geological map of the Danube Lowland – Trnavská pahorkatina Upland (1:50.000). ME SR and SGIDŠ, Bratislava (in Slovak and English)

Maglay J (ed), Pristaš J, Kučera M, Ábelová M (2008a) Quaternary geological map of Slovakia—Quaternary cover thickness (1:500.000). Publ. Ministry of the Environment of the Slovak Republic and State geological Institute of Dionýz Štúr, Bratislava

Maglay J (ed), Pristaš J, Kučera M, Ábelová M (2008b) Quaternary geological map of Slovakia – genetical (deposits) types (1:500.000). Publ. Ministry of the Environment of the Slovak Republic and State geological Institute of Dionýz Štúr, Bratislava (in Slovak)

Mati R, Ivančo J, Kupčo M (1995) Podzemné vody. In: Šútor J (ed) Hydrológia Východoslovenskej nížiny, 1. vyd., Media Group, Michalovce, pp 201–280 (in Slovak, English Summary)

Mazúr E, Lukniš M (1980) Geomorfologické členenie. In: Mazúr E (ed), Atlas SSR. Veda, Bratislava, 54–55 (in Slovak)

Miroššayová E (2019) Laténske osídlenie na Východoslovenskej nížine. Studia Historica Nitriensia 23 (Suppl. – Sedem kruhov Jozefa Bujnu):237–253 (in Slovak, English abstract)

Novák V, Benetín J, Šoltész A, Štekauerová V, Šútor J, Radčenko I (1986) Regulácia vodného režimu pôd Východoslovenskej nížiny. In: Ekologická optimalizácia využívania Východoslovenskej nížiny – II. diel. Zborník z vedeckého sympózia v dňoch 13–16. 5. 1986) Zemplínska Šírava, ÚEBE – KPVS, Bratislava, pp 5–71 (in Slovak)

Rudinec R (1989) New view onto the paleogeographic development of the Transcarpathian depression during the Neogene. Mineralia Slovaca 21:27–42

Šály R, Bedrna Z, Bublinec E, Čurlík J, Fulajtár E, Gregor J, Hanes J, Juráni B, Kukla J, Račko J, Sobocká J, Šurina B (2000) Morphogenetic soil classification system of Slovakia – Basal reference taxonomy. VÚPOP, Societas pedologica slovaca, Bratislava 74 p. (in Slovak)

Šťastný P, Ivančo J, Horáková A (1998) Groundwater table levels regime of Ondava Lowland during the period 1962–1994. Acta facultatis studiorum humanitatis et naturae Universitatis Prešoviensis, Prírodné vedy XXIX – Folia geographica (1):341–362 (in Slovak, English Summary)

Šútor J (1985) Hydrologické aspekty nenasýtenej zóny v podmienkach VSN. Vodohospodársky Časopis 33:458–467

Šútor J, Ivančo J, Kupčo M (1995) Voda v zóne aerácie. In: Šútor J (ed) Hydrológia Východoslovenskej nížiny, 1. vyd., Media Group, Michalovce, pp 281–392 (in Slovak, English Summary)

Vanko J (1988) A rectified map of recent vertical surface movements in the West Carpathians in Slovakia. J Geodyn 10(2–4):147–155. https://doi.org/10.1016/0264-3707(88)90021-X

Vass D (2002) Lithostratigraphy of Western Carpathians: Neogene and Buda Paleogene. SGIDŠ, Bratislava 202 p. (in Slovak, English Summary)

Vass D, Elečko M, Janočko J, Karoli S, Pereszlenyi M, Slávik J, Kaličiak M (2000) Paleogeography of the East-Slovakian Basin. Slovak Geological Magazine 6(4):377–407

Vilček J (2004) Geography of the East Slovakian Lowland Farmland. Acta facultatis studiorum humanitatis et naturae Universitatis Prešoviensis, Prírodné vedy XLII – Folia geographica (7):220–246 (in Slovak, English Summary)

Vojtko R, Petro Ľ, Benová A, Bóna J, Hók J (2012) Neotectonic evolution of the northern Laborec drainage basin (northeastern part of Slovakia). Geomorphology 138(1):276–294. https://doi.org/10.1016/j.geomorph.2011.09.012

Dušan Barabas is an Assistant Professor at the Institute of Geography, Faculty of Natural Sciences, Pavol Jozef Šafárik University in Košice. He focuses on the landscape water balance, climate change, fluvial geomorphology and landscape ecology. In the past, he focused on the evaluation of erosion and the effectiveness of hydromelioration in the East Slovakian Lowland.

Ján Bóna is a research assistant at the Institute of Geography, Faculty of Natural Sciences, Pavol Jozef Šafárik University in Košice. His research concentrates on the geology of the Carpathian Flysch belt with the current interest in (neo)tectonics and identification of tectonic structures from lidar data by geomorphometric analysis.

A Unique Braided-Wandering River in Slovakia: Recent Development and Future of the Belá River

15

Anna Kidová, Milan Lehotský, Miloš Rusnák, and Peter Labaš

Abstract

The Belá River represents a reference braided-wandering river system to observe natural, or semi-natural, forms and processes in the mountain environment. In this chapter, human impact and spatio-temporal bio-morphological evolution of the Belá River from the second half of the twentieth century are described. For the high-energy Belá River, new gravel bar formation as well as their re-formation due to frequent channel avulsion is typical. An overall trend of simplification of the braided and wandering river planform, narrowing of the river active zone and vegetation succession on in-channel landforms with an increased island area was registered. However, in-channel landforms and processes typical for the braided rivers are still prevailing in some river reaches, and these represent unique natural entities. Additionally, the Belá River is included in the network of protected areas Natura 2000, with habitats of European importance.

Keywords

Braided river • Morphology • River active zone • Floods • Human impact • Vegetation dynamics

A. Kidová (✉) · M. Lehotský · M. Rusnák · P. Labaš
Department of Physical Geography, Geomorphology and Natural Hazards, Institute of Geography of the Slovak Academy of Sciences, Štefánikova 49, 814 73 Bratislava, Slovakia
e-mail: geogkido@savba.sk

M. Lehotský
e-mail: geogleho@savba.sk

M. Rusnák
e-mail: geogmilo@savba.sk

P. Labaš
e-mail: geoglaba@savba.sk

15.1 Introduction

Natural processes and processes caused by human activity affect river corridors and can act in isolation, but more often they do so simultaneously, in mutually reinforcing interactions. The disruption of the original structure of the river systems is accompanied by a causal chaining effect that permanently weakens their basic ecological functions. Ongoing processes in the river system can be seen as changes at two levels, at the floodplain and river channel level. High-energy braided rivers are characterized by frequent changes in the riverbed position. It can be stated that lateral channel migration is a measure of the dynamics of processes taking place in braided and wandering river systems. The morphological changes are caused by disruption of the river banks and bars by erosion processes (Charlton 2008). At the same time, during channel migration, numerous bars are formed, and abandoned channels are filled with river sediment (Nordseth 1973). Channel migration or shifting within the river active zone according to Bertoldi et al. (2009) represents a specific change in the position, size and arrangement of an individual part (channel element) of a river system over time while maintaining a constant number of channels. High-energy multi-thread river systems often re-occupy abandoned channels, where bar formation or transformation is supported in the river active zone. Channel avulsion such as relocation or shifting of the channel to the lower part of the floodplain is considered to be a sign of lateral channel instability (Schumm 1985) and exceedance of the equilibrium thresholds. Natural processes mostly operate within the range of states of dynamic equilibrium of the river system and have some self-regulating potential for restoration and return to the original form, while changes caused by human activity very often require revitalizing respective renaturation measures. Over the last decades to hundreds of years, river dynamics in many fluvial systems worldwide have been greatly influenced by human intervention in the form of land-use change and urbanization (Surian and Rinaldi 2003).

M. Lehotský and M. Boltižiar (eds.), *Landscapes and Landforms of Slovakia*, World Geomorphological Landscapes,
https://doi.org/10.1007/978-3-030-89293-7_15

Understanding these changes requires data collection to explain local geographic and historical impacts. When disrupting the natural development of a river active zone as a result of anthropogenic influences, there is a certain sequence of morphological changes in the river channel in the process of adaptation to changed conditions, according to which we can define its current state of development and predict its development in the future (Halaj 2004). In terms of hydromorphological evaluation, the Water Framework Directive proclaims indicators and significance criteria for their impact (Pedersen et al. 2004), which interpret the disturbance of the lateral and longitudinal continuity of a river system. The study of geomorphological processes has an irreplaceable role in such an evaluation, including the determination of the current state of the river system. It needs to be acknowledged that landscape systems do not evolve gradually and linearly, but non-linearly, i.e. through fluctuations manifested by new system properties (emergencies) causing the system to go into a new state. The theory of complexity (Nicolis and Prigogin 1977; Jantsch 1980) points to a higher level of organization, with “special attractors” as influences and impulses playing an important role.

Several authors pointed to changes in river behaviour at the end of the nineteenth century, associated with the end of the cooler period, a decrease in the humidity and intensity of meteorological and hydrological processes, and thus to a change in morphological processes in the river systems (Kotarba 1989; Surian 1999; Liébault and Piégay 2002). Bauch and Hickin (2011) emphasized the correlation between climate change, hydrological regime, and geomorphology of the river channels. The increase in the size and duration of floods is a major factor in the acceleration of riverbed changes. Pekarová et al. (2003, 2010) pointed to the cyclical nature of flood discharges, their occurrence in periods with more intense floods. Gurnell and Petts (2002) pointed to the shift of several river systems in Europe from gravel-bed, braided and wandering rivers to a stable single-thread channel as a result of climate change (the end of the Little Ice Age) and intensive human intervention in the river landscape. Wyżga (1991, 1993, 1996, 2001), Lach and Wyżga (2002), and Zawiejska and Wyżga (2010) explained riverbed degradation and channel incision in the Polish Carpathians as a consequence of river basin reforestation, river channel engineering, and gravel mining. In the North-Eastern Italy, there is a visible trend of narrowing and incision of the originally wide braided rivers and its conversion into a wandering or single-thread channel due to anthropogenic interventions and river sediment extraction (Surian et al. 2009; Rinaldi et al. 2005; Surian and Rinaldi 2003). In France (Liébault and Piégay 2002), the transformation of rivers (the conversion trend from braided rivers through wandering to meandering ones accompanied by narrowing of the riverbed) is explained by climate change at the end of the Little Ice Age when the afforestation of river basins and construction of dams led to sediment input reduction. In the Czech Republic, changes of river systems were identified in the Moravian-Silesian Beskydy (Hradecký 2002, 2007; Hradecký et al. 2012; Škarpich et al. 2012, 2013, 2016; Galia et al. 2012). Škarpich et al. (2013) pointed to significant channel incision and degradation of the former multi-thread Morávka River due to anthropogenic interventions and river regulation, resulting in channel incision rate achieving 8 m over the past 40 years. Škarpich et al. (2016) reported the lithological structure (flysch), anthropogenic impacts (river regulation, construction of stony grade-control structures, dam construction), absence of regular floods as well as acceleration of channel incision by concentration of the flow into a narrow and deep channel be the main factors of the Ostravice River transformation.

The occurrence of river systems with braided and wandering pattern in Slovakia in the period 2006–2009 represented approximately 0.2% of the total length of national river network (49,774.8 km, Fig. 15.1). Recognizing the fact that most of the major Slovak rivers (Danube, Váh, Hron, Ondava, Laborec, etc.) had a former multi-thread pattern (documented since 1949 on the first aerial photos of Slovakia), we are entitled to consider the current 95.1 km of such river systems as unique and rare (Kidová and Lehotský 2012). At the same time, they provide opportunities for seeking and interpreting the environmental causes of their changes, which naturally implies not only the need for their research but also their protection. Only several of them, e.g. Mútňanka River, Jakubianka River, Torysa River, Topľa River or Sveržovka River exceed 5 km in their multi-thread pattern length. The longest river (14.3 km) with the multi-thread river pattern is distinctly the Belá River.

15.2 The Belá River as a Study Area

The Belá River represents a unique morphological and ecological water body in Slovakia. Its multi-thread river pattern and typical gravel-bed character represent a near-natural braided and wandering river system (Kidová and Lehotský 2012; Kidová et al. 2016, 2017; Lehotský et al. 2017; Rusnák and Kidová 2018). The Belá River is the largest right tributary of the upper Váh River and flows through the Liptov Basin (Fig. 15.2), whose origin can be dated back to the Sáva orogenesis that followed the Palaeogene. During this period, layers of flysh slate and sandstone were deposited. Through the mountain-forming processes, the strata were deformed into megatinclinals (Fatra Mountains, Tatras Mountains, Chočské vrchy Mountains) and intervening megasynclinals (Liptov and Spiš Basin) formed (Kontriš 1981). The depression of the Liptov Basin is filled with Tertiary flysch layers of cumulative

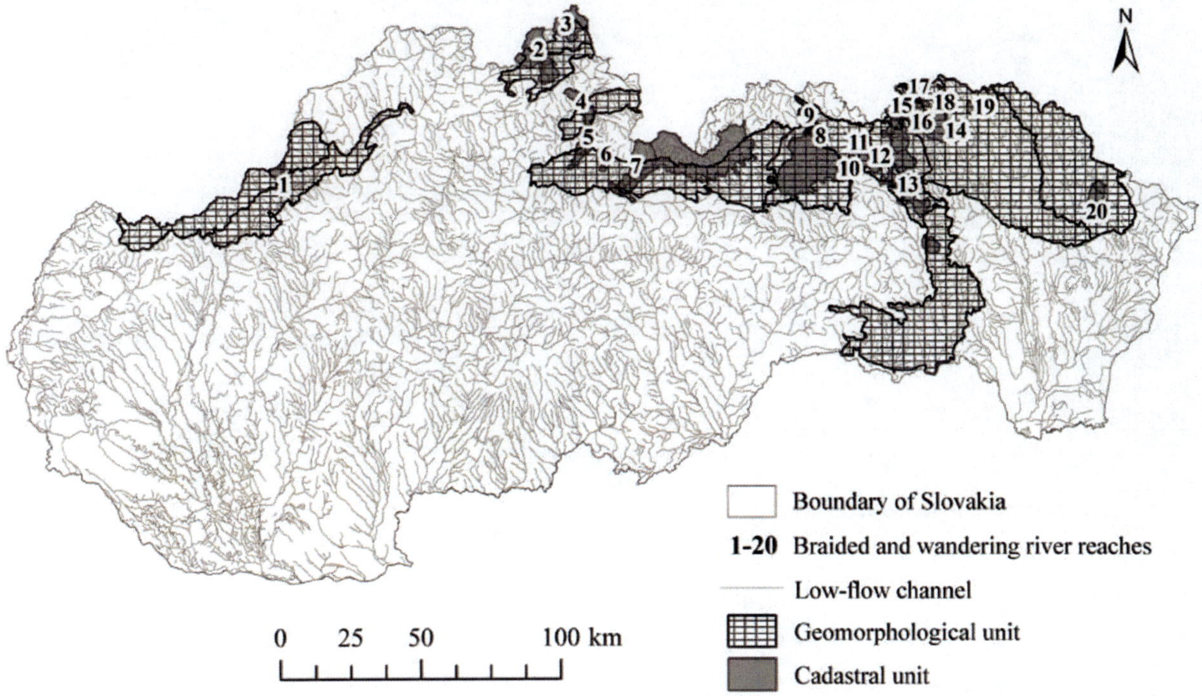

Fig. 15.1 The map illustrates the occurrence of river systems with braided and wandering pattern in Slovakia within its geomorphological (GU) and cadastral (CU) unit (modified according to Kidová and Lehotský 2012). The length of braided and wandering river reaches was identified based on the orthophotomaps of Slovakia in GoogleEarth at a scale of approximately 1:1000 from 2006 and 2009. The criterion for the classification of river sections was the width of the stream greater than 2 m and the exclusion of areas with forest cover. Published with permission of Geografický časopis (Kidová and Lehotský 2012). (Legend: 1—Vlára River in GU Biele Karpaty, Považské Podolie and CU Horné Srnie; 2—Mútňanka River in GU Podbeskydská vrchovina, Podbeskydská brázda and CU Mútne, Novoť, Beňaďovo, Krušetnica, Breza; 3—Suchý creek in GU Podbeskydská brázda and CU Oravská Polhora; 4—Studený creek in GU Skorušinské vrchy and CU Nižná, Oravský Biely Potok; 5—Suchý creek in GU Podtatranská kotlina and CU Kvačany, Liptovské Matiašovce, Liptovská Sielnica; 6—Smrečianka River in GU Podtatranská kotlina and CU Smrečany; 7—Belá River in GU Podtatranská kotlina and CU Podtureň, Liptovský Hrádok, Liptovský Peter, Vavrišovo, Pribylina, Liptovská Kokava, Vysoké Tatry; 8—Jakubianka River in GU Spišsko-šarišské medzihorie, Levočské vrchy and CU Nová Ľubovňa, Jakubany, Javorina; 9—Kolačkovský creek in GU Spišsko-šarišské medzihorie, Levočské vrchy and CU Kolačkov, Nová Ľubovňa; 10—Torysa River in GU Levočské vrchy, Spišsko-šarišské medzihorie, Košická kotlina and CU Tichý potok, Brezovica, Torysa, Krivany, Lipany, Jakubova Voľa, Pečovská Nová Ves, Sabinov, Šarišské Michalany, Ostrovany, Veľký Šariš, Kendice, Petrovany; 11—Kučmanovský creek in GU Spišsko-šarišské medzihorie and CU Šarišské Dravce, Torysa; 12—Lúčanka River in GU Spišsko-šarišské medzihorie and CU Lipany; 13—Ľutinka River in GU Čergov, Spišsko-šarišské medzihorie and CU Pečovská Nová Ves, Ľutina, Olejníkov; 14—Topľa River in GU Čergov, Ondavská vrchovina and CU Gerlachov, Malcov, Lukov, Livov, Bardejov, Komárov, Hrabovec; 15—Kamenec River in GU Ondavská vrchovina and CU Petrová, Gaboltov, Sveržov; 16—Sveržovka River in GU Ondavská vrchovina, Busov and CU Nižný Tvarožec, Sveržov, Tarnov; 17—Oľchovec River in GU Ondavská vrchovina and CUPetrová; 18—Kamenec River in GU Ondavská vrchovina and CU Chmeľová, Zborov, Bardejov; 19—Ondava River in GU Ondavská vrchovina and CU Dubová, Nižný Mirošov; 20—Udava River in GU Laborecká vrchovina and CU Nižná Jablonka, Papín, Zubné)

thickness in excess of 1500 m. Below the flysh are fragments of carbonate nappes and deeper beneath them are Paleozoic crystalline rocks. Within the geotectonic units, the bedrock formation at the bottom of the Liptov Basin is the youngest and least resistant. Faults trending in two directions occur in the vicinity of the basin: the older one's trend East–West and the younger ones follow North–South. The most famous representative of the East–West system is the Podtatranský Fault that stretches along the foot of the High Tatras and its activity is responsible for uplift of the Tatras to their present form. The formation of floodplains and their subsequent re-division to a system of river terraces and alluvial fans is evident in the Quaternary (Kontriš 1981). The original depositional morphology of fans and terraces was remodelled through subsequent fluvial incision. Quaternary sediments in the form of fluvioglacial gravels extend throughout the fore-mountain part of the Belá drainage basin. For the most part, they are non-carbonate. The material of river sediments is generally well sorted (Kontriš 1981).

According to the River Morphology Hierarchical Classification (RMHC) approach (Lehotský 2004), the Belá River is divided into the source, transfer, and the response zone (Fig. 15.3). The Belá River originates as a confluence

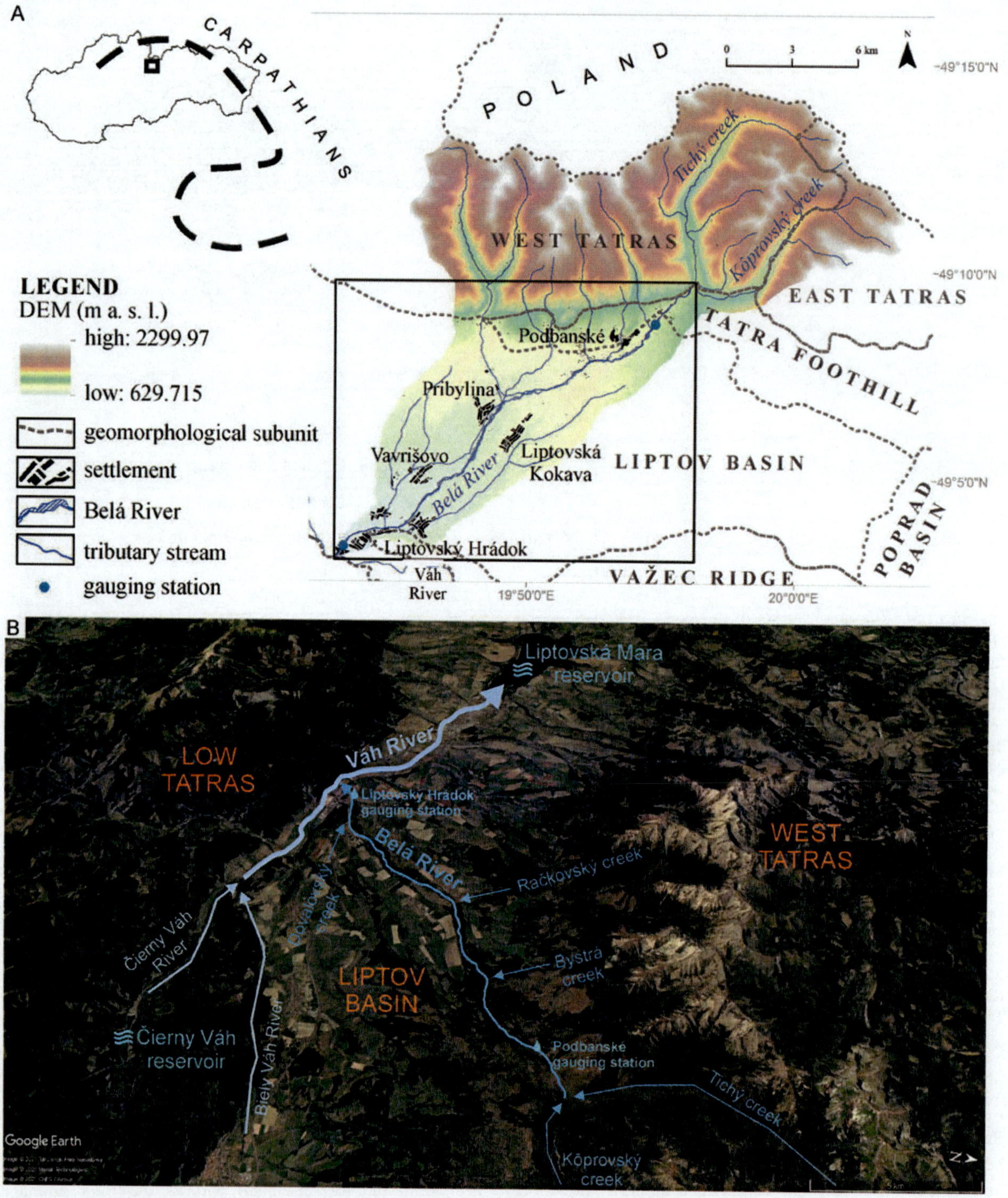

Fig. 15.2 The Belá River location with wider geomorpological division (**a**) and basic river network ordering (**b**). The (**a**) section published and coloured with permission of Acta Scientarium Polonorum (Kidová et al. 2017)

Fig. 15.3 The specific examples of fluvial landforms along the Belá River drainage basin: the source zone - A, the transfer zone - B, and the response zone - C. *Photos* A. Kidová, M. Lehotský

of the Kôprovský creek and Tichý creek in the source zone (A), situated in the High Tatras mountain area. Its higher segments are found in the alpine and subnival environment, characterized by many elements of fossil glacial relief, especially in the form of cirques with several mountain lakes. Geomorphologically, the high-mountain area is characterized by a fluvially dissected highland and moraine and glacial-fluvial piedmont. From a geological point of view, its peaks are built by granitoids and crystalline slates. The low-lying segments of the source zone have the character of a typical glacier valley (trog), with a U-shaped cross-section (Kotarba 2004; Hreško et al. 2005). The bottom of the valleys is usually without floodplain, with channel-bed and bank erosion on some river reaches.

In terms of recent fluvial processes, the transfer zone (B) is the most dynamic. It is represented by the high-mountain, glacial relief, and the fluvially dissected highlands of the Kamenistá, Bystrá, Račková, and Jamnická valleys as a part of the Western Tatras, and the proluvial cones and the proluvial undulate plain (Liptovské nivy) of the Liptov Basin. The presence of the simple-thread, meandering as well as braided-wandering planform of the Belá River is a typical sign for transfer zone. The relief of this zone is a complex of slightly undulating, hilly terrain landforms in Tatras foreland as well as dissected in the mountainous terrain, which is built by granitoids and crystalline slates. The typical flysch of the intra-Carpathian Palaeogene (claystones, siltstones and sandstones) crops out in several places in the Liptov Basin part. However, younger glacifluvial and fluvial sediments dominate (Gross et al. 1979). In the upper part, below the confluence of the Kôprovský and Tichý creeks, there are outcrops of granitoids and crystalline slates of Palaeozoic age. The middle and upper parts of the Belá River from Podbanské to Dovalovo are bordered from the right by the Würm-age terrace, which is connected to proluvial and eluvial-deluvial sediments higher upslope. On the left side, the Belá River is bordered by the Mindel Terrace, with numerous landslides in the flysch lithofacies of a typical intra-Carpathian Palaeogene.

The extent of the response zone (C) is limited to the alluvial fan developed above the confluence of the Belá and Váh River. The massive terraced alluvial fan of Belá spread from Važec to Jamník and Podtureň (Lucerna 1972) overlaid on glacifluvial sediments originating from upper parts of the river catchment. At present, we can identify its right-side border by a Pleistocene Riss terrace created by glacifluvial sediments. Its left-side border, we can find in the interface of basal transgressive lithophace and Hronicum nappe in Liptovský Hrádok settlement. Channel regulation for flood protection purposes is characteristic for the response zone as well.

The hydrological regime is influenced by spring snow melting in higher source zone and heavy rainfall during the summer period. On the Belá River, there are currently two gaging stations (Podbanské and Liptovský Hrádok) and one precipitation station (Liptovský Hrádok). The Belá River reaches an average annual discharge at Podbanské gauging station of 3.5 $m^3\ s^{-1}$ (Majerčáková et al. 2007), and 6.8 $m^3\ s^{-1}$ at the mouth in Liptovský Hrádok gauging station for the period 1964–2006 (Šipikalová 2006), respectively, 6.56 $m^3\ s^{-1}$ for the period 1931–1974 (Hlubocký 1974). Minimum discharges are recorded mainly in winter (February), when precipitation occurs in the form of snow. Flood discharges are related to spring snow melting (May), when they are up to 8–9 times higher than in the months with low discharges (Majerčáková et al. 2007), and to summer storms (June, July) reflecting the mesoclimatic conditions of the river basin. The average annual rainfall is 1590 mm (at 1544 m above sea level) near the Podbanské gauging station. At Liptovský Hrádok precipitation station, the average rainfall achieved is 680–685 mm. From the geomorphological point of view, it is interesting that according to Majerčáková et al. (2007), the hydrological regime of the Belá River has not changed significantly over the last 80 years, i.e. it appears homogeneous (the average annual discharge in the period 1931–1980, it was 3.54 $m^3\ s^{-1}$, and in the second monitored period 1961–2006, it was 3.47 $m^3\ s^{-1}$).

The Belá River together with its riparian zone creates a biocorridor of supra-regional importance and thanks to its habitats of European importance (SKUEV0141), it is included in the Natura 2000 network. The following habitats are subject to protection: floodplain willow-poplar and alder forests, mountain streams and their woody vegetation with *Myricaria germanica* (Fig. 15.4), mountain streams and their woody vegetation with *Salix eleagnos*, hygrophilous

Fig. 15.4 Exposed gravel bars of the Belá River disturbed by frequent flood events preferred by *Myricaria germanica*. *Photo* A. Kidová

marginal communities on floodplains, lowland and piedmont meadows, beech and fir flowering forests as well as animal species of otter (*Lutra lutra*) and common bat (*Myotis myotis*).

15.3 Anthropogenic Impact

Anthropogenic processes associated with channelization cause disturbance of the dynamic structure of in-channel landforms and lead to degradation of river systems. On the Belá River, we can observe anthropogenic interventions in the form of changes in channel direction, which disrupts the hydrodynamic uniformity of the flow, accelerating erosion processes of the bottom and river banks. These interventions also include fragmentation of the channel and its separation from the surrounding area through the construction of bypass channels of small hydropower plants and transverse structures disrupting the connectivity between the channels, the banks and the floodplain. In addition, it leads to reduced sediment transport and disruption of the homogeneity and structure of the river system. Brierley and Fryirs (2005) classified direct changes affecting the river bed character, distinguishing regulation, river sediment extraction, removal of large wood debris and riparian vegetation, construction of dams and reservoirs, water drainage to other systems (e.g. irrigation). Indirect changes are linked to land cover structure changes, techniques applied in agriculture, forestry, urbanization, construction of buildings, infrastructure and navigation systems.

The Belá River was affected by human activity in several places. The river mouth into the Váh River is regulated in the length of 4 km, where the channel is led in the inter-dam area to protect the urban areas of Liptovský Hrádok and Dovalovo. This downstream area is regulated by five stony grade control structures and in the period from 1950 to 2000, more or less 140,000 m^3 of gravel in total was extracted as the flood protection measure for Liptovský Hrádok and Dovalovo settlements and for the protection of road and railway bridges (Kidová et al. 2016). In other parts of the Belá stream from 7th river km, the channel is regulated only at the four bridges over the Belá River and at the river reach where the small hydropower plant (SHP) was built. There are four small hydropower plants between 5,5 and 11,5 river km (one SHP in Vavrišovo settlement and four SHPs in Pribylina settlement), using abandoned side channels of the Belá River.

Kidová (2010) distinguished three specific phases (from 1925 to 2010) in the evolution of anthropogenetic influence on river channel morphology, where six attractors were identified: flood protection measures, agriculture, building-up, leisure activities, mobility of people, and nature protection policy.

The first evolutionary phase (1925–1948) was characterized by water energy utilization as well as the commencement of the flood dam system, thus reducing the area of the river active zone. The urban area of Liptovský Hrádok and other villages near the Belá River were attacked by several significant floods, where the maximum discharge achieved $Q_{max} = 180\ m^3\ s^{-1}$ with recurrence interval (RI) 50 years in Podbanské gauging station in 1934, respectively $Q_{max} = 60–100\ m^3\ s^{-1}$ in 1925, 1930, 1931, 1948 with RI 5–10 years. Due to the protection of this area, human activities were concerned with reducing the degradation of banks by lateral erosion, the admixture of the river gravel in the agricultural soils and channel avulsions in the inundation area. The great energy potential of the river was exploited as early as the first half of the twentieth century for the benefit of those who lived nearby. However, structures (a mill, groynes, a saw) introduced in this period were also aimed at the elimination of unfavourable runoff conditions, which were supposed to fulfil the function of protection of agriculturally utilized landscape. In the State District Archives in Liptovský Mikuláš, five disused water structures are registered in total. The main reasons for anthropogenic interventions in the first period were flood protection measures and improvement of agriculture.

In the second evolutionary phase (1949–1975), channel modifications due to human intervention focused mainly on how to ensure riverbed stability during further floods (RI 10 years in 1949, 1951, RI 50 years in 1958, and RI 5–10 years in 1960, 1965, 1968, 1970, 1973). Changes in the degree of protection of the surrounding area associated with the establishment of the Tatra National Park (in 1948) reflected decreasing human intervention related to river regulation. Among the attractors, flood protection measures, agriculture, building-up and nature protection policy were dominated.

According to the history of measurements of the Slovak Hydro-Metorological Institute (SHMI), the 20-year period from 1976 to 1996 was extremely dry at the Belá River gauging stations, and big floods with RI more than 5 years did not occur. The first major flood in the third evolutionary phase (1976–2010) occurred only in 1997 (RI 7 years). The series of maximum annual flow rates Q_{max}, despite the occurrence of some floods (RI < 5 years) in the Belá River in 2001, 2006, and 2008, had a downward trend. The Belá drainage basin was least affected by human activity in the upstream part of the Podbanské area. For these reasons, the Belá River in this area was, according to Pekárová et al. (2010), particularly suitable for studying the natural hydrological regime. In addition, in 2002, the Belá River was established as an area of European importance within Nature 2000 network (§27 Act No. 543/2002). It is assumed that after 1989 (change of political background), there was also a change in the social way of life in the studied area. Attention was focused on leisure activities and recreation. Flood

Fig. 15.5 The concrete grade control construction as a solution for the decreased groundwater level in 2003. The water marks on bridge pillar indicate the increased water level after channel regulation in 2006 during the third evolutionary phase. *Photos* A. Kidová, M. Lehotský

protection aimed at eliminating flood damage leads to an improvement in the level of sanitary, aesthetic, urban and cultural requirements of the population. A decrease in the depth of groundwater level necessitated the regulation of an incised channel beneath the bridge between the settlements of Liptovský Peter and Liptovský Hrádok in 2006, resulting in an increase in groundwater reserves by building a grade control structure on the river bottom, whereby the morphological diversity of the channel was lost and the hydraulic flow cross-section changed (Fig. 15.5). The exploitation of water energy potential remains attractive, as proved by active operation of four SHPs. In this phase, flood-protection measures, building-up, leisure activities, mobility of people, and nature protection policy are considered as the main attractors.

The period from 2011 to 2020 can be additionally distinguished as the fourth evolutionary phase. From the hydrological point of view, flood events with RI < 5 years are typical for this period, except floods in 2014 and 2018, with 5–10 years RI. After the flood event during summer 2018 (RI 5–7 years), gravel extraction (46,600 m^3 planned in total), relocation of river sediment (24,000 m^3 planned in total), and redirection of the channel course of the Belá River were carried out within the river training (Kidová et al. 2021). Specific morphological processes typical for the multi-thread Belá River were suppressed to a minimum on several river reaches. The new 2 m high artificial banks created by heavy machinery caused a very recent isolation of the floodplain from low-flow channels (Radecki-Pawlik et al. 2019). Although the Belá River belongs to the protection area of Natura 2000, these river training represent the only ones carried out to this extent, except of river regulation within the residential area in the third phase. Thus, the fourth phase could be undoubtedly characterized by contradiction between river training measures and nature protection policy.

15.4 Recent Bio-morphological Evolution of the Belá River's Active Zone

The recent evolution of the Belá River is linked to floodplain creation, i.e. accumulation (accretion) of the sediments and floodplain transformation, i.e. channel lateral migration, avulsions, and channel widening, respectively. To investigate spatio-temporal geomorphic changes resulting from these recent processes, the Post-flood Period Serial Geomorphic Analysis (POPSEGA) approach developed by Kidová et al. (2016) was applied. The whole river active zone was divided into 227 river segments, 100 m long each, analyzed in seven-time horizons for the study period 1949–2009. The decreasing trend in geomorphic diversity and the results of erosion–deposition index (comparison of the areas of erosion and deposition in two successive time series of floods) calculation reveal that the contraction phase (channel narrowing, straightening, incision, mid-channel bar stabilization, island development) currently prevails for the whole river active zone of the Belá River. It is linked with the decreasing long-temporal trend in the magnitude of flood events occurrence (from 1974 with prevailing only with RI 2–5 years) as well as with anthropogenic interventions (flood protection, gravel mining) and reduction in catchment sediment supply due to expansion of forest cover in headwater river reaches.

Despite the general tendency to a decrease in braiding intensity, the opposite process (bank erosion, avulsion, chute cut-offs, lateral and vertical accretion), i.e. expansion phase, in the downstream river reaches was registered. The node density analyses confirmed the very highly dynamic cores (mid-channel forms change constantly under the domination of braiding processes) in the downstream river reaches. Although there is a continuation of braiding processes in some river reaches, the progressive reduction of braidplain width (by about 44% from 1949 to 2009) shows that the Belá River is in a threshold phase, characterized by channel narrowing and incision and increase in island area (Fig. 15.6). The transformation from the type with prevailing bars (braided) to the one typified by bars and islands (braided-wandering) was recognized as an outcome of the study (Fig. 15.7).

The distribution of the observed in-channel landforms (perennial channels, lateral bars, mid-channel bars, islands), which form the channel platform monitored by Kidová et al. (2017), led to the compilation of the spatio-temporal matrix of channel planform types (Fig. 15.8). It includes the channel planform typology based on the number of perennial channels, islands, and mid-channel bars. It consists of single-thread (S), wandering (W), and braided (B) channel planform. The matrix accounts for longitudinal channel planform variability within one-time horizon as well as for channel planform temporal variability of the individual channel segments during the whole study period. According to the authors' findings, the largest proportion of the braided pattern with a well-developed wandering one was identified in 1949, due to high magnitude flood event in 1934 (RI 50 years). In 1958, the next extreme flood event caused probably many channel avulsions, which decreased the number of channel segments with braided pattern and the wandering pattern prevailed. Transformation of the single-thread channel planform to the multi-thread one was observed as well. The Belá River maintained its multi-thread channel planform in 1973. The in-channel landform stabilization as well as channel narrowing between years 1973 and 1992 led to the predominant single-thread channel planform. In 2003, the number of channel segments with the multi-thread channel planform increased due to a flood event with RI 7 years that occurred in 1997. Vice-versa, the mid-channel bar stabilization and their transformation into islands represented the last study horizon (2009). The results confirmed the findings in Kidová et al. (2016) where the simplification of the Belá River′s braided pattern was declared. These conclusions were also confirmed by the application of the original methodology of Postflood Period Sediment Connectivity Assessment (POPSECA) on the connectivity of the coarse sediments of the Belá River (Lehotský et al. 2018). Eight types of potential functional connectivity were identified by interpreting balance connectivity indices at the floodplain–channel, channel–bar, and bar–bar levels. By linear trend analysis of the Integral Connectivity Index (IIC) and the flood periods, it was found that all river reaches show a decreasing trend of IIC values, i.e. a decrease in transport of coarse sediments and thus a decrease in the formation of bars and reduced geodiversity.

The vegetated patches within the river active zone (gravel bars and islands) with a different state of succession phases affect the morphological channel change (stabilization/rejuvenation). The bank and in-channel landforms stabilization by vegetation support the planform simplification process. The size and structure of the vegetated in-channel landform react dynamically by positive feedback to flow discharge changes and directly affect the local variability of sedimentation rate as well as the in-channel landform formation. These reciprocal processes markedly affect the multi-thread planform of the Belá River. Clarification of the role of vegetation cover evolution is essential for detailed detection of vegetation on in-channel forms and was attempted using Braun-Blanquet (1921) approach, based on both abundance and dominance of plant species. An assessment of vegetated patches from remote sensed data (1949–2009) of the Belá River was based on a combined scale of coverage and abundance. Bar and island areas involved in analyses were evaluated in terms of

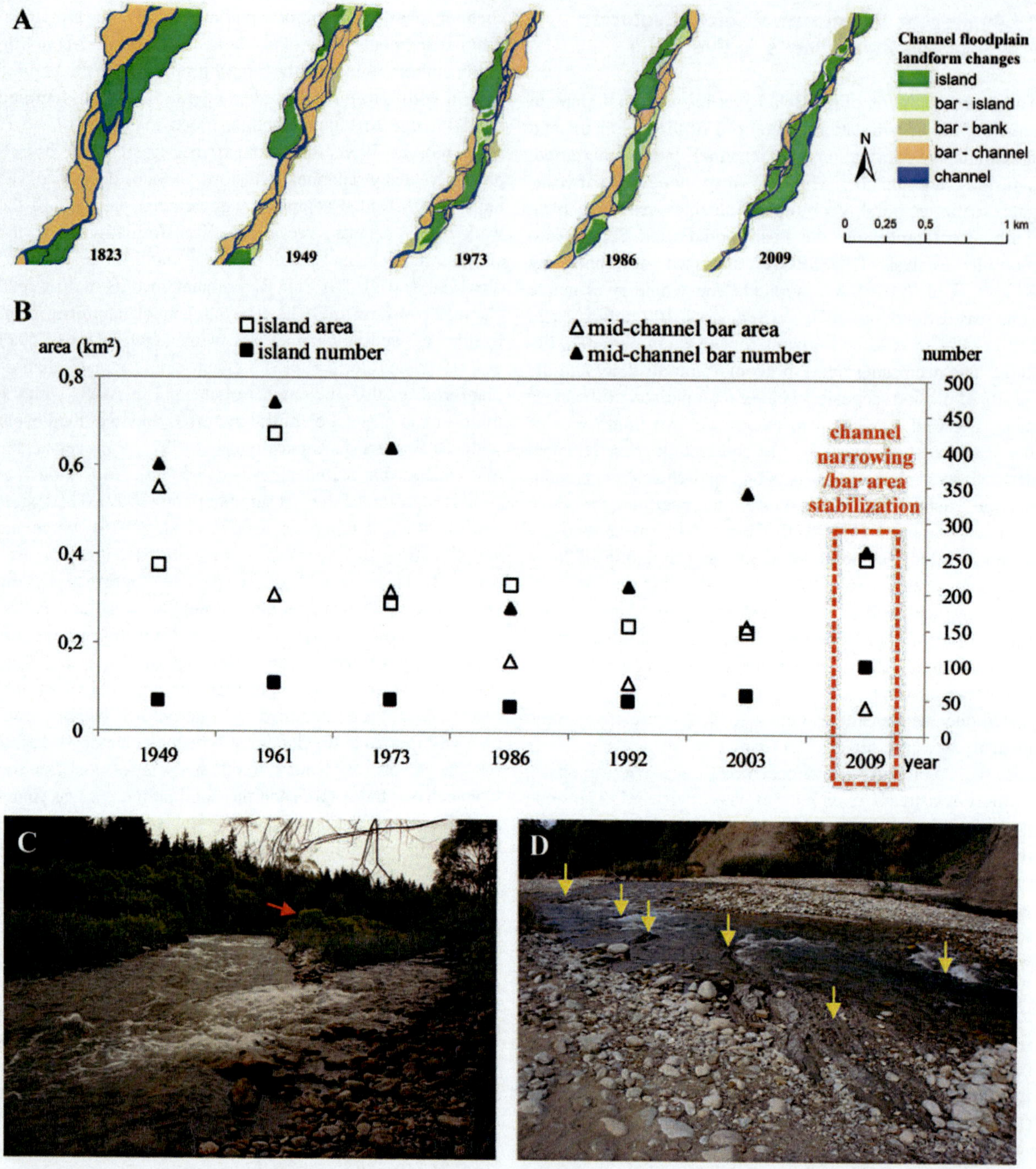

Fig. 15.6 In-channel landform changes from 1823 to 2009 within the river active zone of the Belá River (**a**). Differences between island and mid-channel bar number/area changes during the timespan 1949–2009, where islands represent the most stable in-channel landforms, while the mid-channel bars represent the most unstable ones, are presented on (**b**) section. The *Salix eleagnos* mixed up with *Myricaria germanica* self-sowing in front of the mid-channel bar. Behind the shrub, very well established *Salix* trees (marked with a red arrow) indicate ongoing mid-channel bar stabilization, and its gradual modification to the island (**c**). Channel incision due to progressing backward erosion has created in-channel flysh outcrops, mainly in the downstream river reaches of the Belá River (**d**). *Photo* A. Kidová

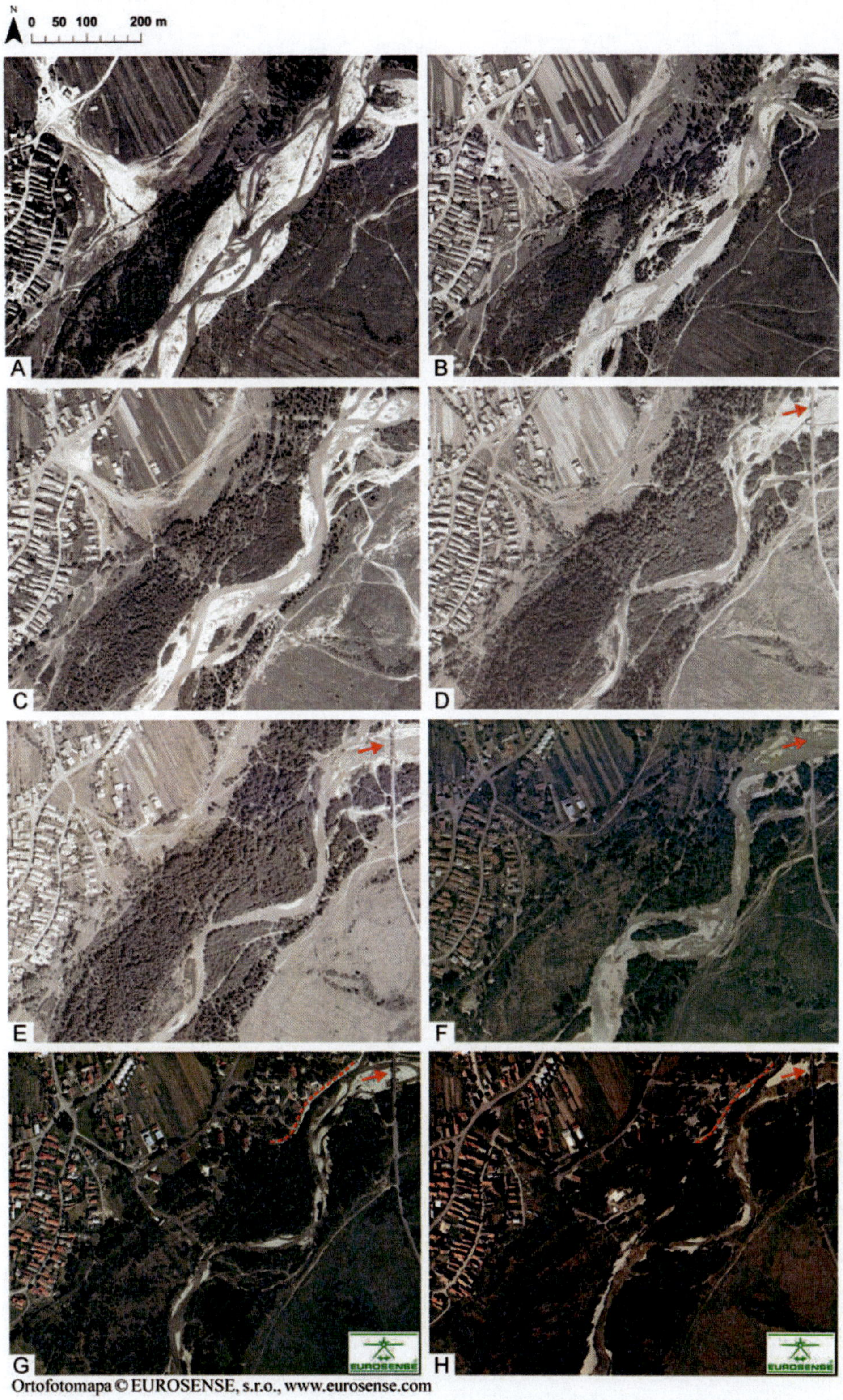

Fig. 15.7 The decreasing trend of the braidplain area from 1949 to 2018 (A: 1949; B: 1961; C: 1973; D: 1986; E: 1992; F: 2003; G: 2012; H: 2018) represented on the river reach near the Pribylina settlement. The aspect of the infrastructure development (the bridge construction in the upper left corner of the image indicated with a red arrow from D to H) in interaction with riverbank strengthening (red dashed line in G and H) due to expansion of the built-up area, stabilized this river reach in the lateral direction. *Source* BW aerial images Topographic Institute in Banská Bystrica; coloured orthophoto ©EUROSENSE, s.r.o.

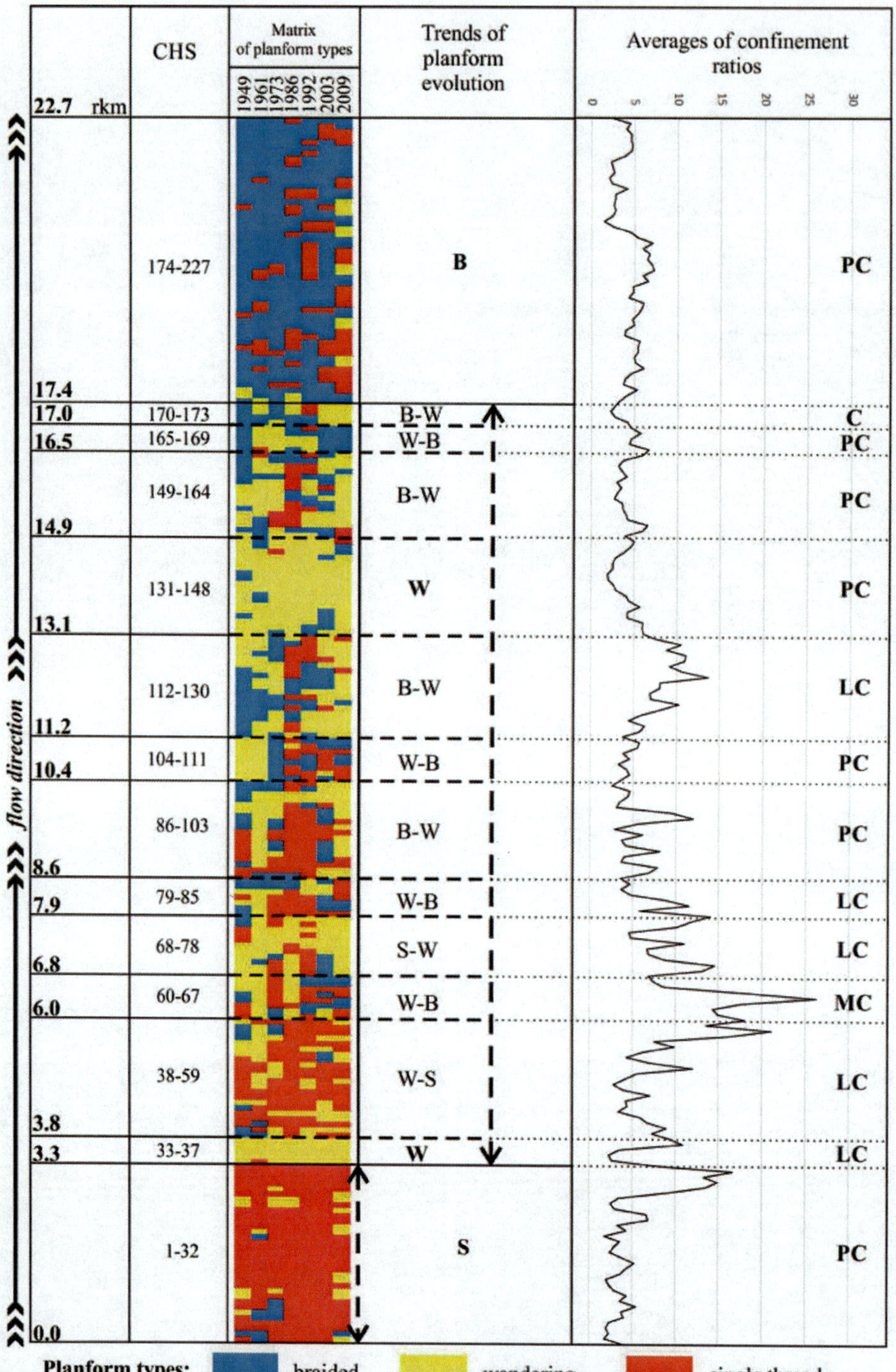

Fig. 15.8 Spatio-temporal matrix of channel planform types, evolutionary trends and averages of valley confinement ratios of the Belá River in study period 1949–2009 (Kidová et al. 2017). The matrix presents the planform type arrangement within the frame of the whole river length in one-time horizon as well as the channel planform within the frame of one channel segment (CHS) in each time horizon. Diversity of planform evolutionary trends allows one to classify CHS into seven categories. Their relation to the valley setting is illustrated by averages of the confinement ratio. Published with permission of Acta Scientarium Polonorum

Table 15.1 Type and numerical code of vegetation patches on individual in-channel landforms in ArcGIS environment

Morphological type	Number code	Vegetation type	Characteristics	Number code	Final type/code
Island	10	xxx	>90% tree vegetation	0	10
Lateral bar (island attached)	20	No vegetation	Vegetation cover < 10%	1	21
		Sparse vegetation	Herb and shrub vegetation < 50%	2	22
		Dense vegetation	Herb and shrub vegetation > 50%	3	23
Lateral bar (bank attached)	30	No vegetation	Vegetation cover < 10%	1	31
		Sparse vegetation	Herb and shrub vegetation < 50%	2	32
		Dense vegetation	Herb and shrub vegetation > 50%	3	33
Mid-channel bar (low flow channel)	40	No vegetation	Vegetation cover < 10%	1	41
		Sparse vegetation	Herb and shrub vegetation < 50%	2	42
		Dense vegetation	Herb and shrub vegetation > 50%	3	43
Low-flow channel	99	xxx	xxx	99	99

changes in their size. A classification key was designed to identify the vegetation structure (Table 15.1).

The areal representation of vegetation types on the Belá River for individual categories of in-channel landforms is quite diverse and varies from year to year. Island-attached lateral bars have the smallest proportion of areas (bars) without vegetation (with vegetation cover up to 10%) during the whole study period. However, this is not surprising, given the natural vegetation succession from islands to their close surroundings. A similar trend is observed for bank-attached lateral bars, where the formation of new bars without a vegetation cover is characteristic for river reaches with the single-thread planform with a sinuous channel. On the opposite side of such river bends (upstream river reaches) point bar formation occurred, but from the point of whole river length, they represent only a small part of this type of bars. The dominant bank-attached lateral bars (code 30) originated mainly due to avulsion processes. This bar type is less influenced by flow discharge changes, compared with the mid-channel bars. This is evidenced by their prevailing total area with a vegetation cover above 50% (code 33), which dominates throughout the whole study period. The most exposed in-channel landforms within the river active zone are mid-channel bars (code 40), whose stabilization by vegetation in multi-thread river systems rarely has a long-term character.

In the case of the Belá River, the area of mid-channel bars without vegetation (code 41) is significantly represented mainly in the period 1949–1973, which was characterized by several extreme floods (in 1948, 1958, 1960s, 1973). After 1973, another large flood (7 years RI) was recorded only in 1997, i.e. 24 years later. During this period, stabilization of the mid-channel bars with vegetation up to 50% (code 42) was registered. In 1992, the mid-channel bar area with vegetation above 50% prevailed. After a flood event with lower magnitude (2–5 years RI in 2001), the mid-channel bars were flooded and remodelled, whereas most of the developed vegetation cover was destroyed (situation in 2003). The area of the mid-channel bars with all three types of vegetation was equalized. During this period, morphologically most significant changes as the channel incision into bedrock and narrowing of the river active zone occurred. These changes resulted in a decrease in the area of mid-channel bars with all types of vegetation cover. On the other hand, the significant prevailing of the island and bank attached lateral bars with vegetation cover above 50% in 2009 was registered.

Both the area and the number of bars with individual types of vegetation cover were analyzed. The number of island-attached lateral bars (21) and mid-channel bars (41) without vegetation (with vegetation up to 10%) prevailed (among the individual types of vegetation cover for 21, 22, 23 and for 41, 42, 43, respectively) during the whole study period 1949–2009. As in other monitored morphometric parameters, the number of bars without vegetation decreased due to stable flow discharges in 1986–1992. A more balanced number of individual types of vegetation cover are observed on the bank-attached lateral bars (code 31, 32, and 33). As expected, the number of bars with vegetation cover above 50% increased during the study

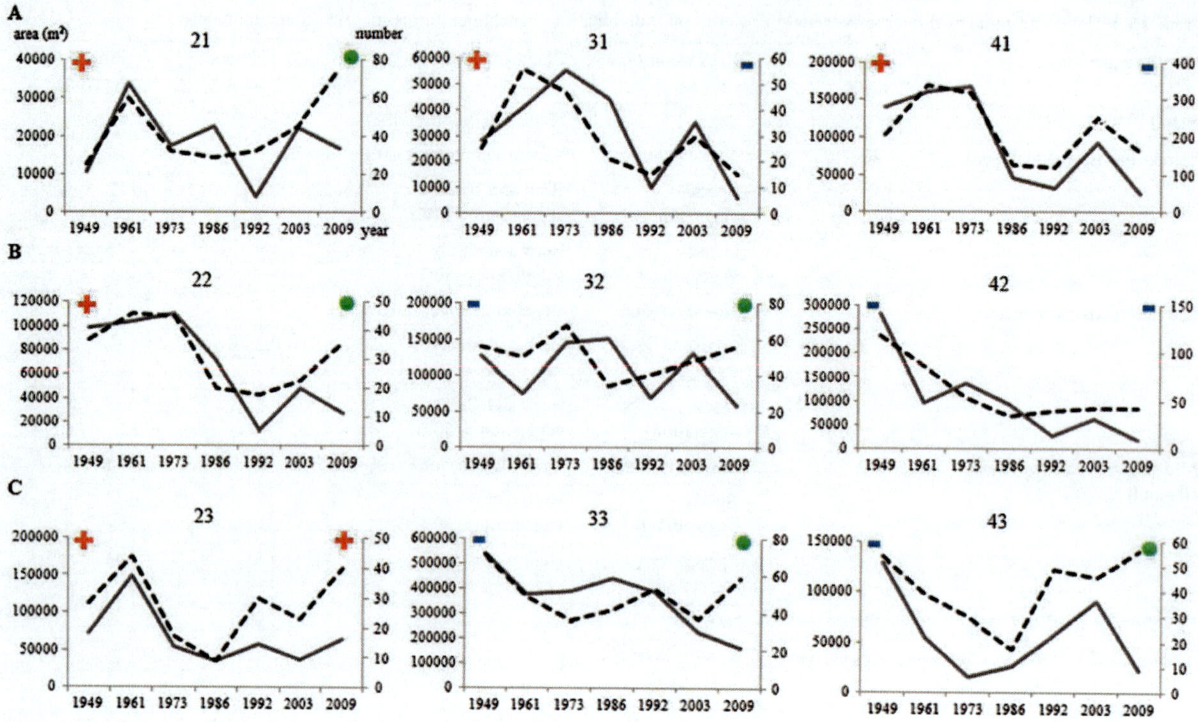

Fig. 15.9 The interdependence of area (full line, m^2) and number (dashed line) of individual categories of three monitored in-channel landforms: island-attached lateral bar (20); bank-attached lateral bar (30); mid-channel bar (40) with vegetation cover <50% (01, A), with herb and shrub vegetation <50% (02, B), with herb and shrub vegetation >50% (03, C) in the study period 1949–2009. Indicated symbols (+, −, •) refer to trends in the first and last period of observation

period. On the contrary, the dominant number of mid-channel bars without vegetation (code 41) shows the presence of ongoing braiding processes.

However, the question remains, which conditions favour changes in the river bed morphology, whether a smaller number of bars with a larger area or a larger number of bars with smaller area within the river active zone? We try to find the answer in the graphs of the interdependence between the number and area of the three types of vegetation cover on the individual in-channel landforms (Fig. 15.9). In the mutual variability of these two parameters (the area and the number), it is possible to find a certain similarity in the progress of individual curves. The same beginning in 1949, which have a rising (+) (codes 21, 31, 41, 22, 23) or a decreasing (−) character (codes 32, 42, 33, 44), we registered for the curves of the number and area. As a result of flood events in the 1950s and 1960s, the arrangement of in-channel landforms has changed, which is reflected in their number as well as in the size of their areas. The curves for all forms with individual types of vegetation are sinusoids. The exception is the mid-channel bar with vegetation up to 50%, whose dependence curves have a decreasing trend over the whole study period. Vegetation on island-attached lateral bars in the period 1961–1973 decreases uniformly for 21, 23 and increases for 22. The mutual deviation of the direction of the curves occurs in 1986–1992. The vegetation cover of the longitudinal benches near the islands increases and their area is shrinking. A similar variability in this period was observed for the vegetation cover on island-attached lateral bars and mid-channel bars. The trend of the number and area curves is towards equilibrium in 2003 for almost all the categories. Smaller flood events (2–5 years RI in 2001 and 2008) probably influenced the morphology of the Belá River in a much more specific way than the larger floods in the past, resulting in the ending of the dependency curves in 2009. The curves are representing decreasing for both area and number (code 31, 41, 42), mutual ascending (code 23) or increasing numbers and decreasing area (code 21, 22, 32, 33, 43) what reflects the fact that these landforms occur more often in the channel but have a smaller area (•). Each vegetated bar area is differently transformed by the internal dynamics of the river, developing and responding to changing hydrological and climatic conditions in the river basin in a different way.

The downward trend of vegetated areas up to 10% and 50% on mid-channel bars and the expansion of small areas with vegetation cover above 50% in terms of the stabilizing function of vegetation in the active river zone have a

Table 15.2 Identification of the driving forces influencing the Belá River as a non-linear river system

Force	Form	Process/Influence
Human disturbances	Flood protection dikes and bank stabilization	Active river zone reduction, channel narrowing
	Stony grade control structure	Groundwater depth changing, loss of the channel morphological segmentation, change of discharge profile, and sediment connectivity
	Gravel mining	Bed erosion processes acceleration, channel incision to bedrock, abort dynamic structure of fluvial landforms, channel narrowing and straightening, island area increasing, loss ability for further lateral bank erosion
	Small hydropower plant	Channel fragmentation, disrupting of the flow continuity, and hydrodynamic uniformity
	Forest landcover change	Erosion base descending, sediment input reduction from source zone and tributaries
Environmental conditions	Physical-geographical	Slope, valley confinement
	Tectonic	Holocene, Würm
	Litologic	Sandstone, claystone, conglomerate
	Hydrologic-climatic	Maximum annual discharge
Inner dynamics of the river	Discharge	Average daily discharge
	Sediments	Sediment size and source, sediment base thickness, fluvial landforms formation
	Vegetation structure	Flow and velocity of sedimentation rates, in-channel landform stabilization

negative impact on the Belá River. The predominant areas with vegetation above 50% indicate degradation of the braided planform and the development of the wandering one. Generally, a change in the vegetation cover on bars primary reflects the internal response of the river system to changes caused by erosion-accretion processes (such as channel incision into bedrock). However, the variability in vegetation cover changes and its stabilization function on the Belá River in the longer term regulate the dynamics of channel transformation.

15.5 Notes on River Management

The findings of the bio-morphological development of the braided-wandering Belá River over the last 60 years, as well as changes in environmental conditions and human disturbances, clarify the driving forces (Table 15.2) that influence the inner dynamics of morphological changes and even help to predict the evolutionary trend. The results have clearly shown a decreasing trend in the area of the river active zone as well as in the magnitude of floods with a maximum of 5–10 years recurrence interval. It also points to the long-term degradation of the river system in the form of simplifying the braided-wandering planform to the wandering-meandering one, increasing island area, narrowing due to channel incision and vegetation succession, as well as channel straightening. Although it is possible to characterize the Belá River as being in a threshold state of metastable equilibrium, it can be expected that the system will continue to be simplified.

Preservation of the natural characteristics of riverbed morphology is closely related to the preservation of its ecological importance and function. Thus, one of the most important input variables in optimizing river management is accepting Water Framework Directive (2000/60/CE) and preventing further irreversible interventions in the river active zone. The understanding of this dynamic non-linear river system, where the lateral/ vertical erosion, bar (re)formation and avulsions are part of autoregulation mechanism of the river, requires a specific attention in decision sphere. Preference given to nature-based solutions and integrated management approach in line with Natura 2000 management recommendation is desired.

Acknowledgements The research was supported by the Science Grant Agency (VEGA) of the Ministry of Education of the Slovak Republic and the Slovak Academy of Sciences; 02/0086/21.

References

Act No. 543/2002 Z. z. Zákon o Ochrane Prírody a Krajiny. https://www.zakonypreludi.sk/zz/2002-543. Accessed 10 Mar 2021

Bauch GD, Hickin EJ (2011) Rate of floodplain reworking in response to increasing storm-induced floods, Squamish River, south-western British Columbia, Canada. Earth Surf Proc Land 36:872–884. https://doi.org/10.1002/esp.2115

Bertoldi W, Zanoni L, Tubino M (2009) Planform dynamics of braided streams. Earth Surf Proc Land 34(4):547–557. https://doi.org/10.1002/esp.1755

Braun-blanquet J (1921) Principien einer Systematik der Pflanzengesellschaften auf floristischer Grundlage. Jb. St. Gall. Naturw. Ges. 57

Brierley GJ, Fryirs KA (2005) Geomorphology and river management. In: Applications of the river styles framework. Blackwell Publishing, Malden, 398 p.

Galia T, Škarpich V, Hradecký J (2012) Dnový transport sedimentů v souvislosti s transformací geomorfologického režimu štěrkonosných toků Moravskoslezskych Beskyd. Geografie 117(1):95–109

Gross P, Vaškovský I, Halouzka R (1979) Geologická mapa Liptovskej kotliny (1:50 000). GÚDŠ, Bratislava.

Gurnell AM, Petts GE (2002) Island-dominated landscapes of large floodplain rivers, a European perspective. Freshwater Biol 47:581–600. https://doi.org/10.1046/j.1365-2427.2002.00923.x

Halaj P (2004) Revitalizácia vodných tokov. Slovenská poľnohospodárska univerzita v Nitre, 200 p.

Hlubocký B (1974) Reprezentatívne povodie Belej – predbežné vyhodnotenie režimu povrchových vôd. Slovenský hydrometeorologický ústav, Bratislava, 79 p.

Hradecký J (2002) Hodnotení časových změn morfodynamiky beskydských toků za využití historických map a leteckých snímků. Geomorphologia Slovaca 2(2):31–39

Hradecký J (2007) Štěrkonosný tok Morávka – vývojové trendy korytové morfologie a revitalizační opatření. In: Hradecký J, Pánek T (eds) Geomorfologický sborník 6, Ostrava (Ostravská univerzita v Ostravě), pp. 65–67

Hradecký J, Škarpich V, Galia T, Dušek R (2012) Complex transformation of the channels in the foothills of the Moravskoslezské Beskydy Mts. In: Blahůt J, Klimeš J, Štěpančíková P, Hartvich F (eds) Geomorfologický sborník 10, Praha (Ústav štruktúry a mechaniky hornín AV ČR), pp. 11–12

Hreško J, Boltižiar M, Bugár G (2005) The present-day development of landforms and landcover in alpine environment—Tatra Mts (Slovakia). Studia Geomorphologica Carpatho-Balcanica 39:23–38

Charlton R (2008) Fundamentals of fluvial geomorphology. Routledge, New York, p 234

Jantsch E (1980) The self-organizing universe: scientific and human implications of the emerging paradigm of evolution, 1st edn. Pergamon Press, Oxford and New York, p 343

Kidová A (2010) Vývoj antropogénneho vplyvu na morfológiu koryta vodného toku - príklad rieky Belej. In: Liga J, Pavlovič V, Strelková M, Ríčiová D, Cápay M (eds) Mladí vedci 2010: zborník vedeckých prác doktorandov, mladých vedeckých a pedagogických pracovníkov. Fakulta prírodných vied Univerzity Konštantína Filozofa, Nitra, pp 485–491

Kidová A, Lehotský M (2012) Časovo-priestorová variabilita morfológie divočiaceho a migrujúceho vodného toku Belá = Spatio-temporal morphological variability of the braided-wandering River Belá. Geografický Časopis 64(4):311–333

Kidová A, Lehotský M, Rusnák M (2016) Geomorphic diversity in the braided-wandering Belá River, Slovak Carpathians, as a response to flood variability and environmental changes. Geomorphology 272:137–149. https://doi.org/10.1016/j.geomorph.2016.01.002

Kidová A, Lehotský M, Rusnák M (2017) Recent channel planform evolution of a braided-wandering river using multitemporal data and GIS (case study of the Belá River, Slovak Carpathians). Acta Scientiarum Polonorum. Formatio Circumiectus 16(1):247–259. https://doi.org/10.15576/ASP.FC/2017.16.1.247

Kidová A, Radecki-pawlik A, Rusnák M, Plesiński K (2021) Hydromorphological evaluation of the river training impact on a multi-thread river system (Belá River, Carpathians, Slovakia). Sci Rep 11:6289. https://doi.org/10.1038/s41598-021-85805-2

Kontriš J (1981) Pôdnoekologické a fytocenologické pomery lužných lesov Liptovskej kotliny. Veda, Bratislava 3(27):164

Kotarba A (1989) On the age of debris flows in the Tatra Mountains. Studia Geomorphologica Carpatho-Balcanica 23:139–152

Kotarba A (2004) Zdarzenia geomorfologiczne w Tatrach Wysokich podczas małej epoki lodowej. Prace Geograficzne IG i PZ PAN 197:9–55

Lach J, Wyżga B (2002) Channel incision and flow increase of the upper Wisłoka River, southern Poland, subsequent to the reafforestation of its catchment. Earth Surf Proc Land 27:445–462. https://doi.org/10.1002/esp.329

Lehotský M (2004) River morphology hierarchical classification (RMHC). Acta Universitatis Carolinae: Geographica 39:33–45

Lehotský M, Rusnák M, Kidová A (2017) Application of remote sensing and the GIS in interpretation of River Geomorphic response to floods. In: Radecki-Pawlik A, Pagliara S, Hradecký J (eds) Open channel hydraulics, river hydraulics structures and fluvial geomorphology, Boca Raton. CRC Press, Taylor & Francis Group, pp 388–399

Lehotský M, Rusnák M, Kidová A, Dudžák J (2018) Multitemporal assessment of coarse sediment connectivity along a braided-wandering river. Land Degrad Dev 29(4):1249–1261. https://doi.org/10.1002/ldr.2870

Liébault F, Piégay H (2002) Causes of 20th century channel narrowing in mountain and piedmont rivers and streams of Southeastern France. Earth Surf Proc Land 27:425–444. https://doi.org/10.1002/esp.328

Lucerna R (ed) (1972) Slovensko II. (Príroda). Kol. autorov, Obzor, Bratislava

Majerčáková O, Škoda P, Danáčová Z (2007) Vývoj vybraných hydrologických a zrážkových charakteristík za obdobia 1961–2000 a 2001–2006 v oblasti Vysokých Tatier. Meteorologický Časopis 10(4):205–210

Nicolis G, Prigogine I (1977) Self-organisation in non-equilibrium systems: from dissipative structures to order through fluctuations. Wiley, 491 p.

Nordseth K (1973) Floodplain construction on a braided river: the islands of Koppangoyene on the River Glomma. Nor Geogr Tidsskr 27:109–126

Pedersen ML, Ovesen NB, Friber GN, Clausen B, Lehotský M, Grešková A (2004) Hydromorphological assessment protokol for the Slovak republic. http://www.shmu.sk/File/implementacia_rsv/twinning/a1_Protocol_final.pdf

Pekárová P, Miklánek P, Pekár J (2003) Spatial and temporal runoff oscillation analysis of the main rivers of the world during the 19th–20th centuries. J Hydrol 274:62–79. https://doi.org/10.1016/S0022-1694(02)00397-9

Pekarová P, Pacl J, Miklánek P, Škoda P, Pekár J (2009) Long-term discharge trend and multi-annual variability analysis of the mountainous Bela River (Slovakia) in 1895–2006. In Herrmann A, Schumann S (eds) Landschaftsökologie und Umweltforschung, 50, Brunschweig, 201–204

Pekárová P, Miklánek P, Pekár J (2010) Možnosti dlhodobej predikcie prietokov slovenských tokov na základe indexu Severoatlantickej oscilácie. Acta Hydrologia Slovaca 11(2):282–290

Radecki-pawlik A, Kidová A, Lehotský M, Rusnák M, Manson R, Radecki-pawlik B (2019) Gravel and boulders mining from mountain stream beds. In: E3S Web of conferences: conference of Civil Enginnering-infrastructure-mining—EDP sciences, vol 106, p 01005. https://doi.org/10.1051/e3sconf/201910601005

Rinaldi M, Wyżga B, Surian N (2005) Sediment mining in alluvial channels: physical effects and management perspectives. River Res Appl 21:805–828. https://doi.org/10.1002/rra.884

Rusnák M, Kidová A (2018) Recent morphological changes of the Slovak rivers Belá and Ondava in Western Carpathians since the second half of the 20th century. Geographia Cassoviensis 12 (2):195–211

Schumm SA (1985) Patterns of alluvial rivers. Annu Rev Earth Pl Sc 13:5–27

Surian N (1999) Channel changes due to river regulation: the case of the Piave River, Italy. Earth Surf Proc Land 24:1135–1151. https://doi.org/10.1002/(SICI)1096-9837(199911)24:12%3c1135::AID-ESP40%3e3.0.CO;2-F

Surian N, Rinaldi M (2003) Morphological response to river engineering and management in alluvial channels in Italy. Geomorphology 50:307–326. https://doi.org/10.1016/S0169-555X(02)00219-2

Surian N, Mao L, Giacomin M, Ziliani L (2009) Morphological effects of different channel-forming discharges in a gravel-bed river. Earth Surf Proc Land 34(8):1093–1107. https://doi.org/10.1002/esp.1798

Šipikalová H (ed) (2006) Spracovanie hydrologických charakteristík za obdobie 1961–2000. Záverečné správy výskumnej úlohy SHMÚ, Slovenský hydrometeorologický ústav, Bratislava

Škarpich V, Galia T, Hradecký J (2012) Podmínky transportu a akumulace sediment v člověkem ovlivněných korytech beskydských toků: příkladová studie soutoku řeky Morávky a Mohelnice. Geologické Výzkumy Na Moravě a Ve Slezsku 19(1–2):53–58. https://doi.org/10.5817/GVMS2012-1-2-053

Škarpich V, Hradecký J, Dušek R (2013) Complex transformation of the geomorphic regime of channels in the forefield of the Moravskoslezské Beskydy Mts.: case study of the Morávka River (Czech Republic). CATENA 111:25–40. https://doi.org/10.1016/j.catena.2013.06.028

Škarpich V, Kašpárek Z, Galia T, Hradecký J (2016) Antropogenní impact a jeho odezva v morfologii koryt beskydských štěrkonosných toků: příkladová studie řeky Ostravice. Česko. Geografie 121 (1):99–120

Wyżga B (1991) Present day downcutting of the Raba River channel and its environmental effects. CATENA 18:551–566. https://doi.org/10.1016/0341-8162(91)90038-Y

Wyżga B (1993) River response to channel regulation: case study of the Raba River, Carpathians, Poland. Earth Surf Proc Land 18:541–556. https://doi.org/10.1002/esp.3290180607

Wyżga B (1996) Changes in the magnitude and transformation of flood waves subsequent to channelization of the Raba River, Polish Carpathians. Earth Surf Proc Land 21:749–763. https://doi.org/10.1002/(SICI)1096-9837(199608)21:8%3c749::AID-ESP675%3e3.0.CO;2-5

Wyżga B (2001) Impact of the channelization-induced incision of the Skawa and Wisloka Rivers, southern Poland, on the conditions of overbank deposition. Regul River 17:85–100. https://doi.org/10.1002/1099-1646(200101/02)17:1%3c85::AID-RRR605%3e3.0.CO;2-U

Zawiejska J, Wyżga B (2010) Twentieth-century channel change on the Dunajec River, southern Poland: patterns, causes and controls. Geomorphology 117:234–246. https://doi.org/10.1016/j.geomorph.2009.01.014

Anna Kidová is a fluvial geomorphologist at the Institute of Geography of the Slovak Academy of Sciences. Her area of research focuses on the behavior and response of river systems to extreme floods and environmental changes, study ongoing hydro-morphological processes, and identification of its driving forces as well as issues related to measures for river management. Her research topics also relate to remote sensing data processing and spatial analysis in GIS. Recently, she is a head of the Department of Physical Geography, Geomorphology and Natural Hazards, and since 2018 she is the president of the Association of Slovak Geomorphologists.

Milan Lehotský is a physical geographer and fluvial geomorphologist at the Institute of Geography of the Slovak Academy of Sciences. He was many years head of the Department of Physical Geography, Geomorphology and Natural Hazards. His research topics are responses of fluvial systems to environmental changes, sedimentological connectivity, evolution trajectories, hydromorphology and GIS and remote sensing applications in rivers and landforms research. He is also working as an external lecturer at the Department of Physical Geography and Geoecology, the Faculty of Natural Sciences of the Comenius University in Bratislava.

Miloš Rusnák is a fluvial geomorphologist at the Institute of Geography of the Slovak Academy of Sciences (Department of Physical Geography, Geomorphology and Natural Hazards). His research topics are fluvial geomorphology, spatial data processing in GIS, UAV data acquisition and processing, fluvial processes and sediment connections in gravel-bed rivers and remote sensing applications in rivers and landforms research. He is the author and co-author of several papers dealing with fluvial system evolution in the Outer Western Carpathians.

Peter Labaš is a Ph.D. student at the Institute of Geography, Slovak Academy of Sciences, graduated from the Comenius University in Bratislava. He is interested in fluvial geomorphology, especially in lateral migration, channel incision, and floodplain vertical discrimination of meandering and braided river systems. He also focuses on question of anthropogenic intervention and river management issues. His point of interest is the Belá and Hornád River.

16 Fluvial Imprints in Flysh Valley Bottoms—Topľa and Ondava Valleys

Miloš Rusnák, Anna Kidová, Milan Lehotský, and Ján Sládek

Abstract

The central part of Ondavská vrchovina upland represents a medium-altitude and moderately dissected relief with narrow elongated valleys. The Topľa and Ondava rivers are the main axis of flysch valleys and their flat bottoms are modelled by dynamic river processes acting upon thin layers of Quaternary sediments. They bear a record of all historical changes of the channel pattern, from a wide wandering gravel-bed wild river to a narrow and sinuous channel. Both rivers are characterized by distinct bank erosion, resulting in bank movement and bank retreat within meander bends. From an economic point of view, erosion of arable land and grassland is a negative consequence of channel migration. Vice versa, from an ecological point of view we can consider bank erosion as a natural process that leads to the increase of geo- and bio-diversity of the riparian landscape. Nowadays, 'green approaches' are applicable in the river management, aiming to eliminate technical interventions in the river basins and allowing for free channel migration. All these processes create valuable ecosystems with natural flood regime and floodplain forest.

Keywords

Flysch • Gravel-bed rivers • The Ondava River • The Topľa River • Bank erosion

M. Rusnák (✉) · A. Kidová · M. Lehotský · J. Sládek
Department of Physical Geography, Geomorphology and Natural Hazards, Institute of Geography, Slovak Academy of Sciences, Štefánikova 49, 814 73 Bratislava, Slovakia
e-mail: geogmilo@savba.sk

A. Kidová
e-mail: geogkido@savba.sk

M. Lehotský
e-mail: geogleho@savba.sk

J. Sládek
e-mail: geogslad@savba.sk

16.1 Introduction

The north-eastern part of Slovakia is formed by the Palaeogene flysch sequence of sedimentary rock (sandstones, claystones and conglomerates), cropping out to the north parts the Klippen Belt. Geological setting of this area conditioned typical upland relief developed upon less or medium-resistant flysch rock formation, moderately to sharply dissected with incised river valleys filled by Quaternary deposits, and reflecting monotonous bedrock lithology. Valleys bottom are reshaped by fluvial processes and create interesting fluvial and cultural landscapes modelled by river systems. River morphology is modified at different spatial taxonomy levels (from basin to channel unit) and sensitive to dynamic events caused by intrinsic variability or external human pressure and flood impact. River behaviour adapts to the complex interaction of erosion and accretion mechanisms (Brierley and Fryirs 2005). Freely migrating and dynamic rivers are strongly affected by changing natural conditions in industrial Europe and altered with increasing human pressure on the landscape. Anthropogenic modification, grade-control structures and channelization resulted in channel narrowing, transformation and incision along many rivers in Europe (Brierley and Mum 1997; James 1997; Rădoane et al. 2013; Scorpio et al. 2018; Škarpich et al. 2013; Wyżga et al. 2016b; Ziliani and Surian 2012). It was pointed out that alteration of sediment supply by gravel extraction (Hajdukiewicz et al. 2017; Korpak 2007; Marston et al. 2003; Surian and Rinaldi 2003; Wyżga et al. 2016a) and afforestation or deforestation (Jefferson and McGee 2013; Liébault et al. 2005; Price and Leigh 2006) are key factors resulting in channel transformation, narrowing and expansion rather than flow regime changes (Comiti et al. 2011). Extraordinary floods and their geomorphological effectiveness are influenced by the actual state of the channel (Gorczyca et al. 2013; Hajdukiewicz et al. 2016; Hooke 2015; Kijowska-Strugała et al. 2017) and affect short-term rejuvenation and channel modification.

M. Lehotský and M. Boltižiar (eds.), *Landscapes and Landforms of Slovakia*, World Geomorphological Landscapes,
https://doi.org/10.1007/978-3-030-89293-7_16

Bank erosion of meandering and sinuous channels are demonstrated by lateral movement and migration of river bends (Bertalan et al. 2019; Hooke 2003). Interaction between erosional and depositional processes lead to channel modification, floodplain creation and construction of riparian landscape. In this system, high magnitude events lead to floodplain reworking and restoration of new habitats. The geomorphological effect of a flood depends on its magnitude, frequency and timing, as well as physical characteristics of the river bed and the floodplain (Miller 1990). It is important to distinguish between the effects of inundation and those of flood-related erosion (Fuller 2008). Flood events rework the riparian zone, modify floodplain topography and increase landscape heterogeneity. Dynamic behaviour of sinuous channels and lateral migration in a form of bank erosion is a natural process (Hooke 2003). Understanding and preserving this dynamic river system, laterally unconfined and capable of bank erosion is a challenge for a society that wishes to protect natural behaviour of rivers. This is because stream bank erosion on the one side represents natural hazard that threatens artificial constructions and human activities in the landscape, but on the other it increases ecological variability and spatial geodiversity of the riparian zone (Bertalan et al. 2018). Preferred 'green approaches' and understanding of river behaviour is essential to understanding river processes and application of proper management strategies. Restrictive interventions are too expensive, lead to increasing river degradation and support irreversible channel changes (Piégay et al. 1997).

16.2 Regional Settings

The Topľa and Ondava River drain the north-eastern part of Slovakia (Fig. 16.1). Both streams are located in the Outer Western Carpathians, in the central part of Ondavská vrchovina upland, which belongs to the area of the Nízke Beskydy mountains (Mazúr and Lukniš 1978). Geologically, the area is the result of flysch sedimentation of the Outer Carpathians, whereas this succession was arranged into three tectono-lithostratigraphic units of the Magura Group during the deformation phase (Kováčik et al. 2011; Nemčok et al. 1990). The southernmost part of the study area is marginally formed by the Krynica unit of proximal deep-marine environment, with prevailingly medium to coarse-grained sandstones and conglomerates. The right slope of the Topľa valley is covered by a narrow strip of the Bystrica unit and the Zlín Formation in "sandstone development", whereas in northern parts of the catchment the Beloveža Formation occurs, formed by thin-bedded claystones and siltstones with layers of sandstones (fine-rhythmical flysch). Central parts of the Ondavská vrchovina upland and slopes of the Ondava valley are formed by the Rača unit with complicated geological settings and variegated facies content. The Beloveža formation creates cores of anticlinal structures (thick-bedded claystones and sandstones) and main central parts of the catchment are formed by the Zlín Formation with claystone and sandstone facial developments. In the central section of the Topľa valley, along the south-western margin of the Rača unit, the Malcov Formation occurs (grey calcareous claystones to siltstones with layers of the quartz-carbonate sandstone). The valleys of both rivers are filled with Holocene fluvial sediments (gravels and sandy gravels) in close genetic relation with proluvial sediments of alluvial fans. Their thickness ranges between 2 and 8 m (Harčár 1995). The shape of the basins is significantly extended in the north–south direction with trellis drainage pattern (main flow with smaller tributaries), where the river creates the main axis of the landscape in prevailing north–south direction.

The river landscape was evaluated in two sections of the eastern Slovakian rivers of Ondava and Topľa. Both represent in their middle part (6th order stream) gravel-bed channels, less affected by technical intervention and regulation. The Topľa River springs at the foothill of the Minčol peak at an altitude 1,015 m a.s.l. in the Čergov Mountains. With the length of 115 km and 1,506 km^2 basin area, it is the largest tributary of the Ondava River. The long term daily average discharge at the Bardejov gauging station is 3.018 m^3 s^{-1} with maximal culmination Q_{max} = 350 m^3 s^{-1} during a flood event with 100-year recurrence interval (17th May 2010). Flow regime is characterised by high variability of discharge due to snow melting and intensive rainfall in summer period. The Ondava River springs at an altitude 546 m a.s.l. at the Slovak-Polish borders and flows for 144.4 km to the confluence with the Latorica River. Its catchment area is 3,355 km^2, and the long term average discharge at Stropkov gauging station is 5.73 m^3 s^{-1}, with maximum 550 m^3 s^{-1} (19th July 1974).

16.3 Fluvial Dynamics on Valleys Floor: Channel Planform Evolution by Old Aerial Images

The last decades of channel evolution are recorded by historical maps and aerial photographs taken from the beginning of twentieth century, which show significant channel pattern modifications (Fig. 16.2). Both rivers flow within narrow valleys floors and valley slopes create a confined environment for river processes and landform reorganisation. The Ondava River valley floor is approximately 1 km wide, and the floodplain is structured to different vertical levels, from 0.5 to 2 m height difference. River terraces are preserved only locally and lateral tributaries created a system of alluvial fans emerging from side valleys. The Topľa River

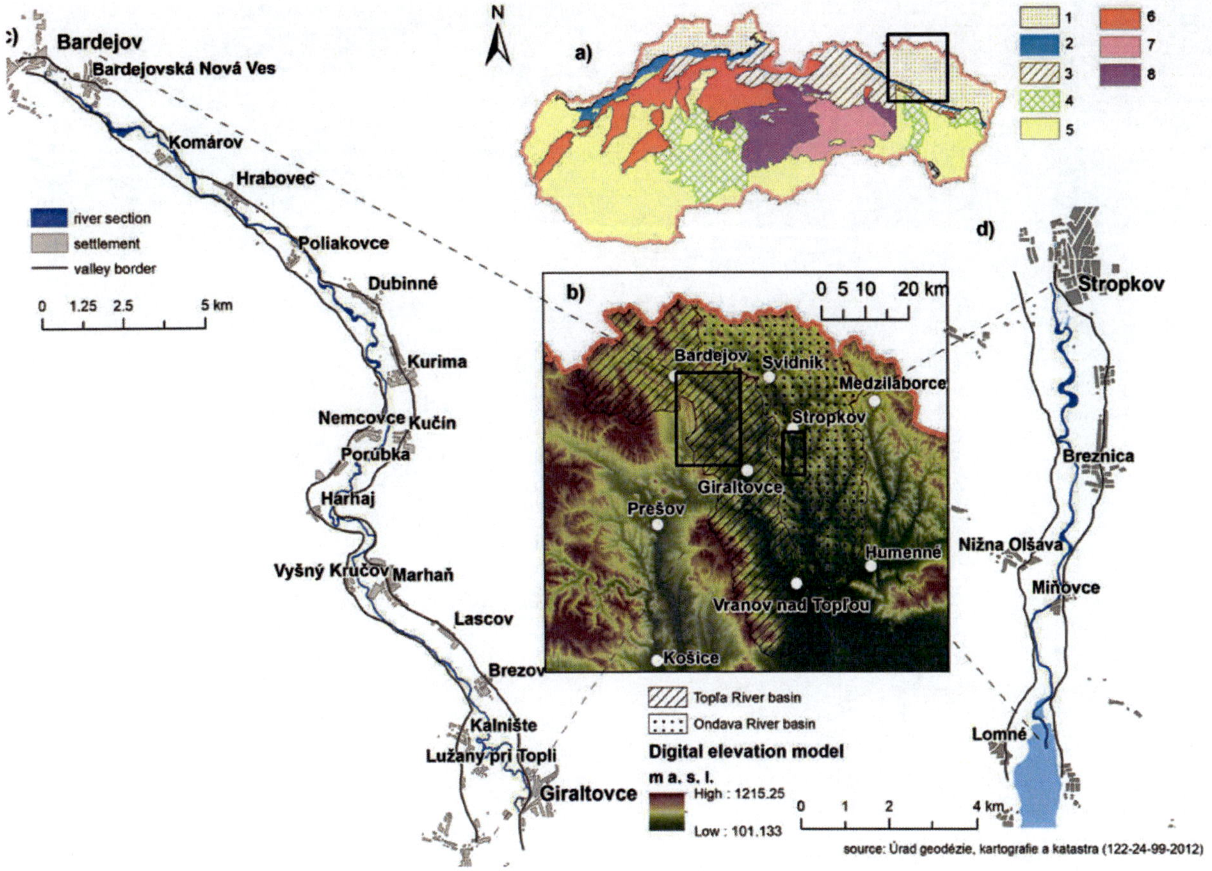

Fig. 16.1 Topľa and Ondava valleys with their sinuous channel patterns of the dynamic river system. Location of the study area in Slovakia (**a**); main tectonic units of the Western Carpathians in Slovakia: 1—Flysch belt, 2—Pieniny Klippen belt, 3—Inner-Carpathian Paleogen, 4—Tatra-Fatra belt of core mountains, 5—Vepor belt, 6—Gemer belt, 7—Neogene volcanics and 8—Neogene basins; topography of the Topľa and Ondava River basins (**b**) and general situation of channels in the valley reaches analysed in detail (**c**, **d**)

floodplain width is approximately 700–1000 m and on systems of lower river terraces are preserved in the valley floor.

Historical maps dating back to the middle of the nineteenth century and aerial photographs point to significant channel modification and pattern changes (Lehotský et al. 2013; Rusnák 2010; Rusnák and Kidová 2018; Rusnák and Lehotský 2014a, b). During the last seventy years lateral dynamics and the numbers of branch channels within the wide gravel plains of maximum width of 226.4 m in the Ondava and 456.7 m in the Topľa River decreased. Gravel-bed rivers have gradually changed from lateral unstable systems with high extent of gravel bar accumulation to simple meandering and sinuous channels with concentrated flow (Fig. 16.3). An average channel width in studied section decreased from 62.1 m in 1949 to 37.2 m in 2009 (Topľa) and from 87.6 m to 39.9 m (Ondava). Overall, the Ondava channel area decreased from 116 ha (1949) to 54 ha (2009), whereas the respective values for the Topľa River are 258 ha and 147 ha. The most significant changes from the point of view of geomorphology are those in gravel bar area, which drastically dropped in the studied period from 76 to 17 ha (Ondava) and from 149 to 53 ha (Topľa). Historical maps and aerial photographs provide direct and indirect evidence of landscape evolution and human impact. Relatively sparsely vegetated area around the river channel was used for agricultural purposes or for grazing. After channel modification and lateral movement accompanied by channel narrowing, newly deposited parts in the form of gravel bars were transformed to a floodplain covered solely with woodland. Riparian forest stabilisation was supported by the process of agricultural collectivisation and subsequent land management resulted in gradual abandonment of small plots in the close vicinity of river channels (Fig. 16.4.).

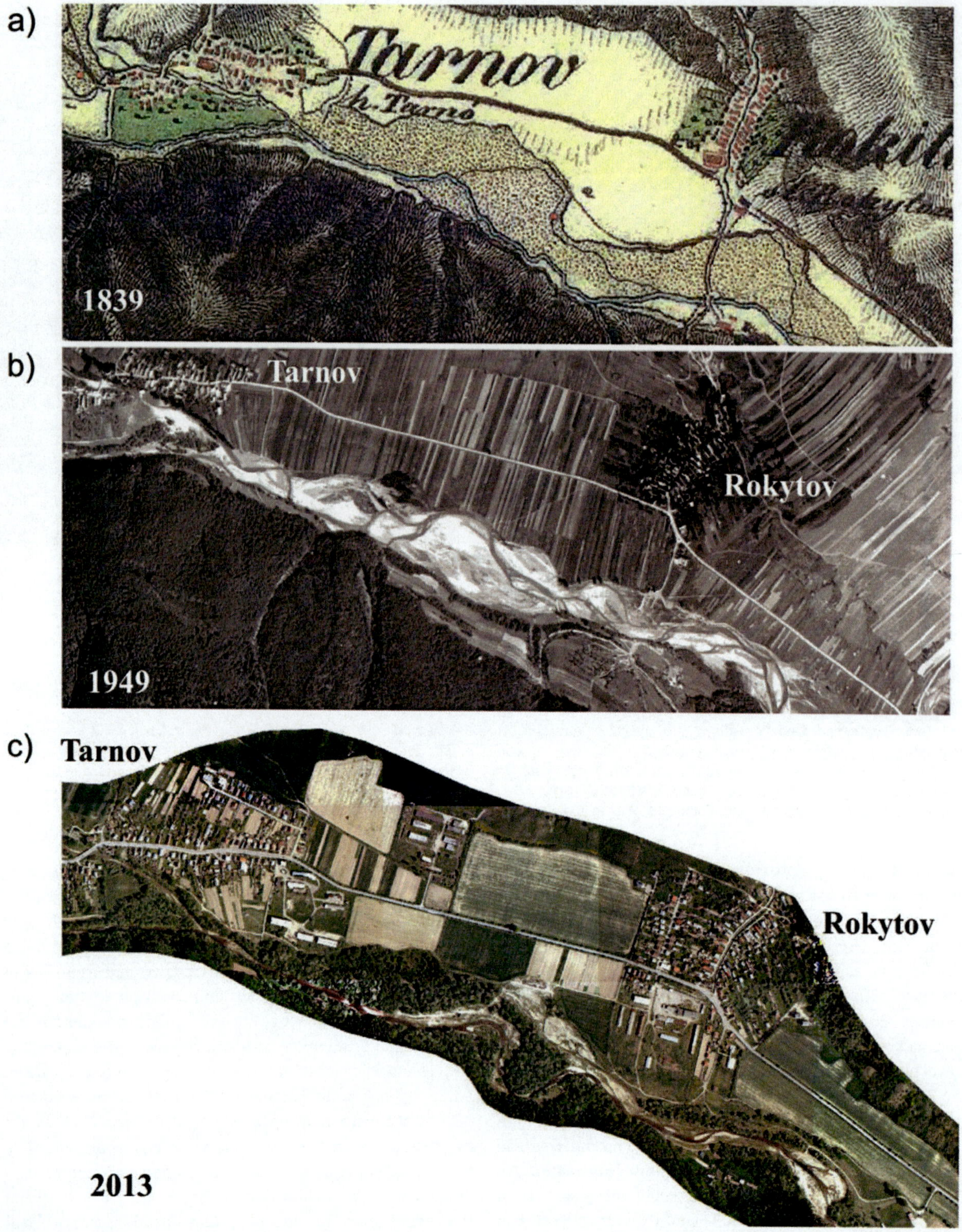

Fig. 16.2 Various sources record the channel pattern evolution of the Topľa River over more than two centuries. An old map produced by the Second Austrian Military Surveys conducted in the second half of the eighteenth century (**a**, 1839), historical aerial photographs from 1949 (**b**) and current channel position in 2013 (**c**) (orthophoto: EUROSENSE Slovakia, Historical orthophoto: GEODIS SLOVAKIA, s r. o., EUROSENSE, s r. o. and Topographic institute Banská Bystrica)

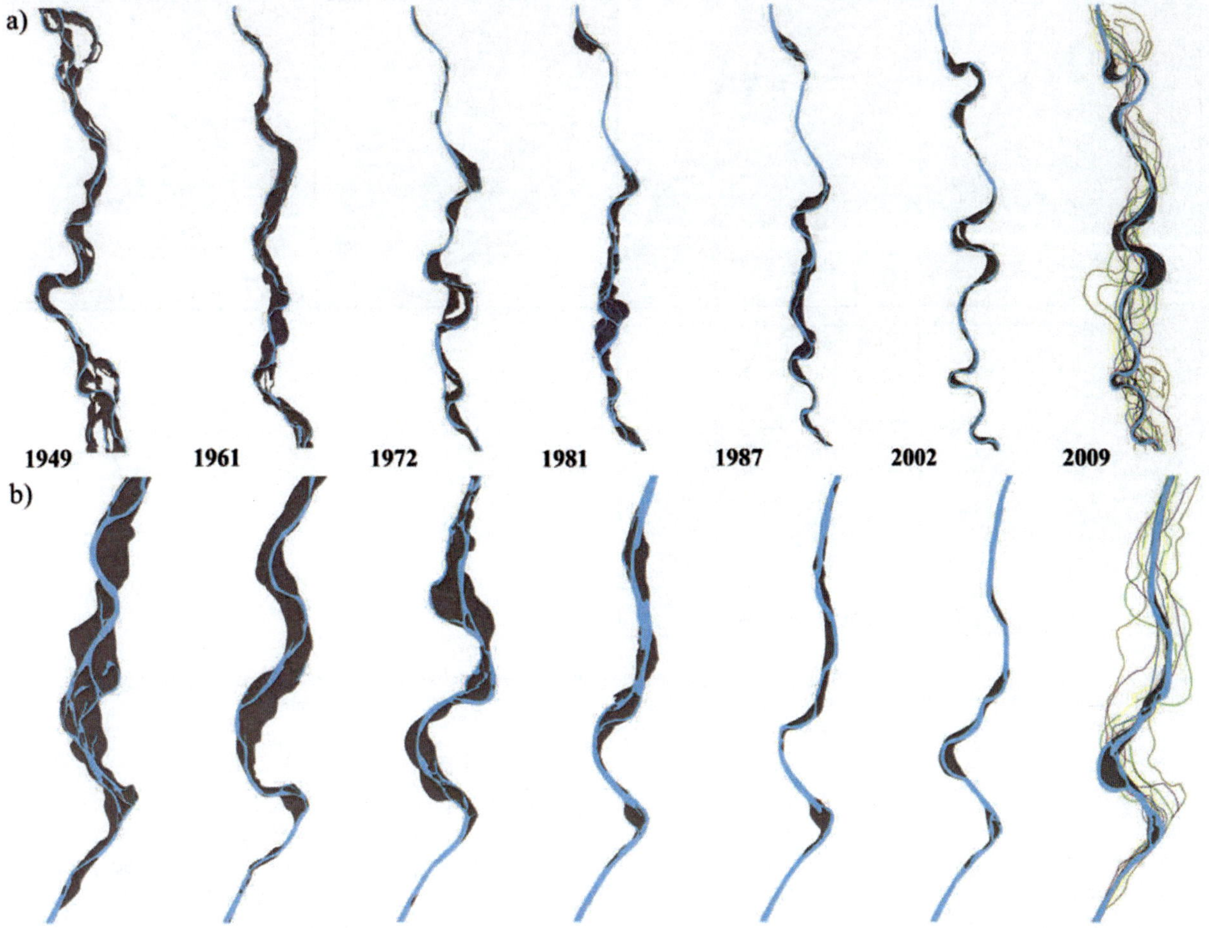

Fig. 16.3 Planform evolution of the Ondava river from 1949 to 2009 in a meandering **(a)** and sinuous **(b)** channel section with historical position of channel plotted as for 2009

16.4 Bank Erosion Hazards and Riverbank Undercuts

North-eastern Slovak rivers have remained unaffected by human intervention and technical bank stabilisation, which still allows for lateral migration, bank erosion and investigation of dynamic behaviour (Fig. 16.5.). Erosion hazard was evaluated by digitalisation of the channel position from the end of 1949. An active channel was defined based on the location of permanent vegetation that marks the boundary between the stabilised floodplain and active channel with flow fluctuations and ongoing in-channel processes (gravel accumulation and bank undercutting). In GIS environment it was possible to calculate the processes of erosion and deposition by superposition of channel positions in different time snapshots (Lehotský et al. 2013). Erosion is understood as replacement of former floodplain areas by river channels, and deposition is recorded by areas of the former channel incorporated into the floodplain by sediment accumulation and vegetation stabilisation (continuous vegetation cover).

From the middle of the twentieth century both rivers eroded an area of 351 ha, as measured in reaches between Bardejov to Giraltovce (length of 40 km) and from Stropkov to Domaša waterwork reservoir (13 km). The intensity of this process is 5.85 ha y^{-1}, which translates into erosion of 1104 m^2 of valleys floor in one year for every kilometre of the river. In both rivers during the second half of the twentieth century the process of deposition prevailed, being 1.5 times higher that process of erosion (accumulation of 524 ha).

Results of bank erosion and deposition processes include the lateral shift of a river channel (Fig. 16.6). In meandering and sinuous rivers this led to gradual movement of bends by cliff erosion during flood events. The average shift of the Ondava river channel in the period from 1949 is 2.68 m per year and 1.91 m y^{-1} for Topľa River. Moreover, during flood events the lateral displacement of a channel can have

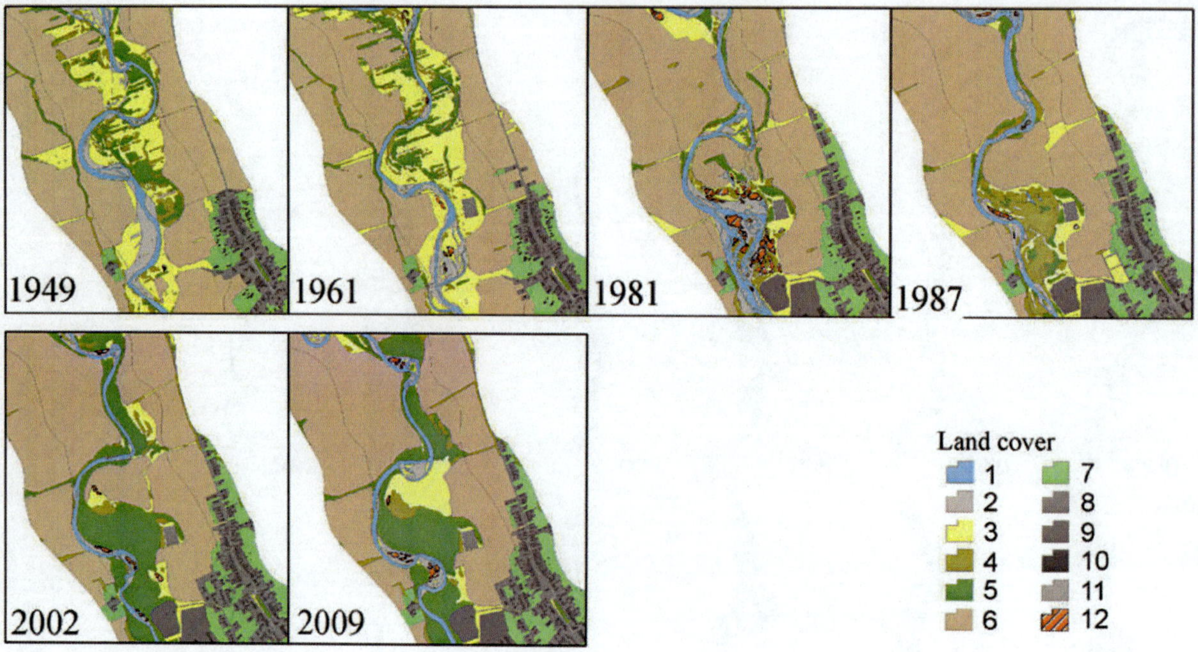

Fig. 16.4 Digitised aerial photographs from the second half of the twentieth century point to the creation of a riparian belt along the Topľa River. Land cover classes: 1—water, 2—gravel bar, 3—meadow, 4—shrubs, 5—forest, 6—arable land, 7—garden, 8, 9—urban area, 10—building, 11—road, 12—gravel mining

reach up to 400 m. In these cases, the main process of channel re-location is avulsion, often occurring at meander necks and leading to meander cut-off. The dynamics of lateral migration is influenced mainly by the distribution and variability of discharge. In the monitored section an increase of bank erosion was recorded during intensive floods in the period 2002–2010. During the period of more frequent floods floodplain erosion dramatically increased, from 13,619 $m^2\ y^{-1}$ to 21,740 $m^2\ y^{-1}$ (Rusnák and Lehotský 2014b). The flood events of 1–2-year recurrence interval do not have destruction effect, but lead to the stabilisation of the system and support vegetation succession (Corenblit et al. 2007; Gurnell et al. 2016; Opperman et al. 2010). On the contrary, frequent floods (high magnitude floods in Topľa and Ondava River in 2004, 2005, 2006, 2008 and 2010) led to intensive destruction and collapse of river banks and in-channel gravel bar accumulation.

These dynamic changes are demonstrated in the section near the village of Hrabovec, where the gravel bar area in a 2 km long reach increased from 5.7 ha in 2002 to 16.2 ha in 2009 due to intensive flood events (see Fig. 16.5I.). Frequent flood occurrence has led to significant channel pattern changes and river could not stabilize themselves by introducing vegetation. The meandering river pattern and its tendency towards lateral migration is controlled by the magnitude of discharge. On the Ondava river one can observe stabilisation of the meander belt, individual meander migration and point bar deposition with an increase of sinuosity index (from 1.34 to 1.57) in the period with low magnitude discharges in the 1990s (Rusnák and Lehotský 2014b). Floods in 2004, 2006 and 2008 led to the straightening of the channel, meander cut-off and abandonment of channel arms (Fig. 16.5b). Consequently, sinuosity decreased to 1.41.

Several authors (Bertoldi et al. 2009; Hickin and Sichingabula 1988) emphasized the role of bank full discharge (relatively low but long duration floods) in bank erosion. In highly dynamic gravel-bed rivers of the flysch Carpathians, as pointed out by Lehotský et al. (2013), low magnitude floods are responsible for the formation of river bed, its planform and relatively slow bank erosion associated with overall behaviour of the channel and helicoidal flow in the river bends. By contrast, high magnitude flows lead to sudden and disastrous bank erosion, channel pattern destruction and partial rejuvenation of stream with gravel bar expansion, destruction of old habitats and creation of new ones.

The combination of dynamic river behaviour and active use of landscape generates problems related to floodplain destruction. From economic point of view destruction of farmland areas is a negative phenomenon. According to the Decree No. 38/2005 the monetary loss resultant from floodplain erosion along the Topľa River was estimated for € 29,924.02 (Rusnák et al. 2016). Identifying unstable reaches

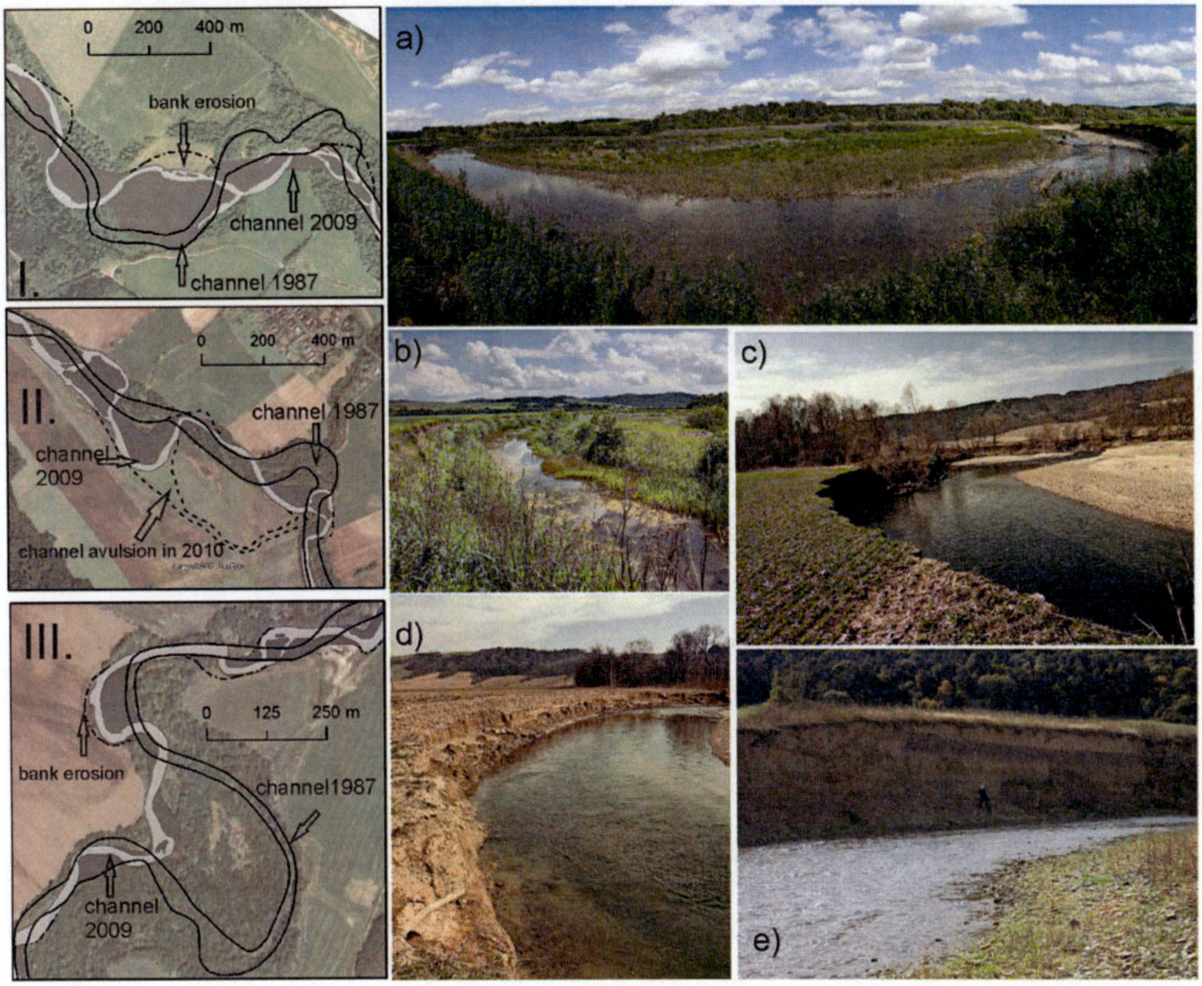

Fig. 16.5 Dynamic changes of channel planform reflected by bank erosion and changes of channel planform of the Topľa River south from Bardejov (I.); avulsion channel creation on the Topľa River near the village of Dubinné during flood events in 2010 (II.) and meander cutoff near the village Porúbka (III.). Example of avulsion channel formation in the meander neck during the flood in 2010 near Stropkov (the Ondava River) (**a**); abandoned channel in the meander loop near the village Breznica (Ondava) after meander neck cutoff (**b**); erosion of arable land with part of winter crops planted before winter 2010/2011 (**c**; Topľa, photo: April 2012); erosion of arable land in the meander loop above the village Lomné (**d**; Ondava) and erosion of Pleistocene terraces near the village of Kurima with height 6 m and 580 m length of eroded bank (**e**; Topľa). Reproduced from Rusnák and Lehotský (2014a)

and potentially at risk is important for effective river management. Bank erosion is a naturally occurring process and 'green approaches' are nowadays applied in management, leading to elimination of technical interventions in river basins and allowing for free channel migration in a certain area where it does not threaten human activity. As pointed out by Piégay et al. (1997), active interventions into the channels are expensive and create spiral effect that leads to degradation and increased flood risk, whereas bank erosion is in most cases a psychological effect, affecting river sections in less valued areas than settlements or farmland.

16.5 Valleys Floor Reworking: Power of River

The energy of river during a flood event is transformed to valley floor re-modelling by erosion-accumulation phases of channel pattern evolution. A characteristic feature of meandering rivers are cutoffs of meander bends. Archive aerial photographs show long-term meander bend evolution near the village of Stropkov from a narrow and straight channel, where township flood defence walls and embankments built

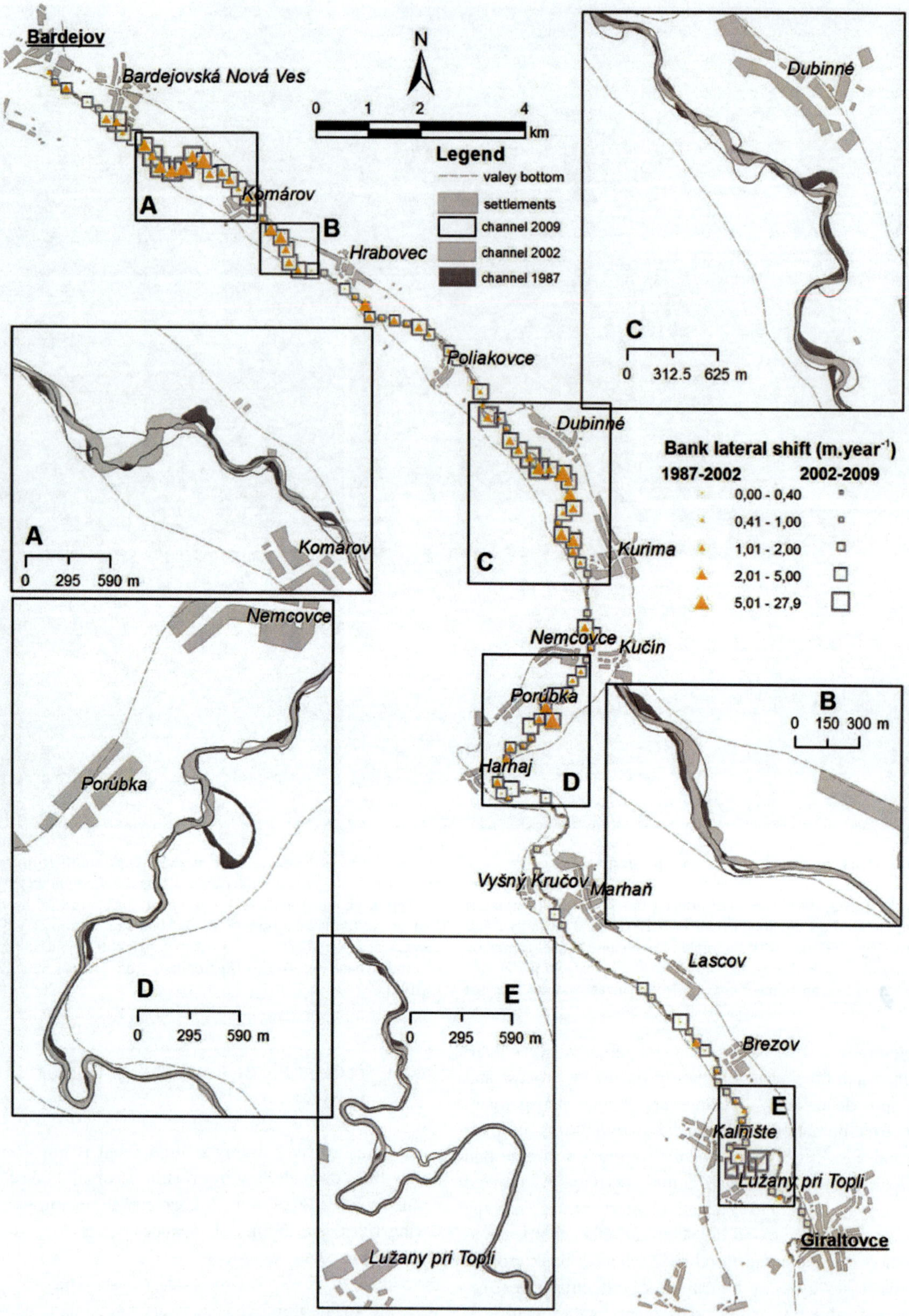

Fig. 16.6 Longitudinal distribution and amount of lateral erosion on the Topľa River in the period 1987–2009, inferred from analyses of historical aerial images. Reproduced from Rusnák et al. (2016)

in the 1960s. During flood events in 2010 a chute cutoff channel was formed inside the meander neck, with the length of 450 m and 150 m wide (Fig. 16.7). The first flood culminated on 17 May (254 $m^3 s^{-1}$) and was followed 18 days later with 220 $m^3 s^{-1}$ (Rusnák et al. 2019). The evolution of the chute cutoff was monitored by drones and UAV-photogrammetry (Unmanned aerial vehicle). After initial reconnaissance by means of field work the first monitoring campaign started on 15 June 2012, followed by the collection of second set of images on 1 April 2014 and the final imaging was performed on 18 July 2014. Overall, 712 images were processed during campaigns. Meander neck cutoff was enforced by headcut gully formation on the floodplain and erosion of topsoil cover. It was also demonstrated that a high-accuracy digital elevation model is important for precise evaluation of the morphological effects of floods.

UAV technology enables one to produce high resolution models with 1–2 cm resolution, that capture chute channel morphology in great detail (Fig. 16.8.). A part of the former meander loop in its position from 2009 is clearly identifiable, with a gradual extension in the direction of water flow. In the central part the original floodplain surface is well-preserved between two main headcut gullies, buried by a thin sandy layer several centimetres in thickness and progressively destroyed by anthropogenic gravel mining. The gravel pits are identifiable throughout the all study area and retreat bank cliff was observed. The avulsion channel bend was slowly shifted laterally by 7.5 m in the monitored period of 24 months.

Low magnitude discharges after 2010 allowed for vegetation stabilisation and rapid vegetation succession with grass formation and shrubs, almost exclusively Salix spp., covered the central part of the chute channel up to the height

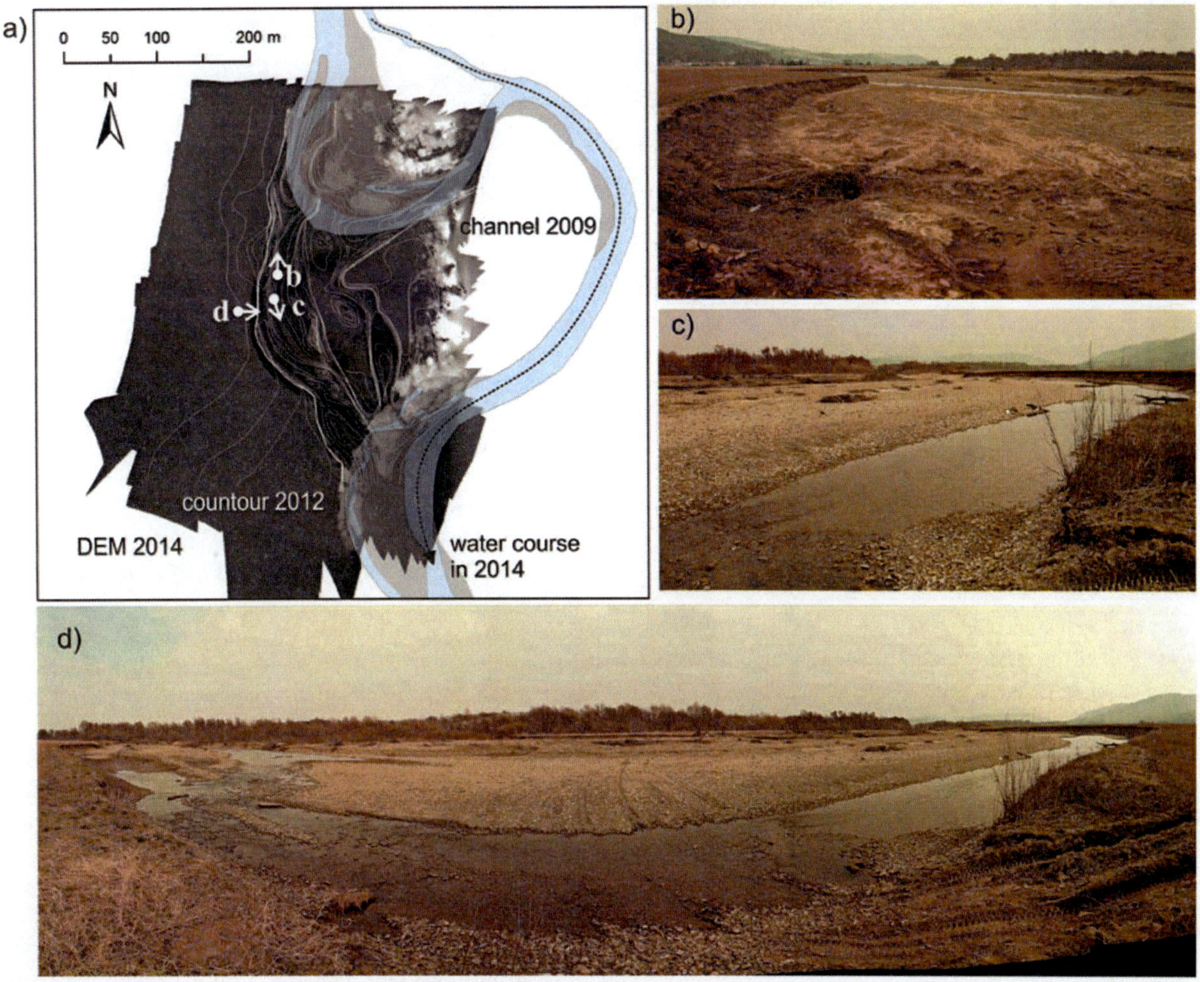

Fig. 16.7 Chute cut-off avulsion channel formation in the meander neck during a flood event in 2010 (**a**) and geomorphological effect of the flood in the upper part (**b**), lower part (**c**) and panoramic view of general situation (**d**)

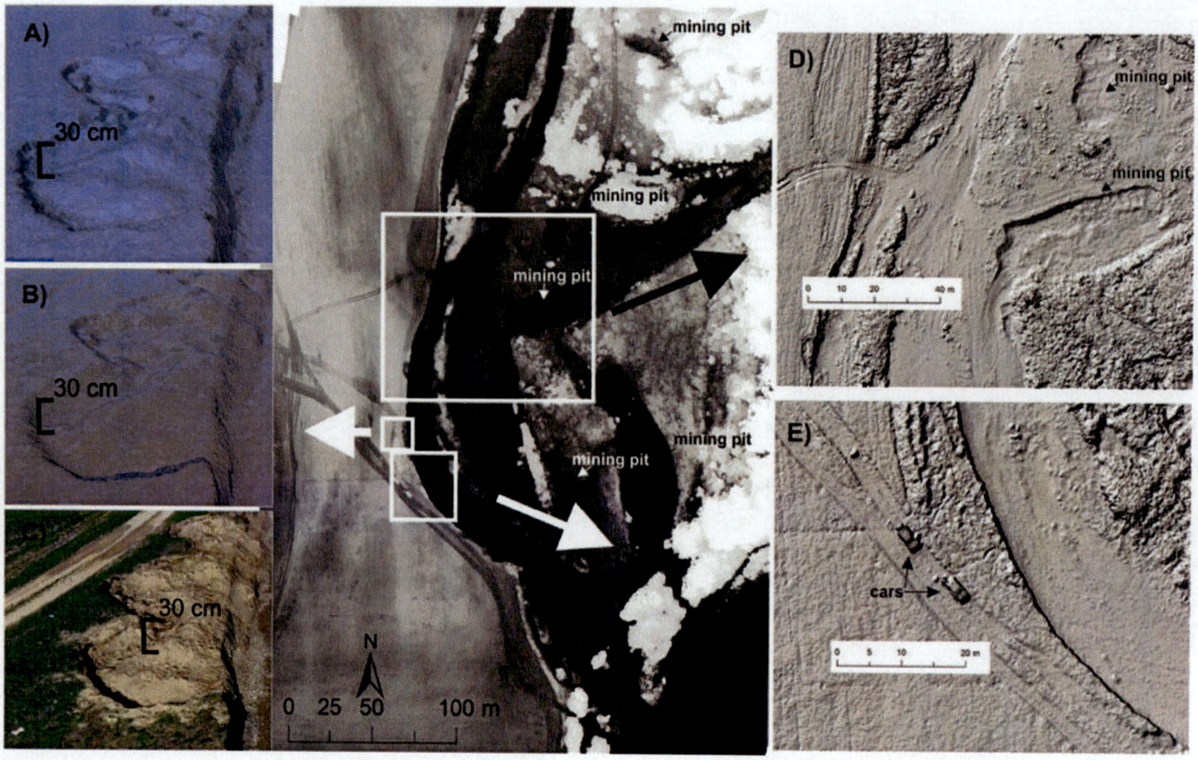

Fig. 16.8 Digital elevation model (DEM from July 2014) of chute cutoff avulsion channel and details of TIN model accuracy (**a**), solid TIN model (**b**), texture 3D model (**c**) and shaded relief of DEM (**d**, **e**)

of 4 m (Fig. 16.9). Vegetation has increased erosion resistance and stabilized the river system by sediment trapping, slowly leading to floodplain transformation. Finally, the flat floodplain was transformed to a new highly structured surface with higher geodiversity and former agriculture plots were replaced riparian forest.

16.6 River and People: Questions in Rivers Management

Since the prehistoric times, the Nízke Beskydy Mountains were typified by a forested hilly landscape, with sporadic evidence of settlements and hence, archaeological findings (Beňko 1985). The oldest traces of settlements are from the Stone Age and point to the transport function of this territory, being associated with the movement of rare hunting groups at the end of the Ice Age. Archaeological findings point to settlements in this region from the late Stone Age (Eneolithic—2500—1800 years BC) onward, but archaeological discoveries are only sporadic. Important development of this area started in the Middle Ages, due to German colonisation associated with the development of trade. Rivers valleys had a transport function, acting as corridors connecting Poland and Hungary. In the fourteenth century, the town of Bardejov became a centre of trade with Poland and gained the status of a royal town in 1376 as a centre of north-eastern Slovakia. The surrounding areas still preserved the natural character and were later affected by the so-called Wallachian colonisation at the beginning of sixteenth century, leading to livestock breeding, grazing and deforestation with agriculture activities confined to valley floor.

The main landscape changes are related to industrialisation and collectivisation during the second half of the twentieth century. Floodplain reworking and replacement of eroded arable land by gravel deposition with rapid vegetation succession in this period led to extensive afforestation in distinct contact with river. A continuous complex of floodplain forest spreading along both flood-affected channels was included in the network of protected sites the NATURA 2000 in 2004. Both rivers create axis of the protected area as an important migratory biocorridors for birds and the nesting places for avifauna. The varying geodiversity is affected by river morphological changes and floods, directly connecting with high biodiversity. Floodplain-channel connection is controlled by water stage fluctuations and discharge regime unaffected by technical infrastructure. Flooding with maximal discharges results from spring snowmelt and may also occur in summer

Fig. 16.9 Vegetation succession in chute cut-off avulsion channel in its central part (photo taken 13 October 2014)

months due to extreme rainfall events. Protected areas are accompanied by the willow, ash, poplar and alder (so-called soft-wood floodplain forest), complemented by wet meadows and in-channel vegetation, especially *Myricaria germanica.* The development of floodplain forest is dependent on effective distribution of floods spilled over the floodplain and active in-channel bar deposition. Unique species of European significance have been recorded in the area, including birds such as *Alcedo atthis, Ciconia ciconia, Ciconia, nigra, Picus canus, Egretta alba,* mammals, especially beavers and otters (*Castor fiber, Lutra lutra*), amphibians (*Bombina variegate*), invertebrates (*Vertigo angustior*) and fish. The nest cavities were usually found on vertical cliffs of undercut banks (Fig. 16.10.), reflecting the importance of geomorphological processes in the riparian landscape.

Most economic activities that take place in the territory of the Ondavská vrchovina upland are linked to the narrow valley floors typified by intensive agriculture, relatively dense population, water demand and the presence of main transportation lines (roads). River water quality and bank vegetation habitats are under pressure from urbanized and agricultural landscape. In-channel bar accumulations are intensive extracted directly from the main channel and trucks destroy armoured gravels bed (Fig. 16.11.). Vegetated banks are excavated by local residents for heating purposes or local river management authorities to prevent flood risk. Recently, the main challenge is to manage interventions aimed to eliminate effects of bank erosion and flooding by "green" approaches, to minimize negative consequences of direct technical solutions.

16.7 Conclusions

Rivers are the most dynamic element of the Ondavská vrchovina upland. Nowadays, active behaviour is the main imprint that creates and reorganises the landscape with harmonic interaction throughout villages, farms and agricultural plots. Understanding the dynamics of rivers in their natural environment allows for predicting further development within nature-friendly management. Well-preserved natural fluvial processes despite human intervention create an extraordinary aquatic and semi-terrestrial system and the value of the landscape lies in its high geo- and bio-diversity.

The eastern Slovak gravel-bed rivers are still one of the less regulated channels that enable for monitoring of

Fig. 16.10 Nest cavities along vertical cliff of undercut banks of the Topľa River

hydrogemorphological processes and bank erosion. Gradual erosion and removal of floodplain material, as well as deposition in the form of gravel bars, generate ever-changing conditions on the valley floors. However, the most dynamic reaches achieve lateral shift several hundred metres, and the average movement is approximately 1 m per year. The geodiversity of the landscape is enriched by the formation of vertical cliffs, bank undercutting and abandoned arm formation. This system is controlled by the occurrence of floods that have a significant destructive effect perceived as a natural hazard, but on the other hand, they are essential for preservation of a dynamic, natural fluvial system.

Acknowledgements The research was supported by Science Grant Agency (VEGA) of the Ministry of Education of the Slovak Republic and the Slovak Academy of Sciences; 02/0086/21.

Fig. 16.11 Gravel extraction from a point bar in the Ondava River channel

References

Beňko J (1985) Osídlenie severného Slovenska. Východoslovenské vydavateľstvo, Košice

Bertalan L, Novák T, Németh Z, Rodrigo-Comino J, Kertész Á, Szabó S (2018) Issues of meander development: land degradation or ecological value? The example of the Sajó River. Hungary. Water 10:1613. https://doi.org/10.3390/w10111613

Bertalan L, Rodrigo-Comino J, Surian N, Šulc Michalková M, Kovács Z, Szabó S, Szabó G, Hooke J (2019) Detailed assessment of spatial and temporal variations in river channel changes and meander evolution as a preliminary work for effective floodplain management. The example of Sajó River, Hungary. J Environ Manage 248:109277. https://doi.org/10.1016/j.jenvman.2019.109277

Bertoldi W, Gurnell A, Surian N, Tockner K, Zanoni L, Ziliani L, Zolezzi G (2009) Understanding reference processes: linkages between river flows, sediment dynamics and vegetated landforms along the Tagliamento River, Italy. River Res Appl 25:501–516. https://doi.org/10.1002/rra.1233

Brierley GJ, Fryirs KA (2005) Geomorphology and river management : applications of the river styles framework. Blackwell Publications

Brierley GJ, Mum CP (1997) European impacts on downstream sediment transfer and bank erosion in Cobargo catchment, New South Wales, Australia. CATENA 31:119–136. https://doi.org/10.1016/S0341-8162(97)00025-8

Comiti F, Da Canal M, Surian N, Mao L, Picco L, Lenzi MA (2011) Channel adjustments and vegetation cover dynamics in a large gravel bed river over the last 200years. Geomorphology 125:147–159. https://doi.org/10.1016/j.geomorph.2010.09.011

Corenblit D, Tabacchi E, Steiger J, Gurnell AM (2007) Reciprocal interactions and adjustments between fluvial landforms and vegetation dynamics in river corridors: a review of complementary approaches. Earth-Sci Rev 84:56–86. https://doi.org/10.1016/j.earscirev.2007.05.004

Fuller IC (2008) Geomorphic impacts of a 100-year flood: Kiwitea Stream, Manawatu catchment, New Zealand. Geomorphology 98:84–95. https://doi.org/10.1016/j.geomorph.2007.02.026

Gorczyca E, Krzemień K, Wrońska-Wałach D, Sobucki M (2013) Channel changes due to extreme rainfalls in the Polish Carpathians. In: Geomorphological impacts of extreme weather, pp 23–35. Springer Netherlands, Dordrecht. https://doi.org/10.1007/978-94-007-6301-2_2

Gurnell AM, Corenblit D, García de Jalón D, González del Tánago M, Grabowski RC, O'Hare MT, Szewczyk M (2016) A conceptual model of vegetation-hydrogeomorphology interactions within river corridors. River Res Appl 32:142–163. https://doi.org/10.1002/rra.2928

Hajdukiewicz H, Wyżga B, Mikuś P, Zawiejska J, Radecki-Pawlik A (2016) Impact of a large flood on mountain river habitats, channel morphology, and valley infrastructure. Geomorphology 272:55–67. https://doi.org/10.1016/j.geomorph.2015.09.003

Hajdukiewicz H, Wyżga B, Zawiejska J (2017) Twentieth-century hydromorphological degradation of Polish Carpathian rivers. Q Int 504(10):181–194. https://doi.org/10.1016/j.quaint.2017.12.011

Harčár J (1995) Reliéf Nízkych Beskýd. Geografický ústav SAV, Bratislava

Hickin EJ, Sichingabula HM (1988) The geomorphic impact of the catastrophic October 1984 flood on the planform of Squamish River, southwestern British Columbia. Can J Earth Sci 25:1078–1087. https://doi.org/10.1139/e88-105

Hooke J (2003) River meander behaviour and instability: a framework for analysis. Trans Inst Br Geogr 28:238–253. https://doi.org/10.1111/1475-5661.00089

Hooke JM (2015) Variations in flood magnitude–effect relations and the implications for flood risk assessment and river management. Geomorphology 251:91–107. https://doi.org/10.1016/j.geomorph.2015.05.014

James LA (1997) Channel incision on the Lower American River, California, from streamflow gage records. Water Resour Res 33:485–490. https://doi.org/10.1029/96WR03685

Jefferson AJ, McGee RW (2013) Channel network extent in the context of historical land use, flow generation processes, and landscape evolution in the North Carolina Piedmont. Earth Surf Proc Land 38:601–613. https://doi.org/10.1002/esp.3308

Kijowska-Strugała M, Wiejaczka Ł, Gil E, Bochenek W, Kiszka K (2017) The impact of extreme hydro-meteorological events on the transformation of mountain river channels (Polish Flysch Carpathians). Z Geomorphol 61(1):75–89. https://doi.org/10.1127/zfg/2017/0434

Korpak J (2007) The influence of river training on mountain channel changes (Polish Carpathian Mountains). Geomorphology 92:166–181. https://doi.org/10.1016/j.geomorph.2006.07.037

Kováčik M, Bóna J, Gazdačko Ľ, Kobulský J, Maglay J, Kučera M (2011) Geologická mapa Nízkych Beskýd - Západná časť [Geological map of the Nízke Beskydy Mts.—Western Part]. 1: 50 000. Bratislava, ŠGÚDŠ

Lehotský M, Frandofer M, Novotný J, Rusnák M, Szmańda JB (2013) Geomorphic/sedimentary responses of rivers to floods: case studies from Slovakia. In: geomorphological impacts of extreme weather, pp 37–52. Springer Netherlands, Dordrecht. https://doi.org/10.1007/978-94-007-6301-2_3

Liébault F, Gomez B, Page M, Marden M, Peacock D, Richard D, Trotter CM (2005) Land-use change, sediment production and channel response in upland regions. River Res Appl 21:739–756. https://doi.org/10.1002/rra.880

Marston RA, Bravard JP, Green T (2003) Impacts of reforestation and gravel mining on the Malnant River, Haute-Savoie, French Alps. Geomorphology 55:65–74. https://doi.org/10.1016/S0169-555X(03)00132-6

Mazúr E, Lukniš M (1978) Regionálne Geomorfologické Členenie SSR [Regional geomorphic division of Slovakia]. Geografický Časopis 30(1):101–125

Miller AJ (1990) Flood hydrology and geomorphic effectiveness in the central Appalachians. Earth Surf Proc Land 15:119–134. https://doi.org/10.1002/esp.3290150203

Nemčok J, Zakovič M, Gašpariková V, Ďurkovič T, Snopková P, Vrana K, Hanzel V (1990) Geologická mapa Pienin, Čergova, Ľubovnianskej a Ondavskej vrchoviny [Geological map of Pieniny and Čergov Mts., Ľubovnianska vrchovina and Ondavská vrchovina uplands], 1: 50 000. Bratislava, ŠGÚDŠ

Opperman JJ, Luster R, McKenney BA, Roberts M, Meadows AW (2010) Ecologically functional floodplains: connectivity, flow regime, and scale 1. J Am Water Resour Assoc 46:211–226. https://doi.org/10.1111/j.1752-1688.2010.00426.x

Piégay H, Cuaz M, Javelle E, Mandier P (1997) Bank erosion management based on geomorphological, ecological and economic criteria on the Galaure River, France. Regul Rivers Res Manage 13:433–448. https://doi.org/10.1002/(SICI)1099-1646(199709/10)13:5%3c433::AID-RRR467%3e3.0.CO;2-L

Price K, Leigh DS (2006) Morphological and sedimentological responses of streams to human impact in the southern Blue Ridge Mountains, USA. Geomorphology 78:142–160. https://doi.org/10.1016/j.geomorph.2006.01.022

Rădoane M, Obreja F, Cristea I, Mihailă D (2013) Changes in the channel-bed level of the eastern Carpathian rivers: climatic vs. human control over the last 50years. Geomorphology 193:91–111. https://doi.org/10.1016/j.geomorph.2013.04.008

Rusnák M (2010) Zmeny pôdorysnej vzorky a presúvanie koryta laterálne čiastočne uzavretej rieky. Geographia Cassoviensis 4 (1):154–158

Rusnák M, Kidová A (2018) Recentné morfologické zmeny slovenských vodných tokov Belá a Ondava v Západných Karpatoch od druhej polovice 20. storočia. Geographia Cassoviensis 12(2):195–211

Rusnák M, Lehotský M (2014a) Povodne, brehová erózia a laterálne presúvanie koryta štrkonosných kľukatiacich vodných tokov (Prípadová štúdia tokov Topľa a Ondava). Acta Hydrologica Slovaca 15:424–433

Rusnák M, Lehotský M (2014b) Time-focused investigation of river channel morphological changes due to extreme floods. Z Geomorphol 58(2):251–266. https://doi.org/10.1127/0372-8854/2013/0124

Rusnák M, Lehotský M, Kidová A (2016) Channel migration inferred from aerial photographs, its timing and environmental consequences as responses to floods: a case study of the meandering Topľa River Slovak Carpathians. Morav Geogr Reports 24(3):32–43. https://doi.org/10.1515/mgr-2016-0015

Rusnák M, Sládek J, Pacina J, Kidová A (2019) Monitoring of avulsion channel evolution and river morphology changes using UAV photogrammetry: case study of the gravel bed Ondava River in Outer Western Carpathians. Area 51(3):549–560. https://doi.org/10.1111/area.12508

Scorpio V, Zen S, Bertoldi W, Surian N, Mastronunzio M, Dai Prá E, Zolezzi G, Comiti F (2018) Channelization of a large Alpine river: what is left of its original morphodynamics? Earth Surf Proc Land 43:1044–1062. https://doi.org/10.1002/esp.4303

Škarpich V, Hradecký J, Dušek R (2013) Complex transformation of the geomorphic regime of channels in the forefield of the Moravskoslezské Beskydy Mts.: case study of the Morávka River (Czech Republic). CATENA 111:25–40. https://doi.org/10.1016/j.catena.2013.06.028

Surian N, Rinaldi M (2003) Morphological response to river engineering and management in alluvial channels in Italy. Geomorphology 50:307–326. https://doi.org/10.1016/S0169-555X(02)00219-2

Wyżga B, Zawiejska J, Hajdukiewicz H (2016a) Multi-thread rivers in the Polish Carpathians: occurrence, decline and possibilities of restoration. Q Int 415:344–356. https://doi.org/10.1016/j.quaint.2015.05.015

Wyżga B, Zawiejska J, Radecki-Pawlik A (2016b) Impact of channel incision on the hydraulics of flood flows: examples from Polish Carpathian rivers. Geomorphology 272:10–20. https://doi.org/10.1016/j.geomorph.2015.05.017

Ziliani L, Surian N (2012) Evolutionary trajectory of channel morphology and controlling factors in a large gravel-bed river. Geomorphology 173–174:104–117. https://doi.org/10.1016/j.geomorph.2012.06.001

Miloš Rusnák is a fluvial geomorphologist at the Institute of Geography of the Slovak Academy of Sciences (Department of Physical Geography, Geomorphology and Natural Hazards). His research topics are fluvial geomorphology, spatial data processing in GIS, UAV data acquisition and processing, fluvial processes and sediment connections in gravel-bed rivers and remote sensing applications in rivers and landforms research. He is the author and co-author of several papers dealing with fluvial system evolution in the Outer Western Carpathians.

Anna Kidová is fluvial geomorphologist at the Institute of Geography of the Slovak Academy of Sciences. Her area of research focuses on the response of river systems to environmental changes, the study of hydro-morphological processes as well as issues related to measures for the river management. Since 2018 she is a president of Association of Slovak Geomorphologists.

Milan Lehotský is a physical geographer and fluvial geomorphologist at the Institute of Geography of the Slovak Academy of Sciences. He was many years head of the Department of Physical Geography, Geomorphology and Natural Hazards. His research topics are responses of fluvial systems to environmental changes, sedimentological connectivity, evolution trajectories, hydromorphology and GIS and remote sensing applications in rivers and landforms research. He is also working as an external lecturer at the Department of Physical Geography and Geoecology, the Faculty of Natural Sciences of the Comenius University in Bratislava.

Ján Sládek is a geomorphologist at the Institute of Geography of Slovak Academy of Sciences in Bratislava as well as he works as a professional UAV operator and beta tester of laser scanner products in private sector. His research is focused on morphotectonics, morphostructure analysis and fluvial landscape monitoring by using of UAVs. The region of his interest covers Slovak part of Carpathian Mountain Range mainly surroundings of Turčianska kotlina Basin and other core mountains.

Slovak Karst: Surface and Subsurface Geodiversity of the Karst Plateau in the Temperate Climate Zone

17

Ľudovít Gaál, Pavel Bella, and Jozef Jakál

Abstract

Slovak Karst is the largest karst area in Slovakia and is considered as one of the most distinctive karst plateaus in Europe. The Slovak Karst consists of several carbonate block slices of the Silica Nappe of Silicicum Unit, lying on deep-marine rocks of the Turnaicum and Meliaticum units. Each of these blocks has an impermeable Werfen shale at the base, therefore, they form hydrologically individual, disconnected structures. The Slovak Karst is featured by a complete range of surface and subsurface karst landforms. Abundant karren, dolines, uvalas, blind and dry valleys, gorges, canyons as well as rare inactive poljes occur within the karst landscape. Subsurface karst phenomena are represented mostly by vadose vertical shafts and stepped abysses. Moreover, several epiphreatic horizontal and subhorizontal caves were developed due to the corrosive activity of autochthonous and allochthonous underground streams. At present, 1184 caves are known in the Slovak Karst. The vast majority of them is originated in the Pliocene and Quaternary. The oldest phase of karstification is associated with residues of the upper Cretaceous paleokarst phenomena. The subsurface karst phenomena of the Slovak Karst and the adjacent Aggtelek Karst in Hungary represent an unusual example of high variability of cave morphology and origin, carbonate speleothems as well as rare cave fauna and archaeological findings in the plateau karst of the temperate climatic zone. Therefore, they were declared as a part of the UNESCO World Heritage in 1995.

Keywords

Slovak Karst • Karst landscape • Plateau karst • Karren • Doline • Blind valley • Paleo-polje • Cave • Abyss • Paleokarst

17.1 Introduction

The Slovak Karst area is a typical example of plateau karst of the temperate climate zone, expressing almost a complete range of surface and subsurface karst landforms. The surface karst landforms are represented by karren, doline, uvala, blind and dry valley, gorges, canyons, several rare inactive poljes, whereas as many as 1184 caves occur in the subsurface. The subsurface karst phenomena of the Slovak Karst and adjacent Aggtelek Karst in Hungary, representative for the temperate climatic zone, represent an unusual example of high variability of cave morphology and origin, carbonate speleothems as well as rare cave fauna and abundance of archaeological findings. Many abysses in extraordinary high density developed on the top of karst plateaus. Based on the bilateral Hungarian-Slovak nomination project, the caves of Slovak and Aggtelek karsts were declared as a part of the UNESCO World Heritage in 1995 as an outstanding example of significant on-going geological processes in the development of landforms and geomorphological features. Besides karst phenomena, rare and endemic plants and animals occur here and are noteworthy.

The Slovak Karst was declared as a protected landscape area in 1973 and as a national park in 2002. In 1977, the most precious part of the Slovak Karst has been included in the international network of biospheric reserves within the

Ľ. Gaál (✉) · P. Bella
Slovak Cave Administration, Hodžova 11, 031 01 Liptovský Mikuláš, Slovakia
e-mail: ludovit.gaal@ssj.sk

P. Bella
e-mail: pavel.bella@ku.sk; ; pavel.bella@ssj.sk

P. Bella
Department of Geography, Faculty of Education, Catholic University in Ružomberok, Hrabovská cesta 1, 031 04 Ružomberok, Slovakia

J. Jakál
Department of Physical Geography, Geomorphology and Natural Hazards, Institute of Geography, Slovak Academy of Sciences, Štefánikova 49, 814 73 Bratislava, Slovakia

M. Lehotský and M. Boltižiar (eds.), *Landscapes and Landforms of Slovakia*, World Geomorphological Landscapes,
https://doi.org/10.1007/978-3-030-89293-7_17

program of Man and Biosphere (MaB). Several national nature reserves and national nature monuments within the national park are of European importance and represent various natural units and phenomena both on the surface and underground spaces, protected by European network Natura 2000. In 2001, the Domica and Baradla caves were inscribed on the list of Ramsar sites as wetlands of international importance, with many endemic fauna species.

Such diversity of surface and subsurface karst phenomena of the Slovak Karst resulted from a specific geological and geomorphological development. The first karstification period in the Cretaceous was enacted in tropical climatic conditions, then surface planation was expressive in the Miocene, and at last, the large planation surface was dissected by fluvial valleys during updoming of the area in the Late Miocene.

The richness of karst phenomena has attracted several notable explorers and scientists since the eighteenth century. The Silická Ice Cave was described as early as 1723 by Matthias Belius, then the first list of caves was given by Karl Siegmeth (1887, 1891). The karst landforms were first studied by a significant Polish geomorphologist Ludomir S. Sawicki (1908) (Fig. 17.1).

17.2 Location and General Characteristics

The Slovak Karst is located in the southern part of the Western Carpathians, in the south-eastern part of the Slovenské rudohorie Mountains. It consists of several karst plateaus (Koniarska, Plešivská, Silická, Horný vrch, Dolný vrch, Bôrčanská, Zádielska and Jasovská plateaus) separated by deep fluvial valleys. Within the regional geomorphological division (Mazúr and Lukniš 1978), in addition to karst plateaus, the Slovak Karst also includes the slightly rugged Jelšava Karst and Turnianska Basin (Fig. 17.2). The Slovak Karst neighbours the Rimavská Basin and Bodvianska Hilly Land in the south-west, the Revúcka Mountains in the west, the Rožňavská Basin and Volovské Mountains in the north and the Košická Basin in the east. The Dolný vrch Plateau and Silická Plateau continue southward into the territory of Hungary, where they form the Alsó-hegy Plateau and Aggtelek Karst. The Slovak Karst territory extends more than 361 km^2.

The surface inclination shows a general tilt of the Slovak Karst plateaus from the north to the south, at an angle of 2°–4° (Jakál 2008). While the northern edge of the Plešivská Plateau has an elevation of 720 m above sea level, its southern edge reaches only 520 m. The central part of the Silická Plateau is situated at 500–600 m a.s.l., Horný vrch Plateau and Zádielska Plateau in the eastern part of the Slovak Karst at 700–800 m a.s.l. The highest place of the Slovak Karst is the Matesová skala Rock on the Bôrčanská Plateau, with an elevation of 925 m a.s.l. The plateaus are bordered by steep slopes connecting to the bottoms of adjacent canyons of Slaná and Štítnik rivers, Zádielska and Hájska gorges and the Rožňavská and Turnianska basins. The Turnianska Basin is located between the Horný vrch and Dolný vrch plateaus. Slope inclination within marginal escarpments of the plateaus reach as much as 45°, while frequently having the character of vertical rock slopes in their upper parts. The height of slopes is 200–550 m (Jakál

Fig. 17.1 The first researchers of Slovak Karst: **a** Matthias Belius (1684–1749); **b** Karl Siegmeth (1845–1912); **c** Ludomir S. Sawicki (1884–1928)

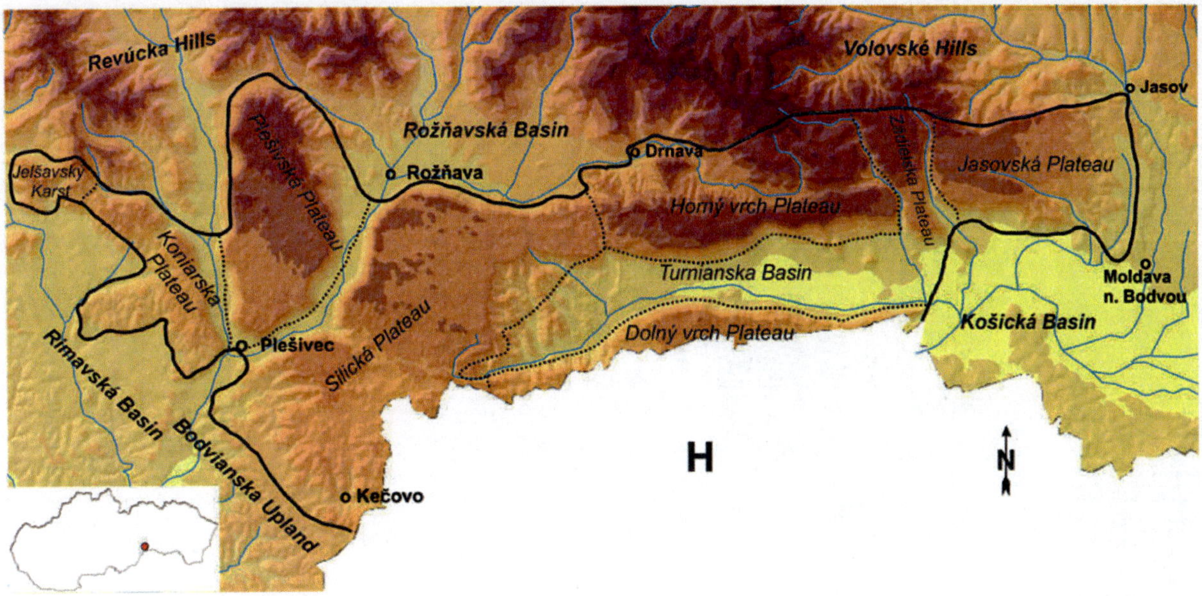

Fig. 17.2 Geomorphological subdivision of the Slovak Karst (according to Mazúr and Lukniš 1978)

2008). Karren, dolines, uvalas, abysses, inactive poljes, as well as dry valleys and blind valleys occur on the plateaus, within a slightly dissected surface. Only cone hills on the Plešivská and Koniarska plateaus protrude more distinctly from the planation surface as remains of the Late Cretaceous karstification.

The Slovak Karst lies in the temperate climate zone, in the transition area between oceanic and continental climate. The average annual air temperature reaches 5.7–8.5 °C, and the average annual precipitation is 630–990 mm. Climatic inversions, also causing vegetation inversions, occur mainly in the deep gorges of the Zádielska and Hájska gorges. The distinctive natural conditions of the Slovak Karst are also seen in the variety of plants and animals. The richest region from floristic point of view is the district of the Pra-Matra xerothermous flora (Matricum) in the region of the Pannonian flora (Pannonicum). There are xerothermous species, calciphytes, as well as cryophilic and hygrophilous mountain species preserved in inversion locations of deep and narrow gorges (Háberová 1994). Mainly endemic, subendemic and relict species of flora are important. Also, several endemic animal species are known from the caves, underground water streams and springs with typical ecological conditions.

17.3 Geological and Geographical Settings

The Slovak Karst area consists of six great block slices of the Silica Nappe structure of Silicicum unit. They originated from shallow-water marine sediments in a large carbonate platform and are lying on the deep-water Turnaicum and Meliaticum units. They were defined as Bôrka, Hačava-Jasov, Silica-Turňa, Plešivec-Brezová, Kečovo and Domica Block (Bystrický 1964; Mello et al. 1997; Gaál 2008). The blocks lie in west–east direction, are monoclinal with a moderate dip to the south, or form shallow synclinal structures. Each of the blocks has impermeable Werfen shales in background. The clastic rocks of Werfen Formation are denuded more quickly than limestone, and therefore, they underlie topographic depressions such as the Rožňava and Turna basins and the slope valley above the Gombasek settlement. Impermeable beds of Werfen shales are also exposed on the surface on both sides of the Fabianka Hill in the southern part of Silická Plateau and have conditioned the development of contact karst. However, two carbonate blocks lie in contact in the Plešivská Plateau, without intervening Werfenian shales, supporting the development of a homogenous karst plateau (Mello 1977). Besides shales and sandstones (Bodvaszilas Beds) also the thin marly sandstones and limestones (Szin Beds) occur in the upper part of Werfen Formation (Mello et al. 1997) (Fig. 17.3).

The carbonate parts of slide blocks consist of several types of limestones and dolomites of Middle or Late Triassic age (Fig. 17.4). The dark-grey, bedded Gutenstein limestone and dolomite lie in the lower part, passing upward into the light-grey, massive Steinalm limestone. Both limestone types originated in the Middle Triassic (Anisian) times and form about 200–600 m wide belts at the surface. The Jasovská and Krásnohorská caves have developed in the Steinalm limestone and Gutenstein dolomite.

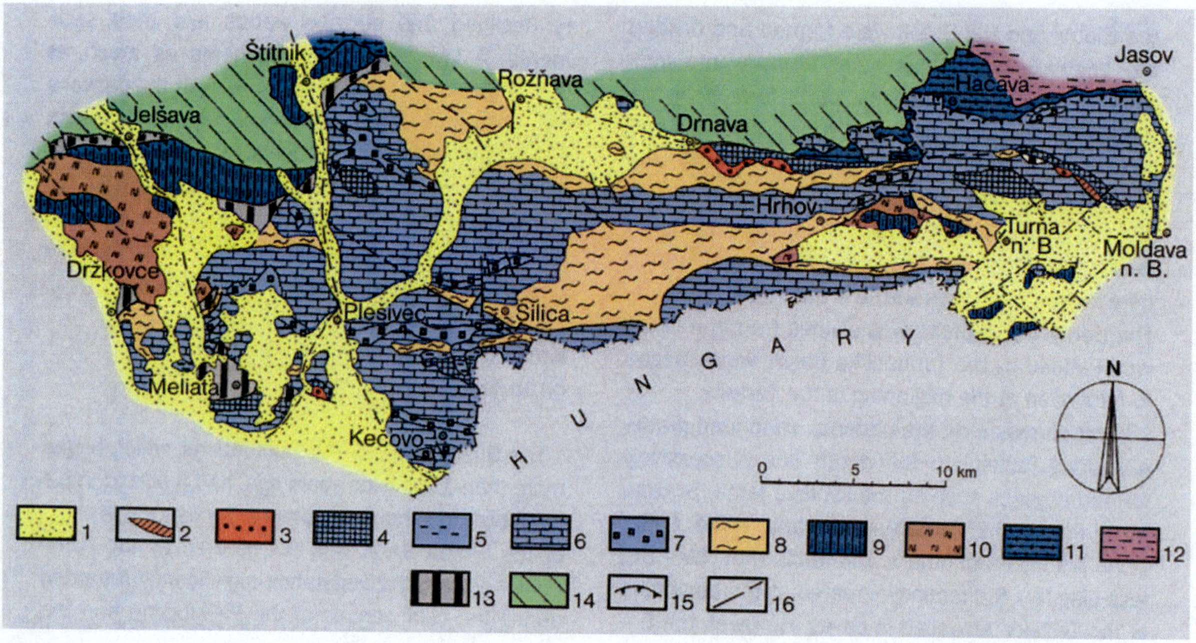

Fig. 17.3 Geological map of the Slovak Karst (according to Mello et al. 1997, simplified). Legend: 1—Neogene and Quaternary gravel and sand; 2—Upper Cretaceous limestone; **Silica Nappe**: 3—Jurassic limestone, breccia and radiolarite; 4—Upper Triassic Tisovec and Dachstein limestones; 5—Upper Triassic Hallstatt limestone and Allgäu Beds; 6—Middle and Upper Triassic Gutenstein, Steinalm and Wetterstein limestones, Middle Triassic Schreyeralm, Nádaska, Reifling basin limestones; 8—Lower Triassic shale and sandstone of Werfen Formation, **Turnaicum**: Middle and Upper Triassic limestone and shale; 10—Lower Triassic shale; **Meliaticum**: 11—Triassic and Jurassic phyllite, limestone and metabasic rock of Bôrka Nappe; 12—Lower Triassic shale, sandstone, rhyolite, 13—Triassic and Jurassic radiolarite, shale, limestone and metabasalt; **Gemericum** (basement): 14—Paleozoic sandstone, phyllite, lydite and metarhyolite tuff; 15—nappe lines; 16—fault lines

Fig. 17.4 Basic rock sequence of the Slovak Karst: **a** Hallstatt limestone, **b** Wetterstein limestone, **c** Reifling limestone, **d** Werfenian shale (compiled by Ľ. Gaál)

The prevalent part of the Slovak Karst plateaus is built of the Wetterstein limestone. This limestone is light grey, generally massive, locally grading to dolomite. They occur in two main facies: reef and lagoon. The calcispongiae of reefs or calcareous algae of lagoons are oftentimes visible on the weathered limestone surface. The age of Wetterstein limestone covers the period since Ladinian up to Cordevolian (Bystrický 1964; Mello et al. 1997). The 800–1200 m thick Wetterstein limestone was deposited in a slightly inclined continental shelf area (Mello et al. 1997). The Wetterstein and Steinalm limestones are chemically very pure, the content of $CaCO_3$ can reach 97–99%. Many large caves of the Slovak Karst were formed in the Wetterstein limestone (Gombasecká, Domica, Hrušovská caves and Silická Ice Cave).

Of minor extension are some basin facies that occur within the Middle Triassic succession. These are dark-grey, bedded Reifling and Raming limestone, red bedded Schreyeralm limestone and variously coloured Nádaska limestones. They originated from pelagic basins and only locally crop out at the surface, as narrow belts.

Also, some Upper Triassic shallow water and basin facies limestones participate in the geological structure of the area. Carnian light-grey Tisovec (or Waxeneck) lagoon limestone with lamellibranches occurs in smaller areas on the Silická and Jasovská plateaus as well as at the Ostré vŕšky Hill on the Plešivská Plateau (Kochanová 1987). The Carnian-Norian grey or reddish, bedded cherty Hallstatt basin limestone comes out only locally near the villages of Silická Brezová and Bohúňovo (Balogh 1940; Mišík and Borza 1976; Mock 1980). In the upper part of the Hallstatt limestone, the Rhaetian marly Zlambach Beds are known to occur near the Bohúňovo and Silická Brezová villages (Mock 1973). The Norian light-grey massive Dachstein limestone is uncovered sporadically in the southern part of the Jasovská Plateau and at the Drienkova Hill near the village of Drnava. The Skalistý potok Cave in the Jasovská Plateau has been developed in the Dachstein limestone with some accumulation of lamellibranches. The stratigraphically highest beds of Triassic limestone in the Slovak Karst are present at Rhaetian Bleskový prameň limestone in the northern part of karst, near Drnava. The locality contains rich fauna of brachiopoda, lamellibranchiate and corals (Stürzenbaum 1879; Kollárová-Andrusovová and Kochanová 1973; Siblík 1967 and others).

Some deep-water Jurassic sediments have been preserved near the villages of Bohúňovo and Drnava, as small outcrops of radiolarites, marly shales (Allgäu Beds) and reddish Adnet and Hierlatz limestones with ammonites. Shallow water limestone of Late Jurassic age is known only from pebbles in conglomerate (Mišík and Sýkora 1980). After termination of Triassic and Jurassic sedimentation, uplift occurred, with sliding of nappes.

Upper Cretaceous sediments are represented by shales and sandstones in paleokarst cavity fillings in the Gombasek, Hosťovce and Včeláre quarries (Mello and Snopková 1973; Marschalko and Mello 1993; Cílek and Svobodová 1999; Gaál et al. 2007a) and remain of shallow marine limestones in the Miglinc valley (Mello and Salaj 1982).

The younger neoalpine sediments are discovered only sporadically in the Slovak Karst. Locally the Eocene–Oligocene coal beds, clays and freshwater limestone of Šomod Formation and Drienovec conglomerate are preserved in the Turnianska Basin (Vass et al. 1994). Also, a small bauxite deposit is covered by conglomerate (Matějka 1958). Fluvial Sarmatian-Pannonian angular gravel has been transported into the abysses and caves. Its larger accumulation occurs in the Zúgo Cave in the western foot of Plešivská Plateau and in the paleokarstic cavity of Cseppkő quarry near the village of Plešivec (Gaál 2008).

The Late Miocene-Pliocene fluvial gravel and sand of Poltár Formation survived in the deep valleys of Slaná, Štítnik and Bodva rivers as well as on the former fluvial plain in front of the Domica Cave.

The Pleistocene and Holocene cave sediments were studied in several caves, e.g. in the Jasovská, Domica, Maštaľná, Leontina, Zbojnícka, Ardovská, Silická ľadnica and Moldavská caves, mostly during archaeological researches (e.g. Bárta 1955; Ložek et al. 1956; Lichardus 1968; Ložek and Horáček 1992; Soják 2007). The Holocene tufa cascades were created in front of several cave entrances and springs (Krásnohorská Cave, Gombasecká Cave, Hájska Gorge, Hrhov and Eveteš springs—e.g. Ložek 1958; Gradziński et al. 2013).

The present structure of the Slovak Karst, with slice blocks of the Silica Nappe, has important implications from the hydrological point of view. Each carbonate slice of the nappe has impermeable Werfenian shales at its base and, therefore, they represent individual hydrogeological structures (Orvan 1999; Gaál 2009). Rainfall—and snowmelt-derived water as well as that from sinking allogenic streams circulate and accumulate in karst aquifers and reappear at the surface at the foot of karst plateaus or at the contact of carbonate rocks with the underlying impermeable shales, in some cases through deep drainage routes. The most significant springs are Zúgo and Brzotínska in the foot of Plešivská Plateau; Buzgo, Hradná, Pstružia, Bohúňovo, Biela and Čierna vyvieračka beneath the Silická Plateau and Eveteš in the south-western foot of Horný vrch Plateau (Orvan 1999; Haviarová and Gruber 2014). The flooded, collapsed shaft Morské oko in the Rimavská Basin near the town of Tornaľa represents an artesian spring of deep-circulation waters along the south-western fault in the marginal part of the Slovak Karst (Orvan 1973; Gaál et al. 2007b; Bella and Gaál 2017).

17.4 Karst Evolution

There were several phases in the alpine geodynamical evolution of the Slovak Karst, during which some thick carbonate deposits were created together with paleokarst and karst phenomena.

17.4.1 Beginnings of Sedimentary Basin Origin of the Slovak Karst

The beginnings of Alpine development of the territory reach back to the Late Permian, when continental basins were formed, into which disintegrated clastic rocks were transported. They are preserved as conglomerate and sandstone beds in the Rožňava Formation in the southern part of Gemericum unit in the Slovenské rudohorie Mountains (Vozárová and Vozár 1988). These basins have already been in connection with slowly transgressing sea from the south in the latest Permian. Sandstones and carbonates of this environment are preserved in the Štítnik Formation. The concentration of salts and sulphates in lagoons during periods of warm and dry climate with extensive evaporation led to the accumulation of evaporates (Perkupa Evaporite Formation in Hungary).

A shallow sea character of the environment persisted also in the Early Triassic. The layer of fine clastic sediments coming from the nearby coast managed to compensate slow subsidence of basins. Thick masses of clastic sediments were deposited, which are now known to occur in the form of greenish and violet siltstones, slates and sandstones, which sporadically evaporates in the lower part of the Werfen Formation (Bodvaszilas Beds). Later on, in the upper part of Early Triassic, the sedimentation had a carbonate character with marly sandstones and limestones (Szin Beds).

17.4.2 Rifting of Carbonate Platform and the Origin of the Meliata Ocean

Carbonate character of sedimentation, without the contribution of clastic components, was strengthened in the early Anisian when Gutenstein, Steinalm and Honce limestones were formed (Mello et al. 1997). This sedimentation was interrupted by rifting of the carbonate platform in Pelsonian (circa 237 Ma ago) (Fig. 17.5). Re-melted masses of basic (alkaline) rocks from deeper parts of the crust and upper mantle invaded into the rift. Due to this rifting, a new narrow and long ocean basin was born in NW–SE direction, named as Meliata Ocean or Meliata-Hallstatt Ocean (Kozur and Mock 1973; Kozur 1991; Mock et al. 1998). An ophiolitic sequence, classified as Meliaticum unit, was created on this ocean bottom (l.c.).

The widening of the floor of the Meliata Ocean continued until the Carnian in the Late Triassic. Sedimentation on the ocean floor continued, represented mainly by radiolarites, dark clays and marls (at present shales) in the Jurassic, until the end of Callovian. The Meliata Ocean was definitively closed by subduction in the Oxfordian (Mock 1978; Kozur 1991, 1997; Mock et al. 1998).

The process of ocean crust subduction from the south to the north had many consequences. The pressure of submerging crust metamorphosed the turbidite sediments, previously accumulated in the oceanic trench, under low temperature and high-pressure conditions and obducted them partially on the continental crust of the southern part of Gemericum in an intensively overfolded form of the Bôrka Nappe. On the basis of radiometric dating of fengite using $^{40}Ar/^{39}Ar$ method, the age of high-pressure metamorphism was determined to 172–152 Ma (Mello et al. 1983; Maluski et al. 1993; Dallmeyer et al. 1993), which corresponds to the Callovian until the Oxfordian.

Prevailing parts of oceanic deposits of the Meliaticum unit and Bôrka Nappe were absorbed by subduction, and only their remains were left in the highest parts of the Earth crust adhered to the continental crust or preserved close to it. At present, these remains occur near the villages of Meliata, Držkovce, Štítnik, Bôrka and Hačava.

Sediments deposited on the continental slope or nearby are classified into the Turnaicum. They were developed under deep-water turbidity conditions, often with volcanic intercalations, and are represented mainly by shales, phyllites with intercalations of sandstone and silicite, and dark cherty limestones. Their bulk part is weakly metamorphosed and they contain allodapic intercalations (Mello 1979; Mello et al. 1997). They are preserved on the surface mainly in the Turnianska Basin.

17.4.3 Carbonate Platform on Triassic Shelf

While rifting and the creation of the oceanic crust occurred in the deep Meliata Ocean, the formation of limestone continued in the shallow parts of the shelf sea. Limestones were formed either from reef or from algal stromatolites of a lagoon, which spread behind the reefs (e.g. Mello 1974). In this way, around 2 km thick mass of limestones was deposited on slowly subsiding shelf during the Triassic.

Probably as a reflection of extension rifting in the Anisian, some neritic intraplatform basins were created with limy mud sedimentation in the Early Ladinian. Similar basins occurred also in the Upper Triassic.

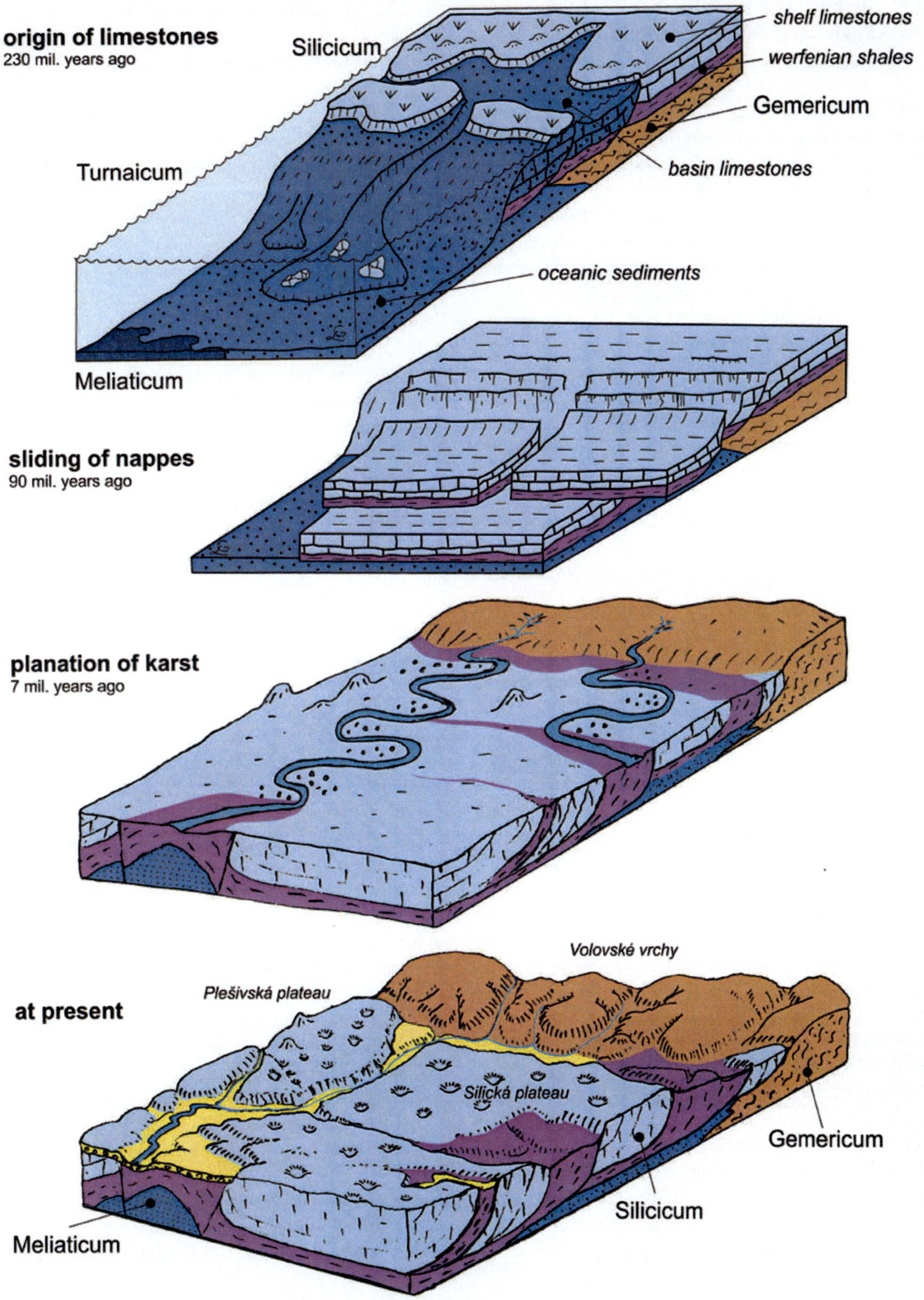

Fig. 17.5 Geological and geomorphological development of the Slovak Karst (Ľ. Gaál)

17.4.4 Jurassic Sea and the Origin of the Oldest Paleokarst Phenomenon in the Silica Nappe

Locally, between the Triassic and the Jurassic, a short emersion and tilting of carbonate blocks occurred. Karstic weathering and gravitational movements after emergence led to the creation of crevices (neptunic dykes) and abrasion cavities. A cavity in the Late Triassic Dachstein limestone with the dimension of 44 × 35 m, filled with Jurassic Hierlatz limestone, has been documented in the western vicinity of the Slovak Karst, between the villages of Bretka and Meliata (Gaál 2008).

This emergence ended with the transgression of the Early Jurrasic Sea. After the Hierlatz limestone, the dark spotty marl (Allgäu Beds), and later also deep-water radiolarites sedimented in the Middle Jurassic. Shallowing of the sea happened again only at the end of the Jurrasic (Titonian). However, these limestones are not preserved to the present and we know about their existence only from the pebble content of Tertiary deposits (Mišík and Sýkora 1980).

17.4.5 Movement of the Silica Nappe

Early Cretaceous sediments are absent in the Slovak Karst; therefore, we can assume the rise of the carbonate masses at that time. Due to this, uplift listric faults originated in the brittle carbonate complex overlying the ductile Early Triassic shales with evaporites. These layers were suitable environments for the movement of thick carbonate blocks along the listric faults. Probably due to vaulting of Veporic and Gemeric Palaeozoic domain in the Turonian, the transport of slice blocks to the south occurred, forming the Silica Nappe (Mock 1978; Bajaník and Reichwalder 1979; Mello et al. 1997; Gaál 2008). The Silica Nappe moved into the deep water sediments of Turnaicum and Meliaticum units, probably in the southern neighbourhood of the carbonate platform of the Silicicum unit. However, another concept of location of the Silica Nappe has appeared, mainly to the south of Meliaticum (e.g. Csontos 1992; Hók et al. 1995; Rakús 1996; Csontos and Vörös 2004). The origin of the Silica Nappe in the domain of Gemericum is supported by the similarity of limestone successions with the Muráň and Stratená nappes, as well as with the Hronic Unit, the absence of the Mesozoic cover formations of Gemericum, and Austroalpine character of Conodonta association of the Silica Nappe, unlike the Dinaride Conodonta features of Meliata, Turna and Bükk units (Mock 1978).

17.4.6 The Cretaceous Paleokarst

Despite the probable emergence of carbonate masses in the Early Cretaceous, we have no evidence of any karstic phenomena from that time. Karstic processes are only indirectly suggested by the lack of the prevailing part of 400–500 m thick Dachstein limestone, which is preserved only in small areas in the Slovak Karst.

The best evidence of paleokarstic phenomena from the Late Cretaceous postdates the shift of the Silica Nappe to its current place. Phenomena from this phase are observable in paleokarstic cavities in several quarries. The cavity in the Gombasek quarry, with a length of 40 m and a height of 20 m, is filled with dark-grey claystone and shale which alternates with laminated sandstone (Fig. 17.6). There is a turbidity flyshoid cave sediment with cruciform and convolute texture, as well as with collapsed boulders (Marshalko and Mello 1993). The cavity walls are covered by thin lateritic brim with Fe-hydroxide. The Campanian age of sediment was determined by palinomorphs (Mello and Snopková 1973). Other paleokarstic cavities of Late Cretaceous ("Senonian") age were recognized in the Gombasek and Hosťovce quarry by Cílek and Svobodová(1999). Moreover, two paleokarstic cavities in the Včeláre quarry in the eastern part of Slovak Karst were found. One of them can be clearly distinguished, having a length of 5 m and a height of 3.5 m, lying 89 m beneath the karst surface of Dolý vrch Plateau. It is filled by shale and light-grey bedded marly sandstone with thin Fe-hydroxide brim and a layer of collapsed limestone fragments. Its Late Cretaceous age has been determined by the presence of palinomorphs, however, a spicule of freshwater silicispongiae was found too (Gaál et al. 2007a). Ferric sediments of caves in redox conditions were created in occasionally present water bodies. Sources of iron could be from humic substances in swamps, marshes and lakes. Even these days, in many places and in caves of the Slovak Karst, mainly in the Plešivská Plateau, residual iron pisolites and crusts can be found (Kováčik 1955; Cílek and Bednářová 1994).

Remains of Cretaceous karstification have been preserved on the surface too. Some conical hills in the Plešivská Plateau like Berc, Vlčí štít, Ostré vŕšky and Malý vrch may be the remnants of tropical conical karst (Lukniš 1972; Jakál 1975; Činčura 2002). The Late Cretaceous karstic processes occurred under warm and humid tropical conditions. Accordingly, there were well developed and dissected karst areas in certain parts of the present-day Slovak Karst, with around 200–300 m thick vadose zones, with deep cockpits, shafts, mighty vadose and epiphreatic fluviokarst

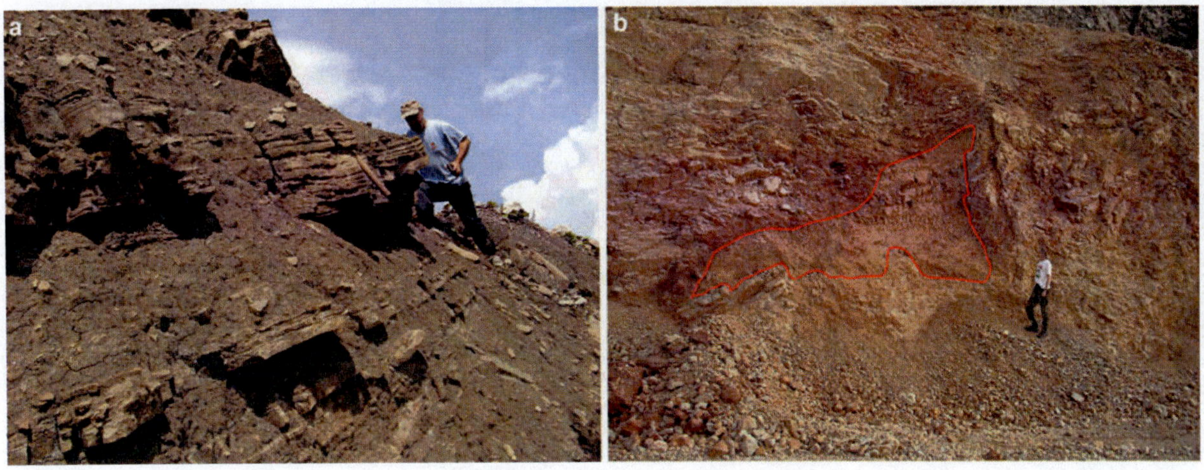

Fig. 17.6 Upper Cretaceous fillings of paleokarstic cavities: **a** Gombasek quarry (*photo* J. Kilik), **b** Včeláre quarry. The filling is outlined in red (*photo* Ľ. Gaál)

underground passages (Gaál 2008). According to the occurrences of limestone fragments and boulders in the fill of paleokarst cavities, mechanical roof flaking also took place. Larger underground systems were probably not created in every limestone block with sufficiently vadose zone. Moreover, younger karst processes erased traces of older karstification. Therefore, currently known evidence of Cretaceous karst comes from the last period before the termination of the karst cycle.

The karstic processes in the Late Cretaceous have finished as a consequence of slightly subsidence of the area with the entry of lacustric and brackish water, later (in the Late Santonian or Campanian) also a shallow sea (Mello and Salaj 1982).

17.4.7 Blocks Movement in the Cenozoic and Its Consequence

At the beginning of the Paleogene, the Alcapa microplate (including the Eastern Alps, Western Carpathians and North Pannonian Pelso unit) moved to the neighbourhood of Tisia unit (Csontos and Vörös 2004). Because the Slovak Karst is located near this connection, a very strong tensional field arose. The direction of compression was mainly from SE to NW (Vass et al. 1993), which resulted in thrusting of carbonate blocks towards the north. Due to strong pressure, the plastic Werfenian shales were deformed and mostly reduced, while the solid carbonate blocks were deformed only slightly to form shallow synclines or monoclinal structures. Several blocks have got northern orientation (vergence). The filling of Cretaceous paleokarstic cavities was also slightly deformed (Sasvári et al. 2006). Probably, the whole territory of the Slovak Karst was a large circumlittoral plain with the shallow vadose zone and without significant karst phenomena in the Paleogene.

Due to the subduction of outer Carpathian, flysch oceanic crust in the extension regime with thinning of continental crust was created in many places of Western Carpathians. In the Late Oligocene and Early Miocene, several blocks have subsided along extension faults, which created the inner Carpathians basins with NW–SE orientation (Kováč et al. 1989; Kováč 2000). The Rimavská and Košická basins were originated in the neighbourhood of the Slovak Karst. In the karstic area between these basins, many faults were created, mainly in the same direction (NW–SE or WNW-ESE), which importantly influenced the direction of karst water circulation and the formation of underground spaces and doline chains.

In consideration to extensional tectonic regime, in the Neogene the prevailing part of the Slovak Karst was still in depression, with the narrow vadose zone. Only the south-eastern part of Jasovská Plateau was moderately uplifted in consequence of tectonic pressure of the Darnó line. The uplifted carbonate block provided material for the Drienovec conglomerates, which accumulated in the form of rain-wash cones in the Turnianska Basin (Vass et al. 1994; Mello et al. 1997).

The Middle Miocene (Badenian Sarmatian) volcanism influenced the evolution of relief only to the west of the Slovak Karst. However, most likely some of underground parts of the Slovak Karst were affected by postvolcanic hydrothermal activity. Hydrothermal metasomatic dykes and impregnations of Pb–Zn ore in carbonate rocks of the Silica Nappe near the village of Ardovo can originate from this period (Maderspach 1877; Slavkay and Beňka 1995). We assume that warm solutions, from which the crystals in the Drienka Cave in the Silická Plateau precipitated, were also the outcomes of the post-volcanic activity.

17.4.8 Planation of Karst Relief

In the Sarmatian–Pannonian, the planation surface in the Western Carpathians, as a tectoplain, was formed during prevailing extension (subsidence) tectonics by several exogenous processes depending on changing climatic conditions. The tectonic stability has prevailed only during the final phase of its formation (Minár 2003; following the concept of the initial polygenetic planation surface given by Lukniš 1964). In the Slovak Karst, Sawicki (1908) was the first to write about the formation of planation surface by lateral erosion of rivers in the Miocene. Mazúr and Jakál (1971), as well as Jakál (1975), assumed that the Pannonian planation surface originated through pediplanation due to the lateral erosion of allogenic streams flowing from the Slovenské rudohorie Mountains. The riverbeds of these allogenic streams were covered by gravel.

In relation to the formation of this planation surface, some horizontal fluviokarst caves were formed in protruding limestone elevations in the Pannonian. Their remains are preserved to the present, e.g. the Ortováň Cave (Fig. 17.7) and Silická kvapľová Cave on the Silická Plateau (Bella and Gaál 2005; Gaál 2008).

17.4.9 Uplift of the Slovak Karst

After surface planation, the Slovenské rudohorie Mountains, including the Slovak Karst as its southern marginal part, were uplifted in the end of the Pannonian. The uplift was probably caused by isostatic movement of the Slovenské rudohorie Mountains as a consequence of geodynamical events in the Pannonian back-arc basin in the Late Miocene.

According to Jakál (1975), the uplift occurred in two stages. The first stage is more distinctively shown only along the edges of the Slovak Karst, where around 70 m under the plateau level horizontal caves were formed like the Hámorská Cave and Prastarý vývery. In the northern part of the Silická Plateau, only the stepwise descending old ponor cave of Drienka was formed.

In the second stage, the vertical extent of the uplift of central part of the Slovak Karst reached 500 m. Due to the vault character of uplift, the karst surface was inclined slightly to the southward respectively to west and eastward. As a result, several deep canyons and other valleys were incised by allogenic rivers and streams. The originally continuous karst planation surface was divided into several plateaus. The Koniarska, Plešivská and Silická plateaus in the central part of Slovak Karst are divided by 500 m deep canyons incised by the allogenic Štítnik and Slaná rivers, while the Zádielska and Hájska gorges in the eastern part of the area are incised only by 100 m. As a result of relatively quick uplift vertical direction of karstification dominated in this phase, many dolines, shafts and stepped abysses originated below the karst plateaus such as Zvonivá jama, Malá Žomboj, Brázda, Obrovská abyss and others. Remains of allochthonous gravel of the Sarmatian–Pannonian planation surface were displaced into abysses and vertical caves.

The age of uplift of the Slovak Karst at the end of Pannonian is testified by some boreholes in the Turnianska Basin, which revealed rock fragments transported from the Slovenské rudohorie Mountains in the Pannonian, whereas these sediments are missing in the Pontian. It means cessation of their transport in the Pontian (Gaál 2008).

17.4.9.1 Gravel Deposition and the Late Tertiary and Quaternary Tectonic Movements

The uplift of Slovenské rudohorie Mountains generated high kinetic energy of surface flows, which caused increased erosion of the mountains. The washed down clastic material was deposited in the Rimavská, Rožňavská, Turnianska and Košická basins as coarse-grained gravel, less abundant sand, and in a few isolated cases also clay of the Poltár Formation. The sedimentation had begun in the Pontian and lasted until the Pliocene (Vass et al. 1989; Gaál 2008). The canyons of Štítnik and Slaná rivers were partly filled by fluvial deposits. Due to the aggradation of canyon bottoms, the erosion base has been elevated, while the vertical extent of vadose zone was reduced to 100–200 m (Jakál 2001). Old caves along the edges of the deepest part of canyons were fully filled by allochthonous deposits (Skřivánek 1966). Also, the subsided tectonic blocks of Triassic carbonates at the southwestern and southern margins of the Silická Plateau were covered by these fluvial deposits (Jakál 1975; Móga 1999).

At the end of the Pliocene, the southern part of the Slovak Karst was raised along the Plešivec fault, which passes in the eastern direction towards the village of Silická Brezová (Gaál 2008). This uplift had an amplitude of 70–80 m, therefore, in this area (to the south of Plešivec and Silická Brezová villages), the abysses are shallower than those in the northern part of Slovak Karst (e.g. the depth of Bezodná ľadnica and Malá ľadnica is only 69 m respectively 56 m). The uplift also caused the displacement of Štítnik River from the Plešivec–Dlhá Ves–Trizs (Hungary) line to the current line Plešivec–Čoltovo–Lenártovce. Large accumulation plain, with a width of 2–5 km, originated at the south edge of Silická Plateau. Occasional streams, flowing on the sediments of the Poltár Formation, sunk underground at the contact with Triassic limestones and formed the initial corrosion channels. Because the Jósva brook in Hungary lays lower than the Slaná River, the Jósva began to cut down quicker during the uplift phase than the Slaná River. A part of accumulation plain to the north-west from the village of Dlhá Ves began to be drained by Slaná River, while the south-eastern part, with a higher erosion potential, by the Jósva brook. The incision of the valley of the Jósva brook

Fig. 17.7 The Ortováň Cave developed during the Pannonian planation of karst plateau (*photo* P. Bella)

was also facilitated by the presence of less resistant Lower Triassic shales in the Jósvafő–Szinpetri–Szin zone in Hungary.

Some horizontal caves, including the levels of the Baradla–Domica cave system, originated in the Middle–Late Pliocene and Early Pleistocene during the interruption of tectonic uplift, when the erosion base was stabilized. Based on ^{10}Be and ^{26}Al cosmogenic nuclide dating, allochthonous fluvial sediments (mostly angular quartz gravels) were deposited in the upper level of this cave system in the Middle–Late Pliocene and in the Early Pleistocene in the main evolution level (Bella et al. 2014, 2019). Within the denudation chronology of the Western Carpathians, these cave levels probably correspond to the "foothill" (Late Pliocene) and "river-side" (Early Pleistocene) levels (l.c.).

The Quaternary development of the caves was markedly influenced by uplift in the interglacial between the Günz and Mindel glacials and also in the Mindel, with amplitude of more than 50 m. The erosional potential of Styx stream and its branches in the Domica-Baradla cave system was influenced by the occurrence of quartz gravels of the Poltár Formation (e.g. Kessler 1938), however, the corrosive potential of underground stream also increased since in the period of glacial the water kept its aggressivity for a longer time. During the Pleistocene, slight uplift and rhythmic climate changes influenced the development of multi-level caves, e.g. the Krásnohorská cave, as well as the lower horizontal parts of Jasovská and Gombasecká caves.

The majority of crevice caves in the Slovak Karst were formed in the Quaternary. Jointed zones formed in response to high tensile stresses, with a width of several tens to one hundred meters. They were particularly frequent along plateau borders as a consequence of their destabilization due to erosional deepening of the valley. Tensile failures, usually parallel to the edges of the plateaus, occurred, leading to the opening of several crevice caves like Snežná diera in the Bôrčianska Plateau.

17.5 Surface Landforms

17.5.1 Karren

Although karren are small-sized solution forms, they are among the most significant features of the Slovak Karst. They are especially concentrated on steep and deforested slopes of karst plateaus, on doline slopes and rock walls (Jakál 1975, 2008). High chemical purity of bedrock, mainly of the Wetterstein and Steinalm limestones, results in the formation of only a thin layer of weathering residue remaining after its dissolution. Karren occur in several parts of the Slovak Karst and originated during the Holocene, mainly in the Atlantic interval (5,000–4,000 years ago), when precipitation was from 60 to 70% and temperature 3 °C higher than the present. After deforestation of some karstified areas, karrenfields were uncovered by soil erosion, caused mostly by pasture of domestic animals (Mazúr and Jakál 1971; Jakál 1975; Ložek 1994).

The morphology and genesis of karren are influenced by the lithological and tectonic attributes of carbonates. The most typical karrenfield occurs near the village of Kečovo in the southern part of Silická Plateau (Bella et al. 2009). The most common *rundkarren* (subsoil rounded runnels) are formed by subsoil corrosion. The widespread occurrence of *fissure karren* is conditioned by intensive tectonic disintegration of limestones. Several *tube-like karren* (solution tubes) and *shaft (well) karren* occur in typical appearance. After deforestation and uncovering of subsoil karren, some types of younger karren (*solution pits* and *pans*, *rinnenkarren*) were shaped by corrosion on bare limestone surfaces (l.c.) due to rock interactions with precipitation water (Fig. 17.8).

The steep, upper part of the southern slope of Plešivská Plateau is a remarkable site of *rillenkarren*, *rinnenkarren* and *wandkarren* (wall solution runnels) (Bella and Gaál 2010). Dominant morphological and genetic karren types of Plešivská Plateau are unlike those within the Kečovo karrenfield on deforested gentler slopes in the southern part of Silická Plateau. Some karrenfields cover the south-facing slopes of the Zádielska and Jasovská plateaus.

17.5.2 Dolines

Solution dolines are closed depressions, rounded, oval or longitudinal in ground plan, deepened into the limestone surface (Fig. 17.9). Their dimensions are mostly around 50–100 m, the biggest are 200–250 m in diameter and 10–15 m deep, while the deepest ones reach 30–40 m (Jakál 1975, 2008). The slopes reach various inclinations and are bare or covered only by a thin layer of soil. The Plešivská and Silická plateaus have the density of 40 dolines per 1 km^2, while the Jasovská Plateau has much lower density. The most common shapes are funnels, bowls and kettles. Doline asymmetry in cross-section results mostly from bedding attitude of stratified limestone or a different lithology of carbonates (Jakál 1975, 2008). The dolines within the layered, monoclinally laid Gutenstein limestones are distinctly asymmetrical. Dolines are most frequently arranged in lines being controlled by faults.

A different process leads to the origin of *collapse dolines,* which originated by collapses of cave roofs into the voids beneath, to form a good shape. The well-known collapse doline in the Slovak Karst is the depression in front of the entrance to the Silická ľadnica Cave on the Silická Plateau.

Alluvial dolines (funnel depressions on the sediment surface) were formed by suffosion processes on fluvial or unconsolidated sediments as a result of dissolution of the limestone which lies under these sediments. They are created on the bottoms of larger karst depressions covered by sediments (Jakál 2008).

17.5.3 Uvalas

They represent compound enclosed depressions, resulting from the connection of several neighbouring solution dolines after the separating ridges were mostly denuded. Large, several 100 m long depressions are formed in this way. The floors of uvalas are dissected by bottom parts of former dolines. The origin of uvalas is also connected with the bottoms of paleo-valleys within the karst plateaus.

17.5.4 Inactive Poljes

In the Slovak Karst, typical active poljes are absent. Larger closed depressions about 60 m deep, with a flat bottom occupying an area of about 1 km^2 on the Silická and Plešivská plateaus, are considered as paleo-poljes (Jakál 1984, 2008). They were developed during the period after a minor river incision into the karst plateau, probably after the first stage of uplift. Their bottoms were bound to the erosion base of water courses draining the karst territory. They were occasionally flooded and widened by lateral corrosion. Then, in the second stage of uplift, they were elevated 500–600 m upwards, and the process of deep karstification was renewed. Therefore, the depressions lost their hydrological function and they ceased to be occasionally flooded. Likewise, the possibility of further enlargement was severely reduced. A typical example of such a depression occurs in the NW part of Silická Plateau and was named Bučina.

Another paleopolje, named Skalické polje, is situated at the contact of the Koniarska Plateau and Rimavská Basin

Fig. 17.8 Karrenfield near the village of Kečovo: **a** general view, **b** kamenitza, **c** well karren, **d** rundkarren, **e** fissure karren (*photo* Ľ. Gaál)

(Fig. 17.10). Its origin was enabled by tilting of the marginal block of the limestone-dolomite complex towards the surrounding relief (Kemény 1961; Jakál 1984). In active stage, it behaved as a border polje, but later, during uplift of the Rimavská Basin in the late Pliocene it lost this function. A probable remnant of a hum hill occurs in the central part of the polje, hosting the small Skalická Cave, which can behave as a ponor or an estavella, depending on hydrological conditions (Hochmuth 1996; Gaál 2008).

17.5.5 Blind Valleys

Blind valleys occur in the Slovak Karst mainly in its marginal parts, at the contact of karstic and non-karstic rocks. They are closed in the direction of the watercourse by steep rock slopes, under which the waters are sinking into the underground through ponors. They are most common on the Silická Plateau, for example to the NW from the village of Silická Brezová, where an occasional stream of about 1 km

Fig. 17.9 Dolines on the Silická Plateau (*photo* Ľ. Gaál)

Fig. 17.10 The Skalické inactive polje in the southern foot of Koniarska Plateau (*photo* Ľ. Gaál)

long flows through a valley deepened into Werfenian impermeable shales, and at the contact with limestones, it sinks into the underground through a ponor under the Červená skala hill (Fig. 17.11.). Similar examples occur on both sides of Fabianka Hill built by Werfenian shales, to the east of Silica village. In the southern part of this shale belt occasional streams flow, sinking into the underground after several hundred meters through ponors in the limestone of Vápenný Hill. Some caves at the site of older ponors are known in this area, the largest of them being the Majková Cave, with the length of 560 m. One of these ponors, named Farárova jama near Silica, was choked up by clay minerals derived from weathering of Werfenian shales and the karst marginal lake was created. The same situation is known in the northern part of Fabianka Hill, where the Jašteričie Lake was created (Kunský 1939).

A basin-like depression near Silická Brezová on the Silická Plateau presents a set of blind valleys bounded by low limestone ridges. The floors of these valleys are flat, as a result of aggradation of fine-grained and less permeable sediments. Occasionally flowing streams are sinking into the underground through invasion ponors (Milada and Matilda inflow caves) in the terminal parts of blind valleys. They are flooded during rainy periods and at the time of snow melting. As a consequence of a hydraulic gradient between invasion ponors and the underground river bed in the Milada Cave, steep drawdown cave passages originated in the inflow parts of the Milada and Matilda caves. The spring of a permanent stream, which flows through the Milada Cave, occurs in the middle part of the Silická Plateau at the narrow belt of Early Triassic shales (Bella and Gaál 2013).

The same situation occurs in the southern border part of the Silická Plateau, in front of the Domica-Baradla cave system, although non-karstic rocks are different. An inactive accumulation surface, 2–5 km wide, originated on gravel, sand and clay of Poltár Formation. Intermittent streams, flowing on the sediments of the plateau, sunk underground in contact with limestones and formed the initial corrosion channels. Ponors were functioning on the accumulation platform with side branches from the Smradľavé Lake, Domický Brook, Demikov Ponor and others. These ponors already enabled for inflow of larger amounts of torrential waters into cave spaces. They are drained by the Jósva spring in the Hungarian part. On the Slovak side, the Dlhoveská blind valley was formed, with the ponor of Ardovská Cave which itself is drained by the Slaná River.

17.5.6 Dry Valleys

Dry valleys were formed by occasional surface flow, mainly in the lower-lying marginal part of karst plateaus, e.g. between the village of Ardovo and the Domica Cave (Jakál 1975). Other slope dry valleys are observed in the Aggtelek Karst, Hungary.

Fig. 17.11 Blind valley located the NW from the village of Silická Brezová (*photo* P. Bella)

17.5.7 Fluviokarst Valleys

There are two main types of classically formed valleys in the Slovak Karst: gorges and canyons, which differ by size, morphology and age (Fig. 17.12). Gorges are deep valleys with a "V" shaped cross-section and steep up to vertical slopes with cliff relief. They are characterized by irregular gradient of the river bed, without forming a floodplain on the valley bottom (Jakál 2008). The most beautiful gorges are the Zádielska and Hájska in the eastern part of the Slovak Karst. They are incised by 200–250 m into the karst plain and due to considerable gradient of the river bed they strongly erode into the underlying limestone. The valley bottoms reach widths of only 6–10 m. Cascades, rock steps with waterfalls, potholes and springs are frequent in the river bed. Travertine cascades are formed at the mouth of the Hájska Gorge. The occurrence of caves within the valley sides indicates the possibility of occasional influence of underground branch watercourses. The valley shape and cave sediments testify to young Quaternary uplift of the eastern part of the Slovak Karst.

Canyons are mighty valleys characterized by great depth, steep rocky slopes, regular gradient of the river bed and a wide aggraded floodplain without river terraces. There are two canyons in the Slovak Karst formed along the Slaná and Štítnik rivers. They are 350–450 m deep in respect to the level of the upland, with the floodplain width of 250–700 m (Jakál 2008). Slope inclinations reach up to 40°, with the upper slopes being increasingly steep upward, up to vertical, and terminating with a sharp edge at the slope/plateau junction. Lower slopes are bordered by a belt of Pleistocene debris cones.

17.5.8 Karst Ridges and Cone Hills

The surface of uplifted plateaus is dissected by irregularly arranged many karst depressions, mostly dolines and uvalas. Elevated ridges and hillocks of 30–60 m high represent remains of an original planation surface of the mid-mountain level from the Pannonian period. Noticeable conical hills with a more relative height of 50–120 m above the general level of the planation surface of the karst plateaus are considered as remains of the tropical Cretaceous cone karst (Lukniš 1972; Jakál 1975; Gaál 2008), e.g. Ostré vŕšky Hills (775 m a.s.l.) on the Plešivská Plateau and Veterník Hill (609 m a.s.l.) on the Koniarska Plateau.

17.6 Underground Geodiversity

17.6.1 Cave Morphology and Origin

Various morphological and genetic types of caves occur in the Slovak Karst, resulting from the development of the plateau karst in the temperate climatic zone (Tables 17.1 and 17.2). Vertical shafts and stepped abysses are dominant. They originated in the vadose zone during and after the uplift of the carbonate massif by corrosion of infiltrating meteoric water. The Veľká Žomboj cave in the Silická Plateau represents a typical simple vertical abyss with the depth of 147 m. The historically remarkable 100.5 m deep Zvonivá jama in the Plešivská Plateau is featured by voluminous chambers at its bottom. It is the oldest surveyed abyss in the Slovak Karst, first mentioned by Windish in 1780. The Brázda, Veľká Bikfa and Diviačia abysses have a

Fig. 17.12 Valleys in the Slovak Karst: **a** Štítnik canyon, **b** Zádielska gorge (*photo* Ľ. Gaál)

Table 17.1 The longest caves of the Slovak Karst and adjacent Medzevská Hilly Land

Name of cave	Location	Length (m)
Skalistý potok	Jasovská Plateau	8,215
Domica	Silická Plateau	8,127 (30,154 m with Baradla)
Jasovská jaskyňa	Medzevská Hilly Land	3,924
Moldavská jaskyňa	Medzevská Hilly Land	3,070
Gombasecká jaskyňa	Silická Plateau	3,057
Silická ľadnica	Silická Plateau	2,300
Drienovská jaskyňa	Jasovská Plateau	1,558
Krásnohorská jaskyňa	Silická Plateau	1,550
Ardovská jaskyňa	Silická Plateau	1,510
Milada	Silická Plateau	1,308
Jaskyňa v ponore Jašteričieho jazera	Silická Plateau	1,189
Hrušovská jaskyňa	Horný vrch	1,139
Veľká Bikfa	Silická Plateau	1,101
Brázda	Silická Plateau	1,100
Nová brzotínska jaskyňa	Plešivská Plateau	800

Table 17.2 The deepest caves of the Slovak Karst

Name of cave	Location	Length (m)
Skalistý potok	Jasovská Plateau	376
Kunia priepasť	Jasovská Plateau	203
Čertova diera	Horný vrch	186
Brázda	Silická Plateau	181
Malá Žomboj	Silická Plateau	147
Veľká Bikfa	Silická Plateau	147
Ponorová priepasť	Silická Plateau	135
Čertov dych	Horný vrch	130
Diviačia priepasť	Plešivská Plateau	123
Veterná priepasť	Horný vrch	120
Silická ľadnica	Silická Plateau	117
Natrhnutá priepasť	Dolný vrch	112
Hlinoš	Dolný vrch	110
Attilova priepasť	Silická Plateau	109
Gombasecká jaskyňa	Silická Plateau	104

stair-step vertical morphology (Fig. 17.13). The lower parts of the Kunia Abyss in the Jasovská Plateau are sculpted by an occasional stream. In several places, the abysses are concentrated in a very high density—up to 20 abysses occur within an area of 0.5 km^2 on the Dolný vrch Plateau (Bella et al. 2021), where the best known and most typical is the Obrovská abyss, 100 m deep. Several abysses or cave entrances are the result of cave ceiling collapse, so-called "light holes", i.e. the entrance part of Silická Ice Cave in the Silická Plateau and Zombor Abyss in the Plešivská Plateau (Roth 1939, 1940a; Mitter 1988; and others).

Horizontal and subhorizontal caves have been formed by underground streams, which emerge to the surface at the foot of karst plateaus in the wide canyons or adjacent basins. The Brzotínska Cave and Zúgo Cave are typical active fluviokarstic caves beneath the Plešivská Plateau. At the foot of Silická Plateau, the outflow multi-level Gombasecká and Krásnohorská caves occur. The Hrušovská Cave is a large fluviokarstic cave of the Horný vrch Plateau, and the Drienovská Cave is the largest one beneath the Jasovská Plateau. The upper part of the Drienovská Cave, with a cupola-shaped morphology and large calcite crystals, has some features of hypogene origin (Zacharov 2013; Filipčíková et al. 2020). Senile horizontal caves close to the surface of karst plateaus were formed at the time of the formation of the Sarmatian–Pannonian planation surface or during its

Fig. 17.13 Diviačia Abyss in the Plešivská Plateau (*photo* J. Stankovič)

initial dissection by younger landforms, e.g. Ortováň Cave (Bella and Gaál 2005). The so-called "unroofed caves" with ancient sedimentary fill are all that remained from senile caves (Mitter 1988; Lešinský 2006; Gaál 2008).

Several occasional allogenic streams sinking at the edge of karst plateaus, also permeable gravel formations deposited on the subsided tectonic blocks of carbonates at the southern margin of Silická Plateau are drained into the Domica Cave. The Domica-Baradla cave system (Fig. 17.14), with the length of 29.9 km (of it 8.1 km in Domica and 21.8 km in Baradla), was gradually created mostly in the shallow phreatic zone or along the piezometric surface of

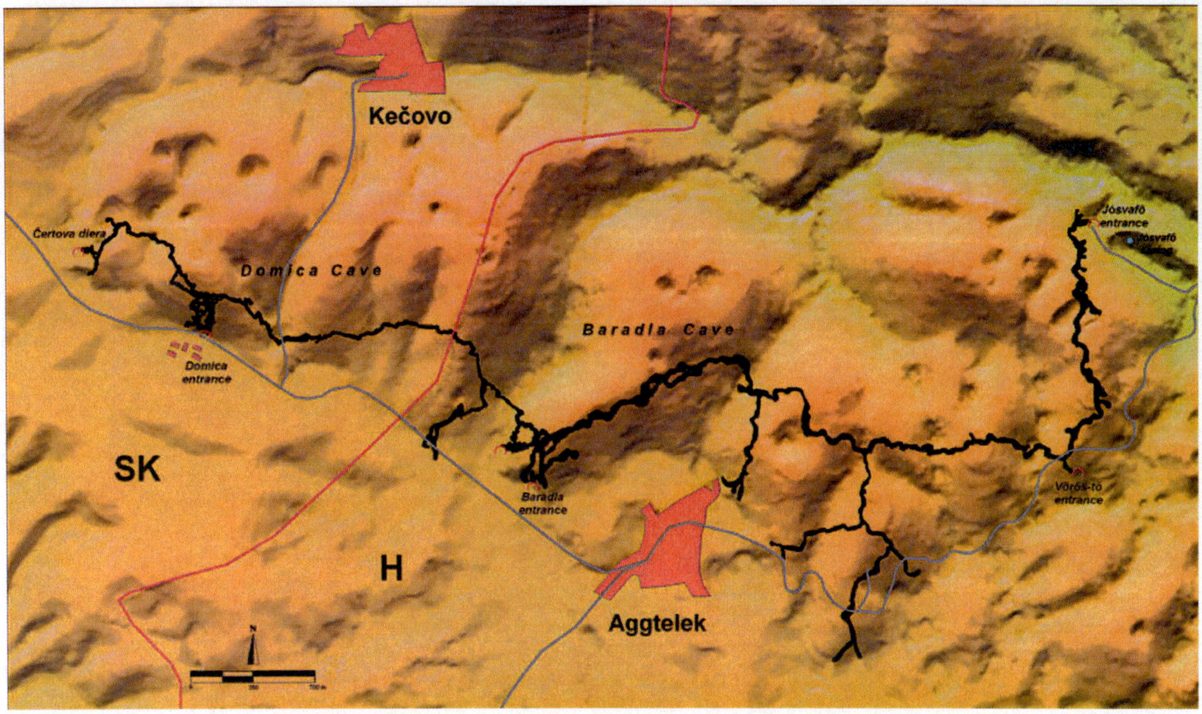

Fig. 17.14 The map of Domica-Baradla Cave System (according to Kessler 1938 and Droppa 1961)

underground karst waters. Therefore, it has several evolution levels of reflecting phases of Jósva valley incision (see Kessler 1938; Roth 1940b; Droppa 1961, 1971, 1972; Jakucs 1975 and others). Slightly inclined passages leading from the Domica Cave through the state boundary and continued by the main part of the Baradla Cave to Jósvafő are an excellent example of an "ideal watertable cave" (Ford 2000; Ford and Williams 2007). They have been formed in relation to the formation of large pediments in the Jósva River valley (Bella et al. 2019). Many passages of the cave system are featured by ceiling and wall channels (Fig. 17.15). For the first time in the world, the morphology and evolution of ceiling channels were described by Roth (1937) in Domica Cave (see Bella and Bosák 2015). Ján Majko, with fellow cavers, discovered the Domica Cave in 1926 and connected it to the Čertova diera Hole in 1929. In 1932, Hubert Kessler and József Sandrik first went from the Baradla Cave through the siphon to the Domica Cave (Lalkovič 2014; Székely 2014).

Some vadose drawdown inflow caves originated on the Silická Plateau, at the contact of impermeable Werfenian shales with limestones. The Majková Cave and the Jaskyňa v ponore Jašteričieho jazera Cave developed on both sides of Fabianka Hill built of Werfenian shales near the village of Silica. The Milada Cave is an underground continuation of blind valleys near the village of Silická Brezová.

The lower part of the Skalistý potok Cave, located in the southern part of Jasovská Plateau, consists of subhorizontal channels with numerous loops and shallow lakes linked with 25 siphons from a few meters to 100 m long (Hochmuth 2013). It is a typical example of "a cave with mixture of phreatic and watertable-levelled components" (referred also by Ford and Williams 2007).

In the hilly land between the eastern part of Jasovská Plateau and the western part of Košická Basin, which consists of subsided carbonate blocks covered by gravels, the Jasovská Cave was created, consisting of five levels (Droppa 1971). There are some remarkable smaller rock-cut morphologies such as meandering ceiling channels and flat ceilings. The lower parts of the cave are characterized by a maze pattern. A more representative example of maze cave is the Moldavská Cave, which is a complicated labyrinth of phreatic and epiphreatic passages and halls with a total length of 3,070 m (Hochmuth 2000). It originated mainly due to the diffuse penetration and repeated flood injections of allochthonous waters from the Bodva River, mixed with seeping meteoric waters (Bella et al. 2007, 2018).

Crevice abysses and caves developed by gravity block movements occur along the steep edges of the plateaus and through valleys, e.g. Snežná diera and Čertova diera abysses in the Horný vrch Plateau, the segment of Nová Brzotínska Cave parallel with the edge of Plešivská Plateau (Skřivánek

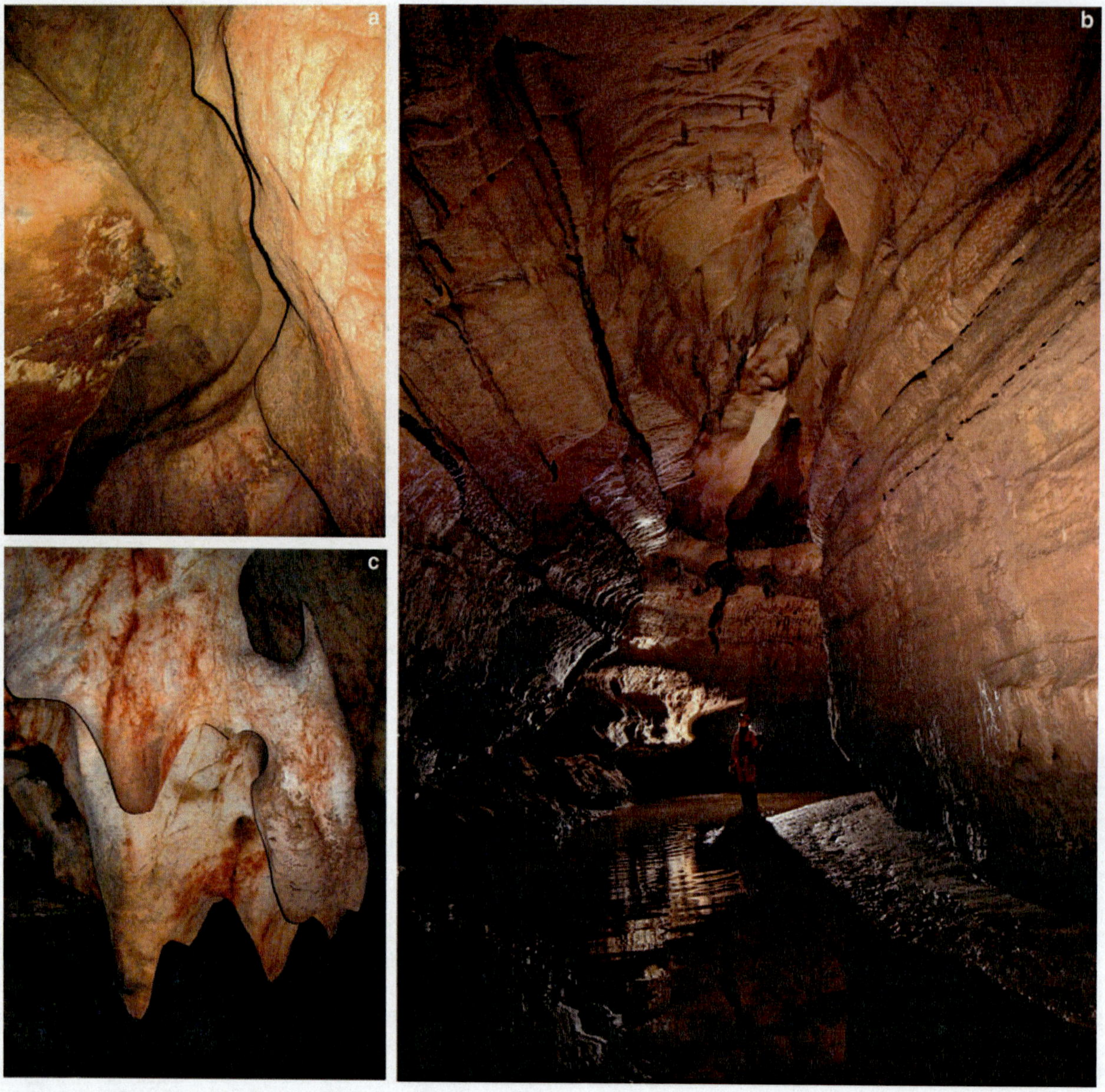

Fig. 17.15 Morphology of the Domica Cave: **a** meandering ceiling channel (*photo* P. Bella), **b** canyon-like passage of the Styx stream (*photo* P. Staník), **c** paragenetic pendants (*photo* P. Bella)

and Stárka 1956; Kučera 1976; Mitter 1988). According to Baroň (2002), crevice abysses formed also in the gravity fractured edges of old collapse dolines, partially filled by limestone boulders and debris.

Shelter caves are formed by weathering, rock breakdown and gravity-driven sediment transport, e.g. caves in the Gerlašské Rocks on the northern slope of Plešivská Plateau (Mitter 1988). In the Hájska Valley, where travertine cascades occur, a small constructional progradational (syngenetic) cave within a constructive travertine waterfall is known (Gradziński et al. 2018 and others).

17.6.2 Variability of Cave Deposits

The highest variability and richness of carbonate speleothems occur in the Domica Cave. From the 38 main types of carbonate speleothem forms described in the world (see Hill and Forti 1997), 17 types are known in the Baradla–Domica cave system (Gaál et al. 2014). There are various shapes of stalactites, including spherical form and helictites, stalagmites in pagoda-, palm- and stick-like forms, columns, draperies, flowstones, rimstone dams, shelfstones and coralloids, sporadically also cave pearls (Fig. 17.16). The

Fig. 17.16 Various types of carbonate speleothems: **a** soda-straws, Gombasecká Cave; **b** column, Krásnohorská Cave; **c** shield, Domica Cave; **d** rimstone dams, Domica Cave (*photo* P. Staník)

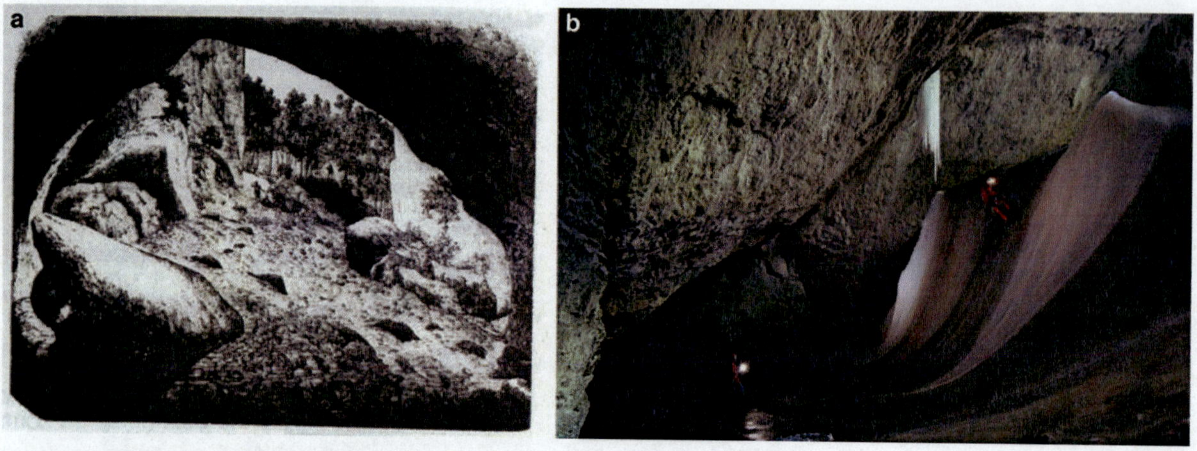

Fig. 17.17 Silická Ice Cave: **a** the entrance in 1858, **b** the ice fall at present (*photo* J. Stankovič)

Domica Cave is one of the most significant caves in the world for the occurrence of shields (two plates of concentrically layered calcite separated by a thin, capillary-sized crack, probably originated when calcite-rich water under hydrostatic pressure moves through the joints within the bedrock). More than 450 shields and drums in many morphological forms occur, mostly on walls and ceilings (Bella 1999). They also sporadically occur in the form of "stegamites", growing up from the floor. The Domica Cave is long known for the presence of brushite (Kašpar 1934). Subsequently, Kettner (1948) studied the destruction of dripstones and flowstones by guano corrosion. Powder viséite was found in the clayey cave sediment at the place of a former ritual cave sanctuary of Neolithic (Bükk) culture, as the third presence in a karst cavity in the world (Cílek 1999).

White soda straws with a length up to 3 m, and reddish flowstones occur in the Gombasecká Cave. The Column of Rožňava Cavers in the Krásnohorská Cave is 34 m high and belongs to the largest columns in Europe (Stankovič et al. 2005). Spectacular helictites and various eccentric stalactites, mostly in bizarre shapes, were precipitated in the Hrušovská Cave. Subaeric coralloids (popcorn) are observed in Brázda and Diviačia abysses, as well as in the Krásnohorská Cave, together with a manganese crust (Stankovič et al. 2005). White moonmilk coatings are deposited on rock walls of the Gombasecká jaskyňa and some other caves. Subaqueous flowstones, recently growing in turbulent stream water in the cave parts with elevated CO_2 content, were studied in the Drienovská and Krásnohorská caves (Wróblewski et al. 2017).

The 1,100 m long Silická ľadnica Ice Cave in the Silická Plateau is the lowest-elevation perennial ice cave up to the latitude of 50° north, located in the temperate climatic zone (Bella and Zelinka 2018; Fig. 17.17). Its entrance lies at an elevation of 470 m. Morphologically, it consists of two different parts. The upper part represents a steeply inclined spacious cavity, with a large opening to the surface (corrosive-collapsed abyss, *light hole*), while its lower, mostly subhorizontal and fluvially modelled passages were formed by an underground stream. The upper abyss part of the cave (from the entrance to the depth of 50 m) is partially filled with ice (at 425–465 m a.s.l.). Roda et al. (1974) reported an ice surface of 710–970 m^2 and ice volume of 213–340 m^3. The surface and volume of ice are varied depending on temperature and precipitation conditions during individual years. During 2016–2018, the highest difference in ice volume in the range of about 70 m^3 was observed between summer 2016 and spring 2017, as well as between spring 2017 and autumn 2017 (Šupinský et al. 2019). Floor ice mass is moving downwards on the inclined rock floor, but in the lower part of the abyss, ice is melting under the influence of warmer air circulating through debris plug (Rajman et al. 1985). Based on archaeological findings, the origin of ice in the upper part of the cave started 400–0 years BC (Roth 1939; Bárta 1995).

17.7 Conclusion

The Slovak Karst is an outstanding example of plateau karst containing almost a complete range of surface and subsurface karst landforms. One can observe the legacy of karstic processes occurring since the Late Cretaceous through the Neogene up to the present. Distinctive values represented by a genetic and morphological variability of subsurface karst phenomena enabled one to declare them as a part of the UNESCO World Heritage in 1995. In 2001, the Domica and Baradla caves were inscribed to the list of Ramsar sites (Fig. 17.18). The caves of Slovak Karst contain abundant

Fig. 17.18 The Styx stream in the Domica Cave is a Ramsar site (*photo* P. Staník)

sedimentary evidence of human development in prehistoric times.

The values of Slovak Karst are presented to the public in the field through several touristic paths and educational trails in the national park as well as in several show caves. The most significant educational trails occur in the Zádielska Gorge and around the Domica and Krásnohorská caves. Four caves are open for visitors in the Slovak Karst (Domica, Gombasecká, Krásnohorská and Jasovská caves), and these are annually visited by 65–70 thousand people.

Acknowledgements The geomorphological study of the Slovak Karst and preparation of this chapter were supported by the VEGA grant 1/0146/19. The authors are grateful to Jana Nováková and Igor Balciar for their help with drawing of graphic figures.

References

Bajaník Š, Reichwalder P (1979) Outline of the paleotectonic development of the Gemeric and its relation to adjected tectonic units. In: Maheľ M, Reichwalder P (eds) Czechoslovak geology and global tectonics. Veda, Bratislava, pp 141–154

Balogh K (1940) Adatok Pelsőcardó Környékének Földtani Ismeretéhez. Tisia 4:152–182

Baroň I (2002) Speleogenesis along sub-vertical joints: a model of plateau karst shaft development: a case study: The Dolný vrch Plateau. Cave Karst Sci 29(1):5–12

Bárta J (1955) Praveké Osídlenie Juhoslovenského Krasu. Krásy Slovenska 32(10):382–390

Bánesz L (1962) Prieskumy v Juhoslovenskom krase pri Rožňave. Študijné Zvesti Aú SAV 9:237–240

Bárta J (1965) Príspevok k pravekému osídleniu jaskýň Domickej sústavy. Slovenský Kras 5:58–73

Bárta J (1995) Pomoc archeológie pri datovaní zaľadnenia Silickej ľadnice. In: Bella P (ed) Ochrana ľadových jaskýň. Zborník referátov, Správa slovenských jaskýň, Liptovský Mikuláš, pp 81–84

Bárta J (1958) Majda-Hraškova jaskyňa a jej kultová funkcia v dobe halštatskej. Slovenská Archeológia 6:347–360

Belius PM (1723) Hungariae antiquae et novae prodromus, cum specimine, quomodo in singulis operis partibus elaborandius, versari constitverit. Norimbergae, p 204

Bella P (1999) Typy sintrových štítov a bubnov v jaskyni Domica. Aragonit 4:3–6

Bella P, Gruber P, Gaál Ľ, Papáč V, Bárány I (2021) Caves of Aggtelek and Slovak karsts, Northern Hungary and Southern Slovakia. Zeitschrift für Geomorphologie 62(Suppl. 3):19–47. https://doi.org/10.1127/zfg_suppl/2021/0689

Bella P, Bosák P (2015) Ceiling erosion in caves: early studies and Zdeněk Roth as author of the concept. Acta Carsologica 44(1):139–144. https://doi.org/10.3986/ac.v44i1.768

Bella P, Bosák P, Braucher R, Pruner P, Hercman H, Minár J, Veselský M, Holec J, Léanni L (2019) Multi-level Domica-Baradla cave system (Slovakia, Hungary): Middle Pliocene-Pleistocene evolution and implications for the denudation chronology of the Western Carpathians. Geomorphology 237:62–79. https://doi.org/10.1016/j.geomorph.2018.10.002

Bella P, Bosák P, Pruner P, Hochmuth Z, Hercman H (2007) Magnetostratigrafia jaskynných sedimentov a speleogenéza Moldavskej a Jasovskej jaskyne. Slovenský Kras 45:15–42

Bella P, Braucher R, Holec J, Veselský M (2014) Datovanie pochovania kremitého štrku na vrchnej vývojovej úrovni jaskyne Domica pomocou kozmogénnych nuklidov. Slovenský Kras 52 (1):15–24

Bella P, Gaál Ľ (2005) Jaskyňa Ortováň pri Silickej Brezovej – jedna z najstarších jaskýň Slovenského krasu. Aragonit 10:3–6

Bella P, Gaál Ľ (2010) Škrapy na južnom svahu Plešivskej planiny v Slovenskom krase. Slovenský Kras 48(2):209–222

Bella P, Gaál Ľ (2013) Reliéf a hydrografia ponorovej zóny brezovsko-kečovského podzemného hydrologického systému, Silická planina. Slovenský Kras. Slovenský Kras 51(1):15–29

Bella P, Gaál Ľ (2017) Hypogene caves in Slovakia. In: Klimchouk A, Palmer A, De Waele J, Aurel A, Audra P (eds) Hypogene Karst regions and caves of the world. Cave and Karst systems of the world. Springer, Cham, pp 299–311. https://dx.doi.org/10.1007/978-3-319-53348-3_19

Bella P, Gaál Ľ, Kilík J (2009) Kečovské škrapové pole v Slovenskom krase. Slovenský Kras 47(2):187–200

Bella P, Gaál Ľ, Littva J, Hochmuth Z, Haviarová D, Soták J (2018) Moldavská jaskyňa – horizontálny labyrint vytvorený alochtónnymi vodami Bodvy v kontaktnom krase Medzevskej pahorkatiny. Mineralia Slovaca 50(2):157–176

Bella P, Zelinka J (2018) Ice caves in Slovakia. In Perşoiu A, Lauritzen SE (eds) Ice caves. Elsevier, Amsterdam – Oxford – Cambridge, pp 657–689. https://doi.org/10.1016/B978-0-12-811739-2.00029-2.

Böhm J, Kunský J (1933) Domica, jeskyně neolitického člověka. Československý státní archeologický ústav, Praha, 2–8

Böhm J, Kunský J (1938) Silické Lednice. Sborník Československé Společnosti Zeměpisné 44:129–133

Bystrický J (1964) Slovenský kras. Stratigrafia a Dasycladaceae mezozoika Slovenského krasu. Ústredný ústav geologický, Bratislava, p 204

Cílek V (1999) Mineralogické Výzkumy v Jeskyni Domica. Aragonit 4:9–11

Cílek V, Bednářová J (1994) Bahenní železné rudy v krasových výplních lomu Gombasek ve Slovenském krasu. Slovenský Kras 32:41–46

Cílek V, Svobodová M (1999) Svrchnokřídové výplně závrtů v lomu Hosťovce a Gombasek ve Slovenském krasu. In: Šmídt J (ed) Výskum a ochrana prírody Slovenského krasu. Zborník referátov, Slovenská agentúra životného prostredia, Správa CHKO Slovenský kras, Brzotín, pp 41–48

Činčura J (2002) Krasový reliéf pripovrchových príkrovov centrálnych Západných Karpát: vývoj a paleotektonické korene. Slovenský Kras 40:7–17

Csontos L (1992) Mesozoic geologic and geodynamic problems of the Gemer-Bükk region. Terra Nova 4:12–13

Csontos L, Vörös A (2004) Mesozoic plate tectonic recontruction of the Carpathian region. Palaeo 210, Elsevier, pp 1–56

Dallmeyer RD, Neubauer E, Putiš M (1993) 40Ar/39Ar mineral age controls for the pre-Alpine tectonic evolution of nappe complexes in the Western Carpathians. In: Pitoňák P, Spišiak J (eds) Pre-Alpine events in the Western Carpathians realm. Excursion Guide, Stará Lesná, pp 13–20

Droppa A (1961) Domica–Baradla – jaskyne predhistorického človeka. Šport, Bratislava

Droppa A (1971) Vzťah horizontálnych chodieb Jasovskej jaskyne k terasám Bodvy. In: Kvitkovič J (ed) Problémy geomorfologického výskumu. Zborník referátov, Vydavateľstvo Slovenskej akadémie vied, Bratislava, pp 99–106

Droppa A (1972) Príspevok k vývoju jaskyne Domica. Československý Kras 22:65–72

Filipčíková N, Koltai G, Dublyansky Y, Spötl C (2020) Morphological and isotopic evidence of a hypogene origin of the Drienovská Cave. Slovak Karst. Aragonit 25(1):59–60

Ford DC (2000) Caves Branch, Belize, and the Baradla-Domica system, Hungary and Slovakia. In: Klimchouk AB, Ford DC, Palmer AN, Dreybrodt W (eds) Speleogenesis. National Speleological Society, Huntsville, Evolution of Karst Aquifers, pp 391–396

Ford DC, Williams PW (2007) Karst hydrogeology and geomorphology. Wiley, Chichester. https://doi.org/10.1002/9781118684986

Gaál Ľ (2008) Geodynamika a vývoj jaskýň Slovenského krasu. Správa slovenských jaskýň, Liptovský Mikuláš, 166 p

Gaál Ľ (2009). Vývoj hydrogeologických štruktúr Slovenského krasu. In: Fľaková R, Ženišová Z, Jašová I, Ďuričková A (eds) Hydrogeochémia ´09. Zborník z medzinárodnej vedeckej konferencie, Slovenská asociácia hydrogeológov, Bratislava, pp 33–40

Gaál Ľ, Bella P (2005) Vplyv tektonických pohybov na geomorfologický vývoj západnej časti Slovenského krasu. Slovenský Kras 43:17–36

Gaál Ľ, Vaněková H, Sýkora M (2007a) Nový nález vrchnokriedovej výplne krasových dutín v Slovenskom krase (lokalita Včeláre). Slovenský Kras 45:143–149

Gaál Ľ, Balciar I, Belanová E, Megela M, Papáč V, Vaněková H (2007b) Zatopená priepasť Morské oko v Rimavskej kotline. Aragonit 12:4–9

Gaál Ľ, Gruber P, Móga J (2014) Chemická a mechanická výplň jaskynného systému. In: Gaál Ľ, Gruber P (eds.) Jaskynný systém Domica-Baradla. Aggteleki Nemzeti Park Igazgatóság, Jósvafő, pp 117–158

Gradziński M, Bella P, Holúbek P (2018) Constructional caves in freshwater limestone: a review of their origin, classification, significance and global occurrence. Earth-Sci Rev 185:179–201. https://doi.org/10.1016/j.earscirev.2018.05.018

Gradziński M, Hercman H, Jaśkiewicz M, Szczurek S (2013) Holocene tufa in the Slovak karst: Facies, sedimentary environments and depositional history. Geol Q 57(4):769–788. https://doi.org/10.7306/gq.1131

Haviarová D, Gruber P (2014) Povrchové vody. In: Gaál Ľ, Gruber P (eds.) Jaskynný systém Domica-Baradla. Aggteleki Nemzeti Park Igazgatóság, Jósvafő, pp 161–184

Hill CA, Forti P (1997) Cave minerals of the World, 2nd edn. National Speleological Society, Huntsville, p 463

Hochmuth Z (1996) Geomorfologické pomery centrálnej časti Revúckej vrchoviny a priľahlých častí Rimavskej kotliny a Slovenského krasu. Geografické Práce 6(1):1–110

Hochmuth Z (2000) Moldavská jaskyňa vo vzťahu ku geomorfologickému vývoju doliny Bodvy. In: Bella P (ed) Výskum, využívanie a ochrana jaskýň. Zborník referátov z 2. vedeckej konferencie, Správa slovenských jaskýň, Liptovský Mikuláš, pp 52–58

Hochmuth Z (2013) Atlas jaskyne Skalistý potok. Slovenská speleologická spoločnosť, Liptovský Mikuláš, p 80

Hók J, Kováč P, Rakús M (1995) Výsledky štruktúrneho výskumu Vnútorných Karpát a ich interpretácia. Mineralia Slovaca 27:231–235

Jakál J (1975) Kras Silickej planiny. Osveta, Martin, p 146

Jakál J (1984) Príspevok k poznaniu polií v Západných Karpatoch. Geografický Časopis 36(2):108–119

Jakál J (2001) Vývoj reliéfu Slovenského krasu v etape neotektonického vyzdvihnutia územia. Slovenský Kras 39:7–14

Jakál J (2008) The surface morphology of karst landscape. In: Jakál J, Bella P (eds) Caves of the world heritage in Slovakia. Správa slovenských jaskýň, Liptovský Mikuláš, pp 37–46

Jakucs L (1975) A Baradla-barlang. In: Jakucs L (ed) Aggteleki karsztvidék. Sport Kiadó, Budapest, pp 78–185

Kašpar J (1934) Genese guanových mineralů v jeskyni Domica. Věstník Státního Geologického Ústavu Československé Republiky 10(3–4):104–111

Kemény A (1961) Geomorfologické Pomery Planiny Koniar. Geografický Časopis 13(2):104–134

Kessler H (1938) Az Aggteleki-Barlangrendszer Hidrográfiája. Földrajzi Közlemények 66(1–3):1–30

Kettner J (1948) O netopýřím guanu a guanových korosích v jeskyni Domici. Věstník Státního Geologického Ústavu Československé Republiky 15:41–64

Kochanová M (1987) Obertriassiche muscheln von Ostré vŕšky (Slowakishen Karst). Západné Karpaty, Paleontológia 12:31–46

Kollárová-Andrusovová V, Kochanová M (1973) Molluskenfauna des Bleskový prameň bei Drnava (Nor, Westkarpaten). Slovenská akadémia vied, Bratislava, p 235

Kováč M (2000) Geodynamický, paleogeografický a štruktúrny vývoj Karpatsko-panónskeho regiónu v miocéne. Nový pohľad na neogénne panvy Slovenska. Slovenská akadémia vied, Bratislava, p 202

Kováč M, Baráth I, Holický I, Marko F, Túnyi I (1989) Basin opening in the lower miocene strike-slip zone in the SW part of the Western Carpathians. Geologický Zborník Geol Carpath 40(1):37–62

Kováčik J (1955) Reziduálne železné rudy z Plešiveckej planiny. Geologický Sborník 6(3–4):232–251

Kozur H (1991) The evolution of the Meliata-Hallstatt ocean and its significance for the early evolution of the Easter Alps and Western Carpathians. Palaeogeogr Palaeoclimatol Palaeoecol 87(1–4):109–135. https://doi.org/10.1016/0031-0182(91)90132-B

Kozur H (1997) The evolution and continuation of the Western Carpathian Meliata-Hallstatt ocean (MHO). In: Plašienka et al (eds) Alpine evolution of the Western Carpathians and related areas. Abstracts, Geologická služba Slovenskej republiky, Bratislava, pp 25–26

Kozur H, Mock R (1973) Zum alter und zur tektonischen stellung der Meliata-Serie des Slowakischen Karstes. Geologický Zborník Geologia Carpathica 24(2):365–374

Kučera B (1976) Propast Čertova Diera. Československý Kras 27:79–84

Kunský J (1939) Jezera Slovenského krasu. Rozpravy II. Tř. Česká Akademie 49(25):1–17

Lalkovič M (1988) Príspevok k histórii Jasovskej jaskyne. Slovenský Kras 26:47–72

Lalkovič M (2014) Objav a poznávanie Domice. In: Gaál Ľ, Gruber P (eds.) Jaskynný systém Domica-Baradla. Aggteleki Nemzeti Park Igazgatóság, Jósvafő, pp 425–449

Lešinský G (2006) Príspevok k najstarším jaskyniam v Slovenskom krase. Sinter 14:22–23

Lichardus J (1968) Jaskyňa Domica – najvýznamnejšie sídlisko ľudu bukovohorskej kultúry. Pamätníky našej minulosti, 5, Vydavateľstvo Slovenskej akadémie vied, Bratislava, p 120

Ložek V (1958) Stratigrafie a měkkýši holocénních travertinů v Háji u Turni. Anthropozoikum 7:27–36

Ložek V (1994) Vývoj prírody v najmladšej geologickej minulosti. In: Rozložník M, Karasová E (eds) Chránená krajinná oblasť – Biosférická rezervácia Slovenský kras. Osveta, Martin, pp 77–87

Ložek V, Sekyra J, Kukla J, Fejfar O (1956) Výzkum Velké Jasovské Jeskyně. Anthropozoikum 6:197–282

Ložek V, Horáček I (1992) Slovenský kras ve světle kvartérní geologie. Slovenský Kras 30:29–56

Lukniš M (1964) Pozostatky po starších povrchoch zarovnania reliéfu v československých Karpatoch. Geografický Časopis 16(3):289–298

Lukniš M (1972) Reliéf. In: Lukniš M (ed) Slovensko 2—Príroda. Obzor Bratislava, pp 124–202

Maderspach L (1877) A Pelsőcz-Ardói czink és gálmafekhelyek. Földtani Közlöny 7:121–124

Maluski H, Railich P, Matte Ph (1993) 40 Ar/39 Ar dating of the inner carpathian variscan basement and alpine mylonitic overprinting. Tectonophysics 223:313–337. https://doi.org/10.1016/0040-1951(93)90143-8

Marschalko R, Mello J (1993) Turbidites as filling of cavities in Triassic limestones of the Silica nappe (Western Carpathians, Plešivec Karst Plateau). Geol Carpath 44(1):35–42

Matějka A (1958) Výskyt bauxitických hornin u Drienovce na jižním Slovensku. Věstník Ústředního Ústavu Geologického 33(4):279–281

Mazúr E, Jakál J (1971) Podklad a reliéf. In: Mazúr E (ed) Slovenský kras. Regionálna fyzickogeografická analýza. Geografické práce 2 (1–2):6–22

Mazúr E, Lukiš M (1978) Regionálne Geomorfologické Členenie SSR. Geografický Časopis 30(2):101–125

Mello J (1974) Facial developments and facial relations of the Slovak Karst Middle and Upper Triassic (West Carpathians, Southern part

of Gemerids). Österreichische Akademie Der Wissenschaften, Schriftenreihe Der Erdwissenschaftlichen Kommissionen 2:147–155
Mello J (1977) Aplikácia litofaciálnej analýzy triasových vápencov v tektonike (južná časť Plešivskej planiny). Geologické Práce, Správy 67:175–180
Mello J (1979) Meliatska séria v Turnianskom tektonickom okne (Slovenský kras, Západ. Karpaty). Geologické Práce, Správy 72:61–76
Mello J, Snopková P (1973) Vrchnokriedový vek výplní v dutinách triasových vápencov gombaseckého lomu. Geologické Práce, Správy 61:239–253
Mello J, Salaj J (1982) Nález vápencov gosauskej kriedy v údolí Miglinc (Slovenský kras). Geologické Práce, Správy 77:49–54
Mello J, Mock R, Planderová E, Gaál Ľ (1983) Nové stratigrafické poznatky o meliatskej skupine. Geologické Práce, Správy 79:61–76
Mello J, Elečko M, Pristaš J, Reichwalder P, Snopko L, Vass D, Vozárová A, Gaál Ľ, Hanzel V, Hók J, Kováč P, Slavkay M, Steiner A (1997) Vysvetlivky ku geologickej mape Slovenského krasu 1 : 50 000. ŠGÚDŠ, Bratislava, p 255
Minár J (2003) Stredohorská roveň v Západných Karpatoch ako tektoplén: náčrt pracovnej hypotézy. Geografický Časopis 55 (2):141–158
Mišík M, Borza K (1976) Obere Trias bei Silická Brezová (Westkarpaten). Acta Geologica Et Geographica Universitasis Comenianae 30:5–49
Mišík M, Sýkora M (1980) Jura der Silica-Einheit, rekonstruiert aus Geröllen, und Oberkretazische Süsswasserkalke des Gemerikums. Geologický Zborník Geologica Carpathica 31(3):239–261
Mitter (1988) Speleologický výskum krasových javov Plešivskej planiny. Výskumné Práce z Ochrany Prírody 6A:75–95
Mock R (1973) Nález zlambašských vrstiev v Slovenskom krase. Geologické Práce, Správy 60:221–224
Mock R (1978) Nové poznatky o južných častiach Západných Karpát. In: Vozár et al (ed) Paleogeografický vývoj Západných Karpát. Geologický ústav Dionýza Štúra, Bratislava, pp 321–341
Mock R (1980) Triassic of the West Carpathians. In: Guidebook of the 2nd European Conodont Symposium (ECOS II). Abhandlungen Der Geologischen Bundesanstalt in Wien 35:69–98
Mock R, Sýkora M, Aubrecht R, Ožvoldová L, Kronome B, Reichwalder P, Jablonský J (1998) Petrology and stratigraphy of the Meliaticum near the Meliata and Jaklovce Village. Slovakia. Slovak Geol Mag 4(4):223–260
Móga J (1999) Reconstruction of the development history of karstic water networks on the southern part of the Gömör-Torna Karst on the basis of ruined caves and landforms. Acta Carsologica 28 (2):159–174
Orvan J (1973) Hydrogeologické Pomery Rimavskej Kotliny. Mineralia Slovaca 5(3):271–278
Orvan J (1999) Podzemné vody Slovenského krasu. In: Šmídt J (ed) Výskum a ochrana prírody Slovenského krasu. Zborník referátov, Slovenská agentúra životného prostredia, Správa CHKO Slovenský kras, Brzotín, pp 51–59
Rajman L, Roda Š, Ščuka J (1985) Výskum dynamiky ľadu v Silickej ľadnici. Slovenský Kras 23:253–260
Rakús M (1996) Jurassic of the innermost Western Carpathian zones—its importance and influence on the geodynamic evolution of the area. Slovak Geol Mag 3–4(96):311–317
Roda Š, Rajman L, Erdös M (1974) Výskum mikroklímy a dynamiky zaľadnenia v Silickej ľadnici. Slovenský Kras 12:157–174
Roth Z (1937) Vývoj Jeskyně Domice. Bratislava 11:129–163
Roth Z (1939) Několik geomorfologických poznámek o Jihoslovenském Krasu a o Silické Lednici. Rozpravy České Akademie Věd a Umění, Třída II 8:1–24
Roth Z (1940a) Charakteristika a vývoj šachtovitých závrtů a propastí typu *Light holes* Cvijić a novo-tvořených portálů vyšších pater říčních jeskyň. Věstník Královské české společnosti nauk, Třída mat.-přír. 6: 1–11
Roth Z (1940b) Vývojový vztah jeskyně Baradel k jeskyni Domici v Jihoslovenském Krase. Věstník Královské české společnosti nauk, třída matematicko-přírodovědecká, 1–9
Sasvári T, Jacko S, Kondela J (2006) Structural-tectonic architercure of limestone quarry Gombasek on Plešivec Plateau of Silica Nappe. Acta Montanistica Slovaca 11(1):70–85
Sawicki L (1908) Szkic krasu slowackiego z pogladem na cykl geograficny w krasie w ogóle. Kosmos 33(6–7):395–444
Siblík M (1967) Ramenonožci noru z lokality Drnava (Slovenský kras). Geologické Práce, Správy 43:81–97
Siegmeth K (1887) Az Abauj-torna-gömöri barlangvidék I. A Magyar Kárpátegyesület Évkönyve 14:1–48
Siegmeth K (1891) Az Abauj-torna-gömöri barlangvidék II. A Magyar Kárpátegyesület Évkönyve 18:33–52
Skřivánek F (1958) Výzkum propastí severní části Silické planiny v Jihoslovenském krasu. Československý Kras 11:115–129
Skřivánek F (1966) Vývoj krasu Plešivecké planiny v Jihoslovenském krasu. Československý Kras 17(1965):42–58
Skřivánek F, Stárka V (1956) Krasové zjevy státní přírodní reservace Zádielska dolina. Ochrana Přírody 11(10):289–295
Slavkay M, Beňka J (1995) Geológia, mineralógia a metalogenéza oloveno-zinkových rúd pri Ardove. Mineralia Slovaca 27(2):79–88
Soják M (2007) Osídlenie blízkeho okolia Moldavy nad Bodvou. In: Soják M, Terray M (eds) Moldavská jaskyňa v zrkadle dejín. A Szepsi-barlang a történelem tükrében. Mestský úrad, Moldava nad Bodvou, pp 50–72
Soják M (2014) Domica v praveku. In:Gaál Ľ, Gruber P (eds) Jaskynný systém Domica–Baradla. Aggteleki Nemzeti Park Igazgatóság, Jósvafő, pp 353–364
Stankovič J, Cílek V, Bruthans J, Gaál Ľ, Višňovská Z, Kovács Á, Rozložník M, Schmelzová R, Zeman O, Kováč Ľ, Mock A, Ľuptáčik P, Hudec I, Nováková A, Košel V, Fenďa P (2005) Krásnohorská jaskyňa Buzgó. Regionálna rozvojová agentúra, Rožňava, p 151
Stürzenbaum J (1879) A Dernői Kösseni Rétegekről. Földtani Közlöny 4:217
Székely K (2014) Dejiny jaskyne Baradla. In: Gaál Ľ, Gruber P (eds) Jaskynný systém Domica-Baradla. Aggteleki Nemzeti Park Igazgatóság, Jósvafő, pp 365–423
Šupinský J, Kaňuk J, Hochmuth Z, Gallay M (2019) Detecting dynamics of cave floor ice with selective cloud-to-cloud approach. Cryosphere 13(11):2835–2851. https://doi.org/10.5194/tc-13-2835-2019
Vass D, Kraus I, Elečko M (1989) Výplň Rožňavskej kotliny a údolia rieky Slaná pri Slavci. Mineralia Slovaca 21(1):71–75
Vass D, Hók J, Kováč P, Elečko M (1993) Sled paleogénnych a neogénnych tektonických udalostí v juhoslovenských kotlinách vo svetle napäťových analýz. Mineralia Slovaca 25(2):79–92
Vass D, Elečko M, Horská A, Petrík F, Barkač Z, Mello J, Vozárová A, Dubéci B (1994) Základné črty geológie turnianskej depresie. Geologické Práce, Správy 99:7–22
Veress M, Unger Z (2015) Baradla–Domica: Large Cave system on the Hungarian-Slovak border. In: Lóczy F (ed) Landscapes and

Landforms of Hungary. Springer International Publishing, pp 167–175
Vozárová A, Vozár J (1988) Late Paleozoic in West Carpathians. GÚDŠ, Bratislava, p 314
Windisch KG (1780) Geographie des Königreichs Ungarn II. Pressburg, p 322
Wróblewski W, Gradziński M, Motyka J, Stankovič J (2017) Recently growing subaqueous flowstones: Occurrence, petrography, and growth conditions. Quatern Int 437, Part A: 84–87. https://doi.org/10.1016/j.quaint.2016.10.006
Zacharov M (2013) Nové poznatky z výskumu Drienovskej jaskyne. Slovenský Kras 51(2):111–120

Ľudovít Gaál is a geologist of State Nature Conversation of Slovak Republic—Slovak Caves Administration in Liptovský Mikuláš. His main scientific interest is in geodynamic influence of cave development in Slovak Karst, preserving of cave geosystems, management of World Heritage caves of Slovak Karst and the genetic development of caves in the volcanic structures of Slovakia.

Pavel Bella is Associate Professor of Physical Geography and Geoecology at the Catholic University in Ružomberok and head of the cave protection department at the State Nature Conservancy of the Slovak Republic, Slovak Caves Administration in Liptovský Mikuláš. His research is focused mainly on karst geomorphology and speleology, especially on cave morphology and genesis, hypogene speleogenesis, and the reconstruction of cave development linked with landscape evolution. He also deals with cave geoecology and environmental problems in karst landscape and caves.

Jozef Jakál is a retired geomorphologist. He worked at the Institute of Geography of the Slovak Academy of Science nevertheless he gave lectures of Quaternary geology at the Faculty of Nature Science of Comenius University in Bratislava. He was interested in the development of karst processes and karst phenomenon in Slovakia, mainly Slovak Karst. He as the first defined the complexity and unsteadiness of cave system in Slovakia and warn to the necessity of complex understanding of cave preservation taking into account not only underground elements but also the surface processes influencing cave system.

18 Slovenský Raj—The Land of Karst Plateaus, Gorges and Waterfalls

Piotr Migoń, Wioleta Porębna, and Kacper Jancewicz

Abstract

Slovenský raj is a dissected highland in the Inner Western Carpathians, at the border of central and east Slovakia. Geologically it is predominantly built of limestones and dolomites of Triassic age, which have supported the development of karst phenomena. Characteristic landforms include remnants of karst plateaus, with uvalas and dolines, incised valleys of Hnilec and Hornád rivers, numerous gorges deeply cut into limestone bedrock, with impressive waterfalls, slot canyon reaches and bedrock channels. The Dobšiná Ice Cave in the southern part of the region is a large ice-filled cavern, inscribed as a UNESCO World Heritage property. The history of karst relief evolution may be traced back to the beginning of the Cenozoic, with some landforms inherited from the times prior to the Late Eocene marine transgression. However, the origin of spectacular gorges is much younger and related to the Pliocene–Quaternary uplift, relative base-level lowering and headward erosion in the tributary valleys.

Keywords

Karst • Plateaus • Exhumed landforms • Gorges • Fluvial erosion • Caves

18.1 Introduction

Slovenský raj (= Slovak Paradise), located in the Inner Western Carpathians at the border between central and eastern Slovakia, represents a geomorphological landscape that is mainly distinctive through the large number of impressive gorges incised into a forested limestone plateau. Since more than a century, the area has been valued for its scenery and popular among visitors, being considered as one of geomorphologically most attractive parts of the country. At the same time, it is easily accessible, particularly from the north, where it borders with a long intra-Carpathian basin corridor of Hornádska kotlina, and is well-developed for tourism. Whereas karstified limestone plateaus are fairly common in the Inner Slovak Carpathians (Jakál 2001) and are classically developed in the Slovakian Karst region (see Chap. 17), the gorges, famous due to the occurrence of narrow slots, bedrock steps and waterfalls, give specific identity to Slovenský raj. In addition, the epigenetic valley of the Hornád River, cut into the northern limb of the plateau, represents an excellent and most scenic example of a water gap, with few parallels elsewhere in the country. Another highlight of the area is the UNESCO World Heritage Dobšinská l'adová jaskyňa (= Dobšiná Ice Cave), recognized for its scenic beauty and scientific importance, also as a representative ice-filled cave in a temperate region. The cave is accessible to visitors and remains a site of considerable geomorphological interest.

This chapter adopts the following structure. Wider geographical context and geological structure are presented first, providing the necessary background to discuss landscape diversity and karst morphology of the area, with particular focus on high-elevation karst plateaus. In the next part valley landforms from the northern part of Slovenský raj are the central theme, starting with the Hornád River Gorge, which sets the boundary conditions for subsequent examination of various tributary gorges. The Dobšiná Ice Cave, a section of a large cave system in the southern part of the area, is

P. Migoń (✉) · W. Porębna · K. Jancewicz
Institute of Geography and Regional Development, University of Wrocław, pl. Uniwersytecki 1, 50–137 Wrocław, Poland
e-mail: piotr.migon@uwr.edu.pl

W. Porębna
e-mail: wioleta.porebna@uwr.edu.pl

K. Jancewicz
e-mail: kacper.jancewicz@uwr.edu.pl

M. Lehotský and M. Boltižiar (eds.), *Landscapes and Landforms of Slovakia*, World Geomorphological Landscapes,
https://doi.org/10.1007/978-3-030-89293-7_18

presented separately. Interrelated issues of conservation and tourism management conclude the chapter.

Unfortunately, very little of the fundamental work about geomorphology of Slovenský raj has been published in English. Hence, this chapter fills an evident gap in this respect. Among literature sources available in Slovak language, the following are of particular use. Lukniš (1945) provided the first comprehensive presentation of geomorphology of the area and many of his observations remain valid until today. Mičian (1962) developed an interpretation of the Hornád River gorge, which is an important reference landform for the region. Novotný and Tulis were long involved in karst and cave research in Slovenský raj, with a benchmark study "Karst of Slovenský raj" (Engl. translation) published in 2005. The same authors produced a highly detailed monograph of the Stratenská Cave in the southern part of the area, which is the third longest known cave in Slovakia (Tulis and Novotný 1989). Hochmuth (2008) paid considerable attention to Slovenský raj in his overview of karst and caves of Slovakia.

18.2 Geographical Setting

Slovenský raj constitutes the north-eastern part of a larger geomorphic unit in central Slovakia, known as the Spiš–Gemer Karst, which itself is a part of the Slovak Ore Mountains (Slovenské rudohorie). Being part of a larger karst area, its conventional boundaries are primarily defined by the extent of the dominant limestone series which support distinctive relief of the area. Some of these boundaries correspond with a clear change in topography, such as along the northern margin, where the inclined and dissected limestone upland abruptly gives way to the nearly flat floor of the intra-Carpathian basin of Hornádska kotlina. Other boundaries, however, run inside the mountainous terrain and follow major fault-controlled valleys (e.g., in the NW) or have rather poor geomorphic expression (in the south) (Fig. 18.1).

In terms of general topography, Slovenský raj represents a medium-altitude mountain relief that may be considered typical for Central Europe. Its characteristic features include relative heights in the order of 500–1000 m, absolute elevation not reaching the timberline, dominance of moderately steep slopes, and the presence of low-gradient surfaces in water-divide positions. The highest peaks of Slovenský raj include Mt. Ondrejisko (1272 m), Mt. Remiaška (1168 m) and Mt. Havrania skala (1156 m), whereas locally conspicuous tracts of low-relief plateaus occur in the 900–1100 m a.s.l. altitude belt. Deeply incised valleys of Hnilec and Hornád rivers cut through the mountain block in the south and north, respectively.

In terms of climate, Slovenský raj enjoys moderately warm to moderately cool climate, depending on altitude (Rosič et al. 2011). Consequently, the mean annual temperature is between 4 and 6 °C, with July being the warmest month (15–17 °C) and January being the coldest (−5 to −6 °C). The number of days with snow cover ranges from c. 80 in the least elevated parts to c. 120 on high-elevation plateaus, whereas there may be up to 170 days annually with temperature falling below zero. Temperature inversions are common, especially within deeply incised gorges, which are therefore very foggy. These circumstances allow to suspect that physical weathering is particularly effective in shaping limestone outcrops. The annual precipitation is 800–900 mm, and it prevails over evaporation, making the shady gorges persistently damp. Of special note is the number of days with thunderstorms per year (50–65), which is among the highest in Slovakia (Rosič et al. 2011). They are important for runoff generation and enhancement of streamflow, and it is very likely that numerous examples of coarse woody debris deposited in the gorges result from these extreme hydrometeorological events.

In these climatic conditions forests can thrive and indeed, forest communities occupy about 90% of the area of Slovenský raj. Beech communities, typical for limestone bedrock in this part of Europe, are most represented, but in least hospitable, rocky places relict pine and larch stands have survived. Unstable scree slopes in the gorges, in turn, provide a habitat for lime and sycamore maple. Meadows, which occur on some of the karst plateaus and occupy c. 5% of the area in total, are largely due to human intervention in the past and forest clearance for animal husbandry.

18.3 Geology

Slovenský raj is predominantly a carbonate area (Mello et al. 2000; Novotný and Tulis 2005) (Fig. 18.2). Limestones and dolomites are of Middle and Late Triassic age, representing different facies. They include darker Guttenstein limestones, typified by rather thin bedding (10–40 cm), and the overlying Wetterstein reef limestones, more thickly bedded or massive, which are most widespread. The latter support the most impressive rock outcrops in the region, including precipitous escarpments around the high-elevation plateaus and rugged sides of fluvial gorges. Tectonically, the Triassic sediments occur within two nappes, the Vernar Nappe in the west and the Stratená Nappe in the east, separated by the SW–NE trending Muráň fault zone. The Stratená Nappe consists of two minor nappe units, Geravy unit in the south and Glac unit in the north, separated by the Tomašovský thrust. In the Vernár Nappe dolomites prevail over limestones, whereas the opposite is true for the Stratená Nappe.

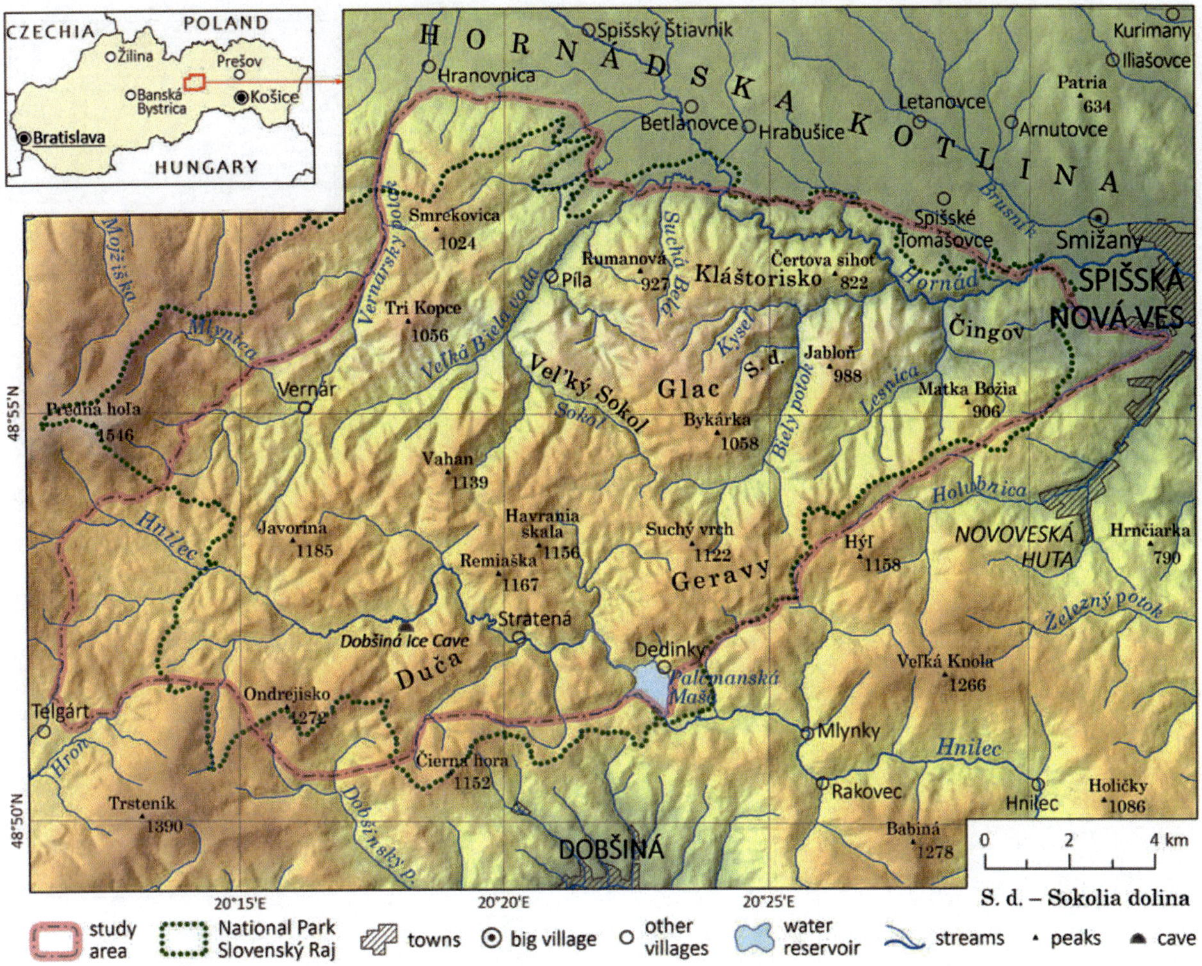

Fig. 18.1 Slovenský raj—location and general relief (data source: digital elevation model DMR3.5 of 10 m resolution. The DMR3.5 dataset was provided by GKU Bratislava under the CC-BY 4.0 licence)

Therefore, karst relief is much better developed and more widespread in the latter. Dolomites appear less resistant to denudation and underlie topographic depressions, such as the headwater parts of Biely potok between the Geravy and Glac plateaus.

Post-Triassic sediments have only localized occurrence today, but these are all but remnants of much more widespread covers that existed in the past. In several places along the Hnilec River valley Upper Cretaceous clastic and carbonate sediments occur, probably remnants of a fill of an intramontane basin. Bauxites found on the karst plateaus are considered as products of early Palaeogene weathering in humid tropical conditions. Marine transgression in the Late Eocene eventually covered the entire area of Slovenský raj, resulting in the deposition of lithologically diverse complex of conglomerates, sandstones, and calcarenites, followed by fine-grained sandstones and siltstones up to 150 m thick. They occur to the north of the Hornád River Gorge and underlie the Čingov plateau in the north-east, but an isolated patch is present to the east of Kláštorisko plateau (Čertova sihot') and scattered gravels on the Glac and Geravy plateaus show the former extent of this series. Quaternary sediments are represented by tufa accumulations, in situ weathering products—eluvia, fluvial deposits, including high terrace levels up to 65 m above the valley floors (Novotný and Tulis 2005), and various hillslope deposits, depending on slope steepness and the transport mechanism involved.

18.4 Geomorphological Landscapes—An Overview

18.4.1 Subdivision

Although limestone scenery is an overarching theme in Slovenský raj, the region may be nevertheless informally

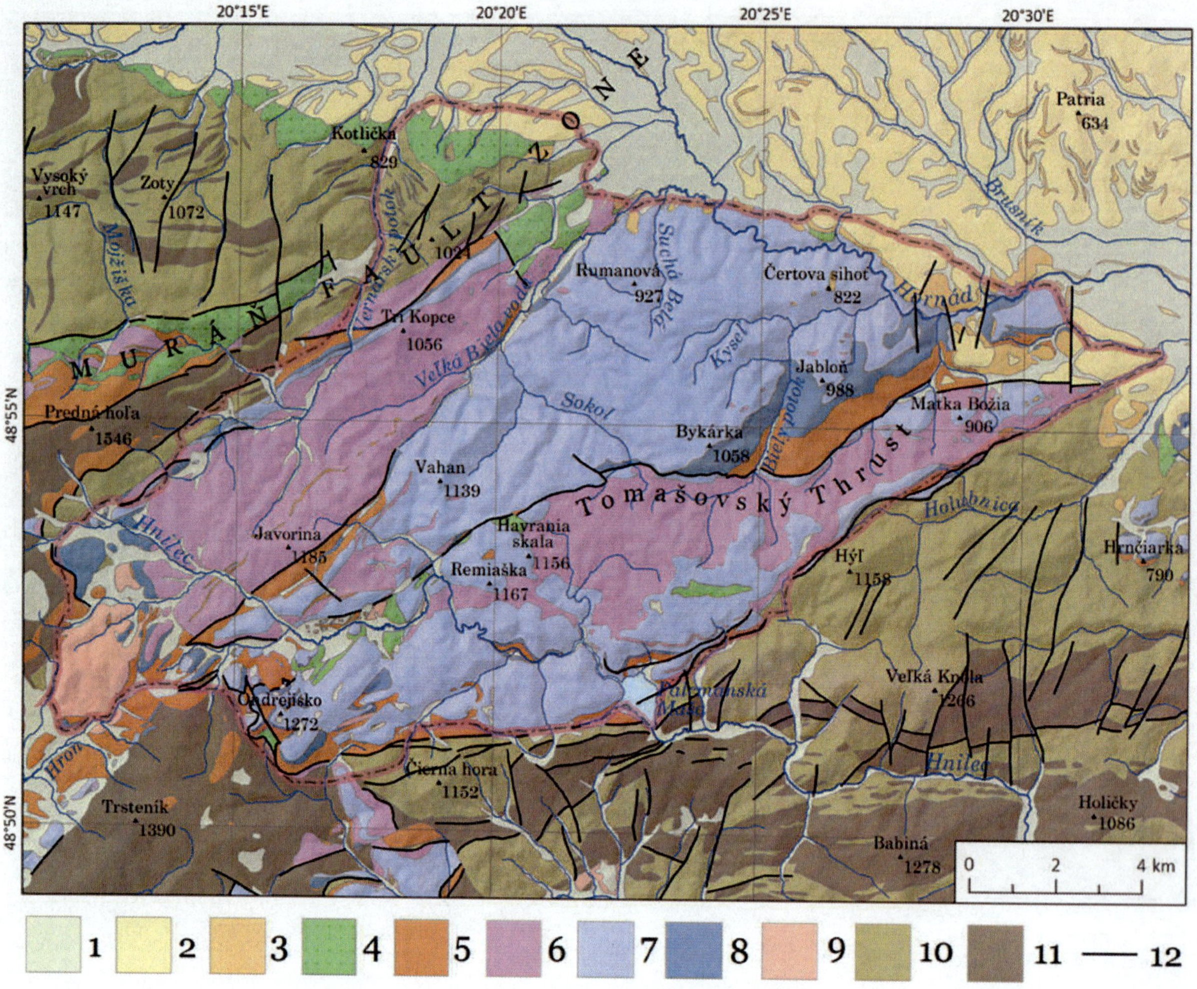

Fig. 18.2 Geological map of Slovenský raj and its surroundings. Cenozoic: 1—unconsolidated Quaternary deposits; 2—sandstones, shales; 3—basal conglomerates; Jurassic–Cretaceous: 4—conglomerates, shales, sandstones, limestones; Triassic: 5—sandstones, shales; 6—dolomites; 7—Wetterstein limestones; 8—Guttenstein limestones; 9—quartzites and rhyolites; Palaeozoic: 10—phyllites, shales, conglomerates (Carboniferous–Permian); 11—basement rocks (igneous and metamorphic); 12—faults. Based on: Geologická mapa Slovenska M 1:50 000 [online]. Bratislava: Štátny geologický ústav Dionýza Štúra, 2013. Dostupné na internete: http://apl.geology.sk/gm50js

divided into several parts, which are morphologically different from one another (Fig. 18.3). In the south, the Hnilec River valley separates the southernmost part, with the karst plateaus of Duča, Pelc and Skala, from the central, more dissected region. The valley itself is nearly 300 m deep and follows a sinuous course, with an impressive entrenched meander to the north of the Duča plateau, followed in the downstream direction by the narrow gorge of Stratenský kaňon, composed of a few small incised meanders. The latter is 1.3 km long in total, whereas the straight line connecting the entrance and exit is only 0.4 km long. The Hnilec River gorge is considered as an example of epigenesis (superimposition), and it is suspected that the river may have partly followed a subterranean course prior to the Quaternary, carving the galleries of the Stratenská Cave (Stratenská jaskyňa) (Novotný and Tulis 2005).

The central part of Slovenský raj has complicated topography. Dissection of its western sector, with Mt. Havrania skala, appears more advanced and low-relief surfaces in water-divide position are limited in extent. The considerable extent of gorges of Vel'ký Sokol and Malý Sokol is notable, as both have cut deeply into the core part of the highland. By contrast, the eastern sector includes two large high-elevation karst plateaus in the region—Geravy in the south and Glac in the north. The latter is at slightly lower elevation (~950–1000 m a.s.l. versus 1000–1050 m a.s.l. for Geravy), demonstrating very modest tilting to the north. Going northward, the density of erosional dissection

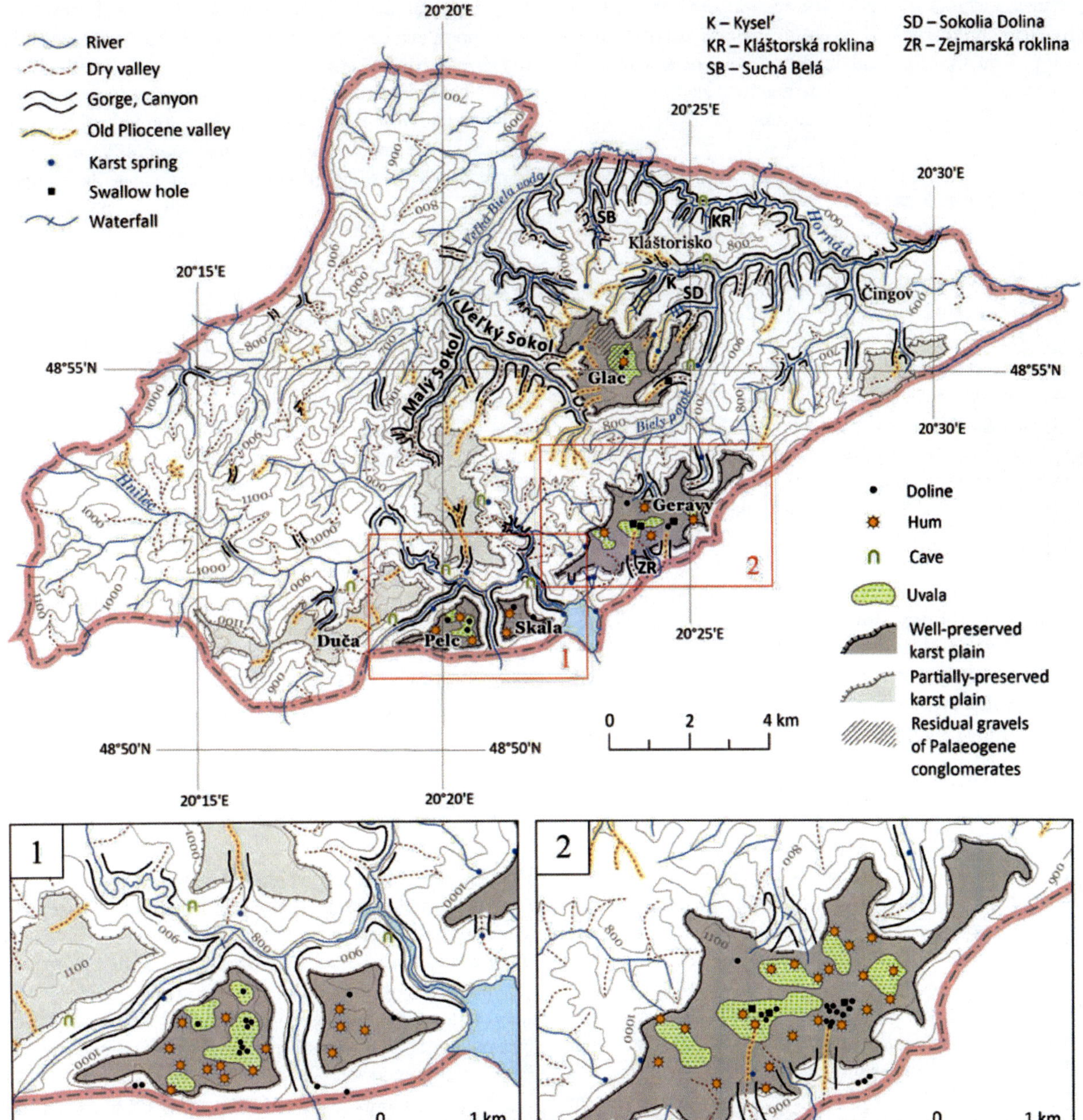

Fig. 18.3 Geomorphological map of Slovenský raj based on Lukniš (1945), supplemented by landform recognition and morphometric analysis (slope, profile curvature) carried out by the authors, with the use of digital elevation model DMR3.5 of 10 m resolution. The DMR3.5 dataset was provided by GKU Bratislava under the CC-BY 4.0 licence

increases and the valley morphology changes. Not only are the valleys deeply incised, but they become extremely rocky, with high rock slopes and slot canyon reaches. The northern part of the sloping highland, with plateau remnants much limited in extent in comparison to the central part, is incised by the Hornád River Gorge and its tributary gorges. To the north of the gorge, a narrow belt of elevations below 700 m a.s.l. runs east–west, marking the northern topographic margin of Slovenský raj.

The north-west part of the region, between the valleys of Vel'ká Biela voda and Vernársky potok, represents a different type of mountainous topography, lacking extensive plateaus and rocky gorges. The highest peak is Tri kopce (1056 m). Similar relief occurs to the east of Biely potok valley (Tomášovská Belá brook), where plateau relief is reduced and rocky gorges are largely replaced by V-shaped valleys with occasional rock outcrops.

18.4.2 Karst Plateaus

In this section, three karst plateaus will be presented in more detail. These are Duča in the south, Geravy in the central part of Slovenský raj, and Glac further to the north (Fig. 18.3).

Among karst plateaus to the south of the Hnilec River valley the Duča plateau is the most elevated, reaching 1142 m a.s.l. It is nearly 3 km long in WSW–ENE direction and up to 1 km wide. In contrast to the other two described below, it is rather poor in surface landforms, but hides an extensive system of Stratenská Cave. The marginal slopes of the plateau are rugged and rocky, and include the biggest example of a collapse doline in the area (Duča doline). It is 115 × 56 m in ground plan and 33 m deep, and originated due to roof collapse above a large underground chamber. The Stratenská Cave is about 24 km long and consists of chambers and galleries developed at five different levels from 995 m a.s.l. to 855 m a.s.l. (Novotný and Tulis 2005). The largest chamber is 192 m long and 11 m high on average, and the volume is calculated as 79,000 m^3. The Dobšiná Ice Cave, although physically disconnected by now, is also a part of this system.

The Geravy plateau is the largest in Slovenský raj and forms part of the water divide between the drainage basins of Hnilec in the south and Hornád rivers in the north. Most of its surface lies in the altitude belt 1000–1050 m a.s.l., but isolated elevations rise above this general level to more than 1100 m a.s.l. (Suchý vrch—1121 m, Červená skala—1109 m, Holý kameň—1107 m). The plateau is elongated from WSW to ENE, approximately 5 km long and up to 2 km wide. Caves under the plateau are rare, with the total length of 18 separate localities being only 153 m (Novotný and Tulis 2005). The deepest shaft, the Kira Cave (28 m deep) owes its origin to deep-seated gravitational displacements affecting the marginal parts of the plateau rather than dissolution. By contrast, surface karst landforms are more abundant and include residual hills—hums, several uvalas drained entirely into underground, open uvala depressions that continue as very shallow trough valleys, more than 40 solution dolines and, locally, karren fields. Scattered remnants of gravels of Palaeogene age occur on the plateau (Novotný and Tulis 2005).

The Glac plateau occurs at 950–1000 m a.s.l. on average, with the highest elevations around the central flat, reaching 1058 m a.s.l. (Mt. Bykárka). The outline of the plateau is very irregular due to extension of valley heads from the west, north and east. Novotný and Tulis (2005) indicated the total area of the plateau as 4.5 km^2 and interpreted narrow ridges between these valley heads as markers of a formerly more extensive plateau level. Karst morphology is most evident in the central part of the plateau (Vel'ká pol'ana = Large Meadow), being also easiest to appreciate due to open terrain (Fig. 18.4). It may be resolved into two residual hills (hums) in the middle and two large uvalas in the north and south, respectively. The hums are 15 m high, whereas the uvalas are up to 400 m long and 150–300 m wide. Their floors are punctuated by numerous solution dolines, with the biggest one being 23 × 13 m (Novotný and Tulis 2005). Subterranean karst appears poorly developed or has not yet been discovered. No caves below the plateau are known and none of those found within steep marginal slopes is particularly long. The most famous one is Medvedia jaskyňa (= Bear Cave), which yielded abundant bones of bear (*Ursus spelaeus*) and is an important bat habitat. The length of the cave is c. 500 m. Of particular significance for deciphering the evolution of karst relief is the occurrence of remnants of Palaeogene gravels on the plateau, including limonite and goethite concretions and bauxitic residuals. These gravels have also been found in the Medvedia Cave. They demonstrate considerable role of the early Cenozoic inheritance in both the present-day morphology of the Glac plateau and the long-term evolution of its caves (Novotný and Tulis 2005).

18.4.3 Landscape Evolution

The history of landform evolution, mainly derived from the examination of spatial relationships between Palaeogene sediments preserved in Slovenský raj and its various karst features, was summarized by Novotný and Tulis (2005) and was not challenged by subsequent research. Marine submergence during the Palaeogene divides the history into "palaeoalpine" and "neoalpine" period, although each of these actually represented a complex series of events.

The key observation is that remnants of Palaeogene (late Eocene) gravels occur on the karst plateaus (Geravy, Glac), within shallow trough valleys in the lowered parts of these plateaus, as well as in several caves located a few tens of metres below the plateau level (Medvedia jaskyňa, Skalné okno in Piecky gorge). Their presence indicates the existence of karstic topography of medium relief in the Eocene, which had formed during the Late Cretaceous and part of the Palaeogene. Consequently, the contemporary plateaus of Geravy and Glac are exhumed (re-exposed) features, although subsequent modelling obviously occurred and equating their relief with the Eocene landscape would be incorrect. It is possible that the pre-late Eocene relief included even older, Early Cretaceous facets and Novotný

Fig. 18.4 Glac plateau. **a** General view, with a hum in the middle of the bowl-shaped depression (right) and dolines incised into grassy surface (left), **b** close view of two dolines

and Tulis (2005) connected scattered bauxite occurrences with that very distant period.

Marine transgression in the late Eocene resulted in the burial of karst relief and the sedimentary cover was probably sealing the pre-existing topography until the Badenian ($\sim$16.3–12.8 Ma), when regional denudation commenced and re-exposure was initiated. The most elevated plateaus and trough valleys were exposed first, whereas some gravels were re-deposited into the caves (e.g., Medvedia jaskyňa). Late Miocene–Pliocene uplift of the area was initially slow, causing selective lowering of plateau surfaces and integration of uvalas into an open drainage network, but accelerated towards the end of Pliocene. Large-scale tilting to the north occurred, helping longer preservation of Palaeogene sediments. Incision into these cover deposits gave rise to the Hornád River Gorge, which in turn initiated the phase of headward erosion in the tributaries and set the local base level. However, a period of lesser uplift is hypothesized to occur sometime during the Pliocene, as deduced from the presence of discontinuous mid-slope benches along the Hnilec and Hornád rivers, some 100–180 m above the present-day channels. In the cave environment it is recorded by an extensive development of horizontal galleries in the Stratenská Cave at the IV. level, c. 100 m above the valley floor of Hnilec River.

18.5 Hornád Gorge

The gorge of Hornád in the northernmost part of Slovenský raj appears as a geomorphological anomaly. The river, instead of continuing to flow eastwards within the wide intramontane basin of Hornádska kotlina (Fig. 18.5), turns to the south, enters the marginal part of the limestone highland, and cuts through it along the next 12 km, until leaving back to the trough near the town of Spišská Nová Ves. The river course within the gorge is highly sinuous, with a few large incised meander bends, whereas the straight line connecting the upstream gate and the downstream exit of the gorge is only c. 7.5 km long. The paradox of the gorge is exacerbated by the fact that the highland belt to the north of it is very narrow, 1 km wide at most, passing into the basin of Hornádska kotlina without any clear break of slope. Consequently, there is evident altitude asymmetry of the opposite sides of the gorge. Elevations within the northern shoulder do not exceed 670 m a.s.l. (Ludmanka, 667 m; Zelená hora, 654 m), whereas altitudes in the south rise to 800–900 m a.s.l. Nonetheless, both valley sides are steep and rocky, with rock cliffs up to 100 m high in total descending from the plateau edge down to the river on the northern side (Fig. 18.6A). Locally, limestone pinnacles have been isolated by selective weathering of jointed limestones. The gorge floor is very narrow for most of its length, with hardly any floodplain and limestone walls directly overlooking the channel (Fig. 18.6B). Small caves are present within the valley sides, but they do not reach large dimensions.

For elucidating the probable origin of the gorge, the relationships to geological structure are important. The gorge itself is cut into limestones of Triassic age, dipping to the north. The northern dip accounts for the subtle cross-sectional asymmetry of the gorge, with the northern side being steeper. To the south of the gorge no other rock series than the limestones occur. However, in the north conglomerates of Palaeogene age occur beyond the local water divide, increasing in thickness towards the Hornádska kotlina basin. This onlap suggests the former flow of the Hornád River over the conglomerates, whose area of occurrence extended further south. At the time of asymmetric uplift of the Slovenský raj block, less intense in the north, fluvial incision was able to keep pace with elevation gain, although after the conglomerates were eroded away and the resistant limestones were reached, lateral erosion was severely suppressed and only vertical downcutting was possible. Local discontinuities probably dictated the specific course of the gorge. The presence of a few minor streams originating in the Hornádska kotlina and cutting through the highland belt to join the Hornád River within the gorge is consistent with this scenario. Thus, in terms of origin the Hornád Gorge has been created mainly by epigenesis (continuous downcutting into more resistant beds after removal of cover deposits) combined with antecedence (incision in response to uplift).

Along with the incision of the Hornád River, its large right-hand tributaries (Biely potok brook in the Tomašovská Belá valley, Lesnica brook) followed, although their valleys generally lack the gorge morphology and show V-shaped cross sections, with only localized more rocky reaches. Likewise, their longitudinal profiles appear fairly graded. By contrast, several minor right-hand tributaries in the upstream part of the gorge (Hlboká dolina, Kláštorská roklina, Zelená dolina) have their long profiles very steep and uneven, with waterfalls as high as 8.5 m (Duchový vodopád in Kláštorská roklina).

18.6 Tributary Gorges of the Northern Part of Slovenský Raj

18.6.1 Occurrence and Characteristic Features

A distinctive geomorphic feature of the northern part of Slovenský raj is the presence of rocky valleys (gorges), deeply incised into the marginal, but locally also the inner parts of the Glac Plateau (Fig. 18.5). They vary in length, from more than 6 km (Vel'ký Sokol) to less than 1 km (a series of valleys descending directly to the Hornád River

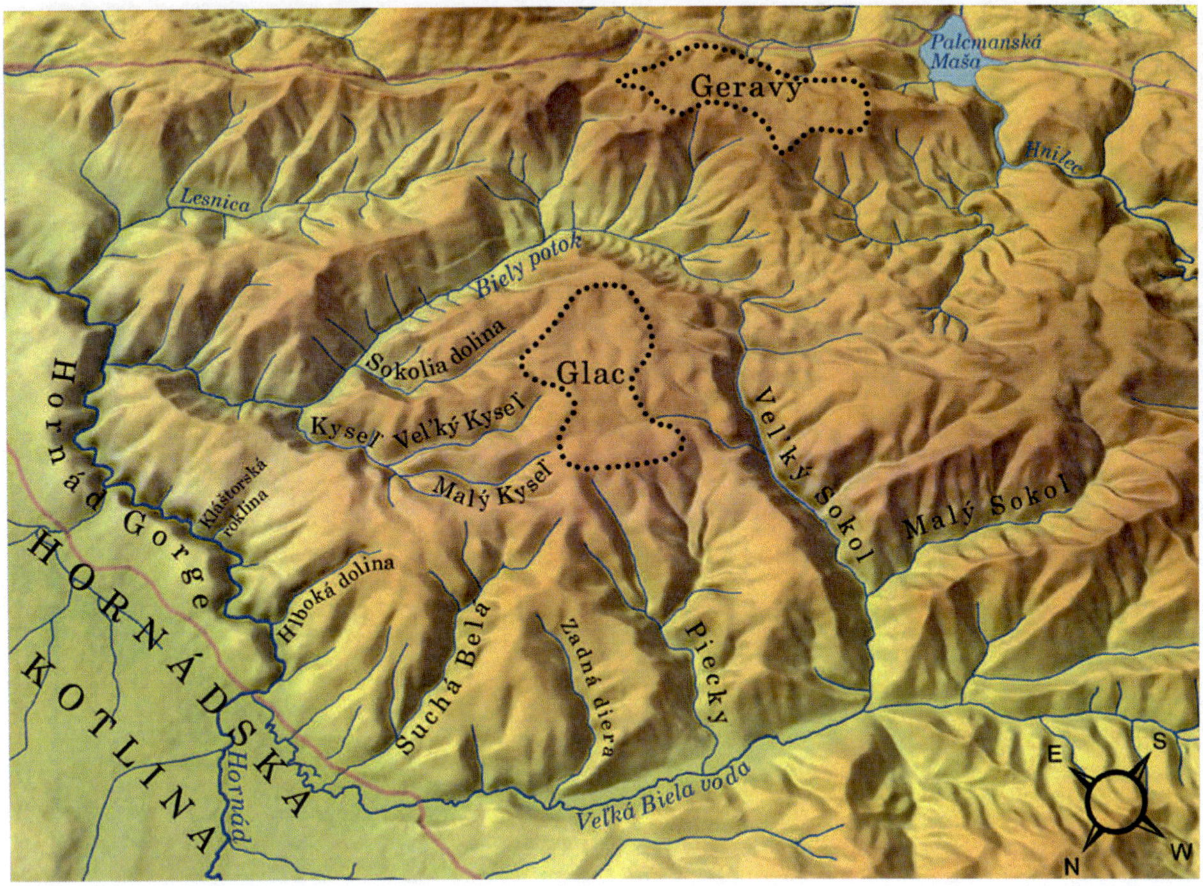

Fig. 18.5 Three-dimensional view of the northern part of Slovenský raj, with the radial network of gorges. Dotted lines show the extent of karst plateaus of Geravy and Glac (data source: digital elevation model DMR3.5 of 10 m resolution. The DMR3.5 dataset was provided by GKU Bratislava under the CC-BY 4.0 licence)

Gorge), and in places attain more than 250 m deep. They radiate in different directions, from WNW (Veľký Sokol, Piecky) to E (Kyseľ). The short ones are simple incisions, whereas the longer ones may be joined by tributary valleys of the second order, some being as long as the trunk gorge (e.g., Veľký Sokol versus Malý Sokol). Most gorges have tripartite morphology, with a relatively shallow upper reach a few tens of metres below the plateau level, the steep middle reach with abundant bedrock channel sections and waterfalls, and a slightly wider lower reach (Fig. 18.3). This subdivision corresponds with changing longitudinal gradient along the watercourse, which may be as high as 300‰ in the middle reach. The regional context allows one to interpret this succession as the evidence of rejuvenation in progress, where the wave of headward erosion, progressing upstream from the trunk Hornád River, has not yet reached the headwater parts. Interestingly, similar gorges do not exist further east (east of the Tomašovská Belá brook valley), despite considerable fluvial dissection, whereas only one similar and rather short gorge—Zejmarská roklina, exists in the southern part of Slovenský raj, drained to the Hnilec River. This pattern of occurrence is best interpreted as bedrock-controlled, with gorges being predominantly associated with the Wetterstein limestone series.

Steep rock slopes, bedrock channels, slot canyon sections and waterfalls are the most characteristic landmarks of these gorges. Composite rock slopes may be as high as 200 m, with individual rock walls not uncommonly in excess of 50 m. They descend straight to the channels in the most incised reaches, but may be separated from the channel by voluminous scree cones and aprons. These, in turn, are further modelled by debris flows and dry debris avalanches. Bedrock channels are ubiquitous, have highly irregular longitudinal profiles, and abound in minor landforms of fluvial erosion. More than 20 sizeable waterfalls occur within the gorges, with the Závojový waterfall in Sokolia dolina being the highest one ($\sim$75 m). Both single-fall and multi-step waterfalls occur. Tufa deposition occurs widely,

Fig. 18.6 Hornád River Gorge. **a** General view from the northern rim (the river flows away from the viewer), **b** limestone cliffs above the river bed. Note technical facilities along the hiking trail through the gorge

both on valley sides, where it is related to emerging groundwater, and in channels, where it is caused by changes in flow turbulence.

Despite the presence of limestones, caves are rather infrequent within the gorges and do not have impressive dimensions. Most are short, structure-controlled rock shelters and simple passages. Horizontal overhangs close to the valley floors may partly owe their origin to fluvial undercutting. One accessible cave is Biela Cave, within a limestone crag in the Kláštorská roklina gorge. It consists of two widely open, separate chambers, with the total length of 26 m. It may be hypothesized that fluvial incision that produced the gorges was so rapid and recent, that it outpaced the progress of deep-reaching karstification.

18.6.2 Suchá Belá

Suchá Belá is arguably the finest and certainly the most visited limestone gorge in Slovenský raj, which is due to both its superb scenic qualities and easiest access (Fig. 18.7). The valley is 3.6 km long and generally retains a meridional course up to the mouth, although local sinousity, almost certainly structure-controlled, typifies the middle reach. The overall gradient is c. 120‰, with the valley depth up to 250 m. The inconspicuous valley head is located at 930–950 m a.s.l., just below the plateau level. Further down extends the upper reach of the valley, which is broadly V-shaped, with a flat, depositional floor and occasional bedrock cliffs protruding from moderately steep slopes. A sudden right-angle turn signifies the beginning of the middle reach, which is the steepest one and represents a deep slot along most of its length, with near-vertical rock walls more than 50 m high. Nevertheless, the width of the slot varies, from sections where the channel bed equals the valley floor to local widenings, allowing for temporary storage of coarse alluvial sediment. Notable is the series of waterfalls, occurring close to one another. Korytový vodopád (= Channel Waterfall) is 8.5 m high in total and consists of two steps of similar height, separated by a large regular pothole. Okienkový vodopád (= Window Waterfall) is 12.5 m high and borrows its name from a natural rock opening, 1–1.5 m in diameter, at the upper rim of the fall. Former streamflow through the opening is likely, whereas nowadays the window is used by the tourist trail. The highest one (29.5 m) is the multi-step Misové vodopády (= Bowl Falls), with potholes of different depth at the base of each step. Minor waterfalls are located at the outlets of minor tributary valleys and reach up to 8 m. Bedrock erosional features in the Suchá Belá valley are not restricted to waterfalls and are in fact ubiquitous, represented by circular potholes, elongated troughs and connecting chutes. The lower valley reach, below the Misové vodopády, has fewer bedrock channel reaches and lacks high waterfalls, but the erosional nature of the gorge is still evident. The gravel-bed channel typically occupies the entire width of the valley floor, which is less than 10 m and just 2–3 m in many places, whereas rock gates less than 2 m wide are not uncommon. The gorge opens to the intramontane basin of Hornádska kotlina. Thus, Suchá Belá is an excellent example of a tripartite valley, with the shallow upper reach, the steep middle reach with waterfalls, and the slightly wider lower reach.

18.6.3 Sokolia Dolina

Sokolia dolina is a left-side tributary valley of the trunk valley of Tomašovská Belá brook, primarily known to include the highest waterfall in Slovenský raj—Závojový vodopád (= Veil Waterfall) (Fig. 18.8). The rather shallow valley head is located to the north of Mt. Bykárka (1058 m) and the valley below consists of three sections, as many others. The upper reach is c. 1.5 km long and follows straight SSW–NNE course. It is moderately incised, lacking high waterfalls and long impressive bedrock reaches, although minor bedrock steps do occur. The middle reach, although the shortest one (~0.5 km), is the most spectacular, with a series of waterfalls allowing the stream to descend for more than 150 m over a short distance. It also signifies a clear turn of the stream towards ENE. The lowest section is trending W–E and is c. 0.8 km long, joining the trunk brook of Biely potok at 573 m a.s.l. The valley shoulders in this reach attain 300 m high and are very steep and rocky, but the valley floor locally widens and contains only occasional low in-channel bedrock steps, with the Skalný waterfall as the highest one (3.5 m). Thus, the steepest middle section marks the transition between the upper, non-rejuvenated reach and the lower reach, which has already adjusted to the falling base level in the Hornád River valley.

The geomorphic highlight of the valley is the waterfall of Závojový vodopád, reaching the total height of 75 m. However, it consists of several individual steps, with the final one being the highest. They are separated by narrow benches, punctuated by potholes. Notable are tufa accumulations on some steps, especially in the middle part. Above the Závojový vodopád a narrow slot continues upstream, with alternating potholes and chutes, terminating with the Vyšný waterfall, the second highest in the valley (15.5 m). Its morphology is complex, with a steeply inclined (~50°) slab in the lower part and a vertical, partly overhanging face above. In turn, below the Závojový vodopád the stream flows over a series of bedrock steps, following the base of an impressive limestone wall more than 50 m high.

Fig. 18.7 Suchá Belá gorge. **a** depositional lower reach, **b** one of many rock steps in the gorge floor, **c** series of potholes, **d** Misové vodopády, **e** one of several slots in the middle reach of the gorge

Fig. 18.8 Sokolia dolina gorge. **a** lower reach with minor bedrock steps, **b** one of narrowings of the gorge, **c** high limestone cliffs in the steepest part of the gorge, **d** Závojový vodopád. Note considerable volume of woody debris in the gorge floor, transported by floods during extreme events

18.6.4 Vel'ký Sokol

Vel'ký Sokol, a tributary valley of Vel'ká Biela voda brook valley, is the longest among limestone gorges in the area (Fig. 18.9). It is more than 6 km long, descending from c. 940 m a.s.l. to 620 m a.s.l. at the mouth. For most of its length, it follows ESE–WNW direction, although in detail the stream takes a sinuous course, likely adjusted to local zones of structural weakness. As the other valleys, Vel'ký Sokol consists of the upper, non-rejuvenated reach above 870 m a.s.l., the middle reach with slots, bedrock cascades and waterfalls, and the lower reach, where the valley floor is

Fig. 18.9 Vel'ký Sokol gorge. **a** gravel bars in the lower reach, **b** widening of the gorge in the middle reach, **c** Rothova roklina, **d** Vel'ký vodopád

relatively wide and the gradient much lower than in the middle reach (30‰ versus 95‰). In contrast to the two gorges presented above, the lower reach of Vel'ký Sokol is depositional rather than bedrock-erosional and the channel is gravel-bed along its length, with several sections where the streamflow can be entirely diverted into the gravel fill. In terms of channel characteristics, braided sections with gravel bars alternate with anabranching reaches, where parallel channels are separated by narrow, persistent and partly vegetated islands.

The most impressive part of Vel'ký Sokol is the middle reach, which begins (on the upstream side) with a 200 m long and very narrow slot known as Rothova roklina. It is 10–15 m high and only 1–2 m wide, with numerous potholes, chutes and substantial amount of coarse woody debris. The latter likely comes from two sources. One is direct delivery to the channel from partly vegetated rock walls of the slot, but the majority seems to be transported from the wider upper reach during occasional extreme events, choking the narrow slot, insufficiently wide to allow the logs to pass down. Downstream of Rothova roklina are two waterfalls. The upper one is Vel'ký vodopád (= Large Waterfall), a simple one-step fall c. 7 m high, with a pothole below. Not far below is Malý vodopád (= Small Waterfall), similar in shape and height. Both are negotiated by tourists using steel ladders. The reach downstream of waterfalls is the deepest within the entire valley, up to 250 m. Jagged rock slopes, with vertical faces and overhangs, line the valley floor and are locally buffered by coarse scree accumulations, forming regular cones and less regular aprons. Shallow caverns dot the rock walls, following discontinuities in the rock mass. Locally large tufa accumulations cover the lowermost valley sides, with interesting microrelief of terraces, cascades and dams.

18.7 Dobšiná Ice Cave

Dobšinská l'adová jaskyňa (= Dobšiná Ice Cave) is one of the greatest highlights of the area (Tulis and Novotný 1989, 2007; Jakál 2005). It owes its uniqueness to the location in the temperate climate zone, outside the Alpine region, while its considerable part is permanently filled with ice. As a part of the system of Stratenská Cave (Fig. 18.10), its genesis is connected with an underground flow of the Hnilec River. The original relief related to the modelling of cave walls and ceilings by flowing water has been preserved in the upper, non-iced chambers of the cave. The remaining parts were changed as a result of frost weathering.

The conditions allowing for the formation of ice in the cave occurred after ceiling collapses the Stratenská Cave system in the middle Quaternary, about 400,000 to 200,000 years ago (Tulis and Novotný 1989, 2007; Jakál 2005; Tulis and Novotný 1989, 2007). As a result of this event, the cave took the form of large, steeply descending corridors and separate chambers developed at three different levels, with only one opening in their upper part, which is the current entrance to the cave. It is located in the northern slopes of the Duča Plateau, at 969 m a.s.l. Ice accumulation in the Dobšiná Ice Cave probably started in the Riss period (Glacial 3., about 300,000 to 140,000 years ago) as a result of stagnation of cold air in the lower parts of the cave, it is ongoing and permanent until now (Jakál 2005; Bella 2007; Pflitsch et al. 2007). The ice monolith fills most of the cave, locally up to the ceiling (maximum thickness of 26.5 m), occupying an area of ca. 9772 m^2 with an estimated volume of 110,132 m^3 (Bella 2007). In the upper parts, the ice monolith expands, whereas at the contact with bedrock it loses mass by melting. These processes are in balance—loss is compensated by growth of successive layers of ice on its upper surface. The ice fill occurs in the form of floor ice, icefalls, stalactites and stalagmites up to 10 m high and up to 4 m in diameter. The latter are concentrated at the intersection with tectonic faults and other discontinuities in the karst massif, which allow for more efficient infiltration of rainwater. The intensive supply of water occasionally leads to flow over the ice, forming channels incised into the surface of ice monolith. During the year the air temperature in the cave ranges from −5.2 to +0.2 °C.

The Dobšiná Ice Cave is an example of a rare natural phenomenon, especially in terms of the occurrence of cave ice at an altitude below 1000 m. The ice monolith is located at only 920 to 950 m a.s.l., being therefore of interest to scientists of different specializations. It is the most researched and described cave phenomenon in Slovakia (Jakál 2005).

Fig. 18.10 Interior of the Dobšiná Ice Cave, with its ice monolith in Great Hall (situation in 2012) (*photo* courtesy of Tymoteusz Sawiński)

18.8 Conservation and Tourism

The value of Slovenský raj was not immediately recognized and the protracted period of forest exploitation in the 19th and early twentieth century, beside land cover change, brought about modifications of stream channels to allow for log rafting and construction of many forest tracks. However, in the late 1920s conservationists succeeded in saving several gorges from track building, which would have damaged natural rock-cut channel landforms (Novotný and Tulis 2005). Territorial protection was introduced in 1964, and the area was declared a Protected Landscape Area (Chránená krajinná oblast'). In 1988 its protection status was raised and the Slovak Paradise National Park (Národný park Slovenský raj) established. It covers 198 km^2, including areas outside the geographically defined region. Within the national park, several areas received higher form of protection, as state nature reserves (11 areas) or nature reserves (8 areas). All limestone gorges are state nature reserves and visitations are only possible using signposted hiking trails. In addition, as elsewhere in Slovakia, all caves are automatically protected by law and the Dobšiná Ice Cave is a UNESCO World Heritage Site since 2000.

Geomorphological curiosities of the region have been visited since the second half of the nineteenth century. As early as 1871 the Dobšiná Ice Cave was made accessible to the public. However, the key assets of Slovenský raj today—the gorges—remained unexplored until the early twentieth century. Vel'ký Sokol gorge was first negotiated in 1898, followed by the lower parts of Suchá Belá (1900), Kysel' (1907), Sokolia dolina (1910) and Piecky (1911). Nonetheless, for long the gorges could have only be visited by climbers, whereas the history of their opening to the recreational tourism started much later, in the second half of the twentieth century. The path through the Hornád River Gorge was completed in 1974. Nowadays most of gorges are crossed by hiking trails, although these may be locally demanding due to the necessity to use steel ladders, rock-attached steps and wooden bridges, some in very exposed places. Therefore, almost all trails in the gorges are one-directional (in upstream direction). The overall density of hiking trails in Slovenský raj is fairly high and they allow an interested geotourist to visit all sites of prime geoscientific interest, including the trunk valleys (Hnilec, Hornád rivers) and tributary gorges, karst plateaus, several excellent viewing points and the ice cave. Some trails also play the role of educational paths, with interpretation panels and/or printed leaflets available.

18.9 Conclusions

Slovenský raj is arguably one of the finest geomorphological landscapes of Slovakia, appreciated by hundreds of thousands of visitors each year. Its most spectacular landforms, easily accessible to a casual tourist, include the gorge of Hornád River, numerous tributary gorges turning into rock canyons with series of impressive waterfalls, limestone escarpments serving as excellent viewing points, and the Dobšiná Ice Cave. However, the geomorphological story behind this scenery is no less exciting. Landform and rock record in Slovenský raj allows one to reconstruct a protracted history that probably goes back to the Cretaceous and some features of contemporary relief may even be of that distant age, owing this remarkable survival to long-term burial and recent exhumation. Characteristic karst plateaus, with their numerous solution dolines, are in gross inherited pre-Late Eocene landscape facets, subject to burial by Eocene marine deposits and gradually uncovered since the Middle Miocene. Some caves in the region may also date back to the Eocene, although the main phase of development of the more than 20 km long Stratenská Cave occurred in the Pliocene. Late Cenozoic uplift, particularly intense since the Pliocene, is responsible for an increase in relative relief, incision of the trunk valleys, headward erosion in the tributaries, and consequently, the origin of picturesque gorges in the northern part of the dissected plateau. Their rejuvenation is ongoing, as attested by impressive waterfalls, some of which are more than 50 m high. The entire area is protected as a National Park, established in 1988, which ensures proper safeguarding of its natural heritage, including outstanding geomorphology.

Acknowledgements The authors are grateful to Milan Lehotský for providing an opportunity to join this publication project and supplying relevant literature. Tymoteusz Sawiński kindly let us use the photograph of ice monolith in the Dobšiná Ice Cave. Participation of Milena Różycka in fieldwork, which was carried out in August 2019, is also acknowledged.

References

Bella P (2007) Morphology of ice surface in the Dobšiná ice cave. In: Zelinká J (ed) 2nd International Workshop on Ice Caves, Demänovská Dolina, Slovak Republic, May 8 – 12, 2006. Proceedings, Liptovský Mikuláš, pp 15–23

Hochmuth Z (2008) Krasové územia a jaskyne Slovenska. Geographia Cassoviensis II(2):1–210

Jakál J (2001) Porovnávacia analýza krasových planín Západných Karpát. Geografický Časopis 53(1):3–20

Jakál J (ed) (2005) Jaskyne Svetového Dedičstva na Slovensku. Správa Slovenských Jaskýň, Liptovský Mikuláš, p 160

Lukniš M (1945) Príspevok ku geomorfológii povrchového krasu Ztratenskej vrchoviny (Slov. raja) Sborník prác Prírodovedeckej fakulty Slovenskej Univerzity v Bratislave 12: 5–45

Mello J, Filo I, Havrila M, Ivan P, Ivanička J, Madarás J, Németh Z, Polák M, Pristaš J, Vozár J et al (2000) Vysvetlivky ku geologickej mape Slovenského raja, Galmusu a Hornádskej kotliny. GÚDŠ Bratislava, p 303

Mičian L (1962) Niekoľko poznámok k prelomu Hornádu v Stratenskej hornatine a morfológii jeho okolia. Geografický Časopis 14(1):57–65

Novotný L, Tulis J (2005) Kras Slovenského raja. Správa slovenských jaskýň a Slovenská speleologická spoločnost', Liptovský Mikuláš, p 175

Pflitsch A, Piasecki J, Sawiński T, Strug K, Zelinka J (2007) Development and degradation of ice crystal sediment in Dobšinská Ice Cave (Slovakia). In: Zelinká J (ed) 2nd International Workshop on Ice Caves, Demänovská Dolina, Slovak Republic, May 8–12, 2006. Proceedings, Liptovský Mikuláš, pp 38–49

Rosič M, Mitríková J, Klamár R, Kovaľ M, Orlovská A (2011) Predpoklady cestovného ruchu na území Národného parku Slovenský raj s empirickým príkladom komparácie spokojnosti návštevníkov počas zimnej a letnej sezóny v jeho južnej časti. Folia Geographica (prešov) 17:323–351

Tulis J, Novotný L (1989) Jaskynný systém Stratenskej jaskyne. Osveta Martin pre SSS Liptovský Mikuláš, p 456

Tulis J, Novotný L (2007) Dobšinská l'adová jaskyňa Ice Cave – 136 years from its discovery. In: Zelinká J (ed) 2nd International Workshop on Ice Caves, Demänovská Dolina, Slovak Republic, May 8–12, 2006. Proceedings, Liptovský Mikuláš, pp 8–14

Piotr Migoń is Professor of Geography at the University of Wrocław, Poland, and Director of Institute of Geography and Regional Development. His main subjects of research are rock control in geomorphology, especially in granite and sandstone areas, weathering, mass movement in mountain terrains and long-term landform evolution. The Sudetes Mountains in Central Europe is his main research area, but he was also involved in projects in other European countries (Czech Republic, Sweden, Great Britain, Portugal), in the United States, Mexico, China, Jordan and Namibia. During 1997–2001 he was the Secretary of the International Association of Geomorphologists, and its Vice-President in 2009–2013. He is also the series editor of 'World Geomorphological Landscapes'.

Wioleta Porębna is a Ph.D. student at the Institute of Geogrpahy and Regional Development, University of Wrocław, Poland. She is interested in the development of sandstone areas, studying the origin, controls and evolutionary pathways of the valley network in sandstone-dominated tablelands, with geographical focus on the Bohemian Cretaceous Basin.

Kacper Jancewicz works as a specialist in the Institute of Geography and Regional Development, University of Wrocław, Poland. His main research interests focus on various LiDAR DEM applications in geomorphological studies, geomorphometric analyses and digital geomorphological mapping.

Permanent Gullies as Important Indicators of Past Environmental Changes

19

Miloš Stankoviansky, Štefan Koco, Pavol Papčo, and Libor Burian

Abstract

Geomorphic investigation of permanent gullies in Slovakia shows three milestones. The first one was an elaboration of the unique map of spatial distribution and density of gully network in Slovakia at the scale of 1: 400 000 in the 1950s. The second milestone is represented by an assessment of gullies in the extensive area of the Myjava Hilly Land (370 km^2) in Western Slovakia (the 1990s and early 2000s), directed to elucidate regularities of spatial distribution of gullies, their age, causes and course of their evolution. It was proved that gully formation was controlled mostly by historical artificial linear landscape elements. Two periods of possible gully formation were suggested, the younger one sometime between the mid-16th and the mid-nineteenth centuries, and an older one in the Late Middle Age. The fact that significant human interventions in this region overlapped temporally with the influences of the Little Ice Age emphasizes the role of permanent gullies as important indicators of geomorphic response to past environmental changes. The third and the last milestone represents a detailed study of five gully sites (the late 2000s and early 2010s), directed to an assessment of their historical development, and to modelling and simulation of gully erosion in two cases.

M. Stankoviansky (✉) · L. Burian
Department of Physical Geography and Geoinformatics, Faculty of Natural Sciences, Comenius University in Bratislava, Bratislava, Slovakia

Š. Koco
Department of Geography and Applied Geoinformatics, Faculty of Humanities and Natural Sciences, University of Presov, Presov, Slovakia
e-mail: stefan.koco@unipo.sk

P. Papčo
Department of Geography, Faculty of Education, Catholic University in Ružomberok, Ružomberok, Slovakia
e-mail: pavol.papco@ku.sk

Keywords

Slovakia • Gully development • Indicator of environmental changes • Phases of gullying • Little Ice Age • Myjava Hilly Land

19.1 Introduction: Brief Outline of Major Topics in Geomorphic Research of Gullies in Slovakia

An increased interest to study permanent gullies stems from the growing awareness that these geomorphic features provide important insights into the impacts of both climate and land use changes on erosion processes (e.g. Vanwalleghem et al. 2005). In fact, permanent gullies represent indicators of these environmental changes in the past. This assumption was confirmed also by research in Slovakia.

The Slovak territory provides ideal conditions for the study of permanent gullies due to their considerable density and diversity. Their formation is a result of high geomorphic effectiveness of gully erosion, influenced by both natural and socio-economic conditions. Among natural conditions, gully formation is supported mostly by noticeable vertical and horizontal dissection of relief and large spatial extent of geological complexes of high erodibility. Conditions for intense gullying began, however, only in association with historical transformation of original woodland to cultural land. Extensive agricultural areas, but also devastated forests, were affected by severe gully erosion, resulting in the formation of dense network of gullies.

In the mountainous part of the country (the Carpathians) gullies occur mostly in uplands, lower piedmonts and erosion-denudation furrows. They relate mostly to little resistant rock complexes of flysch and volcanic belts. In lowland (Pannonian) regions, gully formation has affected mostly higher parts of hilly lands, especially in the contact

M. Lehotský and M. Boltižiar (eds.), *Landscapes and Landforms of Slovakia*, World Geomorphological Landscapes,
https://doi.org/10.1007/978-3-030-89293-7_19

zones with the mountains. There they occur mostly in loess covers and Neogene sediments.

The detailed research of gullies started only after the 2nd World War. Leaders of gully research were mainly geomorphologists, but foresters and geologists were also involved. In the period of seven post-war decades geomorphic investigation of gullies was focused on three main topics. The aim of chronologically the first, Slovakia-wide directed phase of research (the 1950s and 1960s) was to assess the density of gully network. It enabled one to obtain not only a general picture on spatial distribution of these features in Slovakia, but also to process an important database to serve subsequent, more detailed research on lower hierarchical levels. The second, regional phase of gully research (the 1990s and early 2000s) was carried out in the framework of an assessment of geomorphic response to environmental changes in the territory of the Myjava Hilly Land. It was directed to elucidate regularities of spatial distribution of gullies, their age, causes and course of their formation. The third, local phase (the late 2000s and early 2010s) involved detailed studies of selected gullies and gully systems to assess their historical development (sites Prašice, Babikovce and Rybník) and modelling and simulation of gully erosion (sites Bardejov and Veľký Šariš) (Fig. 19.1).

19.2 The Country-Wide Level: Spatial Organization and Density of Gullies in Slovakia

In 1958 Š. Bučko and V. Mazúrová published a unique map of density of gully network in Slovakia at the scale of 1: 400 000. They processed it using topographic map sections at the scale of 1:25 000 from the 1950s. The authors used a square network with an area of the quadrate 4 km^2. The constructed frequency curve enabled them to delimit six-degree scale of density of gullies and sunken lanes, whereas the highest degree represents >3 km/km^2 (Fig. 19.1). Since the 1960s, map sections at the scale 1:10 000 with more precisely depicted network of gullies were available. Confrontation of gully networks as shown by these maps with the situation in the field at selected sites enabled the authors to establish the upper limit of gully density that exceptionally reached up to 9 km/km^2 (Bučko 1980). Šimurková (2012), using the map by Bučko and Mazúrová (1958), computed the weighted average of gully density for all 84 Slovak geomorphic units. The highest average values are shown by the Kysucká vrchovina Upland (1.08 km/km^2), Myjava Hill Land (1.06 km/km^2) and Podbeskydská vrchovina Upland (0.88 km/km^2).

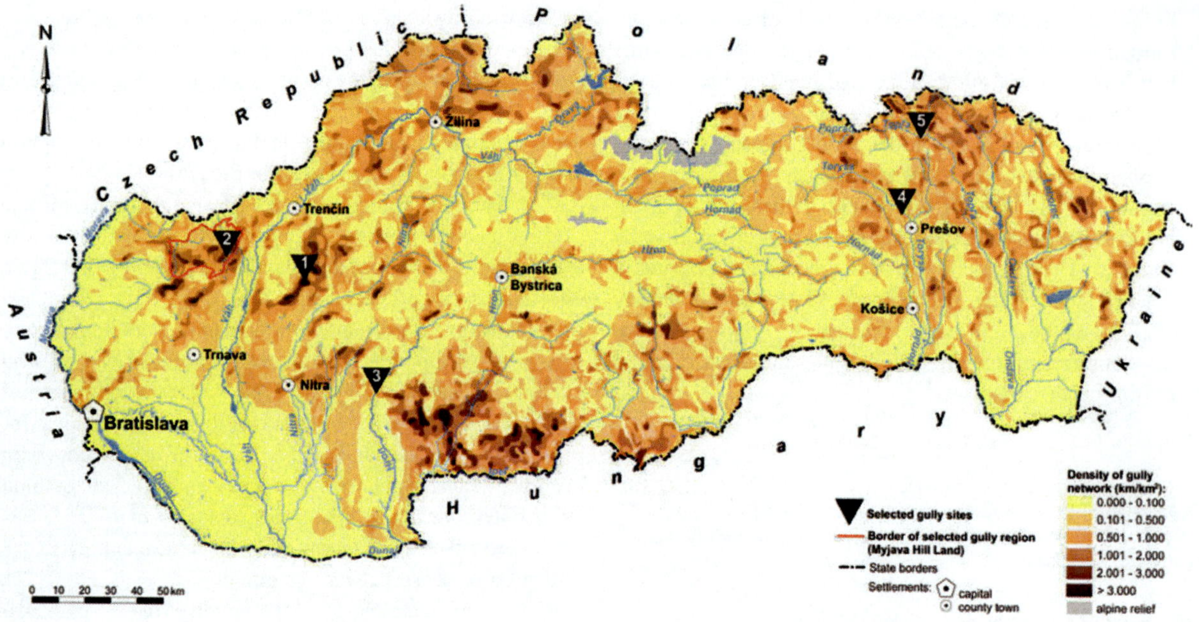

Fig. 19.1 Map of density of gully network in Slovakia (elaborated by Š. Koco on the basis of the 1:400 000 map by Bučko and Mazúrová 1958) with location of studied gully region and sites (1 Prašice, 2 Babikovce, 3 Rybník, 4 Veľký Šariš, 5 Bardejov)

19.3 The Regional Level: The Myjava Hilly Land

Though the spatial organization and density of permanent gullies in Slovakia are known since the late 1950s, the question remained when were they formed and what were the main causes of their origin. Some historical factors triggering gully formation were already indicated by Bučko and Mazúrová (1958), who suggested that forest clearance and overgrazing, associated mostly with the youngest colonization waves (Walachian and „kopanitse " ones), resulted in the formation of dense road and path network that controlled gully formation on deforested slopes. However, they did not date gully formation itself and other authors did not attempt dating either.

Four decades later a campaign to date gullies was carried out in the Myjava Hilly Land, western Slovakia (Fig. 19.1) (cf. Stankoviansky 2003a, b, c). There are at least two reasons why this region is particularly suitable for a study of gully network evolution. Firstly, it belongs to geomorphic units with the highest gully density. Secondly, relatively late human occupation helped to narrow the time interval of their possible formation. For better understanding of gully evolution in time and space in this area it is useful to outline briefly its natural conditions, as well as the settlement and land use history.

19.3.1 Natural Conditions, Settlement and Land Use History

The Myjava Hilly Land (Fig. 19.2) represents a lowered zone between two higher mountain ranges, the Biele Karpaty Mts (White Carpathians) and the Malé Karpaty Mts (Little Carpathians), with an area of ca 370 km^2. It is a low (maximum elevation—Bradlo Hill, 543 m), flat-topped geomorphic unit, with relief of the order of 40–130 m. It is built predominantly of medium- to low-resistant Senonian, Paleogene and Neogene sedimentary rocks, covered with a considerably thick, mostly fine-textured regolith. Small islands of loess loams occur locally. Cambisols and Luvisols are the most frequent soil types. The mean annual precipitation is 650–700 mm. The natural vegetation was represented mostly by oak and oak-hornbeam forests.

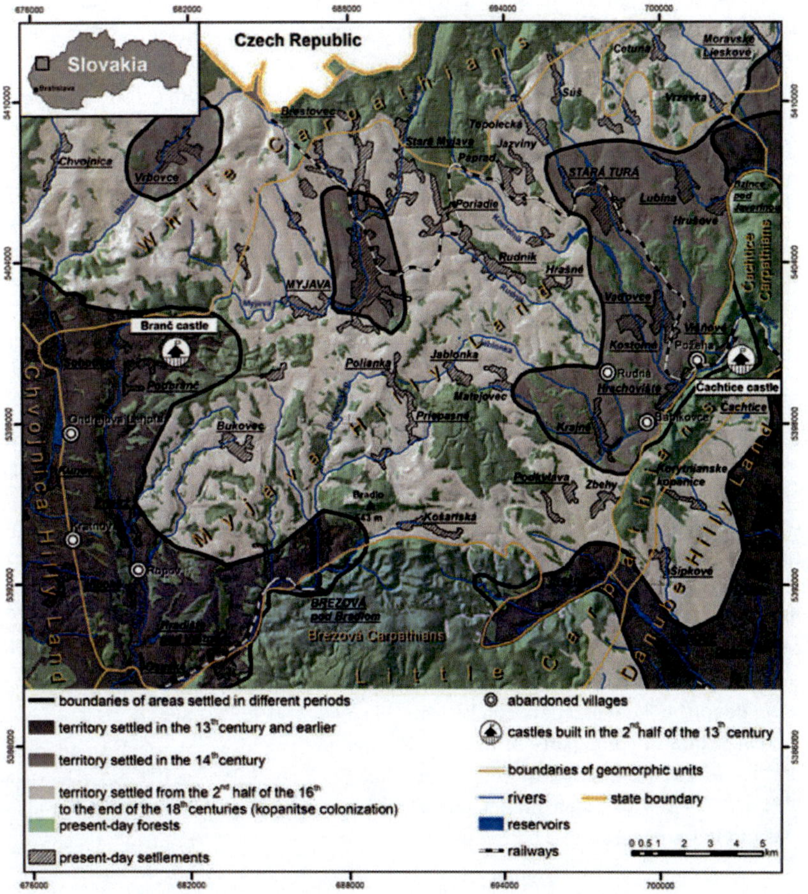

Fig. 19.2 Map of settlement history in the Myjava Hilly Land (modified from Dotterweich et al. 2013)

Most of the area at the end of the thirteenth century was still unsettled. Only its marginal parts, especially in the west, had been discontinuously occupied since prehistoric times and more villages existed there as early as the High Middle Age (the first half of the thirteenth century). Significant change in the history of human occupation is connected with the erection of the nearby Čachtice Castle in the second half of the thirteenth century. The owners of the Čachtice domain founded numerous villages in the eastern part of the area until the mid-fourteenth century. The main wave of the settlement, however, was associated with the so-called kopanitse colonization that started in the second half of the sixteenth century and culminated at the transition from the 18th to the nineteenth century. An important stream of settlers came also in the framework of the Walachian colonization. This overlapped considerably with the kopanitse settling, as it was performed in this area mostly in the second half of the 16th and in the seventeenth centuries. Activity of kopanitse farmers and Walachian shepherds in the newly settled territories resulted in forest clearance almost in the whole central part of the area and in the transformation of acquired ground into pastures, fields and meadows. The settlers established a large number of small settlements—hamlets ("kopanitse" in Slovak) dispersed irregularly in the new farm land. With time, population growth resulted in the gradual extension of arable fields at the expense of originally more widespread pastures and in hereditary division of existing fields into ever smaller plots. An important part of this land was also a dense network of artificial linear landscape elements, mostly access roads. It appears that these linear features represented key controlling factor of gully formation.

After the Second World War, the kopanitse landscape changed because of collectivization in agriculture that resulted in merging of the former small private plots into large cooperative fields and removal of the network of artificial linear landscape elements.

19.3.2 Shape and Dimensions of Gullies

Gullies in this area have mostly V-shaped cross profiles, depths commonly to 10 m, less frequently to 15 m and rarely to 20 m. Their maximum length is up to 800 m on slopes, 700 m along divides, and usually up to 1700 m in the dry valley bottoms, rarely reaching even 2–3 km. Most of gullies are straight, but some meander because of bends of the original roads. Gullies are either singular or occur in sets, both dendritic and parallel. Bučko and Mazúrová (1958) indicated that in some parts of the area the maximum gully density reached 2–3 km/km^2 (Fig. 19.1), but field research revealed sites with much higher concentrations of gullies, exceptionally even up to 11 km/km^2.

19.3.3 Spatial Organization of Gullies in Relation to Topography and Land Use

The field reconnaissance together with an analysis of comtemporary topographic maps and old cadastral maps at the scale of 1:2880 suggests that the pattern and density of gullies have been controlled mostly by historical artificial linear landscape elements, such as access roads, paths, field boundaries, banks, lynchets and exceptionally also by drainage furrows and furrows indicating cadastral boundaries. Only a few gully sites are linked with original pastures. From this viewpoint it is possible to distinguish five main types of gullies (Fig. 19.3).

The first type, quite frequent in the area, occurs on hillslopes previously cultivated along contours (Fig. 19.3a). Parallel gullies were formed along access roads or paths in inter-parcel positions, following gradient lines. Much rarer is the second type that occurs on hillslopes with previous up-and-down slope cultivation (Fig. 19.3b) and originated at various locations along parcel boundaries. Both these diametrically different types of gullies are very well developed on the southern slope of the Bradlo Hill (Figs. 19.4 and 19.5). The third type occurs in hillslope sections, where previous parcels were cultivated subhorizontally, i.e. where the lynchets separating the plots dip at a low angle (Fig. 19.3c). Parallel gullies were created along these linear features by deepening of drainage furrows excavated along the foot of each lynchet. A subtype of the third type (Fig. 19.3c1) is represented by two subhorizontal gullies on the dry valley side, created by erosional deepening of the drainage ditch dug along the border of two sets of parcels with different tillage directions. Later the upper of these gullies was captured at the right angle by a gradient gully emptying into the valley floor gully. The fourth and the most common gully type is linked to field roads and paths (Fig. 19.3d). They follow gradient lines, run aslant the contour lines or even follow the divide lines and often form closely arranged parallel sets. Many gullies originated by deepening of sunken lanes. The fifth type represents the former pasture, where parallel gullies have developed below spots of overgrazed grassland (Fig. 19.3e). A very specific linear arrangement of gullies was found between the villages of Vaďovce and Bzince pod Javorinou. It is very probably controlled by a furrow dug by special plough for marking out cadastral boundaries (Figs. 19.6 and 19.7).

Gullies linked to linear elements of an old landscape occur mostly on slopes (valley-side gullies) and, as it was already stated, even along ridges in water-divide positions. A much smaller group of gullies is situated in the bottoms of dells and dry valleys (valley-floor gullies), and these are controlled topographically. However, this is not a universal control, as some of these gullies do not directly follow

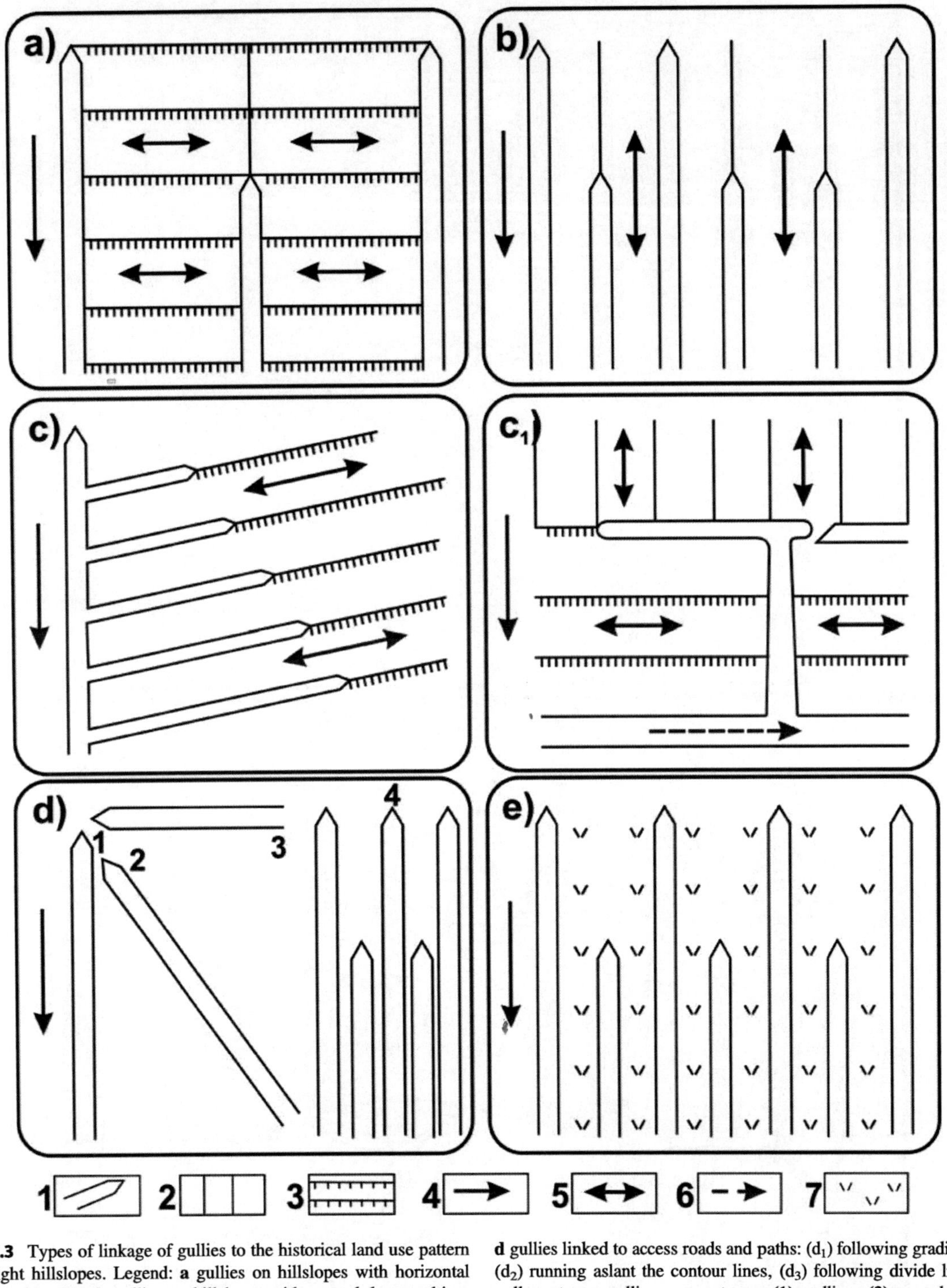

Fig. 19.3 Types of linkage of gullies to the historical land use pattern on straight hillslopes. Legend: **a** gullies on hillslopes with horizontal contour cultivation, **b** gullies on hillslopes with up-and-down cultivation, **c** lateral gullies on hillslopes with subhorizontal cultivation entering to the main gradient gully, (c_1) example of the gully linked to the boundary between two field blocks of different tillage direction, captured by gradient gully emptying into the valley floor gully, **d** gullies linked to access roads and paths: (d_1) following gradient lines, (d_2) running aslant the contour lines, (d_3) following divide lines, (d_4) gully sets, **e** gullies on pastures. (1) gullies, (2) parcel borders, (3) lynchets (steps of terraced fields), (4) direction of hillslope gradient, (5) direction of cultivation, (6) direction of thalweg gradient in cut along dry valley bottom, (7) pastures (after Stankoviansky 2003b)

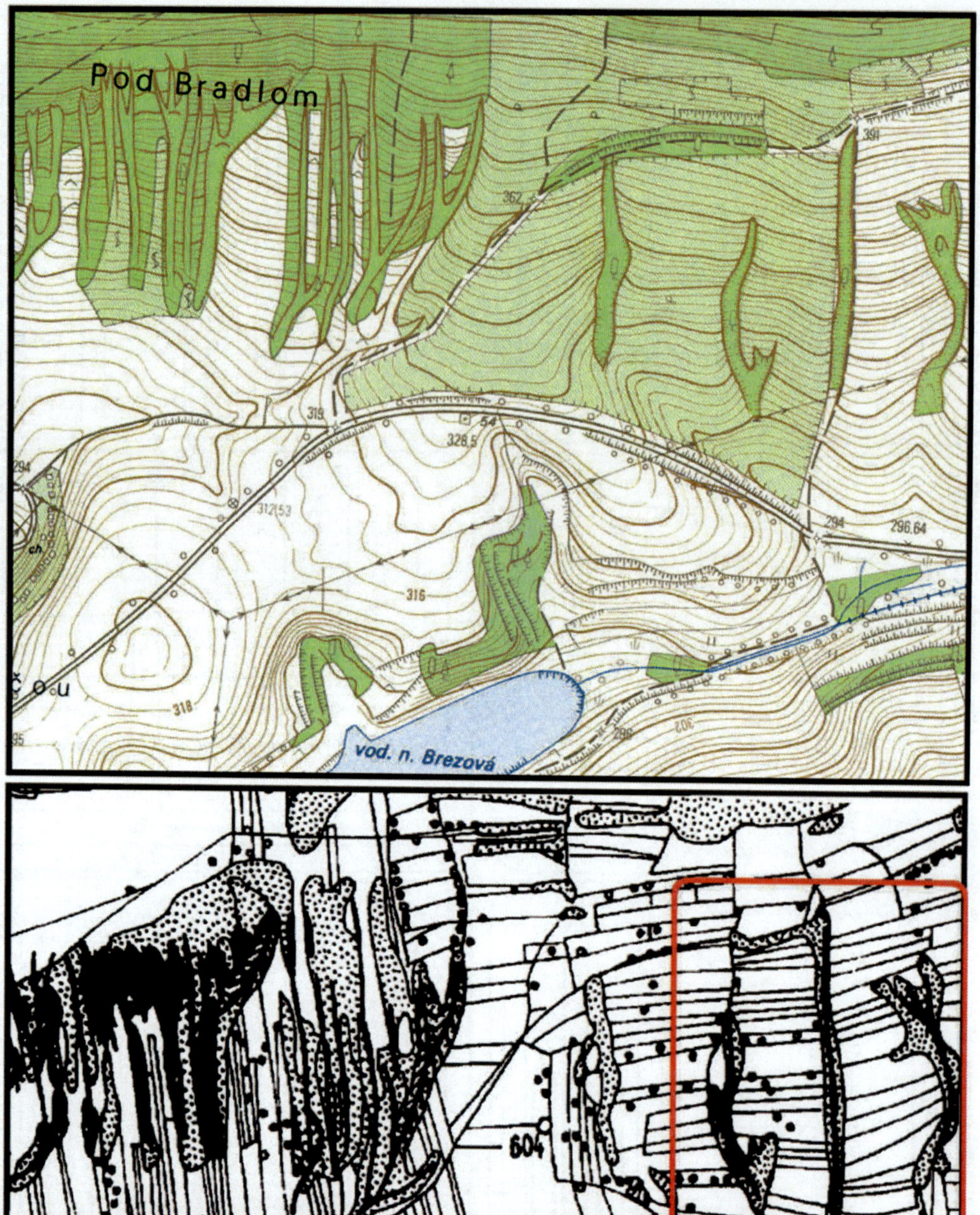

Fig. 19.4 The relationship of gullies on the southern slope of the Bradlo Hill, indicated on the topographic map from 1991 (top), to land use pattern that existed before collectivization, as indicated on the cadastral map from 1900 (bottom) (2003b modified from Zachar 1970, p. 151 and Stankoviansky 2003b); gullies follow borders of parcels with both up-and-down cultivation (left) and contour cultivation (right)

Fig. 19.5 The contemporary land use pattern on the southern slope of the Bradlo Hill, with forested gullies (cf. red rectangle in Fig. 19.4)

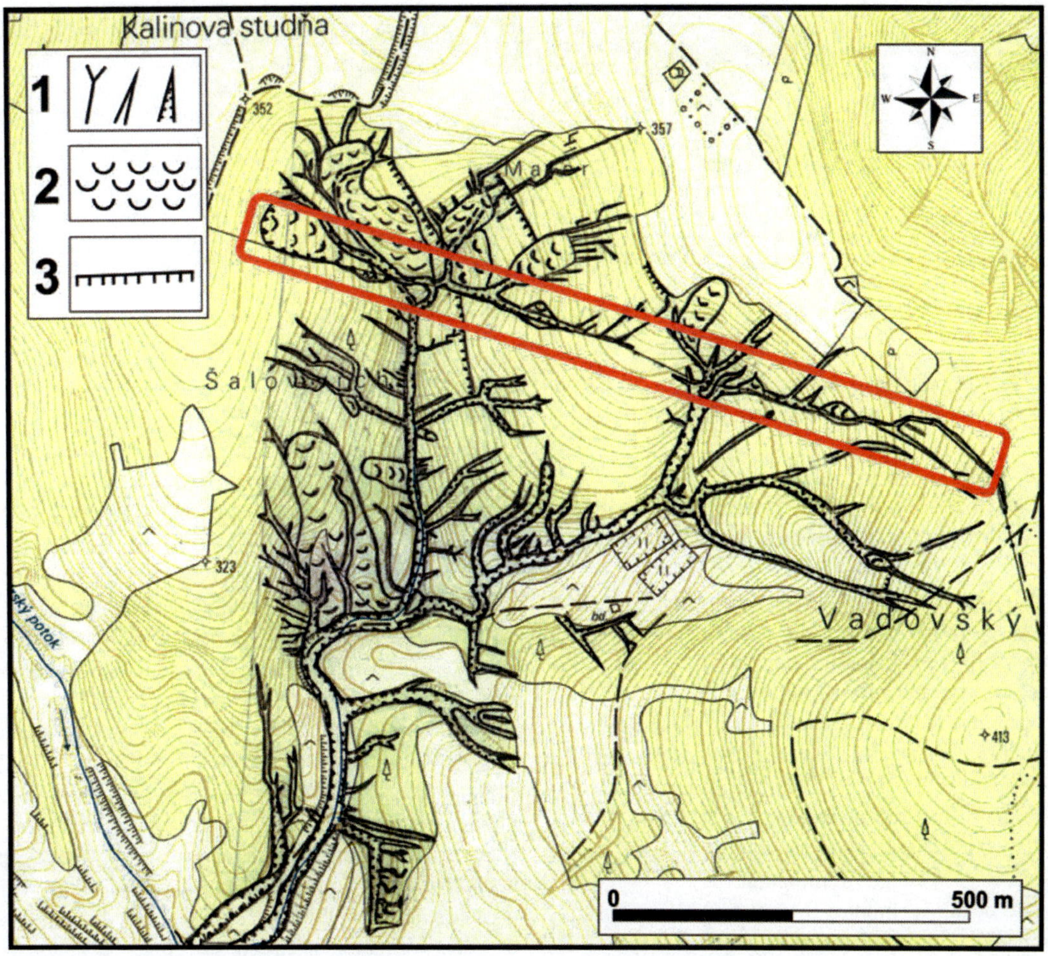

Fig. 19.6 Gully system between the villages of Vaďovce and Bzince pod Javorinou, formed mostly on former pastures, represents the highest gully density in the area (11 km/km^2); linear arrangement of gullies in the red rectangle is controlled by a cadastral boundary. Legend: (1) gullies, (2) landslides, (3) remnants of terraced fields (after Stankoviansky 2003b)

Fig. 19.7 Typical V-shaped gully as a part of linear gully system controlled by a cadastral boundary (see Fig. 19.6)

thalwegs. Topographically-conditioned gullies occur also on landslides that are older than the onset of settlement.

19.3.4 Dating of Gully Formation and Outline of Their Spatial Development in Time

Given that most of the Myjava Hilly Land was settled so late, it is possible to constrain the onset of possible gully formation for the predominant part of its territory. For clarification of its termination, an analysis of historical maps and both regional and local written documents can be used. From among old cartographic materials, the maps of the 1st (1782), the 2nd (1838) and the 3rd (1882) military surveys were used, the former two at the scale of 1:28 800, and the latter at 1:25 000. These historical sources indicate at least two periods of possible gully formation.

The maps of the 1st military survey depict a surprisingly high number of gullies. However, while the maps of the western part show almost all large present gullies, the maps of the eastern part show a maximum of one third, although the present-day density of these features is similar in both parts of the area. It is thus obvious that most of western gullies originated before 1782, whereas most of eastern ones were formed later (Fig. 19.8). However, the exact age of pre-1782 gullies remains unknown. Nevertheless, written sources relevant to two villages in the west (Bukovec and Prietrž) indicate that part of them could have existed as early as the 1730s. It suggests that the first marked period of gully formation in the "kopanitse" area in the New Age had to take place sometime between the mid-sixteenth century and the 1730s.

The maps of the 2nd military survey depict almost all large gullies known at present. The density of gullies on these maps is roughly similar for both the western and eastern part, of course, it does not fully reach the present state. A comparison with the maps of the 1st survey indicates a slight increase in the density of gullies in the west and a greater increase in the east. This suggests that another period of gully formation occurred sometime between the 1780s and the 1840s, and that it had greater effect in the east.

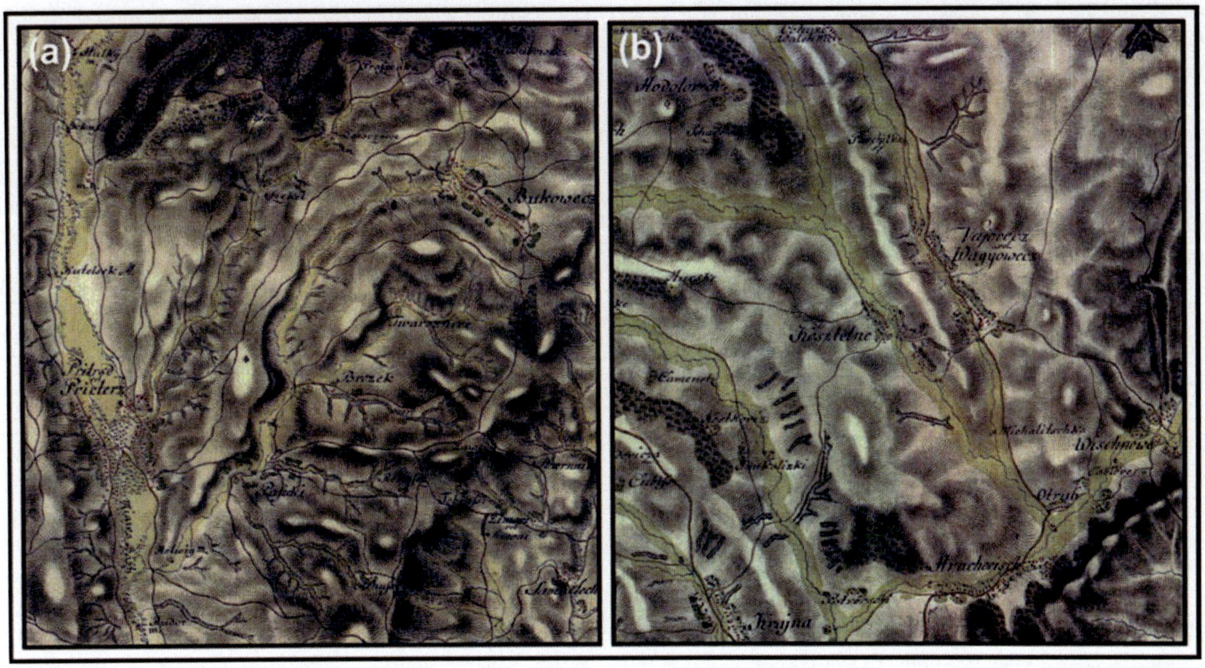

Fig. 19.8 Illustration of different density of gully network in the western (**a**) and eastern (**b**) part of the Myjava Hilly Land in maps of the 1st military survey (1782)

The distribution of gullies on maps of the 3rd military survey is locally virtually identical to the contemporary pattern. A comparison with the maps of the 2nd survey shows that some smaller gullies were formed or the older, larger gullies expanded slightly. A minor increase in the network of gullies after 1882 is indicated by comparing these maps with the maps of military surveys produced in the 1950s and 1960s.

Naturally, these findings are valid only for the central part of the Myjava Hilly Land, settled during the youngest colonization waves. In its eastern part, colonized between the building of the Čachtice Castle and the mid-fourteenth century, gullies could have developed earlier. Suitable conditions promoting gully formation existed particularly in the western marginal part of the area, where the history of human occupation is longer. The abandonment of some villages in this area in the Late Middle Age may indirectly point to the effects of such an older phase of gully erosion. For example, in the case of the villages of Hlboké and Kratnov it is likely that they have been abandoned due to disastrous gullying and accompanying muddy floods at that time because they were situated in the bottoms of dry valleys, incised later by large gullies in thalweg positions (Fig. 19.2). By the way, Hlboké was resettled later.

It is supposed, in conformity with Bork (1989) and other authors, that a repeated gully formation occurred in the parts of the area settled earlier. It seems that most of the Late Middle Age gullies were later either in part or totally filled, but re-appeared in the course of New Age phases of gullying. Relatively sharp outlines and edges of permanent gullies, pointing to their young age, are consistent with this hypothesis. Thus, it may be said that at least a part of contemporary gullies can be the result of alternating cutting and filling, which can be traced back to the Late Middle Age.

19.3.5 Influence of Land Use and Climate Changes on Gully Formation

It was proved that the origin of most gullies in the area was conditioned by the socio-economic activity of humans as these gullies could be formed only during or after the transformation of woodland into farmland. The gullies literally follow the farmland pattern. They were formed above all by vertical incision of ephemeral flows generated during extreme rainfall and sudden snowmelt events along the linear landscape features, both topographical and artificial, with the latter playing a much more important role. The majority of gullies was formed in stages, and they expanded during individual extreme events.

The period of the kopanitse colonization (ca 1550–1800), with accompanying land use changes, overlaps temporally with the period of the Little Ice Age (LIA) sensu stricto (1550–1850 by Lamb 1984). The fact that the most of larger gullies in this area were formed between the mid-16th and the mid-nineteenth century suggests that the decisive role in gully formation was played by climatic factors. In other words, extensive forest clearance and expansion of the farmland could have predisposed the land to gullying, but the triggering mechanism of disastrous gully erosion may have been represented by extreme events during this cold climatic fluctuation. This assumption is in conformity with the opinion of Starkel (2000) that of all climatic fluctuations, the wetter and at the same time colder periods are typified by increased frequency of extreme meteorological-hydrological events.

Gully formation in the early settled parts of the area, coincident with the Late Middle Age, can be also linked with the LIA, though in this case sensu lato (since 1250 by Porter 1986) (cf. Stankoviansky and Pišút 2011). Thus, the recognition of synchronous occurrence of land use and climate changes as factors triggering gully erosion and exacerbating geomorphic response is valid for the whole period of LIA, but particularly for the LIA sensu stricto.

Since the mid-nineteenth century there has been a decrease in gully growth because of the afforestation of gullied areas and climatic improvements since the termination of the LIA. In the post-collectivization period the filling of some gullies was recorded, mostly of those with large catchment areas, as well as the complete infilling of numerous smaller gullies due to land levelling. However, during extreme rainfall events of particularly high magnitude considerable gully deepening can happen also under contemporary conditions, as confirmed by Stankoviansky and Ondrčka (2011) in the villages of Kunov and Prietrž in the western part of the area in June 2009.

Research carried out in this region suggests that permanent gullies, formed as one of the most profound phenomena associated with the geomorphic response to environmental changes, represent significant indicators of both land use and climatic changes (Stankoviansky 2003c). Especially, they can serve as a tool for understanding the geomorphic significance of the Little Ice Age.

19.4 The Site Level: Selected Gullies at Prašice, Babikovce, Rybník, Bardejov and Veľký Šariš

19.4.1 The Prašice Gully Site

19.4.1.1 Location, Natural Conditions, Land Use History

The site is situated in the northern part of the Nitra Hilly Land, Western Slovakia, which is a geomorphic subunit of the Danube Lowland. It lies ca 800 m to NE from the village of Prašice (Fig. 19.1). The gully is cut in the right slope of the Chotina Brook valley (Fig. 19.9). The forked gully is ca 250 m long, to 5–6 m deep and to 15–20 m wide. The gully is incised in the 5–15 m thick bed of Late Pleistocene loess, overlying the Lower Pleistocene fossil soils. The flat colluvial fan is superimposed on the floodplain of the Chotina Brook, covered with the Holocene loams. Haplic and Albic Luvisols predominate. The mean annual precipitation is 550–700 mm. Original vegetation was oak-hornbeam forest on the slope and alder forest on the floodplain.

The first historical document about the village of Prašice is from 1245. Farming in its surroundings significantly expanded and intensified mostly during the Late Middle Ages, whereas forests almost totally disappeared. The result of historical land use development lasting several centuries was a mosaic of ever smaller fields. This pattern disappeared due to large-scale land use changes connected with collectivization after the Second World War. Today the gully represents an island of secondary deciduous forest in the farmland.

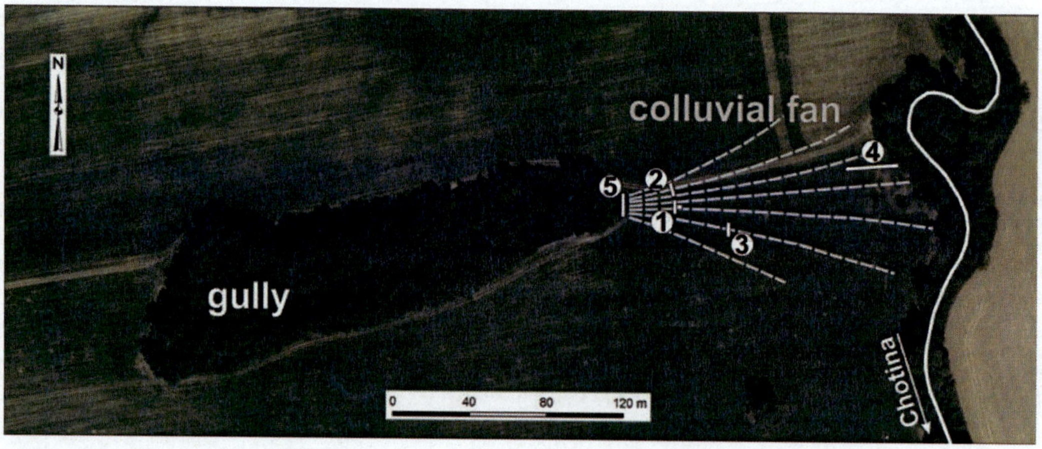

Fig. 19.9 Aerial image of the Prašice gully site with location of opened test profiles (modified from Papčo 2011)

19.4.1.2 Development of Gully System

The assessment of gully development was based on an analysis of historical and contemporary maps and an interpretation of correlative sediments. For the purpose of dating gully formation, the maps of the 1st (1783), 2nd (1838) and 3rd (1882) military surveys and recent topographical map (2001) were analysed. Maps were georeferenced with an exception of the oldest one. The gully appears for the first time on the map dated for 1838, while an increase of its length was recorded at each of the following surveys, roughly by 20% and 35%, respectively. For the purpose of the study of sediments correlative to the historical erosional events five test profiles of big dimensions, excavated by backhoe, were opened and analysed (Fig. 19.9). Their lengths ranged from 10 to 25 m, widths were 1–1.5 m and the depths were 2–2.5 m. Field work was directed mostly to identification of sediment layers and soil formation. The timing of sedimentation was determined by ^{14}C-dating of charcoal and archaeological age estimations of pottery.

According to the old maps, the gully was formed sometime in the period 1783–1838. However, accumulation of the colluvial fan as a consequence of gradual gully incision had to start much earlier. This is suggested by interpretation of the profile No. 4 (Fig. 19.10). There, the buried floodplain facies were detected in the depth of 135–185 cm below the body of the fan. Such a discovery succeeded in this profile only as the fan thickness was, naturally, the lowest in its distal part. ^{14}C analysis of charcoal sample from the floodplain surface yielded an age with years 1332, 1339 and 1399. It would suggest the possible cessation of the formation of floodplain facies as a consequence of ensuing colluvial fan accumulation as early as the fourteenth century.

In the profile No. 5, excavated in the gully mouth, an incision of an older gully, buried below the sediment, was discovered (Fig. 19.11). It suggests that this site was a place of repeated gully cutting and filling. The results of dating of buried floodplain surface testify that the oldest gully in this place could not have been formed earlier than in the fourteenth century. The age of archaeological artefacts (small pieces of broken pottery), found in the fan body at the depth between 87 and 163 cm, as well as in the fill of an older gully between 120 and 200 cm, expertly estimated to come from the sixteenth-nineteenth centuries, is in conformity with dating of the gully system development. Unfortunately, the so-called uniform ceramics did not permit to narrow this rather broad time interval (Papčo 2010, 2011).

19.4.2 The Babikovce Gully Site

19.4.2.1 Location, Natural Conditions, Settlement and Land Use History

The gully site is named after an abandoned medieval village of Babikovce, lying at the eastern margin of the Myjava Hilly Land, somewhere between the villages of Hrachovište, Krajné and Kostolné (Figs. 19.1 and 19.2). The gully system consists of the gully itself, its catchment and colluvial fan (Fig. 19.12). The gully runs along the bottom of the dry

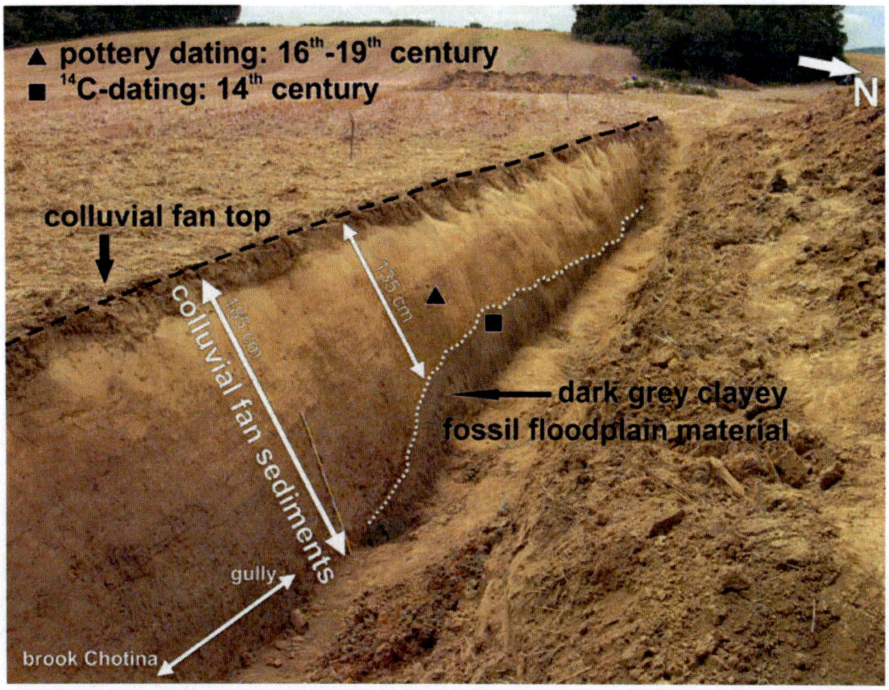

Fig. 19.10 The Prašice gully site, in the foreground is the colluvial fan, in the background one can see the forested gully; the opened profile No. 4 shows the contact of buried floodplain facies with colluvial sediments (modified from Papčo 2011)

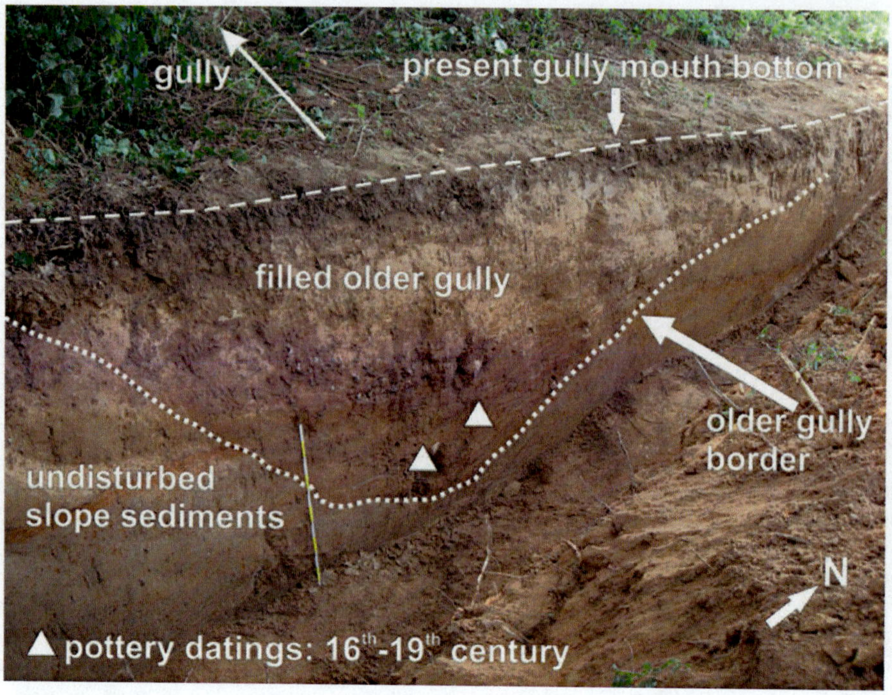

Fig. 19.11 Profile No. 5 at the gully mouth with buried older gully (modified from Papčo 2011)

valley incised into the left (NW) slope of the Jablonka River valley. Its depth reaches up to 7.5 m, with the width up to 20 m. Its flat colluvial fan with an area of about 3.0 ha is superimposed on the Jablonka River floodplain. The catchment covers an area of about 50 ha and its elevation ranges from 222 to 323 m. From the north to south, it is built by silty clays (Karpathian), carbonate conglomerates (Eggenburgian), marls and shales (Santonian). Almost the whole catchment is covered by sandy-loamy deluvial deposits. The predominant soils range from Albic Luvisols to Stagnic Glossisols. The mean annual precipitation is 650–680 mm. The natural vegetation was oak-hornbeam forest.

The first written mention of the villages of Babikovce, Hrachovište and Krajné occurs in a letter of donation from 1392, but they are certainly older, originating prior to the mid-fourteenth century in connection with the construction of the Čachtice Castle. However, Babikovce was abandoned sometime at the end of the Late Middle Age as the last reference to it as a village is from 1452. It is supposed that the studied gully occurred just in the territory of this village. After its abandonment, the area of Babikovce was divided between the neighbouring villages. Almost in the whole catchment (with an exception of the forest in its western part), the mosaic of small fields gradually expanded. Such a situation lasted until collectivization when the transformation of about 150 small parcels into five large blocks occurred. Today, the steeper slopes of the catchment are under grassland, whereas flatter areas and the fan are used as arable fields. The gully itself is covered by secondary forest (cf. Dotterweich et al. 2013).

19.4.2.2 Development of Gully System

The study used sedimentological, pedological, geoarchaeological and historical data. Field work was focused on surveying the gully system and colluvial deposits. Two sites within the fan were investigated at naturally exposed outcrops along the Jablonka River (E1, E2). The remaining nine exposures were excavated by backhoe in various places within the gully system (E3-E11), some of them with a length of 20 m and a depth of about 2.5 m. In addition, drillings with a linear percussion system were carried out on the fan (L2-L4). An analysis of historical sources helped to reconstruct the settlement history. The timing of sedimentation was determined by ^{14}C-dating of charcoal and archaeological estimates of the age of pottery.

^{14}C-dating indicated that the first clear-cuts in the gully catchment took place sometime in the period between the eleventh and thirteenth centuries. The blocky and turbulent sediments in the exposure E7 indicate deposition by muddy flood sometime in the thirteenth-fourteenth century (Fig. 19.13). In the dry valley bottom, extreme gullying caused sudden vertical incision through the Luvisol, underlying loess beds, solifluction deposits, down to the conglomeratic bedrock (Fig. 19.14). The eroded sediments led to the formation of a significant gully fan, very rich in stones, with a volume of about 16,500 m^3 at the outlet of the

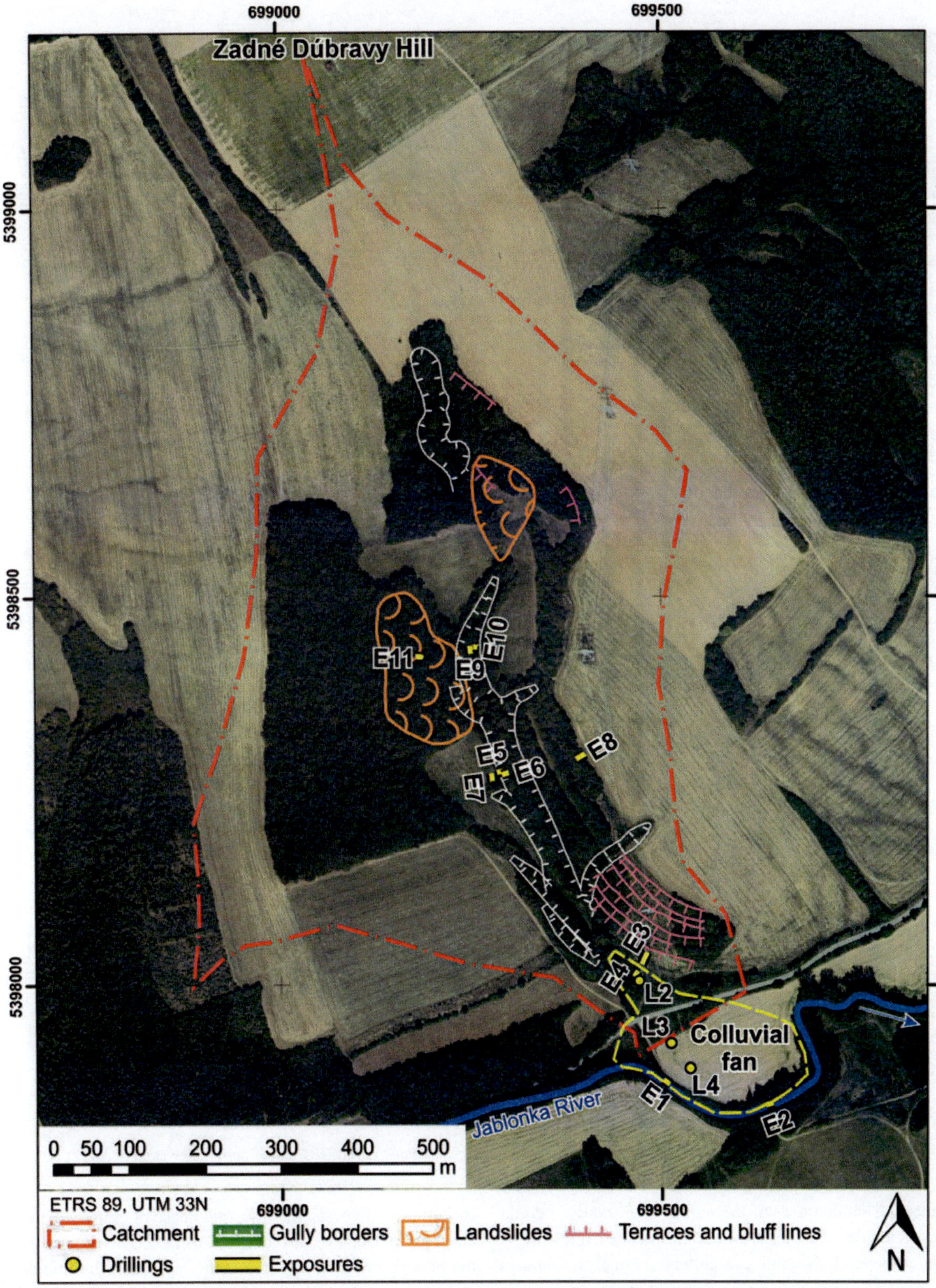

Fig. 19.12 Aerial image of the Babikovce gully site with location of exposures and drillings (after Dotterweich et al. 2013)

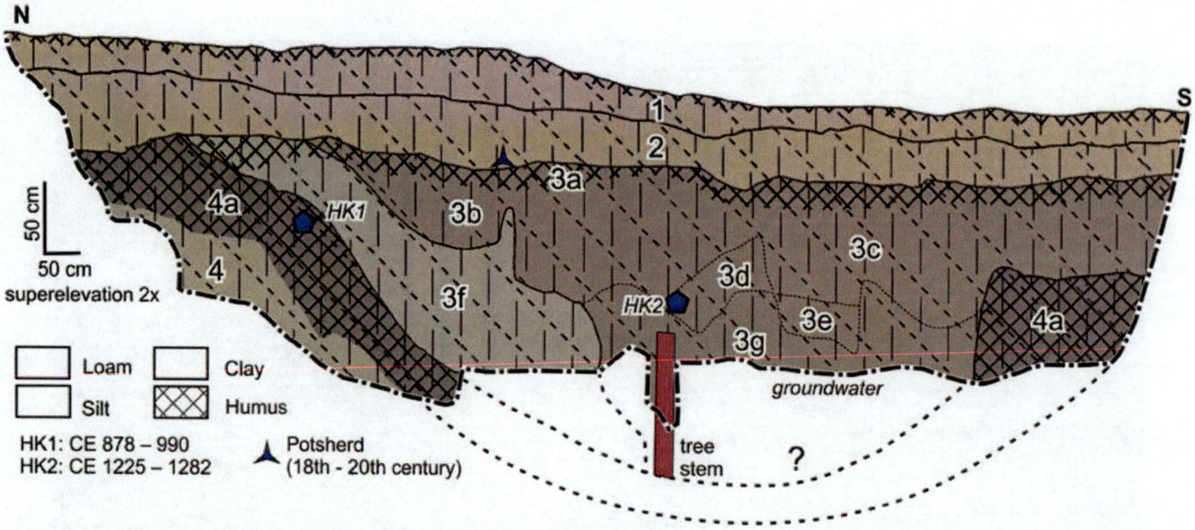

Fig. 19.13 Soil-sediment structure and location of the 14C samples in exposure E7; label "4" here means unit "4" other than subunit 4a (after Dotterweich et al. 2013)

Fig. 19.14 Exposures E5 and E6 showing 7.5 m thick sequence of geological complexes incised by the gully (after Dotterweich et al. 2013)

catchment. Abandonment of the village of Babikovce could indicate reaction of settlers to these disastrous events.

The embedded charcoal and pottery fragments found in the upper 2 m of fan sediments show that the material could not have been deposited before the 16th/seventeenth centuries. It suggests that a second phase of extreme gullying and muddy flood occurrence had to take place sometime in the period of following three centuries, when the influences of both land use (kopanitse settlement) and climate changes (the LIA sensu stricto) temporally overlapped and possible geomorphic effects multiplied. The gully system in the catchment extended in this period and split into several tributaries further upslope. Gullying and sheet wash accumulated about 60,000 m^3 of silty-clayey colluvial material in the fan (cf. Dotterweich et al. 2013).

19.4.3 The Rybník Gully Site

19.4.3.1 Location, Natural Conditions, Settlement and Land Use History

The gully site lies in the NW part of the village of Rybník cadastral area, close to the border with the village of Kozárovce (Fig. 19.1). It is situated on the left side of the so-called Slovenská brána (Slovak Gate), by which the Hron River separates the Kozmálovské vŕšky Hills from the volcanic body of the Štiavnické vrchy Mts (48°20′7″ N 18°33′ 32″ E). The site represents a gully system consisting of the forked gully and two colluvial fans (Fig. 19.15). Its altitudes range between 145 and 230 m. The length of the gully is 250 m, and the depth varies from 4 to 8 m. Detailed morphometric parameters are presented in Fig. 19.16. The gully runs along the bottom of the dell, incised into the marginal slope of the Štiavnické vrchy Mts, with an average inclination of 14°. The slope is built of weathered Sarmatian andesites covered by a loess bed with an average thickness of two meters. Colluvial fans are deposited on the irregularly inundated floodplain terrace of the Hron River. Dominant soil type is Eutric Cambisols. The mean annual precipitation is about 700 mm. Original vegetation was oak and hornbeam forest. Currently the colluvial fans are used as arable land. The gullied slope is covered by secondary deciduous forest with predominant black locust.

The first written mention on the village of Rybník is from 1075, but the settlement is, however, much older, as testified by the results of archaeological research. The main socio-economical activities of the village since its beginnings were viticulture and agriculture. Vineyards and arable fields extended gradually at the expense of original forests. A significant growth of the village occurred especially in the eighteenth century, after the threat of Turkish aggression on the territory of the present Slovakia was definitely over. At

Fig. 19.15 Aerial image of the Rybník gully system; labels (a) and (b) indicate position of photos, arrows direction of their taking

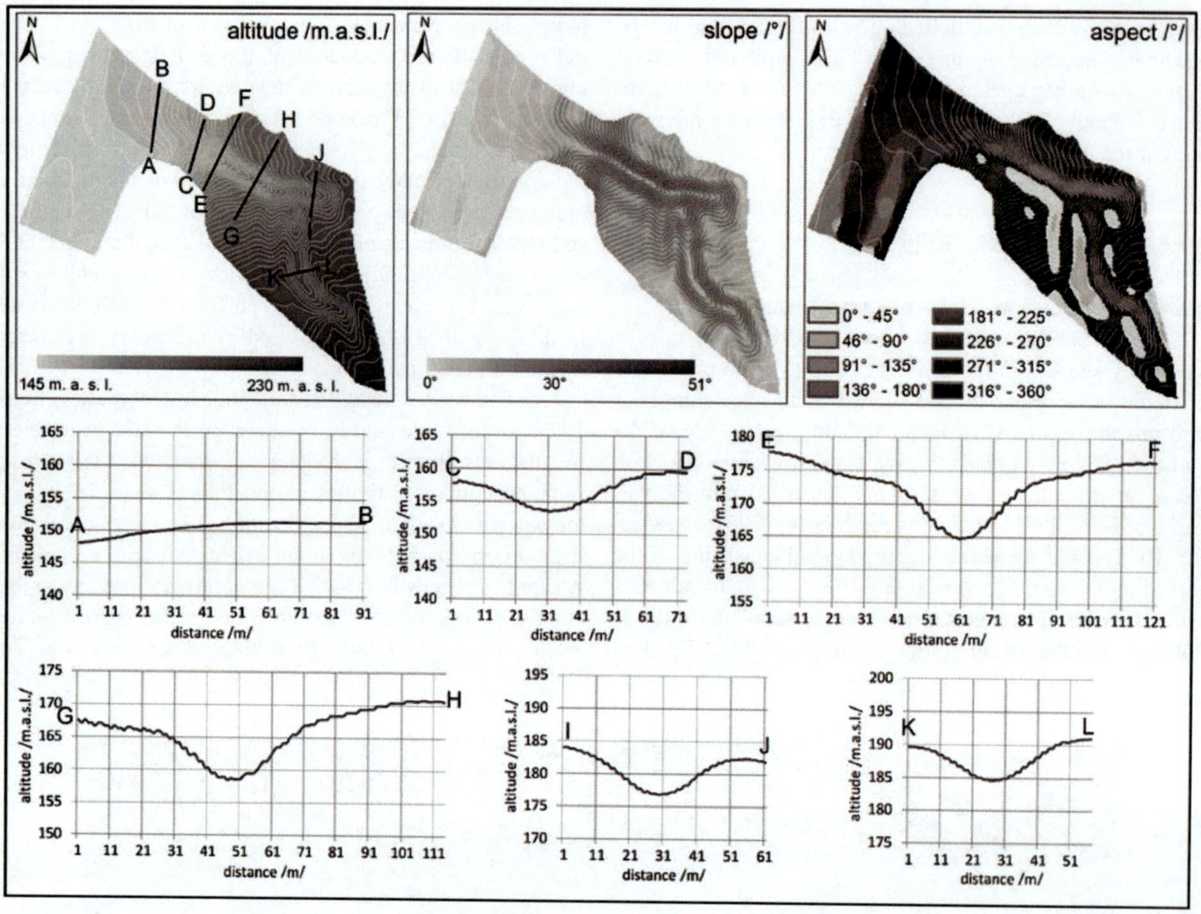

Fig. 19.16 Digital elevation model, derived morphometric variables, and selected transversal profiles across the gully (in the case of aspect, 0° represents north)

the end of the eighteenth century only the steepest slopes remained forested, including also the part of the slope studied here (Fig. 19.17).

19.4.3.2 Development of Gully System

For the reconstruction of gully system evolution an analysis of old maps, analysis of correlative sediments within colluvial fans (using three opened profiles with depth up to 280 cm, excavated by backhoe), dendrochronological analysis, XRF analysis of samples and ERT profiling were used.

The gully appears for the first time on the map of the 2nd military survey (1843) and thus its beginnings reach the period since the late 18th until the mid-nineteenth centuries (Fig. 19.17). Although the gully was in fact controlled by topography, its formation could be also conditioned by the path or access road running along the bottom of the dell. It was a part of broader network of such linear artificial landscape elements that expanded after the end of the Turkish threat. Thus, gully formation can be considered as the response to the climate fluctuation of the LIA sensu stricto.

An important step in the gully development was a landslide event, which occurred between 1875 and 1950. The landslide displaced an outflow from the gully and initiated the formation of the second colluvial fan (Fig. 19.15). This event was confirmed by ERT profiling and dendrochronological research. Nowadays, there is no connection between the gully and the older colluvial fan, with all sediments deposited in the area of younger fan. Confirmed thickness of older and younger fans is 180 cm and 70 cm, respectively.

The last stage of a gully system development was influenced by collectivization, resulting in increased erosion rates. It is testified by results of an analysis of correlative sediments in the younger colluvial fan, where a presence of heavy metals was recognized in the upper 50 cm of the profile. The source of these metals was the nearby energy machine works in Tlmače, founded in 1951. Current

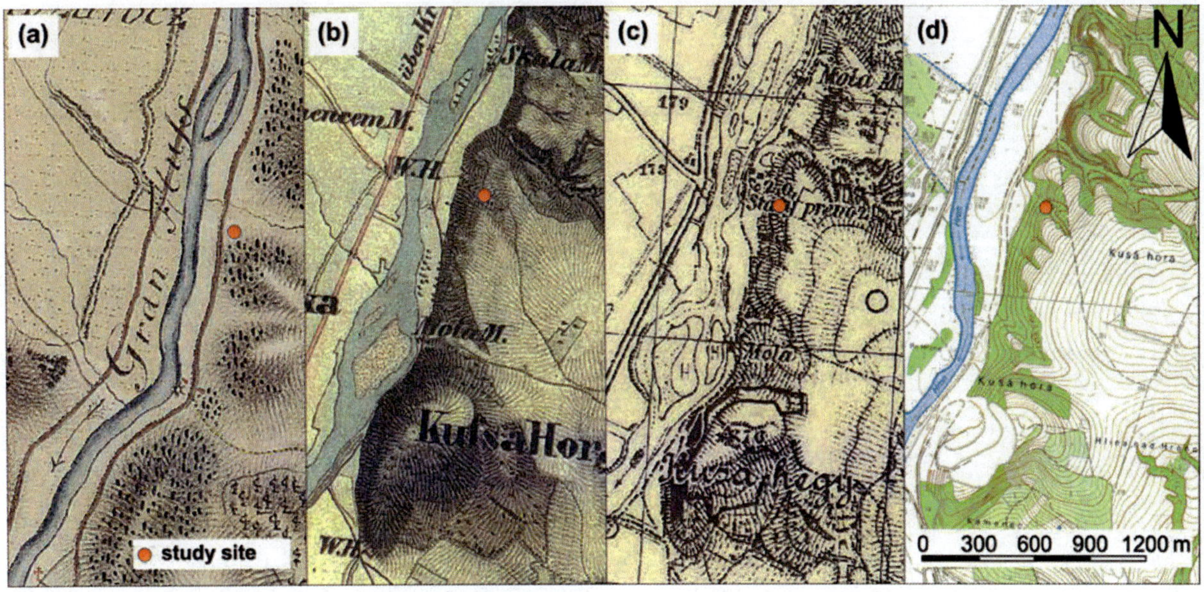

Fig. 19.17 Land use and gully development at the gully site of Rybník, depicted on the maps of **a** the 1st military survey—1782, **b** the 2nd military survey—1843, **c** the 3rd military survey—1875 and **d** recent map—2002

headward erosion of the gully is negligible, as confirmed by 3-years long measurements using installed erosion pins (Burian 2016).

19.4.4 The Bardejov and Veľký Šariš Gully Sites: Modelling and Simulation of Gully Erosion

19.4.4.1 Location, Natural Conditions, Land Use History and Age of Gullies

Both gully sites were named according to well-known east-Slovakian historical towns of Bardejov and Veľký Šariš. Bardejov is situated in the Topľa river valley in the NW part of the Ondavská vrchovina Upland (Fig. 19.1). The gully site represents a system of six gullies around the town (Fig. 19.18), with a total area of 20 km^2. They were formed within the bottoms of dry valleys. Their length ranges from 500 to 1200 m. The depth of gullies is mostly 5–10 m. The gullied slopes are built primarily of Paleogene flysch rocks, covered in the lower sections by river terraces. Prevailing soils types are Cambisols. Currently none of the gullies is active, and almost all of them are completely covered by forest or shrub vegetation.

Veľký Šariš lies in the SE part of the Spišsko-šarišské medzihorie Mts, with a hilly relief. The site itself represents the gully and its catchment (0.26 km^2). The length of the gully is 550 m, and the depth reaches 2–5 m. The catchment is created by a wide, shallow dell within the lower slope of the Torysa river valley (Fig. 19.19), namely in its terraces overlying the Paleogene flysch rocks. Soil types are Haplic Luvisols. The mean annual precipitation (660 cm) and natural vegetation (oak–hornbeam forest) are the same for both gully sites. At present, the forested gully is in the suburban zone, intensely used for agriculture.

A common denominator of both towns is at least eight centuries long human intervention in the landscape, as the first written mention on Bardejov is from 1241 and on Veľký Šariš even from 1217. Bardejov was an important medieval trade and weaving craft centre, achieving the greatest economic success in the fourteenth and fifteenth centuries. The Šariš Castle above Veľký Šariš was the seat of the historical Saros County. The town had a prime economic position in the region and served as the economic support of the castle.

Extensive agriculture in the wider surroundings of both towns in the Late Middle Ages resulted in the significant expansion of farmland at the expense of woodland. Thus, historical land use changes in both towns, together with natural conditions, created preconditions for accelerated sheet wash and gully erosion.

Analysis of old maps showed, however, that permanent gullies observed today at both sites are relatively young. Some of six gullies at the Bardejov site originated before the 1st military survey (1782), whereas the remaining ones formed between the 1st and 2nd survey (1820). The Veľký Šariš gully was also formed within the time interval between these two oldest mappings. The map of the 1st military survey depicts only a road running along the dell bottom.

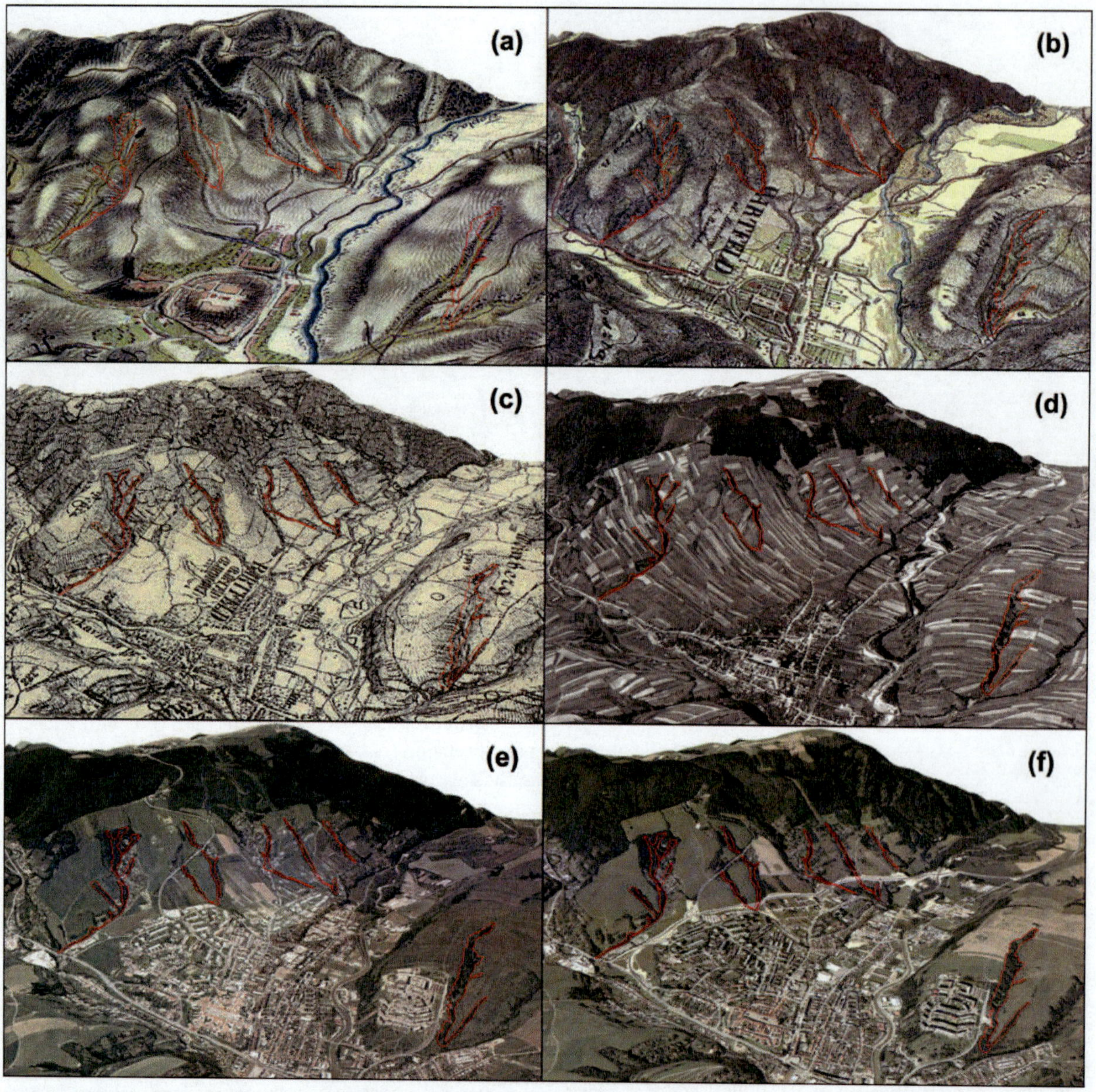

Fig. 19.18 Evolution of gullies and landscape structure at the Bardejov site depicted on historic maps and aerial images: **a** 1782—the 1st military survey, **b** 1820—the 2nd military survey, **c** 1880—the 3rd military survey, **d** 1950, **e** 2002 and **f** 2016. The red line represents current boundaries of gullies

Naturally, in the case of both sites and similarly as in the Myjava Hilly Land, it is not possible to exclude the formation of gullies also in older times though such gullies were not preserved.

19.4.4.2 Modelling and Simulation of Gully Erosion Using Geographic Information Systems

Tools of Geographic Information Systems were used to predict gully evolution based on input parameters conditioning initiation and geomorphic effectiveness of gully erosion. At the Bardejov site, static gully erosion models were used, focused on spatial prediction of gully inception, while at the Veľký Šariš site a dynamic simulation of gully development was used.

It turned out that the model for ephemeral gullies by Desmet et al. (1999) is the most suitable for evaluating the potential for gully formation at the Bardejov site. Though the model is focused on ephemeral gullies, it was assumed that it is usable also for permanent gullies that were often created

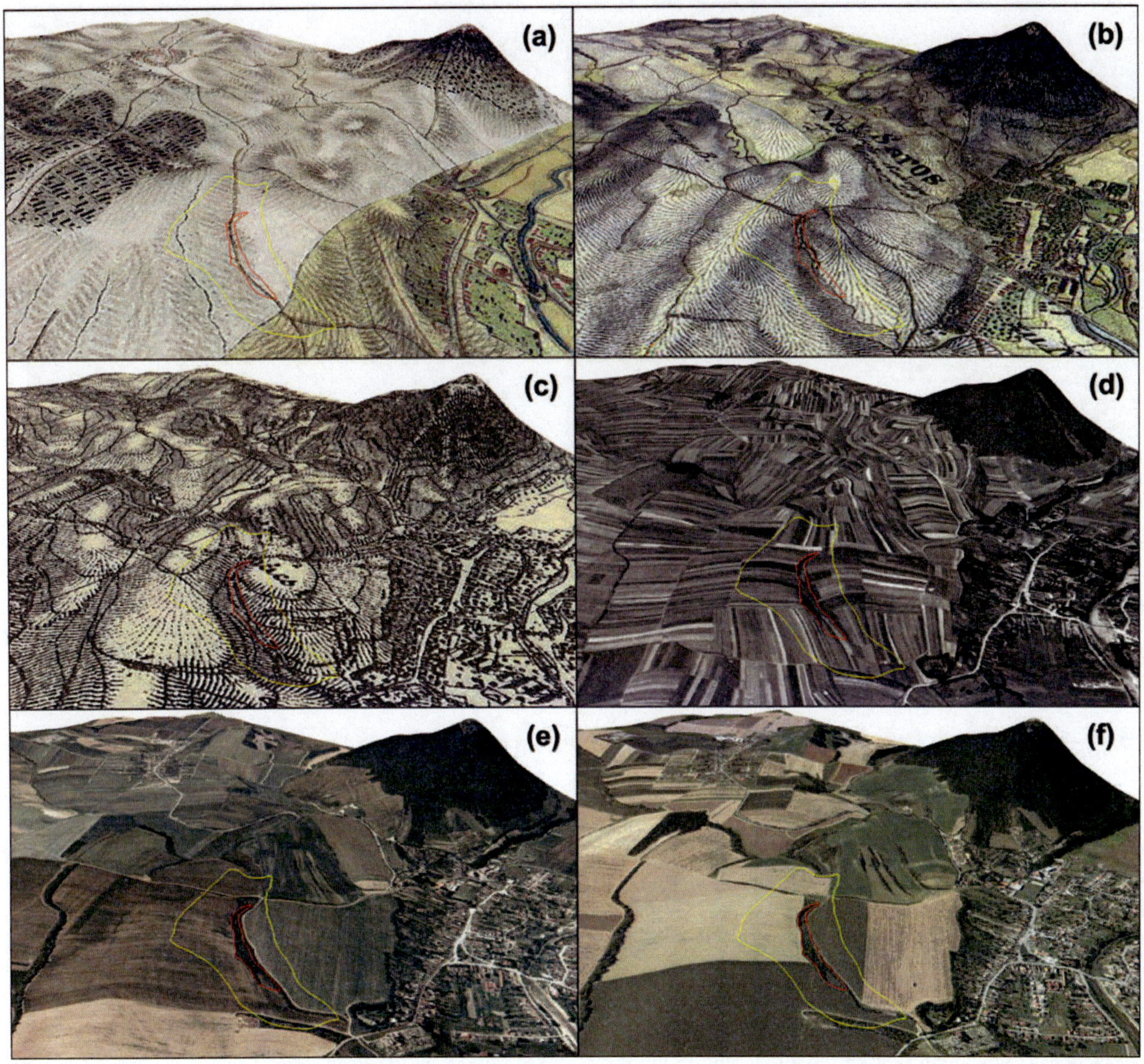

Fig. 19.19 Evolution of gully and landscape structure at the Veľký Šariš site depicted on historic maps and aerial images: **a** 1782—the 1st military survey, **b** 1838—the 2nd military survey, **c** 1880—the 3rd military survey, **d** 1950, **e** 2002 and **f** 2016. The red line represents the current gully boundary, the yellow one the boundary of the gully catchment area

by deepening of ephemeral gullies in the past. Methodological concept of gully erosion simulation at the Veľký Šariš site is based on distributed process-based model SIMWE (Mitas and Mitasova 1998) and GRASS module r.landscape.evol (Mitasova et al. 2013). The module uses a raster data model and creates a new map, where each raster cell carries a numerical value, which represents the simulated values of erosion or deposition in meters estimated for that cell. This value is then added to (deposition) or subtracted from (erosion) the topographic map (DEM) of the previous temporal step. The input parameters of both approaches were optimized for possible conditions of the study sites at the time of gully inception. At the Veľký Šariš site combination of several tested scenarios was used to determine initial conditions for gully formation and its development to reach the current state.

The application of the static model to the Bardejov site provided spatially-distributed values of gully formation potential before the incision of gullies (Fig. 19.20). The model has been applied only to arable land as this land use class is most susceptible to gullying due to minimal protective function of land cover. The results showed the highest values of the gully formation potential along the valley bottoms, very close to the location of current gullies.

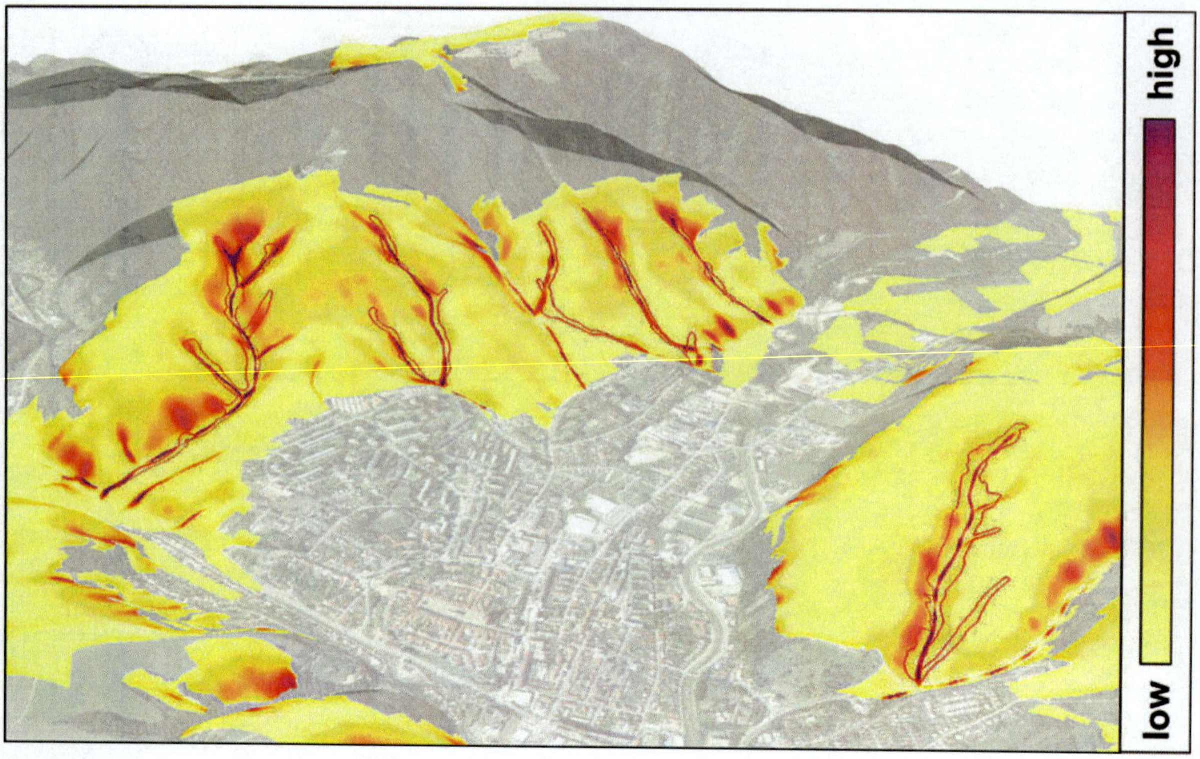

Fig. 19.20 Model of gully erosion at the Bardejov site according to Desmet et al. (1999). The red line represents the current gully boundaries

It is obvious that topography and especially upslope contributing areas play a very important role in gully formation at this site. The application of the model has shown that the initiation of gully erosion is primarily determined by land use. Naturally, the topographic factor also plays an important role because it determines the location of gullies via concentration of flowing water (Hofierka and Koco 2009).

The results of simulation within the assigned conditions at the Veľký Šariš site referred to gully development, which approximates the current state of the gully. During the simulation the gully showed gradual backward increase. The final maps of simulated erosion/deposition (Fig. 19.21) show that the values of erosion/deposition in the place of flow concentration along the bottom of the catchment are growing together with an increasing number of iterations. At the same time, continuous spread of concentrated lines of higher values of erosion/deposition in reverse direction is visible. The simulation shows gully formation in the lower part of its catchment already after the third iteration (Fig. 19.22) and gradually reaches the current state of the gully shape. The simulation proved that this kind of model is able not only to identify the areas with higher potential for gully formation, but that its results are also applicable in the simulation of landform changes due to erosion and deposition. Several scenarios of initial conditions were simulated, according to which the probable cause of gully inception in the topographically-controlled position (i.e. in thalweg of dell) was the combination of extreme rainfalls (3.5 mm.min^{-1}) and land cover class of arable land (Koco 2011).

19.5 Conclusions

Revealing secrets of development of permanent gullies in Slovakia are a fascinating and exciting story. In the period of seven post-war decades three mutually complementing phases of their research took place, namely (1) Slovakia-wide, (2) regional and (3) local. The key result of the first phase (the 1950s and 1960s) was the unique map of the gully network in Slovakia at the scale of 1: 400 000, enabling one to obtain a general picture of spatial distribution and density of these features in the country.

The second phase of research (the 1990s and early 2000s) was executed in the region of the Myjava Hilly Land, Western Slovakia. The research was directed to elucidate regularities of spatial distribution of gullies, their age, causes and course of their formation. Field reconnaissance and analysis of old cadastral maps suggested that the pattern of gullies has been controlled mostly by historical artificial linear landscape elements and yielded a typology of gullies, linked to land use and spatial distribution of the linear elements. An analysis of old military maps and written

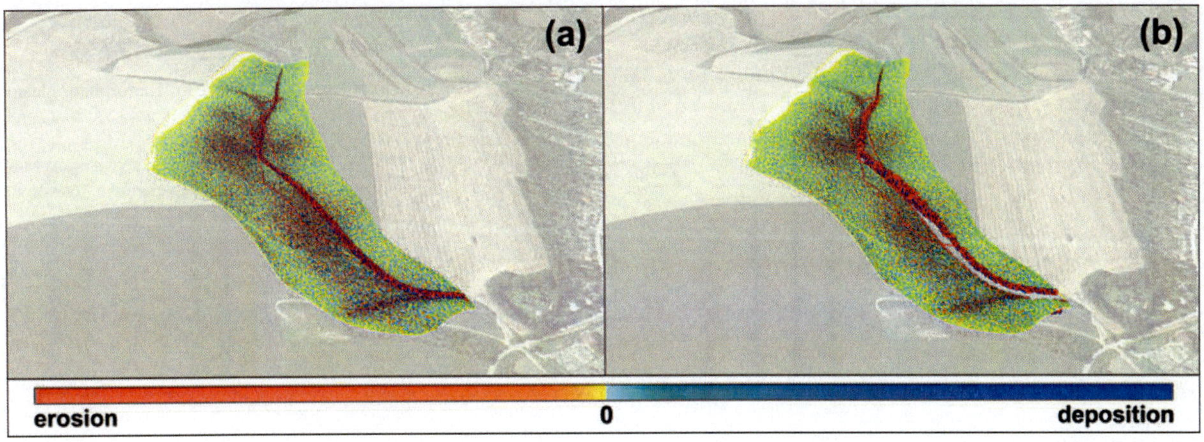

Fig. 19.21 Erosion/deposition development at the Veľký Šariš site by SIMWE model for **a** 3th, and **b** 7th iteration of gully erosion simulation. The red line represents the current gully boundary

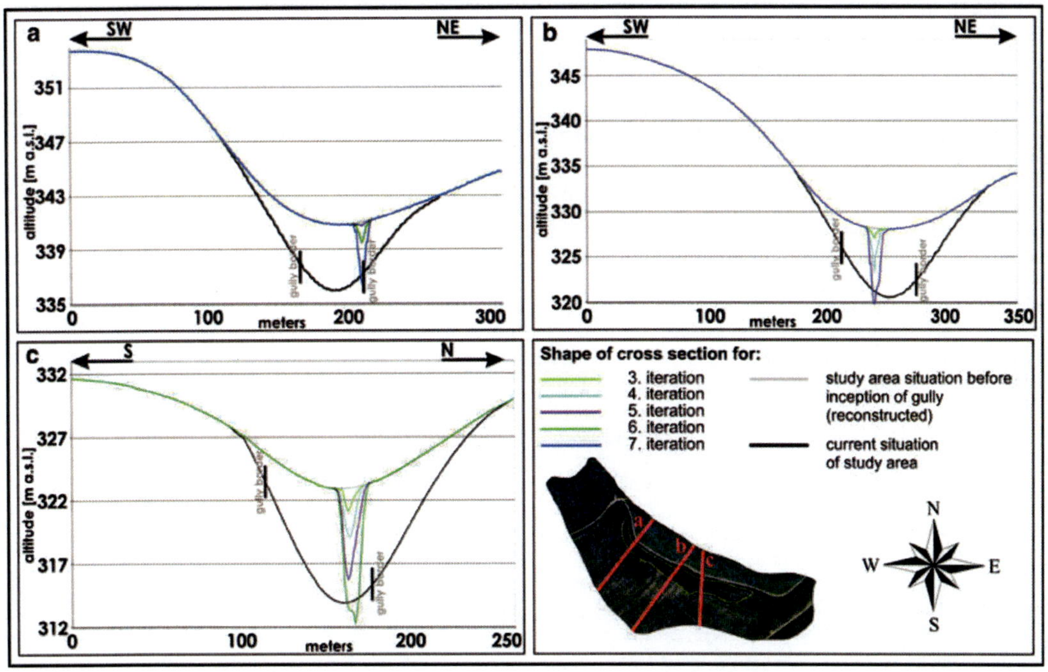

Fig. 19.22 Simulation of gully erosion at the Veľký Šariš site by SIMWE model and changes in vertical profile. The red line represents the cross section

historical sources enabled one to identify the New Age phase of possible gully formation. This statement, however, is valid only to the central part of this region, settled since the mid-16th to the turn of the nineteenth century. In its marginal parts, colonized earlier, gullies could have developed already in the medieval times. The abandonment of some villages in these parts in the Late Middle Age may indirectly point to the adverse effects of an older phase of gullying. The fact that significant human interventions in this region overlapped temporally with the influences of climatic fluctuation of the Little Ice Age emphasizes the role of permanent gullies as important indicators of geomorphic response to past environmental changes, in both land use and climate.

The third phase (the late 2000s and early 2010s) was directed to a detailed study of five selected gully sites to assess their historical development and also involved modelling and simulation of gully erosion. The research confirmed both the New Age and Late Middle age phases of gully formation at the Babikovce and Prašice sites, whereas at Bardejov, Veľký Šariš and Rybnik sites only the younger of them was recognized. Modelling of gully erosion at the Bardejov site provided spatially-distributed values of gully formation potential before the incision of gullies. Simulation of gullying at the Veľký Šariš site revealed gully development approximating the current state of the gully.

Acknowledgements Miloš Stankoviansky, Pavol Papčo and Štefan Koco acknowledge distinguished experts in soil erosion Prof. Hans-Rudolf Bork and Dr. Markus Dotterweich (Germany) for their significant help in an assessment of gully evolution in the Prašice and Babikovce sites.

References

Bork HR (1989) Soil erosion during the past millennium in central Europe and its significance within the geomorphodynamics of the Holocene. Catena Suppl 15:121–131

Bučko Š (1980) Vznik a vývoj eróznych procesov v ČSSR. In: Protierózna ochrana – zborník z konferencie, Dom techniky ČSVTS, Banská Bystrica, 8–10 October 1980

Bučko Š, Mazúrová V (1958) Výmoľová erózia na Slovensku. In: Zachar D (ed), Vodná erózia na Slovensku. Slovak Academy of Sciences, Bratislava, p 68–101

Burian L (2016) Assessment of the historical trends of gully system development and its causes. Case study from Hronská pahorkatina Hills. Dissertation, Comenius University

Desmet PJJ, Poesen J, Govers G, Vandaele K (1999) Importance of slope gradient and contributing area for optimal prediction of the initiation and trajectory of ephemeral gullies. CATENA 37:377–392. https://doi.org/10.1016/S0341-8162(99)00027-2

Dotterweich M, Stankoviansky M, Minár J, Koco Š, Papčo P (2013) Human induced soil erosion and gully system development in the Late Holocene and future perspectives on landscape evolution. Geomorphology 201:227–245. https://doi.org/10.1016/j.geomorph.2013.06.023

Hofierka J, Koco Š (2009) Modelling the inception of gully erosion around the town of Bardejov using geographic information systems. In: DeDapper M, Vermeulen F, Deprez S, Taelman D (eds) Ol'Man River - Geo-Archaelogical Aspects of Rivers and River Plains. Meeting on Ol'Man River - Geo-Archaelogical Aspects of Rivers and River Plains, Gent, September 2006. Archaeological Reports Ghent University., vol 5. Ghent University, Gent, p 627

Koco Š (2011) Simulation of gully erosion using the SIMWE model and GIS. Landform Anal 17:81–86

Lamb HH (1984) Climate in the last thousand years: natural climatic fluctuations and change. In: Flohn H, Fantechi R (eds) The climate of europe: past, present and future. D Reidel Publishing Company, Dordrecht, pp 25–64

Mitas L, Mitasova H (1998) Distributed soil erosion simulation for effective erosion prevention. Water Resour Res 34(3):505–516. https://doi.org/10.1029/97WR03347

Mitasova H, Barton CM, Ullah II, Hofierka J, Harmon RS (2013) GIS-based soil erosion modeling. In: Shroder J, Bishop MP (eds) Remote Sensing and GIScience in Geomorphology. Academic Press, San Diego p, pp 228–258

Papčo P (2010) Geomorfologická odozva zmien využívania krajiny vo vybranom území Nitrianskej pahorkatiny. Dissertation, Comenius University

Papčo P (2011) Výmoľová erózia v čase – mapové podklady versus korelátne sedimenty (Príkladová štúdia). Geografický Časopis 63 (3):287–298

Porter SC (1986) Pattern and forcing of Northern Hemisphere glacier variations during the last millennium. Quaternary Res 26:27–48. https://doi.org/10.1016/0033-5894(86)90082-7

Stankoviansky M (2003a) Geomorfologická odozva environmentálnych zmien na území Myjavskej pahorkatiny. Univerzita Komenského, Bratislava

Stankoviansky M (2003b) Historical evolution of permanent gullies in the Myjava Hill Land. Slovakia. Catena 51(3–4):223–239. https://doi.org/10.1016/S0341-8162(02)00167-4

Stankoviansky M (2003c) Gully evolution in the Myjava Hill Land in the second half of the last millenium in the context of the Central European area. Geogr Pol 76(2):89–107

Stankoviansky M, Ondrčka J (2011) Current and historical gully erosion and accompanying muddy floods in Slovakia. Landform Anal 17:199–204

Stankoviansky M, Pišút P (2011) Geomorphic response to the Little ice age in Slovakia. Geogr Pol 84:127–146

Starkel L (2000) Heavy rains and floods in Europe during last millennium. In: Obrębska-Starkel B (ed) Reconstructions of climate and its modelling, prace geograficzne 107. Institute of Geography of the Jagellonian University, Cracow, pp 55–62

Šimurková M (2012) Hodnotenie vzťahu výmoľovej siete na Slovensku ku geologickému podkladu. Comenius University, Thesis

Vanwalleghem T, Poesen J, Van Den Eeckhaut M, Nachtergaele J, Deckers J (2005) Reconstructing rainfall and land-use conditions leading to the development of old gullies. The Holocene 15(3):378–386. https://doi.org/10.1191/0959683605hl807rp

Zachar D (1970) Erózia pôdy. Vydavateľstvo SAV, Bratislava

Miloš Stankoviansky is a retired Associate Professor in the field of Physical Geography and Geoecology at the Faculty of Natural Sciences, Comenius University in Bratislava. His research activities were concentrated mostly on geomorphic response to environmental changes in the landscape. His key topics were regularities of operation of gully erosion in the pre-collectization landscape and reconstruction of historical evolution of permanent gullies in the Myjava Hilly Land, Western Slovakia.

Štefan Koco is Assistant Professor at the Department of Geography and Applied Geoinformatics of University of Prešov in Prešov and Researcher at the Soil Science and Conservation Research Institute of National Agriculture and Food Centre, with specialization on cartography, geoinformatics, landscape changes and evaluation of soil properties and degradation processes.

Pavol Papčo is Assistant Professor at the Department of Geography of the Catholic University in Ružomberok, Slovakia. His research is focused on environmental history and natural hazards with a special interest in long-term gully erosion.

Libor Burian has been geomorphologist at the Department of Physical Geography and Geoecology, Comenius University in Bratislava with specialization on Geomorphology and morphometry. Nowadays he is working in commercial sphere.

20 Landslides in Slovakia—Spatial Diversity, Activity and Impacts on Society

Pavel Liščák, Juraj Holec, and Peter Pauditš

Abstract

Based upon inventories and engineering geological mapping at various scales, in the years 1997–2006 the geological project titled "Atlas of Slope Stability Maps SR at 1: 50,000" was being completed. For each slope deformation registered in the Atlas, an inventory sheet—the so-called passport—was drawn. The passport of a slope failure includes: serial number of slope failure, territory administrative division (district, geomorphological unit, engineering-geological region), data on exploration and visualization scale, slope deformation description, slope failure dimensions, endangered objects, causes of slope failure generation and corrective measures. The majority of the registered slope deformations belong to the group of potential sliding activity. In some regions of Slovakia (mainly the flysch areas of north-western and eastern Slovakia—Region of Carpathian Flysch, but also in the territories of Neogene volcanic fields) the share of affected land is often greater than 10%. In the scope of the Atlas, 21,190 slope deformation occupying the total area of 257,591.2 ha, which constitutes 5.25% of the total area of Slovakia, were registered.

Keywords

Atlas • Landslides • Slope deformation • Inventory • Slope stability maps

P. Liščák (✉) · J. Holec · P. Pauditš
Slovak Hydrometeorological Institute, Bratislava, Slovakia
e-mail: pavel.liscak@geology.sk

J. Holec
e-mail: juraj.holec@shmu.sk

P. Pauditš
e-mail: peter.paudits@geology.sk

20.1 Introduction

The development and consequences of the catastrophic Handlová Landslide at the break of 1960/1961 meant undeniable landmark in the perception of the importance of slope movements in Slovakia and the need of their study not only by professional and lay circles, but also by competent bodies of the state administration. This disastrous event has given a momentum to the systematic research of slope deformations in the former Czechoslovakia. The Czechoslovak school of engineering geologists supervised by Professor Milan Matula from the Department of Engineering Geology FNS CU, Bratislava, and, in particular, the "landslide specialists" from the Department of Geotechnics, Faculty of Civil Engineering, Slovak University of Technology, Bratislava, led by Arnold Nemčok focused on slope failures inventory throughout the country, with emphasis on investigating perspective areas. This attitude was based on the assumption that new landslides would be preferentially generated in areas that have already been affected by slope failures. The concept of regional distribution of slope deformations has allowed one to analyse the patterns of their formation and evolution, and to derive other facts leading to the understanding of the phenomenon, its forecasting and timely adoption of the necessary stabilization measures. The inventory of slope movements was carried out in three stages, during which the method of registering individual slope deformation (using record sheets) was updated Kováčik and Suchánková (1993). Based upon these inventories and engineering geological mapping at various scales, in the years 1997–2006 the geological project titled "Atlas of Slope Stability Maps SR at 1: 50,000" (hereinafter the Atlas; Šimeková, Martinčeková et al. 2006) was carried out. Since the issuing of the Atlas, further work has been carried out on the register of slope failures at the Slovak Geological Institute of Dionýz Štúr (ŠGÚDŠ). To date, the ŠGÚDŠ register contains 24,222 records of slope deformations.

M. Lehotský and M. Boltižiar (eds.), *Landscapes and Landforms of Slovakia*, World Geomorphological Landscapes,
https://doi.org/10.1007/978-3-030-89293-7_20

20.2 Geomorphological and Tectonic Conditions for Mass Wasting Development in Slovakia

The Western Carpathians are a mountain range characterized by very complicated structural tectonic and geomorphological evolution. They are component of the Alpine-Carpathian mountain system. In the Carpathian mountain range rocks are present, which originated in an immense time span of geological time, from the oldest, dating back over 600 millions years, until the youngest, which have been formed recently (river deposits, weathering scree, various debris, loams and other).

The geological structure of Slovakian territory is made up of the Outer and Inner Carpathians separated by the Klippen Belt. The oldest rock series in the territory of Slovakia are metamorphosed rocks. Originally, they had been several kilometres thick sediments, deposited at the bottom of the primeval ocean, mainly in the older Palaeozoic and maybe even earlier (roughly in the period some 600–400 million years ago). In the wake of primeval Carpathians, these sediments had sunk to the depths of the Earth's crust; in the zone of the increased pressure and temperature they turned into schists. At the same time, some of these rocks in the zones with the highest temperature melted, creating magma, which after cooling and solidification created a colourful mosaic of granite (granitoid) varieties of rocks. In the subsequent movements of the crust these rocks were uplifted and denuded as a result of deep weathering/erosion processes. To date, they form the central (core) parts of the Malé Karpaty, Považský Inovec, Tribeč, Strážovské vrchy, Vysoké and Nízke Tatry, Malá and Veľká Fatra, Žiar and Branisko Mts.; therefore we call them Core Mountains. The magmatic rocks were formed during the whole Palaeozoic, but mainly in the late Palaeozoic (350–300 million years). In the next period, the territory of this primeval mountain range was peneplained and later submerged in the ocean. Carbonatic rocks of Mesozoic age (limestones and dolomites) were the dominant deposits. These rocks were later folded due to pressure from the African Plate upon the European Platform and the secondary Carpathian Mountains evolved. After uplifts and subsidence of crustal blocks in the Tertiary period (Palaeogene) these rocks were, along with the older ones, uncovered. After partial subsidence and partial peneplanation, sandy, gravelly and clayey sediments deposited upon them during the Neogene in a marine and freshwater lake environment.

The Neogene sediments have been preserved in depressions (in lowlands and intermontane basins). Movements of blocks along faults were accompanied by an intense volcanic activity, the maximum of which fell within a period of approximately 10–13 million years ago. The fundamental geomorphic features of the Slovak Carpathians according to Lukniš (1962) reflect the Pliocene development phase, during which individual ranges developed as geomorphological units (macrorelief). The geomorphological development during the Quaternary is connected to the Pliocene phase of development of the topography and is characterized by the formation of smaller morphological forms (meso- and microrelief). The total uplift of the Carpathian mega-anticlinorium during the Pliocene and the Quaternary was not uniform, but highly differentiated in space, with the relative sinking of the adjacent inner Carpathian throughs. Lukniš (1959) suggested that the Tatras were uplifted by 300–400 m against the Liptovská and Popradská kotlina basins during the Pliocene and Pleistocene. Similarly, according to Matějka and Kodym (1935), the Malá Fatra massif was uplifted against the Turčianska basin. The Nízke Tatry Mts. were uplifted against the Liptovská kotlina Basin according to Droppa (1955), who compared the cave levels of the Demänovská cave with the terraces of the river Váh.

The total downcutting of the river network during the Quaternary in the Slovak Carpathians is about 100 m. The result of the geological-geomorphological development in the Slovak Carpathians is the richly articulated, young topography, in which conditions favouring the development of slope movements have existed since the Neogene sea and lakes receded. Most important are the various uplift movements and both vertical and lateral erosion, which continuously create unbalanced conditions within the slopes. However, the topography-forming processes and slope movements operate differently in various geomorphic units and reflect regional peculiarities of various areas. It is therefore necessary to link the types of slope movements with areas that have similar geologic-tectonic structure and similar geologic underpinning of the topography (Nemčok 1982).

The mega-anticlinorium system of the Slovak Carpathians became dry land at the end of the Neogene. The sea receded, lakes disappeared and volcanic activity also ceased. Some volcanoes were extinct merely one million years ago, the youngest had become extinct approximately 120,000 years ago (Putikov vŕšok). The tropical Neogene climate was replaced in the Quaternary by initially moderate, then for a time glacial, and again by moderate climate which lasts until the present.

At the beginning of the Quaternary period (approx. 2.5 million years ago), a variable thick sheet of Quaternary deposits of different genetic and lithological types evolved on the underlying solid rocks, in the exclusively terrestrial environment. They are formed of weathering scree, in particular in areas built of granitoid and carbonate rocks of the Slovak Core Mountains. Their flanks are covered by colluvial deposits. The southwestern part of Slovakia is typified by aeolian Quaternary sediments of loess, loess loams and sands. Along the streams alluvial sediments—sands and gravels—have been deposited in the form of fluvial plains, alluvial fans and river terraces. In the Vysoké and Nízke Tatry, Veľká and Malá Fatra Mts., during the glaciations glacial sediments evolved. In several areas of Slovakia travertines have been formed along faults.

Considering the criteria of uniformity of geologic-tectonic structures and of macrorelief, Matula (1969) delineated four engineering-geological regions (units of the first order) and nine areas (units of the second order) within the engineering geological zoning of the Carpathians. The development and form of slope movements can be connected to the above-mentioned engineering geological units. The regional analysis of slope movements should, therefore, be made within their framework.

20.3 Slope Failure Distribution Within Geomorphological Units of Slovakia

Recent geomorphological division in Slovakia was established by Mazúr and Lukniš (1978). The hierarchically highest geomorphological boundary in Slovakia divides the Carpathian Mountains and the Pannonian Basin subsystems. The Carpathian Mountains are further divided into provinces of Western and Eastern Carpathians, and subsequently split into sub-provinces of Inner and Outer Western/Eastern Carpathians. The Pannonian Basin is divided into provinces of Western and Eastern Pannonian Basin, and subsequently split into sub-provinces of Vienna Basin, Little and Large Danube Basin. Geomorphological sub-provinces are divided into 17 geomorphological areas and, finally, into 84 geomorphological units which represent the basic level of geomorphological division used for this chapter (Fig. 20.1, Table 20.1).

Figure 20.1 shows the slope failures area share (in %) within geomorphological units of Slovakia. Several important clusters of slope failure occurrence can be observed. The biggest cluster, comprising six from seven geomorphological units with percentage of slope failures higher than 20%, is present in the northern part of Slovak territory. These units and respective slope failure percentages are: Oravská vrchovina (33.6%), with the highest slope failure percentage in the whole territory of Slovakia, Skorušinské vrchy (26%), Jablunkovské medzihorie (24.2%), Kysucké Beskydy (22.8%), Kysucká vrchovina (21%) and Oravská Magura (20%). The cluster continues to the west, where the units with smaller but still significant percentages between 12 and 18% can be found—Javorníky, Turzovská vrchovina, Moravsko-Sliezske Beskydy and Biele Karpaty. All the units within this cluster belong to the sub-province of Outer Western Carpathians. The second important cluster is represented by four geomorphological units—Vtáčnik (23.3%), Hornonitrianska kotlina (17.3%), Kremnické vrchy (15.1%) and Žiarska kotlina (14.1%). The third cluster extends on a larger territory of eastern Slovakia, with four geomorphological units over 15% of slope failure percentage —Ľubovnianska vrchovina (almost 20%) and Busov (18.8%) along the border with Poland, Beskydské predhorie (16.5%) and Slanské vrchy (16.3%). It is extended by another four geomorphological units covered by more than 10% of slope failures (Bukovské vrchy, Laborecká vrchovina, Ondavská vrchovina and Košická kotlina).

On the contrary, the southwestern part of Slovakia has the majority of geomorphological units with very low slope failure percentage (<1%). Besides those lying in the sub-province of the Western Pannonian Basin, comprising plains with very small vertical dissection of land surface, it also includes the geomorphic units of the Inner Western Carpathians sub-province such as Malé Karpaty, Považský Inovec and Tribeč. An area of very low slope failure percentage continues through the southern part of Central Slovakia to Krupinská planina, Ostrôžky, Veporské vrchy, Stolické vrchy and Volovské vrchy. The eastern part of the territory of Východoslovenská rovina does not have registered slope failures due to insufficient vertical dissection of land surface.

20.4 Slope Failure Distribution Within Categories of Altitude and Slope

Several workers dealt with slope failure distribution within morphometric categories, e.g. altitude, slope, aspect or curvatures, usually for the purposes of slope failure susceptibility assessment. Bednarik and Liščák (2010) provided an analysis for the whole territory of Slovakia, whereas Holec et al. (2013) did so for the territory of Western Carpathians. Several case studies were published for smaller areas, e.g. Bednarik and Pauditš (2010); Bednarik et al. (2010); Tornyai et al. (2016) or Buša et al. (2019).

The distribution of slope failures within altitude categories is depicted in Table 20.2. The intervals are based on altitude levels division within lowlands and lower, middle and upper highlands, established for the Slovakian territory by Lukniš (1972). Slope failure data are based on Atlas of slope stability maps of the Slovak Republic (Šimeková et al. 2006, 2014). Digital elevation model created from base maps at the scale of 1: 25,000 provided by Geodetic and Cartographic Institute of Slovak Republic, was used for computation.

The highest percentage of areas affected by slope failures —9.68%, lies within the category of higher uplands. However, this area and slope failures within it occupy the smallest area. It has to be noticed that they belong usually to

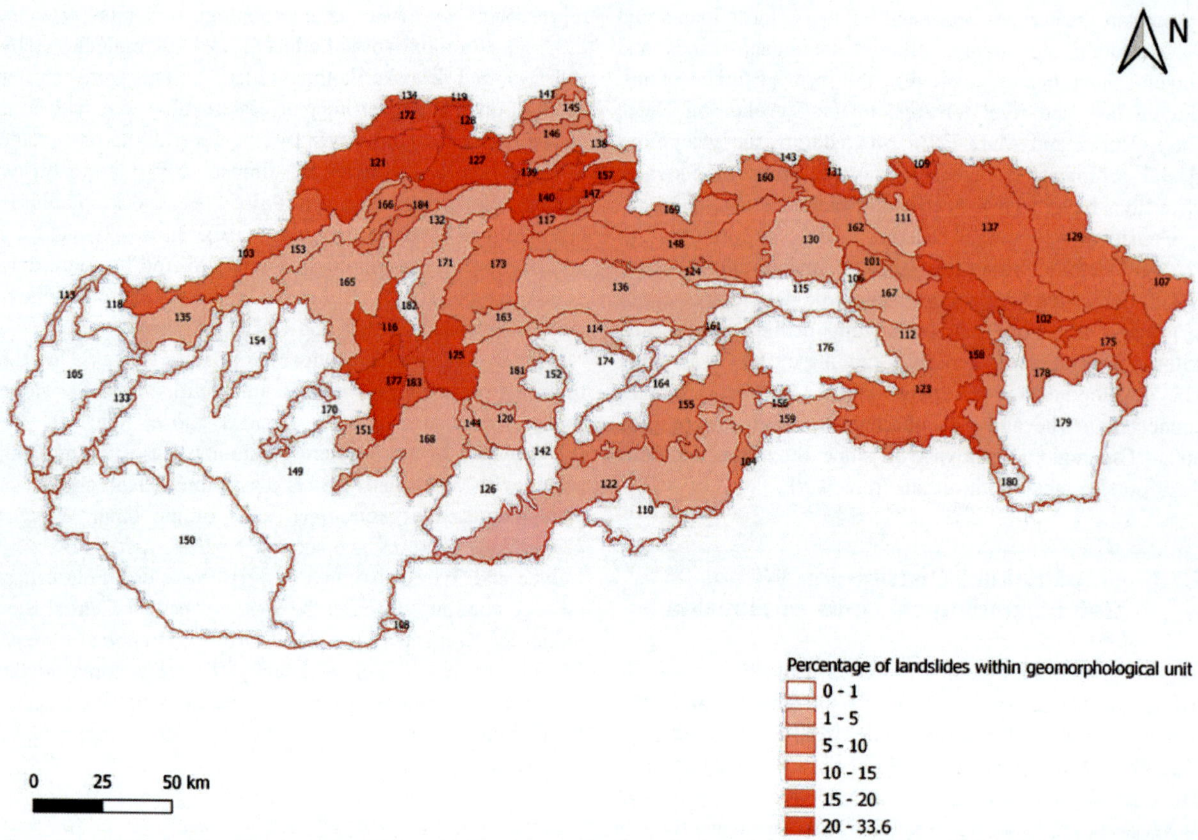

Fig. 20.1 Slope failures area share (in %) within geomorphological units of Slovakia. For code explanation, see Table 1

Table 20.1 Codes of geomorphological units used in Fig. 1

Code	Unit	Code	Unit	Code	Unit	Code	Unit
101	*Bachureň*	122	*Juhoslovenská kotlina*	143	*Pieniny*	164	*Stolické vrchy*
102	*Beskydské predhorie*	123	*Košická kotlina*	144	*Pliešovská kotlina*	165	*Strážovské vrchy*
103	*Biele Karpaty*	124	*Kozie chrbty*	145	*Podbeskydská brázda*	166	*Súľovské vrchy*
104	*Bodvianska pahorkatina*	125	*Kremnické vrchy*	146	*Podbeskydská vrchovina*	167	*Šarišská vrchovina*
105	*Borská nížina*	126	*Krupinská planina*	147	*Podtatranská brázda*	168	*Štiavnické vrchy*
106	*Branisko*	127	*Kysucká vrchovina*	148	*Podtatranská kotlina*	169	*Tatry*
107	*Bukovské vrchy*	128	*Kysucké Beskydy*	149	*Podunajská pahorkatina*	170	*Tríbeč*
108	*Burda*	129	*Laborecká vrchovina*	150	*Podunajská rovina*	171	*Turčianska kotlina*
109	*Busov*	130	*Levočské vrchy*	151	*Pohronský Inovec*	172	*Turzovská vrchovina*
110	*Cerová vrchovina*	131	*Ľubovnianska vrchovina*	152	*Poľana*	173	*Veľká Fatra*
111	*Čergov*	132	*Malá Fatra*	153	*Považské podolie*	174	*Veporské vrchy*
112	*Čierna hora*	133	*Malé Karpaty*	154	*Považský Inovec*	175	*Vihorlatské vrchy*
113	*Dolnomoravský úval*	134	*Moravsko-sliezske Beskydy*	155	*Revúcka vrchovina*	176	*Volovské vrchy*

(continued)

Table 20.1 (continued)

Code	Unit	Code	Unit	Code	Unit	Code	Unit
114	*Horehronské podolie*	135	*Myjavská pahorkatina*	156	*Rožňavská kotlina*	177	*Vtáčnik*
115	*Hornádska kotlina*	136	*Nízke Tatry*	157	*Skorušinské vrchy*	178	*Východoslovenská pahorkatina*
116	*Hornonitrianska kotlina*	137	*Ondavská vrchovina*	158	*Slanské vrchy*	179	*Východoslovenská rovina*
117	*Chočské vrchy*	138	*Oravská kotlina*	159	*Slovenský kras*	180	*Zemplínske vrchy*
118	*Chvojnická pahorkatina*	139	*Oravská Magura*	160	*Spišská Magura*	181	*Zvolenská kotlina*
119	*Jablunkovské medzihorie*	140	*Oravská vrchovina*	161	*Spišsko-gemerský kras*	182	*Žiar*
120	*Javorie*	141	*Oravské Beskydy*	162	*Spišsko-šarišské medzihorie*	183	*Žiarska kotlina*
121	*Javorníky*	142	*Ostrôžky*	163	*Starohorské vrchy*	184	*Žilinská kotlina*

Table 20.2 Slope failures (SF) distribution within altitude intervals

Category	Interval (m a.s.l.)	Area (km^2)	Area with SF (km^2)	Percentage of area covered by SF (%)
1	94–300 (lowland)	20,289	365	1.80
2	300–800 (lower upland)	22,192	1,768	7.97
3	800–1,500 (middle upland)	6,166	308	4.99
4	1,500–2,655 (higher upland)	383	37	9.68

falling, or creeping kinematics present in the Tatry and Nízke Tatry Mountains. The second highest percentage (7.97%) and the biggest area of slope failures (1,768 km^2) occur within the lower uplands. Approximately 71% of the area affected by the registered slope failures in Slovakia lies within this altitude category. The lowest percentage of slope failures (1.8%) is recorded within lowlands, however they occupy a slightly bigger area than the middle uplands, which have 4.99% of their area covered by slope failures.

The distribution of slope failures within slope categories is depicted in Table 20.3. The intervals are based on the methodology of Hrašna (1986).

The highest percentage (10.26%), as well as the biggest area occupied by slope failures (835 km^2) falls within the interval of 7–12°. It is followed by the 12–17° interval (9.07%), with the second biggest area affected by slope failures (575 km^2). The following intervals are 17–25° (6.11%) and 3–7° (5.94%). Slopes with an angle below 3° have infrequent occurrence of slope failures.

Table 20.3 Slope failures (SF) distribution within slope intervals

Category	Interval (°)	Area (km^2)	Area with SF (km^2)	Percentage of area covered by SF (%)
1	0–1	11,887	41	0.34
2	1–3	5,621	58	1.04
3	3–7	7,666	455	5.94
4	7–12	8,137	835	10.26
5	12–17	6,339	575	9.07
6	17–25	6,224	380	6.11
7	25–35	2,693	112	4.15
8	$\geq$ 35	464	22	4.64

Table 20.4 Stages of previous inventory of slope deformations

Inventory stage	Period	Provider	Output document	Territory
1st	1962–1964	GIDŠ Bratislava, FCE SUT Bratislava, FNS CU Bratislava, Central Institute of Geology Prague	Depiction of landslide contours on 1: 25,000 map; description on punch cards	Endangered areas along communication lines and in the vicinity of municipalities
2nd	1974–1978	Department of Geotechnics FCE SUT Bratislava	Depiction of landslide contours on 1: 25,000 map; Monograph (Nemčok 1982)	High mountain areas, selected areas in flysch and volcanic regions and intramontane depressions
3rd	1981–1988	GIDŠ Bratislava	Depiction of landslide contours on 1: 10,000 map; inventory sheets designed for computer processing	Flysch regions, volcanic mountains, intramontane depressions

20.5 Atlas of Slope Stability Maps SR at Scale 1: 50,000

To date, the Atlas represents the most comprehensive nationwide document on the distribution of slope failures across the territory of Slovakia. Another key component of the Atlas is the map of a territory susceptible to slope deformations. The compilation of the Atlas had been preceded by several geological projects focused on, among others, mapping and registration of landslides and other slope deformations, which represented the input materials for documentation of slope deformations and development of their databases in the Atlas. These included the following documents:

- Results of previous inventories of landslides carried out in three stages (Table 20.4), which were processed in the Landslides register of Geofond.
- Thematic engineering-geological maps of scales 1: 10,000, 1: 25,000, 1: 50,000 (multipurpose maps of engineering-geological conditions, engineering-geological zoning maps, special maps of slope deformations and territorial susceptibility to landsliding).
- Geological maps at scales 1: 50,000, 1: 25,000.
- Passportization of slope deformations in relation to the road and rail network and major routes of pipelines of the Central Slovakia Region and selected districts of the East Slovakia Region.
- Map documents with data on slope deformations, developed in the scope of geological exploration tasks for civil engineering structures, for evaluation and optimal selection of alternative sites and routes.
- Results of engineering surveys of investigated landslides.

Since the process of inventory was implemented in different periods and was performed by different organizations and authors, the map documents often lacked correlation and acceptance of the results of previous surveys and inventories. This resulted in numerous ambiguous multiple registrations of the same slope deformations, but with different interpretations of their borders and characteristics, including slope deformation type and activity.

The goal of the Atlas was processing and synthesis of existing data obtained from primary documents, coherent assessment of disturbed areas, verification of insufficient or contradictory data on slope deformations taken from archival materials. Approximate reconnaissance of an unexplored territory in order to detect hazardous slope deformations, posing a threat to the population and existing infrastructure was carried out. The project did not include a more detailed mapping of unexplored areas nationwide.

20.6 Brief Presentation of the Atlas of Slope Stability Maps of the SR at Scale 1: 50,000

Each slope failure in the map is marked with digit and contours or point mark (slope deformations of less than 50 m long). Estimated slope deformations activity is expressed in colour and type of slope failures is shown by hatch (Fig. 20.2). A simplified classification by Nemčok et al. (1974), which distinguishes slope movements based on the rate of the movement—creeping, sliding, flowing and falling, was adopted.

20.6.1 Slope Stability Zoning Maps

Slope stability zoning maps at the scale of 1: 50,000 are the most significant results of the project. They were compiled separately for each map of the layout 1: 50,000 (in total 132 map sheets) and then "welded" (Figs. 20.2 and 20.3). The main zoning criterion is the stability of slopes defined as the

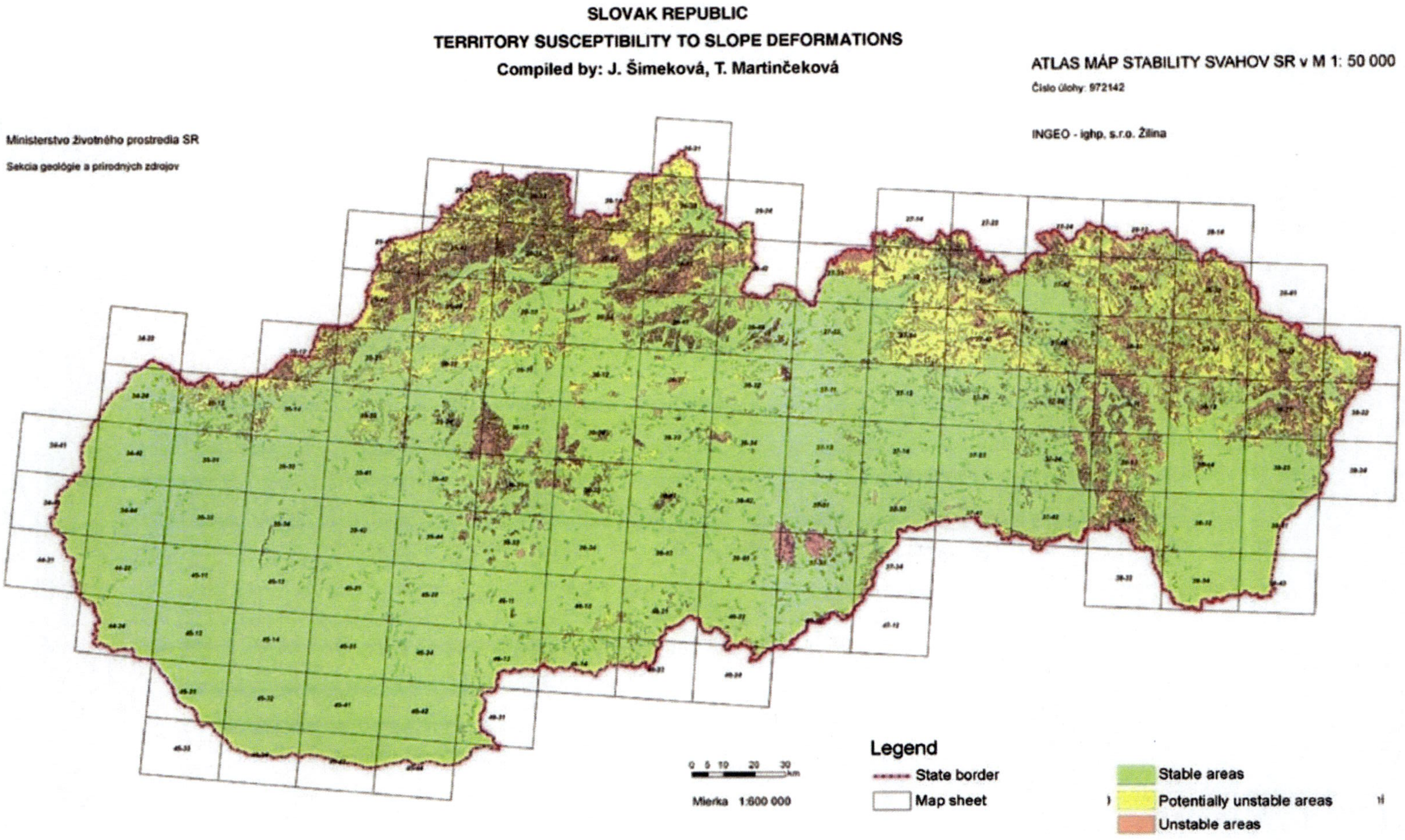

Fig. 20.2 Landslide susceptibility map of the Slovak Republic with depiction of map sheet layout at the scale of 1:50 000 (The map is available in Slovak at the Map Server of ŠGÚDŠ)

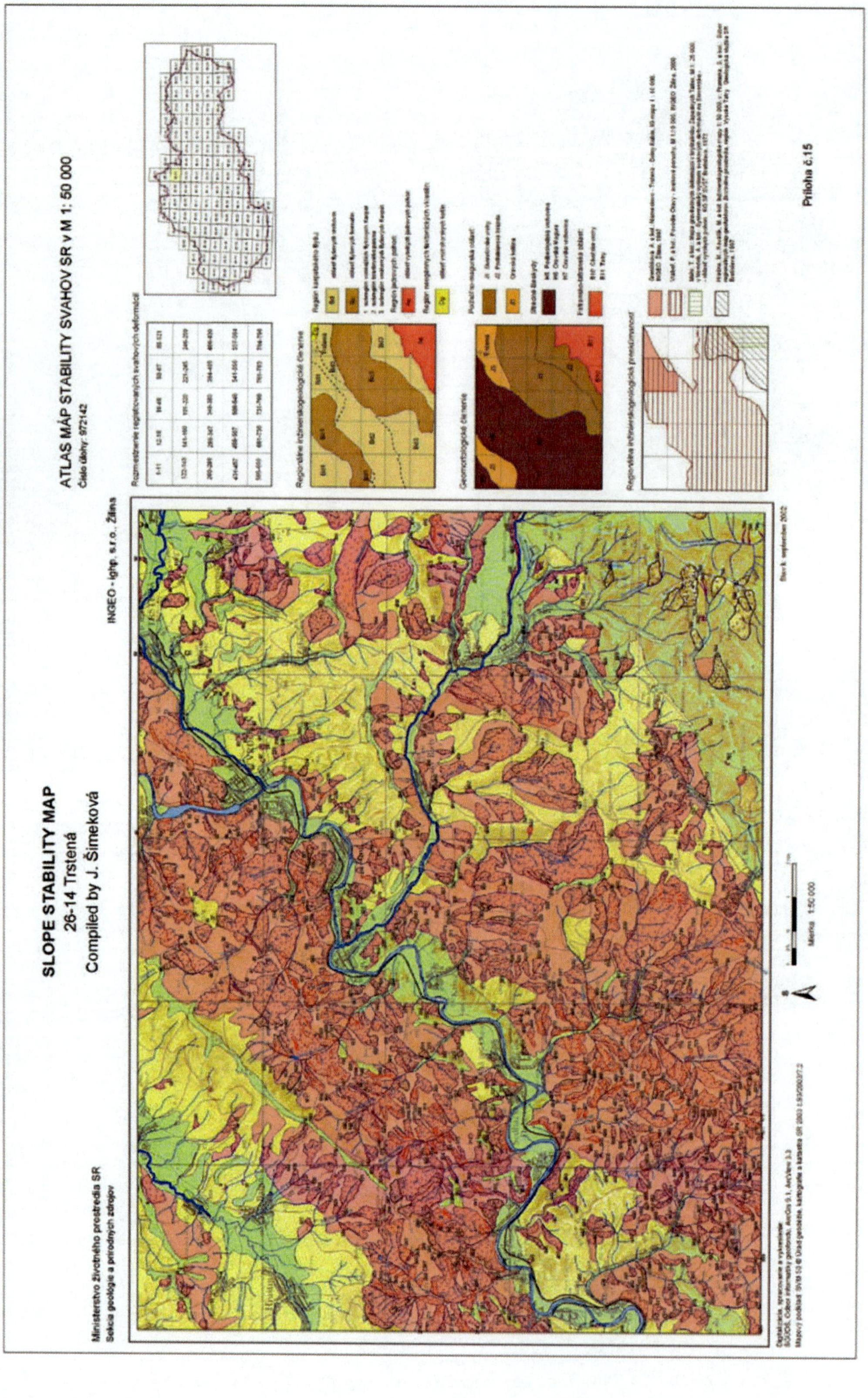

Fig. 20.3 Example of slope stability zoning map, map sheet 26–41 Trstená (to the right are complementary maps of small scale)

resistance of the geological environment to be subject to a gravitational deformation, or territorial susceptibility to the development of slope deformations (Table 20.5).

The map was created by an empirical method. It depicts three zones in colour (the "light" method: shades of green, yellow and red), reflecting the degree of susceptibility of the area to the formation of slope deformations (the database field rajon). In addition to the colour representation, sub-zones are also shown in the map from I. to III.C. (the database field podrajon), while the increasing susceptibility of the sub-zone to sliding is expressed by saturation of the colour.

In order to make the results accessible for the public, the Slovak territory susceptibility to slope deformations has been visualized on the map server SGIDŠ (http://mapserver.geology.sk/zosuvy/).

Components of each 1: 50,000 map sheet are additional maps at the scale of 1:250,000 (Fig. 20.3), which in transparent manner indicate complementary information—location of map sheet in the Slovak Republic map layout, distribution of registered slope failures in the respective map sheet, geomorphological subdivision of the territory, regional engineering-geological division of the territory and existing engineering-geological maps of regional character.

Table 20.5 Susceptibility to slope movements

Zone	Susceptibility to slope deformation	Subzone	Description	Area [km^2]
I	Stable area	I	Areas mostly stable, or areas with a very low degree of susceptibility to slope deformations (in morphologically favourable areas, with insufficient level of investigation, sporadic existence of slope deformations as well as local occurrence of new slope deformations of smaller dimensions cannot be ruled out)	36,789.88
II	Potentially unstable area	II.A	Areas with hitherto unregistered slope deformations, with a favourable geological structure which does not exclude, in the case of favourable morphological conditions, an occasional occurrence of slope deformations (especially landslides and creep) due to natural conditions. Areas are sensitive to negative anthropogenic interventions. In areas with insufficient level of investigation, the existence of as yet unregistered slope deformations is assumed. The district also includes areas affected by intense gully erosion and areas threatened by fall of rock fragments	7,929.93
		II.B	Areas of stabilized slope deformations from the group of creep, with minimal possibility of activation due to natural conditions. In the case of extremely unfavourable climatic factors, depending on engineering-geological conditions and morphology of a particular area, local activation of small-scale slope deformations cannot be ruled out. Areas are sensitive to anthropogenic interference	237.31
III	Unstable area	III.A	Areas in the immediate vicinity of registered slope deformations and areas with as yet unregistered slope deformations, with favourable geological structure for the formation of slope deformations (especially landslides and creep). The degree of susceptibility to slope deformations varies according to the morphology of a particular area. Areas are sensitive to negative anthropogenic interventions. In areas with insufficient level of investigation, the existence of as yet unregistered slope deformations is assumed	1,948.48
		III.B	Areas of slope deformations with a medium degree of susceptibility to activation of slope deformations (slopes with subsurface creep deformations of stabilized character and stabilized forms of rock fall). The possibility of activation due to natural conditions is significantly reduced, local activation of slope deformations of smaller extent cannot be ruled out, depending on the engineering-geological conditions and morphology of the specific area. Areas are sensitive to larger anthropogenic interventions	133.46
		III.C	Slope deformation areas with a medium to high degree of susceptibility to slope deformation activation (slopes with active, potential and stabilized slope deformations, with the exception of stabilized subsurface creep deformations and stabilized rock fall). Activation of slope deformations is possible due to natural conditions or negative anthropogenic factors, or their combination	2,084.85
Lakes, reservoirs				0.63
Total				49,123.91

Note There is obvious inconsistency in the figures of a total area of Slovakia (officially 49,034 km^2) and figures from the Atlas, which is based on topographic work ZM 50

20.6.2 Inventory Sheets (Passports) of Slope Deformations

For each slope deformation registered in the Atlas, an inventory sheet—the so-called passport—was drawn. The passport provides more detailed information about the respective slope deformation (Fig. 20.4). Source information on slope deformation was retrieved from older inventories (Baliak et al. 2014) of landslides stored in Geofond. For those slope deformations, which were not registered in the Geofond, or the information in the registry was inadequate or questionable, the data were obtained from the study of other existing archival materials and additional field mapping.

The passport of a slope failure includes: serial number of slope failure (column 1), territory administrative division (district, geomorphological unit, engineering-geological region, columns 2–4), data on exploration and visualization scale (columns 5–8), slope deformation description (columns 9–14), slope failure dimensions (columns 15–18), endangered objects (columns 19–25), causes of slope failure generation (columns 26–27) and corrective measures (column 28).

Statistical processing of an extensive set of data contained in individual passports of slope deformations provided groundwork for comprehensive analysis of the project results.

20.6.3 Slope Deformations Occurrence

In the Atlas, 21,190 slope deformations were registered by the end of the project in 2006. Of the total number of slope deformations registered:

- 13,012 (61.4%) slope deformations were retrieved from archival documents without field validation;
- Due to nonconformity of archival documents, 6,076 (28.7%) slope deformations were validated in the field (determination of the type, extent, activity, additional data on the endangered objects, etc.);
- 2,102 (9.9%) slope deformations were discovered and registered during field work. Most of them were randomly identified during field reconnaissance of the archived landslides, in rare cases they were detected by targeted mapping of the territories with potential threats to infrastructure based on various indications, e.g. topographic studies.

The number of registered slope deformations gives only an indication of the overall territory of Slovakia affected by slope failures. The actual number of slope deformations is significantly higher. Due to the output scale of 1:50,000 it was often necessary to combine complex and interlinked slope deformations in intensively disturbed areas. Thus, such clusters of individual failures were recorded under one label.

Fair idea of the territory of Slovakia affected by slope failures is illustrated by Table 20.6; the final figure of the Atlas equals to 257,591.2 ha, representing 5.25% of the total land area of the Slovak Republic. For comparison, in the scope of the previous registration, which preceded the compilation of the Atlas, there were estimated 9,194 slope deformations in Slovakia affecting 3.06% of the total territory (Nemčok 1982).

From the data in Table 20.6, which outlines land-use of affected areas (built-up areas, playgrounds, cemeteries), it is evident that agricultural and forest lands are disturbed about equally (50.6% and 46.7%). The share of other land use categories amounts to a mere 2.7%. Some areas of agricultural land disturbed by slope failures have been abandoned because of difficult farming conditions, and are now overgrown with grassland, shrub communities or forest stands. This is confirmed by Liščák et al. (2009), according to whom directly affected by slope failures are 120,855 ha of the forest stands (the analysis was based on Corine Land Cover 2000 input data).

The sloping agricultural land areas are the most affected by landslides, or by their combinations with other types of slope deformations (about 90%). In forested areas landslides are the most abundant (67.4%), but also important (32.6%) are slope deformations of creep, fall and flow types. Other areas, particularly built-up areas (2.7%), are highly susceptible to landslides. The areas above the tree line are dominantly susceptible to failures of creep, fall and flow types.

Of the total number of registered slope deformations (21,190 units) 90.15% are landslides; other types of slope deformations account for less than 10%. In terms of the assessment of disturbed areas acreage (Fig. 20.5), which is more representative than the above numbers, the share of landslides is also significantly higher (78.12%) than that of block fields (15.31%) and other types of slope failures, including the combined ones (total 6.57%).

Statistical analysis of the occurrence of slope deformations in relation to bedrock (geological formations) shows that the most affected are Palaeogene strata representing 60.1%, whereas respective figures for other rock series are as follows: Neogene—18.7%, Cretaceous—9.9%, Palaeozoic—4.2% and Triassic—2.3%. The remaining 4.8% goes to other geological complexes. The occurrence of various types of deformation is often specific to particular bedrock, or a combination of geological complexes. For example, disintegrating massifs are common in rocks of Palaeozoic age (75.0%), block ridges (Fig. 20.6) and block fields typify Neogene volcanites (68.0% and 61.2%, respectively), landslides occur widely in Palaeogene formations (Fig. 20.7, 65.9%), debris flows are associated with Palaeozoic bedrock

Label	District	Geomorphologic unit	Engineering geological area	Investigation Level: Geofond - Landslides register (Nr.)	Investigation Level: Geofond - report (Nr.)	Investigation Level: Present state	Investigation Level: Scale	Slope failure type	Activity	Bedrock	Geological complex	Hydrogeol. cond.	Slope angle (°)	Acreage: In total (ha)	Acreage: Agricultural land (ha)	Acreage: Forest (ha)	Acreage: Other landuse (ha)	Endangered objects: Roads: Motorway and Ist class (m)	Endangered objects: Roads: II. and III. class (m)	Endangered objects: Railway (m)	Endangered objects: Buildings: Housing estates (number)	Endangered objects: Buildings: Others (number)	Endangered objects: Eng. Netw.: Above ground (m)	Endangered objects: Eng. Netw.: Underground (m)	Causes: Natural	Causes: Anthropogenic	Corrective measures
1	2	3	4	5	6	7	8	9	10	11	12	13	14	15	16	17	18	19	20	21	22	23	24	25	26	27	28
267	DK	H6	Bc1	-	83646	-	B	Z	P	Pg	ZF	S	13	1,2		1,2									KE		N
268	DK	H6	Bc1	x	83646	-	B	Z	A	Pg	ZF	P	17	3,5	1,5	2,0									WE		N
269	DK	H6	Bc1	-	83646+	-	B	Z	P	Pg	ZF	P	12	3,5		3,5			200				150		KE		K
270	DK	H6	Bc1	56372	83646	-	B	Z	P	Pg	ZF	P	13	24,2	4,0	20,2			300						KE		N
				54385	58280+				A	Pg	Z	Z	13	4,4		4,4									K		N
271	DK	H7	Bd1	54385	83646	-	B	Z	P	Pg	ZF	P	13	10,0	3,0	7,0									KE		N
					46025+				A	Pg	Z	Z	14	5,2	2,0	3,2									KE		N
272	DK	H7	Bd1	54385	83646+	-	B	Z	P	Pg	ZF	Z	11	11,7		11,7									KE		N
273	DK	H7	Bd2	54384	83646+	-	B	Z	P	K	ZF	Z	13	7,6		7,6									KE		N
274	DK	H7	Bd2	54384	83646+	-	B	Z	P	KJ	ZS	P	12	13,3		13,3									KE		N
275	DK	H7	Bd2	-	83646	-	B	Z	P	JPg	ZF	Z	18	9,9		9,9									KE		N
									A	Pg	Z	Z	11	2,3		2,3									KE		N
276	DK	H7	Bd2	-	83646	-	B	Z	P	K	ZF	P	15	5,4		5,4									KE		N
								P	A	K	Z	P	13	0,8		0,8									KW		N
277	DK	H7	Bd2	54383	83646	-	B	Z	P	Pg	ZF	Z	18	5,1		5,1									KE		N
					46025				A	Pg	Z	Z	13	1,6		1,6									KE		N
					58280			P	P	Pg	ZF	Z	17	2,0		2,0									KE		N
278	DK	H7	Bd2	54382	83646	-	B	Z	P	Pg	ZF	Z	15	11,5	3,0	8,5									KE		N
					46025+				A	Pg	Z	Z	16	0,9		0,9									KE		N
279	DK	H7	Bd2	54378	83646+	-	B	Z	P	K	F	P	15	56,2		56,2									KE		N
280	DK	H7	Bd2	-	83646	-	B	B	S	Pg	F	S		21,4	6,4	15,0									P		N
281	DK	H7	Bd1	54417	83646	-	B	Z	P	Pg	F	P	11	66,0	46,0	20,0			550		7	1	900		WE		N
					58280+				A	Pg	Z	P	13	12,8	4,4	8,3	0,1		45						KW	GM	N
282	DK	H7	Bd2	51943	83646	-	B	L	P	Pg	F	P	17	7,4		7,4									PW		N
				54393	58280			Z	P	PgK	F	P	11	55,5	15,6	39,9			650		6	1			EW		N

Fig. 20.4 Passport of slope deformations 267–282 from the map sheet 26–412 Trstená

Table 20.6 Acreage of area disturbed by slope failures (Šimeková, Martinčeková et al. 2006)

Acreage	Acreage in total	Acreage of slope failures	Disturbance by slope failures [%]	
	[ha]	[ha]	Versus acreage in total	Versus land use
Slovak Republic	4,903,347	257,591.2	5.25	–
Agricultural land use	2,436,876	130,289.9	2.66	50.6
Forest land use	2,004,100	120,243.3	2.45	46.7
Other land use	462,371	7,058.1	0.14	2.7

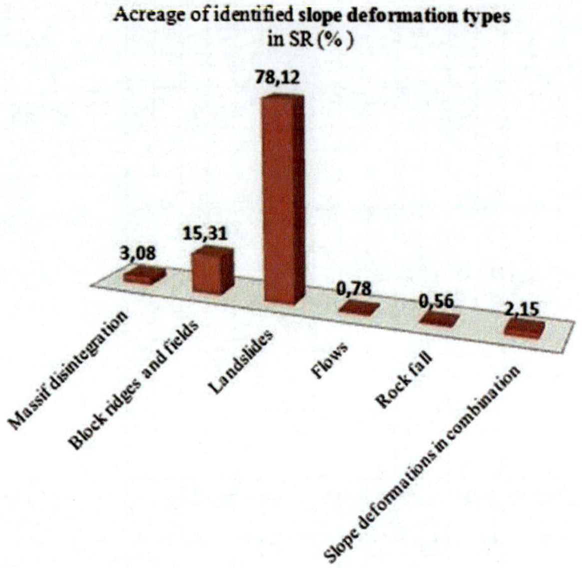

Fig. 20.5 Proportion of slope deformation types in Slovakia (Šimeková et al. 2014)

(55.6%), and rock fall occurs most often in Neogene volcanites (40.7%). For creep deformations of block movements group, besides Neogene volcanites, typical is combination of several rock complexes differing in strength properties. Most often combinations are: Triassic—Cretaceous, Neogene—Palaeogene, whereas Jurassic—Cretaceous and Jurassic—Triassic combinations are less common.

The average angles of slope deformation types are as follows: disintegrating massifs—26.7°, block fields—15.2°, landslides—13.3°, debris flows—25.2°, rock falls—27.6°. On the steepest slopes disintegrating massifs, flows and rock falls are frequently present. Artificial slopes affected by slope failures have an average slope of 21°. Landslides, which make up 90% of all slope deformations, have a mean inclination of 13.3°; to be more accurate, 65.5% of landslides are formed on slopes ranging in inclination from 7° to 17° (Kopecký et al. 2008).

Fig. 20.6 Disintegrating of the edge of the Neogene andesite lava flow, Biely kameň, Vtáčnik Mt., Central Slovakia; **a** Toppling of andesite cliffs at the edge of the lava flow; **b** rock fall accumulation below

Fig. 20.7 A small landslide (originated in 2010) in the Inner Carpathian Flysch of Palaeogene age, Spišské Vlachy, Northern Slovakia

20.6.4 Slope Deformation Distribution in the Engineering-Geological Regions and Areas

The evolution and forms of slope movements have their own characteristics depending on the geological setting and geomorphological evolution. For the purpose of this assessment, an engineering-geological division of the territory of Slovakia according to Matula (Matula and Pašek 1986) was adopted in the Atlas, which takes into account the criteria of uniformity of geological structures and macro-relief. This division was later modified and updated by Hrašna and Klukanová in the Atlas of Landscape SR (2002). The territory of Slovakia is divided into 4 engineering-geological regions and 8 engineering-geological areas as follows:

A – Region of Core Mountains:
  Aa – Area of High Core Mountains
  Ab – Area of Core Highlands

B – Region of Carpathian Flysch:
  Bc – Area of Flysch Highlands
  Bd – Area of Flysch Uplands

C – Region of Neogene Volcanites:
  Ce – Area of Volcanic Highlands
  Cf – Area of Volcanic Uplands

D – Region of Neotectonic Depressions:
  Dg – Area of Inner-Carpathian Depressions
  Dh – Area of Inner-Carpathian Lowlands

The Region of Carpathian Flysch (B) is dominated by various types of landslides. In the Region of Core Mountains (A) most frequent are creep deformations leading to massif disintegration. In the Region of Neogene volcanites (C) deformations of block type are significant and in the Region of Neotectonic Depressions (D) landslides prevail over block deformations (Tables 20.7, 20.8, Fig. 20.8).

Table 20.7 Occurrence of main types of slope deformations in the engineering-geological regions and areas of Slovakia (Šimeková, Martinčeková et al. 2006)

EG Region	EG Area	Massif disintegration	Block ridges and fields	Landslides	Flows	Rock fall	Slope deformations in combination	Total EG Area	Total EG Region
		[ha]	[ha]	[ha]	[ha]	[ha]	[ha]	[ha]	[ha]
A	Aa	6,355.5	2,726.7	5,291.1	862.9	975.0	1,442.2	17,653.4	23,741.9
	Ab	1,301.1	579.5	3,934.1	9.8	75.5	188.6	6,088.5	
B	Bc	134.8	2,266.8	24,732.2	153.8	9.1	231.8	27,528.5	138,169.7
	Bd	25.3	2,150.4	105,956.2	568.7	1.3	1,939.3	110,641.2	
C	Ce	61.0	19,085.4	17,737.7	263.1	373.3	1,045.6	38,566.1	52,214.3
	Cf		10,311.4	3,167.3	14.7	13.2	14.6	13,648.2	
D	Dg	50.6	2,207.0	33,449.7	97.5	3.9	538.8	36,347.5	43,465.5
	Dh		113.1	6,969.7	35.2			7,118.0	
In Total		7,928.3	39,440.3	201,238.0	2,005.7	1,451.3	5,527.9	257,591.4	257,591.4

Table 20.8 Frequency and acreage of main types of slope deformations in the engineering-geological regions of Slovakia (Šimeková, Martinčeková et al. 2006)

EG Region	Massif disintegration		Block ridges and fields		Landslides		Flows		Rock fall		Slope deformations in combination		TOTAL	
	No	area [ha]	No	area [ha]	No	area [ha]	No	area [ha]	No	area [ha]	No	area [ha]	No	area [ha]
A	114	7,656.6	88	3,306.2	1,260	9,225.2	672	872.7	84	1,050.5	41	1,630.8	2,259	23,741.9
B	5	160.1	147	4,417.2	12,624	130,688.4	156	722.5	4	10.4	28	2,171.1	12,964	138,169.7
C	6	61.0	518	29,396.8	1,708	20,905.0	46	277.8	61	386.5	22	1,187.2	2,361	52,214.3
D	3	50.6	51	2,320.1	3,513	40,419.4	34	132.7	3	3.9	4	538.8	3,608	43,465.5
Total	128	7,928.3	804	39,440.3	19,105	201,238.0	908	2,005.7	152	1,451.3	95	5,527.9	21,192	257,591.4

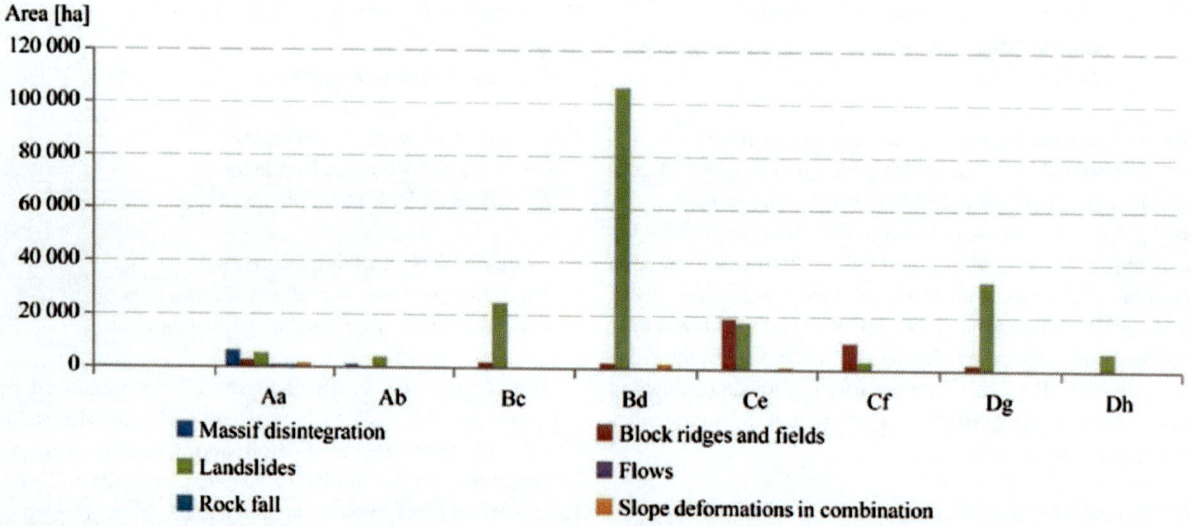

Fig. 20.8 Acreage of the main types of slope deformations in the engineering-geological regions and areas of Slovakia (Šimeková et al. 2014). Aa – High Core Mountains, Ab – Core Highlands, Bc – Flysch Highlands, Bd – Flysch Uplands, Ce – Volcanic Highlands, Cf – Volcanic Uplands, Dg – Inner-Carpathian Depressions, Dh – Inner-Carpathian Lowlands

Table 20.9 Summary disturbance of engineering geological areas of Slovakia (Šimeková, Martinčeková et al. 2006)

EG Area	Share of EG Area in the area of SR	Acreage of EG Area disturbed by slope deformations		Share of slope deformations of EG Area in the area of SR	Share of slope deformations of EG Area to disturbed area of SR
	%	[km^2]	%	%	%
Aa	7.70	176.53	4.79	0.36	6.85
Ab	17.90	60.89	0,71	0.12	2.36
Bc	5.40	275.28	10,58	0.56	10.69
Bd	15.30	1,106.41	15.13	2.26	42.95
Ce	6.90	385.66	11.78	0.79	14.97
Cf	4.00	136.48	7.19	0.28	5.30
Dg	14.40	363.48	5.27	0.74	14.11
Dh	28.40	71.18	0.52	0.15	2.76
Total	100.00	2,575.91		5.25	100.00

In summary presentation of slope deformation acreage in various engineering-geological areas (Tab. 20.9) the most disturbed is the Area of Flysch Uplands, followed by areas of Volcanic Highlands and Flysch Highlands. In relation to the total area of the Slovak Republic the first place is occupied by the Area of Flysch Uplands followed by the areas of Volcanic Highlands and Inner-Carpathian Depressions.

20.6.5 Slope Deformations Distribution In territorial-Administrative Units (Regions)

In terms of territorial-administrative division of the Slovak Republic (8 regions, 79 districts) relatively highest disturbance by slope failures is documented in the Prešov region, followed by Žilina and Banská Bystrica regions (Fig. 20.9). In terms of percentage disturbance to the total area of the region, Žilina is the most affected region, followed by Prešov and Trenčín regions (Fig. 20.10). Bratislava, Nitra and Trnava regions are least affected by slope failures.

20.6.6 Threat for Buildings and Other Objects

The high disturbance of the Slovak territory by landslides also implies considerable threat to buildings and other structures (Table 20.10). The data provided have to be considered as guidelines. In common, they were derived from basic topographic maps at the scale of 1: 10,000, with the last update 10 to 20 years prior to the release of the Atlas. In numerical terms in the table are included objects which had been damaged or set out of operation, as well as objects that were stabilized by remedy works (during the

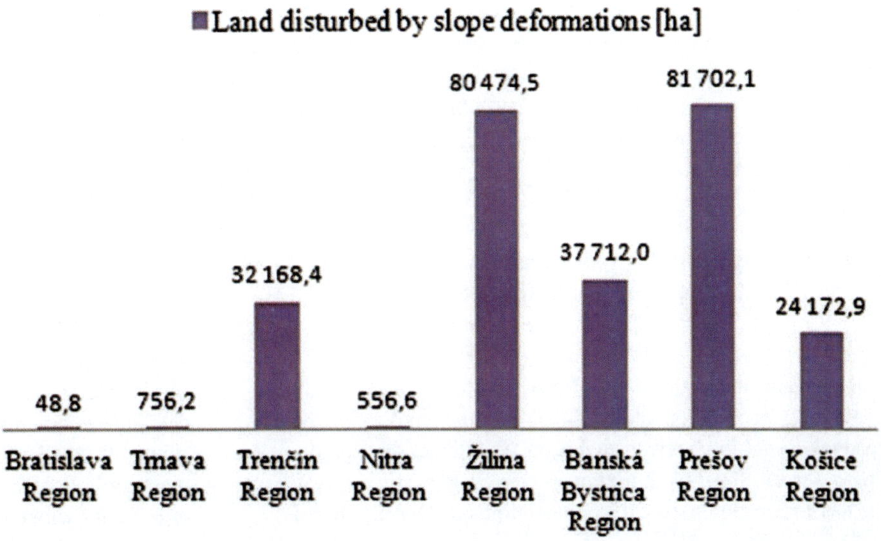

Fig. 20.9 Acreage of slope deformations in the regions of the Slovak Republic (Šimeková et al. 2014)

Fig. 20.10 Disturbance of the regions of the Slovak Republic by slope deformations in % (Šimeková et al. 2014)

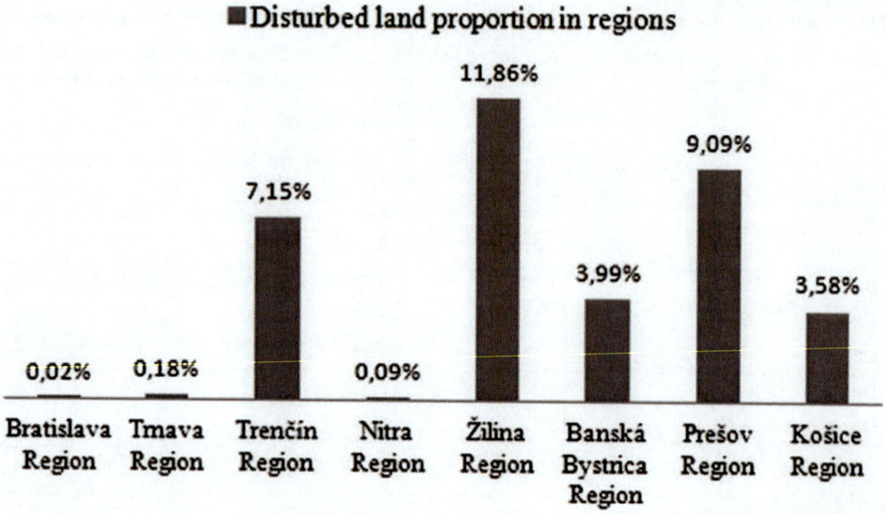

Table 20.10 Objects jeopardized by slope deformations (Šimeková, Martinčeková et al. 2006)

Type of object	Extent
Buildings (housing estates and other buildings)	27,920 objects
Other structures (water-supply tanks, cemeteries, playgrounds, bridges and others)	600 objects
Sites with more than 50 objects endangered*	168 sites
Motorways and roads of the Ist class	98,816 m
Roads of the IInd and IIIrd classes	571,408 m
Railway	67,210 m
Above-ground engineering networks	1,116,056 m
Gas pipelines	101,180 m
Oil pipelines	3,500 m
Water pipelines	290,925 m

*Endangered municipalities, urban suburbs, recreation and garden areas, which are fully or partially located in the territories affected by slope deformations

construction of an object or as a result of its emergency state).

The most frequently threatened "other structures" include water-supply tanks (38.5%) and cemeteries (32%). Degree of threat to objects is dependent on several factors, but mainly on the type and activity of slope failure, its extent and depth, relative position of an object threatened by slope deformation, etc.

20.6.7 Application of the Results of the Atlas of Slope Stability Maps SR at Scale 1:50,000 in Practice

Given that the Atlas is the first complete digital map depicting slope deformations throughout the territory of the Slovak Republic, freely available on the Internet through the map server ŠGÚDŠ (http://mapserver.geology.sk/zosuvy/), it has been often used for various purposes. During almost one decade of its use, a number of shortcomings and examples of its misuse have been identified, for the purposes of which the map scale and the related generalization of landslide areas have not been intended. The insufficient scale of the Atlas is evident especially in experiments, or efforts aimed at extrapolating the landslide areas contained in the Atlas into documents at larger scales, e.g. in the development of spatial plans and the evaluation of specific building sites.

The deficiencies of the Atlas can be divided into several groups.

(a) deficiencies arising from the adopted scale of maps:
- generalization and distortion of mapped boundaries of slope deformations during transformation of the detailed documentation into the scale of 1: 50,000 (Figs. 20.11a–d);
- merging of composite disturbed areas and landslides directly interlinked or lying in close proximity and

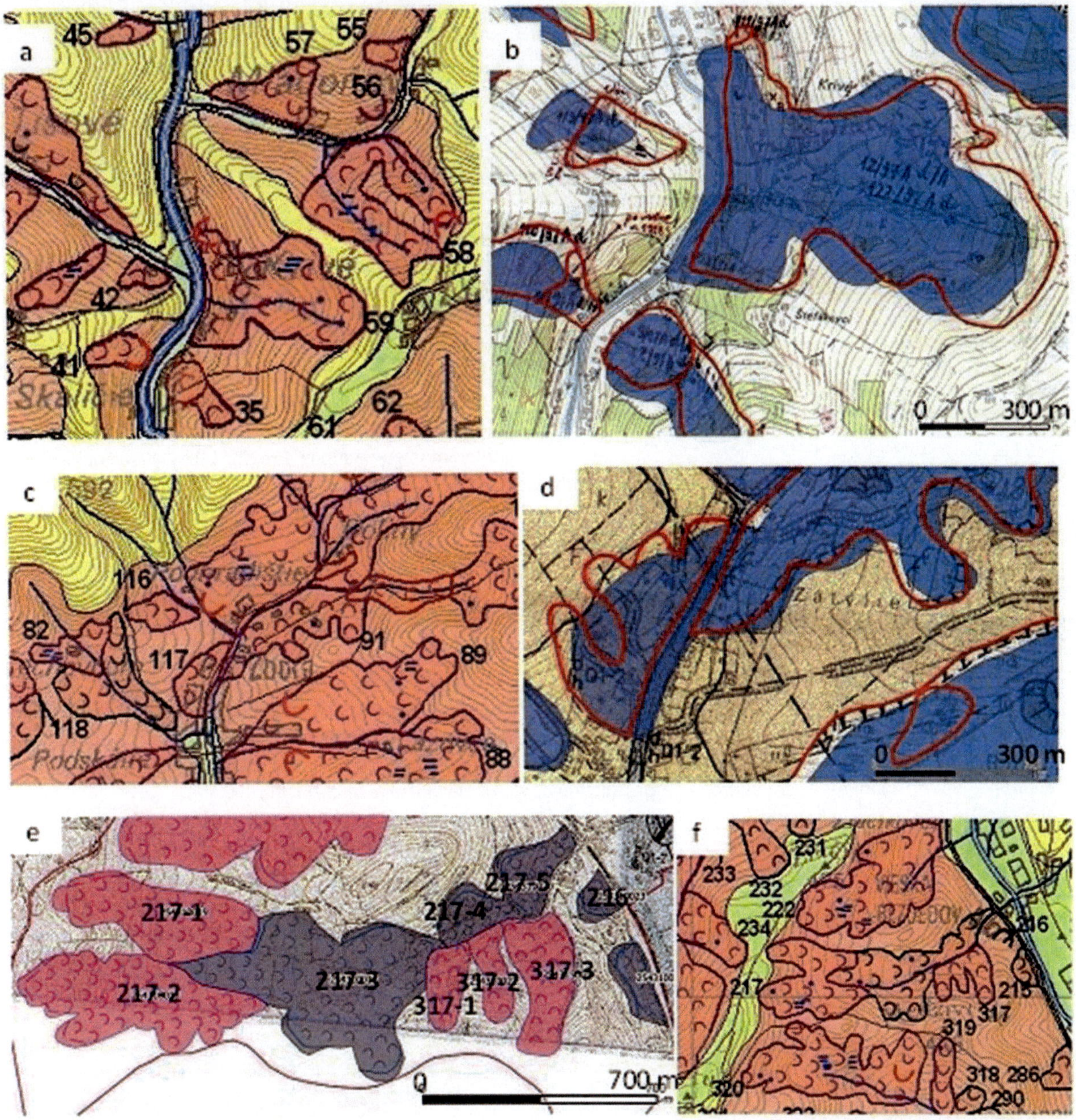

Fig. 20.11 Depiction of deficiencies of the Slope Stability Atlas SR 1:50 000 due to map scale (Šimeková et al. 2014). Explanations: a, c, f —detail from Atlas' zoning map; b, d—detail from working Atlas' groundwork map at scale 1: 10,000, with highlighted landslides boundaries (red line) and area in colour, representing landslides areas transformed from Atlas into topography ZM 10; e – complex sliding area with 8 partial landslides (217—1 to 5, 317—1 to 3) mapped on topography at scale 1: 10,000 and f—its final depiction in the Atlas at scale 1: 50,000 (landslides 217 and 317)

their registration as a singular slope failure (one label; Figs. 20.11e, f);

- visualization of smaller slope deformations (with dimensions < 250 m) only by icons (e.g. Fig. 20.11e, f—landslide no. 216);
- inequalities (distortions) among topographic documents of different scales.

(b) deficiencies arising from the lack of nation-wide mapping:

- existence of previously unregistered slope deformations cannot be excluded, even in relatively stable areas (green areas in the zoning map).

(c) deficiencies arising from aging of map documents (completion in 2006):

- there is no systematic update of the status on newly registered, newly formed and reactivated slope deformations (e.g. in 2010 after massive activation of landslides, especially in eastern Slovakia, ŠGÚDŠ registered 551 new active slope failures using precise GNSS measurements (Liščák et al. 2010)). However, it should be noted that a database of slope deformations in the Slovak Republic is still being systematically built at ŠGÚDŠ, while slope deformations from the Atlas, after their proper topological correction, form the basis of the database.

(d) deficiencies resulting from inaccuracies of source documents:

Objective errors:

- field mapping performed only by validating the facts with available topographic document of appropriate scale (usually 1: 10,000), without accurate registration of contours and morphological elements of slope deformations by modern geodetic GPS devices;
- inaccurate or outdated topographical maps used in mapping work;
- slope deformations without significant morphological manifestations facilitating their unambiguous identification and delineation (disguised by secondary modelling through erosion-denudation and weathering processes or anthropogenic activity—agriculture, construction activities);
- vegetation cover during field mapping (dense vegetation, especially in summer and spring seasons, does not allow the mapping staff to move around freely and veils morphological shapes of slope deformations);
- inaccessibility of the terrain in terms of build-up land or other land use (fenced private facilities, military facilities, etc.).

Subjective errors:

- subjective approach of a mapping geologist.

Despite the above-outlined weaknesses, the Atlas of Slope Stability Maps SR at the scale of 1:50,000 has many advantages and benefits:

- The Atlas of Slope Stability Maps of the SR is a single clear map series, processed in digital form in GIS environment, showing the occurrence of slope deformations, distinguished by type and activity throughout the territory of the Slovak Republic.
- For each slope failure registered in the Atlas, a passport (database) of a slope failure was prepared, with its detailed characteristics, territorial-administrative affiliation, level of investigation, figures on geometric characteristics and objects at risk, and possible corrective measures, which form the basis of an extensive statistical file.
- The data on level of investigation listed in passports of slope failures include a link to the source map documents, provided they are available. Most of these documents are available at Geofond ŠGÚDŠ Bratislava (Geofond Archive stores and makes available results of geological works across Slovakia). Previous multiple registrations of the same landslide area led to a variety of records with different interpretations. Provided they were not validated in the scope of the Atlas field work, the most reliable records were those with a more detailed scale mapping and actual time-date.
- Compilation of the Atlas supplemented the existing Landslide Register of Geofond regarding new slope deformations. Contradictory data on slope deformations produced during previous multiple registrations were eliminated.
- Newly registered slope deformations were validated in the field and spatially corrected, then plotted on the primary working topographic maps at scale 1: 10,000.
- All quantitative data in the passports (acreage of slope deformation, the extent of endangered objects) were derived from existing documents or detailed working maps of the input documentation of the Atlas.

As was already mentioned, the usefulness of the output maps of the Atlas is partially limited due to the used scale of 1: 50,000.

20.7 Slope Failures Research After Publication of the Atlas

The geological task "Engineering geological mapping of slope deformations in the most endangered areas of the flysch zone in M 1: 10,000", was carried out by GeoSlovakia s.r.o., Košice in 2006–2011 and was completed in 2011. As part of this task, purpose-built engineering-geological maps were compiled for an area of 4,042.22 km^2, covered with 285 map sheets at the scale of 1: 10,000. The aim was to evaluate the engineering-geological conditions of the mapped area, with the focus on slope deformations, reassessment of slope deformations in the defined area based on engineering-geological mapping and refinement of older map data, determination of landslide and flood hazards, definition and evaluation of risks, evaluation of the risk of endangering buildings due to floods and slope movements, detailed visual evaluation of selected most dangerous landslide sites in the municipalities of the assessed area, along with a conceptual proposal of technical measures for their remediation and estimation of their financial demands.

In the year 2010, the Eastern Slovakia in particular was heavily hit by extensive precipitation in the spring months, resulting in floods and mass wasting. The geological task "Registration, evaluation and emergency measures on newly formed slope deformations in 2010 in the Prešov and Košice regions" (Liščák et al. 2010) resulted in 577 (551 by ŠGÚDŠ) newly registered landslides in the Prešov and Košice regions, and for the first time in the regional inventory of slope failures in Slovakia GNSS technology was used on a regular basis.

Since 2010 slides, the Section of Geology and Natural Resources of the Ministry of the Environment, in cooperation with ŠGÚDŠ, introduced the practice of inspecting and registering emergency slope failures reported by representatives of municipalities or citizens (2011–22 slope failures; 2012–2; 2013–18; 2014–14, an event in Vrátna dolina Valley (Fig. 20.12), however, consisted of several dozen landslides and debris flows; 2015–4; 2016–12; 2017–12 slope failures, 2018–10, 2019–12 slope failures).

Recently, the use of a high-precision digital relief model derived from aerial laser scanning has been fully established in the preparatory phase of identification of slope failures (Van Den Eeckhaut et al. 2007, Schulz 2007, Migoń et al. 2010, Guzzetti et al. 2012). Improving the landslide prevention by defining the risk arising from slope deformation in geological environment prone to landslides within the Slovak Republic, is the primary objective of the geological task "Identification, Inventory and Engineering Geological Mapping of Slope Deformations" planned for 2018–2023 under the umbrella of the Operational Programme Quality of

Fig. 20.12 Catastrophic slides and debris flows in Vrátna dolina Valley, July 2014

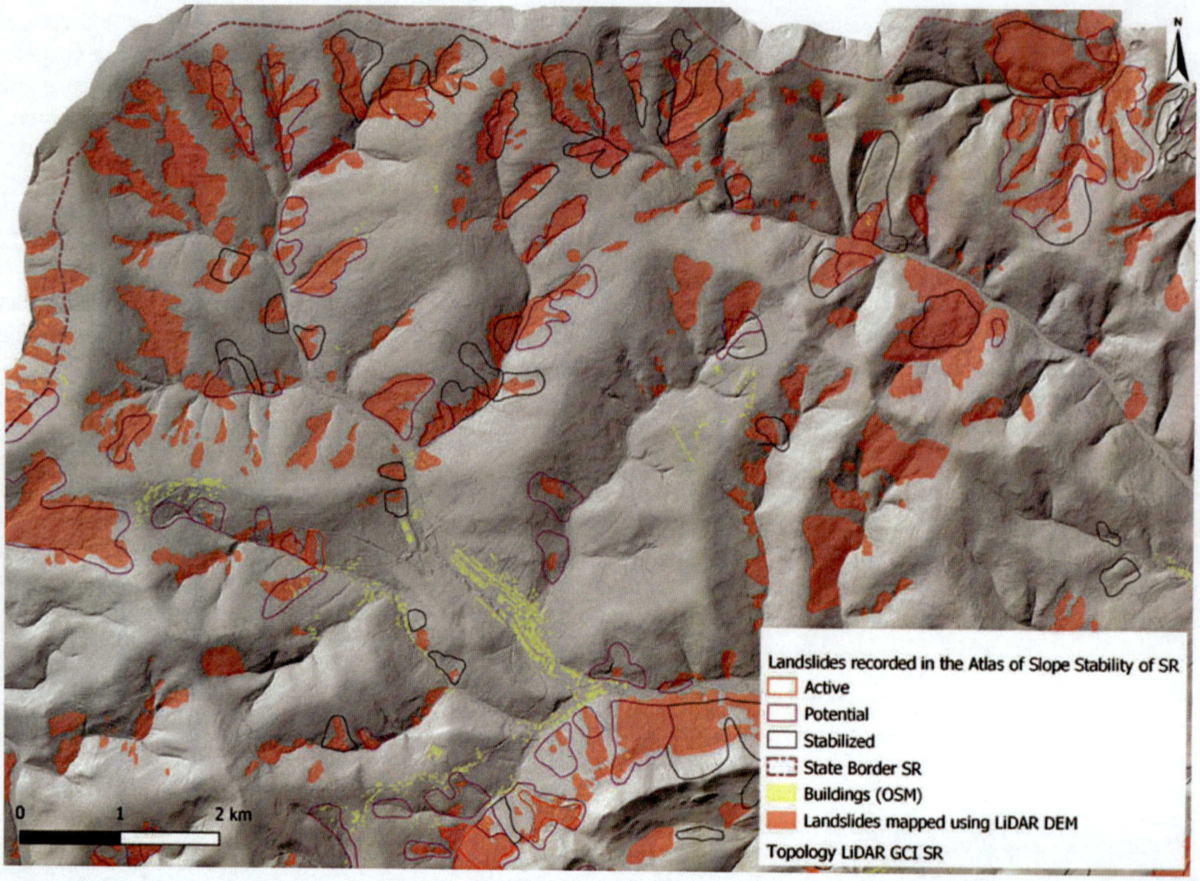

Fig. 20.13 Comparison of currently mapped slope deformations with the Atlas of slope stability maps of SR (Šimeková, Martinčeková et al. 2006). ZM50—Basic Map SR 1: 50,000; OSM—Open Street Map; GCI—Geodetic and Cartographic Institute SR

Environment. It concerns the exploration of the Biele Karpaty, Javorníky, Vtáčnik, Slanské vrchy and Vihorlatské vrchy mountain ranges, all significantly threatened by slope failures with a total mapped area of approximately 3,195 square kilometres. The inventory itself is preceded by the creation of a detailed digital elevation model (DEM, product of Geodetic and Cartographic Institute, Bratislava GCI) generated from airborne LiDAR data, with a detailed capture of various terrain elements, including those originating from slope deformations and other geohazards subject to mapping. The DEM obtained in this way makes it possible to detect slope deformations even in wooded terrain, which is a technological step forward compared to the aerial photogrammetry used so far. Engineering-geological and geological mapping are aimed at refining the position of the cover Quaternary complexes and slope deformations and other geohazards using precise DEM and GNSS technologies. This is followed by updating the database and the register of slope deformations and by the preparation and creation of parametric maps necessary for the development of landslide hazard forecast by numerical methods in the GIS environment. The first field work was successfully launched in the Biele Karpaty Mts. region. Comparison between the product of LiDAR-based mapping and the Atlas is in Fig. 20.13. It is obvious that the accuracy of slope failures depiction using LiDAR-based DEM is significantly higher—it allows to visualize the slope deformation even in cadastral maps.

20.8 Conclusions

Slope deformations are the most important geohazard, which threatens the territory of Slovakia. The process of understanding this phenomenon in recent decades has gone through several stages of development, reflecting the current state of knowledge, but also the degree of development of society and its requirements to eliminate this adverse phenomenon.

In the first half of the twentieth century, the assessment of slope stability issues was associated mainly with human intervention in the natural environment during construction works, and many experts at the time took the view that there was no significant risk of slope movements in our natural conditions. After the catastrophic Handlová landslide from the break of 1960/1961, the perception of the problem of slope movements changed significantly.

Concerns about the possible occurrence of slope movements of a similar extent in other parts of Slovakia and their accompanying adverse consequences have stimulated systematic research into slope deformations, starting with their inventory and mapping of the most vulnerable areas. This long-term purposeful activity resulted in the creation of the Atlas of Slope Stability Maps at a scale of 1: 50,000 for the entire territory of the Slovak Republic in 2006. At the same time, considerable progress in respect to slope failures mitigation has been made in these decades.

The Atlas presents an overview of digital maps developed according to the unified methodology for the whole territory of the Slovak Republic. In the scope of this work, 21,190 slope deformation occupying the total area of 257,591.2 ha, which constitutes 5.25% of the total area of Slovakia, were registered.

The majority of the registered slope deformations belong to the group of potential sliding activity. In some regions of Slovakia (mainly the flysch areas of north-western and eastern Slovakia—Region of Carpathian Flysch, but also in the territories made of Neogene volcanic rocks and overlying sediments in the central and eastern part of Slovakia—Region of Neogene Volcanites) the share of affected land is often greater than 10%. In terms of territorial-administrative division of Slovakia, the Žilina and Prešov regions are most affected. This is reflected in the high number of endangered objects, plenty of emergency situations due to activation of slope movements and considerable financial resources expended for corrective measures.

The Atlas has provided a wide range of uses, however, with certain limitations. The most important issues are too small a scale of 1: 50,000 and related incompatibility of the used topography ZM50 with topography of larger scales, causing relatively large distortions in the transformation of slope deformation boundaries from the Atlas into more detailed map documents.

Despite undeniable progress in the methods of research of slope deformations and a wide spectrum of corrective measures, we still face activation or creation of slope movements, especially in connection with extreme precipitation events. An example from the last period was the weather anomaly of May and June 2010—"the year of landslides" in Slovakia or the Vrátna catastrophic event of July 2014. Since then, we have experienced a handful of slope deformations each year, asking for immediate interventions. It should be emphasized that the development of knowledge about slope deformations in recent decades has had a positive effect in such extreme situations—new landslides have been recorded in existing and functional databases and they are shown in cartographic documents on the susceptibility of specific areas to landslides.

References

Atlas krajiny Slovenskej republiky [Atlas of Landscape of the Slovak Republic] (2002) 1st edition, Bratislava: Ministry of Environment SR; Banská Bystrica: Slovak Environmental Agency, 344 p

Baliak F, Wagner P, Liščák P (2014) History of systematic research of slope failures in Slovakia. Slovak Geol Magaz 14(1):5–18

Bednarik M, Liščák P (2010) Landslide susceptibility assessment in Slovakia. Mineralia Slovaca 42(2):193–204

Bednarik M, Pauditš P (2010) Different ways of landslide geometry interpretation in a process of statistical landslide susceptibility and hazard assessment: Horná Súča (western Slovakia) case study. Environ Earth Sci 61(4):733–739. https://doi.org/10.1007/s12665-009-0387-8

Bednarik M, Magulová B, Matys M, Marschalko M (2010) Landslide susceptibility assessment of the Kraľovany-Liptovský Mikuláš railway case study. Phys Chem Earth, Pt A/B/C 35(3–5):162–171. https://doi.org/10.1016/j.pce.2009.12.002

Buša J, Tornyai R, Bednarik M, Greif V, Rusnák M (2019) Hodnotenie zosuvného hazardu pomocou multivariačnej a bivariačnej štatistickej analýzy v Košickej kotline (Západné Karpaty). Geografický Časopis 71(4):383–405. https://doi.org/10.31577/geogrcas.2019.71.4.20

Droppa A (1955) Výskum Demänovských Jaskýň. Geografický Časopis 7(3–4):133–163

Guzzetti F, Mondini AC, Cardinali M, Fiorucci F, Santangelo M, Chang KT (2012) Landslide inventory maps: New tools for an old problem. Earth-Sci Rev 112:42–66. https://doi.org/10.1016/j.earscirev.2012.02.001

Holec J, Bednarik M, Šabo M, Minár J, Yilmaz I, Marschalko M (2013) A small-scale landslide susceptibility assessment for the territory of Western Carpathians. Nat Hazards 9(1):1081–1107. https://doi.org/10.1007/s11069-013-0751-6

Hrašna M (1986) Mapy inžinierskogeologického rajónovania pre účely urbanizácie. In: Zborník referátov z vedeckej konferencie Progresívne smery v inžinierskogeologickom výskume. PRIF UK Bratislava, Katedra inžinierskej geológie, pp 68–82.

Kopecký M, Martinčeková T, Šimeková J, Ondrášik M (2008) Atlas zosuvov – výsledky riešenia geologickej úlohy [Atlas of landslides – results of the geological project solution]. In: Proceedings of the 6th Conference Geology and Environment. Bratislava: SGIDŠ, 2008, pp 105–110, ISBN 976-80-89343-06-5. (In Slovak).

Kováčik M, Suchánková Z (1993) Systematic inventory of slope deformations in the Slovak Republic, its use and possibilities. In: Novosad S, Wagner P (eds) Landslides. Proceedings of 7th International Conference and Field Workshop, Bratislava. A. A. Balkema. Rotterdam, Brookfield. pp 43–50

Liščák P, Bednarik M, Feranec J (2009) Landslide hazard in afforested territories of Slovakia. In: Proceedings of the 33rd International

Symposium on Remote Sensing of Environment (ISRSE), May 4–8, 2009, Stresa, Italy
Liščák P, Pauditš P, Petro Ľ, Iglárová Ľ, Ondrejka P, Dananaj I, Brček M, Baráth I, Vlačiky M, Németh Z, Záhorová Ľ, Antalík M, Repčiak M, Drotár D (2010) Registration and evaluation of newly evolved slope failures in 2010 in Prešov and Košice regions. Mineralia Slovaca 42(2):393–406
Lukniš M (1959) Reliéf a rozčlenenie kvartérnych útvarov vo Vysokých Tatrách a na ich predpolí. Geol. Zbor. SAV (bratislava) 10(1):233–268
Lukniš M (1962) Die Reliefentwicklung der Westkarpaten. Wiss Z M Luther-Univ Halle-Wittenberg 10(10):1235–1244
Lukniš M (1972) Reliéf. In: Lukniš M (ed). Slovensko 2 – Príroda, Obzor, Bratislava, pp 124–202
Matula M, Pašek J (1986) Regionálna inžinierska geológia ČSSR [*Regional Engineering Geology of Czechoslovakia*). Bratislava: Alfa. 295 p.
Matula M (1969) Regional engineering geology of Czechoslovak Carpathians. Bratislava, Editorial office SAS. 211 p.
Mazúr E, Lukniš M (1978) Regionálne Geomorfologické Členenie SSR. Geografický Časopis 30(2):101–125
Migoń P, Pánek T, Malik I, Hrádecký J, Owczarek P, Šilhán K (2010) Complex landslide terrain in the Kamienne Mountains, Middle Sudetes SW Poland. Geomorphology 124(3–4):200–214. https://doi.org/10.1016/j.geomorph.2010.09.024
Nemčok A, Pašek J, Rybář J (1974) Dělení svahových pohybů [*Division of slope movements*]. Sborník Geol. Věd. ř. HIG 11:77–93 (In Czech)
Nemčok A (1982) Zosuvy v Slovenských Karpatoch [*Landslides in the Slovak Carpathians*]. Bratislava: Slovak Academy of Sciences. 320 p. (In Slovak)
Schulz WH (2007) Landslide susceptibility revealed by LiDAR imagery and historical records, Seattle, Washington. Eng Geol 89:67–87. https://doi.org/10.1016/j.enggeo.2006.09.019
Šimeková J, Liščák P, Jánová V, Martinčeková T (2014) Atlas of slope stability maps of the Slovak Republic at Scale 1:50,000 – Its results and use in practice. Slovak Geol Mag 14(1):19–30
Šimeková J, Martinčeková T, Abrahám P, Gejdoš T, Grenčíková A, Grman D, Hrašna M, Jadroň D, Záthurecký A, Kotrčová E, Liščák P, Malgot J, Masný M, Mokrá M, Petro Ľ, Polaščinová E, Solčiansky R, Kopecký M, Žabková E, Wanieková D, Baliak F, Caudt Ľ, Rusnák M, Sluka V (2006) Atlas máp stability svahov SR v M 1 : 50 000, orientačný IGP [Atlas of Slope Stability Maps SR at Scale 1: 50,000]. MoE SR, INGEO-ighp, Department of Geotechnics FEC SUT, SGIDŠ, GEOKONZULT, FNS CU. (In Slovak)
Tornyai R, Bednarik M, Havlín A (2016) Application of neural network to assess landslide hazard and comparison with bivariate and multivariate statistical analyses. Acta Geologica Slovaca 8:109–118
Van Den Eeckhaut M, Poesen J, Verstraeten G, Vanacker V, Nyssen J, Moeyersons J, Vandekerckhove L (2007) Use of LiDAR-derived images for mapping old landslides under forest. Earth Surf Proc Land 32:754–769. https://doi.org/10.1002/esp.1417

Pavel Liščák is the Head of the Department of Engineering Geology at the State Geological Institute of Dionýz Štúr. He has conducted several research and application projects concerning slope deformation area mapping, recording and rehabilitation as well as monitoring of environmental geological components. He is also working as a lecturer at the Department of Engineering Geology, Faculty of Natural Sciences of the Comenius University in Bratislava.

Juraj Holec is a researcher at Slovak Hydrometeorological Institute, Department of Climatological Service. He has worked in fields of fluvial geomorphology, research of Quaternary, Natural Hazards, Spatial Climatology, Geographical Information Systems and Geostatistics.

Peter Pauditš works at the Department of Engineering Geology at the State Geological Institute of Dionýz Štúr. He is an expert in geological hazards forecasting by relevant database creating and data processing in GIS environments. The compilation of graphical outputs in the form of maps and 3D projections are also his activity at the department.

Part III

Geoheritage and Landscape Protection

21 Geoheritage, Historical and Cultural Landscape and Its Protection in Slovakia

Ľubomír Štrba, Ján Lacika, Mikuláš Huba, Pavel Liščák, and Mário Molokáč

Abstract

Geological history of the territory of the Slovak Republic resulted in the evolution of many features and sites of national and international importance that contribute to the geodiversity of the country. Within the national framework, these locations can be considered as geoheritage of Slovakia. As many of such places have been used by humans for hundreds of years, human activity significantly affected or contributed to changes in recent georelief. Recognition of importance and uniqueness of such places, both geoheritage sites and historical and cultural sites related to landscape and landforms in Slovakia, has led to their protection which has a relatively long tradition from the thirteenth century. However, until the present day, there is no specific protection of this heritage and these sites or locations are protected within a wider framework—Act No. 543/2002 on the Protection of Nature and Landscape enabling to establish protected areas, protect caves, waterfalls, minerals, morphological features, etc. Adopting legal, conceptual and other documents, directly and indirectly, helps conservation and sustainable development of this heritage. Additionally, various civic organizations and geopark management teams contribute to the protection and preservation of geoheritage and related cultural-historical heritage in Slovakia. Thanks to such activities, protection of geological heritage and the cultural and historical landscape are more effective so they can be preserved and appreciated not only by recent but future generations too.

Keywords

Geoheritage • Geoconservation • Cultural and historical landscape • Protection • Slovakia

Ľ. Štrba (✉) · M. Molokáč
Faculty of Mining, Ecology, Process Control and Geotechnologies, Institute of Earth Resources, Technical University of Kosice, Košice, Slovakia
e-mail: lubormir.strba@tuke.sk

M. Molokáč
e-mail: mario.molokac@tuke.sk

J. Lacika
Department of Geography, Geoinformatics and Regional Development, Faculty of Natural Sciences, Constantine the Philosopher University in Nitra, Nitra, Slovakia
e-mail: jan.lacika@gmail.com

M. Huba
Department of Human and Regional Geography, Institute of Geography, Slovak Academy of Sciences, Bratislava, Slovakia

Institute of Management, Slovak University of Technology, Bratislava, Slovakia

P. Liščák
State Geological Institute of Dionýz Štúr, Bratislava, Slovakia
e-mail: pavel.liscak@geology.sk

21.1 Introduction

Geological heritage (geoheritage) has been, for various purposes, used by humans for thousands of years. People used caves as shelters, exploited mineral resources, built settlements near mineral deposits, on fertile plains, or built many castles on hilltops. Various sites representing geological heritage (geosites) have become places of special interest to people due to their natural features (e.g. valleys, cliffs, caves, river meanders, etc.), location of historical heritage at a specific location (e.g. a castle of the top of the hill) or overall attractiveness. The territory of the Slovak Republic includes many of such sites that have been and still are places of increased interest to many people, companies, civic organizations, etc., for various purposes ranging from mineral exploitation to tourism development. In this regard, human-induced and ongoing historical changes of georelief in the Slovak landscape are referred to as historical landscape structures (HLS). Recognition of the importance of these sites has led to their protection to preserve this heritage.

M. Lehotský and M. Boltižiar (eds.), *Landscapes and Landforms of Slovakia*, World Geomorphological Landscapes,
https://doi.org/10.1007/978-3-030-89293-7_21

This chapter presents an overview of geoheritage, focusing on landscape and landforms features and the cultural-historical landscape in Slovakia and their protection.

21.2 Geodiversity, Geoheritage and Geosites

The concept of geodiversity has been introduced as "the diversity of Earth features and systems" (Sharples 1993) in 1993 and lately modified and extended by various authors (e.g. Dixon 1996; Eberhard 1997; Gray 2004, 2013). Geodiversity is defined as "the natural range (diversity) of geological (rocks, minerals, fossils), geomorphological (landforms, topography, physical processes), soil and hydrological features. It includes their assemblages, structures, systems and contributions to landscapes" (Gray 2013). Various aspects of geodiversity have been discussed in several publications by Gray (2004, 2008, 2013, 2018). Additionally, Panizza (2009) introduced the term "geomorphodiversity", referring to the diversity of landforms and processes.

Geoheritage (geological heritage) represents various geological forms, formations, types and features that are significant from a scientific, educational, cultural, economic, or aesthetic perspective. Its definition is based on the definition of natural heritage (UNESCO 1972). As mentioned by Kubalíková (2016), geoheritage comprises examples of geodiversity that are seen as significant according to particular subjective values. However, Brilha (2016) discussed geoheritage as a part of geodiversity justified by scientific value only and consisting of geosites (in situ elements) and geoheritage elements (ex-situ elements). Hose (2016) introduced a slightly modified ProGEO's definition of geoheritage (ProGEO 2012) saying that "geoheritage is an applied scientific discipline which focuses on unique, special and representative geosites and geomorphosites, supporting both modern sciences of geology and ageomorphology and their past development, and their place in modern culture".

The term "geosite" started to appear in publications of various authors (e.g. Cowie 1992, Cowie and Wimbledon 1994) in the first half of the 1990s. Recently, when the idea and concept of geosites are relatively well established and receive attention of a wide range of professionals, these sites have become a subject of research and study interest of many authors (e.g. Wimbledon 1996; Wimbledon et al. 2000; Joyce 2008; Ilieş and Josan 2009; Martínez-Torez et al. 2011; Wimbledon and Smith-Meyer 2012; Gavrilă and Anghel 2013; Brilha 2016; Kubalíková and Kirchner 2016; Migoń and Pijet-Migoń 2017; Cortaza et al. 2019; Nazarauddin 2019; Prosser et al. 2019). Some of these authors have presented their own definitions of the term "geosite", mostly referring to its uniqueness, representativeness, importance and values. However, one of the most understandable and the most appropriate definitions of geosite has been introduced by Reynard (2004) who said that geosites are 'portions of the geosphere that present particular importance for the comprehension of Earth history. More precisely, geosites are defined as geological or geomorphological objects that have acquired a scientific (e.g. sedimentological stratotype, relict moraine representative of a glacier extension), cultural/historical (e.g. religious or mystical value), aesthetic (e.g. some mountainous or coastal landscapes) and/or social/economic (e.g. aesthetic landscapes as tourist destinations) value due to human perception or exploitation'.

21.3 Geoheritage Values

When considering geoheritage as a substantial part of geodiversity, it is obvious that some geoheritage elements are of higher importance than others. As mentioned by Gray (2018), geodiversity is a backbone of geoheritage and geoconservation. Recognizing the role, significance and wider context of various types of geoheritage elements (most commonly referred to as geosites), many authors (Wimbledon et al. 2000; Pereira et al. 2007; Reynard et al. 2007; Zouros 2007; Kubalíková 2009, 2013; Rybár 2010; Baca and Schuster 2011; Bruschi et al. 2011; Poirier and Daigneault 2011; Fassoulas et al. 2012; Kazanci 2012; Tomić and Božić 2014; Štrba and Rybár 2015; Brilha 2016; Kubalíková and Kirchner 2016) published methods to evaluate geoheritage sites for specific purposes, especially in the context of geoconservation, management and tourism activities development. There are two major groups of assessment methods: (1) qualitative assessment method (descriptive), and (2) quantitative assessment methods (numerical). The second group includes a significantly larger number of assessment methodologies. However, no quantitative assessment method is generally accepted yet. According to Brilha (2018), a quantitative assessment is an important step to support the geoconservation strategy. All assessment methods primarily focus on the following values: scientific, educational, economic, aesthetic and added.

In Slovakia, two authors have primarily focused their research interest on geoheritage and its assessment. Rybár (2010) published the first assessment method in this territory. His approach was based on two major assessment categories for each object. However, after years of application of this assessment method, it has been shown that it has several limitations. Therefore, the revised version of the assessment procedure has been introduced by Štrba and Rybár (2015). Additionally, discussing geoheritage assessment in general or focusing on some specific sites or areas, Štrba (2015, 2019), Štrba and Lukáč (2016), Štrba et al. (2015, 2018, 2020) have published papers on this topic.

However, other authors also dealt with this issue. In-depth characteristics of geological heritage located in Slovakia has been presented by Liščák and Antalík (2018). Hronček et al. (2017) presented geoheritage located in the Kysuce region and Jeleň et al (2009) published a guide on geological and geographical sites located in the central part of Slovakia.

21.4 Geoheritage Protection—Geoconservation

It is undeniable fact that land, its shape, type, form and geological structure are the main factors on which all aspects of the natural environment and life in it are based. These factors affect such elements as topography, biodiversity, water distribution, or soil type. However, the initiative to systematically protect geological heritage dates back to the early 1990s only. In June 1991, under the patronage of UNESCO, the 1st International Symposium on the Conservation of our Geological Heritage was held. During this event, the International Declaration of the Rights of the Memory of the Earth has been adopted. This can be considered as the decisive step in the protection of geological heritage. Nevertheless, the first known geoconservation initiative was the protection of a cave in Germany in 1668 (Grube 1994). One of the first documented attempts to preserve a geomorphological site was in the United Kingdom in 1860, when the Cheesewring tor site in Bodmin Moor was protected against the negative effects of mining in the adjacent quarry (Hose 2012).

Geoconservation, as a part of nature conservation (Gray 2013), focuses on protection and preservation of geological, geomorphological and soil elements and processes, as well as on maintaining natural changes of these elements (Sharples 2002). In the last decade, significant progress in geoconservation has been made, especially through the UNESCO Global Geoparks network, protection of geoheritage under the World Heritage Convention, IUCN initiatives (Gordon et al. 2018) and ProGEO initiatives. Additionally, a relatively new concept of "geoethics" (Peppoloni and Di Capua 2017) can be considered as an effective support tool for geoconservation initiatives.

21.5 Geoheritage of Slovakia—Geomorphological Localities

Considering geoheritage value, many internationally significant geosites are located in Slovakia, e.g. Herľany Geyser—a cold water geyser of artificial origin (Fig. 21.1.); Dubnik Opal Mines—a precious opal deposit (Fig. 21.2), where the biggest precious opal mineral in the world, Harlequin, was found and probably the oldest precious opal mine ('*Libanka*') is located in this territory; or the Krásnohorská Cave with one of the highest columns (34 m) in the world—Dripstone of Rožňava Cavers (Fig. 21.3).

Until the last two decades, no complex or unified attention has been paid to the geoheritage of Slovakia. In 2008, the Ministry of the Environment of the Slovak Republic commissioned the State Geological Institute of Dionýz Štúr to execute a geological project 07-08, termed "*Information System of Important Geological Sites SR*" to create an open-access database of geological heritage sites located in Slovakia.

Based on the specific geoheritage features as defined by Brilha (2016), 480 identified geoheritage sites are located in Slovakia (Liščák et al. 2012; VGL 2019). According to Liščák et al. (2012) and based on the primary character of the geoheritage site or location, following geoheritage categories or topics can be distinguished in Slovakia: archeology (7 sites), basic geology (189 sites), geomorphology (44 sites), hydrogeology (37 sites), mineralogy (63 sites), paleontology (20 sites), tectonics (5 sites) and volcanites (88 sites). However, it is obvious that many sites include diverse geological or geomorphological features. From the landscape and landforms perspective, probably the most interesting and important sites are grouped within the category of geomorphology. According to this categorization, based on the Information System of Important Geological Sites of the Slovak Republic (Liščák et al. 2012) and Liščák and Antalík (2018), the following text brings a brief overview of 44 important geomorphological sites located in Slovakia that are included in the online system. These geomorphological sites include landforms of various origin, such as karst, fluvial, glacial, volcanic and eolic.

21.5.1 Karst Landforms

Belianska jaskyňa (cave): The cave is formed in the Mesozoic (Middle Triassic) dark grey Gutenstein limestones of the Krížna Nappe. The origin of the cave was conditioned by stratification, with additional role of tectonic faults, along which the waters penetrated and flowed into deeper parts of the underground. The primary caverns were entirely filled with water which, except corrosion, worked by slow pressure flow. Large cupolas and smaller kettle hollows were modeled on the ceilings. The cave length is 3,640 m, with an elevation range of 160 m. The air temperature reaches 5.0–6.3 °C, relative humidity is 90–97%.

Belianska dolina (valley): Belianska dolina Valley is one of the most famous in the Veľká Fatra Mts. It starts at the village of Belá-Dulice and its headwater section is below Mt. Borišov (distance 15.6 km along a green tourist trail). As one of the most enjoyed touristic attractions of the Veľká Fatra mountains, it represents a complete profile of the

Fig. 21.1 Herlany Geyser (*photo*Ľ Štrba)

Fig. 21.2 Precious opal from the Dubnik Opal Mines (*photo*Ľ Štrba)

Šiprúň Sequence of the Veľká Fatra Tatricum. The lithostratigraphic profile covers a period from Ladinian till Albian-Cenomanian (Ramsau Dolomite, Došná Beds, Carpathian Keuper, crinoidal biosparites, crinoidal limestones, Allgäu Beds, radiolarian limestones, Calpionella limestone, black cherty limestones, Poruba Beds. Presence of Lower Triassic carbonate complexes and flowing water resulted into karstification of the whole area. Several individual karst zones with different level of karstification and richness of sinter decoration are located in the valley. The most important are Suchá massif (1112 m a.s.l.) and Žiarna dolinka valley.

Bešeňovské travertíny (quarry): Bešeňová travertines represent the abandoned quarry, located a few hundred meters north of the village, but there are also several travertine cascades in the vicinity, covering the area of 3,18 hectares. Down the cascade run perennial streams of mineral water, lending the stone cascades, loaves and terraces a wonderful living shine.

Bojnice (travertine heaps): Bojnice travertines are among the largest in Slovakia. The travertines are exposed in terraces in the village of Bojnice, over the alluvium of the Nitra River. Their area of occurrence stretches from the Bojnice Castle up to the edge of the granite mountains of Malá Magura, in the village of Dubnica. Continuity of this massive stripe is interrupted only once by a fluvial incision near Dubnica (Ivan 1943).

Brázda (abyss): It is located on the Silická Plateau, 2.5 km to the northwest from the village of Silica. The depth is 181 m. This corrosion fissure abyss, with several vertical steps or horizons, is formed in Mesozoic (Middle Triassic) pale Wetterstein limestones of the Silica Nappe. Several cave walls are decorated by remarkable pea-like protrusions. A branch cascade passage 150 m long opens to the lower part of the abyss, which leads upwards into a dome with magnificent flowstone decoration.

Demänovská jaskyňa (cave): The cave represents morphologically the most varied part of the Demänovský Cave System. It has formed in the Middle Triassic dark-grey Gutenstein limestones of the Krížna Nappe along the tectonic faults by the ancient ponor flow of Demänovka and its tributaries. The length of the cave is 8,126 m, whereas the vertical distance is 120 m. Among the flowstone fills, the water lilies and other lacustrine forms (sponge, coral, grapes forms), as well as eccentric stalactites, are unique. Mighty flowstone waterfalls and columns, sphaerolitic stalactites and many other forms of stalactites and stalagmites are captivating. A thick layer of white soft sinter—moonmilk can be found in the Great Dome. There are several lakes in the cave, with the biggest one, The Great Lake, which is 52 m long, 5–12 m wide and more than 7 m deep. Air temperature is from 6.1 to 7.0 °C, relative humidity is 94–99%. The bones of the cave bear (*Ursus spelaeus*) were found in the Bear's Passage (Medvedia chodba). The finding of the palpigrade

Fig. 21.3 Dripstone of Rožňava Cavers located in the Krásnohorská Cave (Slovak Karst) (*photo*Ľ Štrba)

Eukoenenia spelaea indicates one of the most northern occurrences of this arachnid in the world, which ranks the cave among the biospeleological localities of European importance.

Dobšinská ľadová jaskyňa (cave): Dobšinská ice cave belongs to the most important ice caves in the world. Since 2000 it is inscribed on the World Heritage list. As compared with the well-known high-mountain ice caves Eisriesenwelt and Dachstein-Rieseneishöhle in the Austrian Alps or with the Romanian Scărişoara Cave in Bihor Mts., which has the glaciated part at elevations of 1,100–1,120 m a.s.l., the monumental glaciation of the Dobšinská Ice Cave has persisted for thousands of years at elevations of only 920–950 m a.s.l.

Dolina Hlboče (rock outcrop): The locality is unique by the depth of pre-Albian (Cretaceous) erosion and by preserved paleokarst phenomena (Aubrecht et al. 2006). It is a typical semi-blind karst valley with the only small waterfall in the mountains (currently mostly dry). In an overturned position, the contact of Upper Bajocian crinoidal limestone (Smolegowa Lst Fm.) and Albian sediments is visible. Along with crinoids and brachiopods, the limestones also contain numerous bivalves. Except the crinoidal limestones, some relics of younger organodetritic limestones with "filamentous" microfacies (Bathonian–Callovian) were found in the basement rocks, mainly as filling of fissures in the crinoidal limestones. The rest of Jurassic sediments is missing. The Albian rocks, as a rule, represent both the organodetritic limestones with crinoids, belemnites, bivalves and siliciclastic admixture (up to the size of small pebbles) and somewhat younger, more pelagic red marlstones to marly limestones containing only planktonic fauna. At the base of the Albian deposits, greenish-yellow, brown to black P-Fe-Mn stromatolites and oncoids occur. Locally, white marly limestones fill the karstic depressions, containing Middle Turonian microfauna. This is further evidence of continuous episodes of non-deposition and/or erosion still after flooding.

Domica (cave): Domica Cave is located on the south-western edge of the Silická Plateau in the Slovak Karst National Park, close to the state border with Hungary. It has been formed in the Middle Triassic pale Wetterstein limestones of the Silica Nappe along the tectonic faults by corrosive and erosive activities of Styx and Domický Brook and smaller underground tributaries draining waters mainly from the non-karstic part of the basin. Horizontal oval passages with ceiling troughs dominate in the cave and are widened into domes and halls in places. The passage of Styx gains a character of an underground canyon with meanders. Three development levels are situated in a relative distance of 8–12 m. The lowest level is filled with gravel and loam. The Domica Cave is connected with the Čertova diera Cave—together they reach the length of 5,358 m. They also form one genetic unit with the Baradla Cave in Hungary, with a total length of about 25 km, from which almost one quarter is in the Slovak territory. The cave is rich in sinter fills, among which the most typical are shields and drums, cascade pools (Roman Spa—Rímske kúpele, Plitvice Lakes—Plitvické jazierka), onion-like stalactites and pagoda-like stalagmites. Also, other sinter forms occur in the cave. Air temperature ranges from 10.2 to 11.4 °C and relative humidity is from 95 to 98%. The cave system is an important wintering place for bats.

Dreveník (cliff): Dreveník hill (Fig. 21.4), formed by precipitation at mineral water springs and composed of multiple travertine mounds, represents the largest travertine territory in Slovakia, with numerous caves and cavities, a rocky city, travertine mounds and tectonic faults. Its current form is one of a table mountain, with striking displaced blocks and often perpendicular edges of rock walls.

Drieňovská jaskyňa (cave): A spring river cave has formed in the Mesozoic (Middle Triassic) pale Wetterstein limestones of the Silica Nappe, locally also in breccia or limestone conglomerates. The underground stream flows through the lower parts of the cave forming lakes, cascades and smaller waterfalls in some places. The cave contains four levels, testifying to successive stage of speleogenesis. The length is 1,348 m and vertical span is 85 m. The upper parts are remarkable, with the occurrence of crystalline gypsum crusts and druses of calcite, with crystals as much as 9–12 cm big, sporadically also 15 cm. Among the most interesting flowstone fills are pisolite, coral and dentritic forms, originally made of transparent aragonite and in places of recrystallized calcite. Archeological discoveries found in the cave point to Neolithic (Bukk-Mountain Culture) and Bronze Age (Piliny Culture) human occupation.

Gombasecká jaskyňa (cave): The cave was formed in the Middle Triassic pale Wetterstein limestones, dark-grey Guttenstein limestones, pale limestones and dolomitic limestones by corrosive and erosive activities of the Čierny Potok (Black Brook) along tectonic faults. A role is also played by its intermittent tributary from the well in the Marble Hall. The cave represents a fluviokarst outlet cave and is 1,525 m long. The cave is unique by thin straw stalactites, which can reach as much as 3 m long. Also, other forms of stalactites, stalagmites, sinter curtains, coatings and crusts can be found here. Air temperature in the cave ranges between 9.0 and 9.4 °C, and the relative humidity is between 95 and 97%. Remarkable is the finding of a blind multipede of *Typhloiulus sp.* genus with 26 mm body length, which represents the biggest troglobite—a genuine cave animal ever found in the Slovakian caves.

Hrušovská jaskyňa (cave): Situated in the eastern foothills of the Silická Plateau close to the village of Hrušov, the cave is formed by an underground stream in the Mesozoic (Middle Triassic) pale Wetterstein limestones of the Silica Nappe in three developmental levels. The cave consists of mainly fluvially-modeled passages, which are in places widened into smaller halls and bigger collapse domes. The cave is 780 m long. The flowstone fill is rich in eccentric large-crystal formations of calcite, as well as it includes a big flowstone shield growing from the floor. Stalactites, stalagmites, flowstone curtains and other layers are also present. Cascades and waterfalls occur along the underground watercourse.

Jaskyňa Skalistého potoka (cave): The cave is located in the southern foothills of the Jasovská Plateau, reaching the length of 5,855 m and the vertical range of 317 m. It is formed in the Mesozoic (Middle Triassic) pale Wetterstein limestones of the Silica Nappe. The cave consists of the

Fig. 21.4 Dreveník—disintegrated rock cliffs of travertine, separated from one another due to mass movements along the edges of the tabular hill

main horizontal lower passage with many water siphons and lakes. Some terminal parts of the cave reach almost under the border of the Háj village. A distinct side branch of the cave was formed by an ascending river that modeled a passage leading closely towards the plateau surface. This cave is the longest and deepest cave of the Jasovská Plateau and whole Slovak Karst.

Jaskyňa Snežná diera (cave): The cave is situated on the Horný vrch Plateau, to the west from the upper part of the Zádiel Gorge. This typical, gravitationally widened crevasse cave reaches the length of 100 m and depth of 25 m. The roof is composed of collapsed boulders. It is formed in the Mesozoic (Middle Triassic) pale Wetterstein limestones of the Silica Nappe. The cave has pea flowstone decoration in some places. There are smaller dripstones and layers of ice in the western part of the cave. The ice partially melts in summer months, yet it persists all year round.

Jazierske travertíny (travertine heap): The right-bank of the Trlenská dolina Valley and the left-bank of the Revúca Brook are covered by travertines (strongly recrystallized and karstified) of Middle to Late Pleistocene (Riss-Würm) age. The younger, Holocene to recent (current) calcareous deposits are the foam sinters and they originated from cold springs and their flows (the contents of calcium and magnesium in them vary from 0.28 to 0.77 $g.l^{-1}$). The system of foam-sinter cascades has developed till the confluence with the Revúca Creek, where the waters had created a waterfall (partly also in the quarry face). The length of the deposits is about 400 m and the width is up to 150 m. Spectacular are two small lakes enclosed by a travertine rim.

Krásnohorská jaskyňa (cave): The cave is located in the northern foothills of the Silická Plateau in the Slovak Karst National Park, on the edge of the Rožňavská Basin and to the south of the village of Krásnohorská Dlhá Lúka. It is known mainly by the mighty dripstone Kvapeľ rožňavských jaskyniarov (Dripstone of Rožňava Cavers) (Fig. 21.3), reaching the height of 32,6 m (among the largest in the world), as well as by mysterious underground riverbed. The cave is formed in the Mesozoic (Middle Triassic) grey to dark-grey Gutenstein dolomites and dolomitic limestones, with the rear part lying in the light grey Steinalm limestones of the Silica Nappe. The origin of underground spaces was preconditioned by a distinct tectonic fault along which corrosive and erosive activity of the underground stream did its work. The total length of the cave is 1,355 m. The water course from the spring to the Marikino Lake is 520 m long. The main part of the cave is a canyon passage widened on crosswise faults into mighty domes as much as 45 m high. There are flow-through lakes within the underground riverbed. A rich flowstone decoration occurs in some side branches of the cave, among which various eccentric helictites are the most interesting. Also, soil pyramids were formed here in a remarkable form. Air temperature is 9 °C, the relative humidity is 98%. The part of the cave, 420 m long, is open to the public.

Kunia priepasť (abyss): Formed in the Mesozoic (Middle Triassic) pale Wetterstein limestones of the Silica Nappe, the abyss reaches the length of 813 m and depth of 203 m. The set of cascading abysses ends with horizontal passages. Underground water course appears in the depth of 150 m under the Easter Estavella and forms several waterfalls and cascades. Waters fall 23 m in the Great Waterfall. Flowstone fills include straw stalactites, stalagmites, curtains, cascade rimstones and lacustrine water lilies.

Malá ľadnica (abyss): Malá ľadnica, a 56 m deep abyss, is located in the cadastre of the village of Silická Brezová, at the altitude 441 m a.s.l. The chasm is of crevice nature, with a roughly oval funnel-shaped entrance, facing NNW–SSE, with a longer axis of 6 m and the shorter one 2 m. A human tooth from the Malá ľadnica sedimentary fill is the first credible finding of the remains of Late Pleistocene man from the territory of Slovakia.

Mažarná (cave): The cave is located in the Middle Triassic (Mesozoic) Gader limestones of the Choč nappe and was formed in a huge cliff at the closure of Mažarná Valley (the right tributary of Vápenná dolina Valley). The total known length of the cave is 130 m. The bottom of the cave consists of mostly sharp-edged debris of cryogenic nature (effect of frost weathering) and poorly rounded clasts, which testify to short transport in the aquatic environment. The decoration consists of stalagnate columns, small sinter lakes and a soft sinter of white, yellowish to beige colour. In the spring months, an occasional water flow flows through the cave. The average humidity of the cave is 92% and the temperature is around 6 °C. The cave was formed approximately 15 million years ago, and currently is in the senile stage of development, as well as its decoration. In winter months, in the portal entry an impressive winter ice-dripstone decoration is formed regularly.

Obrovská priepasť (abyss): A corrosive-collapse abyss, 100 m deep, situated on the Dolný vrch Plateau above the village of Jablonov nad Turňou, has formed in the Mesozoic (Middle Triassic) pale Wetterstein limestones of the Silica Nappe. It is the third deepest abyss in the Slovak part of the Dolný vrch Plateau.

Ohnište (rock window): Ohnište is an important mountain massif made of Gutenstein Limestone and Dolomite of Hronicum. The top part of the hill—a karst plateau—contains numerous abysses and sinkholes. Distinct features include the southern and south-western cliffs on the flank of the hill, reaching several hundreds of meters in length, separated from each other by narrow benches. The hill is split into two peaks—below the higher one is the 125 m deep Veľká ľadová priepasť (Huge Ice Abyss) and not far from the summit one can see an opening in the rock mass—Okno (Window).

Rojkovská travertínová kopa (travertine mound): The spring of Rojkov Lake is located on the southern outskirts of the village of Rojkov, in the slope setting. The spring is also known as the "Rojkov Travertine Mound". The water from the spring has created a travertine mound with a crater of circular shape and a diameter of about 10 m. The lake is artificially modified for swimming. The bottom is uneven, stepped and it reaches a depth of 2–3 m in its central part. The spring is used in summer for swimming, water temperature is around 19 °C. The water source is cold, carbonated, of Ca-Mg-HCO_3-SO_4 type, moderately mineralized (5022.7 mg. l^{-1}), with a CO_2 1340.0 mg.l^{-1}.

Silická ľadnica (cave): It is one of the most important natural phenomena of the Silická Plateau in the Slovak Karst. It is located 2 km westerly from the village of Silica. The lower, mostly horizontal passages were formed by the underground stream of Čierny potok Brook. The cave is 1,100 m long and 110 m deep. It is formed in the Mesozoic (Middle Triassic) pale Wetterstein limestones of the Silica Nappe. The waters of the Čierny Brook flow into the Gombasecká Cave. The upper abyss part of the cave is permanently filled with ice. It is the lowest-lying classical ice cave, located up to 50° of the northern latitude temperate climatic zone. The phenomenon of biological inversion can be studied here. Invertebrates are represented by several rare endemic forms of troglophilic species of beetles like *Duvalius bokori* and *Duvalius hungaricus*, which can be found only on the Silická Plateau in the Slovak Karst. The cave was settled several times before the glaciation. Archaeological findings are dated for the Neolithic, Bronze Age and La Tène culture. Possible is also a young Paleolithic settlement.

Spišský hradný vrch (travertine heap): Spiš Castle Hill is composed of travertines and foam sinters, Late Miocene to Holocene in age. The travertines overlie the claystone of the Huty Formation with very sharp morphological contact. It was declared as National Natural Monument in 1990. The protected area of geological, tourism and archaeological importance has an area of 242 064 m^2. The rocks of the travertine hill are intensively destructed by erosional processes such as karstification and slope movements (Wróblewski et al. 2010) which represent one of the most important denudation processes. The top of the hill is 634 m a.s.l. According to Vlčko (2004), the travertine under the Spiš Castle reaches some 52 m in thickness. Creeping slope movements destructing travertine bodies of the hill cause deformations in the form of block disintegration, shifts and block fields. They also affect the building of the castle. Rock falls on steep to almost vertical rock walls occur along the outer perimeter of the castle hill. These falls result from slow creeping of perimeter blocks and mechanical weathering of travertine. Surface and subsurface karst forms (karrens, caves) are also located in this area, e.g. Podhradská jaskyňa cave located in the southern part of the hill, near the castle entrance gate, under the Perúnova skala rock.

Travertíny na Lúčkach (travertine heap): In the Spa Lúčky, which is located upon the Prosiek Fault dividing the Chočské vrchy Mts. and the Liptovská kotlina Basin, several water outflows with the temperature of 17–33 °C occur, with a total yield of about 30 l.s^{-1}. During the water ascent to the surface, pressure decreases, water cools down and gas escape takes place. Therefore, precipitation of calcium carbonate occurs at the surface, with an additional influence of plants and animals. The plant debris underwent decomposition; however, their casts remained, giving the Lúčky travertine its characteristic appearance. The seasonality of this process is reflected in the travertine in the form of alternation between thick pale layers and thin, dark layers. The rate of precipitation, the colour and the shape of the travertine are affected by the surficial matter contribution, water temperature, the quantity of carbon dioxide, chemical content of the water, organisms in the water and the geological-tectonic setting of the country.

Zádielska tiesňava (canyon/gorge) (Fig. 21.5): Typical karst canyon with a length of about 4 km. At the bottom the width reaches from 20 to 100 m, the depth of the incision is up to 300 m. The canyon cuts through sandstones and dolomites of Triassic age. In the middle section of the valley there is a remarkable rock cliff—Cukrová homoľa (Sugar Loaf), enjoyed by climbers. In the upper part of the canyon, there are several exsurgences and numerous caves (Mišík 1976).

Zvonivá jama (abyss): The abyss is located in the central part of the Plešivská Plateau and is 100,5 m deep. It is formed in the Mesozoic (Middle Triassic) pale Wetterstein limestones of the Silica Nappe. The entrance vertical shaft ends in the depth of 45 m, where a great underground dome with a mighty debris cone on the bottom occurs. Its top is 84 m down from the surface opening. Along the sides of the dome, there are inclined ascending passages 180 m long, which originated in connection with the underground draining of adjacent large dolines. The entrance shaft was formed by corrosion of atmospheric waters percolating along the crossing tectonic faults. There are pagoda stalagmites more than 10 m high in the abyss. The highest one is 26 m.

21.5.2 Fluvial Landforms

Kostelecká tiesňava (canyon/gorge): Kostelec Gorge is wider and more open than the Manín Gorge, but of the same origin. The Manín Brook has cut through the Drienovka Cliff creating two peaks: Kavčia skala (Jackdaw's Rock) and Drienovka. One of the biggest overhangs of the Western Carpathians can be found in the gorge. It is called "*The Kostolec Cathedral* " or also "The Roof of Slovakia ". Large scree cones occur beneath, whereas horizontal folds can be seen on the rock face of Drienovka.

Kvačianska dolina (canyon): The National Nature Reserve Kvačianska dolina Valley is a natural narrow passage of fluviokarstic origin formed within Mesozoic rocks and cliffs. It is about 3.2 km long, till the Oblazy area. The Kvačany Canyon belongs to the most beautiful and very popular canyons in the Chočské Vrchy Mts., together with Prosiek Canyon nearby. Slopes in the Kvačany Canyon are

Fig. 21.5 Zádielska tiesňava gorge in the Slovak Karst area

covered with exposed rocks and rock formations as well as with forests. The forest has never been cut thanks to its inaccessibility. Consequently, nature here is well preserved and unique.

Meandre Hrona (meanders): The protected territory with an area of approx. 1 km^2 is situated in the upper reach of the river Hron, SSW of the village of Švermovo. The site is a classic example of a meandering river. The meanders have low, perpetually modeled banks and many of them are today turned to oxbows without water, overgrown with vegetation. When viewed from the above, from the surrounding slopes, the remnants of abandoned river channels are very clear. Part of the territory is crossed by a creek (left tributary), supplied from a voluminous exsurgence.

Meandre Kamenistého potoka (meanders): Kamenistý potok Creek flows down the picturesque, 25 km long Kamenistá dolina Valley. Its upper reach is protected in the form of "Protected Areal"—meandre Kamenistého potoka (meanders of the Kamenistý potok Creek) over a length of 2.5 km.

Šútovská epigenéza (epigenetic gorge): A unique morphological form of the two-fold epigenetic incision of the River Váh and Šútovský potok Creek originated by the end of the Neogene. The rock substrate is made of the Wetterstein Dolomite. The epigenesis is the process in which the river, after cutting through less resistant rocks, gradually incises into more resistant bedrock and creates a break-through section of the valley while maintaining the original direction of the flow (Jeleň et al. 2009).

Vojtovský prameň (spring): Prameň Vojty (Vojty spring) —is a mofette (Italian from Latin mephitis, a pestilential exhalation), with methane and hydrogen sulphide discharge. This is quite a rare phenomenon in the Western Carpathians—generally, the term mofette is related to volcanic activity. In the case of Vojty Spring, the mother rock environment for methane discharge is the Beloveža Formation of Palaeocene–Eocene age. Discharge of the spring is 0,05 l.s^{-1}, mineralization is 1640,54 mg.l^{-1} and temperature is 9 °C.

21.5.3 Glacial Landforms

Mengusovská dolina (glacial valley): This large glacial valley has a triangular shape, with the base formed by the main ridge of the Tatra Mts. and the peak pointing south, towards the foothills. Bedrock is composed of granitoids. The valley terminates by several glacial cirques. On the eastern side there is the Zlomiská Valley with the smaller valleys Dračia and Rumanova, not accessible for the tourists. Westward the cirque of the Žabie plesá Tarns and the Hincova kotlina cirque are present, divided by the arête of the Mengusovský Volovec Hill.

Mlynická dolina (glacial valley): Climbing the valley above the upper forest boundary, one passes several monadnock-like forms formed by the glacier. Immediately behind them, a massive erratic block is worth noting (named as the fireplace). From both opposite ridges, the debris cones reach this place. The valley is crossed by a distinct mylonite zone, with frequent occurrences of hydrothermal mineralization (mainly chalcopyrite and siderite). Behind the first glacier step, across which the cascades of the Skok Waterfall overflow (height ca 25 m), the small shallow tarn Nad Skokom is situated. The next glacier rock step hides the Malé Kozie plesá and Nižné Kozie plesá tarns and behind the last rock step, situated below the walls of the Mt. Hrubý vrch and Štrbský štít Peak, the Capie pleso tarn is situated. Above it, there is one smaller rock step enclosing the small tarn of Okrúhle pleso. Downstream the Capie pleso, ice-polished surface and roche moutonné (so-called "roundstone") is visible, incised by the Mlynický potok Stream, which exploits joints in the granodiorite.

Vyhne – kamenné more (blockfield): The site represents the material of a shallow intrusive rhyolite body, which is exposed as a secondary accumulation in the form of a block field. The rhyolite is sanidine-plagioclase, with felsitic development of matrix. The rhyolite is affected by processes of silicification and adularization (replacement by or introduction of the mineral adularia). The block field represents colluvial accumulation of rhyolite blocks, which originated during the last glaciation period.

21.5.4 Volcanic Landforms

Fiľakovo, hradný vrch (cliff): The Fiľakovo Castle Hill is an example of less eroded maar rim of the Cerová basalt formation of Early Pleistocene age, with a preserved fill of phreatomagmatic pyroclastics. Volcanic layers can be studied on several outcrops accessible along the educational trail. In several places, especially in the upper part of the castle hill, volcanic bombs consisting of sandstone reaching the size of a fist can be found. Castle hill covers the area of approx. 2,35 ha and its top is the height of 200 m a.s.l.

21.5.5 Eolic Landforms

Chotínske piesky (sand dunes): The layer of aeolian (wind-blown) calcareous sands attains a height of 2–3 m. The surface area of 7 ha provides a habitat for sand-based and xerophyllic plant and animal communities on stabilized sand dunes. The main objective is the protection of rare grasshopper, which is quite abundant here.

21.5.6 Mass Movement Landforms

Chabenec (gravitational slope deformation): The marked features of the Chabenec Hill include several antislope scarps on the south-eastern slope, running parallel to the Lomnistá Valley (1300 m a.s.l.). The Chabenec deformation is one within the series of similar deformations developed between the Chopok and Prašivá Hills (their distance is 25.2 km), within the granitoid massifs forming the main ridge of the Low Tatras Mts. (Nemčok 1982). Almost 38% of the mountain ridge are more or less affected by such deformations. The series of distinct antislope scarps crosses the southern slope of the Chabenec Hill in the Nízke Tatry Mountain range.

Handlovský zosuv (landslide): Massive landslide in the town of Handlová occurred in 1960/1961 and destroyed 150 housing units, a water pipeline, a state road and an electric line. The sliding volumes equaled to 20 mil.m^3 (Nemčok 1982).

Izra (lake): Izra is a lake in the Slanské vrchy Mts. occupying an area of 3.70 ha, with a maximum length of 250 m and width 212 m. The lake was formed by a landslide, which dammed one of the tributaries of Izra Brook–Malá Izra Creek. Later on, this natural dam was anthropogenically reshaped resulting in the enlargement of the lake area and its depth.

Jezerské jazero (landslide lake): The lake developed within one of the most extensive slope deformations of the region "Jezersko - Bukovina", with an area of several km^2. It is a fossil landslide, probably of Pleistocene age, with a very rough morphology. The estimated depth to the slip surface is 10–50 m. This is a significant geomorphological territory with the occurrence of some rare species of sphagnum communities (peat), springs and endemic species of newts. The spruce stands were destroyed by the storm in 2004.

Vrátna dolina (valley): Huge debris flows on 21 July 2014 were triggered in the Vrátna dolina Valley by a rainfall anomaly and caused considerable material losses (Liová et al. 2015). Colluvial deposits are characterized by their large lithological variability, depending on the basement rock present in the source areas. Loamy-stony and clayey slope sediments are developed on marly sediments of the Mráznica Fm. Debris flows were activated in these sediments below the ridge of about 1,600–1,500 m a.s.l. Slope movements along the planar slip surface were conditioned by favourable inclination of bedding/foliation. Within the detachment, areas of these landslides of rather small thickness quite small, but numerous "plates" of sliding material (with an area of several hundred m^2) broke away and generally moved at a speed of several meters per second, preferentially down avalanche chutes, often above extremely wet vegetation cover (grass, blueberries).

21.5.7 Geomorpholohical Heritage Within Another Geoheritage Sites

Geomorphological phenomena occur also within other pre-defined categories of geosites. For example, Morské oko Lake in the eastern part of Slovakia is primarily defined as a hydrogeological site. However, it is of massive landslide origin. More examples can be found in other categories as defined within the Information System of Important Geological Sites in the Slovak Republic (Fig. 21.6). The database is accessible via the following link: http://apl.geology.sk/g_vgl/?jazyk=EN and the interactive map with geosites, available in Slovak language only, can be accessed via http://apl.geology.sk/g_vglg/.

Each geosite entry includes descriptive characteristics such as general information, photo documentation, geological characteristics, protection status, site description and references and record processing. Additionally to this database, supported by the Ministry of Education of the Slovak Republic within projects APVV-0280–07 "Paleodiversity in the area of Western Carpathians in Mesozoic and Cenozoic as a reflection of paleoclimatic, paleogeographic and paleoocanographic changes" and KEGA 3/7226/09 "Implementation of the new technology (E-learning, web testing) in the teaching of paleontology in the Universities, paleontological "adventures" for primary and secondary school students, as well as for the students of the special schools", an online database of important paleontological localities of the Slovak Republic was created. The database, in Slovak language only, is available at http://www.paleolocalities.com. It includes the total number of 52 paleontological sites. Focusing on the central part of Slovakia, Jeleň et al. (2009) published an "Educational guide round the geological geographical localities in the Central Slovakia". In this publication, the authors described 98 sites representing various geoheritage values.

21.6 Protection of Geoheritage in Slovakia

21.6.1 History of Geoheritage Protection

Historically, the protection of natural heritage, at least partially connected to geoheritage, in the territory of Slovakia dates back to the medieval ages when Slovakia was a part of Royal Hungary. As summarized by Škodová and Mazúrek (2012), in the thirteenth century, king Belo IV decreed to protect animals, forests and land near the city of Zvolen. Today, a part of this territory covers one of the oldest legally protected areas in Slovakia, the National Nature Reserve "Primeval Forest of Badín", with its rare primary fir and beech habitat almost untouched by human activity. As the

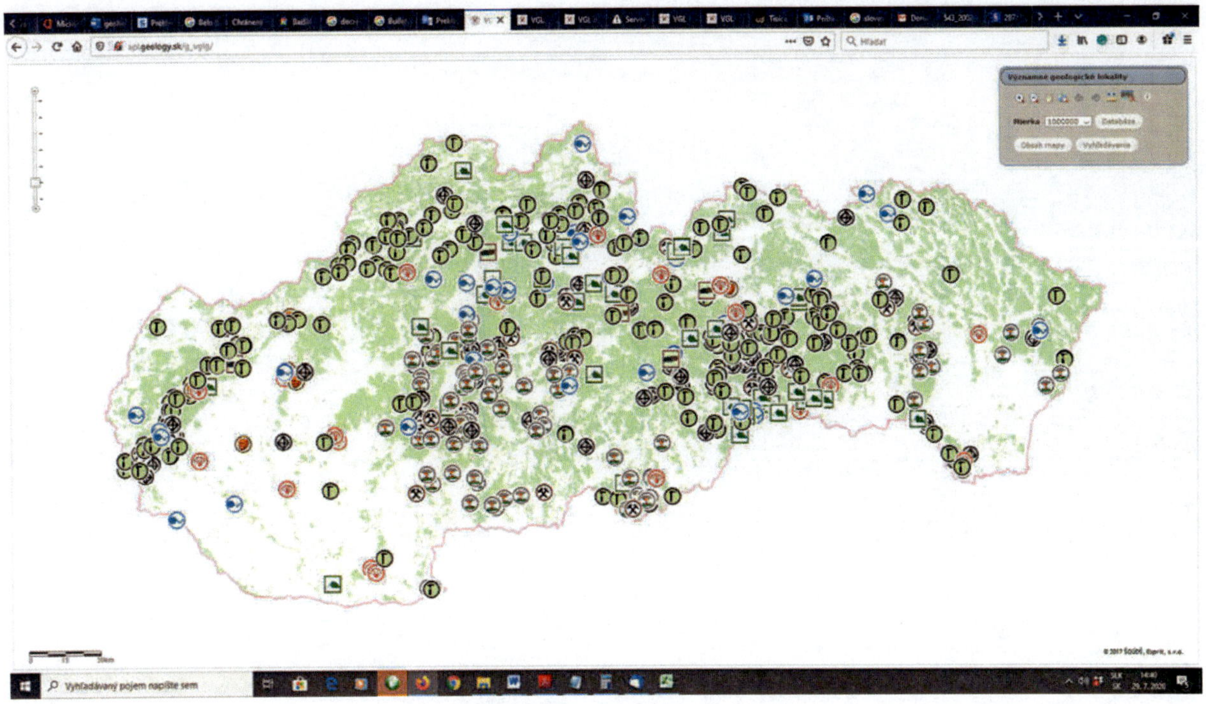

Fig. 21.6 Interactive map of the Important Geological Sites of the Slovak Republic (VGL 2019)

forest is located in the area of Neogene volcanism, besides various important fauna and flora species, this area includes several geosites, e.g. pyroclastics flow remnants. Protection of the Súľovské skaly ("Súľov Rocks") (Fig. 21.7), currently a National Nature Reserve with geo(morpho)logically important, picturesque and touristically attractive locations, predominantly composed of Neogene conglomerates, was declared in 1672 based on the order of Conte Draškovič.

Based on the decree of the king Leopold I from 1682, an island spa in Piešťany, recently the most important and the best-known spa in Slovakia, was protected. In this territory, the warmest natural groundwater spring is located. King Charles III Habsburg declared protection of the Teplička valley and Trenčianske Teplice spa in 1715. In 1769, based on the Theresian Forest Order, sustainable forest management was established. In 1900, the Decree of the Royal Hungarian Ministry of Plowing was issued on the inventory of memorial trees and other natural monuments and their protection. Subsequently, the reserves Szabóova skala (1907), Súľovské skaly (1907), Stužica (1908) and Valné pramene pri Vyhniach (1909, currently the Nature Reserve Kamenné more) were declared. In 1909, the Royal Hungarian Ministry of Plowing approved the first concept of declaring protected areas and natural monuments. In 1923, the Slovak Decree "On the Protection of the Natural Beauty of Mountain Streams" was issued. In 1932, the Slovak Nature Reserve in Pieniny Mountains was opened. In 1934, the first official lists of protected areas were created and published in the Bulletin of the Ministry of Education and National Enlightenment of the Czechoslovak Republic. The from 1945 to 1955, 60 protected areas were declared, including the first national park in Slovakia—Tatra National Park. The first legislation document focusing on nature protection was adopted by the Slovak National Council in 1955—Act of Slovak National Council nr. 1/1955 on National Protection of Nature. In 1959, the official List of State Nature Reserves was approved, with a supplement in 1962. A more comprehensive program of declaring national parks and protected landscape areas was proposed by the Administration on the Protection of the Situation in Nature and Natural Resources in Slovakia and approved in 1964. After 40 years, the original Act on nature protection was amended by Act No. 287/1994 on the Protection of Nature and Landscape. In this Act, several geoheritage elements, including rocks, minerals, relief, fossils, raw materials, landscape, lakes, springs, gorge, blockfield, downsand, are mentioned. Protection of caves, abysses, natural waterfalls, minerals and fossils was defined within individual paragraphs in this Act.

Fig. 21.7 Súľovské skaly ("Súľov Rocks") have been protected as a valuable natural locality since the end of the seventeenth century

21.6.2 Recent Geoheritage Protection

At present, there is no official document aimed at protection of geoheritage or geoconservation in Slovakia. Protection of geoheritage (or its part) is included within a wider context, into overall nature protection in Slovakia. As in many other countries worldwide, the need to protect living nature is generally accepted and relatively well-established, whereas geoheritage (non-living nature) protection is insufficient. But, as mentioned by Lukáč and Baláž (2016), Liščák and Antalík (2018), the majority of geoheritage has no legal protection in Slovakia. In this regard, the protection of Slovak geoheritage should be considered in a broader context within the nature protection, which is regulated under the Act no. 543/2002 Coll. of 25 June 2002 on Protection of Nature and Landscape and Executive Decree to the Act No. 24/2003. In this Act, geological objects are mentioned several times: in § 2 par. 2, letter b, ecosystem rock and minerals (rocks, minerals, fossils, morphological features, soil types, lakes, springs, ponors, geysers) are mentioned; under letter c, significant landscape element or a part of a territory creating a characteristic landscape or contributing to its ecological stability (peatbog, cliff, gorge, block sea, sand dune) is mentioned; § 24 of this act declares caves and waterfalls as natural monuments; and according to § 33 par. 2, minerals and fossils are included among protected species.

Besides legal protection of sites located within protected areas (Natural Parks and Protected Landscape Areas) or declared as natural reserves or monuments, some geoheritage sites with no legal protection are protected as a part of geoparks (Fig. 21.8). These areas group various geosites and cultural-historical sites, informing on geology, geomorphology and other geosciences together with history and culture of the territory, as described in Chap. 22 of this volume.

21.7 Historical and Cultural Landscape of Slovakia

Recent landscape and landform features in Slovakia are the result of long-term geological evolution, geodynamic and geomorphological processes and other factors affecting georelief. Many of such geoheritage sites, e.g. caves or cliffs, have been used by humans for hundreds of years. Human-induced, historical and ongoing changes in the

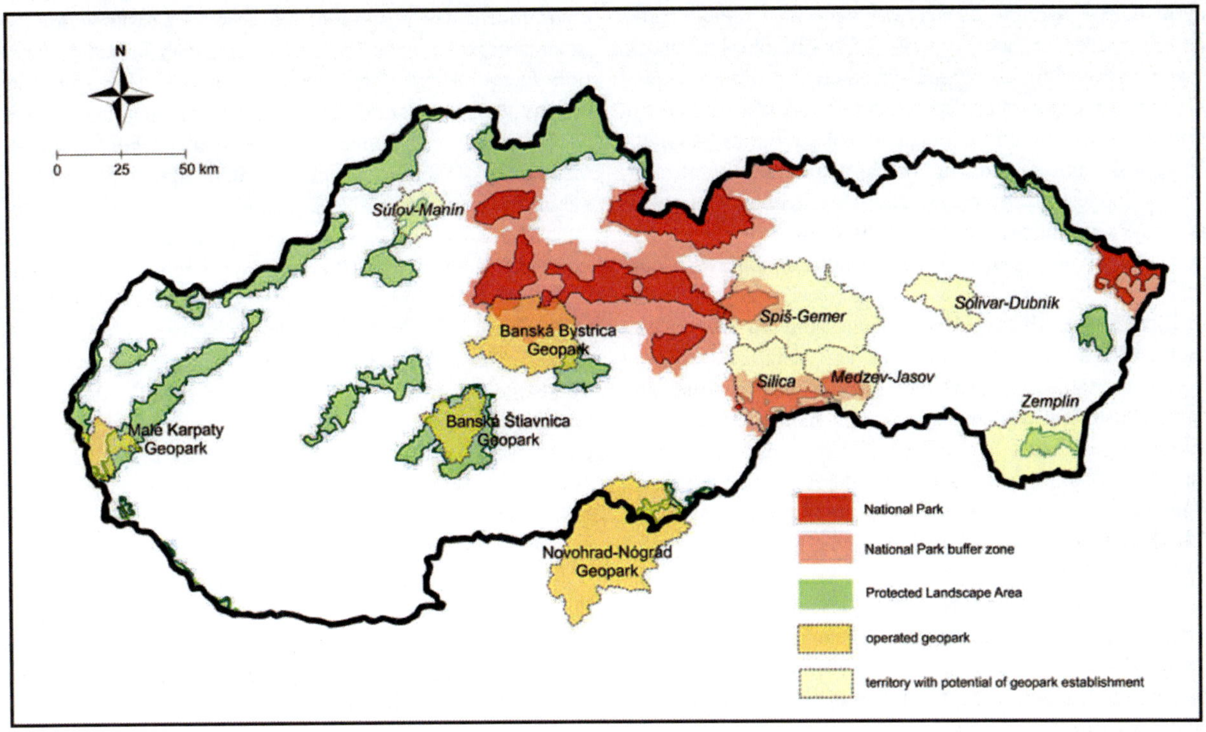

Fig. 21.8 Location of geoparks and territories with the potential of geopark establishment compared to large-scale protected areas in Slovakia

Slovak landscape are referred to as historical landscape structures (HLS).

According to Huba et al. (1988), "historical landscape structures are where the intentional human activity markedly reshaped nature in the course of history and manifestations of such historical transformation were not removed or covered by the present ways of land use (for instance by constructing/destructing interventions of technologies proper to the second half of the twentieth century (and/or beginning of the twenty-first century): industrialized agriculture, industry, constructions, transport, mining and the like. Hence, HLS represents, along with the structures of natural character and the newly emerging structure of the present landscape, a specific, time-delimited and ever-diminishing subtype of landscape structures as a whole. They can be also characterized as a material part of cultural heritage with distinctive application in space".

21.7.1 General Features of Historical Landscape Structures

Types of HLS have become a subject of several studies. A classical one, also focused on historical structures of the vineyard landscape, is the monograph "Geography of Landscape of Jur near Bratislava" (Lukniš 1977). It brings a detailed classification and characteristics of the vineyard landscape asset in the landscape georelief. The founding of strips of vineyards on steeper and higher situated slopes required the building of numerous artificial terraces, also referred to as "shelves", often supported by stone walls. Stones recollected during centuries gradually formed bulky, 3–4 m tall and wide and tens of kilometers long accumulations also called "runes". As a rule, they copy the fall line and are the most remarkable elements or lines of historical structures in the vineyard landscape. The building, protection and permanent maintenance of a network of roads and paths required considerable labor. Water and carts had hollowed deep sunken roads that often became sources of crevice waters. Four decades later, this topic (including the relationship of vineyard landscape to georelief) has been studied in the territory of Slovakia by Lacika (2016a), Hanušin et al. (2013, 2014), Štefunková and Hanušin (2019).

Among those that have invested most effort and time in the study of historical structures of agricultural landscape (HSAL) in Slovakia are (or were) the employees of the Institute of Landscape Ecology, Slovak Academy of Sciences (e.g. Špulerová et al. 2017). HSAL are characterized by the following features: preserved original small-area structures, i.e. the pattern of plots, and/or the landscape where land use was not altered by the collectivization of agriculture, and/or conserved agrarian relief forms (ARF) and/or unchanged traits of traditional husbandry. Agrarian relief forms especially define the relationship of HSAL to the georelief. Six variants of ARF were distinguished: slopes of terraces, stepped balks, mounds, ramparts, loose walls and mounds and ramparts piled into terraces.

The longest history of georelief modification by humans is connected with settlements. Many of the historical towns and villages are situated in strategically or otherwise convenient places, which were, as a rule, settled already in the prehistoric era. Historical settlements suffered the greatest losses of the architectural, cultural and historical values in the years 1948–1989. The most valuable or the best conserved historical settlements are now monument reserves, monument reserves of folk architecture and monument zones. Several Slovak towns (Bardejov, Levoča, Banská Štiavnica, Spišské Podhradie and the rural settlement of Vlkolínec) are now included into the UNESCO List of Cultural and Natural Heritage.

Slovakia as a country rich in castles and their ruins and other medieval monuments attracts the interest of authors involved in this topic (e.g. Lacika 2016b, c). One of the biggest and best-known castle complexes with its surrounding settlements and ecclesiastic monuments in the landscape of the Spiš region has been also inscribed in the UNESCO List of Cultural and Natural Heritage. It is the locality of the Spišský hrad (Spiš Castle) (Fig. 21.9) and the monuments in its environs, inscribed in 1993 and expanded in 2009 by adding the historical town of Levoča.

It can be assumed that geomorphic forms and processes were an important locating factor for the distribution of castles in the territory of present-day Slovakia.

There is a very close relationship between the distribution pattern of Slovak castles and typology of geomorphic environment. The position of the place for a castle within the mountains, basins and lowlands was a very important location factor. These three basic geomorphic environments occupy 29% (lowlands), 16% (intra-Carpathian basins) and 55% (Carpathian Mountains) of the country. One can further distinguish 4 subtypes within the mountainous part of Slovakia: inner mountains (19% of Slovakia), marginal mountains bordered to basins (23%), marginal mountains bordered to lowlands (11%) and intermountain depressions (2%). According to local geomorphic conditions, eight types of the local geomorphic position of Slovak castles can be defined. Flat terrain is characteristic for lowlands and basins. There are 73 castles built within flat relief, without any topographic limits on the building. Non-existent topographic protection is substituted by hydrological protection, provided by lowland river systems. Most of the lowland and basin castles were built in wet alluvial plains, very close to natural rivers. There are 22 valley-type former castles, e.g. Bytča castle, with a similar local geomorphic environment. These castles were built within flat relief of the valley bottoms, on alluvial plains and terraces of the main rivers and in the Carpathian basins. A group of the 36 former Slovak castles was built on hilltops of minor elevations overlooking very flat landscapes. Elevation position on the top of narrow hills improves conditions for castle defense and affords a strategic place for control of the surrounding area. Twelve castles were built on very distinct, isolated elevations jutted from flat surroundings, which offered a good strategic position. For example, they carry the Devín Castle and Beckov Castle in the west of Slovakia, as well as the Orava Castle in Central Slovakia that was built on an extremely steep and rocky calcareous Klippen hill, whereas the medieval builders of the Spiš Castle (Fig. 21.9) in East Slovakia had chosen dominant travertine hills rising from the hilly

Fig. 21.9 The Spiš Castle built on top of an isolated travertine hill

bottom of the Hornádska Kotlina Basin. The isolated volcanic hill not so far from the city of Prešov offered an ideal geomorphic setting for the Šariš Castle (Fig. 21.10). Seven castles were built within the plateau-type of relief. Usually, these historical buildings were located on the border of the upper plateau, on the sharp edge between the flat plateau surface and steep rocky lateral slopes.

Several studies were dealing with the historical landscape structures created or influenced by mining and associated activities (Hudecová 2006). One of the areas demonstrating this type of HLS in Slovakia has been included in the UNESCO List of Cultural and Natural Heritage. It is the locality of the historical town of Banská Štiavnica and the industrial/mining monuments in its environs. The size of this area is over 2000 ha.

Apart from mining, this HLS category also contains preserved historical remnants of metal production and processing, along with industrial works in the landscape that were indispensable for efficient mining and ore processing in historical times (besides extensive earth works and terrain adaptations there were also indigenious water and energy works). A unique system of water reservoirs (referred to as "tajchy") and the interconnecting canals are situated in the valleys of Mlynská dolina in Bratislava or Kvačianska dolina in the Chočské vrchy Mts.

However, georelief and the visual aspect of the landscape were also altered by technical interventions of other kind. Such rather important interventions into the landscape (including its georelief) were related to the construction of roads, railways, counter-flood systems, etc., but also constructions exploiting the energy potential of rivers and water works, which changed the character of river networks, land use and georelief in entire regions (e.g. the Podunajská nížina and Východoslovenská nížina lowlands).

Kozová et al. (2010) presented a methodology of processing the typology of the landscape in Slovakia. According to them, "the variety of the European countries constitutes a vast natural and cultural heritage. For several decades the individual countries produce maps classifying and above all typifying the landscape which also contains the mutual relationships between the land, relief, climate and also the cultural and historical development of land use. Such maps play an important role in, for example, prediction of future changes in land use. Classification of the cultural landscape in a prevailing part of Europe, known as landscape

Fig 21.10 Šariš Castle (*photo* Lacika)

typology, is much more demanding than the classification of the natural landscape".

Landscape typology is the first step to the recognition of its structure and basic properties (potential, carrying capacity, stability). A simpler way is the application of one dominant trait/criterion. A more complicated way is the delimitation of landscape units applying several criteria. Delimitation of potential natural landscape types is methodically simpler than the typology of cultural landscape. Each scientific branch attempts classification/typology of entities that are subjects of its interest; in this case, it is landscape (a complicated, dynamic and unstable object). Division or regionalization of the landscape into smaller quasi-homogeneous units can be individual or typological. The number, size, or rate of homogeneity of delimited units also depends on the chosen precision scale of the division.

Several strategic documents and conventions emphasize the importance of classification (typological) maps that should also include the assessment of landscape appearance, such as the Pan-European Biological and Landscape Diversity Strategy (PEBLDS 1995), European Landscape Convention (European Council 2000), Agenda 2000, the reform program of the European Commission concerning the Common Agricultural Policy, activity adopted by the European Conference of Ministers responsible for spatial planning (CEMAT 2005) and others.

21.7.2 Effects of Historical Activities on Changes of Landscape

Generally, we can distinguish two major groups of changes in landscape based on the direction of the change: vertical and horizontal changes.

The greatest intervention into the vertical dissection has been caused by mining activities and the associated historical waterworks, including the unique water economic system existing in the Štiavnické vrchy Mts. In many places, at the first sight, the most dominant visual features of the landscape are castles (nowadays mostly castle ruins in various stages of degradation or reconstruction or conservation), historical settlements (cores of medieval towns), fortification systems (ramparts, strongholds and the like), hill forts, chateaux, monasteries and churches in open spaces, as well as some transport constructions or industrial buildings (viaducts, factories and chimneys, water-driven facilities, etc.).

Horizontal dissection of the landscape was most influenced by farming (including viticulture and forestry). As far as the land use types are concerned, Špulerová et al. (2017) classified four basic types of such HLS in Slovakia:

- Arable land,
- Permanent grassland,
- Vineyards,
- Building objects.

They estimated, based on their measurement, that these structures spread on approximately 0.9% of the Slovak territory, corresponding to about 1.8% of the total farmland area. As far as other HLS types are concerned, their total area in Slovakia is smaller than that of HSAL, which means that the total HLS area is less than two percent of the total area of Slovakia.

According to Jančura (2004), other HLS influencing georelief include those of technical nature, associated with military activities and defense (hill forts, bulwarks, trenches, bunkers …), related to transportation (roads, railway routes, hollow ways, embankments …), associated with production and processing of raw material (historical queries, terraces of charcoal kilns, historical settlements), historical architectural dominants in an open landscape and historical architectural dominants within the settlements.

21.7.3 Protection of Historical Landscape Structures in Slovakia

In terms of human activities, a landscape with dominant historical structures suffers from diminishing role or even disappearance of traditional users and operation approaches, followed by the abandonment and decay, including inconsiderate ways of transformation or modernization; in other words, from adaptation of HLS to the ideas of the present-day land users. As a rule, the result is the disappearance of cognitive, scientific, cultural, historical, aesthetic, but also utility (or potentially usable) values, with the lost potential of various tourism forms development. However, HLS are also permanently threatened by nature itself. On the one side, there are the effects of destructive or entropic processes of physical decay (due to climatic agents, soil erosion, wind-borne greenery or gravitation), whereas one can also observe the return of natural structures to places temporarily occupied by the HLS (for instance, the secondary succession of shrub and forest associations in historically deforested areas of HLS used for farming, mining, water economic, military and other purposes) on the other one.

The history of efforts put into the protection of HLS in Slovakia is much shorter than that of nature conservation or care for monuments. As a matter of fact, no consistent protection strategy exists. It lacks an efficient legal and organizational framework (despite the existence of the Act on Nature and Landscape Protection) and it is also a conceptual problem. While the aims and tools are more or less clear and their nature is dominantly a conservational one, in case of the cultural landscape and conservation of its historical structures it is mostly not possible or even desirable. For instance, in terms of historical structure of the agricultural landscape, the largest HLS in Slovakia, intervention-free regime may mean a total disappearance of this type of landscape and its assets due to secondary succession, erosion, etc.

The landscaping approach in Slovakia was first of all applied at the moment of the promulgation of large-scale protected areas. The Protected Landscape Areas (PLA), as their name suggests, and some National Parks, concern equally the cultural landscape and nature (the best example is the National Park of Pieniny). Another legislative category of HLS covers the national cultural monuments, town monument reserves (so far 18 in the country), monument reserves of folk architecture (10) and monument zones (82). The value and need for protection of HLS also motivated the inscription of some localities into the UNESCO List of the World Cultural and Natural Heritage (7 so far).

A special category of protection, also relevant for HLS, includes the Biosphere Reserves in the framework of the UNESCO Man and Biosphere Program. There are four such reserves in Slovakia and their mission is to protect the natural and cultural variety, to support and provide for sustainable development of local population, as well as to support science, research and education.

Geoparks, nature trails, etc. also offer the possibility to know and interpret nature and landscape. The document of the European Council, that is the European Landscape Convention (European Council 2000) was meant to provide an important impetus for landscape protection. Slovakia became its contracting partner on 1 January 2006. The document compels the signatories of the convention to identify, classify and efficiently protect, manage and revitalize all types of landscape in its territory (Mišíková et al. 2010). However, in conflict with its commitments with the ELC, the Slovak Republic (the Government followed by the parliament) omitted landscape protection as one of the objectives of Nature and Landscape Protection law at the occasion of the penultimate extensive amendment of this law (2013, effective from 1 January 2014). Nevertheless, the Slovak legal order contains other legal standards concerning landscape (e.g. Huba and Kubišová 2017), but the question remains to what extent are they respected and applied.

Museums (in-door and out-door ones, and the like) can protect only a relatively small part of HLS. The future existence of HLS in Slovakia in their authentic form, if possible, depends on giving these areas and their fragments new functions. The most viable way seems to be their role in recreation and education.

21.8 Future Perspectives of Geoheritage and Landscape Protection in Slovakia

Protection of geoheritage and various landscape features, of both natural and cultural-historical importance, has a relatively long tradition in Slovakia. The country also adopted several legal, conceptual and other documents (of national and international nature) that should, both directly and indirectly, help conservation and sustainable development of these important (geo)sites and traces of historical development for the future. Unfortunately, they are not always and consequently applied and this, under specific unfavourable conditions, may lead to the gradual loss of geoheritage sites and HLS areas and their authenticity.

More addressed support of institutions focusing on the management and protection of this heritage (State Geological Institute of Dionýz Štúr, Slovak Academy of Sciences, administrations of protected areas and geoparks, universities, civic organizations, stakeholders, etc.) and their cooperation will lead to the enhancement of geoheritage and HLS protection and preservation for future generations. Additionally, education at various levels (primary schools, secondary schools, educational trails, thematic camps, etc.), is one of the most effective ways to increase the awareness of the importance of geoheritage and HLS protection. Here, not only educating the general public but also professionals in the field of geoheritage and HLS is crucial, as only specialists in this field, cooperating with scientists from other related disciplines (e.g. history, archeology, biology, environmental protection, etc.), can (1) properly interpret the role and importance of geoheritage and HLS to the general public, (2) identify the needs and requirements for their protection and (3) discuss these issues with the government and local authorities.

Acknowledgements This paper was supported by the VEGA Grant Agency (GP 2/0013/18) as well as ConectGREEN Restoring and managing ecological corridors in mountains as the green infrastructure in the Danube basin, Interreg Project, 2018-2021.

References

Act No. 543/2002 Z. z. Zákon o Ochrane Prírody a Krajiny. Available online: https://www.zakonypreludi.sk/zz/2002-543. Accessed 20 Feb 2020

Aubrecht R, Krobicki M, Sýkora M, Mišík M, Boorová D, Schögl J, Šamajová E, Golonka J (2006) Early Cretaceous hiatus in the Czorsztyn succession (Pieniny Klippen Belt, Western Carpathians): submarine erosion or emersion? Ann Soc Geol Pol 76(2):161–196

Baca I, Schuster E (2011) Listing, evaluation and touristic utilisation of geosites containing archaeological artefacts. case study: Ciceu Ridge (Bistrița-Năsăud County). Roman Rev Geogr Acad 5(1):5–20

Brilha J (2016) Inventory and quantitative assessment of geosites and geodiversity sites: a review. Geoheritage 8:119–134. https://doi.org/10.1007/s12371-014-0139-3
Brilha J (2018) Geoheritage: inventories and evaluation. In E Reynard, J Brilha (Eds) Geoheritage: assessment, protection, and management (pp 69–85), Amsterdam: Elsevier.
Bruschi VM, Cendrero A, Albertos JAC (2011) A statistical approach to the validation and optimization of geoheritage assessment procedures. Geoheritage 3(3):131–149. https://doi.org/10.1007/s12371-011-0038-9
CEMAT (2005) 13th European conference of ministers responsible for regional/spatial planning (Cemat): documents of the conference - Ljubljana (Slovenia), 16–17 September (Territory and Landscape). Council of Europe, Strasbourg
Cortaza P, Vandelli V, Fiorentini L, Paliaga G, Faccini F (2019) Bridging terrestrial and marine geoheritage: assessing geosites in Portofino Natural Park (Italy). Water 11:2112. https://doi.org/10.3390/w11102112
Cowie JW (1992) Report of Task Force Meeting, Paris, France, February 1991, UNESCO World Heritage Convention, Working Group on Geological (inc. Fossil) Sites, Trondheim: IUGS Secretariat.
Cowie JW, Wimbledon WAP (1994) The World Heritage List and its relevance to geology. In D O'Halloran, C Green, M Harley, M Stanley, J Knill (Eds) Geological and Landscape Conservation (pp 71–73), London: Geological Society.
Dixon G (1996) Geoconservation: an international review and strategy for Tasmania. Occasional paper 35, Parks & Wildlife Service, Tasmania, 101 p
Eberhard R (ed) (1997) Pattern and process: towards a regional approach to national estate assessment of geodiversity. Technical series no. 2, Australian heritage commission and environment forest taskforce. Environment Australia, Canberra
European Council (2000) European landscape convention. Available online: https://rm.coe.int/1680080621. Accessed 15 May 2019
Executive Decree No. 24/2003 to the Act No. 543/2002. Available online: https://www.epi.sk/zz/2003-24
Fassoulas C, Mouriki D, Dimitriou-Nikolakis P, Iliopoulos G (2012) Quantitative assessment of geotopes as an effective tool for geoheritage management. Geoheritage 4(3):177–193. https://doi.org/10.1007/s12371-011-0046-9
Gavrilă IG, Anghel T (2013) Geomorphosites inventory in the Măcin Mountains (South-Eastern Romania). GeoJ Tour Geosites 11(1):42–53
Gordon JE, Crofts R, Díaz-Martínez E, Woo KS (2018) Enhancing the role of geoconservation in protected area management and nature conservation. Geoheritage 10:191–203. https://doi.org/10.1007/s12371-017-0240-5
Gray M (2004) Geodiversity: valuing and conserving abiotic nature. Wiley, Chichester, 448 p
Gray M (2008) Geodiversity: the origin and evolution of a paradigm. Geol Soc Lond Spec Publ 300(1):31–36
Gray M (2013) Geodiversity: valuing and conserving abiotic nature, 2nd edn. Wiley Blackwell, Hoboken, 508 p
Gray M (2018) Geodiversity: the backbone of geoheritage and geoconservation. In: Reynard E, Brilha J (eds) Geoheritage: assessment, protection, and management. Elsevier, Amsterdam, pp 13–25
Grube A (1994) The national park system in Germany. In: O'Halloran D, Green C, Harley M, Stanley M, Knill J (eds) Geological and landscape conservation. Geological Society, London, pp 175–180
Hanušin J, Cebecauerová M, Huba M, Ira V, Lacika J, Sládeková Madajová M, Oťaheľ J, Podolák P (2013) Kultúrna krajina podmalokarpatského regiónu. Geografický ústav SAV, Bratislava, 157 p
Hanušin J, Cebecauerová M, Huba M, Ira V, Lacika J, Sládeková Madajová M, Oťaheľ J, Pazúr R, Podolák P, Šeba D, Šveda M (2014) Historical cultural landscapes - problems and reflection: (Sub-Little-Carpathian region, Slovakia). In: Vital landscapes: valorisation and sustainable development of cultural landscapes using innovative participation and visualisation techniques. NEBE s.r.o., České Budějovice, pp 55–66
Hose TA (2012) 3G's for modern geotourism. Geoheritage 4:7–24. https://doi.org/10.1007/s12371-011-0052-y
Hose TA (2016) Introduction. In: Hose TA (ed) Geoheritage and geotourism: a European perspective. The Boydell Press, Woodbridge, pp 1–13
Hronček P, Rybár P, Štrba Ľ, Hvizdák L, Molokáč M, Jesenský M (2017) The identification of geosites as elements of geotourism offer in Kysuce region, Slovakia. Dela 48: 115–128.
Huba M, Čuchor J, Flamík J, Gojdič I, Hrubec I, Kapusta M, Krajčovič R, Križan L, Kubáček L, Lukáčová D, Mlynka L, Pauliniová Z, Podoba J, Procházka K, Slivka M, Šimkovic P, Urbánek J (1988) Historické štruktúry krajiny. MV SZOPK, Bratislava
Huba M, Kubišová K (2017) Legislatívne aspekty ochrany charakteristického vzhľadu (kultúrnej) krajiny v medzinárodnom kontexte (komparatívna štúdia). Geogr Slovaca 33:31–58
Hudecová B (2006) Príspevok k poznaniu historických krajinných štruktúr v stredoslovenskej banskej oblasti. In: Venkovská krajina: sborník příspěvků. Veronika Brno, Brno, pp 64–67
Ilies DC, Josan N (2009) Geosites-geomorphosites and relief. GeoJ Tour Geosites 3:78–86
Ivan Ľ (1943) Výskyty travertínov na Slovensku I. Bratislava, ŠGÚDŠ
Jančura P (2004) Historické krajinné štruktúry vo vzťahu k vývoju poľnohospodárskeho využívania zeme. Partner, Banská Štiavnica
Jeleň S, Galvánek J, Andráš P, Bendík A, Beláček B, Bozalková I, Gaál Ľ Gajdoš A, Háber M, Konečný V., Križáni I, Luptáková J, Mazúrek J, Michal P Soták J, Staňová S, Šimo V, Šurka J, Wetter R (2009) Náučno-poznávací sprievodca po geologických a geografických lokalitách Stredného Slovenska. GÚ SAV, Banská Bystrica, 320 p
Joyce EB (2008) Geosites of Australia: preparing an inventory and framework for the Global Inventory of trhe IUGS. http://earthsci.unimelb.edu.au/Joyce/heritage/GeositesAustraliaJoycems2008.htm. Accessed 10 Aug 2016
Kazancı N (2012) Geological background and three vulnerable geosites of the Kızılcahamam-Çamlıdere Geopark project in Ankara, Turkey. Geoheritage 4(4):249–261. https://doi.org/10.1007/s12371-012-0064-2
Kozová M, Dvořáková V, Kinka M, Huba M, Izakovičová Z, Jančura P, Oťaheľ J, Pauditšová E et al (2010) Metodický postup spracovania typológie krajiny Slovenskej republiky. PRIF UK, MŽP SR, Bratislava
Kubalíková L (2009) Koncepce "geomorphosites" jako jedna z možností hodnocení geomorfologických lokalit. Geomorphol Slovaca Bohem 9(2):28–36
Kubalíková L (2013) Geomorphosite assessment for geotourism purposes. Czech J Tour 2(2):80–104
Kubalíková L (2016) Geomorphological heritage and geoconservation in the Czech Republic. In: Pánek T, Hrádecký T (eds) Landscapes and landforms of the Czech Republic. Springer, Heidelberg, pp 387–398
Kubalíková L, Kirchner K (2016) Geosite and geomorphosite assesment as a tool for geoconservation and geotourism purposes: a case study from Vizovická vrchovina Highland (eastern part of the Czech Republic). Geoheritage 8(1):5–14. https://doi.org/10.1007/s12371-015-0143-2
Lacika J (2016a). Transformation of historical cultural landscape exemplified by Svätý Jur. In: Lněnička L (ed) Central European area

in view of current geography: Proceedings of the 23rd Central European conference. Masaryk University, Brno, pp 202–215

Lacika J (2016b). Kultúrna pamiatka ako prvok historickej kultúrnej krajiny. In: Huba M, Hanušin, Dlouhá (eds) Poznávanie a interpretácia kultúrnej krajiny ako súčasť environmentálnej výchovy a vzdelávania: zborník rozšírených abstraktov zo seminára s medzinárodnou účasťou. Geografický ústav SAV, Bratislava, pp 36–38

Lacika J (2016c). Vybrané geografické aspekty rozšírenia hradov na Slovensku. In: Hrady a hradné panstvá na Slovensku: dejiny, majitelia, prostredie. VEDA, Bratislava, pp 179–205

Liová S, Zvolenský M, Petrášek M (2015) Natural disaster in Vrátna dolina on July 21, 2014. Available online: http://www.shmu.sk/File/Hydrologia/Publikacna_cinnost/2015/2015_MPaPRaHD_Liova_kol_Prirodna_katastrofa_vo_Vratnej_doline.pdf

Liščák P, Antalík M (2018) Information system of important geosites in the Slovak Republic. Slovak Geol Mag 18:5–68

Liščák P et al (2012) Významné geologické lokality. Štátny geologický ústav Dionýza Štúra, Bratislava, 2012. Available online: http://apl.geology.sk/g_vglg.

Lukáč M, Baláž B (2016) Placement of geoparks in the system of territorial protection of natural heritage in Slovakia. In: F Ugolini, V Marchi, S Trampetti, D Pearlmutter, A Raschi (eds) Proceedings of the GEOTOUR 2016—International conference on geotourism, mining tourism, sustainable development, and environmental protection, Florence, Italy, 18–20 October 2016, IBIMET-CNR, Florence, pp 21–26

Lukniš M (1977) Geografia krajiny Jura pri Bratislave. Univerzita Komenského, Bratislava

Martínez-Torez LM, Alonso J, Valle JM (2011) The Upper Aptian-Lower Albian Amber deposit of the Peñacerrada II geosite (Basque–Cantabrian basin, Northern Spain): geological context and protection. Geoheritage 3:55–61. https://doi.org/10.1007/s12371-010-0030-9

Migoń P, Pijet-Migoń E (2017) Viewpoint geosites—values, conservation and management issues. Proc Geol Assoc 128:511–522. https://doi.org/10.1016/j.pgeola.2017.05.007

Mišík M (1976) Geologické exkurzie po Slovensku. SPN, Bratislava, 359 p

Mišíková P, Kozová M, Huba M (2010) The scientific background of the national programme for implementation of the European landscape convention in the Slovak Republic and its main priorities. In: Living landscape: the European landscape convention in research perspective. Conference materials, vol I. UNISCAPE, Bandecchi&Vivaldi, Pontedera (Pisa), pp 330–342

Nazaruddin DA (2019) Selected geosites for geoheritage, geotourism, and geoconservation in Songkhla Province, Southern Thailand. Quaest Geogr 38:161–177. https://doi.org/10.2478/quageo-2019-0011

Nemčok A (1982) Zosuvy v Slovenských Karpatoch. Veda, Bratislava

Panizza M (2009) The geomorphodiversity of the dolomites (Italy): a key of geoheritage assessment. Geoheritage 1(1):33–42. https://doi.org/10.1007/s12371-009-0003-z

PEBLDS (1995) Pan-European biological and landscape diversity strategy. Available online: https://www.cbd.int/doc/nbsap/rbsap/peblds-rbsap.pdf. Accessed 15 May 2019

Peppoloni S, Di Capua G (2017) Geoethics: ethical, social and cultural implications in geosciences. Ann Geophys-Italy 60(7):1–8. https://doi.org/10.4401/ag-7473

Pereira P, Pereira DI, Alves MIC (2007) Geomorphosite assessment in Montesinho Natural Park (Portugal). Geogr Helv 62(3):159–168. https://doi.org/10.5194/gh-62-159-2007

Poirier B, Daigneault RA (2011) La mise en valeur du patrimoine géologique du Sentier national du Québec dans les Laurentides. Le Nat Can 135(1):45–55

ProGEO (2012) Conserving our shared geoheritage. In: Wimbledon WAP, Smith-Meyer S (eds) Geoheritage in Europe and its conservation. ProGEO, Oslo, pp 14–18

Prosser CD, Díaz-Martínez E, Larwood JG (2018) The conservation of geosites: principles and practice. In: Reynard E, Brilha J (eds) Geoheritage: assessment, protection and management. Elsevier, Amsterdam, pp 193–212

Reynard E (2004) Geosite. In: Goudie AS (ed) Encyclopedia of geomorphology. Routledge, London, p 440

Reynard E, Fontana G, Kozlik L, Scapozza C (2007) A method for assessing "scientific" and "additional values" of geomorphosites. Geogr Helv 62(3):148–158. https://doi.org/10.5194/gh-62-148-2007

Rybár P (2010) Assessment of attractiveness (value) of geotouristic objects. Acta Geoturistica 1(2):13–21

Sharples C (1993) A methodology for the identification of significant landforms and geological sites for geoconservation purposes. Report to Forestry Commission, Tasmania, 23 p

Sharples C (2002) Concepts and principles of geoconservation. Available online on the Tasmanian Parks & Wildlife Service website, 79 p. http://dpipwe.tas.gov.au/Documents/geoconservation.pdf. Accessed 20 Apr 2018

Škodová M, Mazúrek J (2012) Chránené územia Slovenska. Fakulta prírodných vied UMB Banská Bystrica, Banská Bystrica

Špulerová J, Štefunková D, Dobrovodská M, Izakovičová Z, Kenderessy P, Vlachovičová M, Lieskovský J, Piscová V, Petrovič F, Kanka R, Bača A, Barančoková M, Bezák P, Bezáková M, Boltižiar M, Mojses M, Dubcová M, Gajdoš P, Gerhátová K, Izsóff M, Kalivoda H, Miklósová V, Degro M, Šatalová B, Krištín A, Dankaninová L, Kalivodová E, Majzlan O, Mihál I, Sašiov S, Šolomeková T, Ambros M, Baláž I, Halabuk A (2017) Historické štruktúry poľnohospodárskej krajiny Slovenska. VEDA, Bratislava

Štefunková D, Hanušin J (2019) Viticultural landscapes: localised transformations over the past 150 years through an analysis of three case studies in Slovakia. Morav Geogr Rep 27(3):155–168. https://doi.org/10.2478/mgr-2019-0012

Štrba Ľ (2015) Identification and evaluation of geosites along existing tourist trail as a primary step of geotourism development: case study from the Spiš region (Slovakia). GeoJ Tour Geosites 16:126–140

Štrba Ľ (2019) Analysis of criteria affecting geosite visits by general public: a case of Slovak (geo)tourists. Geoheritage 11:291–300. https://doi.org/10.1007/s12371-018-0283-2

Štrba Ľ, Lukáč M (2016) Assessment of geological heritage of the Plešivec Plateau (Slovak Karst, Slovakia). In International Multidisciplinary Scientific GeoConference: SGEM, vol. 1 (pp 89–96), Sofia: STEF92 Technology.

Štrba Ľ, Rybár P (2015) Revision of the "assessment of attractiveness (value) of geotouristic objects." Acta Geoturistica 6(1):30–40

Štrba Ľ, Kršák B, Sidor C (2018) Some comments to geosite assessment, visitors, and geotourism sustainability. Sustainability 10(8):2589. https://doi.org/10.3390/su10082589

Štrba Ľ, Rybár P, Baláž B, Molokáč M, Hvizdák L, Kršák B, Lukáč M, Muchová L, Tometzová D, Ferenčíková J (2015) Geosite assessments: comparison of methods and results. Curr Issues Tour 18 (5):496–510. https://doi.org/10.1080/13683500.2014.882885

Štrba Ľ, Kolačkovská J, Kudelas D, Kršák B, Sidor C (2020) Geoheritage and geotourism contribution to tourism development in protected areas of Slovakia—Theoretical considerations. Sustainability 12(7):2979. https://doi.org/10.3390/su12072979

Tomić N, Bozič S (2014) A modified geosite assessment model (M-GAM) and its application on the Lazar Canyon area (Serbia). Int J Environ Res 8:1041–1052. https://doi.org/10.22059/IJER.2014.798

UNESCO (1972) Convention concerning the protection of the world cultural and natural heritage. http://whc.unesco.org/archive/convention-en.pdf. Accessed 10 Aug 2019

VGL Databáza. Available online: http://apl.geology.sk/g_vgl/?jazyk=EN. Accessed 20 Jan 2020

Vlčko J (2004) Extremely slow slope movements influencing the stability of Spis Castle UNESCO Site. Landslides 1(1):67–71. https://doi.org/10.1007/s10346-003-0007-8

Vyhláška č. 24/2003 Z. z. Available online: https://www.epi.sk/zz/2003-24. Accessed 20 Feb 2020

Wimbledon WAP (1996) Geosites — a new conservation initiative. Episodes 9:87–88. https://doi.org/10.18814/epiiugs/1996/v19i3/009

Wimbledon WAP, Ishchenko AA, Gerasimenko NP, Karis LO, Suominen V,Ohansson CE, Freden C (2000) Geosites - an IUGS initiative: science supported by conservation. In: Barettino D, Wimbledon WAP,Gallego E (eds) Geological heritage: its conservation and management. ITGE, Madrid, pp 69–94

Wimbledon WAP, Smith-Meyer S (eds) (2012) Geoheritage in Europe and its conservation. Pro GEO, Oslo, p 405

Wróblewski W, Gradziński M, Hercman H (2010) Suggestions of the allochthonous origin of terra rossa from Dreveník hill (Spiš, Slovakia). Acta Carstologica Slovaca 48(2):153–161

Zouros NC (2007) Geomorphosite assessment and management in protected áreas of Greece, case study of Lesvos Island—coastal geomorphosites. Geogr Helv 62(3):169–180. https://doi.org/10.5194/gh-62-169-2007

Ľubomír Štrba is Associate Professor and a guarantor of the first (Bachelor) level of the study programme 'Geotourism' at the Faculty of Mining, Ecology, Process Control and Geotechnologies, Technical University of Košice, intensively focuses his research activities mainly on identification and characterization of geological heritage and various approaches of its assessment with the potential of its use within specific tourism forms as, e.g., geotourism and mining tourism and their development.

Ján Lacika is an emeritus Associate Professor of Physical Geography. He was researcher at the Department of Physical Geography, Geomorphology and Natural Hazards of the Institute of Geography SAS in Bratislava (1980 – 2019) and teacher and research worker of the Department of Geography, Geoinformatics and Regional Development Faculty of Natural Sciences of the Constantine the Philosopher University in Nitra. Originally his main research interest was structural geomorphology (neotectonics, post-volcanic transformation of landforms and development of the valley networks). Later his research was focused on geographical aspects of the natural and cultural landscape development.

Mikuláš Huba is an environmental geographer - senior research worker at the Institute of Geography of the Slovak Academy of Sciences as well as professor in the field of geography a geoecology. More than 30 years ago he developed the concept of historical landscape structures.

Pavel Liščák is the Head of the Department of Engineering Geology at the State Geological Institute of Dionýz Štúr. He has conducted several research and application projects concerning slope deformation area mapping, recording and rehabilitation as well as monitoring of environmental geological components. He is also working as a lecturer at the Department of Engineering Geology, Faculty of Natural Sciences of the Comenius University in Bratislava.

Mário Molokáč is Associate Professor at the Faculty of Mining, Ecology, Process Control and Geotechnologies, Technical University of Košice. His research in geotourism is primarily focused on geoparks from their preparation, through their implementation to their management and the promotion of the general public. He is a member of the "Interdepartmental Commission of the Network of Geoparks of the Slovak Republic.

Check for updates

22 Geotourism and Geoparks—Promoting Geoheritage and Geodiversity to People

Ľubomír Štrba and Mário Molokáč

Abstract

The promotion of geological heritage and geodiversity via geotourism has become popular in various regions of the world as a result of the growing demand from visitors for such kind of experience in the last two decades. Nowadays, geoparks represent the most effective tool for geotourism-related activities. In Slovakia, geoheritage and geodiversity are promoted mainly by the Slovak Geoparks Network, State Geological Institute of Dionyz Stur, management bodies of protected areas, universities and civic organizations, with the support of Slovak Government. At present, four geoparks are operated in the territory of the Slovak Republic, among which one is a member of the UNESCO Global Geoparks Network. Considerable support of geoheritage promotion, especially karstic, provides the Slovak Caves Administration and private subjects via showcave visits. Future promotion and sustainable use of geoheritage and geodiversity depend on the cooperation of various stakeholders, focusing on the establishment and/or operation of geoparks and other geotourism-related activities. These should be aimed, among other things, at effective geoeducation, which on one hand increases the overall awareness on geological heritage and, on the other hand, contributes to its protection and preservation for future generations.

Keywords

Geotourism • Geoheritage • Promotion • Slovakia • Geopark

Ľ. Štrba (✉) · M. Molokáč
Faculty of Mining, Ecology, Process Control and Geotechnologies, Institute of Earth Resources, Technical University of Kosice, Košice, Slovakia
e-mail: lubomir.strba@tuke.sk

M. Molokáč
e-mail: mario.molokac@tuke.sk

22.1 Introduction

People have been fascinated by various geological phenomena for centuries, not only in the modern history of mankind. Caves, waterfalls, mountains, volcanoes, canyons, etc., serve many purposes, including education, inspiration search or relax, were and still are the subject of interest in any part of our planet (e.g., in Slovakia the Plavecká Cave was opened in 1802, and the Jasovská Cave in 1846). However, in the last decades, the number of people looking for deeper and/or alternative experience closely connected with nature is growing. As a consequence, various nature-based tourism forms have been introduced, defined and practiced. Here, geotourism and ecotourism play a key role. But, as recent expectations of visitors are on relatively high level, such nature-connected tourism forms offer not only strong experience related to specific natural phenomena, but they also require supporting services, including well-addressed interpretation, safe trails, routes and paths, catering and accommodation, etc.

Recently, from the geodiversity and geoheritage perspective, geotourism can be considered as the primary way of promoting geological and geomorphological heritage, its values, and its importance to the general public. Sharples (2002) pointed out the importance of geodiversity geoconservation for future generations. In this context, geotourism plays a crucial role as it supports geoconservation on one hand and promotes understanding of geoheritage among the general public.

22.2 Geotourism and Promotion of Geoheritage

In the last two decades, geotourism has become a global phenomenon (Dowling 2011). In many countries around the world, including Slovakia, this tourism form has been considerably developed and practiced. Although geology and

M. Lehotský and M. Boltižiar (eds.), *Landscapes and Landforms of Slovakia*, World Geomorphological Landscapes,
https://doi.org/10.1007/978-3-030-89293-7_22

tourism were mentioned by some authors (e.g., Maini and Carlisle 1974; Jenkins 1992; De Bastion 1994; Spiteri 1994), the concept of geotourism has not been introduced until 1995. Hose (1995) defined geotourism for the first time as "provision of interpretative and service facilities to enable tourists to acquire knowledge and understanding of the geology and geomorphology of a site (including its contribution to the development of the Earth sciences) beyond the level of mere aesthetic appreciation".

This first geotourism definition was adopted and modified by several authors (e.g., Hose 2000, 2012; Joyce 2006, 2008; Dowling and Newsome 2006, 2010; Sadry 2009; Dowling 2013). According to Slovak authors (Rybar et al. 2010), "understanding geotourism in a broader context, this form of tourism can be understood based on knowledge of geological objects and processes with emphasis on their aesthetic and historical value, and knowledge of technical monuments associated with mining (abandoned mines, mining museums, trade routes with the transport of goods obtained through mining activities) as well as technical and cultural monuments associated with historical mining activities".

An alternative approach to geotourism and its understanding has been introduced by National Geographic saying that "geotourism is defined as tourism that sustains or enhances the distinctive geographical character of a place - its environment, heritage, aesthetics, culture, and the well-being of its residents". Based on this uncertainty in the global understanding of geotourism, the Arouca Declaration (2011) attempted to unify the definition of geotourism as "tourism which sustains and enhances the identity of a territory, taking into consideration its geology, environment, culture, aesthetics, heritage and the well-being of its residents. Geological tourism is one of the multiple components of geotourism". Dowling (2011) emphasized that geotourism includes making "geoproducts" designed to protect the geological heritage, helps to build local associations, informs about the geological heritage and promotes it. Taking into account 3Gs of modern geotourism, as defined by Hose (2012): geoconservation, geointerpretation and geohistory, and respecting principles of geotourism (Dowling 2011), this form of tourism plays a primary role when discussing leisure, tourism and geoheritage promotion. Moreover, according to the majority of geotourism definitions, (geo) education is one of the key features and can be offered in-situ (e.g., educational trails, information panels, guides) and ex-situ (e.g., museums, geoheritage related publications).

In Slovakia, there are several good examples of geoheritage promotion. Guides in showcaves provide information on various, mostly karstic, geoheritage features for centuries. Recently, there are 18 such caves open to the public (Table 22.1), among the total number of more than 7000 caves in Slovakia. Besides 13 showcaves operated by the Slovak Caves Administration, 45 caves located in Slovakia are freely accessible to the public and five caves (Table 22.1) opened to the public are operated by other entities than the Slovak Caves Administration.

Various leaflets, brochures and books are available to people when visiting caves. Jákal and Bella (2008) edited a publication 'Caves of the World Heritage in Slovakia' which can be considered as a scientific-educational book promoting the most important cave-related geoheritage in Slovakia. Many publications, including books, book chapters or papers discussing Slovak geoheritage from various perspectives have been published in the last two decades. However, only a few of them are easily available to the general public. Among these, a book series "Natural Beauties of Slovakia" consisting of 12 individual titles presents, within specific topics, different geoheritage sites. Taking into account books related to geoheritage within this series, Muntág (2007) presented the most beautiful hills and mountains in Slovakia, Bizubová (2008) published a book devoted to the rocks, Lacika (2009) presented the most beautiful valleys, Bella (2011) introduced Slovak caves, Lacika and Ondrejka (2009) described national parks in Slovakia, Hanušin (2009) focused on the most important water features, and Lackovič (2015) described Slovak tarns. Mineral deposits within the historical mining context were described by Marza et al. (2018). Semrád (2012) published a book devoted to the precious opals from the Dubník area, the only opal mines in Europe. Among the online platforms or projects that promote the geological heritage of Slovakia, the most extensive is the database of Important Geological Sites of the Slovak Republic as described in Chap. 21 of this volume. Online map server of the State Geological Institute of the Slovak Republic, in the section "Educational Geology" (https://www.geology.sk/maps-and-data/mapovy-portal/educational-geology/?lang=en), includes four thematic geological/educational maps, namely Tatry Mts., Cerová vrchovina Upland, Vihorlatské vrchy Mts. and Zemplínske vrchy Mts. Moreover, two geopark maps are available here: Banská Bystrica Geopark and Geological-Educational Map of the Sandberg-Pajštún Geopark. Other online sites devoted to this topic are mostly connected with the historical mining in the territory of the Slovak Republic, e.g., The Story of Banská Štiavnica—volcano (http://pribeh.banskastiavnica.travel/en/volcano/) or the Multimedia Geotourism Guide (https://www.montanistika.eu/), which is available only in the Slovak language.

The Slovak Geological Society regularly organizes lectures on various topics related to geology, geomorphology and geoheritage. Twice a year, they also organize a geologically oriented guided field trip. Exhibitions in museums provide promotion of various geoheritage features, most frequently within geological collections where rocks and minerals, but also fossils, raw materials, etc., are presented to visitors. However, considering the most effective way how

Table 22.1 List of showcaves located in Slovakia (SSJ 2020)

*Belianska jaskyňa**
*Bojnická hradná jaskyňa ("Bojnice Castle Cave")***
*Brestovská jaskyňa**
*Bystrianska jaskyňa**
*Jaskyňa mŕtvych netopierov ("Dead Bats Cave")***
*Demänovská jaskyňa slobody**
*Demänovská ľadová jaskyňa (ice cave)**
*Dobšinská ľadová jaskyňa (ice cave)**
*Domica**
*Driny**
*Gombasecká jaskyňa**
*Harmanecká jaskyňa**
*Jasovská jaskyňa**
*Krásnohorská jaskyňa***
*Ochtinská aragonitová jaskyňa**
*Prepoštská jaskyňa***
*Stanišovská jaskyňa***
*Važecká jaskyňa**
*Zlá diera ("Bad Hole Cave")***

* showcave operated by the Slovak Caves Administration
** showcave operated by other entity than the Slovak Caves Administration

to introduce and interpret the significance and role of geoheritage from various perspectives to the general public, geoparks recently play the most important role.

The potential and importance of geotourism development in Slovakia have also been recognized from the education perspective. In Slovakia, at the Technical University of Košice, students have an opportunity to study geotourism within an individual study program "Geotourism". Education in this field, as well as in related geoscientific fields, gives a very strong base for future geotourism development and geoheritage promotion to the general public in Slovakia.

22.3 Geoparks

22.3.1 Geopark System in Slovakia

UNESCO (2020) defines geoparks as unified geographical areas where the landscape and its localities are of international geological significance and are managed for the purpose of conservation, education and sustainable development. The European Geopark is a territory that includes a special geological heritage with a sustainable territorial development strategy, supported by the European Development Program. It must contain a specific number of geological sites of particular importance in terms of their scientific quality, rarity, aesthetic or educational value. Geopark is a new type of geological heritage protection. The Program of geoparks was adopted by UNESCO in 2015 creating a new label—the UNESCO Global Geopark.

Generally, geopark is an area containing several sites of scientific importance not only in geological terms but also in terms of its archaeological, economic, or cultural specificity of international importance. It is consistent with the Sustainable Development Strategy and has a strong governance structure, can be supported by various funding programmes that contribute to its further development. In addition to the potential for environmental research (including education), the geopark is important for local economic development as it contributes to increased employment and new economic activities in the region, while its functionality is autonomous.

Based on the global growth of importance of geoparks and geotourism and in accordance with Agenda 21, geoparks were for the first time included in the Programme Statement of the Government of the Slovak Republic of 2006. Based on the Report of implementation of the Concept of Geoparks in the Slovak Republic of 2012, the commission consisting of members from various fields was established. In 2015, the commission drew an update of the original concept of geoparks.

In terms of the current state, practical needs and proposals for further creation of geoparks, the updated concept of geoparks in Slovakia distinguishes three basic categories of geoparks:

1. geoparks with membership in GGN and EGN,
2. geoparks in operation,
3. areas with the potential of inclusion into the geopark network.

The first category includes UNESCO Global Geoparks and the second category covers national geoparks. These categories of geoparks, if compared to the original categorizations (category A—operated geoparks, category B—geoparks in a process of their implementation and category C—proposed geoparks), reflect the shift in geopark creation, their current status and changes in the philosophy of using geopark status and its name in the country. Since the territory, which is supposed to become a geopark, has to meet certain criteria for at least two years, only geoparks in the first and second categories are considered as official geoparks. The third category includes ideological designs of some areas with a well-known potential and with potential of inclusion into geoparks. These territories are based on the original category C (proposed geoparks), having undergone a revision of the content, area and name in the light of new knowledge and enhancement of the portfolio of phenomena that deserve attention. However, these territories are neither binding in terms of area, content, or future implementation, but are merely an offer for local initiatives.

This is how the four operated geoparks in Slovakia work. The Banská Štiavnica Geopark is operated by the Association of Legal Entities for the Development of Banská Štiavnica and its surroundings—the Sitno Region. The Banská Bystrica Geopark is created and operated by the active cooperation of regional and local self-government, professional institutions in the Banská Bystrica District and representatives of the private sector, under the umbrella of the Banská Bystrica Geo-mining Park Civic Association. The Association of legal entities of Geopark Novohrad—Nógrád is responsible for the operation of the Geopark Novohrad—Nógrád. These associations develop projects, implement them, and carry out all necessary activities for which they receive funding. Malé Karpaty Geopark is created by the bottom-up intiative of several enthusiasts of the non-profit organization Barbora, based on a wide range of activities. It consists of six geomining areas and is an opportunity to connect all tourism actors in the area of geopark.

According to the above-mentioned categorization (Fig. 22.1), the Novohrad-Nógrád Geopark is the only Slovak geopark included in the Global Geoparks Network. The second group of geoparks includes three geoparks in operation: Banská Štiavnica Geopark, Banská Bystrica Geopark and Malé Karpaty Geopark. The potential of inclusion have, according to official documents, six areas:

(1) Dubník Geopark located in the Slanské vrchy Mts., where the world-famous precious opal was mined near Červenica until 1922;
(2) Zemplín Geopark in the Zemplínske vrchy Mts. and its surroundings with examples of rock complexes as well as examples of old mining activity. The world-known Tokaj wine region (Hungary) also extends to this area;
(3) Spiš Geopark, predominantly located in the area of the Slovenský rad NP ("Slovak Paradise National Park"). It will also include demonstrations of old mining activity and metallurgy and some important cultural monuments of the Spiš Region;
(4) Silica Geopark located in the western part of the Slovenský kras NP ("Slovak Karst National Park") (Silica Plateau), with already built educational trails and educational sites (caves: Domica, Gombasecká jaskyňa, Silická ľadnica, Krásnohorská jaskyňa).
(5) Jasov Geopark located in the eastern part of the Slovenský raj NP ("Slovak Karst National Park") (Zádiel and Jasov Plateau) including sites such as the Jasovská jaskyňa Cave and Zádielska tiesňava, Háj Gorge, Turniansky hradný vrch (castle hill), Jasovská skala and Jasov Monastery with its surroundings;
(6) Súľov-Manín Geopark in the Protected Landscape Area of Strážovske vrchy Mts., focusing on Súľovské skaly, Manínska tiesňava, and Kostolecká tiesňava, already with partly built infrastructure of environmental education.

22.3.2 Novohrad-Nógrád Geopark

The Novohrad - Nógrád Geopark (NNG) (Fig. 22.4) is the Slovak-Hungarian cross-border area abundant in unique geological and natural values, cultural and historical legacy, folk traditions, modern culture and gastronomy. The Geopark is managed by the Geopark Novohrad - Nógrád Association of Legal Entities on the Slovak side (based in Fiľakovo) and by the Novohrad - Nógrád Geopark, a non-profit Kft. (based in Salgótarján) on the Hungarian side. The NNG founding organizations were established on the initiative of Fiľakovo after a five-year effort in the year 2008. After signing the cooperation agreement, they started forming a joint management team of the Geopark and prepared for membership into the European Geoparks Network (EGN) and the Global Geoparks Network (GGN). In March 2010, the NNG became the 37th member of the EGN and

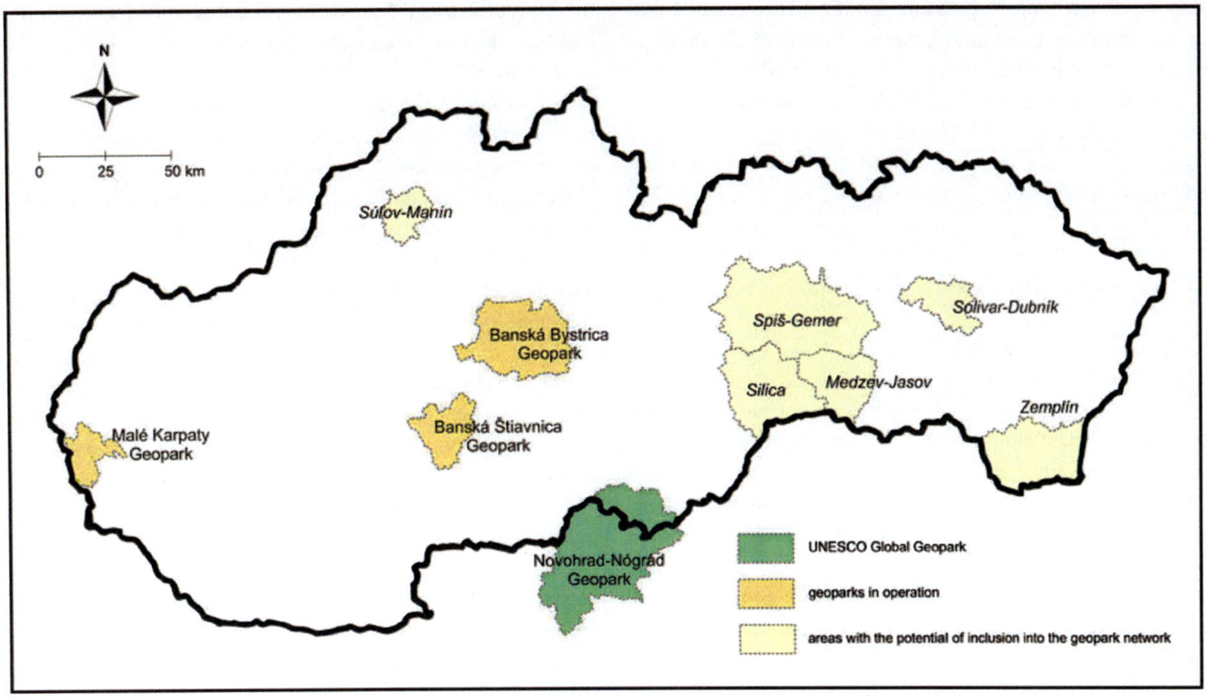

Fig. 22.1 Location of Slovak geoparks and areas with the potential of inclusion into the geopark network in Slovakia

subsequently the 66th member of the GGN as the first one in both countries and as the first cross-border bilateral Geopark in the world. The 1,630 km^2 large NNG (336 km^2 in Slovakia and 1,294 km^2 in Hungary) currently comprises 28 municipalities in the southern districts of Rimavská Sobota and Lučenec and 64 municipalities in the northern part of the Nógrád county in Hungary. By having been granted the Green Card in years 2014 and 2018, the NNG re-affirmed its membership in the EGN and the GGN.

Several outstanding geosites are located in the Slovak part of the NNG. The most famous geological formation of the NNG is the 9 m high Stone Waterfall (Fig. 22.3c), situated below the Šomoška Castle. It was formed by slow solidification of the lava flowing from the volcano about 4 million years ago, which formed today's basaltic neck. It features 15–20 cm wide and 5–6-sided basalt columns, the bending of which is a European rarity. Under the rock outcrop, the Šomoška Stone Sea was formed by debris derived from weathering of basalt and from the waste of broken basalt columns, uniquely used for the construction of the castle. Medveš represents the relic of the largest and highest basaltic volcanic body in the Slovak-Hungarian border area. The flowing lava formed the tabular hill of Bear Heights, covering the area of 12.8 km^2 (7.8 km^2 in Hungary), with the peak on the remnants of the volcanic cone of Medveš (659 m). The sheet thickness reaches from 11 to 107 m. The rock of Hajnáčka Castle Hill represents an erosional remnant of a filled volcano conduit (diatreme), on which the castle was built in the Middle Ages. The nearby impressive Steblová skala (486 m) represents a volcano with distinct columnar separation. One of the most representative examples of volcanic activity in the Cerová vrchovina mountain range is Ragáč (537 m). Almost 2 km long lava flow extended from the volcano at that time, where two basalt layers may be distinguished. A unique geomorphological formation at the top of the flow is the Ragáč Crater (the relic of a cinder cone and lava flow; dominantly formed by agglutinate relics referring to Hawaiian eruption type), about 9 m deep. Below that, the Ebeczky Cave was discovered, also formed by the pressure of emerging volcanic gases.

Geosites of non-volcanic character illustrate mainly the sediments of the Parathetys Sea and younger gravel, tuff and sand layers, which were deposited after its retreat. They contain remains of fossil flora and fauna. For example, the Čakanovský profile (Fig. 22.3d) uncovers the succession of both marine and continental deposits. At the end of the abandoned brown coal mines, which operated between 1901 and 1932, deposits of lakes and marshes have been preserved. In the fomer volcano funnel—the Starobaštianska diatreme—large fragments of sandstone and rhyodacite tuffs with a diameter of up to 5 m were found. On the outskirts of the village of Tachty in the Tachtian diatreme, a stone quarry revealed vertical contact of the volcano with the surrounding sandstones. The Lipovianske sandstones (Fig. 22.3b) have also been uncovered by sand mining. These sandstones contain sand benches and unique sand balls of the diameter of up to 50 cm. One of the rare paleo-zoological sites is Kostná dolina Valley at Hajnáčka, where the bones of animals from the end of the Neogene and the beginning of the Quaternary, such as octopuses, pandas, rhinos, hyenas, tapirs and monkeys, were discovered (Fig. 22.3b).

22.3.3 Geopark Banská Štiavnica

Geopark Banská Štiavnica (Fig. 22.4) is the oldest geopark established in Slovakia. The origins of its establishment date back to years 1998–2000 when the first projects intended to present important sites on the Slovak territory were born in the State Geological Institute of Dionýz Štúr. The purpose of the establishment of this geopark was to ensure harmonic and sustainable development of the territory in the Banská Štiavnica region which, through the economic, demographic and natural potential of the region, would lead to economic growth, increase in employment and better quality of the landscape.

The geopark has an area of 374 km^2 and is situated on the territory of Banská Štiavnica, Žarnovica and Žiar nad Hronom districts. The geopark has been classified as a national geopark of the Slovak Republic and integrated into the Slovak Geoparks Network in 2016. On its territory, there are 156 sites of interest and it partially coincides with the Protected Landscape Area Štiavnické vrchy Mts. Three information centers, two drainage galleries, four nature trails and educational geological exposition are located in the area of the geopark. It also includes objects registered in the UNESCO World Cultural and Natural Heritage List "Historic Town of Banská Štiavnica and Technical Monuments of the Surroundings". The area of the geopark partially extends to the surrounding orographic units. In addition to the Štiavnica Mts., the area of interest includes the Krupina Plateau, Javorie, Kremnické vrchy Mts., Vtáčnik and Pohronský Inovec.

Fig. 22.2 Logo of the Novohrad-Nógrád Geopark

Štiavnické vrchy Mts. (Fig. 22.5a) are part of volcanic mountains representing denudation relics of andesitic stratovolcanoes, whose development took place in the region of Central Slovakia during the younger Neogene in the time interval of Badenian to Sarmatian. The territory of the Banská Štiavnica Geopark was identified based on the above-mentioned denudation relic of an extensive andesitic Štiavnica stratovolcano, whose volcanic rocks cover an area of more than 2,200 km^2. with a caldera of 18 × 22 km diameter. The Štiavnica stratovolcano is characterized by a complicated structure comprising rocks of varied composition, ranging from basalts, basaltic andesites and andesites to rocks of rhyolite composition. From the orographic point of view, these relics of an unusually massive Štiavnica stratovolcano include the area of the Štiavnica Mts. and Pohronský Inovec, in the NW direction they reach into the eastern and southern parts of the Vtáčnik Mts., whereas in the northern direction they include the southern edges of the Kremnické Mts. The products of this stratovolcano deposited in the marine environment in the area of the Krupina Highlands and the Ipeľ Hills continue to the south.

One of the first elements of the system of the Banská Štiavnica Geopark, which was handed over to the public in 2003, is the Educational Geological Trail Paradajs. It focuses on the presentation of natural geological values of the region, describes the origin and development of the Štiavnica volcano (stratovolcano), the rocks involved in its structure, and the development of the most important ore veins. The exposition is a modern textbook of volcanology with many rock exhibits, pictures, geological sections, maps, or describing panoramic views of the widest surroundings. The route is a closed circuit, 3.9 km long.

The Bieber drainage gallery is the original stone portal, which was reconstructed in cooperation with the Banská Štiavnica-Hodruša mining association. The gallery was called the "Mother of All Mines of the World" in the eighteenth century because of the volume of mined silver and gold. In 1627, for the first time in the world, gunpowder was used to disintegrate rocks in the mining field of this gallery. The Glanzenberg drainage gallery is an accessible adit in the center of the historic town of Banská Štiavnica (Fig. 22.5b). Historical exceptionality, but especially the fact that the gallery leads at a low depth below the city center, led in 2001 to the decision to renovate the site. After reconstruction and safety works, the gallery was opened to the public (Fig. 22.5c). Educational geological exposition is built in the area of the Mining Museum in nature, as an open-air museum. It describes the geological development and structure of the Slovak Republic territory. It contains many rock samples, illustrations and geological sections.

Fig. 22.3 Examples of Slovak geosites in the Novohrad-Nógrád Geopark, **a**: Mučínska jaskyňa—12 m long pseudokarst cave, **b**: Neogene shallow marine fossils from Lipovianske pieskovce (sandstones) location, **c**: "Stone waterfall"—an example of columnar jointing in basalt (Šomoška site), **d**: Čakanovský profile

Part of the exposition is also a large billboard geological map of the Slovak Republic.

ŠB-06 Piarg Trail is an educational-tourist route and leads through the southern part of the village of Štiavnické Bane. There are nine stops along the route with educational boards describing the most important mining works—shafts and galleries. The trail can be completed in 2–3 h and allows the visitors to get acquainted with the rich history of mining activities in the village of Štiavnické Bane.

ŠB-08 The Piarg Water Management System Trail with its 16 stops describes the historical water management system (Fig. 22.5d) in the wider surroundings of the village, namely the Evička, Veľká Windšachta, Bakomi, Richňava water reservoirs and the most famous reservoir of Počúvadlo. The

Fig. 22.4 The logo of Banská Štiavnica Geopark

educational boards also pay attention to several head-race canals with a description of their functional connection with mining works. The trail has two variants, a longer one, measuring 15.1 km with an elevation difference of 100 m, and a shorter one, measuring 10.3 km with 85 m difference in elevation. The round trail starts at the Great Richňava water reservoir and can be completed in 6–7 h.

The Trail on the Ecology of the Piarg Water Reservoirs describes biological life and the associated water quality in selected water reservoirs (Richňava, Bakomi, Krechsengrund, Veľká Windšacht and Evička), using five stops. The trail with a total length of 4.4 km and a difference in elevation of 50 m can be walked in 2–3 h.

22.3.4 Banská Bystrica Geopark

The Banská Bystrica Geopark (Fig. 22.6) was established as a civic association in 2006. It became a public partnership only in 2009, when 30 members from the ranks of municipalities, micro-regions, non-governmental organizations, but also entrepreneurs operating in the area of interest joined this association. Since 2016, the Geopark has been a national geopark of the Slovak Republic, included in the Geoparks Network of the Slovak Republic.

The Banská Bystrica Region is very interesting for its complex geological structure, which gave rise to the diverse relief of this part of Slovakia, but also to the wealth of mineral resources, especially ores. In the past, ore mining was the main factor influencing life in this area. The towns of Banská Bystrica and Kremnica were centers of business families of Thurzo and Fugger, who mined precious metals (gold, silver, copper) in the wider area of these two towns and exported them worldwide. Many sites still yield impressive specimens of many rare minerals (e.g., malachite, azurite, realgar, etc.) World-famous was especially mining of copper ores in the Špania Dolina Valley, at Staré Hory and Ľubietová, which, before the discovery of America, were the largest copper deposits in the "old world". Many historical mining monuments have been preserved, especially buildings of shafts and smelters, knockers, galleries and a sophisticated water system with a length of approximately 40 km.

The establishment of the Banská Bystrica Geopark, which now covers an area of 886 km^2, began in 2006 and presupposed the existence and use of geological and mining sites, ecological phenomena and historical monuments of Banská Bystrica and its surroundings. The area of the Geopark includes the district of Banská Bystrica, Žiar nad Hronom District (town Kremnica) and part of Brezno District (municipalities Nemecká and Ráztoka).

From the point of view of tourism and landscape, there are many hiking trails in the area and the surrounding mountains provide significant space for winter as well as summer sports. The deciduous and mixed forests around Banská Bystrica not only provide a sanctuary for many animals but also hide botanically interesting and valuable places. In total, there are 292 localities in the geopark: geological (46), geological and mining (53), geological and natural (40), mining (36), mining-historical (1), cultural-historical (80), natural (29) and archaeological (7). The Vel'ka Fatra National Park, the Nízké Tatry National Park and the Poľana Protected Landscape Area also reach into its territory.

In the Banská Bystrica Geopark there are also many relics of technical monuments, especially representing metallurgical, coal and rope sector, such as hammer mills, smelters, blast furnace, lime kiln, etc. Very important are also archaeological sites in the form of 17 castles situated in difficult mountain terrain and other localities. The surroundings of Banská Bystrica are also interesting from an ethnographic point of view, with preserved original folk architecture, not to mention the burgher architecture of Banská Bystrica and Kremnica themselves. A world-famous rarity is also the lace from Špania Dolina. 16 nature trails and 30 hiking trails are available within the geopark.

The area of the Geopark is divided into geo-mining areas of three categories:

Fig. 22.5 Examples of geoheritage in the Banská Štiavnica Geopark, **a**: Volcanic landscape of the Štiavnické vrchy Mts., **b**: Glazenberg adit portal, **c**: inside the Glanzenberg adit, **d**: Tajchy—artificial water reservoirs built to provide energy for mining

Fig. 22.6 The logo of Banská Bystrica Geopark

1. Category: Staré Hory– Špania Dolina, Poníky—Ľubietová and Kremnica geo-mining areas (Geopark core territory)
2. Category: towns Banská Bystrica and Kremnica (tourist centers)
3. Category: Brusno—Bukovec and Badín—Tajov geo-mining areas (territory of geopark interest)

Špania Dolina, the center of Banská Bystrica Geopark, was established as a mining settlement more than 4,000 years ago. Gold, silver, but above all, copper were first mined on the surface, later extraction was carried out in the underground spaces (Fig. 22.7).

Geopark is a phenomenon bringing people back to roots and protecting values for the future at the same time, while not being only a product, but also an area of exceptional value. It is established and operated by active regional and local autonomy cooperations, professional institutions and representatives of the private sector. The current management of Banská Bystrica geo-mining park unites subjects for common interests, examines, educates, informs and helps building the infrastructure for ecotourism, geotourism and protection of geo-mining elements, objects and sites in the area of Banská Bystrica and its surroundings.

22.3.5 Malé Karpaty Geopark

The Malé Karpaty Geopark (Fig. 22.8) is a relatively new, fourth geopark in Slovakia, opened in 2015 as the Sandberg-Pajštún Geopark. The geopark includes nine scientifically valuable and tourist-attractive locations, important mainly from a geological, speleological and paleontological point of view. Geosites are located in the southern part of the Little Carpathians, especially in the area of the Malé Karpaty Protected Landscape Area. A smaller part extends into the territories of two Protected Landscape Areas: Dunajské luhy and Záhorie. The territory of the geopark represents mainly the area of the Devínska Kobyla hill (514 m), as well as the surroundings of the villages of Borinka, Marianka and the karst valley of Prepadlé.

As summarized by Madarás and Zlinská (2013), the most important sites of the geopark, with individual information panels, include: Slovinec Cliff (south of the village of Devínska Nová Ves), with Upper Jurassic—Lower Cretaceous carbonatic rock sequence in overturned stratigraphic position, Abrasion Cave in the Slovinec Cliff—an example of the Neogene sea activity, Sandberg sand pit (Fig. 22.9a)—the Late Neogene Sea coast on the slopes of Mt. Devínska Kobyla, Wait's quarry and historical quarries between Devín and Dúbravka urban parts exposing Middle Miocene shallow-water marine sedimentary environments, Devín Castle Hill—the historical and geological jewel at the Morava—Danube rivers confluence, with a rock sequence covering the period from the Early Paleozoic to the Neogene (Fig. 22.9b), Morava and Danube rivers with Quaternary sediments and fluvial landforms (meanders, ox-bow lakes, river terraces formation), Marianka Black Shale Adit—medieval mining activities in the Lower–Middle Jurassic black shales, Pajštún Castle Cliff—a huge Middle Triassic carbonate block (megaolistolith) within the Jurassic rock sequence, Manganese ore mining in the Lower to Middle Jurassic carbonates and black shales between the villages of Borinka and Lozorno, Borinka Karst—Limbach spring—the Jurassic limestone karst aquifer in the Prepadlé karst valley with numerous caves, sinkholes, swallow holes, sinking streams and springs.

22.4 Future Perspectives

The promotion of geoheritage in Slovakia at various levels is provided by the State Geological Institute of Dionýz Štúr, Slovak Geoparks Network, Slovak Geological Society, Slovak Caves Administration and universities. Moreover, the management of large-scale protected areas (national parks and protected landscape areas) also contributes to the promotion of geoheritage and geodiversity to people, especially via guided tours, information panels along hiking trails, leaflets, brochures and other publications. However, the current status of nature protection in Slovakia seems to be primarily oriented on biodiversity conservation. And thus, due to the intense focus on biodiversity, the importance of geodiversity is underestimated (Štrba et al. 2020).

Fig. 22.7 Examples of geosites from the Banská Bystrica Geopark, **a**: Jacob-Langitka adit —copper mine (turquoise colour is caused by the occurence of langite—hydrated copper sulfate mineral), **b**: Dolná prašnica—columnar joints in basaltic andesite, **c**: Moštenické travertíny (travertines)—formation of slope calcareous tufa cascades, **d**: Netopieria jaskyňa (cave)

Fig. 22.8 Logo of the Malé Karpaty Geopark

From the future perspective, the most effective way to increase the overall awareness of geoheritage and geodiversity importance is education at various levels and via various forms. In this regard, geopark establishment and operation, with the support of the Slovak government, local authorities, professionals in the field of geosciences, managements of protected areas, universities, civic organizations and individuals, is the most suitable form of geoheritage and geodiversity promotion to people. Therefore, it is necessary to further support these initiatives. Such cooperation, especially of professionals and local authorities, can considerably contribute to the promotion of geoheritage located outside geopark areas. Also, geoheritage and geodiversity have to be more discussed in primary and secondary schools from various perspectives. As concluded by Hovorka (2008), Slovak geoscientists, especially those working in the field of education, will have to prepare a proposal to address thematically broad modern education in the field of inanimate nature at all levels of schools, because it still lacks.

Fig. 22.9 Examples of geosites from the Malé Karpaty Geopark, a: Sandberg sand pit, b: Devín Castle Hill, c: landscape view from Devínska Kobyla

References

Arouca Declaration (2011) Available online: https://www.dropbox.com/s/q41gbd0cp2nt73o/Declaration_Arouca_%5BEN%5D.pdf?dl=0. Accessed 15 Jan 2020

Bella P (2011) Prírodné krásy Slovenska: Jaskyne. Dajama, Bratislava, 120 p

Bizubová M (2008) Prírodné krásy Slovenska: Kamene. Dajama, Bratislava, 119 p

De Bastion R (1994) The private sector-threat of opportunity? In: O'Halloran D, Green C, Harley M, Stanley M, Knill J (eds) Geological and landscape conservation. The Geological Society, Bath, pp 391–395

Dowling R (2011) Geotourism's global growth. Geoheritage 3(1):1–13. https://doi.org/10.1007/s12371-010-0024-7

Dowling R (2013) Global geotourism - an emerging form of sustainable tourism. Czech J Tour 2(2):59–79. https://doi.org/10.2478/cjot-2013-0004

Dowling R, Newsome D (eds) (2006) Geotourism. Elsevier, Amsterdam, 260 p

Dowling R, Newsome D (eds) (2010) Geotourism: the tourism of geology and landscape. Goodfellow Publishers Ltd., Oxford, p 246

Hanušin J (2009) Prírodné krásy Slovenska: Vody. Dajama, Bratislava, 128 p

Hose TA (1995) Selling the story of Britain's stone. Environ Interpret 10(2):16–17

Hose TA (2000) European "geotourism" - geological intepretation and geoconservation promotion for tourists. In: Barettino D, Wimbledon WAP, Gallego E (eds) Geological heritage: its conservation and management. IGME, Madrid, pp 127–146

Hose TA (2012) 3G's for modern geotourism. Geoheritage 4(1–2):7–24. https://doi.org/10.1007/s12371-011-0052-y

Hovorka D (2008) Vedy o Zemi na univerzitách – výchova geovedcov alebo aj súčasť univerzitného vzdelávania? Mineralia Slovaca/geovestník 40(1–2):5–6

Jenkins JM (1992) Fossickers and rockhounds in Northern New South Wales. In: Weiler B, Hall CM (eds) Special interest tourism. Belhaven, London, pp 129–140

Joyce EB (2006) Geomorphological sites and the new geotourism in Australia. http://web.earthsci.unimelb.edu.au/Joyce/heritage/GeotourismReviewebj.htm

Joyce EB (2008) Geosites of Australia: preparing an inventory and framework for the global inventory of the IUGS. Available at: http://earthsci.unimelb.edu.au/Joyce/heritage/GeositesAustraliaJoycems2008.htm. Accessed 20 Oct 2015

Lacika J (2009) Prírodné krásy Slovenska: Najkrajšie doliny. Dajama, Bratislava, 128 p

Lacika J, Kliment O (2009) Prírdoné krásy Slovenska: Národné parky. Dajama, Bratislava, 127 p

Lackovič M (2015) Prírodné krásy Slovenska: Plesá. Dajama, Bratislava, 128 p

Madarás J, Zlinská A (2013) Sandberg-Pajštún Geopark (SAPAG) - a gateway into the Malé Karpaty Mountains geological history. Geol Výzk Mor Slez 20(1–2):54–55

Maini JS, Carlisle A (1974) Conservation in Canada: a conspectus - Publication No. 1340, Department of the Environment/Canadian Forestry Service, Ottawa

Mârza E, Syrný M et al (2018) Baníctvo na Slovensku a v Rumunsku v 15. až 18. storočí/Mineritul in Slovacia si Romania in secolele XV-XVIII. Múzeum SNP, Banská Bystrica, 359 p

Muntág S (2007) Prírodné krásy Slovenska: Najkrajšie vrchy. Dajama, Bratislava, 127 p

Rybár P, Baláž B, Štrba Ľ (2010) Geoturizmus - Identifiácia objektov geoturizmu. Technical University of Kosice, Kosice, p 101

Sadry B (2009) Fundamentals of geotourism: with a special emphasis on Iran. Samt Organization Publishers, Tehran, p 220p

Semrád P (2012) The story of European precious Opal from Dubník. Granit, Praha, 216 p

Sharples C (2002) Concepts and principles of geoconservation. available online on the Tasmanian Parks & Wildlife Service website, 79 p. http://dpipwe.tas.gov.au/Documents/geoconservation.pdf. Accessed 20 Apr 2018

Spiteri A (1994) Malta: a model for the conservatin of limestone regions. In: O'Halloran D, Green C, Harley M, Stanley M, Knill J (eds) Geological and landscape conservation. The Geological Society, Bath, pp 205–208

SSJ (2020) Caves. Available online: http://www.ssj.sk/en/jaskyne. Accessed 23 May 2020

Štrba Ľ, Kolačkovská J, Kudelas D, Kršák B, Sidor C (2020) Geoheritage and geotourism contribution to tourism development in protected areas of Slovakia—Theoretical considerations. Sustainability 12(7):2979. https://doi.org/10.3390/su12072979

UNESCO (2020) UNESCO global geoparks. Available online: http://www.unesco.org/new/en/natural-sciences/environment/earth-sciences/unesco-global-geoparks/. Accessed 10 Apr 2020

Ľubomír Štrba is Associate Professor and a guarantor of the first (Bachelor) level of the study programme 'Geotourism' at the Faculty of Mining, Ecology, Process Control and Geotechnologies, Technical University of Košice, intensively focuses his research activities mainly on identification and characterization of geological heritage and various approaches of its assessment with the potential of its use within specific tourism forms as, e.g., geotourism and mining tourism and their development.

Mário Molokáč is Associate Professor at the Faculty of Mining, Ecology, Process Control and Geotechnologies, Technical University of Košice. His research in geotourism is primarily focused on geoparks from their preparation, through their implementation to their management and the promotion of the general public. He is a member of the "Interdepartmental Commission of the Network of Geoparks of the Slovak Republic.

Author Index

M. Lehotský and M. Boltižiar (eds.), *Landscapes and Landforms of Slovakia*, World Geomorphological Landscapes,
https://doi.org/10.1007/978-3-030-89293-7

Subject Index

Page numbers followed by "f" and "t" refer to figures and tables, respectively.

M. Lehotský and M. Boltižiar (eds.), *Landscapes and Landforms of Slovakia*, World Geomorphological Landscapes,
https://doi.org/10.1007/978-3-030-89293-7

D

T